OXFORD MATHEMATICAL MONOGRAPHS

Series Editors

OXFORD MATHEMATICAL MONOGRAPHS

A. Belleni-Morante: *Applied semigroups and evolution equations*
I. G. Macdonald: *Symmetric functions and Hall polynomials*
J. W. P. Hirschfeld: *Projective geometries over finite fields*
N. Woodhouse: *Geometric quantization*
A. M. Arthurs: *Complementary variational principles* Second edition
P. L. Bhatnagar: *Nonlinear waves in one-dimensional dispersive systems*
N. Aronszajn, T. M. Creese, and L. J. Lipkin: *Polyharmonic functions*
J. A. Goldstein: *Semigroups of linear operators*
M. Rosenblum and J. Rovnyak: *Hardy classes and operator theory*
J. W. P. Hirschfeld: *Finite projective spaces of three dimensions*
J. C. Lennox and S. E. Stonehewer: *Subnormal subgroups of groups*
K. Iwasawa: *Local class field theory*
A. N. Pressley and G. Segal: *Loop groups*
D. E. Edmunds and W. D. Evans: *Spectral theory and differential operators*

Spectral Theory and Differential Operators

D. E. EDMUNDS
Professor of Mathematics,
University of Sussex

W. D. EVANS
Professor of Mathematics,
University College, Cardiff

CLARENDON PRESS · OXFORD
1987

Oxford University Press, Walton Street, Oxford OX2 6DP

Oxford New York Toronto
Delhi Bombay Calcutta Madras Karachi
Petaling Jaya Singapore Hong Kong Tokyo
Nairobi Dar es Salaam Cape Town
Melbourne Auckland

and associated companies in
Beirut Berlin Ibadan Nicosia

Oxford is a trade mark of Oxford University Press

Published in the United States
by Oxford University Press, New York

British Library Cataloguing in Publication Data
Edmunds, D. E.
Spectral theory and differential operators.
— (Oxford mathematical monographs)
1. Differential operators 2. Spectral theory (Mathematics)
I. Title II. Evans, W. D.
515.7′242 QA329.4
ISBN 0-19-853542-2

Library of Congress Cataloging in Publication Data

Edmunds, D. E. (David Eric)
Spectral theory and differential operators.
(Oxford mathematical monographs)
Bibliography: p.
Includes indexes.
1. Spectral theory (Mathematics) 2. Differential operators. I. Evans, W. D. II. Title. III. Series.
QA320.E32 1987 515.7′246 86-21852
ISBN 0-19-853542-2

Set by Macmillan India Ltd, Bangalore 25
Printed in Great Britain by
J. W. Arrowsmith Ltd, Bristol

CB-SHK
SCIMON

To

Rose and Mari

PREFACE

The relationship between the classical theory of compact operators in Banach or Hilbert spaces and the study of boundary-value problems for elliptic differential equations has been a symbiotic one, each having a profound effect on the other. In the L^2 theory of elliptic differential equations with smooth coefficients and on bounded domains in $\mathbb{R}^n$, the problem of eigenfunction expansions rests upon the fact that there is a naturally occurring operator with a compact self-adjoint resolvent to which the abstract theory may be applied with great success. On the other hand, the early work of Fredholm, Hilbert, Riesz and Schmidt, for example, was stimulated by the needs of problems in integral and differential equations. The theory of compact self-adjoint operators in Hilbert space is particularly rich, but when one drops the self-adjointness substantial difficulties appear; the eigenvalues (if any) may be non-real, and, what is very important from the point of view of applications, there is no Max–Min Principle of proven usefulness for the eigenvalues; furthermore the question of whether the eigenfunctions form a basis for the underlying Hilbert space is then much more complex. When we consider compact linear operators acting in a Banach space, as is often necessary in connection with non-linear problems for example, even greater difficulties appear: to obtain information about eigenvalues indirect methods often have to be adopted. In recent years much work has been done in this area, relating eigenvalues to more geometrical quantities such as approximation numbers and entropy numbers. This work is not limited to purely abstract theory: much effort has been put into the estimation of such numbers for embedding maps between Sobolev spaces, the group in the Soviet Union led by Birman and Solomjak being especially active in this area. These embedding maps provide a natural link between the abstract theory and problems in differential (and integral) equations. Boundary-value problems for elliptic differential equations on unbounded domains or with singular coefficients necessitate the study of non-compact operators. In such cases the spectrum does not consist wholly of eigenvalues but also has a non-trivial component called the essential spectrum. In the literature there are many different ways of looking at the essential spectrum but whichever way is followed a study of Fredholm and semi-Fredholm operators is required. A notable result in this area is that due to Nussbaum and (independently) Lebow and Schechter: the radius of the essential spectrum is the same for all the commonly used definitions of essential spectrum. This brings in the notion of the measure of non-compactness of an operator, which is itself related to the entropy numbers mentioned earlier.

In order to apply the abstract theory to boundary-value problems for elliptic differential equations the first task is to determine an appropriate function space and an operator which is a natural realization of the problem. For linear elliptic problems the natural setting is an L^2 space and in this book we concentrate on the L^2 theory for general second order elliptic equations with either Dirichlet or Neumann boundary conditions.

Let

$$\tau\phi = -\sum_{i,j=1}^{n} D_i(a_{ij}D_j\phi) + \sum_{j=1}^{n} b_j D_j\phi + q\phi, \qquad D_j := \frac{\partial}{\partial x_j},$$

in an open set Ω in $\mathbb{R}^n$, with $n \geqslant 1$, and set

$$t[\![\phi,\psi]\!] = \int_\Omega \left(\sum_{i,j=1}^{n} a_{ij} D_j\phi D_i\bar{\psi} + \sum_{j=1}^{n} b_j D_j \phi\bar{\psi} + q\phi\bar{\psi} \right)$$

for ϕ and ψ in $C_0^\infty(\Omega)$ or $C_0^\infty(\mathbb{R}^n)$, the choice depending on the boundary conditions under consideration. If the numerical range of t, namely the set

$$\Theta(t) = \left\{ t[\![\phi,\phi]\!] : \int_\Omega |\phi|^2 = 1 \right\},$$

lies in a sector in the complex plane with angle less than π, one can invoke the theory of sesquilinear forms to obtain an operator T whose spectrum lies within the aforementioned sector which describes the boundary-value problem associated with τ in a weak sense. If $\Theta(t)$ does not lie in a sector other techniques have to be found. In this case we can make use of the powerful methods which have been developed to tackle the problem of determining sufficiency conditions for the operator T_0 defined by a formally symmetric τ on $C_0^\infty(\Omega)$ to have a unique self-adjoint extension in $L^2(\Omega)$, a problem which has attracted a great deal of attention over the years, particularly because of its importance in quantum mechanics. An important example is Kato's distributional inequality which makes it possible to work with coefficients having minimal local requirements. Once the operator has been obtained, the next step is to analyse its spectrum. For non-self-adjoint operators the location of the various essential spectra is often as much as one can realistically hope for in the absence of the powerful tools available when the operators are self-adjoint, notably the Spectral Theorem and Max–Min Principle. Perturbation methods are effective in determining the dependence of the essential spectra on the coefficients of τ, the effect of these methods being to reduce the problem to one involving a simpler differential expression. The geometrical properties of Ω then become prominent and the properties of the embedding maps between Sobolev spaces which occur naturally achieve a special significance. In this the notion of capacity has a central role, a fact highlighted in the work of Molcanov, Maz'ja and others in the Soviet Union. To obtain information about the eigenvalues one usually has to resort to the indirect methods developed in the abstract theory. For instance, knowledge of the singular numbers of T, i.e. the eigenvalues of the non-negative self-adjoint operator $|T|$, provides information about the l^p class of the eigenvalues of T.

Our main objective in this book is to present some of the results which have been obtained during the last decade or so in connection with the problems described in the previous paragraphs. On the abstract side we deal with operators in Banach spaces whenever possible, especially as some of the most notable achievements can only be appreciated in this context. We specialize to Hilbert spaces in the work on elliptic differential equations reported on, chiefly because it is in the framework of the L^2 theory that most of the relevant recent advances have been made. Furthermore, for the L^p theory with $p \neq 2$ we have nothing substantial to add to what is contained in the books by Goldberg [1] and Schechter [2]. Despite this, when we prepare tools like the embedding theorems and results on capacity we work with L^p spaces if this can be done without much additional strain. In an area as broad as this, one is forced to be selective in one's choice of topics and, inevitably, important omissions have to be made. We say very little about eigenfunctions and expansion theorems, for instance, but we have a clear conscience about this because what we could say is adequately covered in the book by Gohberg and Krein [1]. In any case, our book is already long enough.

The book is primarily designed for the mathematician although we hope that other scientists will also find something of interest to them here and we have kept this goal in mind while writing it. The language of the book is functional analysis and a sound basic knowledge of Banach and Hilbert space theory is needed. Some familiarity with the Lebesgue integral and the elements of the theory of differential equations would be helpful but only the barest essentials are assumed. We have dispensed with a chapter of preliminaries in favour of reminders in the body of the text and where necessary we refer to other books for background material.

Most of the abstract theory is developed in the first four chapters. Chapters I and II deal with bounded linear operators in Banach spaces, the main themes being the essential spectra and the properties of various numbers like entropy numbers and approximation numbers associated with the bounded linear operators. In Chapter III closed linear operators are studied, particular emphasis being given to the behaviour of their deficiency indices and Fredholm index when the operators are extended or are perturbed. We illustrate the abstract results with a comprehensive account of general second-order quasi-differential equations and this covers the Weyl limit-point, limit-circle theory for formally symmetric equations and also its extensions by Sims and Zhikhar to formally J-self-adjoint equations. Sesquilinear forms in Hilbert spaces are the subject of Chapter IV. The basic results are the Lax-Milgram Theorem for bounded coercive forms and the representation theorems for sectorial forms. Also there are perturbation results for the forms of general self-adjoint and m-sectorial operators which have an important role to play later in the location of the essential spectra of differential operators. Another result which will be important later is Stampacchia's generalization of the Lax–Milgram theorem to variational inequalities.

In Chapter V we give a treatment of Sobolev spaces. Apart from their intrinsic interest these spaces are an indispensible tool for any work on partial differential equations and much of what is done in subsequent chapters hinges on Chapter V. Furthermore Sobolev spaces are an ideal testing ground for examining some of the abstract notions discussed in the early chapters and accordingly we devote some space to the determination of the measures of non-compactness and the approximation numbers of embedding maps between Sobolev spaces.

The remaining chapters deal mainly with second-order elliptic differential operators. The weak or generalized forms of the Dirichlet and Neumann boundary-value problems are defined and studied in Chapter VI. The material in Chapter VI is mainly relevant to bounded open sets Ω in $\mathbb{R}^n$ when the underlying operators have compact resolvents in $L^2(\Omega)$, in which case the spectra consist wholly of eigenvalues. Also included is Stampacchia's weak maximum principle and this leads naturally to the notion of capacity. Second-order operators on arbitrary open sets Ω are the theme of Chapter VII. Under weak conditions on the coefficients of the differential expression we describe three different techniques for determining the Dirichlet and Neumann operators. The first applies the First Representation Theorem to sectorial forms, the second is one developed by Kato based on his celebrated distributional inequality and the third has its roots in the work of Levinson and Titchmarsh on the essential self-adjointness of the operator defined by $-\Delta+q$ on $C_0^\infty(\Omega)$ when q is real. Schrödinger operators are an important special case, especially of the third class of operators discussed, and some of the results obtained for highly oscillatory potentials are anticipated by the quantum-mechanical interpretation of the problem.

The central result of Chapter VIII is Molcanov's necessary and sufficient condition for the self-adjoint realization of $-\Delta+q$ (q real and bounded below) to have a wholly discrete spectrum. This necessitates the study of capacity and in the wake of the main result we also obtain necessary and sufficient conditions for the embedding $W_0^{1,p}(\Omega) \to L^p(\Omega)$ to be compact and for important integral inequalities (like the Poincaré inequality) to hold.

In Chapter IX we study the essential spectra of closed operators in Banach and Hilbert spaces and then use the abstract theory to locate the various essential spectra of constant coefficient differential operators in $L^2(\mathbb{R}^n)$ and $L^2(0, \infty)$. In the case when the coefficients are not constant a useful tool for ordinary differential operators is the so-called Decomposition Principle which implies that the essential spectra depend only on the behaviour of the coefficients at infinity. For partial differential operators a Decomposition Principle is obtained in Chapter X as a perturbation result and this is then used to locate the essential spectra of the general second-order operators in $L^2(\Omega)$ discussed in Chapter VII. We analyse the dependence of the essential spectra on Ω in two different ways. In the first the results are described in terms of capacity and sequences of cubes which intersect Ω. The second involves the use of a mean distance function $m(x)$, which is a measure of the distance of x to the boundary of Ω, and an integral inequality obtained by E. B. Davies. This enables us to give estimates for the first eigenvalue and the least point of the essential spectrum of the Dirichlet problem for $-\Delta$ on Ω.

The last two chapters are concerned with the eigenvalues and singular values of the Dirichlet and Neumann problems for $-\Delta + q$. The case of q real, and hence self-adjoint operators, is treated in Chapter XI, the main result being a global estimate for $N(\lambda)$, the number of eigenvalues less than λ when λ is below the essential spectrum. From this estimate asymptotic formulae are derived for $N(\lambda)$ when the spectrum is discrete and $\lambda \to \infty$ and when the negative spectrum is discrete and $\lambda \to 0-$. We also obtain the Cwikel–Lieb–Rosenblyjum estimate for $N(\lambda)$ when $q \in L^{n/2}(\mathbb{R}^n)$ with $n \geqslant 3$, and include the elegant Li–Yau proof of the latter result. In Chapter XII q is complex, and global and asymptotic estimates are obtained for $M(\lambda)$, the number of singular values less than λ. From these estimates the l^p-class of the singular numbers and eigenvalues are derived.

Chapters are divided into sections, and some sections into subsections. For example, §I.3.2 means subsection 2 of section 3 of Chapter I; it is simply written §3.2 when referred to within the same chapter and §2 when referred to within the same section. Theorems, Corollaries, Lemmas, Propositions, and Remarks are numbered consecutively within each section. Theorem I.2.3 means Theorem 2.3 in §2 of Chapter I and is referred to simply as Theorem 2.3 within the same chapter. Formulae are numbered consecutively within each section; (I.2.3) means the third equation of §2 of Chapter I and is referred to as (2.3) within the same chapter. The symbol ■ indicates the end of the statement of a result and □ indicates the end of a proof.

There is also a glossary of terms and notation, a bibliography and an index.

We have made no systematic attempt to go into the complicated history of the results presented here, but hope that the references provided will be helpful to the reader interested in the background of the material.

It is a pleasure to acknowledge the help we have received from many colleagues and in particular from Robin Dyer, Edward Fraenkel, and Desmond Harris. We are especially indebted to Hans Triebel, who read the whole manuscript and offered invaluable comments.

Brighton and
Cardiff
June 1986

D. E. E
W. D. E

CONTENTS

Basic notation

$B(x, r)$: open ball in $\mathbb{R}^n$, centre x and radius r.
$\mathbb{C}$: complex plane; $\mathbb{C}_\pm = \{z \in \mathbb{C}: \text{im}\, z \gtrless 0\}$; $\mathbb{C}^n$: n-dimensional complex space; $\mathbb{R}$: real line; $\mathbb{R}^n$: n-dimensional Euclidean space.
$\mathbb{R}^n_+ = \mathbb{R}^n \backslash \{0\}$.
$D_i u = \partial u / \partial x_i$; if $\alpha = (\alpha_1, \ldots, \alpha_n)$ with α_i non-negative integers, $D^\alpha u = \partial^{|\alpha|} u / \partial x_1^{\alpha_1} \ldots \partial x_n^{\alpha_n}$, where $|\alpha| = \alpha_1 + \ldots + \alpha_n$.
Ω: an open set in $\mathbb{R}^n$; Ω is a *domain* if it is also connected.
$\partial\Omega$: boundary of Ω; $\bar{\Omega}$: closure of Ω; $\Omega^c = \mathbb{R}^n \backslash \Omega$.
$\Omega' \subset\subset \Omega$: $\bar{\Omega}'$ is a compact subset of Ω.
dist $(x, \partial\Omega)$: distance from x to Ω^c.
$\mathbb{N}$: positive integers; $\mathbb{N}_0 = \mathbb{N} \cup \{0\}$; $\mathbb{Z}$: all integers.
$f(t) \asymp g(t)$ as $t \to a$: there exist positive constants c_1, c_2 such that $c_1 \leqslant f(t)/g(t) \leqslant c_2$ for $|t - a|$ $(\neq 0)$ small enough, if $a \in \mathbb{R}$; and for large enough $\pm t$ if $a = \pm\infty$.
$T\restriction_G$: restriction of the operator (or function) T to set G.
$f^+ = \max(f, 0)$, $f^- = -\min(f, 0)$.
$A \subset B$ for sets A, B allows for $A = B$.
Embedding: a bounded linear injective map of a Banach space X to another such space Y.
l^p $(1 \leqslant p \leqslant \infty)$: complex sequence space with norm $\|(\xi_j)\|_p = (\Sigma |\xi_j|^p)^{1/p}$ when $1 \leqslant p < \infty$ and $\|(\xi_j)\|_\infty = \sup_j |\xi_j|$ when $p = \infty$.
$c_0 := \{(\xi_j) \in l^\infty : \lim_j \xi_j = 0\}$.
ω_n: volume of the unit ball in $\mathbb{R}^n$, i.e. $\omega_n = \dfrac{\pi^{n/2}}{\Gamma(1 + \frac{1}{2}n)}$.

I

Linear operators in Banach spaces

Three main themes run through this chapter: compact linear operators, measures of non-compactness, and Fredholm and semi-Fredholm maps. Each topic is of considerable intrinsic interest; our object is not only to make this apparent but also to establish connections between the themes so as to derive results which will be of great interest later. One such result is a formula for the radius $\mathrm{r_e}(T)$ of the essential spectrum of a bounded linear map T.

The theory of compact linear operators acting in a Banach space has a classical core which will be familiar to many, and in view of this we pass rather quickly over it. Perhaps less well-known is the concept of the measure of non-compactness of a set and of a map, a notion due to Kuratowski [1], who introduced it in 1930 for subsets of a metric space. The idea lay more or less dormant until 1955, when Darbo[1] showed how it could be used to obtain a significant generalization of Schauder's fixed-point theorem. Since that time, substantial advances have been made both in the theory and in applications, although the bulk of applications have been to ordinary rather than to partial differential equations. We try to redress the balance later on in the book by use of the formula for $\mathrm{r_e}(T)$ in our discussion of the essential spectrum of various partial differential operators. The interaction between measures of non-compactness and semi-Fredholm maps is of crucial importance in the derivation of this formula, and accordingly we devote some time to this interplay.

1. Compact linear maps

All vector spaces which will be mentioned will be assumed to be over the complex field, unless otherwise stated. The norm on a normed vector space X will usually be denoted by $\|\bullet\|_X$, or by $\|\bullet\|$ if no ambiguity is possible.

Given any Banach spaces X and Y, the vector space of all bounded linear maps from X to Y will be denoted by $\mathscr{B}(X,Y)$, or by $\mathscr{B}(X)$ if $X = Y$; with the norm $\|\bullet\|$ defined by $\|T\| = \sup\{\|Tx\|:\|x\| \leqslant 1\}$, $\mathscr{B}(X,Y)$ is a Banach space. It is natural to try to distinguish members of $\mathscr{B}(X,Y)$ which have particularly good properties. Compact linear maps come into this category, since they have properties reminiscent of linear maps acting between finite-dimensional spaces.

Definition 1.1. Let X and Y be Banach spaces and let $T: X \to Y$ be linear. The map T is said to be *compact* if, and only if, for any bounded subset B of X, the closure $\overline{T(B)}$ of $T(B)$ is compact. ■

Evidently T is compact if, and only if, given any bounded sequence (x_n) in X, the sequence (Tx_n) contains a convergent subsequence. Note also that if T is compact it is continuous, since otherwise there would be a sequence (x_n) in X such that $\| x_n \| = 1$ for all $n \in \mathbb{N}$, and $\| Tx_n \| \to \infty$ as $n \to \infty$, which is impossible.

Examples. (i) If $T \in \mathscr{B}(X, Y)$ and the dimension of the range $\mathscr{R}(T) := T(X)$ of T, $\dim \mathscr{R}(T)$, is finite, then T must be compact, since if B is a bounded subset of X then $\overline{T(B)}$ is closed and bounded, and hence compact.

(ii) Not every bounded linear map is compact: take $X = Y = l^2$, for each $n \in \mathbb{N}$ let $e^{(n)}$ be the element of l^2 with mth coordinate δ_{mn} (equal to 1 if $m = n$, and 0 otherwise), and observe that the identity map of l^2 to itself is continuous but not compact, because the sequence $(e^{(n)})$ has no convergent subsequence.

(iii) Let $a, b \in \mathbb{R}$, $b > a$, $J = [a, b]$, and suppose that $k: J \times J \to \mathbb{C}$ is continuous on $J \times J$; define

$$(Kx)(s) = \int_a^b k(s, t)\, x(t)\, dt$$

for all $s \in J$ and for all x in the Banach space $C(J)$ of all continuous complex-valued functions on J (with norm $\| \bullet \|$ given by $\| x \| = \max \{ |x(s)| : s \in J \}$). Then K is a linear map of $C(J)$ into itself, and in fact, K is compact. To see this, set $M = \max \{ |k(s, t)| : s, t \in J \}$; then $\| Kx \| \leqslant M(b-a) \| x \|$ for all $x \in C(J)$, and so if B is a bounded subset of $C(J)$ then $K(B)$ is bounded. Moreover, for all $s_1, s_2 \in J$ and any $x \in C(J)$,

$$|(Kx)(s_1) - (Kx)(s_2)|^2 = \left| \int_a^b [k(s_1, t) - k(s_2, t)]\, x(t)\, dt \right|^2$$
$$\leqslant \int_a^b |k(s_1, t) - k(s_2, t)|^2\, dt \cdot \int_a^b |x(t)|^2\, dt.$$

Thus given any $\varepsilon > 0$, there exists a $\delta > 0$ such that $|(Kx)(s_1) - (Kx)(s_2)| < \varepsilon$ if $\| x \| \leqslant 1$ and $|s_1 - s_2| < \delta$ $(s_1, s_2 \in J)$. Hence $\{ Kx : x \in C(J), \| x \| \leqslant 1 \}$ is equicontinuous and bounded, and thus relatively compact, by the Arzela–Ascoli Theorem (cf. Yosida [1, p. 85]).

Denote by $\mathscr{K}(X, Y)$ the family of all compact linear maps from X to Y, and put $\mathscr{K}(X) = \mathscr{K}(X, X)$. The following proposition is a well-known consequence of the definition of compactness.

Proposition 1.2. Let X,Y,Z be Banach spaces. Then $\mathscr{K}(X,Y)$ is a closed linear subspace of $\mathscr{B}(X,Y)$; if $T_1 \in \mathscr{B}(X,Y)$ and $T_2 \in \mathscr{B}(Y,Z)$ then $T_2 T_1$ is compact if either T_1 or T_2 is compact. ■

This proposition implies that $\mathscr{K}(X)$ is a closed two-sided ideal in the Banach algebra $\mathscr{B}(X)$.

It has already been noted that if $T \in \mathscr{B}(X, Y)$ is of *finite rank*, that is, $\dim \mathscr{R}(T) < \infty$, then $T \in \mathscr{K}(X,Y)$. In particular, $\mathscr{B}(X,Y) = \mathscr{K}(X,Y)$ if either X or Y is finite-dimensional. The following result complements this, and throws new light on Example (ii) above.

Theorem 1.3. Let X be a Banach space and suppose that the identity map of X to itself is compact. Then $\dim X < \infty$. ■

This follows directly from the following lemma.

Lemma 1.4. Let (X_n) be a sequence of finite-dimensional linear subspaces of a Banach space X such that for all $n \in \mathbb{N}$, $X_n \subset X_{n+1}$ and $X_n \neq X_{n+1}$. Then given any $n \in \mathbb{N}$ with $n \geqslant 2$, there exists $x_n \in X_n$, with $\|x_n\| = 1$, such that $\|x_n - x\| \geqslant 1$ for all $x \in X_{n-1}$. In particular, $\| x_n - x_m \| \geqslant 1$ when $m < n$; the sequence (x_n) has no convergent subsequences. ■

Proof. Let $y_n \in X_n \setminus X_{n-1}$. The function $y \mapsto \| y - y_n \|$ is positive and continuous on X_{n-1} and approaches infinity as $\|y\| \to \infty$; hence it has a minimum, at $z_n \in X_{n-1}$, say, and

$$0 < \| z_n - y_n \| \leqslant \| x + z_n - y_n \|$$

for any $x \in X_{n-1}$. The point $x_n := (y_n - z_n)/\| y_n - z_n \|$ then has all the required properties. □

A related result is the following.

Lemma 1.5 (Riesz's Lemma). Let M be a proper, closed, linear subspace of a normed vector space X. Then given any $\theta \in (0, 1)$, there is an element $x_\theta \in X$ such that $\| x_\theta \| = 1$ and $\operatorname{dist}(x_\theta, M) \geqslant \theta$. ■

Proof. Let $x \in X \setminus M$. Since M is closed, $d := \operatorname{dist}(x, M) > 0$. Thus given any $\theta \in (0, 1)$, there exists $m_\theta \in M$ such that $\| x - m_\theta \| \leqslant d/\theta$. The element $x_\theta :=$ $(x - m_\theta)/\| x - m_\theta \|$ has all the properties needed. □

Compactness of a linear map is preserved by the taking of the adjoint. Before this result is given in a formal way, some remarks about notation are desirable. Given any normed vector space X, by the *adjoint space* X^* of X is meant the set of all *conjugate* linear continuous functionals on X; that is, $f \in X^*$ if, and only if, $f: X \to \mathbb{C}$ is continuous and $f(\alpha x + \beta y) = \bar{\alpha} f(x) + \bar{\beta} f(y)$ for all $\alpha, \beta \in \mathbb{C}$ and all $x, y \in X$. Our choice of *conjugate-linear* functionals, rather than the more common *linear* functionals, is dictated solely by the convenience which will result later on in the book. With the usual definitions of addition and multiplication by scalars the adjoint space X^* becomes a Banach space when given the norm $\|\bullet\|$ defined by

$$\|f\| = \sup\{|(f,x)| : \|x\| = 1\},$$

where (f, x) is the value of f at x (often denoted by $f(x)$ as above. Strictly speaking, we should write $(f, x)_X$, but the subscript will be omitted if no ambiguity is possible. The same omission will be made for inner products in a Hilbert space.). Given any $T \in \mathscr{B}(X,Y)$, the adjoint of T is the map $T^* \in \mathscr{B}(Y^*, X^*)$ defined by $(T^* g, x) = (g, Tx)$ for all $g \in Y^*$ and $x \in X$; note that $(\alpha S + \beta T)^* = \bar{\alpha} S^* + \bar{\beta} T^*$ for all $\alpha, \beta \in \mathbb{C}$ and all $S, T \in \mathscr{B}(X,Y)$. These conventions about adjoints will apply even when the underlying spaces are Hilbert spaces, and there will therefore be none of the usual awkwardness about the distinction between Banach-space and Hilbert-space adjoints of a map that has to be made when linear, rather than conjugate-linear, functionals are used. Of course, many of the results to be given below would also hold had X^* been defined to be the space of all continuous linear functionals on X. Note that the Riesz Representation Theorem (cf. Taylor[1, Theorem 4.81-C]) enables any Hilbert space H to be identified with H^*; and that in view of this, given any $T \in \mathscr{B}(H_1, H_2)$ (H_1 and H_2 being Hilbert spaces), $T^* \in \mathscr{B}(H_2, H_1)$. If $H_1 = H_2 = H$, then both T and T^* belong to $\mathscr{B}(H)$: the map T is said to be *self-adjoint* if $T = T^*$.

Theorem 1.6. Let X and Y be Banach spaces and let $T \in \mathscr{B}(X,Y)$. Then $T \in \mathscr{K}(X,Y)$ if, and only if, $T^* \in \mathscr{K}(Y^*, X^*)$. ∎

The well-known proof may be found in Yosida [1, p. 282].

The notion of the adjoint of a map T will also be needed when T is unbounded. Thus let $\mathscr{D}(T)$ be a linear subspace of X which is *dense* in X (i.e. $\overline{\mathscr{D}(T)} = X$) and let $T: \mathscr{D}(T) \to Y$ be linear. Let $\mathscr{D}(T^*) = \{g \in Y^*$: there exists $f \in X^*$ such that $(f,x) = (g, Tx)$ for all $x \in \mathscr{D}(T)\}$; the adjoint T^* of T is the map $T^*: \mathscr{D}(T^*) \to X^*$ defined by $T^* g = f$, i.e., $(T^* g, x) = (g, Tx)$ for all $x \in \mathscr{D}(T)$ and all $g \in \mathscr{D}(T^*)$. Note that it is essential that $\mathscr{D}(T)$ be dense in X for T^* to be well-defined. A more detailed discussion of self-adjoint maps in a Hilbert space will be given in §III.4.

Next, the results of the Fredholm–Riesz–Schauder theory of compact linear maps will be given; this theory extends in a most direct way the theory of linear maps in finite-dimensional spaces. The complete picture follows from a series of auxiliary results, a number of which are of interest in their own right. Throughout the discussion X will stand for a non-trivial (that is $\neq \{0\}$) Banach space, I will be the identity map from X to X; and given any $T \in \mathscr{B}(X)$ and any $\lambda \in \mathbb{C}$, we shall write T_λ for $T - \lambda I$. The notions of the *resolvent set* and the *spectrum* of a linear map will also be needed; these will be explained in terms of a linear map S from a linear subspace $\mathscr{D}(S)$ of X to X. The resolvent set, $\rho(S)$, of S is defined to be $\{\lambda \in \mathbb{C} : (S - \lambda I)^{-1}$ exists and belongs to $\mathscr{B}(X)\}$; $\mathbb{C} \setminus \rho(S)$ is called the spectrum of S and is written as $\sigma(S)$. Three disjoint subsets of $\sigma(S)$ are distinguished: the *point spectrum* $\sigma_p(S) := \{\lambda \in \sigma(S) : S - \lambda I$ is not injective$\}$, the *continuous spectrum* $\sigma_c(S) := \{\lambda \in \sigma(S)$: $S - \lambda I$ is injective, $(S - \lambda I)\, \mathscr{D}(S)$ is dense in X but not equal to $X\}$ and the *residual spectrum* $\sigma_r(S) := \{\lambda \in \sigma(S) : S - \lambda I$ is injective, $(S - \lambda I)\, \mathscr{D}(S)$ is not dense in $X\}$. The elements of $\sigma_p(S)$ are, of course, *eigenvalues* of S. In general the three subsets of the spectrum given above do not exhaust $\sigma(S)$; it is conceivable that $(S - \lambda I)^{-1}$ could exist, have domain X and yet be unbounded. However, this pathology cannot occur in an important case, namely, when S is *closed*: recall that a linear map T from a linear subspace $\mathscr{D}(T)$ of X to a normed vector space Y is said to be closed if its *graph* $\mathscr{G}(T) := \{\{x, Tx\} : x \in \mathscr{D}(T)\}$ is a closed subset of the product space $X \times Y$, when $X \times Y$ is given the norm $\|\bullet\|$ defined by

$$\|\{x, y\}\| = (\|x\|^2 + \|y\|^2)^{\frac{1}{2}}.$$

It is easy to see that T is closed if, and only if, given any sequence (x_n) in $\mathscr{D}(T)$ such that $x_n \to x$ and $Tx_n \to y$, it follows that $x \in \mathscr{D}(T)$ and $Tx = y$. For more details of these topics, Kato[1, III-5, 6] may be consulted. Finally we write $\mathscr{N}(T)$ for the *kernel* of T, i.e. $\{x \in \mathscr{D}(T) : Tx = 0\}$, and set nul $T = \dim \mathscr{N}(T)$; nul T is the *nullity* of T.

Theorem 1.7. Let $T \in \mathscr{K}(X)$ and suppose that $\lambda \in \mathbb{C} \setminus \{0\}$. Then:

(i) if $\mathscr{R}(T_\lambda) = X$ then T_λ is injective and $T_\lambda^{-1} \in \mathscr{B}(X)$;

(ii) $\mathscr{R}(T_\lambda)$ is closed;

(iii) if T_λ is injective, $T_\lambda^{-1} \in \mathscr{B}(X)$;

(iv) either $\lambda \in \rho(T)$, in which case $\mathscr{R}(T_\lambda) = X$, or $\lambda \in \sigma_p(T)$, in which case $\mathscr{R}(T_\lambda)$ is a proper closed linear subspace of X;

(v) nul $T_\lambda < \infty$. ∎

Definition 1.8. Let $T \in \mathscr{B}(X)$ and suppose that $\lambda \in \sigma_p(T)$. The *geometric multiplicity* of λ is defined to be nul T_λ. ∎

Theorem 1.9. Let $T \in \mathscr{K}(X)$. Then $\sigma_p(T)$ is at most countable and has no accumulation point except possibly 0. Each point of $\sigma(T) \setminus \{0\}$ is an eigenvalue of finite geometric multiplicity. ∎

It is worth remarking that if $\dim X = \infty$ then $0 \in \sigma(T)$ for all $T \in \mathscr{K}(X)$. To see this, suppose that $0 \in \rho(T)$ for some $T \in \mathscr{K}(X)$; then $T^{-1} \in \mathscr{B}(X)$ and so $I = TT^{-1}$ is compact. Theorem 1.3 now gives the desired contradiction.

At this stage the adjoint T^* of T makes an important entrance. If $T \in \mathscr{K}(X)$, it has already been observed that $T^* \in \mathscr{K}(X^*)$; and it is easy to see that $\sigma(T) = \overline{\sigma(T^*)} := \{\bar{\lambda} : \lambda \in \sigma(T^*)\}$, compactness not being needed for this latter result. It follows that if $\lambda \in \mathbb{C} \setminus \{0\}$, then $\lambda \in \sigma_p(T)$ if, and only if, $\bar{\lambda} \in \sigma_p(T^*)$.

Theorem 1.10. Let $T \in \mathscr{K}(X)$ and let $\lambda \in \mathbb{C} \setminus \{0\}$. Then nul T_λ = nul $T^*_{\bar{\lambda}}$. ■

These results may be combined to give the following theorem, generally known as the Fredholm Alternative Theorem:

Theorem 1.11. Let $T \in \mathscr{K}(X)$ and let $\lambda \in \mathbb{C} \setminus \{0\}$. Then the non-homogeneous equations

$$T_\lambda x = y, \tag{1.1}$$

$$T^*_{\bar{\lambda}} y^* = x^* \tag{1.2}$$

have unique solutions for any $y \in X$ and any $x^* \in X^*$ if, and only if, the homogeneous equations

$$T_\lambda x = 0, \tag{1.3}$$

$$T^*_{\bar{\lambda}} y^* = 0 \tag{1.4}$$

have only the zero solutions. If one of these homogeneous equations has a non-zero solution then they both have the same finite number of linearly independent solutions, and in this case (1.1) and (1.2) have solutions if, and only if, y and x^* are orthogonal to all the solutions of (1.4) and (1.3) respectively in the sense that $(y^*, y) = 0$ and $(x^*, x) = 0$ for all y^* satisfying (1.4) and all x satisfying (1.3). ■

The proofs of Theorems 1.7, 1.9, 1.10 and 1.11 can be found in Yosida [1, X-§5].

The next important concept is that of a Fredholm map, but before this can be given, some preliminary definitions are required.

Let X be a normed vector space and let $G \subset X$. The *annihilator* G^0 of G in X^* is defined by

$$G^0 = \{x^* \in X^* : (x^*, g) = 0 \text{ for all } g \in G\};$$

if $H \subset X^*$, then ${}^0H$, the annihilator of H in X, is defined by

$${}^0H = \{x \in X : (h, x) = 0 \text{ for all } h \in H\}.$$

If X is reflexive, and in particular if X is a Hilbert space, we shall often write $G^\perp$ in place of G^0; and because of the possibility of identification of X with X^{**} in this case we shall then use the notation $H^\perp$ instead of ${}^0H$, and $G^{\perp\perp}$ for ${}^0(G^0)$. Given any linear subspace G of a normed vector space X, we have $\bar{G} = {}^0(G^0)$: that $\bar{G} \subset {}^0(G^0)$ is obvious, while if $x \in {}^0(G^0) \setminus \bar{G}$, then by the Hahn–Banach Theorem there exists $x^* \in X^* \setminus \{0\}$ such that $(x^*, g) = 0$ for all $g \in \bar{G}$ and $(x^*, x) \neq 0$. Thus $x^* \in G^0$ and so $(x^*, x) = 0$. This contradiction proves our assertion. Similarly, if H is a linear subspace of X^*, it is clear that $\bar{H} \subset ({}^0H)^0$, but this time the inclusion may be proper if X is not reflexive. These results imply that G is dense in X if, and only if, $G^0 = \{0\}$; and H is dense in X^* if, and only if, ${}^0H = \{0\}$.

One way in which annihilators occur naturally is in connection with *quotient spaces*. Recall that if M is a linear subspace of a vector space X, the quotient space X/M of X by M is the set of all cosets $[x] := x+M \quad (x \in X)$; two cosets $[x]$ and $[y]$ coincide if, and only if, $x-y \in M$. The quotient X/M may be made into a vector space on defining the linear operations by

$$\lambda[x]+\mu[y] = [\lambda x+\mu y],$$

the zero element of X/M being M. The *codimension* of M in X is defined to be $\dim(X/M)$ and will sometimes be written as codim M. If M is a closed linear subspace of a normed vector space X, then X/M becomes a normed vector space when provided with the norm defined by

$$\|[x]\| = \inf\{\|y\| : y \in [x]\} = \operatorname{dist}(x, M) := \inf\{\|x-m\| : m \in M\};$$

if, in addition, X is a Banach space, so is X/M. Verification of these assertions will be found in Kato [1, III-§8]; so will the proof of the statement that if M is a closed linear subspace of a normed vector space, then

$$\operatorname{codim} M = \dim M^0, \qquad \operatorname{codim} M^0 = \dim M. \tag{1.5}$$

Let M and N be closed linear subspaces of a Banach space X, and suppose that $M \cap N = \{0\}$ and $X = M+N := \{m+n\colon m \in M,\ n \in N\}$, so that each element x of X has a unique decomposition of the form $x = y+z$, with $y \in M$ and $z \in N$. Then X is said to be the (topological) *direct sum* of M and N, and we write $X = M \dotplus N$. (The notation $X = M \oplus N$ will be reserved for the special case in which X is a Hilbert space and the subspaces M and N are orthogonal.) The subspaces M and N are said to be *complementary*, and $\dim(X/M) = \dim N$ (cf. Lang[1, VIII-§1]). It may be shown (cf. Taylor[1, §4.8]) that given a closed linear subspace M of a Banach space X, there is a complemntary subspace N (so that $X = M \dotplus N$) if, and only if, there is a projection $P \in \mathscr{B}(X)$ (i.e. a linear map P with $P^2 = P$) of X onto M. It is perfectly possible for there to be no closed subspaces complementary to a given closed subspace, but in important cases this is so, as the next theorem shows.

Theorem 1.12. Let X be a Banach space and let M be a closed linear subspace of X with finite dimension or finite codimension. Then M has a complementary closed linear subspace. ■

A proof of this theorem may be found in Lang[1, VIII-§1]. Improvements of the above result can be made by use of the following lemma.

Lemma 1.13. (Auerbach's Lemma) Let X be a Banach space with $\dim X = n < \infty$. Then there exist elements $x_1, x_2, \ldots, x_n$ in X, and elements $x_1^*, x_2^*, \ldots, x_n^*$ of X^*, such that $(x_i^*, x_j) = \delta_{ij}$ $(i, j = 1, 2, \ldots, n)$. ■

The proof of this is given in Pietsch [1, Part 0, B. 4], together with the following improvement of Theorem 1.12.

Theorem 1.14. Let X be a Banach space and let M be a linear subspace of X with $\dim M = m < \infty$. Then there is a projection $P \in \mathscr{B}(X)$ of X onto M such that $\| P \| \leqslant m^{\frac{1}{2}}$. If N is a closed linear subspace of X with $\operatorname{codim} N = n < \infty$, then given any $\varepsilon > 0$, there is a projection $Q \in \mathscr{B}(X)$ of X onto N such that $\| Q \| \leqslant (1+\varepsilon) n$. ■

We shall also need the following notions. A linear subspace M of X is called an *invariant subspace* of a linear operator T acting in X if T maps M into itself. If X_1 and X_2 are invariant subspaces of T and $X = X_1 \dotplus X_2$, then T is said to be *reduced* by the subspaces X_1 and X_2 and to have the *direct sum decomposition* $T = T_1 \dotplus T_2$, where T_i is the restriction of T to X_i $(i = 1, 2)$. If X is a Hilbert space and X_1 and X_2 are orthogonal we write $X = X_1 \oplus X_2$ and $T = T_1 \oplus T_2$, and call the sums *orthogonal sums*.

To round off our discussion of concepts related to the Fredholm maps which will soon be defined we mention the following theorem, which gives connections between annihilators, kernels, adjoints, and ranges. The proof is left as an exercise.

Theorem 1.15. Let X and Y be normed vector spaces and let $T \in \mathscr{B}(X, Y)$. Then:

(i) $[\overline{\mathscr{R}(T)}]^0 = \mathscr{N}(T^*)$;

(ii) $\overline{\mathscr{R}(T)} = {}^0\mathscr{N}(T^*)$;

(iii) $\overline{\mathscr{R}(T)} = Y$ if, and only if, T^* is injective;

(iv) ${}^0[\overline{\mathscr{R}(T^*)}] = \mathscr{N}(T)$;

(v) $\overline{\mathscr{R}(T^*)} \subset \mathscr{N}(T)^0$;

(vi) $\overline{\mathscr{R}(T^*)} = X^*$ only if T is injective;

(vii) if T is injective, $[{}^0\mathscr{R}(T^*)]^0 = X^*$. ■

After all this, it is at last convenient to introduce the notion of a Fredholm map.

Definition 1.16. Let X and Y be Banach spaces and let $T \in \mathscr{B}(X, Y)$. The map T is said to be a *Fredholm map* if, and only if, $\mathscr{R}(T)$ is closed and

$$\operatorname{nul} T < \infty, \qquad \operatorname{def} T := \operatorname{codim} \mathscr{R}(T) < \infty;$$

def T is called the *deficiency of* T. The *index* ind T of T is defined by

$$\operatorname{ind} T = \operatorname{nul} T - \operatorname{def} T.$$ ■

Examples. (i) Let $T\colon\ l^2 \to l^2$ be the left-shift operator defined by $Tx = (x_2, x_3, \ldots)$, where $x = (x_1, x_2, \ldots)$. Then nul $T = 1$, and since T is surjective, def $T = 0$. Hence T is a Fredholm map of index 1.

(ii) Let $S\colon\ l^2 \to l^2$ be the right shift operator given by $Sx = (0, x_1, x_2, \ldots)$ where $x = (x_1, x_2, \ldots)$. Here nul $S = 0$, def $S = 1$, and so S is a Fredholm map of index -1.

(iii) Let X be a Banach space and let $T \in \mathscr{K}(X)$. Then for any $\lambda \in \mathbb{C} \setminus \{0\}$, we have that T_λ ($= T - \lambda I$, I being the identity map of X to itself) is a Fredholm map of index 0. To verify this, note that by Theorem 1.7, $\mathscr{R}(T_\lambda)$ is closed and nul $T_\lambda < \infty$; and by (1.5), Theorem 1.15(i) and Theorem 1.10,

$$\operatorname{def} T_\lambda = \operatorname{codim} \mathscr{R}(T_\lambda) = \dim \mathscr{R}(T_\lambda)^0 = \operatorname{nul}(T_\lambda)^* = \operatorname{nul} T^*_{\bar{\lambda}} = \operatorname{nul} T_\lambda .$$

We shall have a lot more to say about Fredholm maps later on in this chapter.

To complete our discussion of compact linear operators in a Banach space setting the notion of the algebraic multiplicity of an eigenvalue is introduced.

Definition 1.17. Let X be a Banach space, let $T \in \mathscr{B}(X)$ and suppose that $\lambda \in \sigma_p(T)$. The linear subspace M_λ of X defined by

$$M_\lambda = \{x \in X : T^n_\lambda x = 0 \text{ for some } n \in \mathbb{N}\} = \bigcup_{n=1}^{\infty} \mathscr{N}(T^n_\lambda)$$

is called the *algebraic* (or *root*) *eigenspace corresponding to* λ, and non-zero elements of it are called *generalized eigenvectors* (or *root vectors*) *corresponding to* λ. The dimension of M_λ is called the *algebraic multiplicity of* λ. ■

It is clear that the algebraic multiplicity of an eigenvalue is greater than or equal to its geometric multiplicity. If the map is compact, a good deal more can be said about the algebraic multiplicity of eigenvalues, as is shown by the next results.

Lemma 1.18. Let X be a Banach space, let $T \in \mathscr{K}(X)$, and suppose that λ is a non-zero eigenvalue of T. Then there exists $r \in \mathbb{N}$ such that for all $n \in \mathbb{N}$ with $n \geqslant r$,

$$\mathscr{N}(T^n_\lambda) = \mathscr{N}(T^r_\lambda). \qquad ■$$

Proof. It is enough to deal with the case in which $\lambda = 1$, as otherwise we may divide through by λ. Thus suppose that $\lambda = 1$ and that the result is false. Then, there is a sequence $(\mathscr{N}(T^n_1))_{n \in \mathbb{N}}$ of subspaces of X such that each subspace is strictly contained in its successor. By Lemma 1.5, given any $n \in \mathbb{N}$ with $n \geqslant 2$, there exists $x_n \in \mathscr{N}(T^n_1)$ such that $\|x_n\| = 1$ and $\operatorname{dist}[x_n, \mathscr{N}(T^{n-1}_1)] \geqslant \frac{1}{2}$. Hence if $k \in \mathbb{N}$ and $k < n$,

$$\|Tx_n - Tx_k\| = \|x_n + T_1 x_n - x_k - T_1 x_k\| \geqslant 1,$$

because $T_1 x_n$ (and hence $-T_1 x_n + x_k + T_1 x_k$) lies in $\mathscr{N}(T^{n-1}_1)$. This contradicts the compactness of T and completes the proof. □

Remark 1.19. (i) In view of Lemma 1.18 there will be a *smallest* positive integer r such that $\mathscr{N}(T^n_\lambda) = \mathscr{N}(T^r_\lambda)$ for all $n \in \mathbb{N}$ with $n \geqslant r$. This smallest integer is called the *index* of λ (with respect to T). Note that if r is the index of λ, then the algebraic eigenspace corresponding to λ is simply $\mathcal{N}(T_\lambda^r)$. Moreover, since, by the binomial theorem and Proposition 1.2, we may write

$$T_\lambda^r = T' - \lambda^r(-1)^{r+1} I,$$

where $T' \in \mathcal{K}(X)$, it follows from Theorem 1.7 that nul $T_\lambda^r < \infty$. Hence every non-zero eigenvalue λ of a compact linear map T has *finite algebraic multiplicity* equal to nul T_λ^r, where r is the index of λ.

(ii) Under the hypotheses of Lemma 1.18 and with the index of λ denoted by r, it turns out that there is a positive integer s such that $\mathcal{R}(T_\lambda^s) = \mathcal{R}(T_\lambda^n)$ for all $n \geqslant s$, and that the smallest such integer s equals r. The proof of this is similar to that of Lemma 1.18. Suppose that there is no such positive integer s. Then $\mathcal{R}(T_\lambda^{n+1}) \subsetneqq \mathcal{R}(T_\lambda^n)$ for each $n \in \mathbb{N}$; and since, by Theorem 1.7, $\mathcal{R}(T_\lambda^n)$ is closed, it follows from Lemma 1.5 that given any $n \in \mathbb{N}$, there exists $x_n \in \mathcal{R}(T_\lambda^n)$, say $x_n = T_\lambda^n y_n$, such that $\|x_n\| = 1$ and $\|x_n - x\| \geqslant \frac{1}{2}$ for all $x \in \mathcal{R}(T_\lambda^{n+1})$. If $1 \leqslant m < n$, then

$$\begin{aligned} w := {} & x_n + \lambda^{-1} T_\lambda x_n - \lambda^{-1} T_\lambda x_m \\ = {} & T_\lambda^n y_n + \lambda^{-1} T_\lambda^{n+1} y_n - \lambda^{-1} T_\lambda^{m+1} y_m \in \mathcal{R}(T_\lambda^{m+1}); \end{aligned}$$

thus $\|x_m - w\| \geqslant \frac{1}{2}$. However,

$$\|Tx_m - Tx_n\| = |\lambda| \, \|x_m - w\| \geqslant \tfrac{1}{2}|\lambda|,$$

and the compactness of T is contradicted.

This establishes the existence of $s \in \mathbb{N}$, and it remains to prove that $s = r$. First we claim that T_λ is a bijection of $\mathcal{R}(T_\lambda^s)$ onto itself. To establish this, since $T_\lambda \mathcal{R}(T_\lambda^s) = \mathcal{R}(T_\lambda^{s+1}) = \mathcal{R}(T_\lambda^s)$, all we have to do is to prove that T_λ is injective on $\mathcal{R}(T_\lambda^s)$. If not, then there exists $x_1 \in \mathcal{R}(T_\lambda^s) \setminus \{0\}$ such that $T_\lambda x_1 = 0$; and as $T_\lambda \mathcal{R}(T_\lambda^s) = \mathcal{R}(T_\lambda^s)$ there is a sequence (x_n) such that $T_\lambda x_n = x_{n-1}$ for all $n \in \mathbb{N}$ with $n \geqslant 2$. Thus $T_\lambda^{n-1} x_n = x_1 \neq 0$ and $T_\lambda^n x_n = 0$ for all $n \geqslant 2$; hence $x_n \in \mathcal{N}(T_\lambda^n) \setminus \mathcal{N}(T_\lambda^{n-1})$ for all $n \geqslant 2$. This contradicts Lemma 1.18 and establishes our claim.

Now suppose that $x \in \mathcal{N}(T_\lambda^{s+1})$ and put $y = T_\lambda^s x$. Then $y \in \mathcal{R}(T_\lambda^s)$ and $T_\lambda y = 0$. Since T_λ is a bijection of $\mathcal{R}(T_\lambda^s)$ onto itself, this implies that $y = 0$ and hence $x \in \mathcal{N}(T_\lambda^s)$. Thus $\mathcal{N}(T_\lambda^{s+1}) \subset \mathcal{N}(T_\lambda^s)$, and so $\mathcal{N}(T_\lambda^{s+1}) = \mathcal{N}(T_\lambda^s)$. Hence $s \geqslant r$. Finally we prove that $s \leqslant r$. Let $x = T_\lambda^{s-1} y \in \mathcal{R}(T_\lambda^{s-1}) \setminus \mathcal{R}(T_\lambda^s)$ and $z = T_\lambda x = T_\lambda^s y \in \mathcal{R}(T_\lambda^s)$. Since T_λ^s is a bijection of $\mathcal{R}(T_\lambda^s)$ onto itself, there exists a unique $y' \in \mathcal{R}(T_\lambda^s)$ such that $T_\lambda^s y' = z$, and thus $y - y' \in \mathcal{N}(T_\lambda^s)$. Moreover, $T_\lambda^{s-1}(y - y') = x - T_\lambda^{s-1} y' \neq 0$ since $T_\lambda^{s-1} y' \in \mathcal{R}(T_\lambda^s)$ while $x \notin \mathcal{R}(T_\lambda^s)$. Hence $y - y' \in \mathcal{N}(T_\lambda^s) \setminus \mathcal{N}(T_\lambda^{s-1})$, and $s \leqslant r$. The proof is complete.

Theorem 1.20. Let $T \in \mathcal{K}(X)$, suppose that λ is a non-zero eigenvalue of T, and let r be the index of λ. Then

$$X = \mathcal{N}(T_\lambda^r) \dotplus \mathcal{R}(T_\lambda^r),$$

and each of the spaces in this direct sum is closed and invariant under T. If β is another non-zero eigenvalue of T, distinct from λ, and s is its index, then

$$\mathcal{N}(T_\beta^s) \subset \mathcal{R}(T_\lambda^r). \qquad \blacksquare$$

Proof. We recall that a subspace M of X is said to be *invariant* under T if $T(M) \subset M$. The remarks above make it clear that $\mathcal{R}(T_\lambda^r)$ is closed; that $\mathcal{N}(T_\lambda^r)$ is closed, and that $\mathcal{N}(T_\lambda^r)$ and $\mathcal{R}(T_\lambda^r)$ are invariant under T, is obvious. Suppose that $x \in \mathcal{N}(T_\lambda^r) \cap \mathcal{R}(T_\lambda^r)$. Then, since T_λ maps $\mathcal{R}(T_\lambda^r)$ bijectively onto itself, it follows from the fact that $T_\lambda^r x = 0$ that $x = 0$. Thus $\mathcal{N}(T_\lambda^r) \dotplus \mathcal{R}(T_\lambda^r)$ is a direct sum decomposition of some subspace of X, and since def $T_\lambda^r =$ nul T_λ^r (as ind $T_\lambda^r = 0$) it follows that $X = \mathcal{N}(T_\lambda^r) \dotplus \mathcal{R}(T_\lambda^r)$.

Now suppose that $\beta \in \sigma_p(T)$, $\beta \neq \lambda$, $\beta \neq 0$, and observe that, since $T_\lambda^r T_\beta^s = T_\beta^s T_\lambda^r$, the subspaces $\mathcal{N}(T_\lambda^r)$ and $\mathcal{R}(T_\lambda^r)$ are invariant under T_β^s. Let $x \in \mathcal{N}(T_\beta^s)$. Then there is a unique decomposition of x as $x = y + z$, with $y \in \mathcal{N}(T_\lambda^r)$ and $z \in \mathcal{R}(T_\lambda^r)$. Hence $0 = T_\beta^s x = T_\beta^s y + T_\beta^s z$, and since $T_\beta^s y \in \mathcal{N}(T_\lambda^r)$ and $T_\beta^s z \in \mathcal{R}(T_\lambda^r)$, it follows from the uniqueness of the decomposition that $T_\beta^s y = 0$. However, there are polynomials P and Q in T such that $PT_\beta^s + QT_\lambda^r = I$: to see that this is so, observe that $(\beta - \lambda)^{-1}(T_\lambda - T_\beta) = I$, so that for any $n \in \mathbb{N}$,

$$(\beta - \lambda)^{-n} \sum_{j=0}^{n} \binom{n}{j} T_\lambda^j (-T_\beta)^{n-j} = I.$$

Take $n = r + s$: then

$$I = (\beta - \lambda)^{-r-s} \sum_{j=0}^{r-1} \binom{r+s}{j} T_\lambda^j (-1)^{r+s-j} T_\beta^s T_\beta^{r-j}$$
$$+ (\beta - \lambda)^{-r-s} \sum_{j=r}^{r+s} \binom{r+s}{j} T_\lambda^r T_\lambda^{j-r} (-T_\beta)^{n-j},$$

and our assertion is proved. Thus

$$y = Iy = PT_\beta^s y + QT_\lambda^r y = 0.$$

Hence $x = z \in \mathcal{R}(T_\lambda^r)$, and the proof is complete. $\square$

Theorem 1.20 implies that given any $T \in \mathcal{K}(X)$ and any non-zero eigenvalue λ of T, with index r, there is a bounded projection P_λ of X onto $\mathcal{N}(T_\lambda^r)$, the (finite-dimensional) algebraic eigenspace corresponding to λ. In fact, writing $X = X_1 \dotplus X_2$, where $X_1 = \mathcal{N}(T_\lambda^r)$ and $X_2 = \mathcal{R}(T_\lambda^r)$, we see that $T_\lambda(X_i) \subset X_i$ and $T(X_i) \subset X_i$ for $i = 1, 2$; and given any $x \in X$, we have $P_\lambda x \in X_1$, $(I - P_\lambda)x \in X_2$, $P_\lambda T_\lambda P_\lambda x = T_\lambda P_\lambda x$, $P_\lambda T_\lambda (I - P_\lambda)x = 0$, $(I - P_\lambda) T_\lambda P_\lambda x = 0$ and $(I - P_\lambda) T_\lambda (I - P_\lambda) x = T_\lambda (I - P_\lambda) x$, so that

$$P_\lambda T_\lambda x = P_\lambda T_\lambda [P_\lambda x + (I - P_\lambda)x] = P_\lambda T_\lambda P_\lambda x = T_\lambda P_\lambda x;$$

that is, $P_\lambda T_\lambda = T_\lambda P_\lambda$. Also, $(I-P_\lambda)T_\lambda = T_\lambda(I-P_\lambda)$. Thus

$$P_\lambda T = TP_\lambda \quad \text{and} \quad (I-P_\lambda)T = T(I-P_\lambda).$$

Let $T^{(i)} = T \restriction X_i$, the restriction of T to X_i $(i = 1, 2)$. Then it is easy to see that $\sigma(T) = \sigma(T^{(1)}) \cup \sigma(T^{(2)})$; λ obviously belongs to $\sigma_p(T^{(1)})$; if $\mu \in \sigma_p(T)$, $\mu \neq 0$ and $\mu \neq \lambda$, then for some $x \in X \setminus \{0\}$ we have $Tx = \mu x$ and thus, by Theorem 1.20, $x \in \mathcal{N}(T_\mu) \subset \mathcal{R}(T_\lambda^r)$. Hence $x \notin X_1$ and $\mu \notin \sigma_p(T^{(1)}) = \sigma(T^{(1)})$: since dim $X_1 < \infty$ we have $0 \notin \sigma(T_1)$, because otherwise there would exist a non-zero element x of X_1 such that $Tx = 0$, and since $T_\lambda^r x = 0$, this would imply that $\lambda^r x = 0$, a contradiction.

The significance of all this is that $\sigma(T^{(1)}) = \{\lambda\}$ and $\sigma(T^{(2)}) = \sigma(T) \setminus \{\lambda\}$.

We shall return to this *spectral theory* of compact linear operators in Chapter II in a Hilbert-space setting, when a sharpening of the results is possible. Further details of the theory may be found in Gohberg and Krein [1] and Glazman [1].

2. Measures of non-compactness

As we have seen the compact linear maps possess strikingly good properties. It turns out to be fruitful to view arbitrary continuous linear maps from the point of view of their deviation from compact maps; one way in which this can be done is to use the notion of the measure of non-compactness of bounded subsets of a Banach space, and we now set about the description of this process.

Definition 2.1. Let A be a bounded subset of a metric space X. The *Kuratowski measure of non-compactness* $\psi_X(A)$ of A is defined by

$$\psi_X(A) = \inf\{\delta > 0\colon A \text{ can be covered by finitely many sets of diameter } \leqslant \delta\};$$

the *ball measure of non-compactness* $\tilde{\psi}_X(A)$ of A is defined by

$$\tilde{\psi}_X(A) = \inf\{\delta > 0\colon A \text{ can be covered by finitely many open balls of radius } \delta\}.$$

■

We shall often write ψ and $\tilde{\psi}$ in place of ψ_X and $\tilde{\psi}_X$, if X is understood to be the underlying space.

It is easy to see that for every bounded subset A of X, we have $\tilde{\psi}_X(A) \leqslant \psi_X(A) \leqslant 2\tilde{\psi}_X(A)$. However, the two measures of non-compactness are not the same, even when allowance is made for the obvious factor of 2. To illustrate this take $X = l^2$ and $A = \{e^{(i)} \in l^2 \colon i \in \mathbb{N}\}$, where the ith coordinate of $e^{(i)}$ is 1 and all other entries are zero. Since A has diameter equal to $\sqrt{2}$, we have $\psi(A) \leqslant \sqrt{2}$. However, given any open ball $B \subset l^2$ of radius $\frac{1}{2}\sqrt{2} + \varepsilon$, with $\varepsilon > 0$ small enough, $e^{(i)} \in B$ only if the centre of B has ith coordinate $\geqslant 1 - \frac{1}{2}\sqrt{2} - \varepsilon$; and evidently this can happen for finitely many indices i only. It follows that no finite set of balls of radius $\frac{1}{2}\sqrt{2}+\varepsilon$ can cover A, and hence $\tilde{\psi}(A) > \frac{1}{2}\sqrt{2}$. Thus $\psi(A) < 2\tilde{\psi}(A)$.

Despite this, ψ and $\tilde{\psi}$ have many similar properties, and we next establish some of the more important of these.

Lemma 2.2. Let A and B be bounded subsets of a metric space X. Then:
(i) $\psi(A) = 0$ if, and only if, A is precompact;
(ii) if $A \subset B$ then $\psi(A) \leqslant \psi(B)$;
(iii) $\psi(\bar{A}) = \psi(A)$;
(iv) $\psi(A \cup B) = \max\{\psi(A), \psi(B)\}$.
The same properties hold for $\tilde{\psi}$. ■

Proof. Parts (i) and (ii) are obvious. As for (iii), we see from (ii) that $\psi(A) \leqslant \psi(\bar{A})$; while if $A \subset \bigcup_{i=1}^{n} A_i$, then $\bar{A} \subset \bigcup_{i=1}^{n} \bar{A}_i$ and, since each $\bar{A}_i$ has the same diameter as A_i, we have $\psi(\bar{A}) \leqslant \psi(A)$; thus $\psi(\bar{A}) = \psi(A)$.

Finally, (ii) shows that $a := \max\{\psi(A), \psi(B)\} \leqslant \psi(A \cup B)$. Given any $\varepsilon > 0$, there are coverings of A and B by sets $A_1, A_2, \ldots, A_n$ and $B_1, B_2, \ldots, B_m$ respectively, where $\operatorname{diam} A_i \leqslant \psi(A)+\varepsilon \leqslant a+\varepsilon$ and $\operatorname{diam} B_j \leqslant a+\varepsilon$ for $i = 1, 2, \ldots, n$ and $j = 1, 2, \ldots, m$. Since the A_i and the B_j together form a covering of $A \cup B$, it follows that $\psi(A \cup B) \leqslant a+\varepsilon$ for each $\varepsilon > 0$. Hence $\psi(A \cup B) \leqslant a$, from which (iv) follows.

The proofs of (i)–(iv) for $\tilde{\psi}$ are similar. □

Property (i) above justifies the use of the term 'measure of non-compactness'.

The next result, due to Kuratowski [1], generalizes the elementary observation that in a complete metric space a decreasing sequence of compact non-empty sets has non-empty intersection.

Lemma 2.3. Let X be a complete metric space and let (A_n) be a decreasing sequence of non-empty closed subsets of X such that $\psi(A_n) \to 0$ as $n \to \infty$. Then $\bigcap_{n=1}^{\infty} A_n$ is non-empty and compact. ■

Proof. Since $\psi\left(\bigcap_{n=1}^{\infty} A_n\right) \leqslant \psi(A_m)$ for each $m \in \mathbb{N}$, we have $\psi\left(\bigcap_{n=1}^{\infty} A_n\right) = 0$; thus $\bigcap_{n=1}^{\infty} A_n$ is compact, because it is obviously closed. It remains to prove that it is non-empty.

For each $n \in \mathbb{N}$ put $\psi(A_n) = d_n$. We can write $A_n = \bigcup_{i=1}^{k(n)} A_{i,n}$, where $\operatorname{diam} A_{i,n} \leqslant d_n + n^{-1}$ for $1 \leqslant i \leqslant k(n)$. For some i with $1 \leqslant i \leqslant k(1)$ we have

$A_{i,1} \cap A_n \neq \emptyset$ for all $n \in \mathbb{N}$, since otherwise for some $n \in \mathbb{N}$,

$$A_n = A_1 \cap A_n = \left(\bigcup_{i=1}^{k(1)} A_{i,1} \right) \cap A_n = \emptyset.$$

Let B_1 be one of the sets $A_{i,1}$ such that $A_{i,1} \cap A_n \neq \emptyset$ for all $n \in \mathbb{N}$; to be precise, let us take B_1 to be that one of the $A_{i,1}$ with the smallest subscript i. This is the beginning of an inductive process. Suppose that we have sets $B_1, \ldots, B_m$ such that $\left(\bigcap_{i=1}^{m} B_i \right) \cap A_n \neq \emptyset$ for all $n \in \mathbb{N}$, and with $B_i = A_{j,i}$ for some j with $1 \leqslant j \leqslant k(i)$, for $1 \leqslant i \leqslant m$. Put $B = \bigcap_{i=1}^{m} B_i$. Our object is to determine a set $B_{m+1} = A_{j,m+1}$ for some j with $1 \leqslant j \leqslant k(m+1)$, such that for all $n \in \mathbb{N}$ we have $B \cap B_{m+1} \cap A_n \neq \emptyset$. If for each j with $1 \leqslant j \leqslant k(m+1)$ there is an $n_0 \in \mathbb{N}$ such that $B \cap A_{j,m+1} \cap A_{n_0} = \emptyset$, then for large enough n,

$$B \cap A_n = B \cap A_{m+1} \cap A_n = \bigcup_{j=1}^{k(m+1)} B \cap A_{j,m+1} \cap A_n = \emptyset,$$

contrary to the inductive hypothesis. Hence there exists j, with $1 \leqslant j \leqslant k(m+1)$, such that $B \cap A_{j,m+1} \cap A_n \neq \emptyset$ for all $n \in \mathbb{N}$. Taking the smallest such j we define $B_{m+1} = A_{j,m+1}$, and the inductive step is complete.

It follows that for each $n \in \mathbb{N}$ we have $\emptyset \neq \bigcap_{j=1}^{n} B_j \subset A_n$. For each $n \in \mathbb{N}$ take $x_n \in \bigcap_{j=1}^{n} B_j$. Since for each $m \in \mathbb{N}$,

$$\operatorname{diam} \left(\bigcap_{1}^{m} B_j \right) \leqslant \operatorname{diam} B_m \leqslant d_m + m^{-1},$$

we see that $d(x_j, x_k) \leqslant d_m + m^{-1}$ if $j, k \geqslant m$; thus (x_n) is a Cauchy sequence and there exists $x \in X$ such that $x_n \to x$ as $n \to \infty$. Since for all $n \geqslant m$, the element x_n belongs to the closed set A_m, we have $x \in A_m$ for all $m \in \mathbb{N}$. Hence $x \in \bigcap_{1}^{\infty} A_m$, and the proof is complete. □

Corollary 2.4. Lemma 2.3 holds with ψ replaced by $\tilde{\psi}$. ■

Proof. Since $\psi(B) \leqslant 2\tilde{\psi}(B)$ for every bounded subset B of X, the corollary follows immediately. □

Further results follow if we work in a normed vector space, rather than in a metric space. To give these we need a little more of the notation standard in functional analysis. Given any normed vector space X and any subsets A and B of X we write $A + B$ for $\{a + b: a \in A, b \in B\}$, co A for the intersection of all convex sets containing A, and $\overline{\text{co}}\, A$ for the intersection of all closed, convex sets containing A. The sets co A and $\overline{\text{co}}\,A$ are referred to respectively as the *convex hull* and the *closed convex hull* of A; evidently co A ($\overline{\text{co}}\,A$) is the smallest convex (closed convex) set containing A, and co $A = A$ ($\overline{\text{co}}\,A = A$) if, and only if, A is convex (convex and closed). A characterization of co A is easy to give: it is

$$\left\{\sum_{i=1}^{n} \lambda_i x_i : n \in \mathbb{N};\ x_1, \ldots, x_n \in A;\ \lambda_1, \ldots, \lambda_n \geqslant 0,\ \sum_{i=1}^{n} \lambda_i = 1\right\}.$$

Moreover, $\overline{\text{co}}\,A = \overline{\text{co}\,A}$. To prove this last statement simply note that $\overline{\text{co}\,A}$ is plainly closed and convex; thus $\overline{\text{co}\,A} \supset \overline{\text{co}}\,A$. Moreover, $\overline{\text{co}}\,A$ is closed and contains co A; hence $\overline{\text{co}\,A} \subset \overline{\text{co}}\,A$, and thus $\overline{\text{co}}\,A = \overline{\text{co}\,A}$.

Lemma 2.5. Let A and B be bounded subsets of a normed vector space X. Then $\psi(A+B) \leqslant \psi(A) + \psi(B)$ and $\psi(\text{co}\,A) = \psi(\overline{\text{co}}\,A) = \psi(A)$. Similar results hold for $\tilde{\psi}$. ■

Proof. Put $\psi(A) = d_1$ and $\psi(B) = d_2$, and let $\varepsilon > 0$. There are sets $S_1, \ldots, S_m$ and $T_1, \ldots, T_n$, with diam $S_i \leqslant d_1 + \frac{1}{2}\varepsilon$ and diam $T_j \leqslant d_2 + \frac{1}{2}\varepsilon$ ($i = 1, \ldots, m; j = 1, \ldots, n$) such that $A = \bigcup_{i=1}^{m} S_i$ and $B = \bigcup_{j=1}^{n} T_j$. Since $A + B = \bigcup_{i,j} (S_i + T_j)$, if we can prove that diam $(S_i + T_j) \leqslant d_1 + d_2 + \varepsilon$ it will follow that $\psi(A+B) \leqslant d_1 + d_2 + \varepsilon$. Given $x_1, x_2 \in S_i + T_j$ we may write $x_k = a_k + b_k$ ($k = 1, 2$) with $a_k \in S_i$ and $b_k \in T_j$; thus $\|x_1 - x_2\| \leqslant \|a_1 - a_2\| + \|b_1 - b_2\| \leqslant d_1 + d_2 + \varepsilon$. Hence $\psi(A+B) \leqslant d_1 + d_2 + \varepsilon$, and since this holds for all $\varepsilon > 0$, we see that $\psi(A+B) \leqslant \psi(A) + \psi(B)$.

To show that $\psi(\overline{\text{co}}\,A) = \psi(A)$ we first prove that given any bounded convex subsets C_1 and C_2 of X,

$$\psi(\text{co}(C_1 \cup C_2)) \leqslant \max\{\psi(C_1), \psi(C_2)\}. \tag{2.1}$$

To do this, observe that since C_1 and C_2 are convex,

$$\text{co}(C_1 \cup C_2) = \bigcup_{0 \leqslant \lambda \leqslant 1} [\lambda C_1 + (1-\lambda) C_2].$$

Next, since C_1 and C_2 are bounded, there exists $M > 0$ such that $\|x\| \leqslant M$ for all $x \in C_1 + C_2$. Let $\varepsilon > 0$ and let $\lambda_1, \ldots, \lambda_m$ be points of $[0, 1]$ such that if $\lambda \in [0, 1]$, then for some i with $1 \leqslant i \leqslant m$, we have $|\lambda - \lambda_i| < \frac{1}{2}\varepsilon/M$. Hence

$$\text{co}(C_1 \cup C_2) \subset \bigcup_{i=1}^{m} [\lambda_i C_1 + (1-\lambda_i) C_2 + \varepsilon B],$$

where B is the open unit ball in X. Thus by Lemma 2.2 and the first part of our proof,

$$\psi(\operatorname{co}(C_1 \cup C_2)) \leqslant \max_{1 \leqslant i \leqslant m} [\lambda_i \psi(C_1) + (1-\lambda_i)\psi(C_2) + 2\varepsilon]$$
$$\leqslant \max\{\psi(C_1), \psi(C_2)\} + 2\varepsilon.$$

Since ε may be choosen arbitrarily close to zero, (2.1) follows.

Finally, let $\delta > \psi(A)$. Then there exist sets $A_1, \ldots, A_k$, each with diameter $\leqslant \delta$, such that $A \subset \bigcup_{j=1}^{k} A_j$. Since diam A_j = diam co A_j, we may and shall assume that the A_j are convex. Put

$$B_1 = \operatorname{co}(A_1 \cup A_2), \qquad B_n = \operatorname{co}(B_{n-1} \cup A_n) \text{ for } 1 < n \leqslant k.$$

Then by (2.1)

$$\psi(B_k) \leqslant \max\{\psi(A_1), \ldots, \psi(A_k)\};$$

this, together with the observation that co $A \subset \operatorname{co}\left(\bigcup_1^k A_j\right) \subset B_k$, shows that

$$\psi(\operatorname{co} A) \leqslant \psi(B_k) \leqslant \max\{\psi(A_1), \ldots, \psi(A_k)\} \leqslant \delta.$$

Hence $\psi(\operatorname{co} A) \leqslant \psi(A)$, and since $A \subset \operatorname{co} A$ it follows that $\psi(A) = \psi(\operatorname{co} A)$. Since $\overline{\operatorname{co}}\, A = \overline{\operatorname{co} A}$, we finally obtain $\psi(\overline{\operatorname{co}}\, A) = \psi(\operatorname{co} A) = \psi(A)$.

The argument for $\tilde{\psi}$ is similar but easier, and is therefore omitted. □

Notice that if A is compact, the lemma gives the familiar result that $\overline{\operatorname{co}}\, A$ is also compact.

Given a bounded subset A of a normed vector space X there is not, in general, any simple way in which $\psi(A)$ or $\tilde{\psi}(A)$ may be calculated. However, if A is a ball or a sphere the position is clear. For example, let $A = \{x \in X : \|x\| = 1\}$. If dim $X < \infty$, then A is compact and thus $\psi(A) = 0$. If dim $X = \infty$, then clearly $\psi(A) \leqslant 2$. Suppose that $\psi(A) < 2$. Then there are closed sets $A_1, \ldots, A_n$, each having diameter less than 2, such that $A = \bigcup_{i=1}^{n} A_i$. Let Y be an n-dimensional linear subspace of X and consider $A \cap Y = \bigcup_{i=1}^{n} (A_i \cap Y)$. We now invoke the Ljusternick–Schnirel'mann–Borsuk Theorem (cf. Lloyd [1]): if the unit sphere in an n-dimensional normed vector space is covered by n closed sets, then at least one of these sets contains a pair of antipodal points. Thus $A_j \cap Y$ contains a pair of antipodal points for some j, and

$$2 \leqslant \operatorname{diam}(A_j \cap Y) \leqslant \operatorname{diam} A_j < 2.$$

This contradiction shows that $\psi(A) = 2$. The same argument shows that $\tilde{\psi}(A) = 1$ if dim $X = \infty$. □

We now turn to mappings.

Definition 2.6. Let X and Y be Banach spaces and let $k \geqslant 0$. A map $T: X \to Y$ is called a *k-set-contraction* if, and only if, it is continuous and for every bounded set $B \subset X$, we have $\psi_Y(T(B)) \leqslant k\psi_X(B)$; it is called a *k-ball-contraction* if, and only if, it is continuous and $\tilde{\psi}_Y(T(B)) \leqslant k\tilde{\psi}_X(B)$ for every bounded set $B \subset X$. The same terminology is used, with obvious modifications, when T is not everywhere defined in X. ■

Examples of such maps are plentiful. Thus if T is continuous and *compact*, in the sense that whenever $B \subset X$ is bounded, $\overline{T(B)}$ is compact, then evidently T is a 0-set-contraction. If T is such that $\|Tx - Ty\| \leqslant k\|x-y\|$ for all $x, y \in X$, then it is a k-set-contraction and a k-ball-contraction. Linear combinations of these two kinds of maps give further examples of k-set-contractions and k-ball-contractions, as a simple argument shows. The *radial retraction* $L: X \to X$ defined by

$$L(x) = \begin{cases} x \text{ if } \|x\| \leqslant R, \\ Rx/\|x\| \quad \text{if } \|x\| > R, \end{cases}$$

is a 1-set-contraction, since if B is a bounded subset of X, then $L(B) \subset \overline{\text{co}}\,(B \cup \{0\})$, and hence

$$\psi(L(B)) \leqslant \psi(\overline{\text{co}}\,(B \cup \{0\})) = \psi(B \cup \{0\}) = \psi(B).$$

Illustrations of the useful properties of k-set-contractions and k-ball-contractions will be found in the exercises; we mention in particular the analogue of Schauder's fixed point theorem which holds for all such maps. Here, however, our primary interest lies with *linear* maps as we have applications to linear partial differential equations in mind, and it is to these maps that we now turn.

Definition 2.7. Let X and Y be Banach spaces and let $T \in \mathscr{B}(X, Y)$. We define $\beta_{X,Y}(T)$ to be $\inf\{k: T \text{ is a } k\text{-set-contraction}\}$; $\tilde{\beta}_{X,Y}(T)$ is defined to be $\inf\{k: T \text{ is a } k\text{-ball-contraction}\}$. Usually we shall write $\beta(T)$, and $\tilde{\beta}(T)$ in place of $\beta_{X,Y}(T)$ and $\tilde{\beta}_{X,Y}(T)$ respectively. ■

If $T \in \mathscr{B}(X, Y)$, it is clear that $\beta(T), \tilde{\beta}(T) \leqslant \|T\|$. These inequalities may well be strict; for if $T \in \mathscr{K}(X, Y) \setminus \{0\}$, then $\beta(T) = \tilde{\beta}(T) = 0$ but $\|T\| > 0$.

The following lemma gives various elementary properties of $\beta(T)$ and $\tilde{\beta}(T)$.

Lemma 2.8. Let X, Y, Z be Banach spaces and let $T \in \mathscr{B}(X, Y)$. Then

(i) $\frac{1}{2}\beta(T) \leqslant \tilde{\beta}(T) \leqslant 2\beta(T)$; and $T \in \mathscr{K}(X, Y)$ if, and only if, $\beta(T) = 0$;
(ii) $\tilde{\beta}(T) = \tilde{\psi}(T(B))$, where B is the open unit ball in X;
(iii) if $S \in \mathscr{K}(X, Y)$, then $\beta(T+S) = \beta(T)$ and $\tilde{\beta}(T+S) = \tilde{\beta}(T)$;
(iv) if $U \in \mathscr{B}(Y, Z)$, then $\beta(UT) \leqslant \beta(U)\beta(T)$ and $\tilde{\beta}(UT) \leqslant \tilde{\beta}(U)\tilde{\beta}(T)$. ■

Proof. (i) This follows directly from the inequality $\tilde{\psi} \leqslant \psi \leqslant 2\tilde{\psi}$ mentioned earlier.

(ii) If $\dim X < \infty$ the result is obvious. Suppose that $\dim X = \infty$. Then $\tilde{\beta}(T) = \sup\{\tilde{\psi}(T(B')) : B' \text{ bounded}, \tilde{\psi}(B') = 1\} \geqslant \tilde{\psi}(T(B))$. Moreover, if B' is bounded and $\tilde{\psi}(B') = 1$, then, given any $\varepsilon > 0$, there exist $n \in \mathbb{N}$ and points $x_1, x_2, \ldots, x_n \in X$ such that $B' \subset \bigcup_{i=1}^{n} B(x_i, 1+\varepsilon)$. Thus $T(B') \subset \bigcup_{i=1}^{n} T(B(x_i, 1+\varepsilon))$ and

$$\tilde{\psi}(T(B')) \leqslant \max_{1 \leqslant i \leqslant n} \tilde{\psi}(T(B(x_i, 1+\varepsilon))) = (1+\varepsilon)\tilde{\psi}(T(B)).$$

Hence $\tilde{\psi}(T(B')) \leqslant \tilde{\psi}(T(B))$, and it is now immediate that $\tilde{\psi}(T(B)) = \tilde{\beta}(T)$.

Parts (iii) and (iv) are entirely elementary, and are left to the reader. □

Next we deal with the relationships between T and T^* from the standpoint of measures of non-compactness. The first result generalizes Theorem 1.6, which says that T is compact if, and only if, T^* is compact.

Theorem 2.9. Let X and Y be Banach spaces (real or complex) and let $T \in \mathscr{B}(X, Y)$. Then $\beta(T^*) \leqslant \tilde{\beta}(T)$ and $\beta(T) \leqslant \tilde{\beta}(T^*)$. ■

Proof. For the first part it is enough to prove that if T is a k-ball-contraction for some k, then T^* is a k-set-contraction with the same k; and to do this it is sufficient to show that given any set $S \subset Y^*$ with $\operatorname{diam} S \leqslant d$, $d > 0$, and given any $\varepsilon > 0$, then $T^*(S)$ can be covered by finitely many sets, each with diameter $\leqslant kd + \varepsilon$.

Suppose first that X and Y are real, and let $B = \{x \in X : \|x\| \leqslant 1\}$. Since $\tilde{\psi}(B) \leqslant 1$, the set $T(B)$ can be covered by a finite number of balls $B(y_1), \ldots, B(y_n)$, with centres at $y_1, \ldots, y_n$ and with radius $k + \frac{1}{2}\varepsilon/d$. Since S is bounded it follows that for $j = 1, \ldots, n$, the set $\{y^*(y_j) : y^* \in S\} \subset \mathbb{R}$ is relatively compact and can therefore be covered by a finite number of closed intervals $I_{j,1}, \ldots, I_{j,M(j)}$, each of length $\leqslant \frac{1}{2}\varepsilon$. Let $p = (p_1, \ldots, p_n)$, where each $p_j \in \{1, 2, \ldots, M(j)\}$, and put

$$E_p = \{y^* \in S : y^*(y_j) \in I_{j,p_j} \quad \text{for} \quad j = 1, 2, \ldots, n\}.$$

Then $T^*(S) \subset \bigcup_p T^*(E_p)$, and since this is a finite union the proof will be complete if we can show that $\operatorname{diam} T^*(E_p) \leqslant kd + \varepsilon$ for all p.

To do this let p be fixed and take any $y_1^*, y_2^* \in E_p$. Then

$$\begin{aligned}\|T^*(y_1^* - y_2^*)\| &= \sup\{|(y_1^* - y_2^*, Tx)| : x \in B\} \\ &= \sup\{|(y_1^* - y_2^*, y)| : y \in T(B)\}.\end{aligned}$$

If $y \in T(B)$, then $y \in B(y_i)$ for some $i \in \{1, 2, \ldots, n\}$; since $y_1^*, y_2^* \in E_p$ we have that $|(y_1^* - y_2^*, y_i)| < \frac{1}{2}\varepsilon$, and since $\|y_1^* - y_2^*\| \leqslant d$ and $\|y - y_i\| \leqslant k + \frac{1}{2}\varepsilon/d$, it follows that $|(y_1^* - y_2^*, y - y_i)| \leqslant kd + \frac{1}{2}\varepsilon$. Thus for all $y \in T(B)$,

$$\begin{aligned} |(y_1^* - y_2^*, y)| &\leqslant |(y_1^* - y_2^*, y - y_i)| + |(y_1^* - y_2^*, y_i)| \\ &\leqslant kd + \varepsilon. \end{aligned}$$

Hence $\|T^*(y_1^* - y_2^*)\| \leqslant kd + \varepsilon$, which implies that diam $T^*(E_p) \leqslant kd + \varepsilon$, and completes the proof of the first part of the theorem when the spaces are real. When the spaces are complex an obvious modification of the proof suffices; we use rectangles instead of intervals.

To complete the proof, suppose that T^* is a k-ball-contraction. Then by what we have proved, T^{**} is a k-set-contraction. Let J_X and J_Y be the natural embeddings of X and Y in X^{**} and Y^{**} respectively (cf. Taylor [1, §4.31]), and let $A \subset X$ be bounded. Since J_X is an isometry,

$$\psi\big(T^{**}(\mathrm{J}_X(A))\big) \leqslant k\psi(\mathrm{J}_X(A)) = k\psi(A),$$

and since $T^{**}\mathrm{J}_X = \mathrm{J}_Y T$, we see that $\psi\big(\mathrm{J}_Y T(A)\big) \leqslant k\psi(A)$. However, J_Y is also an isometry, and thus $\psi\big(T(A)\big) \leqslant k\psi(A)$. The proof is complete. □

Corollary 2.10. Let X and Y be Banach spaces and let $T \in \mathscr{B}(X, Y)$. Then

$$\tfrac{1}{2}\tilde{\beta}(T) \leqslant \tilde{\beta}(T^*) \leqslant 2\tilde{\beta}(T), \qquad \tfrac{1}{2}\beta(T) \leqslant \beta(T^*) \leqslant 2\beta(T). \qquad \blacksquare$$

Proof. By Theorem 2.9 and Lemma 2.8 (i),

$$\tfrac{1}{2}\tilde{\beta}(T) \leqslant \beta(T) \leqslant \tilde{\beta}(T^*) \leqslant 2\beta(T^*) \leqslant 2\tilde{\beta}(T) \leqslant 4\beta(T). \qquad \square$$

These results can be greatly improved in a Hilbert–space setting: it turns out as a consequence of the next chain of theorems that $\beta(T) = \tilde{\beta}(T) = \beta(T^*) = \tilde{\beta}(T^*)$.

Theorem 2.11. Let H be a Hilbert space and let $T \in \mathscr{B}(H)$. Then $\tilde{\beta}(T) = \tilde{\beta}(T^*)$. ■

Proof. Since $T = T^{**}$ (cf. Taylor [1, §4.9]) it is enough to prove that $\tilde{\beta}(T^*) \leqslant \tilde{\beta}(T)$. Suppose that T is a k-ball-contraction for some $k \geqslant 0$. We follow the same line of argument as in the proof of Theorem 2.9 and first suppose that H is real with inner product $(\bullet, \bullet)$, the complex case being handled similarly. We simply have to prove that given any ball $\mathrm{B}(z, d)$ in H, and any $\varepsilon > 0$, then $T^*\big(\mathrm{B}(z, d)\big)$ can be covered by finitely many balls of radius $\leqslant kd + \varepsilon$. Let B be the closed unit ball in H; the set $T(B)$ can be covered by finitely many open balls, $\mathrm{B}(y_1, k + \frac{1}{2}\varepsilon d^{-1}), \ldots, \mathrm{B}(y_n, k + \frac{1}{2}\varepsilon d^{-1})$, say. As in the proof of Theorem 2.9, we see that $T^*\big(\mathrm{B}(z, d)\big) \subset \bigcup_p T^*(E_p)$, where

$$E_p = \{y \in \mathrm{B}(z, d) : (y, y_j) \in I_{j, p_j} \quad \text{for} \quad j = 1, 2, \ldots, n\},$$

the I_{j,p_j} being closed intervals, each of length $\leqslant \frac{1}{2}\varepsilon$, such that for $j = 1, 2, \ldots, n$ the intervals $I_{j,1}, \ldots, I_{j,M(j)}$ cover $\{(y, y_j): y \in \mathrm{B}(z, d)\}$, and where $p = (p_1, p_2, \ldots, p_n)$, where each p_j belongs to $\{1, 2, \ldots, M(j)\}$.

The sets E_p are closed and convex; given any non-empty set E_p (and we shall discard any empty ones from now on) and $x \in H$, let $L_p x$ be the (unique) point in E_p nearest to x. The map L_p is Lipschitz–continuous, with constant 1, and thus for all $y \in E_p$,

$$\|y - L_p z\| \leqslant \|y - z\| \leqslant d.$$

Now we show that $T^*(E_p) \subset B(T^* L_p z, kd + \varepsilon)$. Given any $y \in E_p$,

$$\|T^*(y - L_p z)\| = \sup\{|(Tx, y - L_p z)| : x \in B\},$$

and since to each $x \in B$ there corresponds y_j $(1 \leqslant j \leqslant n)$ such that $\|Tx - y_j\| \leqslant k + \frac{1}{2}\varepsilon d^{-1}$, we have

$$\begin{aligned} |(Tx, y - L_p z)| &\leqslant |(Tx - y_j, y - L_p z)| + |(y - L_p z, y_j)| \\ &\leqslant (k + \tfrac{1}{2}\varepsilon d^{-1})d + \tfrac{1}{2}\varepsilon = kd + \varepsilon; \end{aligned}$$

thus $\|T^*(y - L_p z)\| \leqslant kd + \varepsilon$. It follows that $T^*(\mathrm{B}(z, d))$ can be covered by a finite number of balls of radius $kd + \varepsilon$, and the proof is complete. □

Corollary 2.12. Let H be a Hilbert space and let $T \in \mathscr{B}(H)$. Then $\beta(T) \leqslant \tilde{\beta}(T)$. ■

Proof. By Theorems 2.9 and 2.11, $\beta(T) \leqslant \tilde{\beta}(T^*) = \tilde{\beta}(T)$. □

Theorem 2.13. Let H be a Hilbert space and let $T \in \mathscr{B}(H)$. Then

$$\tilde{\beta}(T) = [\tilde{\beta}(T^* T)]^{1/2}.$$ ■

Proof. By Lemma 2.8 (iv) and Theorem 2.11,

$$\tilde{\beta}(T^* T) \leqslant \tilde{\beta}(T^*)\tilde{\beta}(T) = [\tilde{\beta}(T)]^2;$$

it is therefore enough to establish the reverse inequality. Let $k = \tilde{\beta}(T^* T)$; we shall prove that T is a $k^{\frac{1}{2}}$-ball-contraction. Let $\mathrm{B}(z, d)$ be an arbitrary ball in H and let $\varepsilon > 0$. There are points $y_1, \ldots, y_n$ in H such that $(T^* T)(B) \subset \bigcup_{j=1}^{n} \mathrm{B}(y_j, k + \frac{1}{2}\varepsilon d^{-1})$, and with exactly the same notation as in the proof of Theorem 2.11, we see that $T(\mathrm{B}(z, d)) \subset \bigcup_{p} T(E_p)$. For any $y \in E_p$,

$$\begin{aligned} \|T(y - L_p z)\|^2 &= (T^* T(y - L_p z), y - L_p z) \\ &\leqslant \|T^* T(y - L_p z)\| d \\ &\leqslant (kd + \varepsilon)d, \end{aligned}$$

the final inequality following just as in the corresponding stage of the proof of Theorem 2.11. Thus $\|T(y-L_p z)\| \leqslant k^{\frac{1}{2}}d+\sqrt{(\varepsilon d)}$, and so $T(\mathrm{B}(z,d))$ can be covered by finitely many balls of radius $k^{\frac{1}{2}}d+\sqrt{(\varepsilon d)}$: T is therefore a $k^{\frac{1}{2}}$-ball-contraction, and the proof is complete. □

To obtain further information it is convenient to establish the following general result.

Theorem 2.14. Let X, Y_1, Y_2, Y_3 be Banach spaces and let $T \in \mathscr{B}(X, Y_1)$, $F \in \mathscr{B}(X, Y_2)$, and $G \in \mathscr{B}(X, Y_3)$. Suppose that there are constants $p > 0$ and $q \geqslant 0$ such that for all $x \in X$,

$$\|Tx\|^p \leqslant \|Fx\|^q \|Gx\|^{p-q}.$$

Then

$$[\beta(T)]^p \leqslant [\beta(F)]^q [\beta(G)]^{p-q}.$$ ■

Proof. Let B be a bounded subset of X and let $\varepsilon > 0$. Evidently we can write $F(B) = \bigcup_{j=1}^{n} A_j$, where $\operatorname{diam} A_j \leqslant \beta(F)\psi(B)+\varepsilon$ for each j. Put $D_j = F^{-1}(A_j) \cap B$; the sets $G(D_1), \ldots, G(D_n)$ cover $G(B)$, and since

$$\psi(G(D_j)) \leqslant \beta(G)\,\psi(D_j) \leqslant \beta(G)\psi(B),$$

there is a covering of $G(B)$ by sets $E_1, \ldots, E_m$ such that each E_k is contained in some $G(D_j)$ and $\operatorname{diam} E_k \leqslant \beta(G)\psi(B)+\varepsilon$. Set $C_k = G^{-1}(E_k) \cap B$, and observe that $C_k \subset D_j$ for some j and $F(C_k) \subset F(D_j) \subset A_j$, while $\operatorname{diam} G(C_k) \leqslant \beta(G)\psi(B)+\varepsilon$ and $\operatorname{diam} F(C_k) \leqslant \beta(F)\psi(B)+\varepsilon$. Hence

$$\begin{aligned}[\operatorname{diam} T(C_k)]^p &\leqslant [\operatorname{diam} F(C_k)]^q [\operatorname{diam} G(C_k)]^{p-q} \\ &\leqslant [\beta(F)\psi(B)+\varepsilon]^q [\beta(G)\psi(B)+\varepsilon]^{p-q}.\end{aligned}$$

Since ε may be chosen arbitrarily close to 0 it follows that

$$[\psi(T(B))]^p \leqslant [\beta(F)\psi(B)]^q [\beta(G)\psi(B)]^{p-q},$$

which concludes the proof. □

Corollary 2.15. Let H be a Hilbert space and let $T \in \mathscr{B}(H)$. Then $\beta(T) = \beta(T^*)$ and $[\beta(T)]^2 = \beta(T^*T)$. ■

Proof. Since $\|Tx\|^2 = (x, T^*Tx) \leqslant \|x\| \; \|T^*Tx\|$ for all $x \in H$, Theorem 2.14 shows that $[\beta(T)]^2 \leqslant \beta(T^*T) \leqslant \beta(T^*)\beta(T)$. Application of this inequality to T^* gives $[\beta(T^*)]^2 \leqslant \beta(TT^*) \leqslant \beta(T)\beta(T^*)$. If either $\beta(T)$ or $\beta(T^*)$ is zero, then both are zero, by Theorem 1.6, and as T^*T is then compact, $\beta(TT^*) = 0$. If $\beta(T)\beta(T^*) \neq 0$, then we have $\beta(T) \leqslant \beta(T^*)$ and $\beta(T^*) \leqslant \beta(T)$. The corollary follows. □

This result is a useful one *en route* to our next goal, which is to show that if $T \in \mathscr{B}(H)$, then $\beta(T) = \tilde{\beta}(T)$. First we need another lemma.

Lemma 2.16. Let X be a Banach space and let $T \in \mathscr{B}(X)$. Then $\lim_{n\to\infty} [\beta(T^n)]^{1/n}$ and $\lim_{n\to\infty} [\tilde{\beta}(T^n)]^{1/n}$ exist and are equal. ■

Proof. Put $r = \inf\{[\beta(T^n)]^{1/n}: n \in \mathbb{N}\}$. Since $r \leqslant \liminf_{n\to\infty} [\beta(T^n)]^{1/n}$, if we can prove that $\limsup_{n\to\infty} [\beta(T^n)]^{1/n} \leqslant r$, it will follow that $\lim_{n\to\infty} [\beta(T^n)]^{1/n} = r$. Given any $\varepsilon > 0$, there exists $m \in \mathbb{N}$ such that $[\beta(T^m)]^{1/m} \leqslant r + \varepsilon$; given any $n \in \mathbb{N}$, there are unique integers p and q such that $n = pm + q$ with $0 \leqslant q \leqslant m-1$. Then

$$[\beta(T^n)]^{1/n} \leqslant [\beta(T^m)]^{p/n} [\beta(T)]^{q/n} \leqslant (r+\varepsilon)^{pm/n} [\beta(T)]^{q/n}.$$

Since $pm/n \to 1$ and $q/n \to 0$ as $n \to \infty$, we see that $\limsup_{n\to\infty} [\beta(T^n)]^{1/n} \leqslant r + \varepsilon$; and as ε may be chosen arbitrarily close to zero, $\limsup_{n\to\infty} [\beta(T^n)]^{1/n} \leqslant r$. Thus $\lim_{n\to\infty} [\beta(T^n)]^{1/n}$ exists and equals r.

In the same way it follows that $\lim_{n\to\infty} [\tilde{\beta}(T^n)]^{1/n}$ exists, and equals r', say. Since, by Lemma 2.8 (i), $\frac{1}{2}\beta(T) \leqslant \tilde{\beta}(T) = 2\beta(T)$, we see that $r = r'$. □

Theorem 2.17. Let H be a Hilbert space and let $T \in \mathscr{B}(H)$ be self-adjoint. Then $\beta(T) = \tilde{\beta}(T) = \lim_{n\to\infty} [\beta(T^n)]^{1/n}$. ■

Proof. By Corollary 2.15, $\beta(T^2) = [\beta(T)]^2$, and by induction we see that $\beta(T^{2^n}) = [\beta(T)]^{2^n}$ for all $n \in \mathbb{N}$. Thus $\beta(T) = \lim_{n\to\infty} [\beta(T^n)]^{1/n}$; and similarly, in virtue of Theorem 2.13, $\tilde{\beta}(T) = \lim_{n\to\infty} [\tilde{\beta}(T^n)]^{1/n}$. Now use Lemma 2.16. □

Theorem 2.18. Let H be a Hilbert space and let $T \in \mathscr{B}(H)$. Then

$$\beta(T) = \tilde{\beta}(T) = \lim_{n\to\infty} [\beta((T^*T)^n)]^{1/2n}. \quad ■$$

Proof. Apply Theorem 2.17 to the self-adjoint map T^*T: we find that

$$\beta(T^*T) = \tilde{\beta}(T^*T) = \lim_{n\to\infty} [\beta((T^*T)^n)]^{1/n}.$$

Since $\beta(T^*T) = [\beta(T)]^2$ and $\tilde{\beta}(T^*T) = [\tilde{\beta}(T)]^2$, by Theorem 2.13 and Corollary 2.15, the result follows. □

Theorem 2.7 can be improved, in that self-adjointness may be replaced by normality: recall that a map T is normal if $TT^* = T^*T$.

Theorem 2.19. Let H be a Hilbert space and let $T \in \mathscr{B}(H)$ be normal. Then $\beta(T) = \tilde{\beta}(T) = \lim_{n\to\infty} [\beta(T^n)]^{1/n}$. ■

Proof. Let $x \in H$. Then

$$\|T^2x\|^2 = (Tx, T^*T^2x) = (Tx, TT^*Tx) = \|T^*Tx\|^2,$$

and $\|Tx\|^2 \leqslant \|T^*Tx\|\,\|x\|$: thus $\|Tx\|^2 \leqslant \|T^2x\|\,\|x\|$. By Theorem 2.14, $[\beta(T)]^2 \leqslant \beta(T^2) \leqslant [\beta(T)]^2$, and consequently $[\beta(T)]^2 = \beta(T^2)$. Since all powers of a normal operator are normal it follows that $\beta(T^{2^n}) = [\beta(T)]^{2^n}$ for all $n \in \mathbb{N}$. The result is now obvious. □

Theorem 2.19 does not apply to all $T \in \mathscr{B}(H)$. For example, if we take $H = l^2$ and define $T: H \to H$ by $Tx = y$, where $x = (x_n)$ and $y = (y_n)$, with $y_{2n+1} = 0$ and $y_{2n} = x_{2n-1}$ for all n, then $T \in \mathscr{B}(H)$ and T is not compact: hence $\beta(T) > 0$. However, $T^2 = 0$, which implies that $\lim_{n\to\infty} [\beta(T^n)]^{1/n} = 0$.

To take stock of the position, we have shown that if $T \in \mathscr{B}(H)$ then $\beta(T) = \beta(T^*) = \tilde{\beta}(T) = \tilde{\beta}(T^*)$. We now extend our result so as to cope with mappings from one Hilbert space to another.

Theorem 2.20. Let H_1 and H_2 be Hilbert spaces over the same field and let $T \in \mathscr{B}(H_1, H_2)$. Then

$$\beta(T^*T) = [\beta(T)]^2, \qquad \tilde{\beta}(T^*T) = [\tilde{\beta}(T)]^2,$$

and

$$\beta(T) = \tilde{\beta}(T) = \beta(T^*) = \tilde{\beta}(T^*) = \lim_{n\to\infty} [\beta((T^*T)^n)]^{1/2n}. \qquad ■$$

Proof. Since T^*T is positive (i.e. $(T^*Tx, x) \geqslant 0$ for all $x \in H_1$) and self-adjoint it has a unique positive self-adjoint square root $G: H_1 \to H_1$ (cf. Kato [1, *V*-§3]). For all $x \in H_1$,.

$$\|Tx\|^2 = (x, T^*Tx) = (x, G^2x) = \|Gx\|^2,$$

and hence the map $U_1 : G(H_1) \to H_2$ defined by $U_1(Gx) = Tx$ is an isometry; let $U_2 : \overline{G(H_1)} \to H_2$ be the extension of U_1 by continuity. Now let $U: H_1 \to H_2$ be the extension of U_2 which is zero on $\overline{G(H_1)}^{\perp}$. Evidently $\|U\| \leqslant 1$ and $T = UG$. Thus $\tilde{\beta}(T) \leqslant \tilde{\beta}(U)\tilde{\beta}(G) \leqslant \tilde{\beta}(G)$, and since by Theorem 2.13, $\tilde{\beta}(G^2) = [\tilde{\beta}(G)]^2$, we obtain the inequality $[\tilde{\beta}(T)]^2 \leqslant \tilde{\beta}(T^*T)$. Also $\tilde{\beta}(T^*T) \leqslant \tilde{\beta}(T)\tilde{\beta}(T^*)$; thus $[\tilde{\beta}(T)]^2 \leqslant \tilde{\beta}(T)\,\tilde{\beta}(T^*)$ and $\tilde{\beta}(T) \leqslant \tilde{\beta}(T^*)$. If we replace

T by T^* in this last inequality we obtain $\tilde{\beta}(T^*) \leqslant \tilde{\beta}(T)$, from which it follows that $\tilde{\beta}(T) = \tilde{\beta}(T^*)$ and $\tilde{\beta}(T^*T) = [\tilde{\beta}(T)]^2$. In the same way we obtain the equality $\beta(T^*T) = [\beta(T)]^2$. By Theorem 2.17,

$$[\beta(T)]^2 = [\tilde{\beta}(T)]^2 = \beta(T^*T) = \lim_{n\to\infty} [\beta((T^*T)^n)]^{1/n}.$$

The proof is complete. □

For later use we introduce another way of estimating the departure from compactness of an operator. Let X and Y be Banach spaces and for any $T \in \mathscr{B}(X, Y)$ put

$$c(T) = \inf\{\varepsilon > 0\text{: there is a linear subspace } M \text{ of } X \text{ with } \operatorname{codim} M < \infty \text{ and such that } \|Tx\| \leqslant \varepsilon\|x\| \text{ for all } x \in M\}.$$

The function $c: \mathscr{B}(X, Y) \to \mathbb{R}$ has all the properties of a norm save that it may have a non-trivial kernel: such a function is called a *semi-norm*. To see that this is so it is enough to establish the triangle inequality, as the remaining property of semi-norms is obvious. Let $T_1, T_2 \in \mathscr{B}(X, Y)$ and let $\varepsilon_i > c(T_i)$ $(i = 1, 2)$. Note that it is sufficient to consider only *closed* subspaces M in the definition of $c(T)$, for if $\operatorname{codim} M < \infty$ and $\|Tx\| \leqslant \varepsilon\|x\|$ for all $x \in M$, then the same holds with $\overline{M}$ in place of M. Thus there are closed linear subspaces M_1 and M_2 of X, each with finite codimension, such that $\|T_i x\| \leqslant \varepsilon_i \|x\|$ for all $x \in M_i$ $(i = 1, 2)$; and by Theorem 1.12 there are finite-dimensional linear subspaces F_1 and F_2 of X such that $X = M_1 \dotplus F_1 = M_2 \dotplus F_2$. Put $F = F_1 + F_2$ and $M = M_1 \cap M_2$; then $\dim F \leqslant \dim F_1 + \dim F_2 < \infty$, $M \cap F = \{0\}$ and $M + F$ is a closed subspace of X. This last statement may be justified as follows. Let P_1 and P_2 be continuous projections of X onto F_1 and F_2 respectively and let $x \in \overline{M + F}$. Then there is a sequence (x_n) in $M + F$, with $x_n = y_n + z_n$ (where $y_n \in M$ and $z_n \in F$) for each $n \in \mathbb{N}$, such that $x_n \to x$. But $z_n = (P_1 + P_2)x_n \to (P_1 + P_2)x = z \in F$ and $y_n = (I - P_1 - P_2)x_n \to (I - P_1 - P_2)x = y \in M$; thus $x = y + z \in M + F$, and so $M + F$ is closed. Suppose that $M + F \neq X$. Then by the Hahn–Banach Theorem, there exists $f \in X^* \backslash \{0\}$ such that $f(M + F) = \{0\}$. Since $f \neq 0$ and $f(F_1) = \{0\}$, there exists $x \in M_1$ with $f(x) \neq 0$. But $x \notin M_2$, for otherwise $x \in M$ and $f(x) = 0$; thus $x \in F_2$ and again $f(x) = 0$. Hence $X = M \dotplus F$ and for all $x \in M$,

$$\|(T_1 + T_2)x\| \leqslant \|T_1 x\| + \|T_2 x\| \leqslant (\varepsilon_1 + \varepsilon_2)\|x\|.$$

It follows that $c(T_1 + T_2) \leqslant \varepsilon_1 + \varepsilon_2$, and from this the triangle inequality is immediate.

A connection between $c(T)$ and $\tilde{\beta}(T)$ is given by the next theorem.

Theorem 2.21. Let X and Y be Banach spaces and let $T \in \mathscr{B}(X, Y)$. Then

$$\tfrac{1}{2}\tilde{\beta}(T) \leqslant c(T) \leqslant 2\tilde{\beta}(T).$$ ■

Proof. Given any $\varepsilon > 0$ there are elements $y_1, \ldots, y_n$ of Y such that

$$\min\{\|Tx - y_k\| : k = 1, \ldots, n\} \leqslant \tilde{\beta}(T) + \varepsilon$$

for all $x \in X$ with $\|x\| \leqslant 1$. By the Hahn–Banach Theorem, there exist $y_1^*, \ldots, y_n^* \in Y^*$ such that for all $k \in \{1, 2, \ldots, n\}$, $\|y_k^*\| = 1$ and $y_k^*(y_k) = \|y_k\|$; let $M = {}^0\{T^*y_1^*, \ldots, T^*y_n^*\}$ and observe that codim $M = \dim M^0 = \dim ({}^0G)^0$, where G is the vector space spanned by the $T^*y_k^*$, of dimension $m \leqslant n$. Let $g_1, g_2, \ldots, g_m$ be a basis for G. If $f \in ({}^0G)^0$, then $f(x) = 0$ for all $x \in X$ such that $g_1(x) = \ldots = g_m(x) = 0$, and thus by Taylor [1, §3.5-C], $f \in G$. As G is plainly contained in $({}^0G)^0$, this shows that $G = ({}^0G)^0$, and hence codim $M = m < \infty$. Let $x \in M$ be such that $\|x\| \leqslant 1$, and let y_k be one of the elements $y_1, \ldots, y_n$ which is closest to Tx. Since $x \in M$, we have $(y_k^*, Tx) = (T^*y_k^*, x) = 0$, and so

$$\|y_k\| = (y_k^*, y_k) = (y_k^*, y_k - Tx) \leqslant \|y_k - Tx\| \leqslant \tilde{\beta}(T) + \varepsilon.$$

Thus

$$\|Tx\| \leqslant \|Tx - y_k\| + \|y_k\| \leqslant 2[\tilde{\beta}(T) + \varepsilon]$$

from which it follows that $c(T) \leqslant 2\tilde{\beta}(T)$.

To establish the remaining inequality, let $\varepsilon > 0$ and observe that there is a closed linear subspace M of X with codim $M < \infty$, such that for all $x \in M$,

$$\|Tx\| \leqslant [c(T) + \varepsilon]\|x\|.$$

Let P be a bounded projection of X onto M (cf. Theorem 1.14), so that $I - P$ is of finite rank. For any $x \in X$,

$$\begin{aligned}\|Tx\| &\leqslant \|TPx\| + \|T(I-P)x\| \\ &\leqslant [c(T) + \varepsilon]\|Px\| + \|T\|\,\|(I-P)x\| \\ &\leqslant [c(T) + \varepsilon]\|x\| + (2\|T\| + \varepsilon)\|(I-P)x\|,\end{aligned}$$

the last inequality following from the fact that $\|Px\| \leqslant \|x\| + \|(I-P)x\|$ and $c(T) \leqslant \|T\|$. By the compactness of $I - P$ there are elements $x_1, \ldots, x_n$ of the closed unit ball B of X such that

$$\min\{\|(I-P)(x - x_k)\| : k = 1, 2, \ldots, n\} < \varepsilon/(2\|T\| + \varepsilon)$$

for all $x \in B$. Given any $x \in B$ let x_k minimize the left-hand side above: then

$$\begin{aligned}\|T(x - x_k)\| &\leqslant [c(T) + \varepsilon]\|x - x_k\| + (2\|T\| + \varepsilon)\|(I-P)(x - x_k)\| \\ &\leqslant 2[c(T) + \varepsilon] + \varepsilon.\end{aligned}$$

Hence $\tilde{\beta}(T) \leqslant 2c(T)$, and the proof is complete. □

Corollary 2.22. Let $T \in \mathscr{B}(X, Y)$. Then $c(T) = 0$ if, and only if, $T \in \mathscr{K}(X, Y)$; and $c(T) = c(T + K)$ for all $K \in \mathscr{K}(X, Y)$. ■

Proposition 2.23. Let X, Y, Z be Banach spaces and let $S \in \mathscr{B}(X,Y)$ and $T \in \mathscr{B}(Y,Z)$. Then $\mathrm{c}(TS) \leqslant \mathrm{c}(T)\mathrm{c}(S)$. ■

Proof. Given any $\varepsilon > 0$, there is a closed linear subspace M of X, with codim $M < \infty$, such that $\|Sx\| \leqslant [\mathrm{c}(S)+\varepsilon]\,\|x\|$ for all $x \in M$; there is also a closed linear subspace N of Y, with codim $N < \infty$, such that $\|Ty\| \leqslant [c(T) + \varepsilon]\,\|y\|$ for all $y \in N$. Let $W = \{x \in M\colon Sx \in N\}$; then W is a closed linear subspace of X with codim $W < \infty$. This is because there is a finite-dimensional linear subspace F of Y such that $Y = N \dotplus F$, by Theorem 1.12; thus $X = S^{-1}(N) \dotplus (S^{-1}(F)\backslash[\mathscr{N}(S)\backslash\{0\}])$, and so codim $S^{-1}(N) < \infty$; hence as in the paragraph preceding Theorem 2.21, it follows that $W = S^{-1}(N) \cap M$ has finite codimension. Moreover,

$$\|TSx\| \leqslant [\mathrm{c}(T)+\varepsilon]\,[\mathrm{c}(S)+\varepsilon]\,\|x\|$$

for all $x \in W$, and thus

$$\mathrm{c}(TS) \leqslant [\mathrm{c}(T)+\varepsilon]\,[\mathrm{c}(S)+\varepsilon],$$

from which the proposition is immediate. □

Remark 2.24. The material in this section is drawn from the basic work of many authors, including Darbo [1], Goldenstein and Markus [1], Lebow and Schechter [1], and Nussbaum [1]; in particular, the results concerning the relationship between the measures of non-compactness of a map and that of its adjoint are based upon the papers of Stuart [1] and Webb [1, 2].

Exercises Throughout these exercises X and Y will stand for Banach spaces.

(1) For $i = 1, 2$, let $T_i\colon X \to Y$ be a k_i-set contraction. Prove that $T_1 + T_2$ is a $(k_1 + k_2)$-set contraction.

(2) Prove Darbo's fixed-point theorem: given any closed non-empty bounded convex subset C of X and any map $T\colon C \to C$ which is a k-set contraction for some $k \in [0, 1)$, then T has a fixed point in C. (Hint: define $C_1 = \overline{\mathrm{co}}\,T(C)$ and $C_{n+1} = \overline{\mathrm{co}}\,T(C_n)$ $(n \in \mathbb{N})$; use Lemma 2.3 and Schauder's fixed-point theorem.)

(3) Given any subset C of X, any map $S\colon C \to Y$ is said to be *condensing* if it is continuous (C is assumed to have the topology induced by that of X) and $\psi(S(B)) < \psi(B)$ for all bounded subsets B of C with $\psi(B) > 0$. Show that if (a) C is closed, non-empty, and bounded, (b) $T\colon C \to X$ is condensing, and (c) I is the identity map from X to itself, then $I - T$ is *proper*; that is, $(I-T)^{-1}(K)$ is compact whenever K is a compact subset of X.

(4) Let C be a non-empty closed bounded convex subset of X and let $T\colon C \to C$ be condensing. Prove that T has a fixed point. (Hint: let $x_0 \in C$; consider $T_t x = tTx + (1-t)x_0$ $(0 \leqslant t < 1)$; use Darbo's Theorem and Exercise 3.)

3. Fredholm and semi-Fredholm maps

Fredholm maps have already been introduced in §1; here we extend the notions and study the matter in more depth. Throughout this section, X and Y will stand for Banach spaces, I_X will be the identity map of X to itself (often written as I if no ambiguity is possible), and $\mathscr{C}(X,Y)$ will denote the family of all closed linear maps T with domain $\mathscr{D}(T)$—a linear subspace of X (differing from map to map in general)—and values in Y.

Definition 3.1. A map $T \in \mathscr{C}(X,Y)$ is said to be *semi-Fredholm* if $\mathscr{R}(T) := T\mathscr{D}(T)$ is closed and at least one of nul $T := \dim\{x \in \mathscr{D}(T) : Tx = 0\}$ and def $T := \dim Y/\mathscr{R}(T)$ is finite; it is called a *Fredholm map* if $\mathscr{R}(T)$ is closed and both nul T and def T are finite. If T is semi-Fredholm the *index* of T, ind T, is defined by

$$\text{ind } T = \text{nul } T - \text{def } T. \qquad \blacksquare$$

This definition of a Fredholm map is, of course, consistent with that given for *bounded* maps in §1. Earlier it was shown that if $K \in \mathscr{K}(X)$, then $I-K$ is a Fredholm map of index zero. Here the main task is to prove that the property of being Fredholm or semi-Fredholm is stable under suitable perturbations; this fact will be used later to link up the essential spectrum of a map with the measures of non-compactness introduced in §2. Before this, some remarks about maps with closed range may be in order. First, not all maps have closed range: the linear map $T \in \mathscr{B}(\mathrm{C}[0,1])$ defined by

$$(Tf)(x) = \int_0^x f(t)\,\mathrm{d}t \qquad (f \in \mathrm{C}[0,1],\ x \in [0,1])$$

is from many points of view an admirable one but its range contains only differentiable functions and so cannot be closed in $\mathrm{C}[0,1]$. In view of this, criteria are needed to enable one to say when a map does have closed range, and the next observations are directed to this question.

Theorem 3.2. Suppose that $T \in \mathscr{C}(X,Y)$ and that def $T < \infty$. Then $\mathscr{R}(T)$ is closed. $\blacksquare$

Proof. First assume that $T \in \mathscr{B}(X,Y)$. We may assume that T is injective, since if not we simply consider instead the map $\tilde{T}: X/\mathscr{N}(T) \to Y$ defined by $\tilde{T}\xi = Tx$, where x is any element in the equivalence class $\xi \in X/\mathscr{N}(T)$. Note that $\tilde{T}$ is bounded, for $\|\tilde{T}\xi\| = \|Tx\| \leqslant \|T\|\,\|x\|$ for all $x \in \xi$, and so $\|\tilde{T}\xi\| \leqslant \|T\|\,\|\xi\|$ for all $\xi \in X/\mathscr{N}(T)$. Let $n = \text{def } T$, let $\{y_1', \ldots, y_n'\}$ be a basis of $Y/\mathscr{R}(T)$ and let $y_1, \ldots, y_n$ be elements of Y such that for $i = 1, 2, \ldots, n$, the element y_i is mapped into y_i' by the natural map $Y \to Y/\mathscr{R}(T)$; put Z equal to $\operatorname{sp}\{y_1, \ldots, y_n\}$, the linear span of $\{y_1, \ldots, y_n\}$. Then $Z \cap \mathscr{R}(T) = \{0\}$ and $Z \dotplus \mathscr{R}(T) = Y$. Let $S: \mathbb{C}^n \to Z$ be a linear bijection and define $T_1: X \times \mathbb{C}^n \to Y$ by $T_1(x, y) = Tx + Sy$. It is clear that T_1 is a continuous bijection, and so by the Closed-Graph Theorem (cf. Taylor [1, §4.2 – I]), T_1 is a homeomorphism. Hence $\mathscr{R}(T) = T_1(X \times \{0\})$ is closed.

Now let $T \in \mathscr{C}(X, Y)$ and let $\mathscr{D}(T)$ be the domain of T. The idea is to give $\mathscr{D}(T)$ the *graph norm*; that is the norm $\|\bullet\|_T$ defined by

$$\|u\|_T = (\|u\|^2 + \|Tu\|^2)^{\frac{1}{2}} \quad (u \in \mathscr{D}(T)).$$

Since T is closed it is easy to see that $\mathscr{D}(T)$ endowed with the norm $\|\bullet\|_T$ is a Banach space; and of course when $\mathscr{D}(T)$ is given this norm, $T \in \mathscr{B}(\mathscr{D}(T), Y)$. Application of what has been proved already now shows that if def $T < \infty$, then $\mathscr{R}(T)$ is closed and the proof is complete. □

This technique of passage to the graph of a closed map is of great utility and will be used quite often in subsequent chapters.

The next theorem gives a characterization of closed maps with closed range. To explain it let $T \in \mathscr{C}(X, Y)$. Since T is closed it follows that $\mathscr{N}(T)$ is a closed linear subspace of X; thus $X/\mathscr{N}(T)$ is a Banach space when given the quotient norm. Let $\mathscr{D}(T)$ be the domain of T and define $\tilde{T}: \mathscr{D}(T)/\mathscr{N}(T) \to Y$ by $\tilde{T}\xi = Tu$, where u is any element of the equivalence class $\xi \in \mathscr{D}(T)/\mathscr{N}(T)$. The map $\tilde{T}$ is clearly linear; it is also closed. To see this, let (ξ_n) be a sequence in $\mathscr{D}(T)/\mathscr{N}(T)$ such that $\xi_n \to \xi$ and $\tilde{T}\xi_n \to y$ as $n \to \infty$. Let $u \in \xi$ and for each $n \in \mathbb{N}$ let $u_n \in \xi_n$; there is a sequence (v_n) in $\mathscr{N}(T)$ such that $u_n - u - v_n \to 0$ as $n \to \infty$, and since $T(u_n - v_n) = Tu_n \to y$ it follows from the closedness of T that $y = Tu = \tilde{T}\xi$. Thus $\tilde{T}$ is closed. It is plain that $\tilde{T}$ is invertible; that is, $\tilde{T}^{-1}: \mathscr{R}(T) \to X/\mathscr{N}(T)$ exists.

Definition 3.3. The *reduced minimum modulus* $\gamma(T)$ of $T \in \mathscr{C}(X, Y)$ is defined to be $\gamma(T) = \|\tilde{T}^{-1}\|^{-1}$, with the understanding that $\gamma(T) = 0$ if $\tilde{T}^{-1}$ is unbounded, and $\gamma(T) = \infty$ if $\tilde{T}^{-1} = 0$. ■

Note that $\gamma(T)$ is the largest number γ such that $\|Tx\| \geqslant \gamma\|\xi\|$ for all $\xi \in X/\mathscr{N}(T)$ and all $x \in \xi$. (The convention that $\infty \times 0 = 0$ is needed to make this apply to the case in which $\gamma(T) = \infty$, for then $T = 0$). Note also that if $\mathscr{N}(T) = \{0\}$, then $\gamma(T) = \inf\{\|Tx\|: x \in X, \|x\| = 1\}$, the so-called *minimum modulus* of T.

We can now give the promised characterization of closed maps with closed range.

Theorem 3.4. Let $T \in \mathscr{C}(X, Y)$. Then T has closed range if, and only if, $\gamma(T) > 0$. ■

Proof. Since $\gamma(T) > 0$ if, and only if, $\tilde{T}^{-1}$ is bounded, and since the domain of $\tilde{T}^{-1}$ is $\mathscr{R}(T)$ we see that if $\mathscr{R}(T)$ is closed then, by the Closed-Graph Theorem, $\tilde{T}^{-1}$ is bounded and $\gamma(T) > 0$. Conversely, if $\tilde{T}^{-1}$ is bounded then $\tilde{T}$ must have closed range; the reason is that if $Tu_n \to y \in Y$, then, with obvious notation, $\tilde{u}_n \to \tilde{z} \in X/\mathcal{N}(T)$, and so by the closedness of $\tilde{T}$, $y \in \mathscr{R}(T)$. □

Next, we examine the relationship between properties of a densely defined linear map T and those of its adjoint T^*. Let $T \in \mathscr{C}(X,Y)$ have domain $\mathscr{D}(T)$ dense in X, let $y \in Y$ and suppose that $Tx = y$ for some $x \in \mathscr{D}(T)$. Then for all $f \in \mathcal{N}(T^*)$,

$$(f, y) = (f, Tx) = (T^*f, x) = 0.$$

Hence the equation $Tx = y$ has a solution x only if $y \in {}^0[\mathcal{N}(T^*)]$. If this condition, that y should belong to ${}^0[\mathcal{N}(T^*)]$, is also *sufficient* for the equation $Tx = y$ to have a solution, T is said to be *normally soluble*. We formalize this in a definition.

Definition 3.5. Let $T \in \mathscr{C}(X,Y)$ have domain dense in X. Then T is said to be *normally soluble* if, and only if, $\mathscr{R}(T) = {}^0[\mathcal{N}(T^*)]$. ■

This terminology goes back to Hausdorff [1]. In fact, the normally soluble maps are nothing more than the densely defined maps with closed range.

Theorem 3.6. Let $\mathscr{D}(T)$ be a dense linear subspace of X, and let $T \in \mathscr{C}(X,Y)$ have domain $\mathscr{D}(T)$. Then T is normally soluble if, and only if, $\mathscr{R}(T)$ is closed. ■

Proof. If T is normally soluble it is clear that $\mathscr{R}(T)$ is closed. Conversely, suppose that $\mathscr{R}(T)$ is closed. We have already seen that $\mathscr{R}(T) \subset {}^0[\mathcal{N}(T^*)]$; to prove equality, let $y_0 \in Y \backslash \mathscr{R}(T)$ and define a functional f on the closed linear space spanned by $\mathscr{R}(T)$ and y_0 by $f(Tx + \lambda y_0) = \lambda \ \ (x \in \mathscr{D}(T))$. Evidently f is continuous and linear, and so by the Hahn–Banach Theorem there is an extension $F \in Y^*$ of f: $F \in \mathcal{N}(T^*)$ and $F(y_0) = 1$. Thus $y_0 \notin {}^0[\mathcal{N}(T^*)]$, and hence $\mathscr{R}(T) \supset {}^0[\mathcal{N}(T^*)]$. The proof is complete. □

This identity, for closed, densely defined linear maps, between normal solubility and the property of having closed range, lies fairly near the surface. Much deeper results connected with these ideas are known; we summarize them in the following theorem and refer the reader to the books of Kato [1, p. 234] and Yosida [1, p. 205] for the rather intricate proofs.

Theorem 3.7. Let $T \in \mathscr{C}(X,Y)$ have domain dense in X. Then the following statements are equivalent:

(i) $\mathscr{R}(T)$ is closed;
(ii) $\mathscr{R}(T^*)$ is closed;
(iii) T is normally soluble;
(iv) T^* is normally soluble.

If any one of these statements holds, then $\mathscr{R}(T)^0 = \mathscr{N}(T^*)$, $[\mathscr{N}(T)]^0 = \mathscr{R}(T^*)$, nul T = def T^*, and nul T^* = def T. The map T is semi-Fredholm if, and only if, T^* is semi-Fredholm; and in this case, ind $T = -$ ind T^*. ■

An important tool in the detection of normally soluble maps is the concept of *approximate inverse* of a map, and it is to this topic that we now turn.

Definition 3.8. Let $\mathscr{D}(T)$ be a linear subspace of X and let $T: \mathscr{D}(T) \to Y$ be linear. The map T is said to have a *left approximate inverse* if, and only if, there are maps $R_\ell \in \mathscr{B}(Y, X)$ and $K_X \in \mathscr{K}(X)$ such that $I_X + K_X$ extends $R_\ell T$. Similarly, T has a *right approximate inverse* if, and only if, there is a map $R_r \in \mathscr{B}(Y, X)$ such that $R_r(Y) \subset \mathscr{D}(T)$ and $TR_r - I_Y \in \mathscr{K}(Y)$. The maps R_ℓ and R_r are called *left and right approximate inverses of* T, respectively. ■

We next develop some properties of maps which have approximate inverses.

Proposition 3.9. Let $\mathscr{D}(T)$ be a dense linear subspace of X and let $T: \mathscr{D}(T) \to Y$ be linear. Then:

(i) if T has a left (right) approximate inverse, T^* has a right (left) approximate inverse;

(ii) if T has a right and a left approximate inverse, their difference is compact. ■

Proof. (i) Suppose that T has a left approximate inverse R_ℓ, so that $R_\ell T - I_X$ is the restriction to $\mathscr{D}(T)$ of a map K_X in $\mathscr{K}(X)$. Then $T^* R_\ell^* = I_X^* + K_X^*$, and so T^* has a right approximate inverse. The rest is now clear.

(ii) Let R_ℓ and R_r be left and right approximate inverses of T. Then there is a map $K_X \in \mathscr{K}(X)$ such that $I_X + K_X$ extends $R_\ell T$, and $TR_r - I_Y := K_Y \in \mathscr{K}(Y)$. Hence $R_r - R_\ell = R_\ell K_Y - K_X R_r \in \mathscr{K}(X, Y)$, by Proposition 1.2 □

Corollary 3.10. Let $T \in \mathscr{B}(X, Y)$ and suppose that T has left and right approximate inverses R_ℓ and R_r respectively. Then both R_ℓ and R_r are left and right approximate inverses of T. ■

Proof. From Proposition 3.9(ii) we see that $R_\ell - R_r$ is compact, and since the addition of a compact map to a left (or right) approximate inverse produces another left (or right) approximate inverse, it follows that $R_r = R_\ell + (R_r - R_\ell)$ is a left approximate inverse, and that $R_\ell = R_r + (R_\ell - R_r)$ is a right approximate inverse. □

We shall refer to a map which is both a left and a right approximate inverse of a map T as a *two-sided approximate inverse of* T.

Theorem 3.11. Let $\mathscr{D}(T)$ be a dense linear subspace of X, let $T: \mathscr{D}(T) \to Y$ be linear, and suppose that T has a left (right) approximate inverse. Then nul T (nul T^*) is finite. ■

Proof. Suppose that T has a left approximate inverse R_ℓ. Evidently $\mathcal{N}(T) \subset \mathcal{N}(R_\ell T)$, and so nul $T \leqslant$ nul $(R_\ell T)$. But $R_\ell T$ is extended by $I_X + K$ for some $K \in \mathcal{K}(X)$, and by Theorem 1.7 (v) we have nul $(I_X + K) < \infty$. Hence nul $T < \infty$.

If T has a right approximate inverse R_r then by Proposition 3.9 (i), T^* has a left approximate inverse, and thus by what has just been proved, nul $T^* < \infty$. Note that the density of $\mathcal{D}(T)$ is needed merely so that T^* is well-defined: it is not necessary that $\mathcal{D}(T^*)$ should be dense in X^*. □

Our object is now to prove that if a map $T \in \mathcal{C}(X,Y)$ has a right or left approximate inverse it has closed range. To do this we need a lemma.

Lemma 3.12. Let $T \in \mathcal{C}(X,Y)$. Then T has closed range and nul $T < \infty$ if, and only if, there exist a Banach space Y_1, a map $K \in \mathcal{K}(X,Y_1)$, and a positive constant C such that for all x in the domain $\mathcal{D}(T)$ of T,

$$\|x\| \leqslant C(\|Tx\| + \|Kx\|). \tag{3.1}$$ ■

Proof. First suppose that the inequality holds. Then for all $x \in \mathcal{N}(T)$, $\|x\| \leqslant C\|Kx\|$, which by the compactness of K implies that $\overline{\{x \in X: \|x\| \leqslant 1\} \cap \mathcal{N}(T)}$ is compact, and thus that $\mathcal{N}(T)$ is finite-dimensional; that is, nul $T < \infty$. To prove that $\mathcal{R}(T)$ is closed let $y \in \overline{\mathcal{R}(T)}$ and let (y_n) be a sequence in $\mathcal{R}(T)$ with $y_n \to y$ as $n \to \infty$. Then $y_n = Tx_n$ for some $x_n \in \mathcal{D}(T)$, and since nul $T < \infty$, the space X may be written as a topological direct sum, $X = \mathcal{N}(T) \dot{+} Z$ say, and we may suppose that $(x_n) \subset Z$. Suppose that (x_n) is unbounded: without loss of generality we may suppose that $\|x_n\| \to \infty$ as $n \to \infty$. Put $z_n = x_n/\|x_n\|$ $(n \in \mathbb{N})$: then $Tz_n \to 0$ as $n \to \infty$; and as $\|z_n\| = 1$ for all $n \in \mathbb{N}$, there is a subsequence of (z_n), denoted again by (z_n) for simplicity, such that (Kz_n) converges. In view of the inequality (3.1), (z_n) is thus a Cauchy sequence, converging to $z \in Z$, say. Clearly $\|z\| = 1$; since T is closed, it follows that $z \in \mathcal{D}(T)$ and $Tz = 0$. But since $z \in Z$ and $z \in \mathcal{N}(T)$, we have $z = 0$, which contradicts the fact that $\|z\| = 1$. Hence (x_n) must be bounded. Thus by the compactness of K and the convergence of (Tx_n), (3.1) shows that there is a subsequence of (x_n), yet again denoted by (x_n), which converges, to $x \in Z$, say. Since T is closed, $x \in \mathcal{D}(T)$ and $y = Tx \in \mathcal{R}(T)$. Hence $\mathcal{R}(T)$ is closed.

Conversely, suppose that $\mathcal{R}(T)$ is closed and that nul $T < \infty$. Write $X = \mathcal{N}(T) \dot{+} Z$ as above, let P be the corresponding projection of X on $\mathcal{N}(T)$, and denote by T_1 the restriction of T to Z: $\mathcal{N}(T_1) = \{0\}$ and $\mathcal{R}(T_1) = \mathcal{R}(T)$. Since T_1^{-1} is closed and has domain the Banach space $\mathcal{R}(T)$ it is continuous, by the Closed-Graph Theorem: thus there exists a constant $c > 0$ such that for all $z \in Z \cap \mathcal{D}(T)$,

$$\|z\| \leqslant c\|T_1 z\|.$$

Now let $x \in \mathscr{D}(T)$. Then $x = Px + (I - P)x = x_1 + x_2$, say, where $x_1 \in \mathscr{N}(T)$ and $x_2 \in Z$. Thus

$$\|x\| \leqslant \|Px\| + \|x_2\| \leqslant \|Px\| + c\|T_1 x_2\| = \|Px\| + c\|Tx\|,$$

and since P is compact, having finite-dimensional range, the proof is complete. □

Theorem 3.13. Let $T \in \mathscr{C}(X,Y)$. Then T is semi-Fredholm if either (i) T has a left approximate inverse, or (ii) the domain of T is dense in X and T has a right approximate inverse. ■

Proof. First suppose that T has a left approximate inverse R_ℓ, so that $R_\ell T = I_X + S$ for some $S \in \mathscr{K}(X)$. Apply Lemma 3.12 to $I_X + S$: there exist $K \in \mathscr{K}(X)$ and $C > 0$ such that for all $x \in X$,

$$\|x\| \leqslant C[\|(I_X + S)x\| + \|Kx\|].$$

Actually Lemma 3.12 shows that $K \in \mathscr{K}(X, Y_1)$ for some Y_1, but the proof of that lemma makes it plain that we may take K to be in $\mathscr{K}(X)$. Thus for all $x \in \mathscr{D}(T)$,

$$\|x\| \leqslant C(\|R_\ell\| \|Tx\| + \|Kx\|),$$

from which, by Lemma 3.12 again, the result follows.

Under condition (ii), T^* has a left approximate inverse by Proposition 3.9, and so by what has just been proved and the evident closedness of T^*, we see that T^* is semi-Fredholm. Now use Theorem 3.7. □

Definition 3.14. The set of all semi-Fredholm maps $T \in \mathscr{B}(X,Y)$ with nul $T < \infty$ (def $T < \infty$) will be denoted by $\mathscr{F}_+(X,Y)$ ($\mathscr{F}_-(X,Y)$), or by $\mathscr{F}_+(X)$ ($\mathscr{F}_-(X)$) if $X = Y$. The set of all Fredholm maps in $\mathscr{B}(X,Y)$ ($\mathscr{B}(X)$) will be denoted by $\mathscr{F}(X,Y)$ ($\mathscr{F}(X)$). ■

Theorem 3.15. Let $T \in \mathscr{B}(X,Y)$. Then the following statements are equivalent:

(i) $T \in \mathscr{F}(X,Y)$;

(ii) T has a two-sided approximate inverse;

(iii) there is a map $S \in \mathscr{B}(Y, X)$ such that $ST - I_X$ and $TS - I_Y$ are finite-dimensional. ■

Proof. First suppose that $T \in \mathscr{F}(X,Y)$. Then there are closed linear subspaces X_1 and Y_1 of X and Y respectively such that $X = \mathscr{N}(T) \dotplus X_1$, $Y = \mathscr{R}(T) \dotplus Y_1$. Take S to be the composite mapping given by

$$Y = \mathscr{R}(T) \dotplus Y_1 \overset{P}{\to} \mathscr{R}(T) \overset{T^{-1}}{\to} X_1 \overset{J}{\to} X,$$

where P is the natural projection and J is the inclusion map; by the Closed-Graph Theorem T^{-1} is continuous and hence $S = JT^{-1}P \in \mathscr{B}(Y, X)$. Then $I_Y - TS$ and $I_X - ST$ are projections on Y_1 and $\mathscr{N}(T)$ respectively, and since $\dim Y_1 = \operatorname{def} T < \infty$ and nul $T < \infty$, it follows that (i) implies (ii) and (iii). That (iii) implies (ii) is obvious, while Theorems 3.7, 3.11 and 3.13 show that (ii) implies (i). □

Theorem 3.16. Let X, Y, Z be Banach spaces, let $S \in \mathscr{F}(X, Y)$ and suppose that $T \in \mathscr{F}(Y, Z)$. Then $TS \in \mathscr{F}(X, Z)$ and

$$\operatorname{ind} TS = \operatorname{ind} T + \operatorname{ind} S. \qquad \blacksquare$$

Proof. By Theorem 3.15, there are maps $\tilde{S} \in \mathscr{B}(Y, X)$ and $\tilde{T} \in \mathscr{B}(Z, Y)$ such that $\tilde{S}S - I_X$, $S\tilde{S} - I_Y$, $\tilde{T}T - I_Y$ and $T\tilde{T} - I_Z$ are finite-dimensional. Hence $\tilde{S}\tilde{T}TS - I_X$ and $TS\tilde{S}\tilde{T} - I_Z$ are finite-dimensional, and so $TS \in \mathscr{F}(X, Z)$, by Theorem 3.15 again.

We now observe that

$$\operatorname{nul}(TS) = \dim[\mathscr{N}(TS)/\mathscr{N}(S)] + \dim \mathscr{N}(S)$$

and

$$\dim[Z/\mathscr{R}(TS)] = \dim[Z/\mathscr{R}(T)] + \dim[\mathscr{R}(T)/\mathscr{R}(TS)].$$

To proceed further we need the purely algebraic result that if W is a linear subspace of a linear space V and f is a linear map with domain V, then

$$\dim[V/W] = \dim[f(V)/f(W)] + \dim[\mathscr{N}(f)/W \cap \mathscr{N}(f)]$$

in the sense that if two of these quantities is finite then so is the third and the given relation holds. The proof of this is left as an exercise; but page 203 of Lang's book [1] may be consulted if necessary.

We now apply this result with $V = Y$, $W = \mathscr{R}(S)$ and $f = T$, giving

$$\dim[Y/\mathscr{R}(S)] = \dim[\mathscr{R}(T)/\mathscr{R}(TS)] + \dim[\mathscr{N}(T)/\mathscr{N}(T) \cap \mathscr{R}(S)].$$

Moreover,

$$\operatorname{nul}(T) = \dim[\mathscr{N}(T)/\mathscr{R}(S) \cap \mathscr{N}(T)] + \dim[\mathscr{R}(S) \cap \mathscr{N}(T)].$$

Use of these relations now gives

$$\begin{aligned}\operatorname{ind}(TS) - \operatorname{ind} T - \operatorname{ind} S &= \operatorname{nul}(TS) - \dim[Z/\mathscr{R}(TS)] - \operatorname{nul} T \\ &\quad + \dim[Z/\mathscr{R}(T)] - \operatorname{nul} S + \dim[Y/\mathscr{R}(S)] \\ &= \dim[\mathscr{N}(TS)/\mathscr{N}(S)] - \dim[\mathscr{R}(S) \cap \mathscr{N}(T)].\end{aligned}$$

There are linear subspaces X_1 and X_2 of X with $\dim X_1 < \infty$, such that

$$\mathscr{N}(TS) = \mathscr{N}(S) \dotplus X_1, \qquad X = \mathscr{N}(TS) \dotplus X_2,$$

and so

$$X = \mathscr{N}(S) \dotplus X_1 \dotplus X_2 \quad \text{and} \quad \mathscr{R}(S) = S(X_1) \dotplus S(X_2).$$

Evidently $S(X_1) \subset \mathcal{N}(T)$; also if $y \in S(X) \cap \mathcal{N}(T)$, so that $y = Sx$ and $TSx = 0$, then $x = u + v$ with $u \in \mathcal{N}(TS)$ and $v \in X_1$, and $Sx = Sv \in S(X_1)$. Thus $S(X) \cap \mathcal{N}(T) = S(X_1)$, and as S is an isomorphism on X_1,

$$\dim[\mathcal{N}(TS)/\mathcal{N}(S)] = \dim X_1 = \dim S(X_1) = \dim[\mathcal{R}(S) \cap \mathcal{N}(T)],$$

which completes the proof. □

With these results available we can now set about the matter of the stability of the index under perturbations of various kinds. We begin with compact perturbations.

Theorem 3.17. Let $T \in \mathcal{F}(X,Y)$ and suppose that $S \in \mathcal{K}(X,Y)$. Then $T+S \in \mathcal{F}(X,Y)$ and $\operatorname{ind}(T+S) = \operatorname{ind} T$. ■

Proof. By Theorem 3.15, T has a two-sided approximate inverse R; and clearly T is a two-sided approximate inverse of R, so that $R \in \mathcal{F}(Y, X)$. By Theorem 3.16, $0 = \operatorname{ind}(RT) = \operatorname{ind} R + \operatorname{ind} T$; that is, $\operatorname{ind} T = -\operatorname{ind} R$. But R is also a two-sided approximate inverse of $T+S$, and so by the same argument, $\operatorname{ind}(T+S) = -\operatorname{ind} R$. The result follows. □

The next result shows that the index of a Fredholm map is stable under perturbations by bounded maps of sufficiently small norm.

Theorem 3.18. Let $T \in \mathcal{F}(X,Y)$. Then there is a positive number δ such that if $S \in \mathcal{B}(X,Y)$ and $\|S\| < \delta$, then $T+S \in \mathcal{F}(X,Y)$ and $\operatorname{ind}(T+S) = \operatorname{ind} T$. ■

Proof. Let $R \neq 0$ be a two-sided approximate inverse of T, so that there are maps $K_1 \in \mathcal{K}(X)$ and $K_2 \in \mathcal{K}(Y)$ such that

$$RT = I_X + K_1, \qquad TR = I_Y + K_2.$$

Put $\delta = \|R\|^{-1}$: we show that δ has the desired properties. Let $S \in \mathcal{B}(X,Y)$ be such that $\|S\| < \delta$. Then $\|RS\| < 1$ and so $(I_X + RS)^{-1}$ exists and is in $\mathcal{B}(X)$. Thus

$$R(T+S) = I_X + K_1 + RS = (I_X + RS)[I_X + (I_X + RS)^{-1}K_1],$$

and

$$(I_X + RS)^{-1}R(T+S) = I_X + (I_X + RS)^{-1}K_1.$$

Since $(I_X + RS)^{-1}K_1 \in \mathcal{K}(X)$ this shows that $T+S$ has a left approximate inverse. In the same way it follows that $R(I_Y + SR)^{-1}$ is a right approximate inverse of $T+S$. Hence $T+S \in \mathcal{F}(X,Y)$, by Theorem 3.15. By Theorem 3.16,

$$\begin{aligned}\operatorname{ind} R + \operatorname{ind}(T+S) &= \operatorname{ind} R(T+S) = \operatorname{ind}(I_X + RS)[I_X + (I_X + RS)^{-1}K_1] \\ &= \operatorname{ind}(I_X + RS) + \operatorname{ind}[I_X + (I_X + RS)^{-1}K_1] = 0.\end{aligned}$$

Hence $\operatorname{ind}(T+S) = -\operatorname{ind} R = \operatorname{ind} T$, the final step following just as in the proof of Theorem 3.17. □

This important result shows that $\mathscr{F}(X,Y)$ is an open subset of the Banach space $\mathscr{B}(X,Y)$, and that the index is constant on connected components of $\mathscr{F}(X,Y)$.

We now turn our attention to the families $\mathscr{F}_+(X,Y)$ and $\mathscr{F}_-(X,Y)$ of semi-Fredholm maps, our ultimate object being again to obtain stability results for the index together with information about the behaviour of the nullity and deficiency under perturbations.

Theorem 3.19. Let X,Y,Z be Banach spaces and suppose that $T \in \mathscr{F}_{\pm}(X,Y)$ and $S \in \mathscr{F}_{\pm}(Y,Z)$. Then $ST \in \mathscr{F}_{\pm}(X,Z)$. ■

Proof. Suppose that $T \in \mathscr{F}_+(X,Y)$ and $S \in \mathscr{F}_+(Y,Z)$. By Lemma 3.12, there exist Banach spaces Y' and Z', maps $K_1 \in \mathscr{K}(X,Y')$ and $K_2 \in \mathscr{K}(Y,Z')$, and positive constants C_1 and C_2 such that for all $x \in X$ and $y \in Y$,

$$\|x\| \leqslant C_1(\|Tx\| + \|K_1 x\|), \qquad \|y\| \leqslant C_2(\|Sy\| + \|K_2 y\|).$$

Writing $y = Tx$ we see that for all $x \in X$,

$$\|x\| \leqslant C_1 C_2(\|STx\| + C_2^{-1}\|K_1 x\| + \|K_2 Tx\|).$$

Since $C_2^{-1}K_1$ and $K_2 T$ are compact, it follows from Lemma 3.12, taking $Y_1 = Y' \times Z'$ with the sum norm and $K \in \mathscr{K}(X,Y_1)$ defined by $Kx = \{K_1 x, K_2 Tx\}$, that $ST \in \mathscr{F}_+(X,Z)$. The case of $\mathscr{F}_-$ is dealt with by taking adjoints and using what has just been proved. □

Theorem 3.20. Let X,Y,Z be Banach spaces, let $T \in \mathscr{B}(X,Y)$ and $S \in \mathscr{B}(Y,Z)$, and suppose that $ST \in \mathscr{F}_+(X,Z)$ $(ST \in \mathscr{F}_-(X,Z))$. Then $T \in \mathscr{F}_+(X,Y)$ $(S \in \mathscr{F}_-(Y,Z))$. ■

Proof. Suppose that $ST \in \mathscr{F}_+(X,Z)$. By Lemma 3.12, there exist a Banach space Y_1, an operator $K \in \mathscr{K}(X,Y_1)$ and a constant $C > 0$ such that for all $x \in X$,

$$\|x\| \leqslant C(\|STx\| + \|Kx\|) \leqslant C(1 + \|S\|)(\|Tx\| + \|Kx\|).$$

That $T \in \mathscr{F}_+(X,Y)$ now follows from Lemma 3.12 again. The $\mathscr{F}_-$ case results from consideration of adjoints. □

Theorem 3.21. Let $T \in \mathscr{F}_{\pm}(X,Y)$ and suppose that $S \in \mathscr{K}(X,Y)$. Then $S + T \in \mathscr{F}_{\pm}(X,Y)$ and $\operatorname{ind}(S+T) = \operatorname{ind} T$. ■

Proof. Suppose that $T \in \mathscr{F}_+(X,Y)$. By Lemma 3.12, there exist a compact map K and a positive constant C such that for all $x \in X$,

$$\|x\| \leqslant C(\|Tx\| + \|Kx\|),$$

and hence

$$\|x\| \leqslant C(\|(T+S)x\| + \|Sx\| + \|Kx\|).$$

Thus from Lemma 3.12 again, just as in the proof of Theorem 3.19, we see that $T+S \in \mathscr{F}_+(X,Y)$. Theorem 3.17 shows that $\operatorname{ind}(S+T) = \operatorname{ind} T$, both when $\operatorname{def} T < \infty$ and when $\operatorname{def} T = \infty$. The result when $T \in \mathscr{F}_-(X,Y)$ follows by the usual considerations of adjoints. □

We are now in a position to deal with perturbations of semi-Fredholm maps by bounded maps of sufficiently small norm.

Theorem 3.22. Let $T \in \mathscr{F}_\pm(X,Y)$. Then there is a positive number $\delta > 0$ such that if $S \in \mathscr{B}(X,Y)$ and $\|S\| < \delta$ then $S+T \in \mathscr{F}_\pm(X,Y)$, $\operatorname{nul}(T+S) \leqslant \operatorname{nul} T$, $\operatorname{def}(T+S) \leqslant \operatorname{def} T$, and $\operatorname{ind}(T+S) = \operatorname{ind} T$. ■

Proof. Suppose that $T \in \mathscr{F}_+(X,Y)$. As in the proof of Lemma 3.12 we see that there is a map $K \in \mathscr{K}(X)$ with $\dim \mathscr{R}(K) = \operatorname{nul} T < \infty$, and for some constant $C > 0$,

$$\|x\| \leqslant C(\|Tx\| + \|Kx\|)$$

for all x in X. We claim that $\delta_1 := C^{-1}$ has some of the desired properties. Let $S \in \mathscr{B}(X,Y)$ with $\|S\| < \delta_1$. Then for all $x \in X$,

$$\|x\| \leqslant C(\|(T+S)x\| + \|S\|\,\|x\| + \|Kx\|),$$

so that

$$\|x\| \leqslant C(1 - \|S\|\delta_1^{-1})^{-1}(\|(T+S)x\| + \|Kx\|).$$

Lemma 3.12 now shows that $T+S \in \mathscr{F}_+(X,Y)$. Moreover, the above inequality shows that $\|x\| \leqslant C(1-\|S\|\delta_1^{-1})^{-1}\|Kx\|$ for all $x \in \mathscr{N}(T+S)$, and so $\mathscr{N}(T+S)$ is isomorphic to the range of the restriction of K to $\mathscr{N}(T+S)$, and this range has dimension $\leqslant \dim \mathscr{R}(K) = \operatorname{nul} T$. Hence $\operatorname{nul}(T+S) \leqslant \operatorname{nul} T$.

If $T \in \mathscr{F}_-(X,Y)$, consideration of adjoints shows that $T+S \in \mathscr{F}_-(X,Y)$ and $\operatorname{def}(T+S) \leqslant \operatorname{def} T$ if $\|S\| < \delta_1$. In view of Theorem 3.18 it is now enough to show that if $\operatorname{ind} T$ is infinite then so is $\operatorname{ind}(T+S)$ if $\|S\|$ is small enough. Suppose that $T \in \mathscr{F}_+(X,Y) \backslash \mathscr{F}(X,Y)$. Then since $\operatorname{nul} T < \infty$, there is a closed linear subspace M of X such that $X = \mathscr{N}(T) \dotplus M$, by Theorem 1.12. Let J_M be the natural embedding of M in X. Then $TJ_M : M \to T(X)$ is invertible; put $\varepsilon = \|(TJ_M)^{-1}\|^{-1}$ and let $S \in \mathscr{B}(X,Y)$ be such that $\|S\| < \varepsilon$. Observe that, with J standing for the natural embedding of $T(X)$ in Y,

$$JTJ_M + SJ_M = [J + SJ_M(TJ_M)^{-1}]TJ_M. \tag{3.2}$$

Moreover, $\|SJ_M(TJ_M)^{-1}\| < 1$, and so for any $\lambda \in [0, 1]$ it follows that $S_\lambda := J + \lambda SJ_M(TJ_M)^{-1}$ is an injective map in $\mathscr{F}_+(T(X), Y)$. We claim that for all $\lambda \in [0, 1]$,

$$\operatorname{codim} [J + \lambda SJ_M(TJ_M)^{-1}]\, T(X) = \operatorname{codim} T(X) \ (= \operatorname{def} T) \ = \infty.$$

For suppose that $\operatorname{codim} S_\lambda(T(X)) = n < \infty$ for some $\lambda \in [0, 1]$. Given any $\mu \in [0, 1]$, S_μ is invertible with

$$\|S_\mu^{-1}\|^{-1} \geq 1 - \mu \|SJ_M(TJ_M)^{-1}\| \geq 1 - \|SJ_M(TJ_M)^{-1}\| =: \eta > 0.$$

Let $\lambda_1, \mu \in [0, 1]$ be such that $|\mu - \lambda_1| < (n+1)^{-1}\eta$; note that

$$\|S_{\lambda_1} - S_\mu\| = |\mu - \lambda_1| \, \|SJ_M(TJ_M)^{-1}\| < (n+1)^{-1}\eta \leq (n+1)^{-1} \|S_{\lambda_1}^{-1}\|^{-1}.$$

We claim that in view of this, $\operatorname{def}(S_\mu T) = \operatorname{def}(S_{\lambda_1} T)$; this claim will be substantiated at the end of the proof. Accepting this for the moment, the possibility of covering $[0, 1]$ with finitely many intervals of length $\frac{1}{2}(n+1)^{-1}\eta$ shows that $\operatorname{def} T = \operatorname{def}(S_\lambda T) = n < \infty$, which contradicts the assumption that $T \in \mathscr{F}_+(X, Y) \backslash \mathscr{F}(X, Y)$. Thus $\operatorname{def}(S_\lambda T) = \infty$ for all $\lambda \in [0, 1]$ and in particular for $\lambda = 1$.

By (3.2),

$$\operatorname{codim} (JTJ_M + SJ_M)(M) \geq \operatorname{codim} [J + SJ_M(TJ_M)^{-1}]\, T(X),$$

and hence

$$\operatorname{codim} (JTJ_M + SJ_M)(M) = \infty.$$

But since

$$(T+S)(M) \subset (T+S)(X) \subset (T+S)(M) + (T+S)\mathscr{N}(T),$$

and $\mathscr{N}(T)$ has finite dimension, it follows that $\operatorname{def}(T+S) = \operatorname{codim} (T+S)(X) = \infty$; thus $T + S \in \mathscr{F}_+(X, Y) \backslash \mathscr{F}(X, Y)$, as required. The corresponding result when $T \in \mathscr{F}_-(X, Y) \backslash \mathscr{F}(X, Y)$ is a consequence of the usual process of consideration of adjoints: $T^* \in \mathscr{F}_+(X, Y) \backslash \mathscr{F}(X, Y)$, $\operatorname{nul} T^* = \operatorname{def} T$ and $\operatorname{def} T^* = \operatorname{nul} T$.

All that remains is to justify the unproved claim made earlier; this follows immediately from the next lemma. □

Lemma 3.23. Let $T \in \mathscr{F}(X, Y)$ be injective, with $\operatorname{def} T = n < \infty$; let $S \in \mathscr{B}(X, Y)$ and suppose that $\|S\| < (n+1)^{-1} \|T^{-1}\|^{-1}$. Then $\operatorname{def}(T+S) = n$. ■

Proof. By Theorem 1.14, there is a projection $P \in \mathscr{B}(Y)$ with $P(Y) = \mathscr{R}(T)$ and $\|P\| \leq n+1$. Let I be the identity map of Y to itself. Then since

$$\|ST^{-1}P\| \leq \|S\| \, \|T^{-1}\| (n+1) < 1,$$

$I+ST^{-1}P$ is bijective:

$$\text{nul}\,(I+ST^{-1}P) = \text{def}\,(I+ST^{-1}P) = 0.$$

Since $T+S = (I+ST^{-1}P)T$, it follows that $T+S$ is injective, and by Theorem 3.16,

$$\begin{aligned}\text{def}\,(T+S) &= -\text{ind}\,(T+S) = -\text{ind}\,[(I+ST^{-1}P)T]\\ &= -\text{ind}\,(I+ST^{-1}P) - \text{ind}\,T\\ &= -\text{ind}\,T = \text{def}\,T = n.\end{aligned}$$ □

These results establish the stability of the *index*, but so far nothing has been said about the stability of the *nullity* or the *deficiency*. To repair this omission a simple lemma is needed.

Lemma 3.24. Let $T \in \mathscr{F}_+(X,Y)$. Then given any closed linear subspace M of X, the subspace $T(M)$ is closed. ■

Proof. Put $\tilde{X} = X/\mathscr{N}(T)$ and define $\tilde{T}: \tilde{X} \to Y$ by $\tilde{T}\xi = Tx$ $(\xi \in \tilde{X})$, where x is any element of the equivalence class ξ. Then $\tilde{T}(\tilde{M}) = T(M)$, where $\tilde{M} = \{x + \mathscr{N}(T): x \in M\}$. Since $\tilde{T}$ has a bounded inverse, it is enough to prove that $\tilde{M}$ is closed in the quotient space $\tilde{X}$. Let (ξ_n) be a sequence in $\tilde{M}$ such that $\xi_n \to \xi \in \tilde{X}$. Then there are sequences (x_n) and (z_n) in M and $\mathscr{N}(T)$ respectively such that $x_n - x - z_n \to 0$, for some $x \in X$. Since $\text{nul}\,T < \infty$, the subspace $M + \mathscr{N}(T)$ is closed as we shall see in a moment; thus $x \in M + \mathscr{N}(T)$, and $\xi \in \tilde{M}$.

To prove that $M + \mathscr{N}(T)$ is closed, write it as $M \dotplus M_1$ for some finite-dimensional M_1. By Theorem 1.14, there is a bounded projection P of X on M_1; thus if $y \in \overline{M + \mathscr{N}(T)}$, there is a sequence (y_n) in $M \dotplus M_1$ with $y_n \to y$, and hence $Py_n \to Py \in M_1$, $(I-P)y_n \to (I-P)y \in M$ (by the closedness of M), and $y \in M \dotplus M_1$. The proof is complete. □

Theorem 3.25. Let $T \in \mathscr{F}_+(X,Y) \cup \mathscr{F}_-(X,Y)$ and $S \in \mathscr{B}(X,Y)$. Then for all small enough $|\tau| > 0$, $T + \tau S \in \mathscr{F}_+(X,Y) \cup \mathscr{F}_-(X,Y)$ and nul $(T+\tau S)$ and def $(T+\tau S)$ are constant. ■

Proof. First suppose that $\text{nul}\,T < \infty$, and define sequences (M_n) and (R_n) of linear subspaces with $M_n \subset X$ and $R_n \subset Y$, by

$$M_0 = X, \quad R_0 = Y, \quad M_n = S^{-1}(R_n), \quad R_{n+1} = T(M_n) \quad \text{for } n = 0, 1, \ldots.$$

Evidently $M_{n+1} \subset M_n$ and $R_{n+1} \subset R_n$ for all n. An inductive argument shows that the M_n and R_n are closed: in fact if R_n is closed so is $M_n = S^{-1}(R_n)$, and thus by Lemma 3.24, R_{n+1} is closed.

Let T' be the restriction of T to the closed linear subspace $X' := \bigcap_{n \in \mathbb{N}} M_n$. If $u \in X'$, then $u \in M_n$ and $T'u = Tu \in T(M_n) = R_{n+1}$ for all n; thus $\mathscr{R}(T') \subset Y'$

$:= \bigcap_{n\in\mathbb{N}} R_n$. We claim that $\mathscr{R}(T') = Y'$. To see this, let $v \in Y'$, so that $v \in T(M_n)$ for all $n \in \mathbb{N}$; that is, $T^{-1}v \cap M_n \neq \varnothing$. Since $T^{-1}v$ is an affine subspace of the form $u + \mathscr{N}(T)$ for some $u \in X$, and since $\text{nul}\, T < \infty$, the $T^{-1}v \cap M_n$ form a decreasing sequence of finite-dimensional affine subspaces, and thus are ultimately constant, coinciding with $T^{-1}v \cap X'$, which must thus be non-empty, containing an element u say. But then $T'u = Tu = v$, and so $\mathscr{R}(T') = Y'$, as claimed.

The map T' may be viewed as an element of $\mathscr{B}(X', Y')$; and so can the restriction S' of S to X', for if $u \in X'$ then $u \in S^{-1}(R_n)$ for all n, which shows that $Su \in R_n$ for all n; that is, $Su \in Y'$. Now apply Theorem 3.22: we see that for all small enough $|\tau|$, we have $\text{def}\,(T' + \tau S') = \text{def}\, T' = 0$, and $\text{nul}\,(T' + \tau S') = \text{ind}\,(T' + \tau S') = \text{ind}\, T' = \text{nul}\, T'$. Hence $\text{nul}\,(T' + \tau S')$ and $\text{def}\,(T' + \tau S')$ are constant for sufficiently small $|\tau|$. However, $\mathscr{N}(T + \tau S) = \mathscr{N}(T' + \tau S')$ if $\tau \neq 0$, because if $u \in \mathscr{N}(T + \tau S)$ then $Tu = -\tau Su$, and by induction $u \in M_n$ for all $n \in \mathbb{N}$, so that $u \in X'$. Thus $\text{nul}\,(T + \tau S) = \text{nul}\,(T' + \tau S')$ is constant for all small enough $|\tau| > 0$, and since $\text{ind}\,(T + \tau S)$ is constant the result concerning $\text{def}\,(T + \tau S)$ also follows.

The case in which $\text{def}\, T < \infty$ can be reduced to that just considered by means of adjoints, as usual. □

Remark 3.26. Much of the material in this section can be found in the book of Kato [1]; the treatment of approximate inverses is based upon that given by Michlin and Prössdorf [1] and the discussion of the perturbation of semi-Fredholm maps takes advantage of the presentation given by Tylli [1] of this topic. Comprehensive lists of references may be found in these works. The standard properties of semi-Fredholm maps are also given in the book of Caradus, Pfaffenberger, and Yood [1].

Remark 3.27. The results proved in this section for Fredholm and semi-Fredholm maps T in $\mathscr{B}(X, Y)$ continue to hold if $T \in \mathscr{C}(X, Y)$. For if $T \in \mathscr{C}(X, Y)$ then $T \in \mathscr{B}(X(T), Y)$, where $X(T)$ denotes the Banach space defined by $\mathscr{D}(T)$ and the graph norm of T; since $\mathscr{N}(T)$ and $\mathscr{R}(T)$ are unchanged by our new view of T as a bounded map, we may reduce propositions about closed maps to corresponding ones about bounded operators.

4. The essential spectrum

Let X be a complex Banach space and let $T \in \mathscr{B}(X)$. There are various definitions of the essential spectrum of T in the literature, and to take account of the main ones we introduce the following subsets of the complex plane $\mathbb{C}$, letting I stand for the identity map of X to itself.

$\Phi_{\pm}(T) = \{\lambda \in \mathbb{C}: T-\lambda I \in \mathscr{F}_{\pm}(X)\}$,
$\Delta_1(T) = \Phi_{+}(T) \cup \Phi_{-}(T)$,
$\Delta_2(T) = \Phi_{+}(T)$,
$\Delta_3(T) = \{\lambda \in \mathbb{C}: T-\lambda I \in \mathscr{F}(X)\}$,
$\Delta_4(T) = \{\lambda \in \mathbb{C}: T-\lambda I \in \mathscr{F}(X)$ and $\operatorname{ind}(T-\lambda I) = 0\}$,
$\Delta_5(T) =$ union of all components of $\Delta_1(T)$ which contain points of the resolvent set of T.

Finally we set $\sigma_{ek}(T) = \mathbb{C}\backslash\Delta_k(T)$ for $k = 1, 2, 3, 4, 5$, so that $\sigma_{ek} \subset \sigma_{el}(T)$ if $k < l$; and put $\mathrm{r}_{ek}(T) = \sup\{|\lambda|: \lambda \in \sigma_{ek}(T)\}$.

Each of the sets $\sigma_{ek}(T)$ has been referred to in the literature as the *essential spectrum* of T, the extremes $\sigma_{e1}(T)$ and $\sigma_{e5}(T)$ being used by Kato [1] and Browder [1] respectively. Fortunately this confusing situation is partially redeemed by the fact that the *radius* $\mathrm{r}_{ek}(T)$ *of the essential spectrum* of T is the same for all k, and we set about the proof of this and of an interesting and useful formula for $\mathrm{r}_{ek}(T)$.

We begin with some remarks about $\sigma_{e1}(T) = \mathbb{C}\backslash\Delta_1(T)$, where $\Delta_1(T) = \{\lambda \in \mathbb{C}: T-\lambda I$ is semi-Fredholm$\}$. By Theorem 3.22, $\Delta_1(T)$ is open and hence $\sigma_{e1}(T)$ is closed: the same is clearly true for $\sigma_{ek}(T)$ $(k = 2, 3, 4, 5)$. The open set $\Delta_1(T)$ can naturally be written as the union of countably many components $\Delta_1^{(n)}$ $(n \in \mathbb{N})$ and by Theorem 3.22, $\operatorname{ind}(T-\lambda I) = \operatorname{nul}(T-\lambda I) - \operatorname{def}(T-\lambda I) = \nu(\lambda) - \mu(\lambda)$, say, is constant in each $\Delta_1^{(n)}$. By Theorem 3.25, with $S = I$, $\tau = \lambda_0 - \lambda$, and T replaced by $T-\lambda_0 I$, both $\nu(\lambda)$ and $\mu(\lambda)$ take constant values ν_n and μ_n respectively in each component $\Delta_1^{(n)}$, save possibly at some isolated values of λ, say λ_{nj}. Thus if $\lambda \in \Delta_1^{(n)}$ and $\lambda \neq \lambda_{nj}$, we have $\nu(\lambda) = \nu_n$, $\mu(\lambda) = \mu_n$, while

$$\nu(\lambda_{nj}) = \nu_n + r_{nj}, \qquad \mu(\lambda_{nj}) = \mu_n + r_{nj}, \qquad 0 < r_{nj} < \infty.$$

If $\nu_n = \mu_n = 0$, all points of $\Delta_1^{(n)}$, with the exception of the λ_{nj}, belong to the resolvent set $\rho(T)$; it is clear that in this case the λ_{nj} are isolated eigenvalues of T (that is, eigenvalues of T which are isolated in $\sigma(T)$). Kato shows in [1, §IV.5] that these eigenvalues have finite algebraic multiplicities and, of course, they have geometric multiplicities r_{nj}. It is worth observing that $\Delta_k(T)$ $(k > 1)$ is the union of selected components of $\Delta_1(T)$.

The following result shows the invariance of $\sigma_{ek}(T)$ $(k = 1, 2, 3, 4)$ under compact perturbations.

Theorem 4.1. Let $T \in \mathscr{B}(X)$ and $S \in \mathscr{K}(X)$. Then $\sigma_{ek}(T) = \sigma_{ek}(T+S)$ for $k = 1, 2, 3, 4$. ■

Proof. The result follows immediately from Theorem 3.21. □

More will be said about the essential spectrum in Chapter IX.

We can now begin to establish a connection between the radius of the essential spectrum of T and measures of non-compactness of T. Let $\pi: \mathscr{B}(X)$

$\to \mathscr{B}(X)/\mathscr{K}(X)$ be the natural map. Since $\mathscr{B}(X)/\mathscr{K}(X)$ is a Banach algebra under the norm

$$\|\pi(T)\| := \|T\|_{\mathscr{K}} := \inf\{\|T+K\| : K \in \mathscr{K}(X)\},$$

the spectral radius $\mathrm{r}(\pi(T))$ of $\pi(T)$ is given by

$$\mathrm{r}(\pi(T)) = \lim_{n\to\infty} \|T^n\|_{\mathscr{K}}^{1/n}.$$

Lemma 4.2. Let $T \in \mathscr{B}(X)$ and suppose that for some $n \in \mathbb{N}$ either $\beta(T^n) < 1$ or $\tilde{\beta}(T^n) < 1$. Then given any closed, bounded $B \subset X$ and any compact $K \subset X$, the set $\{x \in B : (I-T)x \in K\}$ is compact; that is, $I-T$ is proper on closed, bounded sets. ■

Proof. Let B and K be as above and put $M = \{x \in B : (I-T)x \in K\}$. Since M is closed and bounded it is enough to prove that $\psi(M) = 0$ or $\tilde{\psi}(M) = 0$. Let $x \in M$; then $x = Tx + y$ for some $y \in K$. Then $x = T^2x + Ty + y$, and more generally for any $n \in \mathbb{N}$,

$$x = T^n x + \sum_{j=0}^{n-1} T^j y.$$

Let $K_1 = (\sum_{j=0}^{n-1} T^j)(K)$; the set K_1 is evidently compact as T is continuous, and since from our argument above, $M \subset T^n(M) + K_1$, it follows that $\psi(M) \leqslant \psi(T^n(M))$ and $\tilde{\psi}(M) \leqslant \tilde{\psi}(T^n(M))$. Thus for some $k \in [0, 1)$, either $\psi(M) \leqslant k\psi(M)$ or $\tilde{\psi}(M) \leqslant k\tilde{\psi}(M)$, and so either $\psi(M) = 0$ or $\tilde{\psi}(M) = 0$. The proof is complete. □

Lemma 4.3. Let $T \in \mathscr{B}(X)$ be such that given any closed bounded set $B \subset X$ and any compact set $K \subset X$, the subset $\{x \in B : Tx \in K\}$ is compact. Then $T \in \mathscr{F}_+(X)$. ■

Proof. That $\mathrm{nul}\, T < \infty$ follows directly from Theorem 1.3: any bounded subset of $\mathscr{N}(T)$ is compact (take $K = \{0\}$), and so the identity map of the Banach space $\mathscr{N}(T)$ (with the norm induced by that of X) to itself is compact. Hence by Theorem 1.3, $\mathrm{nul}\, T < \infty$.

To finish the proof we suppose that $\mathscr{R}(T)$ is not closed. Then there is a sequence (x_n) in X such that (Tx_n) converges to a point $y \in X \setminus \mathscr{R}(T)$; evidently we may suppose for all $n \in \mathbb{N}$, that $x_n \notin \mathscr{N}(T)$. For each $n \in \mathbb{N}$ put $d_n = \mathrm{dist}(x_n, \mathscr{N}(T))$; clearly $d_n > 0$. We claim that $d_n \to \infty$ as $n \to \infty$. To see this, suppose otherwise: since there is a sequence (y_n) in $\mathscr{N}(T)$ such that $d_n \leqslant \|x_n - y_n\| < 2d_n$ for all $n \in \mathbb{N}$, then $(x_n - y_n)$ would contain a bounded sequence, again denoted by $(x_n - y_n)$ for simplicity. By the hypothesis of the lemma, some subsequence of $(x_n - y_n)$ must converge, to x, say. Hence $T(x_n - y_n) \to Tx$, and so $y = Tx \in \mathscr{R}(T)$ - a contradiction. Thus $d_n \to \infty$ as $n \to \infty$. Now put $z_n = (x_n - y_n)/\|x_n - y_n\|$. Since $\|z_n\| = 1$ for all $n \in \mathbb{N}$, and $Tz_n \to 0$, our hypothesis shows that there is a subsequence of (z_n) which converges, to z, say, where $z \in \mathscr{N}(T)$. But for all $n \in \mathbb{N}$,

$$\|z_n - z\| = \|x_n - y_n - z\|x_n - y_n\|\|/\|x_n - y_n\| \geqslant d_n/(2d_n) = \tfrac{1}{2},$$

and we have a contradiction. This completes the proof. □

Theorem 4.4. Let $T \in \mathscr{B}(X)$ be such that for some $n \in \mathbb{N}$, either $\beta(T^n) < 1$ or $\tilde{\beta}(T^n) < 1$. Then $I - T \in \mathscr{F}(X)$ and $\operatorname{ind}(I - T) = 0$. ■

Proof. Lemmas 4.2 and 4.3 show that $I - \lambda T \in \mathscr{F}_+(X)$ for all $\lambda \in [0, 1]$. By Theorem 3.22, $\operatorname{ind}(I - T) = \operatorname{ind} I = 0$, which immediately shows that $I - T \in \mathscr{F}(X)$. □

At this point it is convenient to mention some more connections between measures of non-compactness and semi-Fredholm maps. Let $T \in \mathscr{B}(X)$ and write

$$\omega(T) = \inf\left[\psi(T(\Omega))/\psi(\Omega)\right], \qquad \tilde{\omega}(T) = \inf\left[\tilde{\psi}(T(\Omega))/\tilde{\psi}(\Omega)\right]$$

where the infima are taken over all bounded subsets Ω of X which are not relatively compact.

Proposition 4.5. Let $T \in \mathscr{B}(X)$ and suppose that either $\omega(T) > 0$ or $\tilde{\omega}(T) > 0$. Then T is proper on closed bounded sets. ■

Proof. Let B and K be respectively a closed bounded and a compact subset of X. Then $A := B \cap T^{-1}(K)$ is closed and bounded. Suppose that A is not compact and that $\omega(T) > 0$. Then $\psi(T(A)) \geqslant \omega(T)\psi(A)$; but as $T(A) \subset K$, we have $\psi(T(A)) = 0$ and so $\psi(A) = 0$; that is, A is compact. This contradiction shows that when $\omega(T) > 0$, the set A is compact and hence that T is proper on B. The same argument clearly works if instead we assume that $\tilde{\omega}(T) > 0$. □

Theorem 4.6. Let $T \in \mathscr{B}(X)$. Then:

(i) $\omega(T) > 0$ if, and only if, $T \in \mathscr{F}_+(X)$;
(ii) $\omega(T^*) > 0$ if, and only if, $T \in \mathscr{F}_-(X)$;
(iii) $T \in \mathscr{F}(X)$ if, and only if, $\omega(T) > 0$ and $\omega(T^*) > 0$.

The same statements hold with ω replaced by $\tilde{\omega}$. ■

Proof. If $\omega(T) > 0$, then by Lemma 4.3 and Proposition 4.5, $T \in \mathscr{F}_+(X)$. Conversely, suppose that $T \in \mathscr{F}_+(X)$. We may write $X = X_0 \dotplus \mathscr{N}(T)$ for some closed linear subspace X_0 of X. Let P be the natural projection of X onto X_0 and write $T_0 = T \upharpoonright X_0$. Since $\mathscr{R}(T) = \mathscr{R}(T_0)$ is closed, the Closed-Graph Theorem shows that T_0 maps X_0 isomorphically onto $\mathscr{R}(T)$. Thus there is a positive constant c such that $\|T_0 x\| \geqslant c\|x\|$ for all $x \in X_0$, and so for any bounded set $B \subset X_0$, we obtain $c\psi(B) \leqslant \psi(T_0(B))$, which shows that $\omega(T_0) \geqslant c > 0$. Also

$$\omega(T_0 P) = \inf[\psi(T_0 P(\Omega))/\psi(\Omega)],$$

where the infimum is taken over all bounded subsets Ω of X which are not relatively compact. Given any such Ω, the compactness of $I - P$ ensures that $(I-P)(\Omega)$ is relatively compact, and hence $P(\Omega)$ is not relatively compact. Thus

$$\begin{aligned}\omega(T_0 P) &= \inf[\psi(T_0 P(\Omega))/\psi(P(\Omega))\cdot\psi(P(\Omega))/\psi(\Omega)] \\ &\geqslant \inf_{\Omega'}[\psi(T_0 \Omega')/\psi(\Omega')]\cdot\inf_{\Omega}[\psi(P(\Omega))/\psi(\Omega)] \\ &= \omega(T_0)\omega(P).\end{aligned}$$

Hence $\omega(T) = \omega(T_0 P) \geqslant \omega(T_0)\omega(P)$. But since $I - P$ is compact, $\omega(P) = \omega(I - (I-P)) = \omega(I) = 1$, and so $\omega(T) \geqslant \omega(T_0) > 0$. The proof of (i) is complete.

As for (ii), suppose that $\omega(T^*) > 0$. Then $T^* \in \mathscr{F}_+(X)$ and hence $T \in \mathscr{F}_-(X)$. Conversely, if $T \in \mathscr{F}_-(X)$, then $T^* \in \mathscr{F}_+(X)$, so that by (i), $\omega(T^*) > 0$. Statement (iii) follows immediately from (i) and (ii). □

Corollary 4.7. Let $T \in \mathscr{B}(X)$. Then $T \notin \mathscr{F}_+(X)$ if, and only if, it has a singular sequence; that is, a sequence (x_n) in X such that $\|x_n\| = 1$ for all $n \in \mathbb{N}$, (x_n) has no convergent subsequence, and $Tx_n \to 0$ as $n \to \infty$. ■

Proof. Suppose that T has a singular sequence (x_n) and let $\Omega = \overline{\{x_n : n \in \mathbb{N}\}}$; the set Ω is bounded but is not relatively compact. However, $\overline{T(\Omega)}$ is compact. Hence by Theorem 4.6 (i), $T \notin \mathscr{F}_+(X)$. Conversely, suppose that $T \notin \mathscr{F}_+(X)$. If nul $T = \infty$, there must be a sequence (x_n) in $\mathscr{N}(T)$ such that $\|x_n\| = 1$ for all $n \in \mathbb{N}$ and $\|x_n - x_m\| \geqslant 1$ for all $m, n \in \mathbb{N}$ with $m \neq n$. Since $Tx_n = 0$ for all $n \in \mathbb{N}$, the sequence (x_n) is singular. If nul $T < \infty$ then $\mathscr{R}(T)$ cannot be closed, since otherwise T would belong to $\mathscr{F}_+(X)$. There is a closed subspace X_0 of X such that $X = \mathscr{N}(T) \dotplus X_0$. By Theorem 3.4, there is a sequence (x_n) in X_0 such that $\|x_n\| = 1$ for all $n \in \mathbb{N}$ and $\lim_{n\to\infty} T(x_n) = 0$. If (x_n) has a convergent subsequence, with limit x, say, then $\|x\| = 1$ and $x \in \mathscr{N}(T) \cap X_0$, which is impossible. Hence (x_n) is a singular sequence, and the proof is complete. □

Given any $T \in \mathscr{B}(X)$ we recall that $\Delta_3(T) = \{\lambda \in \mathbb{C}: T - \lambda I \in \mathscr{F}(X)\}$. We know that $\Delta_3(T)$ is open; and by Theorem 3.15, $\rho(\pi(T)) = \Delta_3(T)$; thus

$$\mathrm{r}(\pi(T)) = \max\{|\lambda| : \lambda \in \sigma_{e3}(T)\} = \mathrm{r}_{e3}(T).$$

By Theorem 4.4, if $\lambda \in \mathbb{C}$ and $|\lambda| > \beta(T)$, then $T - \lambda I \in \mathscr{F}(X)$ and ind $(T - \lambda I) = 0$. Thus $\lambda \in \Delta_4(T) \subset \Delta_3(T)$, and hence $\lambda \in \rho(\pi(T))$. This shows that $\mathrm{r}(\pi(T)) \leqslant \beta(T)$. The following result clarifies the picture.

Theorem 4.8. Let $T \in \mathscr{B}(X)$. Then

$$\mathrm{r}(\pi(T)) = \lim_{n \to \infty} [\tilde{\beta}(T^n)]^{1/n}. \qquad \blacksquare$$

Proof. That the limit exists is a consequence of Lemma 2.16; denote it by r. Since $\tilde{\beta}(T) \leqslant \|T\|$ and $\tilde{\beta}(T) = \tilde{\beta}(T+K)$ for all $K \in \mathscr{K}(X)$ (by Lemma 2.8), it follows that $\tilde{\beta}(T) \leqslant \|T\|_{\mathscr{K}}$. Hence $r \leqslant \mathrm{r}(\pi(T))$. To prove the reverse inequality let $\lambda \in \mathbb{C}$ be such that $|\lambda| > r$. Then there exists $n \in \mathbb{N}$ such that $|\lambda|^n > \tilde{\beta}(T^n)$; thus $I - \lambda^{-1}T \in \mathscr{F}(X)$, by Theorem 4.4, and hence $\lambda \in \Delta_3(T)$. This shows that $|\lambda| > \mathrm{r}(\pi(T))$; that is, $\mathrm{r}(\pi(T)) \leqslant r$. The proof is complete. □

Corollary 4.9. Let $T \in B(X)$. Then

$$\begin{aligned}\mathrm{r}(\pi(T)) &= \lim_{n \to \infty} [\beta(T^n)]^{1/n} = \lim_{n \to \infty} [\tilde{\beta}(T^n)]^{1/n} \\ &= \lim_{n \to \infty} [\mathrm{c}(T^n)]^{1/n} = \lim_{n \to \infty} \|T^n\|_{\mathscr{K}}^{1/n}.\end{aligned} \qquad \blacksquare$$

Theorem 4.10. Let $T \in \mathscr{B}(X)$. Then

$$\mathrm{r}(\pi(T)) = \max\{|\lambda| : \lambda \in \sigma_{e1}(T)\} = \max\{|\lambda| : \lambda \in \sigma_{e5}(T)\}. \qquad \blacksquare$$

Proof. Let Φ_0 be the unbounded component of $\Delta_1(T)$. Since T is bounded, $T - \lambda I$ is invertible for all large enough $|\lambda|$, and hence $\Phi_0 \subset \Delta_5(T) \subset \Delta_3(T) \subset \Delta_1(T)$. Let $\lambda_0 \notin \Phi_0$ be such that $|\lambda_0| = \max\{|\lambda| : \lambda \notin \Phi_0\}$. Since Φ_0 is a component of $\Delta_1(T)$ and λ_0 is a boundary point of Φ_0, we have $\lambda_0 \notin \Delta_1(T)$. Hence $|\lambda_0| = \max\{|\lambda| : \lambda \notin \Delta_1(T)\}$, and so

$$\begin{aligned}|\lambda_0| &= \max\{|\lambda| : \lambda \in \sigma_{e1}(T)\} \leqslant \max\{|\lambda| : \lambda \in \sigma_{e5}(T)\} \\ &\leqslant \max\{|\lambda| : \lambda \notin \Phi_0\} \leqslant |\lambda_0|,\end{aligned}$$

the second inequality following since $\Phi_0 \subset \Delta_5(T)$. □

Corollary 4.11. The radius of the essential spectrum of a map $T \in \mathscr{B}(X)$ is the same for all the definitions given here of the essential spectrum. Denote this radius by $\mathrm{r_e}(T)$; then $\mathrm{r_e}(T) = \mathrm{r}(\pi(T))$. ■

Remark 4.12. This section leans heavily on the work of Nussbaum [1] and Lebow and Schechter [1]. The relationships between measures of non-compactness and semi-Fredholm maps have been intensively studied in recent years and various striking results have been obtained. For example, Tylli [2] has shown that for all $T \in \mathscr{B}(X)$,

$$\inf\{|\lambda| : \lambda \in \sigma(\pi(T))\} = \inf\{|\lambda| : \lambda \in \sigma_{e2}(T)\} = \lim_{n \to \infty} [\tilde{\omega}(T^n)]^{1/n}.$$

Other interesting work will be found in Zemánek [1,2,3,4,5].

II

Entropy numbers, s-numbers, and eigenvalues.

We have already seen in Chapter I that given any compact linear map acting in a Banach space, the only non-zero points in its spectrum are isolated eigenvalues of finite algebraic (and hence geometric) multiplicity; and for continuous linear maps we have a formula for the radius of the essential spectrum, in terms of measures of non-compactness. In recent years, great attention has been paid to the connections between *analytical* entities related to bounded linear maps (such as eigenvalues and the essential spectrum) and more *geometrical* quantities, typified by entropy numbers, approximation numbers, and n-widths. These connections are not only of considerable intrinsic interest but also seem likely to be of definite use in the theory of partial differential equations; and we therefore devote the present chapter to an account of some of the more important results of this theory.

After giving the basic definitions, we focus on compact linear maps and establish inequalities between the entropy numbers, the approximation numbers, and the non-zero eigenvalues of such maps. These lead ultimately to the proof of a celebrated inequality, due to Weyl [4], which shows that if the approximation numbers form a sequence in l^p (with $0 < p < \infty$), then so do the eigenvalues. Weyl's inequality was given originally in a Hilbert-space setting, but we are able to give a very simple proof of it in the context of a general Banach space, and by the use of Lorentz spaces show that if the nth approximation number is $O(n^{-a})$ for some $a > 0$, as $n \to \infty$, then so is the nth eigenvalue. We also give König's remarkable formula in [3] for the absolute value of the eigenvalues of a compact map in terms of approximation numbers.

All this being done for compact maps, we then point out that at little extra cost it is possible to carry out a broadly similar programme of work for general bounded linear maps, provided that, where eigenvalues are concerned, we consider only those eigenvalues which lie outside the smallest disc containing the essential spectrum. The privileged position of compact maps is thus to some extent undermined.

1. Entropy numbers.

Throughout this chapter, unless otherwise stated, X, Y, Z will stand for Banach spaces and all spaces will be assumed to be complex; B_X will here stand for the closed unit ball in X.

Definition 1.1. Let $T \in \mathscr{B}(X, Y)$ and let $n \in \mathbb{N}$. The nth *entropy number* $\mathrm{e}_n(T)$ of T is defined by

$$\mathrm{e}_n(T) = \inf\{\varepsilon > 0\colon T(\mathrm{B}_X) \subset \bigcup_{i=1}^{2^{n-1}} (y_i + \varepsilon \mathrm{B}_Y)$$
$$\text{for some } y_i \in Y \ (i = 1, 2, \ldots, 2^{n-1})\}. \blacksquare$$

These numbers were introduced by Pietsch in his book [1] where they are referred to as *outer entropy numbers* and their fundamental properties are derived. Certain functions inverse to them had earlier been used by Mitjagin and Pelczynski[1] and by Triebel[1]. Since the $\mathrm{e}_n(T)$ are monotonic decreasing as n increases, their limit exists. Clearly

$$\lim_{n \to \infty} \mathrm{e}_n(T) = \inf\{\varepsilon > 0 : T(\mathrm{B}_X) \text{ can be covered by finitely many open balls of radius } \varepsilon\},$$

and so

$$\lim_{n \to \infty} \mathrm{e}_n(T) = \tilde{\beta}(T),$$

in the notation of Chapter I, §2. We now provide some elementary properties of the entropy numbers.

Proposition 1.2. Let $S, T \in \mathscr{B}(X, Y)$ and suppose that $R \in \mathscr{B}(Y, Z)$. Then:

(i) $\|T\| = \mathrm{e}_1(T) \geqslant \mathrm{e}_2(T) \geqslant \ldots \geqslant 0$;
(ii) for all $m, n \in \mathbb{N}$, $\mathrm{e}_{m+n-1}(S+T) \leqslant \mathrm{e}_m(S) + \mathrm{e}_n(T)$;
(iii) for all $m, n \in \mathbb{N}$, $\mathrm{e}_{m+n-1}(RS) \leqslant \mathrm{e}_m(R)\mathrm{e}_n(S)$. $\blacksquare$

Proof. (i) Since $\|T\| = \inf\{\lambda \geqslant 0\colon T(\mathrm{B}_X) \subset \lambda \mathrm{B}_Y\}$, it follows that $\mathrm{e}_1(T) \leqslant \|T\|$. On the other hand, if $T(\mathrm{B}_X) \subset y_0 + \lambda \mathrm{B}_Y$ for some $y_0 \in Y$ and some $\lambda \geqslant 0$, then given any $x \in \mathrm{B}_X$, there are elements $y_1, y_2 \in \mathrm{B}_Y$ such that $Tx = y_0 + \lambda y_1$ and $-Tx = y_0 + \lambda y_2$. Hence $2Tx = \lambda(y_1 - y_2)$ and $\|Tx\| \leqslant \lambda$: thus $\|T\| \leqslant \lambda$, whence $\|T\| \leqslant \mathrm{e}_1(T)$. This shows that $\|T\| = \mathrm{e}_1(T)$; and the rest of (i) is obvious.

(ii) Let $\lambda > \mathrm{e}_n(T)$ and $\mu > \mathrm{e}_m(S)$. Then there are points $y_1, \ldots, y_M$, $y'_1, \ldots, y'_N \in Y$ such that $S(\mathrm{B}_X) \subset \bigcup_{i=1}^{M} (y_i + \lambda B_Y)$ and $T(\mathrm{B}_X) \subset \bigcup_{j=1}^{N} (y'_j + \mu \mathrm{B}_Y)$, where $M \leqslant 2^{m-1}$ and $N \leqslant 2^{n-1}$. Given any $x \in B_X$, there are points y_i and y'_j such that $Sx \in y_i + \lambda B_Y$ and $Tx \in y'_j + \mu B_Y$: hence

$$(S+T)x \in y_i + y'_j + (\lambda + \mu)B_Y,$$

and thus

$$(S+T)(B_X) \subset \bigcup_{i=1}^{M} \bigcup_{j=1}^{N} [y_i + y'_j + (\lambda+\mu)B_Y].$$

The number of points $y_i + y'_j$ with $i = 1, 2, \ldots, M$ and $j = 1, 2, \ldots, N$ is at most $MN \leqslant 2^{m+n-2} = 2^{(m+n-1)-1}$: it follows that

$$e_{m+n-1}(S+T) \leqslant e_m(S) + e_n(T).$$

(iii) The proof of this is similar to that of (ii). □

Given any $T \in \mathscr{B}(X,Y)$, it is clear that $T \in \mathscr{K}(X,Y)$ if, and only if, $e_n(T) \to 0$ as $n \to \infty$. To put a little more flesh on the bones of the definition of $e_n(T)$ it may, however, be helpful to estimate $e_n(T)$ in some simple cases, and we now set about this.

Proposition 1.3. Suppose that X is real with $\dim X = m < \infty$, and let $I: X \to X$ be the identity map. Then for all $n \in \mathbb{N}$,

$$1 \leqslant 2^{(n-1)/m} e_n(I) \leqslant 4.$$ ■

Proof. Since $\dim X < \infty$, the space X may be identified with $\mathbb{R}^m$ via a homeomorphism; thus there is a translation-invariant Borel measure μ on X, and without loss of generality we may, and shall, assume that $\mu(B_X) = 1$. Let $\lambda > e_n(I)$, so that for some positive integer $q \leqslant 2^{n-1}$ and some points x_1, $x_2, \ldots, x_q$ in X,

$$B_X \subset \bigcup_{i=1}^{q} (x_i + \lambda B_X).$$

Hence $\mu(B_X) \leqslant \sum_{i=1}^{q} \mu(x_i + \lambda B_X)$, which implies that $1 \leqslant q\lambda^m$. Thus $\lambda \geqslant q^{-1/m} \geqslant 2^{-(n-1)/m}$, from which we see that $e_n(I) \geqslant 2^{-(n-1)/m}$.

As for the remaining inequality, first suppose that $n-1 \geqslant m$ and define ρ by $(1+\rho)/\rho = 2^{(n-1)/m}$ and observe that since B_X is compact, there is a maximal family of elements $x_1, x_2, \ldots, x_p \in B_X$ such that $\|x_i - x_j\| > 2\rho$ if $i \neq j$. The sets $\bar{B}(x_i, \rho)$ $(i = 1, 2, \ldots, p)$ are disjoint and $\bar{B}(x_i, \rho) \subset \bar{B}(0, 1+\rho)$. Hence

$$p\rho^m = \sum_{i=1}^{p} \mu(B(x_i, \rho)) \leqslant \mu((1+\rho)B_X) = (1+\rho)^m,$$

and so $p \leqslant (1+\rho)^m/\rho^m = 2^{n-1}$. Given any $x \in B_X$, there exists an i, with $1 \leqslant i \leqslant p$, such that $\|x - x_i\| \leqslant 2\rho$: hence

$$B_X \subset \bigcup_{i=1}^{p} (x_i + 2\rho B_X).$$

It follows that $e_n(I) \leqslant 2\rho = 2/(2^{(n-1)/m} - 1) \leqslant 4 \cdot 2^{-(n-1)/m}$, since $n-1 \geqslant m$. The proof is complete, since if $n-1 < m$ then $2^{(n-1)/m} < 2$ and thus

$$2^{(n-1)/m} e_n(I) < 2\|I\| = 2.$$

If X is complex, the conclusion of Proposition 1.3 holds with m replaced by $2m$. □

A notion related to the entropy numbers is the *entropy* of a set.

Definition 1.4. Let $T \in \mathcal{K}(X)$, let $\varepsilon > 0$, and let $K(\varepsilon, T)$ be the least number of closed balls of radius ε needed to cover $T(B_X)$. The *ε-entropy* of $T(B_X)$ is defined to be

$$H(\varepsilon, T) = \log_2 K(\varepsilon, T).$$ ■

In a certain sense, $(e_n(T))$ is inverse to $H(\varepsilon, T)$. Our purpose in mentioning $H(\varepsilon, T)$ is to point out that it has connections with quantities which are of interest from the standpoint of the distribution of the eigenvalues of T.

Let $T \in \mathcal{K}(X)$, so that the spectrum of T, apart from the point 0, consists solely of eigenvalues of finite algebraic multiplicity: let $(\lambda_n(T))$ be the sequence of all non-zero eigenvalues of T, repeated according to their algebraic multiplicities and ordered so that

$$|\lambda_1(T)| \geqslant |\lambda_2(T)| \geqslant \ldots \geqslant 0.$$

If T has only m $(< \infty)$ distinct eigenvalues we put $\lambda_n(T) = 0$ for all $n \in \mathbb{N}$ with $n > m$. Finally, given any $\varepsilon > 0$, put

$$N(\varepsilon, T) = \sum_{|\lambda_j(T)| \geqslant \varepsilon} 1, \qquad M(\varepsilon, T) = \sum_{|\lambda_j(T)| \geqslant \varepsilon} \log_2(|\lambda_j(T)|/\varepsilon).$$

The following result, due to Carl and Triebel [1], is of crucial importance in what follows.

Theorem 1.5. Let $T \in \mathcal{K}(X)$ and let $\varepsilon > 0$. Then

$$N(4\varepsilon, T) \leqslant M(2\varepsilon, T) \leqslant \tfrac{1}{2} H(\varepsilon, T).$$ ■

Proof. First we show that for all $n \in \mathbb{N}$,

$$\prod_{j=1}^{n} |\lambda_j(T)| \leqslant 2^n \varepsilon^n \sqrt{[K(\varepsilon, T)]}. \tag{1.1}$$

To prove this it is clearly enough to deal with the case in which $\lambda_n(T) \neq 0$. For this case we claim that there is an n-dimensional subspace X_n of X such that T maps X_n onto itself and $T \restriction X_n$ has precisely the eigenvalues $\lambda_1(T), \ldots, \lambda_n(T)$. To justify this claim, let Y_λ be the linear space of all generalized eigenvalues of T corresponding to the eigenvalue λ:

$$Y_\lambda = \bigcup_{k=1}^{\infty} \{x \in X : (T - \lambda I)^k x = 0\};$$

and let Y be the linear subspace of X spanned by all the Y_λ with $|\lambda| \geqslant |\lambda_n(T)|$. Then Y is invariant under T and

$$n \leqslant \dim Y = \max\{k \in \mathbb{N} : |\lambda_k(T)| = |\lambda_n(T)|\} < \infty.$$

The claim now follows by use of the Jordan canonical form of T in Y. Moreover, by Herstein [1, Theorem 6.J], we may and shall assume that $T\restriction X_n$ is realized by a triangular $n \times n$ matrix $[a_{ij}]$, where $a_{ii} = \lambda_i(T)$ and $a_{ij} = 0$ if $i < j$. There are $\mathrm{K}(\varepsilon, T)$ closed balls in X_n of radius 2ε which cover $T(\mathrm{B}_X \cap X_n)$: the radius is taken to be 2ε rather than ε because the centres of the balls have to lie in X_n rather than X. Now view X_n as the $2n$-dimensional real Euclidean space $\mathbb{R}^{2n}$ and equip it with the usual Lebesgue measure. The map of $\mathbb{R}^{2n}$ which is induced by T is realized by a real $2n \times 2n$ matrix of the form

$$\begin{bmatrix} \begin{matrix} \operatorname{re}\lambda_1(T) & -\operatorname{im}\lambda_1(T) \\ \operatorname{im}\lambda_1(T) & \operatorname{re}\lambda_1(T) \end{matrix} & & & \mathbf{0} \\ & \square & & \\ & & \ddots & \\ & & & \square \end{bmatrix}$$

Let V be the Lebesgue measure of $\mathrm{B}_X \cap X_n$ in $\mathbb{R}^{2n}$, which we denote by $V = |\mathrm{B}_X \cap X_n|$. Then

$$|T(\mathrm{B}_X \cap X_n)| = V \prod_{j=1}^{n} |\lambda_j(T)|^2,$$

since T maps $\mathrm{B}_X \cap X_n$ onto an ellipsoid with half-axes $|\lambda_1(T)|$, $|\lambda_1(T)|$, $|\lambda_2(T)|$, $|\lambda_2(T)|, \ldots, |\lambda_n(T)|$, $|\lambda_n(T)|$. As the volume of a ball in $\mathbb{R}^{2n}$ of radius 2ε is $(2\varepsilon)^{2n}V$ and $\mathrm{K}(\varepsilon, T)$ such balls cover $T(\mathrm{B}_X \cap X_n)$, it follows that

$$\mathrm{K}(\varepsilon, T)(2\varepsilon)^{2n} V \geqslant V \prod_{j=1}^{n} |\lambda_j(T)|^2,$$

and the proof of (1.1) is complete.

The proof of the theorem is now easy. First note that

$$\mathrm{N}(2\varepsilon, T) = \sum_{|\lambda_j(T)| \geqslant 2\varepsilon} 1 \leqslant \sum_{|\lambda_j(T)| \geqslant 2\varepsilon} \log_2 [|\lambda_j(T)|/\varepsilon] \leqslant \mathrm{M}(\varepsilon, T).$$

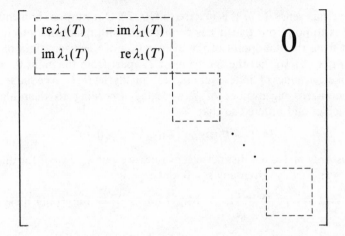

Next, (1.1) shows that

$$\prod_{j=1}^{n} [|\lambda_j(T)|/2\varepsilon] \leqslant \sqrt{K(\varepsilon, T)},$$

and thus for all $n \in \mathbb{N}$,

$$\sum_{j=1}^{n} \log_2 [|\lambda_j(T)|/2\varepsilon] \leqslant \tfrac{1}{2} \log_2 K(\varepsilon, T) = \tfrac{1}{2} \mathrm{H}(\varepsilon, T).$$

This completes the proof. □

Theorem 1.5 may be used to give the following striking connection (due to Carl and Triebel [1]) between the eigenvalues and entropy numbers.

Theorem 1.6. Let $T \in \mathscr{K}(X)$ and let $(\lambda_n(T))$ be as above. Then for all $m, n \in \mathbb{N}$,

$$|\lambda_n(T)| \leqslant \left(\prod_{j=1}^{n} |\lambda_j(T)| \right)^{1/n} \leqslant (\sqrt{2})^{(m-1)/n} \mathrm{e}_m(T). \qquad \blacksquare$$

Proof. Put $\varepsilon = \mathrm{e}_m(T) + \delta$, with $\delta > 0$, in (1.1):

$$\left(\prod_{j=1}^{n} |\lambda_j(T)| \right)^{1/n} \leqslant 2[\mathrm{e}_m(T) + \delta] 2^{(m-1)/2n}.$$

This is true for all $\delta > 0$, and so δ may be replaced by 0. Use this inequality with T replaced by T^N $(N \in \mathbb{N})$, and employ the relation $|\lambda_j(T)|^N \leqslant |\lambda_j(T^N)|$ which follows since (i) $\lambda^N \in \sigma_{\mathrm{p}}(T^N)$ if $\lambda \in \sigma_{\mathrm{p}}(T)$; (ii) $(T - \lambda I)^n x = 0$ implies that $(T^N - \lambda^N I)^n x = 0$, and hence $\dim \bigcup_{n=1}^{\infty} \mathscr{N}(T - \lambda I)^n \leqslant \dim \bigcup_{n=1}^{\infty} \mathscr{N}(T^N - \lambda^N I)^n$. Thus

$$\left(\prod_{j=1}^{n} |\lambda_j(T)| \right)^{1/n} \leqslant \left(\prod_{j=1}^{n} |\lambda_j(T^N)| \right)^{1/nN}$$
$$\leqslant 2^{1/N} [\mathrm{e}_m(T^N)]^{1/N} \cdot 2^{(m-1)/2nN}.$$

Now put $m = N(k-1) + 1$ and use the inequality

$$\mathrm{e}_{N(k-1)+1}(T^N) \leqslant [\mathrm{e}_k(T)]^N$$

which follows from Proposition 1.2(iii). Then

$$\left(\prod_{j=1}^{n} |\lambda_j(T)| \right)^{1/n} \leqslant 2^{1/N} \mathrm{e}_k(T) 2^{(k-1)/2n}$$
$$\to \mathrm{e}_k(T) 2^{(k-1)/2n} \quad \text{as } N \to \infty.$$

This proves the theorem. □

Note that by taking $m = n+1$ the theorem gives the inequality

$$|\lambda_n(T)| \leqslant \sqrt{2}\,\mathrm{e}_{n+1}(T),$$

which is geometrically rather surprising. The theorem also gives rise to the following corollary due to Zemanek [1], which shows that in a sense the entropy numbers may be thought of as deformations of the norm.

Corollary 1.7. Let $T \in \mathscr{K}(X)$. Then for all $n \in \mathbb{N}$,

$$\lim_{k \to \infty} [\mathrm{e}_n(T^k)]^{1/k} = \mathrm{r}(T),$$

the spectral radius of T. ■

Proof. Since $\mathrm{e}_n(T) \leqslant \|T\|$ for all $n \in \mathbb{N}$, we have

$$\limsup_{k \to \infty} [\mathrm{e}_n(T^k)]^{1/k} \leqslant \lim_{k \to \infty} \|T^k\|^{1/k} = \mathrm{r}(T).$$

On the other hand, Theorem 1.6 shows that for all $m \in \mathbb{N}$,

$$|\lambda_1(T)| \leqslant 2^{\frac{1}{2}(m-1)}\mathrm{e}_m(T),$$

and thus for all $k \in \mathbb{N}$,

$$|\lambda_1(T)|^k = |\lambda_1(T^k)| \leqslant 2^{\frac{1}{2}(m-1)}\mathrm{e}_m(T^k).$$

Hence

$$|\lambda_1(T)| \leqslant 2^{(m-1)/2k}[\mathrm{e}_m(T^k)]^{1/k},$$

whence

$$|\lambda_1(T)| \leqslant \liminf_{k \to \infty} [\mathrm{e}_m(T^k)]^{1/k},$$

so that

$$\mathrm{r}(T) \leqslant \liminf_{k \to \infty} [\mathrm{e}_m(T^k)]^{1/k}.$$

The proof is complete. □

We conclude this section by remarking that a problem which has attracted a good deal of attention, and which is still not completely resolved, is to elucidate the relationship between the entropy numbers of a map $T \in \mathscr{B}(X,Y)$ and those of its adjoint T^*. Following work by Carl [3], Gordon, König and Schütt [1] have shown that under certain hypotheses on the spaces X and Y, namely that X and Y^* are of type 2 (a hypothesis satisfied if, for example, X and Y are Hilbert spaces), then there are positive constants c and d, depending only on X and Y, such that for all $n \in \mathbb{N}$,

$$d^{-1}\mathrm{e}_{[nc]}(T) \leqslant \mathrm{e}_n(T^*) \leqslant d\,\mathrm{e}_{[nc]}(T),$$

where $[nc]$ denotes the integer part of nc.

In the absence of hypotheses on X and Y nothing as sharp as this has yet been proved, although Edmunds and Tylli [1] have shown that for all k and n in $\mathbb{N}$,

$$e_k(T^*) \leqslant 2(1+c_{k,2^n})e_n(T) + \|T\|c_{k,2^n}$$

and

$$e_k(T) \leqslant 4(1+c_{k,2^n})e_n(T^*) + 2\,\|T\|c_{k,2^n},$$

where

$$c_{k,m} = \min\{1, 2(2^{(k-1)/m} - 1)^{-1}\}.$$

Even though these inequalities do not seem to perform well in terms of summability (on account of having to make choices of k and n to take care of the correction terms involving $\|T\|$), they still seem appropriate analogues of the formulae of Theorem I.2.10: for example, from the first of these inequalities we obtain, by passage to the limit, the inequality $\tilde{\beta}(T^*) \leqslant 2\tilde{\beta}(T)$.

However, in the Hilbert-space case, much more can be said for compact operators, as the following theorem, due to Edmunds and Edmunds [2], makes plain.

Theorem 1.8. Let $T \in \mathscr{K}(H)$, where H is a Hilbert space. Then

$$e_n(T) = e_n(T^*) = e_n(|T|) \text{ for all } n \in \mathbb{N}. \qquad \blacksquare$$

We shall postpone the proof until §5, when the necessary machinery will have been developed and the definition of $|T|$ given.

2. Approximation numbers

These numbers measure the closeness by which a bounded linear map may be approximated by similar maps but with finite-dimensional range. A comprehensive account is given in the book by Pietsch [1].

Definition 2.1. Let X and Y be normed vector spaces, $T \in \mathscr{B}(X, Y)$ and $n \in \mathbb{N}$. The nth approximation number $a_n(T)$ of T is defined by

$$a_n(T) = \inf\{\|T - L\| : L \in \mathscr{B}(X, Y) \text{ and rank } L < n\}.$$

Here rank $L := \dim \mathscr{R}(L)$. ■

These numbers have various properties similar to those of the entropy numbers and summarized in the following proposition.

Proposition 2.2. Let X, Y, Z be normed vector spaces, let $S, T \in \mathscr{B}(X, Y)$ and suppose that $R \in \mathscr{B}(Y, Z)$. Then:

(i) $\|T\| = a_1(T) \geqslant a_2(T) \geqslant \ldots \geqslant 0$;

(ii) for all $m, n \in \mathbb{N}$, $a_{m+n-1}(S+T) \leqslant a_m(S) + a_n(T)$;

(iii) for all $m, n \in \mathbb{N}$, $a_{m+n-1}(RS) \leqslant a_m(R)a_n(S)$. ■

Proof. The proof of (i) is obvious. To establish (ii), note that given any $\varepsilon > 0$, there exist maps $S', T' \in \mathscr{B}(X,Y)$, with rank $S' < m$ and rank $T' < n$, such that $\|S - S'\| < a_m(S) + \varepsilon$ and $\|T - T'\| < a_n(T) + \varepsilon$. Since rank $(S' + T') < n + m - 1$ we have

$$a_{m+n-1}(S+T) \leqslant \|S+T-S'-T'\| < a_m(S) + a_n(T) + 2\varepsilon,$$

and (ii) follows. As for (iii), given any $\varepsilon > 0$, there are maps $\tilde{R} \in \mathscr{B}(Y, Z)$ and $\tilde{S} \in \mathscr{B}(X, Y)$, with rank $\tilde{R} < m$ and rank $\tilde{S} < n$, such that $\|R - \tilde{R}\| < a_m(R) + \varepsilon$ and $\|S - \tilde{S}\| < a_n(S) + \varepsilon$. Since rank $[\tilde{R}(S - \tilde{S}) + R\tilde{S}] < m + n - 1$, we have

$$\begin{aligned} a_{m+n-1}(RS) &\leqslant \|RS - \tilde{R}(S - \tilde{S}) - R\tilde{S}\| \leqslant \|R - \tilde{R}\| \, \|S - \tilde{S}\| \\ &< [a_m(R) + \varepsilon]\,[a_n(S) + \varepsilon], \end{aligned}$$

and (iii) follows. □

Further useful properties of the approximation numbers are given below.

Proposition 2.3. Let X and Y be normed vector spaces and let $T \in \mathscr{B}(X,Y)$. Then:

(i) $a_n(T) = 0$ if, and only if, rank $T < n$;

(ii) if $\dim Y \geqslant n$ and there is a map $S \in \mathscr{B}(Y, X)$ with $TSy = y$ for all $y \in Y$, then $a_n(T)\|S\| \geqslant 1$;

(iii) if Y is a linear subspace of a normed vector space Z and has the norm induced by that of Z, then (with the natural interpretation of the notation) $a_n^Z(T) \leqslant a_n^Y(T)$ for all $n \in \mathbb{N}$, with equality if Y is dense in Z. ■

Proof. (i) Suppose that $a_n(T) = 0$ for some $n \in \mathbb{N}$, while $\dim \mathscr{R}(T) \geqslant n$. Then there are n linearly independent elements $y_i = Tx_i$ $(i = 1, 2, \ldots, n)$ and n elements $f_i \in Y^*$ $(i = 1, 2, \ldots, n)$ with $f_i(y_j) = \delta_{ij}$. Since $\det[\delta_{ij}] = 1$, there exists $\delta > 0$ such that $\det[\alpha_{ij}] \neq 0$ if $|\delta_{ij} - \alpha_{ij}| < \delta$ for all i and j. Moreover, since $a_n(T) = 0$, there exists $L \in \mathscr{B}(X,Y)$ such that rank $L < n$ and $\|T - L\| < \delta / \max\limits_{i,j} (\|x_i\| \, \|f_j\|)$. Since

$$|\delta_{ij} - f_j(Lx_i)| = |f_j(Tx_i - Lx_i)| \leqslant \|f_j\| \, \|x_i\| \, \|T - L\| < \delta,$$

it follows that $\det[f_j(Lx_i)] \neq 0$. But because $\dim \mathscr{R}(L) < n$, the n elements Lx_i $(i = 1, 2, \ldots, n)$ must be linearly dependent and thus $\det[f_j(Lx_i)] = 0$. This contradiction shows that $\dim \mathscr{R}(T) < n$. The converse is obvious.

(ii) Under the given hypothesis, suppose that $a_n(T)\|S\| < 1$. Then there exists $L \in \mathscr{B}(X,Y)$, with rank $L < n$, such that $\|T - L\| \, \|S\| < 1$. Let I be the identity map of Y to itself: then $I - (T - L)S$ must be invertible; that is, $(LS)^{-1} \in \mathscr{B}(Y)$. This is impossible, since $\dim \mathscr{R}(LS) < n$ and $\dim Y \geqslant n$.

(iii) We deal only with the question of equality when Y is dense in Z, the rest being obvious. Given any $\varepsilon > 0$, there exists $L \in \mathscr{B}(X, Z)$ such that rank $L < n$ and $\|T - L\| < a_n^Z(T) + \varepsilon$. Moreover, for some $f_k \in X^*$ and $z_k \in Z$

$(k = 1, 2, \ldots, n-1)$ we have $Lx = \sum_{k=1}^{n-1} f_k(x) z_k$ for all $x \in X$; and by the density of Y in Z, there exist $y_1, \ldots, y_{n-1} \in Y$ such that $\sum_{k=1}^{n-1} \|f_k\| \, \|z_k - y_k\| < \varepsilon$. Define $S \in \mathscr{B}(X, Y)$ by $Sx = \sum_{k=1}^{n-1} f_k(x) y_k \ (x \in X)$. Then $\|S - L\| < \varepsilon$ and so

$$\mathrm{a}_n^Y(T) \leqslant \|T - S\| \leqslant \|T - L\| + \|L - S\| < \mathrm{a}_n^Z(T) + 2\varepsilon.$$

This completes the proof. □

Corollary 2.4. If $\dim X \geqslant n$ and $I: X \to X$ is the identity map, then $\mathrm{a}_k(I) = 1$ for $k = 1, 2, \ldots, n$. ■

Proof. By Proposition 2.3(ii), $\mathrm{a}_n(I) \geqslant 1$. Since $\mathrm{a}_n(I) \leqslant \|I\| = 1$, it follows that $\mathrm{a}_n(I) = 1$. Hence $1 = \mathrm{a}_1(I) \geqslant \mathrm{a}_2(I) \geqslant \ldots \geqslant \mathrm{a}_n(I) = 1$, and the result is clear. □

Proposition 2.5. Let $T \in \mathscr{K}(X, Y)$ and let $n \in \mathbb{N}$. Then $\mathrm{a}_n(T) = \mathrm{a}_n(T^*)$. ■

Proof. Let $\varepsilon > 0$. Then there is a map $L \in \mathscr{B}(X^{**}, Y^{**})$ such that rank $L < n$ and $\|T^{**} - L\| \leqslant \mathrm{a}_n(T^{**}) + \varepsilon$. Moreover, there are points $y_1, y_2, \ldots, y_m \in Y$ such that $T(\mathrm{B}_X) \subset \bigcup_{i=1}^{m} (y_i + \varepsilon \mathrm{B}_Y)$, as T is compact. Let M be a finite-dimensional subspace of Y^{**} which contains $\mathscr{R}(L)$ and $\mathrm{J}_Y y_1, \ldots, \mathrm{J}_Y y_m$, where J_Y is the canonical map of Y to Y**. By the principle of local reflexivity (Pietsch [1, Part 0, E.3.1]), there exists $R \in \mathscr{B}(M, Y)$ such that $\|R\| \leqslant 1 + \varepsilon$ and $\mathrm{J}_Y R y^{**} = y^{**}$ for all $y^{**} \in \mathscr{R}(\mathrm{J}_Y) \cap M$. Thus $R \mathrm{J}_Y y_i = y_i$ for i $= 1, 2, \ldots, m$.

Define $L_0 \in \mathscr{B}(X, Y)$ by $L_0 x = RL\mathrm{J}_X x \ (x \in X)$; then rank $L_0 < n$. Let $x \in B_X$. Then there exists $i \in \{1, 2, \ldots, m\}$ such that $\|Tx - y_i\| \leqslant \varepsilon$. Hence

$$\begin{aligned}\|Tx - L_0 x\| &\leqslant \|Tx - y_i\| + \|y_i - L_0 x\| \leqslant \varepsilon + \|R\mathrm{J}_Y y_i - RL\mathrm{J}_X x\| \\ &\leqslant \varepsilon + (1+\varepsilon)\|\mathrm{J}_Y y_i - L\mathrm{J}_X x\| \\ &\leqslant \varepsilon + (1+\varepsilon)(\|\mathrm{J}_Y y_i - \mathrm{J}_Y Tx\| + \|\mathrm{J}_Y Tx - L\mathrm{J}_X x\|) \\ &\leqslant \varepsilon + (1+\varepsilon)(\varepsilon + \|T^{**}\mathrm{J}_X x - L\mathrm{J}_X x\|) \\ &\leqslant \varepsilon + (1+\varepsilon)[\mathrm{a}_n(T^{**}) + 2\varepsilon].\end{aligned}$$

Thus

$$\mathrm{a}_n(T) \leqslant \|T - L_0\| \leqslant \varepsilon + (1+\varepsilon)[\mathrm{a}_n(T^{**}) + 2\varepsilon].$$

It follows that $\mathrm{a}_n(T) \leqslant \mathrm{a}_n(T^{**})$. However, given any $S \in \mathscr{B}(X, Y)$, there is a map $F \in \mathscr{B}(X, Y)$ with rank $F < n$ and $\|S - F\| \leqslant (1+\varepsilon)\mathrm{a}_n(S)$; and since rank $F^* < n$ and $\|S^* - F^*\| = \|S - F\| \leqslant (1+\varepsilon)\mathrm{a}_n(S)$, we see that $\mathrm{a}_n(S^*) \leqslant (1+\varepsilon)\mathrm{a}_n(S)$; that is, $\mathrm{a}_n(S^*) \leqslant \mathrm{a}_n(S)$.

Thus $a_n(T) \leqslant a_n(T^{**}) \leqslant a_n(T^*) \leqslant a_n(T)$. The proof is complete. □

Remark 2.6. (i) Proposition 2.5 is false if T is merely required to be in $\mathscr{B}(X,Y)$. For if T and S are the natural embeddings of l^1 in c_0 and of l^1 in l^∞ respectively, then $a_n(T) = 1$ and $a_n(S) = \frac{1}{2}$ for all $n \in \mathbb{N}$ (cf. Pietsch [1, 11.11.9 and 11.11.10]); and $S = T^*$.

(ii) For all $T \in \mathscr{B}(X,Y)$ and all $n \in \mathbb{N}$, we have $a_n(T^*) \leqslant a_n(T)$. This follows from the proof of Proposition 2.5.

(iii) If Y is a Banach space such that there is a projection $P: Y^{**} \to Y$ with $\|P\| = 1$, then for all $T \in \mathscr{B}(X,Y)$ and all $n \in \mathbb{N}$,

$$a_n(T^*) = a_n(T).$$

To see this, let $\lambda > a_n(T^{**})$. Then there exists $K \in \mathscr{B}(X^{**}, Y^{**})$ with rank $K < n$ and $\|T^{**} - K\| < \lambda$. Define $A = PKJ_X$, where J_X is the canonical embedding of X in X^{**}; then rank $A < n$ and

$$\|T - A\| = \|PT^{**}J_X - PKJ_X\| \leqslant \|T^{**} - K\| < \lambda.$$

Thus $a_n(T) \leqslant a_n(T^{**})$; and since by (ii), $a_n(T^{**}) \leqslant a_n(T^*)$, this gives $a_n(T) \leqslant a_n(T^*)$. Together with (ii) this gives the result. It follows that $a_n(T^*) = a_n(T)$ whenever Y is reflexive.

(iv) Even if no conditions are imposed on the spaces, a connection can still be established between $a_n(T)$ and $a_n(T^*)$ when $T \in \mathscr{B}(X,Y)$. For,

$$a_n(T) \leqslant a_n(T^{**}) + 2\beta(T)$$

and hence

$$a_n(T) \leqslant 5\, a_n(T^*).$$

To establish these inequalities, which were proved in Edmunds and Tylli [1], let $\varepsilon > 0$ and $\lambda > \beta(T)$. Then there exist $A \in \mathscr{B}(X^{**}, Y^{**})$, with rank $A < n$ and with $\|T^{**} - A\| < a_n(T^{**}) + \varepsilon$, and $y_1, y_2, \ldots, y_k \in Y$ with $T(B_X) \subset \{y_1, \ldots, y_k\} + \lambda B_Y$. Let M be the linear span of $T(X) \cup \{J_Y y_i : i = 1, 2, \ldots, k\}$. By the principle of local reflexivity there exists $R: M \to Y$ such that $\|R\| \leqslant 1 + \varepsilon$ and $RJ_Y y_i = y_i$ for $i = 1, 2, \ldots, k$. Put $A_0 = RAJ_X \in \mathscr{B}(X,Y)$; then rank $A_0 < n$. Let $x \in B_X$ and choose y_i with $\|Tx - y_i\| \leqslant \lambda$. Thus

$$\begin{aligned}
\|Tx - A_0 x\| &\leqslant \|Tx - y_i\| + \|y_i - A_0 x\| \\
&\leqslant \lambda + \|RJ_Y y_i - RAJ_X x\| \\
&\leqslant \lambda + (1+\varepsilon)(\|J_Y y_i - J_Y Tx\| + \|T^{**}J_X x - AJ_X x\|) \\
&\leqslant \lambda + (1+\varepsilon)[\lambda + a_n(T^{**}) + \varepsilon],
\end{aligned}$$

which shows that $a_n(T) \leqslant a_n(T^{**}) + 2\beta(T)$ as claimed.

To derive the second inequality, note that since $a_n(T^{**}) \leqslant a_n(T^*)$ and, by Corollary I.2.10, $\tilde{\beta}(T) \leqslant 2\tilde{\beta}(T^*)$, we have (cf. Proposition 2.7)

$$\begin{aligned} a_n(T) &\leqslant a_n(T^*) + 4\tilde{\beta}(T^*) \\ &\leqslant 5a_n(T^*). \end{aligned}$$

Let $T \in \mathscr{B}(X,Y)$. We have seen that the sequence $(a_n(T))$ of approximation numbers of T is monotonic decreasing and bounded below by 0; it therefore converges, so that we may define a real number $\alpha(T)$ by

$$\alpha(T) = \lim_{n\to\infty} a_n(T). \tag{2.1}$$

We also set, in line with Chapter I,

$$\|T\|_{\mathscr{K}} = \operatorname{dist}[T, \mathscr{K}(X,Y)] = \inf\{\|T-K\| : K \in \mathscr{K}(X,Y)\}. \tag{2.2}$$

Propositon 2.7. Let $T \in \mathscr{B}(X,Y)$. Then $\tilde{\beta}(T) \leqslant \|T\|_{\mathscr{K}} \leqslant \alpha(T)$. ■

Proof. Since $\alpha(T) = \operatorname{dist}[T, \mathscr{E}(X,Y)]$, where $\mathscr{E}(X,Y) = \{T \in \mathscr{B}(X,Y) : \dim \mathscr{R}(T) < \infty\}$, it is clear that $\|T\|_{\mathscr{K}} \leqslant \alpha(T)$. To establish the remaining inequality, note that for all $K \in \mathscr{K}(X,Y)$, we have $\tilde{\beta}(T) = \tilde{\beta}(T+K) \leqslant \|T+K\|$, and so $\tilde{\beta}(T) \leqslant \|T\|_{\mathscr{K}}$. □

Proposition 2.7 is due to Zemánek [1].

Note that it may well be that $\|T\|_{\mathscr{K}} < \alpha(T)$: this is a consequence of Per Enflo's celebrated work [1] on the basis problem, as a result of which there exist Banach spaces X and Y and a map $K \in \mathscr{K}(X,Y)$ such that K cannot be approximated arbitrarily closely by linear maps of finite rank. This kind of pathology is impossible in a Hilbert-space setting where, as is well known, $\alpha(T) = \|T\|_{\mathscr{K}}$; this follows immediately from Theorem 5.7 below. Note also that if $T \in \mathscr{B}(X)$, then since $\|T\|_{\mathscr{K}}$ is the usual quotient norm in $\mathscr{B}(X)/\mathscr{K}(X)$ of the equivalence class containing T, we see from Corollary I.4.11 that

$$r_e(T) = \lim_{n\to\infty} \|T^n\|_{\mathscr{K}}^{1/n}.$$

The approximation numbers have important connections with eigenvalues, connections which are significant from the point of view of applications to differential equations. The picture is clearest in a Hilbert-space context, when the results are of a classical nature: a detailed account of this will be given in §5. The essence of the matter is that if X is a complex Hilbert space and $T \in \mathscr{K}(X)$, then T^*T has a positive self-adjoint square root $|T| \in \mathscr{K}(X)$, and for all $n \in \mathbb{N}$,

$$a_n(T) = \lambda_n(|T|),$$

the $\lambda_n(|T|)$ being (following the convention of §1) the eigenvalues of $|T|$, repeated according to algebraic multiplicity and ordered so that $\lambda_n(|T|) \geqslant \lambda_{n+1}(|T|)$ for all $n \in \mathbb{N}$. The algebraic multiplicity is also the geometric multiplicity as $|T|$ is self-adjoint: we shall see in §5 that for any self-adjoint compact map the geometric and algebraic multiplicities coincide. The eigenvalues of $|T|$ are often called the *singular values* of T, and seem to have been first considered seriously by E. Schmidt. We observe that if $T \in \mathscr{K}(X)$ is positive and self-adjoint, then $a_n(T) = \lambda_n(T)$ for all $n \in \mathbb{N}$: this relation is of considerable importance in a good deal of work on the asymptotic distribution of eigenvalues of elliptic, self-adjoint operators.

In a general Banach space one could hardly expect to obtain such dramatic results. Nevertheless, a surprising amount can be said, and one of the most interesting developments is that contained in the following theorem, due to König [3]:

Theorem 2.8. Let $T \in \mathscr{K}(X)$. Then for all $n \in \mathbb{N}$,

$$|\lambda_n(T)| = \lim_{k \to \infty} [a_n(T^k)]^{1/k}. \qquad \blacksquare$$

Proof. First we show that for all $n \in \mathbb{N}$,

$$\limsup_{k \to \infty} [a_n(T^k)]^{1/k} \leqslant |\lambda_n(T)|. \tag{2.3}$$

When $n = 1$ this is clear, for

$$[a_1(T^k)]^{1/k} = \|T^k\|^{1/k} \to r(T) = |\lambda_1(T)|.$$

The proof now proceeds by induction. Assume that (2.3) holds with $n = m-1$, for some $m \geqslant 2$. We may suppose that $\lambda_{m-1}(T) \neq 0$, for otherwise $\lambda_m(T) = 0$ and by the inductive hypothesis,

$$\limsup_{k \to \infty} [a_m(T^k)]^{1/k} \leqslant \limsup_{k \to \infty} [a_{m-1}(T^k)]^{1/k}$$
$$\leqslant |\lambda_{m-1}(T)| = 0,$$

and (2.3) holds with $n = m$. Take k to be the least positive integer such that $\lambda_{m-k}(T) \neq \lambda_m(T)$; if no such k exists, then $|\lambda_m(T)| = |\lambda_1(T)| = r(T)$ and (2.3) again holds with $n = m$. Of course, $k = 1$ if $\lambda_{m-1}(T)$ has algebraic multiplicity equal to one. Let P be the spectral projection (see Taylor [1, §5.7]) onto the (finite-dimensional) subspace of X spanned by the generalized eigenvectors of T corresponding to $\lambda_1(T), \ldots, \lambda_{m-k}(T)$ (see the discussion following Theorem I.1.20). Evidently $\dim \mathscr{R}(P) = m-k$ and hence $\dim \mathscr{R}(PT^n) \leqslant m-k$ for all $n \in \mathbb{N}$. Thus

$$a_m(T^n) \leqslant a_{m-k+1}(T^n) \leqslant \|T^n - PT^n\| = \|(I-P)T^n\|$$
$$= \|[(I-P)T]^n\|$$

since P commutes with T and $I-P$ is a projection. As

$$\sigma\big((I-P)T\big) = \sigma(T)\setminus \bigcup_{i=1}^{m-k} \{\lambda_i(T)\} = \bigcup_{i\in\mathbb{N}} \{\lambda_{m+i-k}(T)\},$$

it follows that

$$\begin{aligned}\limsup_{n\to\infty} [\mathrm{a}_m(T^n)]^{1/n} &\leqslant \lim_{n\to\infty} \|\{(I-P)T\}^n\|^{1/n} \\ &= \mathrm{r}\big((I-P)T\big) = |\lambda_{m-k+1}(T)| = |\lambda_m(T)|,\end{aligned}$$

and our inductive proof of (2.3) is complete.

To finish the proof we show that for all $n\in\mathbb{N}$,

$$\liminf_{k\to\infty} [\mathrm{a}_n(T^k)]^{1/k} \geqslant |\lambda_n(T)|. \tag{2.4}$$

This is clear when $n = 1$, and we once more proceed by induction. Assume that (2.4) holds with $n = m-1$ for some $m \geqslant 2$. If $\lambda_m(T) = 0$ there is nothing to prove and we therefore suppose that $\lambda_m(T) \neq 0$. Let P be the spectral projection corresponding to $\Lambda := \{\lambda_1(T),\ \lambda_2(T), \ldots, \lambda_m(T)\}$. Then $k := \dim \mathscr{R}(P) \geqslant m$ and $\lambda_k(T) = \lambda_m(T)$, so that $k = m$ if $\lambda_m(T)$ has algebraic multiplicity equal to one. Since T and P commute, T maps $\mathscr{R}(P)$ onto itself and the restriction $\tilde{T}$ of T to $\mathscr{R}(P)$ is in $\mathscr{B}\big(\mathscr{R}(P)\big)$; since $0 \notin \sigma(\tilde{T}) = \Lambda$, the map $\tilde{T}$ is injective and is therefore a homeomorphism since $\dim \mathscr{R}(P) < \infty$. Let A: $\mathscr{R}(P) \to X$ be the natural injection and set $B_n = (\tilde{T}^{-1})^n P : X \to \mathscr{R}(P)$ $(n\in\mathbb{N})$. Then for each $n\in\mathbb{N}$, the map $B_n T^n A$ is the identity on $\mathscr{R}(P)$, and consequently, by Corollary 2.4 and Proposition 2.2 (iii), since $\dim \mathscr{R}(P) = k$,

$$1 = \mathrm{a}_k(B_n T^n A) \leqslant \mathrm{a}_k(T^n)\|A\|\,\|B_n\|.$$

Hence

$$\begin{aligned}\mathrm{a}_m(T^n) \geqslant \mathrm{a}_k(T^n) &\geqslant \|A\|^{-1}\|B_n\|^{-1} \\ &\geqslant \|P\|^{-1}\,\|(\tilde{T}^{-1})^n\|^{-1}.\end{aligned}$$

The eigenvalue of $\tilde{T}^{-1}$ with the largest modulus is $[\lambda_k(T)]^{-1} = [\lambda_m(T)]^{-1}$. Thus by the formula for the spectral radius of $\tilde{T}^{-1}$,

$$\liminf_{n\to\infty} [\mathrm{a}_m(T^n)]^{1/n} \geqslant [\mathrm{r}(\tilde{T}^{-1})]^{-1} = |\lambda_m(T)|,$$

and our proof of (2.4), and hence of the theorem, is complete. □

We remark that, in general, no inequality of the form $\mathrm{a}_n(T) \geqslant (\leqslant)\, |\lambda_n(T)|$ for all $n\in\mathbb{N}$ can be expected. To illustrate, take $X = \mathbb{C}^2$ and let $T\in\mathscr{B}(\mathbb{C}^2)$ be represented by the matrix

$$\begin{bmatrix} 2 & 0 \\ 1 & 1 \end{bmatrix}.$$

Then $\lambda_1(T) = 2$ and $\lambda_2(T) = 1$; but since T^*T is represented by

$$\begin{bmatrix} 5 & 1 \\ 1 & 1 \end{bmatrix}$$

which has eigenvalues $3 \pm \sqrt{5}$, it follows that since (as will be proved in §5) $a_i(T) = \lambda_i\left((T^*T)^{1/2}\right)$, we have

$$a_1(T) = \sqrt{(3+\sqrt{5})} > \lambda_1(T), \qquad a_2(T) = \sqrt{(3-\sqrt{5})} < \lambda_2(T).$$

However, there is an inequality, due to Carl and Triebel [1], between $|\lambda_n(T)|$ and an appropriate power of $a_n(T)$. To establish this the following lemma is useful.

Lemma 2.9. Let $T \in \mathscr{B}(X,Y)$. Then for all $n \in \mathbb{N}$ and all $m \in \mathbb{N}$ with $m > 1$,

$$e_n(T) \leqslant a_m(T) + 8 \cdot 2^{-(n-1)/2(m-1)} \|T\|. \qquad ■$$

Proof. Let $m, n \in \mathbb{N}$, with $m > 1$, and let $\varepsilon > 0$. Then there exists $L \in \mathscr{B}(X,Y)$ with rank $L < m$, such that $\|T - L\| < (1+\varepsilon)a_m(T)$. Let $Q: X \to X/\mathscr{N}(L)$ be the canonical map, let $J: \mathscr{R}(L) \to Y$ be the natural embedding and observe that there is an injective map $L_0 \in \mathscr{B}\left(X/\mathscr{N}(L),\ \mathscr{R}(L)\right)$, where $L_0\xi = Lx$ ($\xi \in X/\mathscr{N}(L)$, $x \in \xi$), such that $\mathscr{R}(L_0) = \mathscr{R}(L)$, $\|L_0\| = \|L\|$, and $L = LJ_0Q$. Only the assertion that $\|L_0\| = \|L\|$ needs proof here. However, given any $\xi \in X/\mathscr{N}(L)$ and any $x \in \xi$,

$$\|L_0\xi\| = \|Lx\| \leqslant \|L\|\,\|x\|;$$

and since $\|\xi\| = \inf\{\|x\| : x \in \xi\}$, it follows that $\|L_0\xi\| \leqslant \|L\|\,\|\xi\|$, which shows that $\|L_0\| \leqslant \|L\|$. Moreover, for any $\xi \in X/\mathscr{N}(L)$ and any $x \in \xi$,

$$\|Lx\| = \|L_0\xi\| \leqslant \|L_0\|\,\|\xi\| \leqslant \|L_0\|\,\|x\|;$$

thus $\|L\| \leqslant \|L_0\|$ and so $\|L\| = \|L_0\|$.

Hence by Proposition 1.2 (iii) and the remark following Proposition 1.3, and with I as the identity map of $\mathscr{R}(L)$ to itself,

$$e_n(L) \leqslant \|J\| e_n(I) \|L_0\|\,\|Q\| \leqslant 4 \cdot 2^{-(n-1)/2(m-1)} \|L\|.$$

Together with Proposition 1.2 (ii) this gives

$$e_n(T) \leqslant \|T - L\| + e_n(L) \leqslant (1+\varepsilon)a_m(T) + 4 \cdot 2^{-(n-1)/2(m-1)} \|L\|.$$

Since $\|L\| \leqslant \|T\| + \|T - L\| \leqslant \|T\| + (1+\varepsilon)a_m(T) \leqslant (2+\varepsilon)\|T\|$, the result follows. □

Theorem 2.10. Let $T \in \mathscr{K}(X,Y)$ and suppose that $\|T\| \leqslant 1$. Then for all $n \in \mathbb{N}$ and all $m \in \{0, 1, 2, \ldots, n-1\}$,

$$\left(\prod_{i=1}^{n} |\lambda_i(T)|\right)^{1/n} \leqslant 16 a_{m+1}^{(n-m)/n}(T). \qquad ■$$

Proof. Let $n \in \mathbb{N}$ and $m \in \{0, 1, \ldots, n-1\}$. If $m = 0$ the result is obvious, since $a_1(T) = \|T\| \geq |\lambda_1(T)|$. Suppose that $m \geq 1$. By Lemma 2.9, for any $k \in \mathbb{N}$,

$$\mathrm{e}_{k+1}(T) \leq \mathrm{a}_{m+1}(T) + 8 \cdot 2^{-k/2m}.$$

Suppose that $\mathrm{a}_{m+1}(T) \neq 0$. Then we choose $k \in \mathbb{N}$ so that $8 \cdot 2^{-k/2m} \leq \mathrm{a}_{m+1}(T)$: our choice is $k = 6m + 1 + [-\log_2 \mathrm{a}_{m+1}^{2m}(T)]$. Hence $\mathrm{e}_{k+1}(T) \leq 2\mathrm{a}_{m+1}(T)$, and since by Theorem 1.6,

$$\left(\prod_{j=1}^{n} |\lambda_j(T)| \right)^{1/n} \leq (\sqrt{2})^{k/n} \mathrm{e}_{k+1}(T) \leq 2(\sqrt{2})^{k/n} \mathrm{a}_{m+1}(T),$$

the result follows on substitution of our chosen value for k. If $\mathrm{a}_{m+1}(T) = 0$, we take any $\varepsilon > 0$, choose $k \in \mathbb{N}$ so that $8 \cdot 2^{-k/2m} \leq \varepsilon$ and proceed as above to obtain an upper bound for $\left(\prod_{j=1}^{n} |\lambda_j(T)| \right)^{1/n}$ in terms of ε. We may then allow ε to tend to 0 to obtain the desired result. □

Note that if we take $m = n - 1$ in Theorem 2.10 we find that

$$|\lambda_n(T)| \leq 16 \mathrm{a}_n^{1/n}(T) \qquad (n \in \mathbb{N}).$$

The connections between eigenvalues and approximation numbers have naturally been most extensively analysed in a Hilbert-space context. Thus if H is a complex Hilbert space and $T \in \mathscr{K}(H)$, there is a celebrated inequality due to Weyl [4] which states that for all $n \in \mathbb{N}$,

$$\prod_{j=1}^{n} |\lambda_j(T)| \leq \prod_{j=1}^{n} \mathrm{a}_j(T), \tag{2.5}$$

from which it may be deduced that for all $n \in \mathbb{N}$ and all $p \in (0, \infty)$,

$$\sum_{j=1}^{n} |\lambda_j(T)|^p \leq \sum_{j=1}^{n} \mathrm{a}_j^p(T). \tag{2.6}$$

This implies that if $(\mathrm{a}_j(T)) \in l^p$ then $(\lambda_j(T)) \in l^p$, a result which we shall use in Chapter XII to obtain information about the distribution of eigenvalues of certain non-self-adjoint elliptic problems. These classical inequalities (2.5) and (2.6) will be derived at the end of §5 by methods peculiar to Hilbert spaces; but here we shall take advantage of some very recent developments in Banach-space theory to obtain analogues of these inequalities for compact maps acting in a Banach space. An interesting feature of the proof given here, which is essentially that of Carl [1], with modifications due to Teixeira [2], is its heavy dependence on entropy numbers. We begin with some preliminary lemmas.

Lemma 2.11. Let X be a Banach space with $\dim X = n < \infty$, and let $I: X \to X$

be the identity map. Then given any $q \in (0, \infty)$ there is a positive constant c_q such that

$$\left(\sum_{k=1}^{\infty} \mathrm{e}_k^q(I)\right)^{1/q} \leqslant c_q n^{1/q}.$$

The constant c_q can be taken to be

$$2^{5/2}\,(2/q)^{1/q}. \qquad \blacksquare$$

Proof. By the remark following Proposition 1.3, we know that for all $k \in \mathbb{N}$,

$$\mathrm{e}_k(I) \leqslant 4 \cdot 2^{-(k-1)/2n}.$$

Hence

$$\left(\sum_{k=1}^{\infty} \mathrm{e}_k^q(I)\right)^{1/q} \leqslant 4\left(\sum_{k=1}^{\infty} 2^{-q(k-1)/2n}\right)^{1/q}$$

$$= 4\left(\frac{2^{q/2n}}{2^{q/2n}-1}\right)^{1/q}.$$

Since $2^{q/2n} = (1+1)^{q/2n} \geqslant 1 + q/2n$, we obtain

$$\left(\sum_{k=1}^{\infty} \mathrm{e}_k^q(I)\right)^{1/q} \leqslant 4 \cdot 2^{1/2n}\left(\frac{q}{2n}\right)^{-1/q} \leqslant 2^{5/2}(2/q)^{1/q} n^{1/q}. \qquad \square$$

Lemma 2.12. Let $0 < q < p < \infty$, let $T \in \mathscr{B}(X,Y)$, and suppose that rank $T \leqslant n \in \mathbb{N}$. Then there is a positive constant $\rho_{p,q}$, depending only upon p and q, such that

$$\left(\sum_{k=1}^{\infty} \mathrm{e}_k^q(T)\right)^{1/q} \leqslant \rho_{p,q} n^{1/q-1/p} \sup_{1 \leqslant k \leqslant n} k^{1/p} \mathrm{a}_k(T). \qquad \blacksquare$$

Proof. Put $N = [\log_2 n]$, so that $2^N \leqslant n < 2^{N+1}$. Given any $\varepsilon > 0$, there are maps $T_j \in \mathscr{B}(X,Y)$, with rank $T_j < 2^j$, such that for $j = 0, 1, \ldots, N+1$,

$$\|T - T_j\| \leqslant (1+\varepsilon)\, a_{2^j}(T),$$

with $T_0 = 0$ and $T_{N+1} = T$. For $j = 1, 2, \ldots, N+1$ put $S_j = T_j - T_{j-1}$: then $T = \sum_{j=1}^{N+1} S_j$ and rank $S_j < 3 \cdot 2^{j-1}$. Note that for $j = 1, 2, \ldots, N+1$,

$$\|S_j\| \leqslant \|T - T_j\| + \|T - T_{j-1}\| \leqslant 2(1+\varepsilon)\, \mathrm{a}_{2^{j-1}}(T).$$

Using the factorization

$$\begin{array}{ccc} X & \xrightarrow{\quad S_j \quad} & Y \\ {\scriptstyle \tilde{S}_j}\downarrow & & \uparrow{\scriptstyle J} \\ S_j(X) & \xrightarrow{\quad I \quad} & S_j(X) \end{array}$$

where $\tilde{S}_j$ is the map induced by S_j, the map I is the identity, and J is the canonical injection, we thus find, with the help of Proposition 1.2 and Lemma 2.11,

$$\left(\sum_{k=1}^{\infty} \mathrm{e}_k^q(S_j)\right)^{1/q} \leqslant \|\tilde{S}_j\|\left(\sum_{k=1}^{\infty} \mathrm{e}_k^q(I)\right)^{1/q} \leqslant c_q \cdot 3^{1/q} \cdot 2^{(j-1)/q} \cdot 2(1+\varepsilon)\,\mathrm{a}_{2^{j-1}}(T)$$

$$(j = 1, 2, \ldots, N+1). \qquad (2.7)$$

We now claim that, with the convention that $\mathrm{N}_q(S)$ stands for $\left[\sum_{k=1}^{\infty} \mathrm{e}_k^q(S)\right]^{1/q}$,

$$\mathrm{N}_q\left(\sum_{j=1}^{N+1} S_j\right) \leqslant 2^{1/r}\left(\sum_{j=1}^{N+1} \mathrm{N}_q^r(S_j)\right)^{1/r}, \qquad (2.8)$$

where r is given by

$$2^{1/r-1} = 2^{1/q} \max\{1, 2^{1/q-1}\},$$

so that

$$r = q/(q+1) \text{ if } 1 \leqslant q < \infty, \qquad r = \tfrac{1}{2}q \text{ if } 0 < q < 1. \qquad (2.9)$$

In all cases $0 < r < 1$. Accepting this claim for a moment we may proceed to establish the lemma. From (2.7) and (2.8),

$$\mathrm{N}_q(T) \leqslant 2^{1/r}\left(\sum_{j=1}^{N+1} \mathrm{N}_q^r(S_j)\right)^{1/r}$$

$$\leqslant 2^{1+1/r} \cdot 3^{1/q}(1+\varepsilon)c_q\left(\sum_{j=1}^{N+1} 2^{(j-1)r/q}\mathrm{a}_{2^{j-1}}^r(T)\right)^{1/r}$$

$$= 2^{(r+1)/r} \cdot 3^{1/q}(1+\varepsilon)c_q\left(\sum_{j=1}^{N+1} 2^{r(j-1)(1/q-1/p)}\right)^{1/r} \sup_{1 \leqslant j \leqslant N+1} 2^{(j-1)/p}\mathrm{a}_{2^{j-1}}(T).$$

Since

$$\sum_{j=1}^{N+1} 2^{r(j-1)(1/q-1/p)} = (2^{(1/q-1/p)r(N+1)} - 1)/(2^{(1/q-1/p)r} - 1)$$

$$\leqslant 2^{r(1/q-1/p)}n^{r(1/q-1/p)}/(2^{r(1/q-1/p)} - 1),$$

it follows that

$$\mathrm{N}_q(T) \leqslant \rho_{p,q} n^{1/q-1/p} \sup_{1 \leqslant k \leqslant n} k^{1/p} a_k(T),$$

where

$$\rho_{p,q} = 2^{(r+1)/r+1/q-1/p} \cdot 3^{1/q}(1+\varepsilon)c_q/(2^{r(1/q-1/p)} - 1)^{1/r}.$$

To complete the proof it is enough to establish (2.8). To do this we first note that given any $R, S \in \mathscr{B}(X, Y)$,

$$\mathrm{N}_q(R+S) = \left(\sum_{k=1}^{\infty} \mathrm{e}_k^q(R+S)\right)^{1/q} \leqslant \left(2\sum_{k=1}^{\infty} \mathrm{e}_{2k-1}^q(R+S)\right)^{1/q}$$
$$\leqslant 2^{1/q}\left(\sum_{k=1}^{\infty} [\mathrm{e}_k(R)+\mathrm{e}_k(S)]^q\right)^{1/q}$$
$$\leqslant 2^{1/q}\max\{1, 2^{1/q-1}\}[\mathrm{N}_q(R)+\mathrm{N}_q(S)]. \tag{2.10}$$

Now define r by (2.9). For $i = 1, 2, \ldots, m$ let $U_i \in \mathscr{B}(X,Y)$ be such that $\mathrm{N}_q^r(U_i) \leqslant 2^{-k_i}$, where $\sum_{i=1}^{m} 2^{-k_i} \leqslant 1$ and each k_i is in $\mathbb{N}_0$. We claim that under these conditions

$$\mathrm{N}_q\left(\sum_{i=1}^{m} U_i\right) \leqslant 1. \tag{2.11}$$

To prove this, observe that we may assume that $\sum_{i=1}^{m} 2^{-k_i} = 1$, since if not, we simply choose extra points $k_{m+1}, \ldots, k_M$ in $\mathbb{N}_0$ such that $\sum_{i=1}^{M} 2^{-k_i} = 1$ and put $U_i = 0$ for $i = m+1, \ldots, M$. The proof now proceeds by induction. If $\max\{k_1, \ldots, k_m\} = 0$ there is nothing to prove. Suppose that (2.11) is true when $\max\{k_1, \ldots, k_m\} = h$, and assume that $\max\{k_1, \ldots, k_m\} = h+1$. Since $\sum_{i=1}^{m} 2^{-k_i} = 1$, the set $I := \{i : k_i = h+1\}$ must have an even number of elements: $I = \{i_1, i_2, \ldots, i_{2s}\}$, say. To see this, suppose that I had $2t+1$ elements, for some $t \in \mathbb{N}_0$; then

$$1 = \sum_{i=1}^{m} 2^{-k_i} = (2t+1)2^{-h-1} + \sum_{i \in \{1, \ldots, m\} \setminus I} 2^{-k_i}$$
$$= 2^{-h-1} + 2^{-h}M$$

for some $M \in \mathbb{N}$; and this is plainly impossible. Then by (2.10),

$$\mathrm{N}_q^r(U_{i_{2u-1}} + U_{i_{2u}}) \leqslant 2^{1-r}[\mathrm{N}_q(U_{i_{2u-1}}) + \mathrm{N}_q(U_{i_{2u}})]^r \leqslant 2^{-h}$$

for $u = 1, 2, \ldots, s$. Thus $\sum_{i=1}^{m} U_i$ can be represented as $\sum_{j=1}^{l} V_j$, where $\mathrm{N}_q^r(V_j) \leqslant 2^{-h_j}$ and $\sum_{j=1}^{l} 2^{-h_j} = 1$, but $\max\{h_1, \ldots, h_l\} = h$. By the inductive hypothesis it follows that

$$\mathrm{N}_q\left(\sum_{i=1}^{m} U_i\right) = \mathrm{N}_q\left(\sum_{j=1}^{l} V_j\right) \leqslant 1$$

and our claim (2.11) is justified.

Finally, to deal with (2.8), we may, without loss of generality, suppose that $\sum_{j=1}^{N+1} \mathrm{N}_q^r(S_j) = \frac{1}{2}$, since if this were not so, we could simply replace the S_j by λS_j for suitable λ. For $j = 1, 2, \ldots, N+1$ choose $k_j \in \mathbb{N}_0$ such that

$$2^{-k_j-1} \leqslant \mathrm{N}_q^r(S_j) \leqslant 2^{-k_j}.$$

Since $\sum_{j=1}^{N+1} 2^{-k_j-1} \leqslant \frac{1}{2}$, it follows that $\sum_{j=1}^{N+1} 2^{-k_j} \leqslant 1$, and so by (2.11),

$$\mathrm{N}_q\left(\sum_{j=1}^{N+1} S_j\right) \leqslant 1.$$

The proof of the lemma is complete. □

Before we can give the promised generalization of Weyl's inequality (2.6) one more lemma is necessary.

Lemma 2.13. Let $T \in \mathscr{K}(X)$ and let $p \in (0, \infty)$. Then there is a constant C_p, depending only upon p, such that for all $n \in \mathbb{N}$,

$$|\lambda_n(T)|^p \leqslant C_p n^{-1} \sum_{j=1}^{n} \mathrm{a}_j^p(T).$$ ■

Proof. Let $n \in \mathbb{N}$. Then there exists $S \in \mathscr{B}(X)$, with rank $S < n$, such that

$$\|T - S\| \leqslant 2\mathrm{a}_n(T).$$

Hence for $k = 1, 2, \ldots, n$,

$$\mathrm{a}_k(S) \leqslant \mathrm{a}_k(T) + \|T - S\| \leqslant 3\mathrm{a}_k(T); \tag{2.12}$$

and

$$\mathrm{e}_n(T) \leqslant \|T - S\| + \mathrm{e}_n(S) \leqslant 2\mathrm{a}_n(T) + \mathrm{e}_n(S). \tag{2.13}$$

By Lemma 2.12, with $0 < q < p < \infty$, we have

$$n^{1/q}\mathrm{e}_n(S) \leqslant \left(\sum_{k=1}^{\infty} \mathrm{e}_k^q(S)\right)^{1/q} \leqslant \rho_{p,q} n^{1/q-1/p} \sup_{1 \leqslant k \leqslant n} k^{1/p}\mathrm{a}_k(S).$$

Now choose $q = p/(p+1)$ and use (2.12): we find that

$$n^{1/p}\mathrm{e}_n(S) \leqslant 3\rho_{p,\,p/(p+1)} \sup_{1 \leqslant k \leqslant n} k^{1/p}\mathrm{a}_k(T).$$

With the aid of (2.13) this gives

$$n^{1/p}\mathrm{e}_n(T) \leqslant (2 + 3\rho_{p,\,p/(p+1)}) \sup_{1 \leqslant k \leqslant n} k^{1/p}\mathrm{a}_k(T),$$

and hence

$$e_n^p(T) \leqslant (2+3\rho_{p,\,p/(p+1)})^{p_n-1} \sum_{k=1}^{n} a_k^p(T).$$

Since $|\lambda_n(T)| \leqslant \sqrt{2}\, e_n(T)$, by Theorem 1.6, the lemma follows, with

$$C_p = 2^{p/2}(2+3\rho_{p,\,p/(p+1)})^p. \qquad (2.14) \quad \square$$

Theorem 2.14. Let $T \in \mathscr{K}(X)$ and let $p \in (0, \infty)$. Then there is a constant K_p, depending only on p, such that

$$\left(\sum_{k=1}^{\infty} |\lambda_k(T)|^p\right)^{1/p} \leqslant K_p\left(\sum_{k=1}^{\infty} a_k^p(T)\right)^{1/p}. \qquad \blacksquare$$

Proof. We invoke Hardy's inequality (see Hardy, Littlewood, and Polya [1, p. 239]): if $0 < s < p < \infty$ and $\alpha = (\alpha_i) \in l^p$, then

$$\left[\sum_{n=1}^{\infty}\left(n^{-1}\sum_{i=1}^{n}|\alpha_i|^s\right)^{p/s}\right]^{1/p} \leqslant [p/(p-s)]^{1/s}\left(\sum_{i=1}^{\infty}|\alpha_i|^p\right)^{1/p}.$$

By this inequality and Lemma 2.13, applied with p replaced by $\frac{1}{2}p$, we have

$$\left(\sum_{n=1}^{\infty}|\lambda_n(T)|^p\right)^{1/p} \leqslant \left[\sum_{n=1}^{\infty}\left(C_{p/2}\, n^{-1}\sum_{j=1}^{n} a_j^{p/2}(T)\right)^2\right]^{1/p}$$

$$\leqslant (2C_{p/2})^{2/p}\left(\sum_{n=1}^{\infty} a_j^p(T)\right)^{1/p},$$

and the proof is complete. $\square$

Remark 2.15. (i) The constant K_p which appears in Theorem 2.14 is given by

$$K_p = (2C_{p/2})^{2/p} = 2^{1/2+2/p}(2+3\rho_{p/2,\,p/(p+2)})$$
$$\leqslant 2^{3/2+2/p}[1+2^{15/2+6/p}\cdot 3^{2+2/p}(1+2/p)^{1+2/p}\cdot(2^{p/2(p+2)}-1)^{2+4/p}].$$

This bound for K_p goes to infinity as $p \to 0$, but remains bounded as $p \to \infty$. It follows that there is an absolute constant K such that for all (complex) Banach spaces X, all maps $T \in \mathscr{K}(X)$, and all $p \in [1, \infty)$,

$$\left(\sum_{k=1}^{\infty}|\lambda_k(T)|^p\right)^{1/p} \leqslant K\left(\sum_{k=1}^{\infty} a_k^p(T)\right)^{1/p}. \qquad (2.15)$$

(ii) An examination of the proof of Theorem 2.14 makes it plain that it also follows that given any $N \in \mathbb{N}$ and any $p \in (0, \infty)$,

$$\left(\sum_{k=1}^{N}|\lambda_k(T)|^p\right)^{1/p} \leqslant K_p\left(\sum_{k=1}^{N} a_k^p(T)\right)^{1/p}. \qquad (2.16)$$

This observation will be used in the discussion of non-compact operators in §4.

(iii) The methods used here do not give the best constant K_p for which Theorem 2.14 and (2.16) hold, for König [5] has shown that these results are valid with

$$K_p = \begin{cases} 2e/\sqrt{p} & \text{if } 0 < p < 1, \\ 2^{1/p}\sqrt{(2e)} & \text{if } 1 \leqslant p < \infty. \end{cases}$$

His proof relies on the theory of 2-summing operators and does not appear to extend to non-compact operators, unlike the methods given here: we obtain the analogue of (2.16) for merely continuous operators with little extra effort in §4. However, König's striking result, which follows earlier very beautiful work by Pietsch [2] (who obtained $K_p = 2^{1/p}\,\mathrm{e}$), shows that (2.15) holds with $K = 2\sqrt{(2\mathrm{e})}$. Weyl's original inequality (2.6) corresponds to the special case of Theorem 2.14 in which X is a Hilbert space, and of course it has the constant K_p replaced by 1. None of the methods of proof of Theorem 2.14 so far available takes into account any special features which the Banach space X may have, and so it is natural to expect that some improvement in the size of the constant K_p may be possible in particular circumstances.

Weyl's inequality is of clear importance in the theory of non-self-adjoint operators; but it would equally clearly be desirable to have a result to the effect that if $\mathrm{a}_n(T) = O(n^{-a})$ as $n \to \infty$, then also $|\lambda_n(T)| = O(n^{-a})$ as $n \to \infty$. Such results, in a Hilbert-space context, can be found in the well-known book by Gohberg and Krein [1]; here, following König [4], we give a treatment based on Lorentz sequence spaces and valid for arbitrary Banach spaces. We begin with some basic definitions.

Given any bounded sequence $x = (\xi_k)_{k\in\mathbb{N}}$ of real numbers, put

$$\mathrm{t}_n(x) = \inf\{\sigma \geqslant 0\colon\ \#\{k\in\mathbb{N}: |\xi_k| \geqslant \sigma\} < n\} \qquad (n\in\mathbb{N}).$$

If $|\xi_k| \geqslant |\xi_{k+1}|$ for all $k\in\mathbb{N}$, it is easy to see that $\mathrm{t}_n(x) = |\xi_n|$ for all $n\in\mathbb{N}$, and for this reason $(\mathrm{t}_n(x))_{n\in\mathbb{N}}$ is called the *non-increasing rearrangement* of x.

Definition 2.16. Let $p, q \in (0, \infty]$. The *Lorentz space* $l^{p,q}$ is the vector space of all $x = (\xi_k)\in l^\infty$ such that

$$\|x\|_{p,q} := \begin{cases} \left(\sum_{n=1}^{\infty} [n^{1/p-1/q}\,\mathrm{t}_n(x)]^q\right)^{1/q} & \text{if } 0 < q < \infty, \\ \sup_{n\in\mathbb{N}} [n^{1/p}\,\mathrm{t}_n(x)] & \text{if } q = \infty \end{cases}$$

is finite. ■

The Lorentz space $l^{p,q}$ is, of course, simply the familiar sequence space l^p. We leave it as an exercise for the reader to prove that $l^{p_1,q_1} \subset l^{p_2,q_2}$ if $0 < p_1 < p_2 \leqslant \infty$ and $q_1, q_2 \in (0, \infty]$; and that, with $l^{p,\omega}$ standing for the subspace of $l^{p,\infty}$ consisting of all those x with $\lim_{n\to\infty} [n^{1/p}t_n(x)] = 0$, $l^{p,q_1} \subset l^{p,q_2} \subset l^{p,\omega} \subset l^{p,\infty}$ if $p \in (0, \infty)$ and $0 < q_1 < q_2 < \infty$.

At this stage a little interpolation theory is needed. We use the notation of Triebel's book [2], in which a comprehensive account of the subject is given. A pair of Banach spaces X_0, X_1 is said to be an interpolation pair if X_0 and X_1 are continuously embedded in some Hausdorff topological vector space $\mathfrak{X}$; that is, $X_0 \subset \mathfrak{X}$ and $X_1 \subset \mathfrak{X}$ algebraically and topologically. Given such a pair, $X_0 \cap X_1$ is also a Banach space when given the norm

$$\|x\|_{X_0 \cap X_1} = \max\{\|x\|_{X_0}, \|x\|_{X_1}\}.$$

The space

$$X_0 + X_1 := \{x : x \in \mathfrak{X}, \quad x = x_0 + x_1, \text{ where } x_j \in X_j \text{ for } j = 0, 1\}$$

becomes a Banach space when provided with the norm

$$\|x\|_{X_0+X_1} = \inf\{\|x_0\|_{X_0} + \|x_1\|_{X_1} : x = x_0 + x_1, x_j \in X_j \text{ for } j = 0, 1\},$$

and for $j = 0, 1$,

$$X_0 \cap X_1 \subset X_j \subset X_0 + X_1.$$

A Banach space X with $X_0 \cap X_1 \subset X \subset X_0 + X_1$ is called an *interpolation space with respect to* $\{X_0, X_1\}$. Given any two interpolation pairs $\{X_0, X_1\}$ and $\{Y_0, Y_1\}$ we denote by $\mathscr{B}(\{X_0, X_1\}, \{Y_0, Y_1\})$ the set of all linear operators $T: X_0 + X_1 \to Y_0 + Y_1$ such that $T \upharpoonright X_j \in \mathscr{B}(X_j, Y_j)$ for $j = 0, 1$. If X and Y are interpolation spaces with respect to $\{X_0, X_1\}$, $\{Y_0, Y_1\}$ respectively, they are said to be interpolation spaces of exponent θ $(0 < \theta < 1)$ with respect to $\{X_0, X_1\}$ and $\{Y_0, Y_1\}$ if and only if, given any $T \in \mathscr{B}(\{X_0, X_1\}, \{Y_0, Y_1\})$ the restriction of T to X is in $\mathscr{B}(X, Y)$ and

$$\|T\|_{\mathscr{B}(X,Y)} \leqslant \|T\|^{1-\theta}_{\mathscr{B}(X_0, Y_0)} \|T\|^{\theta}_{\mathscr{B}(X_1, Y_1)}.$$

Several methods can be used to obtain interpolation spaces of exponent θ with respect to given interpolation pairs $\{X_0, X_1\}$ and $\{Y_0, Y_1\}$, notably the K-method (cf. Triebel [2, p. 23]) which leads to the spaces $X = (X_0, X_1)_{\theta,p}$ and $Y = (Y_0, Y_1)_{\theta,p}$ $(0 < \theta < 1, 1 \leqslant p \leqslant \infty)$. These spaces are defined as follows:

$$(X_0, X_1)_{\theta,p} := \left\{ x \in X_0 + X_1 : \|x\|_X := \left(\int_0^\infty [t^{-\theta} K(t, x)]^p t^{-1}\, dt \right)^{1/p} < \infty \right\}$$

if $1 \leqslant p < \infty$;

$$(X_0, X_1)_{\theta,\infty} := \left\{ x \in X_0 + X_1 : \|x\|_X := \sup_{0 < t < \infty} t^{-\theta} K(t, x) < \infty \right\}.$$

Here

$$K(t, x) := \inf\{\|x_0\|_{X_0} + t\|x_1\|_{X_1} : x = x_0 + x_1, x_0 \in X_0, x_1 \in X_1\}.$$

It can be shown that if $x \in X_0 \cap X_1$ then

$$\|x\|_{(X_0, X_1)_{\theta,p}} \leqslant \|x\|_{X_0}^{1-\theta} \|x\|_{X_1}^{\theta}.$$

For our purposes the main results we shall need are contained in the following two theorems, proofs of which may be found in Triebel [2, Theorems 1.18.3/1 and 2].

Theorem 2.17. Let $\theta \in (0, 1)$ and $q \in [1, \infty]$. Then

$$(l^1, l^\infty)_{\theta,q} = l^{1/(1-\theta),q}. \qquad \blacksquare$$

This shows that the Lorentz spaces $l^{1/(1-\theta),q}$ may be regarded as Banach spaces as they are interpolation spaces. However, $\|\bullet\|_{1/(1-\theta),q}$ is not, in general, a norm as it does not satisfy the triangle inequality but instead has the property that there is a constant $k \geqslant 1$ such that for all $x, y \in l^{1/(1-\theta),q}$,

$$\|x+y\|_{1/(1-\theta),q} \leqslant k(\|x\|_{1/(1-\theta),q} + \|y\|_{1/(1-\theta),q});$$

this fact is expressed by saying that $\|\bullet\|_{1/(1-\theta),q}$ is a *quasi-norm*. Nevertheless, $\|\bullet\|_{1/(1-\theta),q}$ and $\|\bullet\|_{(l^1,l^\infty)_{\theta,q}}$ are *equivalent* in the sense that there are positive constants c_1 and c_2 such that for all $x \in l^{1/(1-\theta),q}$,

$$c_1 \|x\|_{(l^1,l^\infty)_{\theta,q}} \leqslant \|x\|_{1/(1-\theta),q} \leqslant c_2 \|x\|_{(l^1,l^\infty)_{\theta,q}}.$$

Theorem 2.18. Let $\theta \in (0, 1)$; $p_0, p_1 \in (1, \infty)$ with $p_0 \neq p_1$, and $q_0, q_1, q \in [1, \infty]$. Then

$$(l^{p_0,q_0}, l^{p_1,q_1})_{\theta,q} = l^{p,q},$$

where $p^{-1} = (1-\theta)p_0^{-1} + \theta p_1^{-1}$. This result also holds if l^{p_0,q_0} is replaced by l^1 (taking $p_0 = 1$) or l^{p_1,q_1} is replaced by l^∞ (taking $p_1 = \infty$); both these replacements may be made at the same time. $\blacksquare$

The next lemma is crucial in the proof of the main result relating approximation numbers and eigenvalues in the context of Lorentz spaces.

Lemma 2.19. Let $0 < r < p < \infty$ and $0 < r < q \leqslant \infty$. Then there is a constant c, which depends only on q/r and p/r, such that for every monotone non-increasing sequence $\alpha = (\alpha_n)$ of positive numbers,

$$\|\alpha\|_{p,q} \leqslant \|\beta\|_{p/r,q/r}^{1/r} \leqslant c\|\alpha\|_{p,q},$$

where $\beta = (\beta_n)$ is given by

$$\beta_n = n^{-1} \sum_{j=1}^{n} \alpha_j^r \qquad (n \in \mathbb{N}). \qquad \blacksquare$$

Proof. Define a map $A: l^\infty \to l^\infty$ by $Ax = y$, where $x = (\xi_n)$ and $y = (\eta_n)$, with $\eta_n = n^{-1} \sum_{j=1}^{n} \xi_j$ $(n \in \mathbb{N})$. Evidently $A \in \mathscr{B}(l^\infty)$, and as an element of $\mathscr{B}(l^\infty)$, $\|A\| = 1$. Moreover, if $x = (\xi_n) \in l^1$, then for any $a \in (0, 1)$,

$$\|Ax\|_{a,\infty} = \sup_{n \in \mathbb{N}} \left[n^a \mathrm{t}_n\left(\left(n^{-1} \sum_{j=1}^{n} \xi_j \right)\right)\right] \leqslant \sup_{n \in \mathbb{N}} \left[n \mathrm{t}_n\left(\left(n^{-1} \sum_{j=1}^{n} \mathrm{t}_n(x) \right)\right)\right]$$

$$\leqslant \sum_{n=1}^{\infty} \mathrm{t}_n(x) = \|x\|_1.$$

In the language of interpolation theory we thus have $A \in \mathscr{B}[(l^1, l^\infty), (l^{a,\infty}, l^\infty)]$, and so for any $\theta \in (0, 1)$ and any $s \in (1, \infty]$, it follows from Theorem 2.18 that $A \in \mathscr{B}(l^{1/(1-\theta),s}, l^{a/(1-\theta),s})$. Since $l^{a/(1-\theta),s} \subset l^{1/(1-\theta),s}$, we see that $A \in \mathscr{B}(l^{1/(1-\theta),s})$. Making suitable choices of θ and s this shows that $A \in \mathscr{B}(l^{p/r, q/r})$. The right-hand inequality of the lemma now follows, with c equal to the norm of A as an element of $\mathscr{B}(l^{p/r, q/r})$, from the inequality $\|Ax\|_{p/r, q/r} \leqslant \|A\| \|x\|_{p/r, q/r}$, with $x = (\alpha_n^r)$; the left-hand inequality is obvious. □

We can now give our main result in this direction.

Theorem 2.20. Let $p \in (0, \infty)$ and $q \in (0, \infty]$. Then there is a constant c, dependeing only upon p and q, such that for all $T \in \mathscr{K}(X)$,

$$\|(\lambda_n(T))\|_{p,q} \leqslant c \|(\mathrm{a}_n(T))\|_{p,q}.$$ ■

Proof. Let $r = \frac{1}{2} \min\{p, q\}$. By Lemma 2.13, for all $n \in \mathbb{N}$,

$$|\lambda_n(T)| \leqslant \left(C_r n^{-1} \sum_{j=1}^{n} \mathrm{a}_j^r(T) \right)^{1/r}.$$

Thus by Lemma 2.19,

$$\begin{aligned} \|(\lambda_n(T))\|_{p,q} &\leqslant C_r^{1/r} \left\| \left(\left(n^{-1} \sum_{j=1}^{n} \mathrm{a}_j^r(T) \right)^{1/r} \right) \right\|_{p,q} \\ &\leqslant C_r^{1/r} \left(\left\| \left(n^{-1} \sum_{j=1}^{n} \mathrm{a}_j^r(T) \right) \right\|_{p/r, q/r} \right)^{1/r} \\ &\leqslant c_{p,q} \|\{\mathrm{a}_n(T)\}\|_{p,q}, \end{aligned}$$

as required. □

Corollary 2.21. Let $T \in \mathscr{K}(X)$ and suppose that $a \in (0, \infty)$. Then: (i) if $\mathrm{a}_n(T) = O(n^{-a})$ as $n \to \infty$, also $|\lambda_n(T)| = O(n^{-a})$ as $n \to \infty$; (ii) if $\mathrm{a}_n(T) = o(n^{-a})$ as $n \to \infty$, also $|\lambda_n(T)| = o(n^{-a})$ as $n \to \infty$. ■

Proof. Part (i) is simply the special case $q = \infty$ of Theorem 2.20. As for (ii), let $\varepsilon, \delta > 0$ and let $N \in \mathbb{N}$ be such that $a_n(T) < \varepsilon n^{-a}$ if $n > N$. Application of Lemma 2.13 with $p = 1/(a+\delta)$ gives for $n > N$,

$$n^{a+\delta}|\lambda_n(T)| \leqslant C\left(\sum_{j=1}^{N} a_j(T)^{1/(a+\delta)} + \varepsilon^{1/(a+\delta)} \sum_{j=N+1}^{n} j^{-a/(a+\delta)}\right)^{a+\delta}$$
$$\leqslant C 2^{a+\delta} \max\left\{\left(\sum_{j=1}^{N} a_j(T)^{1/(a+\delta)}\right)^{a+\delta}, \varepsilon O(n^{\delta})\right\}.$$

It now follows immediately that $|\lambda_n(T)| \leqslant \text{const}\cdot\varepsilon n^{-a}$ for all sufficiently large n; the proof is therefore complete. □

3. An axiomatic approach to s-numbers

We have seen in §2 that the approximation numbers have various helpful properties, notably those contained in Propositions 2.2 and 2.3. In this section we follow Pietsch [1] and show that properties similar to those just mentioned may be taken as the basis for an axiomatic approach for *s*-numbers, by which we mean, roughly speaking, real numbers which form a sequence characterizing, in some sense, the compactness properties of an opertor. Various examples of *s*-numbers are given, and the relations between them discussed.

Definition 3.1. A map s which to each bounded linear map T from one Banach space to another such space assigns a sequence $(s_n(T))$ of non-negative real numbers is called an *s-function* if for all Banach spaces W, X, Y, Z:

(i) $\|T\| = s_1(T) \geqslant s_2(T) \geqslant \ldots \geqslant 0$ for all $T \in \mathscr{B}(X,Y)$;

(ii) for all $S, T \in \mathscr{B}(X,Y)$ and all $n \in \mathbb{N}$,

$$s_n(S+T) \leqslant s_n(S) + \|T\|;$$

(iii) for all $T \in \mathscr{B}(X,Y)$, $S \in \mathscr{B}(Y,Z)$, $R \in \mathscr{B}(Z,W)$ and $n \in \mathbb{N}$,

$$s_n(RST) \leqslant \|R\| s_n(S) \|T\|;$$

(iv) for all $T \in \mathscr{B}(X,Y)$ with rank $T < n \in \mathbb{N}$, $s_n(T) = 0$;

(v) $s_n(I_n) = 1$ for all $n \in \mathbb{N}$; here I_n is the identity map of $l_n^2 := \{x \in l^2 : x_i = 0 \text{ if } i > n\}$ to itself.

For all $n \in \mathbb{N}$, we call $s_n(T)$ the nth *s-number* of T. ■

It follows immediately from Propositions 2.2 and 2.3 that the approximation numbers are *s*-numbers. However, in view of Proposition 1.3, the entropy numbers are not *s*-numbers: property (iv) above rules them out.

Remark 3.2. (i) All *s*-numbers depend continuously on their arguments: by axiom (ii), it is easy to see that, $|s_n(S) - s_n(T)| \leqslant \|S - T\|$.

(ii) If $s_n(T) = 0$ for some $n \in \mathbb{N}$ and some $T \in \mathscr{B}(X,Y)$, then rank $T < n$. The reason is that, if not, there are maps $R \in \mathscr{B}(l_n^2, X)$ and $S \in \mathscr{B}(Y, l_n^2)$ such that $STR = I_n$; and thus by axioms (iii) and (v), $1 = s_n(STR) \leqslant \|S\| s_n(T) \|R\| = 0$, which is absurd.

(iii) The approximation numbers are the largest s-numbers. To see this, let $n \in \mathbb{N}$ and $T \in \mathscr{B}(X,Y)$ be given, and let $s_n(T)$ be an s-number. Given any $L \in \mathscr{B}(X,Y)$ with rank $L < n$, it follows from axioms (ii) and (iv) that

$$s_n(T) \leqslant s_n(L) + \|T - L\| = \|T - L\|;$$

thus $s_n(T) \leqslant \inf\{\|T-L\|: L \in \mathscr{B}(X,Y), \text{ rank } L < n\} = \mathrm{a}_n(T)$.

(iv) All s-functions coincide on operators acting between Hilbert spaces. More precisely, given any s-function, any Hilbert spaces H_1 and H_2, any $T \in \mathscr{B}(H_1, H_2)$, and any $n \in \mathbb{N}$, then $s_n(T) = \mathrm{a}_n(T)$. For a proof of this we refer to Pietsch [1, p. 147].

An s-function s is said to be *additive* if

$$s_{m+n-1}(S+T) \leqslant s_m(S) + s_n(T)$$

for all $m, n \in \mathbb{N}$ and all $S, T \in \mathscr{B}(X,Y)$, where X and Y are arbitrary Banach spaces; it is called *multiplicative* if

$$s_{m+n-1}(RS) \leqslant s_m(R) s_n(S)$$

for all $m, n \in \mathbb{N}$ and all $R \in \mathscr{B}(Y,Z)$ and $S \in \mathscr{B}(X,Y)$, where X, Y, Z are arbitrary Banach spaces.

Proposition 2.2 shows that the approximation numbers are both additive and multiplicative.

We now set about the construction of s-functions other than the approximation numbers.

Definition 3.3. Let $T \in \mathscr{B}(X,Y)$ and let $n \in \mathbb{N}$. The nth *Kolmogorov number* (or nth *width*) of T, $\mathrm{d}_n(T)$, is defined by

$$\mathrm{d}_n(T) = \inf\{\|\mathrm{Q}_M^Y T\|: M \text{ is a linear subspace of } Y, \dim M < n\},$$

where Q_M^Y is the canonical map of Y onto Y/M. The nth *Gelfand number* $\mathrm{c}_n(T)$ is defined by

$$\mathrm{c}_n(T) = \inf\{\|T \mathrm{J}_M^X\|: M \text{ is a linear subspace of } X, \operatorname{codim} M < n\},$$

where J_M^X is the embedding map from M to X. ∎

The Kolmogorov numbers have been used extensively in the study of the asymptotic distribution of eigenvalues of elliptic operators (cf. Fleckinger [1], Triebel [2]). Note that

$$\mathrm{d}_n(T) = \inf_{M \subset Y, \dim M < n} \sup_{x \in \mathrm{B}_X} \inf_{y \in M} \|Tx - y\|. \tag{3.1}$$

We leave it to the reader to prove that the Kolmogorov and Gelfand numbers give rise to s-functions which are additive and multiplicative. Moreover, given any $T \in \mathscr{B}(X,Y)$ and any $n \in \mathbb{N}$,

$$c_n(T) = \mathrm{d}_n(T^*), \qquad \mathrm{d}_n(T) \geqslant \mathrm{c}_n(T^*), \tag{3.2}$$

with equality if T is compact. Details of the proofs will be found in Pietsch [1, Part 3, §11]; here we are content to point out that certain characterizations of the Gelfand and Kolmogorov numbers are useful in these proofs. To explain this, some notation is necessary. Let I be an arbitrary index set, let $1 \leqslant p \leqslant \infty$ and let $l^p(I)$ be the Banach space of all $x = (\zeta_i)_{i \in I}$ such that each ζ_i is real or complex, $\{i \in I : \zeta_i \neq 0\}$ is finite or countable if $p < \infty$, and

$$\|x\|_p := \begin{cases} \left(\sum_I |\zeta_i|^p \right)^{1/p} & (1 \leqslant p < \infty), \\ \sup_I |\zeta_i| & (p = \infty), \end{cases}$$

is finite. Define $\mathrm{K}_X : X \to l^\infty(\mathrm{B}_{X^*})$ by $\mathrm{K}_X x = (x^*(x))_{x^* \in \mathrm{B}_{X^*}}$, and define $\mathrm{Q}_X : l^1(\mathrm{B}_X) \to X$ by $\mathrm{Q}_X((\zeta_x)) = \sum_{x \in \mathrm{B}_X} \zeta_x x$. The characterizations mentioned above are given in the following proposition.

Proposition 3.4. Let $T \in \mathscr{B}(X,Y)$ and $n \in \mathbb{N}$. Then

$$\mathrm{c}_n(T) = \mathrm{a}_n(\mathrm{K}_Y T)$$

and

$$\mathrm{d}_n(T) = \mathrm{a}_n(T\,\mathrm{Q}_X). \qquad \blacksquare$$

Proof. Let M be a linear subspace of X, with codim $M < n$. Then there exists $S \in \mathscr{B}(X, l^\infty(\mathrm{B}_{Y^*}))$ such that $SJ_M^X = \mathrm{K}_Y T \mathrm{J}_M^X$ and $\|S\| = \|T \mathrm{J}_M^X\|$. Thus $L := \mathrm{K}_Y T - S \in \mathscr{B}(X, l^\infty(\mathrm{B}_{Y^*}))$ and rank $L < n$, and hence

$$\mathrm{a}_n(\mathrm{K}_Y T) \leqslant \|\mathrm{K}_Y T - L\| = \|S\| = \|T \mathrm{J}_M^X\|.$$

It follows that $\mathrm{a}_n(\mathrm{K}_Y T) \leqslant \mathrm{c}_n(T)$. To prove the reverse inequality, let $\varepsilon > 0$ and let $U \in \mathscr{B}(X, l^\infty(\mathrm{B}_{Y^*}))$ be such that $\|\mathrm{K}_Y T - U\| \leqslant (1+\varepsilon)\mathrm{a}_n(\mathrm{K}_Y T)$ and rank $U < n$. Put $N = \mathscr{N}(U)$. Then since, by Theorem I.3.7, U is normally soluble, it follows from (I.1.5) and Theorem I.3.7 that codim $N = \dim N^0 = \dim \mathscr{R}(U^*) < n$. Thus

$$\begin{aligned} \|T \mathrm{J}_N^X\| = \|\mathrm{K}_Y T \mathrm{J}_N^X\| = \|(\mathrm{K}_Y T - U)\mathrm{J}_N^X\| \leqslant \|\mathrm{K}_Y T - U\| \\ \leqslant (1+\varepsilon)\mathrm{a}_n(\mathrm{K}_Y T). \end{aligned}$$

Hence $\mathrm{c}_n(T) \leqslant \mathrm{a}_n(\mathrm{K}_Y T)$.

The proof for $d_n(T)$ is rather similar. First, let $\varepsilon > 0$ and let N be a linear subspace of Y with $\dim N < n$. Then there exists $S \in \mathscr{B}(l^1(B_X), Y)$ such that $Q_N^Y S = Q_N^Y T Q_X$ and $\|S\| \leqslant (1+\varepsilon)\|Q_N^Y T\|$. Put $L = TQ_X - S$: then $L \in \mathscr{B}(l^1(B_X), Y)$ and rank $L < n$; and hence

$$a_n(TQ_X) \leqslant \|TQ_X - L\| = \|S\| \leqslant (1+\varepsilon)\|Q_N^Y T\|.$$

Thus $a_n(TQ_X) \leqslant d_n(T)$. As for the reverse inequality, let $\varepsilon > 0$ and let $V \in \mathscr{B}(l^1(B_X), Y)$ be such that rank $V < n$ and $\|TQ_X - V\| \leqslant (1+\varepsilon)a_n(TQ_X)$. Put $M = \mathscr{R}(V)$. Then $\dim M < n$ and

$$\|Q_M^Y T\| = \|Q_M^Y T Q_X\| = \|Q_M^Y(TQ_X - V)\|$$
$$\leqslant \|TQ_X - V\| \leqslant (1+\varepsilon)a_n(TQ_X).$$

Hence $d_n(T) \leqslant a_n(TQ_X)$. The proof is complete. □

We now give some more connections between the various numbers.

Proposition 3.5. Let H be a Hilbert space, let $T \in \mathscr{B}(H, Y)$ and let $n \in \mathbb{N}$. Then

$c_n(T) = a_n(T) =$

$$\inf\{\|T - TP\| : P \in \mathscr{B}(H) \text{ is an orthogonal projection with rank } P < n\}.$$ ■

Proof. Let $\varepsilon > 0$, and let $L \in \mathscr{B}(H, l^\infty(B_{Y^*}))$ be such that rank $L < n$ and $\|K_Y T - L\| \leqslant (1+\varepsilon)c_n(T)$. Let $P \in \mathscr{B}(H)$ be an orthogonal projection with $\mathscr{N}(P) = \mathscr{N}(L)$, so that rank $P < n$. Since

$$\|T - TP\| = \|K_Y T(I-P)\| = \|(K_Y T - L)(I-P)\| \leqslant \|K_Y T - L\|,$$

it follows that

$$a_n(T) \leqslant \|T - TP\| \leqslant \|K_Y T - L\| \leqslant (1+\varepsilon)c_n(T).$$

Hence $a_n(T) \leqslant c_n(T)$; and since, by Remark 3.2 (iii), $a_n(T) \geqslant c_n(T)$ the result follows. □

Proposition 3.6. Let H be a Hilbert space, let $T \in \mathscr{B}(X, H)$, and let $n \in \mathbb{N}$. Then

$d_n(T) = a_n(T) =$

$$\inf\{\|T - PT\| : P \in \mathscr{B}(H) \text{ is an orthogonal projection, rank } P < n\}.$$ ■

Proof. Let $\varepsilon > 0$ and let $L \in \mathscr{B}(l^1(B_X), H)$ be such that rank $L < n$ and $\|TQ_X - L\| \leqslant (1+\varepsilon)d_n(T)$. Let $P \in \mathscr{B}(H)$ be the orthogonal projection with $\mathscr{R}(P) = \mathscr{R}(L)$, so that rank $P < n$. Since

$$\|T - PT\| = \|(I-P)TQ_X\| = \|(I-P)(TQ_X - L)\| \leqslant \|TQ_X - L\|,$$

we see that

$$\mathrm{a}_n(T) \leqslant \|T - PT\| \leqslant \|T\mathrm{Q}_X - L\| \leqslant (1+\varepsilon)\mathrm{d}_n(T).$$

Hence $\mathrm{a}_n(T) \leqslant \mathrm{d}_n(T)$; and as, by Remark 3.2(iii), $\mathrm{a}_n(T) \geqslant \mathrm{d}_n(T)$, the proof is complete. □

Proposition 3.7. Let $T \in \mathscr{B}(X, Y)$. Then for all $n \in \mathbb{N}$,

$$\mathrm{a}_n(T) \leqslant 2n^{\frac{1}{2}}\mathrm{c}_n(T), \quad \mathrm{a}_n(T) \leqslant 2n^{\frac{1}{2}}\mathrm{d}_n(T). \qquad \blacksquare$$

Proof. Let $\varepsilon > 0$, and let $L \in \mathscr{B}(l^1(\mathrm{B}_X), Y)$ be such that rank $L < n$ and $\|T\mathrm{Q}_X - L\| \leqslant (1+\varepsilon)\mathrm{d}_n(T)$. By Theorem I.1.14, there is a projection $P \in \mathscr{B}(Y)$ such that $\mathscr{R}(P) = \mathscr{R}(L)$ and $\|P\| \leqslant (n-1)^{\frac{1}{2}}$. Since

$$\begin{aligned}\|T - PT\| &= \|(I-P)T\mathrm{Q}_X\| = \|(I-P)(T\mathrm{Q}_X - L)\| \\ &\leqslant [1 + (n-1)^{\frac{1}{2}}]\, \|T\mathrm{Q}_X - L\|,\end{aligned}$$

we see that

$$\begin{aligned}\mathrm{a}_n(T) \leqslant \|T - PT\| &\leqslant [1 + (n-1)^{\frac{1}{2}}](1+\varepsilon)\mathrm{d}_n(T) \\ &\leqslant 2(1+\varepsilon)n^{\frac{1}{2}}\,\mathrm{d}_n(T),\end{aligned}$$

and hence $\mathrm{a}_n(T) \leqslant 2n^{\frac{1}{2}}\,\mathrm{d}_n(T)$.

A similar procedure establishes the other inequality of the proposition. □

Proposition 3.8. Let $T \in \mathscr{B}(X, Y)$. Then for all $n \in \mathbb{N}$,

$$\mathrm{c}_n(T) \leqslant n\mathrm{e}_n(T), \qquad \mathrm{d}_n(T) \leqslant n\mathrm{e}_n(T). \qquad \blacksquare$$

Proof. Let $n \in \mathbb{N}$. If $\mathrm{c}_n(T) = 0$, the first inequality is obvious. Suppose that $\mathrm{c}_n(T) > 0$, and let $\rho \in (0, \mathrm{c}_n(T))$. An inductive argument shows that there are points $x_1, x_2, \ldots, x_n \in \mathrm{B}_X$ and points $b_1, b_2, \ldots, b_n \in \mathrm{B}_{Y^*}$ such that for $k = 1, 2, \ldots, n$, we have $|b_k(Tx_k)| = \|Tx_k\| > \rho$ and $x_k \in M_k := \{x \in X : b_h(Tx) = 0 \text{ for } h < k\}$. Since $M_k = {}^0\{T^*b_1, \ldots, T^*b_{k-1}\}$, it follows as in the proof of Theorem I.2.21 that codim $M_k < k$. Let $\mathscr{E}^n = \{e = (\varepsilon_1, \varepsilon_2, \ldots, \varepsilon_n) : \varepsilon_i = \pm 1$ for $i = 1, 2, \ldots, n\}$. Then for all $e \in \mathscr{E}^n$ we have $x_e := n^{-1} \sum_{k=1}^{n} \varepsilon_k x_k \in \mathrm{B}_X$. Note that $\# \mathscr{E}^n = 2^n$. Let $e', e'' \in \mathscr{E}^n$, with $e' \neq e''$, and put $h = \min\{k : e'_k \neq e''_k\}$. Then

$$\begin{aligned}|b_h(Tx_{e'} - Tx_{e''})| &= n^{-1}\left|\sum_{k=h}^{n} (\varepsilon'_k - \varepsilon''_k) b_h(Tx_k)\right| \\ &= n^{-1}|\varepsilon'_h - \varepsilon''_h|\,\|Tx_h\| > 2\rho n^{-1}.\end{aligned}$$

Hence $\|Tx_{e'} - Tx_{e''}\| > 2\rho/n$, and consequently $\mathrm{e}_n(T) \geqslant \rho/n$. It follows that $\mathrm{c}_n(T) \leqslant n\mathrm{e}_n(T)$.

The proof of the remainder of the proposition is similar. □

Since both $\mathrm{c}_n(T)$ and $\mathrm{d}_n(T)$ are monotonic decreasing as $n \to \infty$, it follows that $\mathrm{c}(T) := \lim_{n\to\infty} \mathrm{c}_n(T)$ and $\mathrm{d}(T) := \lim_{n\to\infty} \mathrm{d}_n(T)$ both exist; we have, of course, met $\mathrm{c}(T)$ in Chapter I. A connection between these numbers and $\tilde{\beta}(T)$ is given by the following result.

Theorem 3.9. Let $T \in \mathscr{B}(X,Y)$. Then $\mathrm{d}(T) = \tilde{\beta}(T)$ and $\frac{1}{2}\tilde{\beta}(T) \leqslant \mathrm{c}(T) \leqslant 2\tilde{\beta}(T)$. ■

Proof. First we claim that for all $n \in \mathbb{N}$,

$$\mathrm{d}_{2^{n-1}+1}(T) \leqslant \mathrm{e}_n(T). \tag{3.3}$$

To prove this let $\lambda > \mathrm{e}_n(T)$, let $y_1, \ldots, y_{2^{n-1}} \in Y$ with $T(\mathrm{B}_X) \subset \{y_1, \ldots, y_{2^{n-1}}\} + \lambda \mathrm{B}_Y$, and put $M = \operatorname{span}\{y_1, \ldots, y_{2^{n-1}}\}$. Then $\|\mathrm{Q}_M^Y T\| \leqslant \lambda$, and our claim follows.

Next, we assert that for all $k, n \in \mathbb{N}$,

$$\mathrm{e}_k(T) \leqslant (1 + \mathrm{c}_{k,\,2n-2})\mathrm{d}_n(T) + \|T\|\mathrm{c}_{k,\,2n-2}, \tag{3.4}$$

where $\mathrm{c}_{k,n}$ is the upper bound (cf. Proposition 1.3 and its proof) $\min\{1, 2(2^{(k-1)/n} - 1)^{-1}\}$ of $\mathrm{e}_k(\mathrm{id}_G)$, the space G being any n-dimensional real vector space and id_G the identity map of G to itself; the corresponding bound for a complex n-dimensional space is $\mathrm{c}_{k,\,2n}$. To establish this, let $\lambda > \mathrm{d}_n(T)$ and $\mu > \mathrm{c}_{k,\,2n-2}$, and choose a linear subspace N of Y with $\dim N < n$ and $\|Q_N^Y T\| < \lambda$. For a given $x \in B_X$ we have $\|\mathrm{Q}_N^Y\ Tx\| < \lambda$, and hence there exists $z \in N$ with $\|Tx - z\| < \lambda$. Thus $\|z\| \leqslant \|z - Tx\| + \|Tx\| \leqslant \lambda + \|T\|$. On the other hand, $\mathrm{e}_k\big((\lambda + \|T\|)\mathrm{id}_N\big) < (\lambda + \|T\|)\mu$ and there are points $x_1, \ldots, x_{2^{k-1}} \in N$ with

$$(\lambda + \|T\|)\mathrm{B}_N \subset \{x_1, \ldots, x_{2^{k-1}}\} + (\lambda + \|T\|)\mu\, \mathrm{B}_N.$$

Hence we may pick some x_i with $\|z - x_i\| \leqslant (\lambda + \|T\|)\mu$. Thus

$$\|Tx - x_i\| \leqslant \|Tx - z\| + \|z - x_i\| \leqslant (1 + \mu)\lambda + \|S\|\mu,$$

and the proof of (3.4) is complete.

Combination of (3.3) and (3.4), together with appropriate passage to the limit, shows that $\mathrm{d}(T) = \tilde{\beta}(T)$. That the inequality claimed for $\mathrm{c}(T)$ holds is just Theorem I.2.21. □

The result of this theorem dealing with $\mathrm{d}(T)$ is due to Edmunds and Tylli [1]; it improves an earlier result due to Carl [2].

From Theorem 3.9 it follows immediately that the next theorem holds.

Theorem 3.10. Let $T \in \mathscr{B}(X,Y)$. Then

(i) $\mathrm{c}(T) = 0$ if, and only if, T is compact;

(ii) $\mathrm{c}(T) = 0$ if, and only if, $\mathrm{d}(T) = 0$;

(iii) $r_e(T) = \lim_{n\to\infty} [c(T^n)]^{1/n} = \lim_{n\to\infty} [d(T^n)]^{1/n}$;

(iv) if $S \in \mathscr{K}(X,Y)$ then $c(S+T) = c(T)$ and $d(S+T) = d(T)$. ■

We conclude this brief account of s-numbers by mentioning the *Weyl numbers*. Given any $T \in \mathscr{B}(X,Y)$ and any $n \in \mathbb{N}$, the nth Weyl number $x_n(T)$ is defined by

$$x_n(T) = \sup\{a_n(TS): S \in \mathscr{B}(l^2, X),\ \|S\| \leqslant 1\}.$$

It follows easily that the Weyl numbers are additive, multiplicative s-numbers; moreover, by Proposition 3.5,

$$x_n(T) = \sup\{c_n(TS): S \in \mathscr{B}(l^2, X),\ \|S\| \leqslant 1\}.$$

Interesting results can be proved by means of the Weyl numbers: for example, Pietsch [2] has used them to give sufficient conditions for the eigenvalues of a map $T \in \mathscr{K}(X)$ to form a sequence in a Lorentz space; and he identifies a class of maps $T \in \mathscr{B}(X)$, the so-called absolutely $(p, 2)$-summing maps $(2 < p < \infty)$, which have the happy property that $\sigma(T) \setminus \{0\}$ consists of an at most countable set of eigenvalues $\lambda_n(T)$. Arranged in the usual way it turns out that $(\lambda_n(T)) \in l^{p,\infty}$, so that $\lambda_n(T) = O(n^{-1/p})$ as $n \to \infty$. For this and various other interesting results we refer the reader to Pietsch [2]. In König [5] it is shown that Weyl's inequality (2.16) holds with the approximation numbers $a_n(T)$ replaced by the Weyl numbers $x_n(T)$.

4. Non-compact maps

Let X be a complex Banach space and let $T \in \mathscr{B}(X)$. If D is a connected component of $\mathbb{C} \setminus \sigma_{e1}(T)$, then by the discussion at the beginning of §I.4 it follows that $\sigma(T)$ contains either all of D or only an at most countable set of points of D with no point of accumulation in D. In this second case the points in question are eigenvalues of T and the corresponding projections (see §I.4) are finite-dimensional, the dimension being the algebraic multiplicity of the eigenvalue. Let us call these eigenvalues *Riesz points* of $\sigma(T)$. Since $\sigma(T)$ is bounded it cannot contain the unbounded component of $\mathbb{C} \setminus \sigma_{e1}(T)$: thus the set

$$\Lambda(T) := \{\lambda \in \sigma(T): |\lambda| > r_e(T)\}$$

is at most countable and consists solely of Riesz points. We order these points, denoted by $\lambda_n(T)$, in such a way that $|\lambda_1(T)| \geqslant |\lambda_2(T)| \geqslant \ldots > r_e(T) \geqslant 0$, where each eigenvalue is repeated according to its algebraic multiplicity: if there are only n such points $(n = 0, 1, 2, \ldots)$, including multiplicities, we put $|\lambda_{n+1}(T)| = |\lambda_{n+2}(T)| = \ldots = r_e(T)$. Thus $\Lambda(T) = \{\lambda_n(T): n \in \mathbb{N}\}$. It turns out that it is this part $\Lambda(T)$ of $\sigma(T)$ which can be regarded as the appropriate analogue of the non-zero part of the spectrum of a *compact* map. This is because various of the results concerning the eigenvalues of a compact map which were mentioned earlier hold for the eigenvalues in $\Lambda(T)$, where T is merely bounded and not necessarily compact; this observation is due to Zemánek [1]. We give some of these results below, beginning with the analogue of Theorem 1.6.

Theorem 4.1. Let $T \in \mathscr{B}(X)$ and let $\lambda_n(T)$ $(n \in \mathbb{N})$ be as above. Then for all $m, n \in \mathbb{N}$,

$$|\lambda_n(T)| \leqslant \left(\prod_{j=1}^{n} |\lambda_j(T)| \right)^{1/n} \leqslant (\sqrt{2})^{(m-1)/n}\, \mathrm{e}_m(T). \qquad \blacksquare$$

Proof. If $\mathrm{r_e}(T) = 0$ or $|\lambda_n(T)| > \mathrm{r_e}(T)$, the proof is essentially the same as that of Theorem 1.6. Since $|\lambda_k(T)| \geqslant |\lambda_{k+1}(T)| \geqslant \ldots \geqslant \mathrm{r_e}(T)$ for all $k \in \mathbb{N}$, and $\mathrm{r_e}(T) \leqslant \beta(T) \leqslant \mathrm{e}_m(T)$ for all $m \in \mathbb{N}$, the result is also obvious if $|\lambda_1(T)| = \mathrm{r_e}(T)$. To deal with the remaining possibilities we may therefore assume, without loss of generality, that $\mathrm{r_e}(T) = 1$ and that $n+1$ is the least number $k \in \mathbb{N}$ such that $|\lambda_k(T)| = |\lambda_{k+1}(T)| = \ldots = \mathrm{r_e}(T)$. Then $|\lambda_n(T)| > \mathrm{r_e}(T)$, and thus for all $m \in \mathbb{N}$,

$$\left(\prod_{j=1}^{n} |\lambda_j(T)| \right)^{1/n} \leqslant (\sqrt{2})^{(m-1)/n}\, \mathrm{e}_m(T).$$

Since $|\lambda_{n+1}(T)| = 1$ the theorem now follows immediately. $\square$

From this result it can be shown, just as in Corollary 1.7, that if $T \in \mathscr{B}(X)$, then for all $n \in \mathbb{N}$,

$$\lim_{k \to \infty} [\mathrm{e}_n(T^k)]^{1/k} = \mathrm{r}(T). \tag{4.1}$$

The analogue of Theorem 2.8 also holds:

Theorem 4.2. Let $T \in \mathscr{B}(X)$. Then for all $n \in \mathbb{N}$,

$$|\lambda_n(T)| = \lim_{k \to \infty} [\mathrm{a}_n(T^k)]^{1/k}. \qquad \blacksquare$$

Proof. As in the proof of Theorem 2.8 it follows that for all $n \in \mathbb{N}$,

$$\limsup_{k \to \infty} [\mathrm{a}_n(T^k)]^{1/k} \leqslant |\lambda_n(T)|.$$

To prove that for all $n \in \mathbb{N}$,

$$\liminf_{k \to \infty} [\mathrm{a}_n(T^k)]^{1/k} \geqslant |\lambda_n(T)| \tag{4.2}$$

we first observe that if $|\lambda_n(T)| = \mathrm{r_e}(T)$, then since, with the aid of Proposition 2.7, we have

$$\mathrm{r_e}(T) = \lim_{k\to\infty} [\tilde{\beta}(T^k)]^{1/k} \leqslant \liminf_{k\to\infty} [\alpha(T^k)]^{1/k}$$

$$\leqslant \liminf_{k\to\infty} [\mathrm{a}_n(T^k)]^{1/k},$$

(4.2) holds. On the other hand, if $|\lambda_n(T)| > \mathrm{r_e}(T)$ then we can proceed as in the proof of Theorem 2.8. This completes the proof. □

Teixeira [2] has pointed out that Weyl's inequality also holds in the form (2.16): if $T \in \mathscr{B}(X)$ and $p \in (0, \infty)$, then there is a constant K_p, depending only on p, such that for all $N \in \mathbb{N}$,

$$\left(\sum_{k=1}^{N} |\lambda_k(T)|^p\right)^{1/p} \leqslant K_p\left(\sum_{k=1}^{N} \mathrm{a}_k^p(T)\right)^{1/p}.$$

The proof of this is the same as that given for the case when T is compact.

For maps $T \in \mathscr{K}(X)$, much interest has been shown in the manner in which $\lambda_n(T) \to 0$ as $n \to \infty$, a famous example of such matters being the question raised by Lorentz and solved, in a form, by Weyl, about the asymptotic behaviour of eigenvalues of elliptic equations on a bounded open set with zero Dirichlet boundary conditions. For such maps Weyl's inequality shows that $(\lambda_n(T)) \in l^p$ if $(\mathrm{a}_n(T)) \in l^p$; the inequality $|\lambda_n(T)| \leqslant \sqrt{2}\mathrm{e}_{n+1}(T)$ which follows from Theorem 1.6 also shows that $(\lambda_n(T)) \in l^p$ if $(\mathrm{e}_n(T)) \in l^p$. For a non-compact map T it is natural to enquire into the way in which $|\lambda_n(T)| \to \mathrm{r_e}(T)$, and to do this, more delicacy seems to be required. A preliminary result in this direction is given by the following theorem, due to Edmunds and Triebel [1].

Theorem 4.3. Let H be a complex Hilbert space, let $T \in \mathscr{B}(H)$ be self-adjoint and suppose that $0 < p < \infty$. Then if $\sum_{n=1}^{\infty} n^{-1} |\mathrm{e}_n(T) - \mathrm{r_e}(T)|^p$ is finite, so is $\sum_{n=1}^{\infty} n^{-1} [|\lambda_n(T)| - \mathrm{r_e}(T)]^p$. ■

Proof. In view of Theorem 4.1 the result is clear if $\mathrm{r_e}(T) = 0$. If $\mathrm{r_e}(T) > 0$ it is, plainly, enough to deal with the case in which $\mathrm{r_e}(T) = 1$. For each $n \in \mathbb{N}$ we put $\Lambda_n = \lambda_{4^n}(T)$ and $E_n = \mathrm{e}_{2^n}(T)$. Since

$$\sum_{j=0}^{\infty} \sum_{n=4^j}^{4^{j+1}} n^{-1} [|\lambda_n(T)| - 1]^p \leqslant 3 \sum_{j=0}^{\infty} 4^{-j} (\Lambda_j - 1)^p \cdot 4^j$$

and

$$\sum_{n=1}^{\infty} n^{-1} [\mathrm{e}_n(T) - 1]^p \geqslant \tfrac{1}{2} \sum_{j=0}^{\infty} (E_{j+1} - 1)^p,$$

if we can prove that for all $n \in \mathbb{N}$,

$$\Lambda_n - 1 \leqslant c(E_n - 1) + 2^{-n}c, \tag{4.3}$$

then it is clear that the theorem will hold. To establish (4.3), observe that by Theorem 4.1,

$$|\lambda_n(T)| \leqslant (\sqrt{2})^{(m-1)/n}\,\mathrm{e}_m(T),$$

and hence

$$|\lambda_{4^n}(T)| \leqslant (\sqrt{2})^{(2^n-1)/4^n}\,\mathrm{e}_{2^n}(T);$$

that is,

$$\Lambda_n \leqslant (\sqrt{2})^{2^{-n}} E_n \qquad (n \in \mathbb{N}).$$

Thus

$$\begin{aligned}\Lambda_n - 1 \leqslant (\sqrt{2})^{2^{-n}} E_n - 1 &\leqslant (\sqrt{2})^{2^{-n}}(E_n - 1) + [(\sqrt{2})^{2^{-n}} - 1] \\ &\leqslant (E_n - 1) + 2^{-n}\sqrt{2}.\end{aligned}$$

The proof is complete. □

5. Compact linear operators in Hilbert spaces

We have already seen that, in the setting of a general Banach space, a great deal can be said about compact linear maps. In the context of Hilbert spaces the results can be sharpened quite considerably, and here we shall set about the description of the more fundamental points of this classical theory. Occasionally we shall prove results already known, simply to illustrate how the extra structure that Hilbert spaces possess enables a more direct method of proof to be employed. Throughout this section H, possibly adorned with a subscript, will stand for a Hilbert space. In Chapter IV more details about self-adjoint operators will be found.

We begin with a discussion of bounded self-adjoint maps.

Lemma 5.1. Let $T \in \mathscr{B}(H)$ be self-adjoint. Then

$$\|T\| = \sup\{|(Tx, x)| : x \in H, \|x\| = 1\}. \tag{5.1}$$ ■

Proof. Denote the right-hand side of (5.1) by $\mathrm{N}(T)$. Evidently $\mathrm{N}(T) \leqslant \|T\|$. Moreover, if $\lambda > 0$, then

$$4\|Tx\|^2 = (T(\lambda x + \lambda^{-1}Tx), \lambda x + \lambda^{-1}Tx) - (T(\lambda x - \lambda^{-1}Tx), \lambda x - \lambda^{-1}Tx)$$

for all $x \in H$. Thus

$$\begin{aligned}4\|Tx\|^2 &\leqslant \mathrm{N}(T)(\|\lambda x + \lambda^{-1}Tx\|^2 + \|\lambda x - \lambda^{-1}Tx\|^2) \\ &= 2\mathrm{N}(T)(\lambda^2\|x\|^2 + \lambda^{-2}\|Tx\|^2).\end{aligned}$$

If $Tx \neq 0$, the right-hand side above attains its minimum, as a function of the positive variable λ, when $\lambda^2 = \|Tx\|/\|x\|$, and hence

$$\|Tx\|^2 \leqslant \mathrm{N}(T)\|Tx\|\,\|x\|,$$

which shows that $\|Tx\| \leqslant \mathrm{N}(T)\|x\|$. This obviously holds also when $Tx = 0$, and consequently $\|T\| \leqslant \mathrm{N}(T)$. Thus $\|T\| = \mathrm{N}(T)$ as required. □

Theorem 5.2. Let $\dim H = \infty$ and let $T \in \mathscr{K}(H)\backslash\{0\}$ be self-adjoint. Then T has a finite or countably infinite set of eigenvalues, according to whether T is of finite rank or not. Each eigenvalue has finite geometric multiplicity. Denote the eigenvalues by λ_n, each repeated according to its geometric multiplicity and arranged so that $|\lambda_n| \geqslant |\lambda_{n+1}|$ for all n; then $\lim_{n\to\infty} \lambda_n = 0$ if T is not of finite rank. Moreover for all $x \in H$,

$$Tx = \sum_{n=1}^{\infty} \lambda_n (x, \phi_n)\, \phi_n,$$

where each ϕ_n is an eigenvector of T corresponding to the eigenvalue λ_n and (ϕ_n) is an orthonormal sequence. (Here and later we adopt the convention that $\lambda_n = 0$ for $n > N$ if T is of rank $N < \infty$). ■

Proof. Some of these results follow from those given earlier in the context of Banach spaces; we give details here only because the methods used are quite different from those employed before.

By Lemma 5.1, there is a sequence (x_n) in H such that $\|x_n\| = 1$ for all $n \in \mathbb{N}$, and $\lim_{n\to\infty} (Tx_n, x_n) = \lambda_1$, where $|\lambda_1| = \|T\| > 0$. Hence

$$\begin{aligned}\|Tx_n - \lambda_1 x_n\|^2 &= \|Tx_n\|^2 - 2\lambda_1 (Tx_n, x_n) + |\lambda_1|^2\, \|x_n\|^2 \\ &\leqslant |\lambda_1|^2 - 2\lambda_1 (Tx_n, x_n) + |\lambda_1|^2 \to 0 \end{aligned} \tag{5.2}$$

as $n \to \infty$. Since T is compact, (Tx_n) must contain a convergent subsequence, $(Tx_{n(k)})$, say. Put $\phi_1 = \lambda_1^{-1} \lim_{k\to\infty} Tx_{n(k)}$. By (5.2), $x_{n(k)} \to \phi_1$; and as T is continuous, $Tx_{n(k)} \to T\phi_1$. Hence $T\phi_1 = \lambda_1 \phi_1$ and $\|\phi_1\| = 1$; so λ_1 is an eigenvalue of T with corresponding eigenvector ϕ_1.

Let $H_2 = (\mathrm{sp}\,\{\phi_1\})^{\perp}$, the orthogonal complement of the linear span of ϕ_1, and let T_2 be the restriction of T to H_2. Then H_2 is a Hilbert space and T_2 is a self-adjoint map in $\mathscr{K}(H_2)$. Thus if $T_2 \neq 0$, it has an eigenvalue λ_2 such that

$$|\lambda_2| = \|T_2\| \leqslant \|T\| = |\lambda_1|,$$

and an eigenvector $\phi_2 \in H_2$ such that $\|\phi_2\| = 1$. Hence λ_2 is an eigenvalue of T with associated eigenvector ϕ_2; and $\phi_2 \perp \phi_1$.

This process can be repeated indefinitely, unless at some stage, with $H_{m+1} := (\text{sp}\{\phi_1, \phi_2, \ldots, \phi_m\})^\perp$, $T_{m+1} := T \upharpoonright H_{m+1}$ is the zero operator. If the process does not terminate we obtain an infinite sequence (λ_n) of eigenvalues and an orthonormal sequence (ϕ_n) of associated eigenvectors. In this latter case we must have $\lambda_n \to 0$, because otherwise $(\lambda_n^{-1}\phi_n)$ would be bounded and the compactness of T would imply that, since $\phi_n = T(\lambda_n^{-1}\phi_n)$, the sequence (ϕ_n) contains a convergent subsequence, contrary to the orthonormality of (ϕ_n). The same reasoning shows that the geometric multiplicity of each eigenvalue λ_n is finite.

Let m be the number of eigenvectors in the sequence (ϕ_n) if the sequence is finite, that is, if $T_{m+1} = 0$; and let m be an arbitrary positive integer otherwise. Let $x \in H$ and put $y_m = x - \sum_{n=1}^{m} (x, \phi_n)\phi_n$. Since $(y_m, \phi_n) = 0$ $(n = 1, 2, \ldots, m)$, it follows that $y_m \in H_{m+1}$, and so

$$\|Ty_m\| = \|T_{m+1} y_m\| \leqslant \|T_{m+1}\| \, \|y_m\| = |\lambda_{m+1}| \, \|y_m\| \tag{5.3}$$

and

$$Ty_m = Tx - \sum_{n=1}^{m} \lambda_n (x, \phi_n)\phi_n. \tag{5.4}$$

If $T_{m+1} = 0$, then $Tx = \sum_{n=1}^{m} \lambda_n (x, \phi_n)\phi_n$ and so T is of finite rank. Otherwise, since $\lambda_{m+1} \to 0$ as $m \to \infty$, and (y_m) is clearly bounded, we see from (5.3) that $Ty_m \to 0$ and so, from (5.4),

$$Tx = \sum_{n=1}^{\infty} \lambda_n (x, \phi_n)\phi_n. \qquad \square$$

Corollary 5.3. Let $T \in \mathcal{K}(H)$ be self-adjoint. Then the set (ϕ_n) of eigenvectors of T is a complete orthonormal set in $\mathcal{N}(T)^\perp$. ■

Proof. Let $x \in H$ and define M by

$$M = \begin{cases} \infty & \text{if } T \text{ is of infinite rank,} \\ \min\{m \in \mathbb{N} : T_{m+1} = 0\} & \text{otherwise.} \end{cases}$$

Then $\sum_{n=1}^{M} (x, \phi_n)\phi_n$ converges in H, and since T is continuous,

$$T\left(x - \sum_{n=1}^{M} (x, \phi_n)\phi_n\right) = Tx - \sum_{n=1}^{M} \lambda_n (x, \phi_n)\phi_n = 0,$$

by Theorem 5.2. Since T is self-adjoint, each ϕ_n lies in $\mathcal{N}(T)^\perp$, and thus if $x \in \mathcal{N}(T)^\perp$,

$$x - \sum_{n=1}^{M} (x, \phi_n)\, \phi_n \in \mathcal{N}(T) \cap \mathcal{N}(T)^\perp = \{0\}.$$

The result follows. □

Corollary 5.4. Let H and T be as in Theorem 5.2. Then

$$\sigma(T) = \{\lambda_n : n = 1, 2, \ldots\} \cup \{0\}$$ ■

Proof. If T is of finite rank, Theorem 5.2 and Corollary 5.3 imply that $\mathcal{N}(T) \neq \{0\}$; in fact, nul $T = \infty$ since dim $H = \infty$. Hence $\{\lambda_n : n \in \mathbb{N}\} \cup \{0\} \subset \sigma(T)$. If T is not of finite rank, $0 \in \sigma(T)$ since $\lambda_n \to 0$ and $\sigma(T)$ is closed. Thus we need only prove that every λ not in $\{\lambda_n : n \in \mathbb{N}\} \cup \{0\}$ is in $\rho(T)$; and since T is self-adjoint and $\sigma(T)$ is accordingly a subset of $\mathbb{R}$, we can focus our attention on real λ.

Let $\lambda \in \mathbb{R}$, with $\lambda \notin \{\lambda_n : n \in \mathbb{N}\} \cup \{0\}$. Then there exists $\delta > 0$ such that, with $\lambda_0 := 0$, $\inf\{|\lambda - \lambda_n| : n \in \mathbb{N}_0\} \geqslant \delta > 0$. From Theorem 5.2 and Corollary 5.3, if $x \in \mathcal{N}(T)^\perp$ then

$$(T - \lambda I)x = \sum_{n=1}^{\infty} (\lambda_n - \lambda)(x, \phi_n)\, \phi_n,$$

and so

$$\|(T - \lambda I)x\| \geqslant \delta \|x\| \qquad (x \in \mathcal{N}(T)^\perp).$$

If $x \in \mathcal{N}(T)$ then $\|(T - \lambda I)x\| = \|\lambda x\| \geqslant \delta \|x\|$. Since $(T - \lambda I)\mathcal{N}(T) \subset \mathcal{N}(T)$ we have $\|(T - \lambda I)x\| \geqslant \delta \|x\|$ for all $x \in H$. Thus $\mathcal{R}(T - \lambda I)$ is closed and def $(T - \lambda I) =$ nul $(T - \lambda I) = 0$. Hence $\lambda \in \rho(T)$. □

Our next object of study is the *min–max principle*, a most important formula for the eigenvalues of a non-negative self-adjoint compact linear operator.

Theorem 5.5. Let dim $H = \infty$, let $T \in \mathcal{K}(H)$ be non-negative and self-adjoint, and let $\{\lambda_n : n \in \mathbb{N}\}$ be the set of all eigenvalues of T, each repeated according to geometric multiplicity and arranged in descending order. (As usual, if T has finite rank we define λ_n to be zero for all large enough n). Then for each $n \in \mathbb{N}$,

$$\lambda_{n+1} = \min_{\dim G \leqslant n} \; \max_{x \in G^\perp,\, x \neq 0} \; [(Tx, x)/\|x\|^2], \tag{5.5}$$

where it is to be understood that, in taking the minimum, G runs through all linear subspaces of H with dimension $\leqslant n$. ■

Proof. We have already proved in Theorem 5.2 that if $n < N := \text{rank } T$, then

$$\lambda_{n+1} = \max_{x \in M_n^{\perp},\, x \neq 0} (Tx, x)/\|x\|^2,$$

where $M_n = \text{sp}\{\phi_1, \phi_2, \ldots, \phi_n\}$. Denote the right-hand side of (5.5) by μ_{n+1}. Since dim $M_n = n$ we have $\lambda_{n+1} \geqslant \mu_{n+1}$. Let G be an m-dimensional linear subspace of H, with $m \leqslant n$, and let $\{\theta_1, \theta_2, \ldots, \theta_m\}$ be a basis of G. Then the system of m equations in $n+1$ unknowns $\alpha_1, \alpha_2, \ldots, \alpha_{n+1} \in \mathbb{C}$,

$$\sum_{i=1}^{n+1} \alpha_i (\theta_j, \phi_i) = 0 \qquad (j = 1, 2, \ldots, m), \tag{5.6}$$

has a non-trivial solution. Put $\psi = \sum_{i=1}^{n+1} \alpha_i \phi_i$, where $\alpha_1, \alpha_2, \ldots, \alpha_{n+1}$ satisfy (5.6). Thus $\psi \in M_{n+1} \cap G^{\perp}$ and

$$(T\psi, \psi) = \sum_{i=1}^{n+1} \lambda_i |(\psi, \phi_i)|^2 \geqslant \lambda_{n+1} \sum_{i=1}^{n+1} |(\psi, \phi_i)|^2$$

$$= \lambda_{n+1} \sum_{i=1}^{n+1} |\alpha_i|^2 = \lambda_{n+1} \|\psi\|^2.$$

It follows that for all linear subspaces G of H with dim $G \leqslant n$,

$$\max_{x \in G^{\perp},\, x \neq 0} (Tx, x)/\|x\|^2 \geqslant \lambda_{n+1},$$

which shows that $\mu_{n+1} \geqslant \lambda_{n+1}$. Hence $\mu_{n+1} = \lambda_{n+1}$, as required, provided that $n < N$. However, if $n \geqslant N$, (5.5) is immediate. The proof is complete. □

Now, given any $\lambda > 0$, let $K(\lambda)$ be the set of all closed linear subspaces R of H such that $(Tx, x) \geqslant \lambda \|x\|^2$ for all $x \in R$.

Theorem 5.6. Let $T \in \mathscr{K}(H)$ be non-negative and self-adjoint, and let $N(\lambda)$ be the number of eigenvalues of T which are not less than λ; that is, $N(\lambda) = \sum_{\lambda_n \geqslant \lambda} 1$. Then

$$N(\lambda) = \max_{R \in K(\lambda)} \dim R. \tag{5.7}$$ ■

Proof. Let $\lambda > 0$ and let $m \in \mathbb{N}$ be such that $\lambda_m \geqslant \lambda > \lambda_{m+1}$: for the moment, we assume that such an m exists and discuss what happens if it does not later. Then $N(\lambda) = m$; and with $M_m := \text{sp}\{\phi_1, \phi_2, \ldots, \phi_m\}$ we see that if $x \in M_m \setminus \{0\}$, then

$$(Tx, x)/\|x\|^2 = \sum_{i=1}^{m} \lambda_i |(x, \phi_i)|^2 \Big/ \sum_{i=1}^{m} |(x, \phi_i)|^2$$

$$\geqslant \lambda_m \geqslant \lambda.$$

Hence $M_m \in K(\lambda)$ and $\max\{\dim R : R \in K(\lambda)\} \geqslant m$.

Now suppose that $R \in K(\lambda)$ and $\dim R > m$. Then proceeding exactly as in Theorem 5.5, we see that there is an element $\psi \neq 0$ of $R \cap M_m^\perp$. From Theorem 5.2, and since $\psi \in M_m^\perp$,

$$\begin{aligned}(T\psi, \psi) &= \sum_{i=1}^{\infty} \lambda_i |(\psi, \phi_i)|^2 = \sum_{i=m+1}^{\infty} \lambda_i |(\psi, \phi_i)|^2 \\ &\leqslant \lambda_{m+1} \sum_{i=m+1}^{\infty} |(\psi, \phi_i)|^2 \\ &< \lambda \|\psi\|^2.\end{aligned}$$

This contradicts the assumption that $R \in K(\lambda)$, and so $\dim R \leqslant m$ for all $R \in K(\lambda)$. Hence the result follows.

If λ is so large that there is no $m \in N$ with $\lambda_m \geqslant \lambda > \lambda_{m+1}$ then $N(\lambda) = 0$; moreover, if $\phi \in H$, then for all $\phi \in H \setminus \{0\}$,

$$(T\phi, \phi) = \sum_{i=1}^{\infty} \lambda_i |(\phi, \phi_i)|^2 < \lambda \sum_{i=1}^{\infty} |(\phi, \phi_i)|^2 = \lambda \|\phi\|^2,$$

and hence the only element of $K(\lambda)$ is $\{0\}$. The proof of the theorem is now complete. □

We now turn to compact maps which need not be self-adjoint. If $T \in \mathscr{K}(H, H_1)$, then T^*T is a non-negative self-adjoint map in $\mathscr{K}(H)$ and therefore has a non-negative, self-adjoint square root $|T| := (T^*T)^{\frac{1}{2}}$. Since $T = U|T|$, where U is a partial isometry (see Chapter IV, § 3), it follows that $|T| = U^*T$ and hence $|T| \in \mathscr{K}(H)$ by Proposition I.1.2. The eigenvalues of $|T|$ are called the *singular values* of T; they will be denoted by $\mu_1, \mu_2, \ldots$, arranged so that $\mu_1 \geqslant \mu_2 \geqslant \ldots \geqslant 0$, and repeated according to geometric multiplicity. With the convention that μ_n is defined to be zero for all sufficiently large n if T (and hence $|T|$) is of finite rank, we see that in all cases, $\mu_n \to 0$ as $n \to \infty$.

Theorem 5.7. Let $T \in \mathscr{K}(H, H_1)$. Then

$$Tx = \sum_{n=1}^{\infty} \mu_n (x, \phi_n)_H \psi_n \qquad (x \in H), \tag{5.8}$$

$$T^*y = \sum_{n=1}^{\infty} \mu_n (y, \psi_n)_{H_1} \phi_n \qquad (y \in H_1), \tag{5.9}$$

where the ϕ_n are orthonormal eigenvectors of $|T|$ corresponding to the eigenvalues μ_n, and $\psi_n = \mu_n^{-1} T\phi_n$ $(\mu_n \neq 0)$. The series in (5.8) and (5.9) are finite if T is of finite rank. Moreover, T and T^* have the same singular values. ■

Proof. From Theorem 5.2,

$$|T|x = \sum_{n=1}^{\infty} \mu_n (x, \phi_n)_H \phi_n \qquad (x \in H),$$

and hence

$$T^*Tx = |T|^2 x = \sum_{n=1}^{\infty} \mu_n^2 (x, \phi_n)_H \phi_n.$$

Since T^*T is of finite rank if T is, the series is finite if T is of finite rank. With $\mu_n \psi_n = T\phi_n$,

$$T^*Tx = \sum_{n=1}^{\infty} (x, T^*T\phi_n)_H \phi_n = \sum_{n=1}^{\infty} (Tx, T\phi_n)_{H_1} \phi_n$$

$$= \sum_{n=1}^{\infty} \mu_n (Tx, \psi_n)_{H_1} \phi_n \qquad (x \in H). \tag{5.10}$$

We also have

$$\mu_n \mu_m (\psi_n, \psi_m)_{H_1} = (T\phi_n, T\phi_m)_{H_1} = (\phi_n, T^*T\phi_m)_H = \mu_m^2 (\phi_n, \phi_m)_H,$$

which shows that the sequence (ψ_n) is orthonormal in H_1. Hence

$$\left\| \sum_{n=1}^{\infty} \mu_n (Tx, \psi_n)_{H_1} \phi_n \right\|^2 = \sum_{n=1}^{\infty} \mu_n^2 |(Tx, \psi_n)_{H_1}|^2$$

$$\leqslant \mu_1^2 \sum_{n=1}^{\infty} |(Tx, \psi_n)_{H_1}|^2 \leqslant \mu_1^2 \|Tx\|^2. \tag{5.11}$$

It follows from (5.10) and (5.11) that for any $y \in \overline{\mathscr{R}(T)}$,

$$T^*y = \sum_{n=1}^{\infty} \mu_n (y, \psi_n)_{H_1} \phi_n,$$

since both sides of (5.10) depend continuously upon $Tx \in \mathscr{R}(T)$. If $z \perp \mathscr{R}(T)$, then $T^*z = 0$ and also

$$(z, \psi_n)_{H_1} = \mu_n^{-1} (z, T\phi_n)_{H_1} = \mu_n^{-1} (T^*z, \phi_n)_H = 0 \qquad (\mu_n \neq 0).$$

Hence (5.9) holds, for all $y \in H_1$; (5.8) now follows from the relationship $(Tx, y)_{H_1} = (x, T^*y)_H$.

Finally we have from (5.8) and (5.9),

$$TT^*y = \sum_{n=1}^{\infty} \mu_n^2 (y, \psi_n)_{H_1} \psi_n \qquad (y \in H_1), \tag{5.12}$$

and so TT^* has eigenvalues μ_n^2 and eigenvectors ψ_n in H_1. This means, in particular, that the singular values of T^* are the μ_n. The proof is complete. □

At this point we can give the proof of Theorem 1.8, which for convenience we state again below.

Theorem 1.8. Let $T \in \mathscr{K}(H)$, where H is a Hilbert space. Then

$$\mathrm{e}_n(T) = \mathrm{e}_n(T^*) = \mathrm{e}_n(|T|) \quad \text{for all } n \in \mathbb{N}.$$ ■

Proof. Let B be the closed unit ball in H. With the same notation as in Theorem 5.7, define a map $V: T(B) \to T^*(B)$ by $V(Tx) = T^*y$, where $y = \sum_{n=1}^{\infty} (x, \phi_n)\psi_n$. The function V is well-defined since $V(z)$ is independent of the choice of x such that $Tx = z$. Also note that

$$\|y\|^2 = \sum_{n=1}^{\infty} |(x, \phi_n)|^2 \leqslant \|x\|^2 \leqslant 1.$$

An immediate consequence is that $V(T(B)) = T^*(B)$. To see this, let $y \in B$ and define x by $x = \sum_{n=1}^{\infty} (y, \psi_n)\phi_n$. Then

$$\|x\|^2 = \sum_{n=1}^{\infty} |(y, \psi_n)|^2 \leqslant \|y\|^2 \leqslant 1,$$

and $V(Tx) = T^*(y)$.

Moreover, for all $x_1, x_2 \in B$,

$$\|V(Tx_1) - V(Tx_2)\|^2 = \|T^*(y_1 - y_2)\|^2 = \sum_{n=1}^{\infty} \mu_n^2 |(y_1 - y_2, \psi_n)|^2$$

$$= \sum_{n=1}^{\infty} \mu_n^2 |(x_1 - x_2, \phi_n)|^2 = \|Tx_1 - Tx_2\|^2.$$

By Valentine's extension of Kirszbraun's Theorem (see Valentine [1]), there exists $W: H \to H$ such that $\|Wx_1 - Wx_2\| \leqslant \|x_1 - x_2\|$ for all $x_1, x_2 \in H$ and $W \upharpoonright T(B) = V$.

Now suppose that $\varepsilon > 0$ and $n \in \mathbb{N}$. Then there exist $x_1, x_2, \ldots, x_{2^{n-1}} \in H$ such that $T(B)$ can be covered by balls $B_1, B_2, \ldots, B_{2^{n-1}}$, where each B_j has centre x_j and radius $e_n(T) + \varepsilon$. Given any $y \in T^*(B)$, there exist $x \in T(B)$ and $j \in \{1, 2, \ldots, 2^{n-1}\}$ such that $Vx = y$ and $x \in B_j$. Since

$$\|y - Wx_j\| = \|Wx - Wx_j\| \leqslant \|x - x_j\| \leqslant e_n(T) + \varepsilon,$$

it follows that $T^*(B)$ can be covered by 2^{n-1} balls of radius $e_n(T) + \varepsilon$; hence $e_n(T^*) \leqslant e_n(T)$. Similarly, $e_n(T) = e_n(T^{**}) \leqslant e_n(T^*)$, and thus $e_n(T) = e_n(T^*)$.

Define V_1: $T(B) \to |T|(B)$ by $V_1(Tx) = |T|(x)$. Then for all $x_1, x_2 \in B$ we have

$$\|V_1(Tx_1) - V_1(Tx_2)\|^2 = \||T|(x_1 - x_2)\|^2 = \sum_{n=1}^{\infty} \mu_n^2 |(x_1 - x_2, \phi_n)|^2$$

$$= \|Tx_1 - Tx_2\|^2.$$

As before, there is a map $W_1: H \to H$ with $\|W_1 x_1 - W_1 x_2\| \leqslant \|x_1 - x_2\|$ for all $x_1, x_2 \in H$ and such that $W_1 \upharpoonright T(B) = V_1$, from which it follows that $e_n(|T|)$

$\leqslant e_n(T)$. Similarly, it can easily be seen that $e_n(T) \leqslant e_n(|T|)$. The proof is complete. □

Let us define U to be the continuous linear map on $\overline{\mathscr{R}(|T|)}$ such that $U\phi_n = \psi_n$ for all n (that is, extend from the basis elements ϕ_n by linearity and continuity). From Theorems 5.2 and 5.7 it follows that (ϕ_n) and (ψ_n) are orthonormal bases for $\overline{\mathscr{R}(|T|)}$ and $\overline{\mathscr{R}(T)}$ respectively, and therefore U is a unitary map of $\overline{\mathscr{R}(|T|)}$ onto $\overline{\mathscr{R}(T)}$. Moreover, for all $x \in H$,

$$U|T|x = \sum_{n=1}^{\infty} \mu_n (x, \phi_n)_H \psi_n = Tx;$$

thus $T = U|T|$. This is the so-called *polar decomposition* of T (cf. Chapter IV, § 3).

Theorem 5.8. Let $\dim H = \infty$ and $T \in \mathscr{K}(H)$ be normal; that is, $TT^* = T^*T$. Then there is a scalar sequence (λ_n), with $|\lambda_n| = \mu_n$ (the nth singular value of T) and $Te_n = \lambda_n e_n$, $\|e_n\| = 1$, for each n, such that for all $x \in H$,

$$Tx = \sum_{n=1}^{\infty} \lambda_n (x, e_n) e_n.$$

Moreover, a scalar λ is an eigenvalue of T if, and only if, $\bar{\lambda}$ is an eigenvalue of T^*; and T and T^* have the same eigenvectors. The set $\{e_n : n \in \mathbb{N}\}$ of the eigenvectors of T corresponding to the set $\{\lambda_n : n \in \mathbb{N}\}$ of eigenvalues of T is a complete orthonormal set in $\mathscr{N}(T)^{\perp}$ $(= \mathscr{N}(T^*)^{\perp})$, and

$$\sigma(T) = \{\lambda_n : n \in \mathbb{N}\} \cup \{0\}, \qquad \sigma(T^*) = \{\bar{\lambda}_n : n \in \mathbb{N}\} \cup \{0\}.$$ ■

Proof. From Theorem 5.7,

$$T^*T\phi_n = \mu_n^2 \phi_n, \qquad TT^*\psi_n = \mu_n^2 \psi_n, \qquad (n \in \mathbb{N})$$

where $\mu_n \psi_n = T\phi_n$ and (ϕ_n) and (ψ_n) are orthonormal sequences in H. Let E_n be the eigenspace of T^*T corresponding to the eigenvalue μ_n^2 $(\neq 0)$. Since $T^*T = TT^*$, the space E_n is spanned by both the finite-dimensional sets $\{\phi_k : \mu_k^2 = \mu_n^2\}$ and $\{\psi_j : \mu_j^2 = \mu_n^2\}$. Since $T\phi_n = \mu_n \psi_n$ we see that T maps E_n onto itself. Also if $x \in E_n$, then $x = \Sigma c_i \phi_i$ and

$$\|Tx\|^2 = (T^*Tx, x) = \mu_n^2 \Sigma |c_i|^2 = \mu_n^2 \|x\|^2.$$

It follows that $\mu_n^{-1} T$ is a unitary map of E_n onto itself. It is therefore diagonalizable and has eigenvalues $\alpha_1^{(n)}, \alpha_2^{(n)}, \ldots, \alpha_j^{(n)}$ $(j = \dim E_n)$ such that for all $x \in E_n$,

$$\mu_n^{-1} Tx = \sum_{i=1}^{j} \alpha_i^{(n)} (x, e_i) e_i,$$

where $\mu_n^{-1} Te_i = \alpha_i^{(n)} e_i$, $|\alpha_i^{(n)}| = 1$ for $i = 1, 2, \ldots, j$, and $\{e_i : i = 1, 2, \ldots, j\}$ is orthonormal in H. Writing $\lambda_i^{(n)} = \mu_n \alpha_i^{(n)}$ we thus have for all $x \in E_n$,

$$Tx = \sum_{i=1}^{j} \lambda_i^{(n)} (x, e_i) e_i, \qquad |\lambda_i^{(n)}| = \mu_n,$$

and

$$Te_i = \lambda_i^{(n)} e_i.$$

Since $H = \mathcal{N}(T) \bigoplus_{n=1}^{\infty} E_n$, by Corollary 5.3, we have for all $x \in H$,

$$Tx = \sum_{n=1}^{\infty} \lambda_n (x, e_n) e_n, \qquad |\lambda_n| = \mu_n.$$

As T is normal so is $B := T - \lambda I$ for every scalar λ. Hence for all $x \in H$,

$$\begin{aligned} \|(T-\lambda I)x\|^2 &= (Bx, Bx) = (B^*Bx, x) = (BB^*x, x) \\ &= \|B^*x\|^2 = \|(T^* - \bar{\lambda} I)x\|^2. \end{aligned}$$

This shows that λ is an eigenvalue of T if, and only if, $\bar{\lambda}$ is an eigenvalue of T^*; and also T and T^* have the same eigenvectors. Note that $\lambda = 0$ is allowed here so that T and T^* have the same kernel.

We have already seen that $\{e_n: n = 1, 2, \ldots\}$ is a complete orthonormal set in $\mathcal{N}(T)^\perp$, and hence any $x \in X$ can be written as

$$x = \sum_{n=0}^{\infty} (x, e_n) e_n,$$

where $e_0 \in \mathcal{N}(T)$ (e_0 will, in general, depend on x). Hence

$$\|(T-\lambda I)x\|^2 = \sum_{n=0}^{\infty} |\lambda_n - \lambda|^2 |(x, e_n)|^2.$$

If $\lambda \notin \{\lambda_n: n = 1, 2, \ldots\} \cup \{0\}$, then there exists $\delta > 0$ such that

$$\|(T-\lambda I)x\|^2 \geqslant \delta^2 \sum_{n=0}^{\infty} |(x, e_n)|^2 = \delta^2 \|x\|^2,$$

and thus $(T-\lambda I)^{-1}$ exists and is bounded on $\mathcal{R}(T-\lambda I)$. Since $(T-\lambda I)^{-1}$ is also closed, $\mathcal{R}(T-\lambda I)$ must be closed. Moreover, $\mathcal{R}(T-\lambda I)^\perp = \mathcal{N}(T^* - \bar{\lambda} I)$ $= \mathcal{N}(T-\lambda I) = \{0\}$. Thus $\lambda \in \rho(T)$. Finally, $0 \in \sigma(T)$. For if not, then T^{-1} would exist and belong to $\mathcal{B}(H)$. Since $\dim H = \infty$, there is an orthonormal sequence (x_n) in H; and thus $(T^{-1}x_n)$ is bounded. But this implies that $(TT^{-1}x_n) = (x_n)$ has a convergent subsequence, which is impossible. The proof is complete. □

Theorem 5.8 shows that if $T \in \mathcal{K}(H)$ is *normal* the singular values of T are the moduli of the eigenvalues of T. This is not so in general as the following example shows.

Suppose that $\dim H \geqslant 2$; let $e, f \in H$ be such that $\|e\| = \|f\| = 1$ with $|(e,f)| \neq 0, 1$; and define $T \in \mathscr{B}(H)$ by

$$Tx = (x, e)f \qquad (x \in H).$$

Then T is of rank 1 and is thus compact. Since

$$(x, T^*y) = (Tx, y) = (x, e)(f, y) = (x, \overline{(f, y)}\, e),$$

it follows that $T^*y = (y, f)e$ $(y \in H)$, and hence

$$TT^*y = (y, f)f, \quad T^*Tx = (x, e)e. \tag{5.13}$$

Thus T is not normal; and also from (5.13) we obtain

$$\sigma(T^*T) = \{0, 1\}.$$

To see this, let $\{e_i : i \in I\}$ be a complete orthonormal set in H containing e, and let $x \in \mathrm{H}$; then $x = \sum_{i \in I_0} c_i e_i$ for some (at most) countable set $I_0 \subset I$, and hence if $\lambda \neq 0, 1$, and $d_i = -\lambda^{-1} c_i$ if $e_i \neq e$ and $d_i = (1-\lambda)^{-1} c_i$ if $e_i = e$, we have, with $y = \sum_{I_0} d_i e_i$, that

$$\|x\| \geqslant \min\{1, |\lambda|, |1-\lambda|\} \cdot \|y\|$$

and

$$(T^*T - \lambda I)y = (y, e)e - \lambda \sum_{I_0} d_i e_i = \sum_{I_0} c_i e_i = x.$$

This establishes our claim that $\sigma(T^*T) = \{0, 1\}$, as evidently 0 and 1 belong to $\sigma(T^*T)$. In a similar way it follows that $\sigma(T) = \{0, (f, e)\}$. However, $|(e, f)| \neq 0, 1$, by construction, and hence the singular values of T, namely 0 and 1, are not given by the moduli of the eigenvalues of T.

The next result is a converse to Theorem 5.2.

Theorem 5.9. Let H and H_1 be infinite-dimensional Hilbert spaces and let (ϕ_n) and (ψ_n) be orthonormal sequences in H and H_1 respectively. Let (β_n) be a scalar sequence such that $|\beta_1| \geqslant |\beta_2| \geqslant \ldots$ and $\lim_{n \to \infty} \beta_n = 0$. Then the map $T: H \to H_1$ defined by

$$Tx = \sum_{n=1}^{\infty} \beta_n (x, \phi_n)_H \psi_n \qquad (x \in H)$$

is in $\mathscr{K}(H, H_1)$ and $|\beta_n| = \mu_n(T) = \mu_n(T^*)$, the nth singular value of T and T^*. ■

Proof. Let $x \in H$ and $y \in H_1$. Since

$$(x, T^*y)_H = (Tx, y)_{H_1} = \sum_{n=1}^{\infty} \beta_n (x, \phi_n)_H (\psi_n, y)_{H_1}$$

$$= \Big(x, \sum_{n=1}^{\infty} \bar{\beta}_n (y, \psi_n)_{H_1} \phi_n\Big)_H,$$

it follows that

$$T^*y = \sum_{n=1}^{\infty} \bar{\beta}_n (y, \psi_n)_{H_1} \phi_n$$

and

$$T^*Tx = \sum_{n=1}^{\infty} |\beta_n|^2 (x, \phi_n)_H \phi_n. \tag{5.14}$$

The operators F_m defined by

$$F_m x = \sum_{n=1}^{m} |\beta_n|^2 (x, \phi_n)_H \phi_n \qquad (m \in \mathbb{N}, x \in H)$$

evidently belong to $\mathscr{B}(H)$ and are of finite rank. Also

$$\|(T^*T - F_m)x\|_H^2 = \sum_{n=m+1}^{\infty} |\beta_n|^2 |(x, \phi_n)_H|^2$$

$$\leqslant |\beta_{m+1}|^2 \|x\|_H^2 \to 0$$

as $m \to \infty$. Hence $T^*T \in \mathscr{K}(H)$ by Proposition I.1.2. As in the proof of Corollary 5.4 it follows from (5.14) that $\sigma(T^*T) = \{|\beta_n|^2 : n = 1, 2, \ldots\} \cup \{0\}$, and thus $|\beta_n| = \mu_n(T)$. The proof is complete. □

We can now turn to the connection between the approximation numbers $a_n(T)$ of a map T and its singular values.

Theorem 5.10. Let $T \in \mathscr{K}(H, H_1)$. Then for all $n \in \mathbb{N}$ we have $a_n(T) = \mu_n(T)$, the nth singular value of T. ■

Proof. By Theorem 5.5,

$$\mu_n(T) = \min_{\dim G \leqslant n-1} \max_{x \in G^\perp, x \neq 0} \frac{(|T|x, x)_H}{\|x\|_H^2} = \min_{\dim G \leqslant n-1} \max_{x \in G^\perp, x \neq 0} \frac{\|Tx\|_{H_1}}{\|x\|_H},$$

the final equality following from

$$\||T|x\|_H^2 = (T^*Tx, x)_H = \|Tx\|_{H_1}^2$$

and Lemma 5.1.

Let G be a linear subspace of H with $\dim G := r \leqslant n-1$ and let P be the orthogonal projection of H onto G. Then $TP : H \to H_1$ is of rank r and

$$\max_{x\in G^{\perp},x\neq 0}(\|Tx\|_{H_1}/\|x\|_H)=\max_{x\in H,x\neq 0}[\|Tx-TPx\|_{H_1}/\|(I-P)x\|_H]$$

$$\geqslant \max_{x\in H,x\neq 0}(\|Tx-TPx\|_{H_1}/\|x\|_H)$$

$$\geqslant \mathrm{a}_n(T).$$

Hence $\mu_n(T)\geqslant \mathrm{a}_n(T)$.

Now let $F\in\mathscr{B}(H,H_1)$, with $\dim\mathscr{R}(F)=r\leqslant n-1$. There are scalars α_1, $\alpha_2,\ldots,\alpha_n$, not all zero, such that if $\psi=\sum_{i=1}^{n}\alpha_i\phi_i$, then $T\psi\in\mathscr{R}(F)^{\perp}$, where the ϕ_i are the eigenvectors of $|T|$ corresponding to the eigenvalues $\mu_i(T)$. In the notation of Theorem 5.7, $T\psi=\sum_{i=1}^{n}\mu_i\alpha_i\psi_i$, where $\mu_i=\mu_i(T)$ and, as shown in the proof of Theorem 5.7, the ψ_i are orthonormal in H_1. Thus

$$\|(T-F)\psi\|_{H_1}^2=\|T\psi\|_{H_1}^2+\|F\psi\|_{H_1}^2\geqslant\|T\psi\|_{H_1}^2=\sum_{i=1}^{n}|\mu_i\alpha_i|^2$$

$$\geqslant|\mu_n|^2\sum_{i=1}^{n}|\alpha_i|^2=|\mu_n|^2\|\psi\|_H^2.$$

Thus $\|T-F\|^2\geqslant|\mu_n|^2$, which shows that $\mathrm{a}_n(T)\geqslant\mu_n(T)$. □

To conclude this section some further remarks about eigenvalues may be helpful. Let $T\in\mathscr{K}(H)$ and let $\lambda\in\mathbb{C}\backslash\{0\}$. We have seen that $\lambda\in\sigma_p(T)$ if, and only if, $\bar{\lambda}\in\sigma_p(T^*)$; moreover, if $\lambda\in\sigma_p(T)$ then the geometric multiplicity of λ (in $\sigma(T)$) is the same as that of $\bar{\lambda}$ (in $\sigma(T^*)$), in view of Theorem I.1.10: nul $T_\lambda=\mathrm{nul}\,T^*_{\bar{\lambda}}$. Since Theorem I.1.10 also shows that for all $k\in\mathbb{N}$, nul $T^k_\lambda=$ nul $(T^*_{\bar{\lambda}})^k$, it follows that λ and $\bar{\lambda}$ have the same algebraic multiplicities. We still lack, however, a connection between the algebraic and geometric multiplicities of eigenvalues, other than the obvious inequality. The next theorem provides some information about this.

Theorem 5.11. Let $T\in\mathscr{B}(H)$ be normal. Then each eigenvalue of T has index 1, and hence its algebraic and geometric multiplicities are the same. Moreover, eigenvectors associated with distinct eigenvalues are orthogonal. ■

Proof. Let $\lambda\in\sigma_p(T)$ and suppose that $x\in H\backslash\{0\}$ is such that $(T-\lambda I)^2x=0$. Put $y=(T-\lambda I)x$; then $(T-\lambda I)y=0$ and so $(T^*-\bar{\lambda}I)y=0$ since $\|(T-\lambda I)y\|=\|(T^*-\bar{\lambda}I)y\|$. Hence

$$\|(T-\lambda I)x\|^2=((T-\lambda I)x,(T-\lambda I)x)=(y,(T-\lambda I)x)$$

$$=((T^*-\bar{\lambda}I)y,x)=0,$$

and thus $\mathcal{N}((T-\lambda I)^2) \subset \mathcal{N}(T-\lambda I)$. This shows that $\mathcal{N}((T-\lambda I)^2) = \mathcal{N}(T-\lambda I)$, which immediately tells us that λ has index 1.

Finally, let $\lambda, \mu \in \sigma_p(T)$, with $\lambda \neq \mu$, and let x and y be eigenvectors corresponding to λ and μ respectively. Then

$$\lambda(y,x) = (y, \bar{\lambda}x) = (y, T^*x) = (Ty, x) = \mu(y,x),$$

and consequently $(x, y) = 0$. □

We conclude this section with a proof of Weyl's inequality in a Hilbert-space setting by methods totally different from those employed earlier in the context of general Banach spaces.

Lemma 5.12. Let H be a Hilbert space, let $T \in \mathcal{K}(H)$ and let $\nu(T)$ be the sum of the algebraic multiplicities of all the non-zero eigenvalues $\lambda_j(T)$ of T; as usual the $\lambda_j = \lambda_j(T)$ are assumed to be arranged in descending order of magnitude and repeated according to algebraic multiplicity; to be precise we adopt the convention that if $|\lambda_j(T)| = |\lambda_{j+1}(T)|$ then $\arg \lambda_{j+1}(T) \geqslant \arg \lambda_j(T)$, where arg is the principal argument, with values in $(-\pi, \pi]$. Then there is an orthonormal set $\{e_j : j = 1, 2, \ldots, \nu(T)\}$ such that, for suitable complex numbers a_{jk},

$$Te_j = \lambda_j(T)e_j + \sum_{k=1}^{j-1} a_{jk}\, e_k \qquad (j = 1, 2, \ldots, \nu(T));$$

thus

$$(e_j, Te_j) = \lambda_j(T).$$ ■

Proof. For $j = 1, 2, \ldots, \nu(T)$ let P_j be the spectral projection

$$P_j := -(2\pi i)^{-1} \int_{\gamma_j} (T-\lambda I)^{-1}\, d\lambda,$$

where γ_j is the positively oriented circle with centre $\lambda_j(T)$ and radius ε so small that no eigenvalue of T, save $\lambda_j(T)$, lies within or on γ_j. The range of P_j is given by $\mathcal{R}(P_j) = \{x \in H : (T-\lambda_j I)^n x = 0 \text{ for some } n \in \mathbb{N}\}$, and $\dim \mathcal{R}(P_j)$ is the algebraic multiplicity of $\lambda_j(T)$. For each j choose a Jordan normal form for $T \restriction \mathcal{R}(P_j)$; this gives a sequence $(x_j)_{j=1}^{\nu(T)}$ of linearly independent vectors such that

$$Tx_j = \lambda_j(T)x_j + \beta_j x_{j-1}, \tag{5.15}$$

where each β_j is either 0 or 1. Application of the Gram-Schmidt orthonormalization process to the x_j then gives an orthonormal sequence $(e_j)_{j=1}^{\nu(T)}$ such that

$$e_j = \sum_{k=1}^{j} b_{jk} x_k, \tag{5.16}$$

with $b_{jj} \neq 0$. The lemma now follows directly from (5.15) and (5.16). □

The orthonormal set $\{e_j : j = 1, 2, \ldots, \nu(T)\}$ is called a *Schur basis* for T.

Theorem 5.13. Let $T \in \mathscr{K}(H)$ and let $\lambda_n(T)$ and $\mu_n(T)$ as usual be the eigenvalues and singular values respectively of T. Then given any $p \in [1, \infty)$ and any $N \in \mathbb{N}$,

$$\sum_{n=1}^{N} |\lambda_n(T)|^p \leqslant \sum_{n=1}^{N} \mu_n(T)^p. \qquad (5.17) \blacksquare$$

Proof. By Theorem 5.7,

$$Tx = \sum_{n=1}^{\infty} \mu_n(T)(x, \phi_n)\psi_n \qquad (x \in H)$$

where $\{\phi_n\}$ and $\{\psi_n\}$ are orthonormal sets. Let $\{e_n : n = 1, 2, \ldots, v(T)\}$ be a Schur basis for T. Then, by Lemma 5.12,

$$\lambda_m(T) = (e_m, Te_m) = \sum_{n=1}^{\infty} \mu_n(T)(e_m, \psi_n)(\phi_n, e_m).$$

Put $\alpha_{mn} = (e_m, \psi_n)(\phi_n, e_m)$. Then

$$\sum_{m=1}^{\infty} |\alpha_{mn}| \leqslant \left(\sum_{m=1}^{\infty} |(e_m, \psi_n)|^2\right)^{\frac{1}{2}} \left(\sum_{m=1}^{\infty} |(\phi_n, e_m)|^2\right)^{\frac{1}{2}}$$
$$\leqslant \|\psi_n\| \|\phi_n\| = 1;$$

similarly it can be seen that $\sum_{n=1}^{\infty} |\alpha_{mn}| \leqslant 1$. Now let $N \in \mathbb{N}$ and $p \in [1, \infty)$ and write $p' = p/(p-1)$. Use of Hölder's inequality gives

$$\sum_{m=1}^{N} |\lambda_m(T)|^p = \sum_{m=1}^{N} \left(|\lambda_m(T)|^{p-1} \left|\sum_{n=1}^{\infty} \alpha_{mn}\mu_n(T)\right|\right)$$
$$\leqslant \sum_{m=1}^{N} \left(|\lambda_m(T)|^{p-1} \sum_{n=1}^{\infty} |\alpha_{mn}|\mu_n(T)\right)$$
$$\leqslant \left(\sum_{n,m} |\alpha_{mn}|\mu_n(T)^p\right)^{1/p} \left(\sum_{n,m} |\alpha_{mn}||\lambda_m(T)|^p\right)^{1/p'}$$
$$\leqslant \left(\sum_{n} |\mu_n(T)|^p\right)^{1/p} \left(\sum_{m=1}^{N} |\lambda_m(T)|^p\right)^{1/p'}.$$

The result needed follows immediately. □

Remark 5.14. This quick proof of Weyl's inequality is that given by Reed and Simon [2, Theorem XIII.103]; it will be noted that the hypothesis that $p \geqslant 1$ is essential for the use of Hölder's inequality made in the proof to be justified. With a little extra work, however, this condition can be weakened to $p > 0$ while still retaining the conclusion (5.17). See Simon [3, Chapter I] and Gohberg and Krein [1, Chapter 2] for details of this.

III
Unbounded linear operators

Our main concern in this chapter will be with closable and closed operators and especially with special classes in Hilbert spaces such as the symmetric, J-symmetric, and accretive operators. In subsequent chapters the differential operators we define will fall within one of the last three categories. A typical situation in practice is to have available a closable operator readily defined on some minimal domain and the outstanding problem is to find closed extensions which are well determined in some sense. Ideally one would like an extension with a non-empty resolvent set as is the case with self-adjoint and m-accretive extensions of symmetric and accretive operators respectively. Failing that we would like an operator which is regularly solvable in the sense defined in §3, that is, it satisfies the Fredholm alternative. The theory of such extensions in §3 includes the Stone–von Neumann theory of extensions of symmetric operators and also its adaption by Zhikhar to extensions of J-symmetric operators. The stability of self-adjointness, J-self-adjointness, and m-accretiveness under perturbations by suitable operators is investigated in §8. In §10 we illustrate the earlier abstract ideas by means of operators defined by the so-called quasi-differential expressions of second order, and, *inter alia*, generalize Glazman's proof of the Weyl limit-point, limit-circle characterization for formally symmetric expressions and also the extension of the Weyl characterization to the case of formally J-symmetric expressions obtained by Sims.

1. Closed and closable operators

Let X and Y be Banach spaces and let T be a linear operator whose domain $\mathscr{D}(T)$ is a linear subspace of X and whose range $\mathscr{R}(T)$ is in Y. If T is closed, it is bounded on $\mathscr{D}(T)$ if, and only if, $\mathscr{D}(T)$ is a closed subspace of X; for if T is bounded on $\mathscr{D}(T)$, as well as being closed, it follows that $\mathscr{D}(T)$ must be closed, while the converse is true on account of the Closed-Graph Theorem. In general the domain of T is not a closed subspace of X: a typical and important example is given by the differential operator $Tf = f'$ when $X = Y = L^2(0, 1)$. The largest set on which T can be defined is $AC[0, 1]$, the set of functions which are absolutely continuous on $[0, 1]$, and with this as domain, T is in fact a closed operator. However AC $[0, 1]$ is not a closed subspace of $L^2(0, 1)$ and T is not bounded; if $f_n(t) = t^n$ $(n = 1, 2, \ldots)$, then $\|Tf_n\|/\|f_n\| \to \infty$. The set of all closed linear operators with domain in X and range in Y is denoted by $\mathscr{C}(X, Y)$. If $X = Y$ we write $\mathscr{C}(X, X)$ as $\mathscr{C}(X)$.

We recall that $T \in \mathscr{C}(X, Y)$ if its graph $\mathscr{G}(T) = \{\{x, Tx\} : x \in \mathscr{D}(T)\}$ is a closed subspace of the Cartesian product $X \times Y$ whose topology is determined by the norm

$$\|\{x, y\}\| = (\|x\|_X^2 + \|y\|_Y^2)^{\frac{1}{2}}. \tag{1.1}$$

Equivalently $T \in \mathscr{C}(X, Y)$ if $x_n \in \mathscr{D}(T)$, $x_n \to x$ in X, and $Tx_n \to y$ in Y together imply that $x \in \mathscr{D}(T)$ and $y = Tx$. Clearly, T is closed if, and only if, $\mathscr{D}(T)$ is complete with respect to the norm

$$\|x\|_T = (\|x\|_X^2 + \|Tx\|_Y^2)^{\frac{1}{2}} \qquad (x \in \mathscr{D}(T)). \tag{1.2}$$

We shall call $\|\bullet\|_T$ the graph norm of T and denote $\mathscr{D}(T)$ equipped with $\|\bullet\|_T$ by $\mathrm{X}(T)$. Thus $T \in \mathscr{C}(X, Y)$ if, and only if, $\mathrm{X}(T)$ is a Banach space. If X and Y are Hilbert spaces, $\mathrm{X}(T)$ is a Hilbert space with graph inner product

$$(x_1, x_2)_T = (x_1, x_2)_X + (Tx_1, Tx_2)_Y \qquad (x_1, x_2 \in \mathscr{D}(T)). \tag{1.3}$$

The identification map $x \mapsto x$ is a continuous embedding of $\mathrm{X}(T)$ into X with norm $\leqslant 1$ and $T \in \mathscr{B}(\mathrm{X}(T), Y)$. This of course was the basis of the extension of the results on bounded Fredholm and semi-Fredholm operators in §I.3 to closed operators; see Remark I.3.27.

The set $\mathscr{G}'(T) = \{\{Tx, x\} : x \in \mathscr{D}(T)\}$ is called the *inverse graph* of T. It is a linear subspace of $Y \times X$ which is closed if, and only if, $T \in \mathscr{C}(X, Y)$. If T^{-1} exists,

$$\mathscr{G}(T^{-1}) = \mathscr{G}'(T) \tag{1.4}$$

and consequently T^{-1} is closed if T is closed.

We recall that the adjoint T^* of T is the operator with domain in Y^* and range in X^* defined as follows: $y^* \in \mathscr{D}(T^*)$ if, and only if, there exists a unique $x^* \in X^*$ such that

$$(y^*, Tx) = (x^*, x) \qquad (x \in \mathscr{D}(T)),$$

and we set $x^* = T^*y^*$. Thus T^* exists if, and only if, $\mathscr{D}(T)$ is dense in X, and

$$(-T^*y^*, x) + (y^*, Tx) = 0 \qquad (x \in \mathscr{D}(T),\ y^* \in \mathscr{D}(T^*)). \tag{1.5}$$

With the choice of norm (1.1) for $X \times Y$, it follows that $(X \times Y)^* = X^* \times Y^*$ (see Kato [1,III-§5.2, footnote 2]) so that for any $F \in (\mathrm{X} \times Y)^*$ there exists a unique $\{f, g\} \in X^* \times Y^*$ such that for all $\{x, y\} \in X \times Y$,

$$(F, \{x, y\}) = (f, x) + (g, y).$$

Consequently (1.5) expresses the fact that each $\{x, Tx\}$ is annihilated by $\{-T^*y^*, y^*\}$ and so

$$\mathscr{G}'(-T^*) = \mathscr{G}(T)^0 \tag{1.6}$$

and similarly

$$\mathscr{G}(-T^*) = \mathscr{G}'(T)^0. \tag{1.7}$$

Since $\mathscr{G}(T)^0$ is closed in $X \times Y$, even if $\mathscr{G}(T)$ is not, we conclude that $\mathscr{G}'(-T^*)$ is always closed and hence $T^* \in \mathscr{C}(Y^*, Y^*)$ irrespective of whether or not $T \in \mathscr{C}(X, Y)$. We also have the following important result.

Theorem 1.1. If $\mathscr{D}(T)$ is dense in X, and T^{-1} exists with $\mathscr{D}(T^{-1})$ dense in Y, then $(T^*)^{-1}$ exists and $(T^*)^{-1} = (T^{-1})^*$. ■

Proof. The assumed denseness of $\mathscr{D}(T)$ and $\mathscr{D}(T^{-1})$ guarantee the existence of T^* and $(T^{-1})^*$. From (1.4–7),

$$\begin{aligned}\mathscr{G}(-(T^{-1})^*) &= \mathscr{G}'(T^{-1})^0 \\ &= \mathscr{G}(T)^0 \\ &= \mathscr{G}'(-T^*)\end{aligned}$$

and hence

$$\mathscr{G}((T^{-1})^*) = \mathscr{G}'(T^*).$$

This means that $\mathscr{G}'(T^*)$ is a graph and hence for every $y^* \in \mathscr{D}(T^*)$, the element y^* is uniquely determined by T^*y. In other words, T^* is injective and $(T^*)^{-1}$ exists. From (1.4), $\mathscr{G}((T^*)^{-1}) = \mathscr{G}((T^{-1})^*)$ and this gives $(T^*)^{-1} = (T^{-1})^*$. □

If T is not closed, the closure $\overline{\mathscr{G}(T)}$ of its graph in $X \times Y$ may or may not be a graph; for $\overline{\mathscr{G}(T)}$ to be a graph any $\{x, y\} \in \overline{\mathscr{G}(T)}$ must be such that y is uniquely determined by x, or equivalently, $\{0, y\} \in \overline{\mathscr{G}}(T)$ must imply $y = 0$.

Definition 1.2. The map T is said to be *closable* if $\overline{\mathscr{G}(T)}$ is a graph. If T is closable the operator whose graph is $\overline{\mathscr{G}(T)}$ is called the *closure of* T and denoted by $\bar{T}$. ■

An operator S is an *extension* of an operator T (and T a *restriction* of S) if $\mathscr{D}(S) \supset \mathscr{D}(T)$ and $Sx = Tx$ for all $x \in \mathscr{D}(T)$; we write $T \subset S$ or $S \supset T$. Thus $T \subset S$ if, and only if, $\mathscr{G}(T) \subset \mathscr{G}(S)$. The closure of T is the minimal closed extension of T in that for any closed extension S of T we have $\mathscr{G}(T) \subset \overline{\mathscr{G}(T)} \subset \mathscr{G}(S)$ and hence $\bar{T} \subset S$. Also $x \in \mathscr{D}(\bar{T})$ if and only if there exists a sequence (x_n) in $\mathscr{D}(T)$ such that $\{x_n, Tx_n\} \to \{x, \bar{T}x\}$ in $X \times Y$, that is $x_n \to x$ in X and $Tx_n \to \bar{T}x$ in Y.

If (x_n) is a Cauchy sequence with respect to the graph norm $\|\bullet\|_T$ in (1.2) then it is also a Cauchy sequence in X. Hence, for each $z \in \hat{X}(T)$, the completion of

$\mathscr{D}(T)$ with norm $\|\bullet\|_T$, there corresponds a unique $x \in X$ and this correspondence is injective if, and only if, T is closable. Hence T is closable if, and only if, the completion $\hat{X}(T)$ is embedded in this canonical way in X. Moreover this embedding is continuous, with norm not exceeding 1, and $\hat{\mathrm{X}}(T) = \mathrm{X}(\bar{T})$.

Definition 1.3. A subspace E of X is said to be *a core of* $T \in \mathscr{C}(X,Y)$ if E is dense in $\mathrm{X}(T)$. Equivalently E is a core if $T \restriction E$, the restriction of T to E, has closure T. ■

Theorem 1.4. Let S and T be linear operators with domains in X and ranges in Y.

(i) If $S \subset T$ and T is closed, S is closable and $\bar{S} \subset T$.
(ii) If $\mathscr{D}(T)$ is dense in X and T is closable, $(\bar{T})^* = T^*$.
(iii) If T is bounded on $\mathscr{D}(T)$, it is closable, $\mathscr{D}(\bar{T})$ is the closure of $\mathscr{D}(T)$ in X and $\bar{T}$ is bounded on $\mathscr{D}(\bar{T})$ with $\|\bar{T}\| = \|T\|$. ■

The proof is left as an exercise.

The next theorem provides a useful criterion for an operator to be closable. In it $T \in \mathscr{C}(X,Y)$ where X and Y are reflexive Banach spaces and T^{**}, the adjoint of T^*, is regarded as a map from X into Y; in other words X^{**} and Y^{**} are identified with X and Y respectively.

Theorem 1.5. Let X and Y be reflexive Banach spaces and let $\mathscr{D}(T)$ be a dense subspace of X. Then T is closable if, and only if, $\mathscr{D}(T^*)$ is dense in Y^*, in which case $T^{**} = \bar{T}$. ■

Proof. If $\mathscr{D}(T^*)$ is dense in Y^*, the map T^{**} exists; $x \in \mathscr{D}(T^{**})$ if, and only if, there exists a unique $y \in Y$ such that

$$(T^* y^*, x) = (y^*, y) \quad \text{for all } y^* \in \mathscr{D}(T^*).$$

It follows that $T \subset T^{**}$ and since T^{**} is closed, T is closable. Conversely, if T is closable, we see from (1.6) that

$$\mathscr{G}(\bar{T}) = \overline{\mathscr{G}(T)} = \mathscr{G}(T)^{\perp\perp} = \mathscr{G}'(-T^*)^{\perp}. \tag{1.8}$$

If $\mathscr{D}(T^*)$ is not dense in Y^* there exists a non-zero $y \in Y$ such that $y \in \mathscr{D}(T^*)^{\perp}$ (by the Hahn–Banach Theorem). This implies that $\{0,y\} \in \mathscr{G}'(-T^*)^{\perp} = \mathscr{G}(\bar{T})$. Since $\mathscr{G}(\bar{T})$ is a graph this is impossible and hence $\mathscr{D}(T^*)$ must be dense in Y^*; the map T^{**} is therefore defined and, on using (1.7) and (1.8),

$$\mathscr{G}(T^{**}) = \mathscr{G}'(-T^*)^{\perp} = \mathscr{G}(\bar{T}),$$

whence $T^{**} = \bar{T}$. □

2. Numerical range and field of regularity

In this section we consider only operators with domain and range in a complex Hilbert space H.

Definition 2.1. The *numerical range* $\Theta(T)$ of a linear operator T in H is the set of complex numbers $\Theta(T) = \{(Tu, u) : u \in \mathscr{D}(T),\ \|u\| = 1\}$. ■

In general $\Theta(T)$ is neither open nor closed even when T is a closed or bounded operator. It is, however, convex: this was first proved by Hausdorff for finite-dimensional spaces H and then in complete generality by Stone.

Theorem 2.2. $\Theta(T)$ is a convex subset of $\mathbb{C}$. ■

Proof. We exclude the trivial case when $\Theta(T)$ contains only one member. Let l_1 and l_2 be distinct points in $\Theta(T)$, in which case there are $u_1, u_2 \in \mathscr{D}(T)$ with $\|u_1\| = \|u_2\| = 1$, $l_1 = (Tu_1, u_1)$, and $l_2 = (Tu_2, u_2)$. Our task is to show that given any point l on the line segment joining l_1 and l_2, there exists a $u \in \mathscr{D}(T)$ such that $\|u\| = 1$ and $l = (Tu, u)$. We shall obtain u as a linear combination of u_1 and u_2. For $\lambda_1, \lambda_2 \in \mathbb{C}$ define

$$\begin{aligned} A_1 \equiv A_1(\lambda_1, \lambda_2) &:= (T(\lambda_1 u_1 + \lambda_2 u_2), (\lambda_1 u_1 + \lambda_2 u_2)) \\ &= |\lambda_1|^2 l_1 + 2\mathrm{re}((Tu_1, u_2)\lambda_1\bar{\lambda}_2) + |\lambda_2|^2 l_2 \end{aligned}$$

and

$$\begin{aligned} A_2 \equiv A_2(\lambda_1, \lambda_2) &:= \|\lambda_1 u_1 + \lambda_2 u_2\|^2 \\ &= |\lambda_1|^2 + 2\,\mathrm{re}((u_1, u_2)\lambda_1\bar{\lambda}_2) + |\lambda_2|^2. \end{aligned}$$

We shall prove that when $A_2 = 1$, the complex number A_1 can take any value on the line segment between l_1 and l_2 by choice of λ_1 and λ_2. Equivalently

$$A \equiv A(\lambda_1, \lambda_2) := \frac{A_1 - l_2 A_2}{l_1 - l_2} =: |\lambda_1|^2 + a_{12}\bar{\lambda}_1\lambda_2 + a_{21}\lambda_1\bar{\lambda}_2$$

say, takes every value between 0 and 1. Let x and y be real numbers and choose $\lambda_1 = x$ and $\lambda_2 = \gamma y$, where

$$\gamma = \begin{cases} \pm 1 & \text{when } \bar{a}_{12} = a_{21} \\[2ex] \pm \dfrac{\bar{a}_{12} - a_{21}}{|\bar{a}_{12} - a_{21}|} & \text{when } \bar{a}_{12} \neq a_{21} \end{cases}$$

the sign being chosen such that $\beta := \mathrm{re}(\bar{\gamma}(u_1, u_2)) > 0$. We then have

$$A = x^2 + \alpha xy, \qquad A_2 = x^2 + 2\beta xy + y^2$$

where

$$\alpha = \bar{\alpha} = \gamma a_{12} + \bar{\gamma} a_{21} = \pm(a_{12} + a_{21}) \quad \text{or} \quad \pm[(|a_{12}|^2 - |a_{21}|^2)/|\bar{a}_{12} - a_{21}|].$$

Moreover,

$$|\beta| \leqslant |\bar{\gamma}(u_1, u_2)| \leqslant \|u_1\| \, \|u_2\| = 1$$

and $A_2 = 1$ when

$$y = -\beta x + \sqrt{[1-(1-\beta^2)x^2]}, \qquad |x| \leqslant 1.$$

On substituting this in A we obtain

$$A = (1-\alpha\beta)x^2 + \alpha x\sqrt{[1-(1-\beta^2)x^2]}.$$

This is a continuous function for $x \in [0, 1]$ which takes the values 0 and 1 at the end points. It therefore attains every intermediate value as we set out to prove. □

The closure $\Gamma(T)$ of $\Theta(T)$ in $\mathbb{C}$ is therefore a closed convex set and if it does not fill out the whole of $\mathbb{C}$, its complement $\tilde{\Delta}(T)$ has either one or two connected components. The latter possibility occurs when $\Gamma(T)$ is an infinite strip, in which case the two connected components $\tilde{\Delta}_1(T)$ and $\tilde{\Delta}_2(T)$ of $\tilde{\Delta}(T)$ are half-planes. If $T \in \mathscr{B}(H)$, the set $\Theta(T)$ is bounded and $\tilde{\Delta}(T)$ is connected.

Theorem 2.3. Let $T \in \mathscr{C}(H)$ and $\lambda \in \tilde{\Delta}(T) \equiv \mathbb{C} \setminus \Gamma(T)$. Then $T - \lambda I$ is semi-Fredholm with nul $(T-\lambda I) = 0$ and def $(T-\lambda I)$ constant in each connected component of $\tilde{\Delta}(T)$. If def $(T-\lambda I) = 0$ for $\lambda \in \tilde{\Delta}(T)$ ($\tilde{\Delta}_1(T)$ or $\tilde{\Delta}_2(T)$) then $\tilde{\Delta}(T)$ ($\tilde{\Delta}_1(T)$ or $\tilde{\Delta}_2(T)$) is a subset of the resolvent set $\rho(T)$ of T and

$$\|(T-\lambda I)^{-1}\| \leqslant 1/\text{dist}[\lambda, \Gamma(T)]. \tag{2.1}$$ ■

Proof. Let $\delta = \text{dist}(\lambda, \Gamma(T)) > 0$. For $u \in \mathscr{D}(T)$ and $\|u\| = 1$,

$$\delta \leqslant |(Tu, u) - \lambda| = |((T-\lambda I)u, u)| \leqslant \|(T-\lambda I)u\|. \tag{2.2}$$

Hence $\text{nul}(T-\lambda I) = 0$ and $(T-\lambda I)^{-1}$ exists and is a bounded closed operator on $\mathscr{R}(T-\lambda I)$. Therefore $\mathscr{R}(T-\lambda I)$ is closed and $T-\lambda I$ is semi-Fredholm with

$$\text{ind}(T-\lambda I) = -\text{def}(T-\lambda I). \tag{2.3}$$

We now invoke the result in Theorem I.3.22, adapted for closed maps, that the index of $T-\lambda I$ is constant in any connected subset of its semi-Fredholm domain $\Delta_1(T) = \{\lambda \in \mathbb{C}\colon T-\lambda I \text{ semi-Fredholm}\}$. From (2.3) we see that $\text{def}(T-\lambda I)$ is constant in any connected component of $\tilde{\Delta}(T)$. If $\text{def}(T-\lambda I) = 0$ we have $\mathscr{R}(T-\lambda I) = H$ and so $(T-\lambda I)^{-1} \in \mathscr{B}(H)$. Thus $\lambda \in \rho(T)$. The inequality (2.1) follows from (2.2). □

Theorem 2.3 prompts the following definition.

Definition 2.4. If $T \in \mathscr{C}(H)$ and $\tilde{\Delta}(T)$ is simply connected, the constant $\mathrm{m}(T) := \text{def}(T-\lambda I)$, with $\lambda \in \tilde{\Delta}(T)$, is called the *deficiency index* of T. If $\tilde{\Delta}(T)$ has two components, the constants $m_i(T) = \text{def}(T-\lambda I)$, with $\lambda \in \tilde{\Delta}_i(T)$, $(i = 1, 2)$ are called the *deficiency indices* of T; $m_1(T)$ and $m_2(T)$ are unequal in general. ■

The set $\tilde{\Delta}(T)$ is often called the *external field of regularity* of T. When $T \in \mathscr{C}(H)$, the set $\tilde{\Delta}(T)$ is a subset of the so-called field of regularity $\Pi(T)$ of T which is defined as follows.

Definition 2.5. The *field of regularity* $\Pi(T)$ of T is the set of values $\lambda \in \mathbb{C}$ for which there exist positive constants $k(\lambda)$ such that

$$\|(T-\lambda I)u\| \geqslant k(\lambda)\|u\| \quad \text{for all } u \in \mathscr{D}(T). \qquad \blacksquare$$

The set $\Pi(T)$ is easily seen to be open in $\mathbb{C}$ and if $T \in \mathscr{C}(H)$, the inclusion $\tilde{\Delta}(T) \subset \Pi(T)$ follows from (2.2). Notice also that for $\lambda \in \Pi(T)$, the map $(T-\lambda I)^{-1}$ exists and is bounded on $\mathscr{R}(T-\lambda I)$. Hence, if $T \in \mathscr{C}(H)$ then $(T-\lambda I)^{-1}$ is both bounded and closed on $\mathscr{R}(T-\lambda I)$ with the result that $\mathscr{R}(T-\lambda I)$ is closed. Therefore in the notation of §I.4, $\Pi(T) \subset \Phi_+(T)$ when $T \in \mathscr{C}(H)$. By the Closed-Graph Theorem, $\Pi(T) = \{\lambda \in \Phi_+(T) : \text{nul}(T-\lambda I) = 0\}$ if $T \in \mathscr{C}(H)$.

If $T \in \mathscr{C}(H)$, $\text{def}(T-\lambda I)$ is constant in any connected component of $\Pi(T)$ as well as $\tilde{\Delta}(T)$. For $\text{def}(T-\lambda I) = -\text{ind}(T-\lambda I)$ in $\Pi(T)$ and so our assertion follows from Theorem I.3.22.

Theorem 2.6. If T is densely defined and $\Theta(T) \neq \mathbb{C}$ then T is closable. $\blacksquare$

Proof. Since $\Theta(T)$ is convex it must lie in some half-plane and, without loss of generality, we may suppose that it lies in $\{\lambda : \text{re}\,\lambda \geqslant 0\}$, so that $\text{re}(Tu, u) \geqslant 0$ for all $u \in \mathscr{D}(T)$. In order to prove the assertion it suffices to show that if $(u_n) \subset \mathscr{D}(T)$ is such that $u_n \to 0$ and $Tu_n \to u$ then $u = 0$. For any $v \in \mathscr{D}(T)$ and any positive real number λ,

$$\begin{aligned} 0 &\leqslant \text{re}\,(T(u_n + \lambda v), (u_n + \lambda v)) \\ &= \text{re}[(Tu_n, u_n) + \lambda(Tu_n, v) + \lambda(Tv, u_n) + \lambda^2(Tv, v)]. \end{aligned}$$

On allowing $n \to \infty$ and dividing by λ we obtain

$$0 \leqslant \text{re}[(u, v) + \lambda(Tv, v)];$$

letting $\lambda \to 0$, this gives $\text{re}(u, v) \geqslant 0$. Since $v \in \mathscr{D}(T)$ is arbitrary we conclude that $(u, v) = 0$. This in turn gives $u = 0$ since $\mathscr{D}(T)$ is dense. $\square$

3. Extensions of closed operators and adjoint pairs

The next results will have far-reaching consequences in subsequent sections dealing with the characterization of extensions of special closed operators; H is a complex Hilbert space throughout.

Theorem 3.1. Let $T \in \mathscr{C}(H)$ be densely defined and $\Pi(T) \neq \varnothing$. Then for any closed extension S of T and $\lambda \in \Pi(T)$,

$$\mathscr{D}(S) = \mathscr{D}(T) \dotplus \mathscr{N}((T^* - \bar{\lambda} I)(S - \lambda I)). \tag{3.1}$$

If there exists a $\lambda \in \Pi(T)$ such that $\operatorname{def}(T-\lambda I) < \infty$ and $\lambda \in \Delta_3(S) = \{\lambda \in \mathbb{C}: S-\lambda I \in \mathscr{F}(H)\}$, the Fredholm domain of S, then

$$\dim \mathscr{D}(S)/\mathscr{D}(T) = \operatorname{nul}(S-\lambda I) + \operatorname{def}(T-\lambda I) - \operatorname{def}(S-\lambda I). \quad (3.2) \blacksquare$$

Proof. If $\lambda \in \Pi(T)$ then $\operatorname{nul}(T-\lambda I) = 0$ and $\mathscr{R}(T-\lambda I)$ is closed. Therefore we have the orthogonal sum decomposition $H = \mathscr{R}(T-\lambda I) \oplus \mathscr{N}(T^* - \bar{\lambda} I)$, since $\mathscr{N}(T^*-\bar{\lambda}I) = \mathscr{R}(T-\lambda I)^{\perp}$. Hence, if $u \in \mathscr{D}(S)$,

$$(S-\lambda I)u = (T-\lambda I)v + w, \qquad v \in \mathscr{D}(T), \quad w \in \mathscr{N}(T^*-\bar{\lambda}I),$$

whence $(S-\lambda I)(u-v) = w$ and $(T^*-\bar{\lambda}I)(S-\lambda I)(u-v) = 0$. Since the two subspaces on the right-hand side of (3.1) are contained in $\mathscr{D}(S)$ we have therefore proved that

$$\mathscr{D}(S) = \mathscr{D}(T) + \mathscr{N}\big((T^*-\bar{\lambda}I)(S-\lambda I)\big).$$

To establish (3.1) we must show that $\mathscr{D}(T) \cap \mathscr{N}\big((T^*-\bar{\lambda}I)(S-\lambda I)\big) = \{0\}$. If we assume the contrary, there exists a $\phi \neq 0$ such that $\phi \in \mathscr{D}(T)$ and $(T^*-\bar{\lambda}I)(T-\lambda I)\phi = (T^*-\bar{\lambda}I)(S-\lambda I)\phi = 0$. In this case $(T-\lambda I)\phi \in \mathscr{N}(T^*-\bar{\lambda}I) = \mathscr{R}(T-\lambda I)^{\perp}$ from which we infer that $(T-\lambda I)\phi = 0$. But this is a contradiction since $\operatorname{nul}(T-\lambda I) = 0$.

To prove the second part, let $\{\phi_1, \ldots, \phi_m\}$, $\{\psi_1, \ldots, \psi_n\}$, and $\{\theta_1, \ldots, \theta_p\}$ be bases for $\mathscr{N}(S^*-\bar{\lambda}I)$, $\mathscr{N}(T^*-\bar{\lambda}I) \ominus \mathscr{N}(S^*-\bar{\lambda}I)$, and $\mathscr{N}(S-\lambda I)$ respectively where $\lambda \in \Pi(T)$ is such that $\operatorname{def}(T-\lambda I) = m+n$ and $\lambda \in \Delta_3(S)$. Note that from $T \subset S$, we have $S^* \subset T^*$ and hence $\operatorname{def}(S-\lambda I) = \operatorname{nul}(S^*-\bar{\lambda}I) \leqslant \operatorname{nul}(T^*-\bar{\lambda}I) = \operatorname{def}(T-\lambda I)$. We now decompose H as $H = \mathscr{R}(S-\lambda I) \oplus \mathscr{N}(S^*-\bar{\lambda}I)$. For $i = 1, 2, \ldots, n$,

$$\psi_i = (S-\lambda I)u_i + v_i, \qquad u_i \in \mathscr{D}(S), \qquad v_i \in \mathscr{N}(S^*-\bar{\lambda}I). \quad (3.3)$$

If $w \in \mathscr{N}\big((T^*-\bar{\lambda}I)(S-\lambda I)\big)$ then $(S-\lambda I)w \in \mathscr{N}(T^*-\bar{\lambda}I)$ and for some constants c_i and d_i,

$$\begin{aligned}(S-\lambda I)w &= \sum_{i=1}^{m} c_i\phi_i + \sum_{i=1}^{n} d_i\psi_i \\ &= \sum_{i=1}^{m} c_i\phi_i + (S-\lambda I)\sum_{i=1}^{n} d_i u_i + \sum_{i=1}^{n} d_i v_i,\end{aligned}$$

from (3.3). Consequently

$$(S-\lambda I)\left(w - \sum_{i=1}^{n} d_i u_i\right) = \sum_{i=1}^{m} c_i\phi_i + \sum_{i=1}^{n} d_i v_i \in \mathscr{N}(S^*-\bar{\lambda}I) = \mathscr{R}(S-\lambda I)^{\perp}$$

and so $w - \sum_{i=1}^{n} d_i u_i \in \mathscr{N}(S-\lambda I)$. Since $\mathscr{N}(S-\lambda I)$ and the linear span U of $\{u_1, \ldots, u_n\}$ are both subspaces of $\mathscr{N}\big((T^*-\bar{\lambda}I)(S-\lambda I)\big)$ we have proved that

$$\mathscr{N}\big((T^*-\bar{\lambda}I)(S-\lambda I)\big) = \mathscr{N}(S-\lambda I) + U. \quad (3.4)$$

Let $\phi \in \mathcal{N}(S-\lambda I) \cap U$. Then if $\phi = \sum_{i=1}^{n} c_i u_i$ we have from (3.3)

$$\begin{aligned} 0 &= (S-\lambda I) \sum_{i=1}^{n} c_i u_i \\ &= \sum_{i=1}^{n} c_i \psi_i - \sum_{i=1}^{n} c_i v_i. \end{aligned}$$

But $\psi_i \in \mathcal{N}(S^* - \bar{\lambda} I)^{\perp}$ and $v_i \in \mathcal{N}(S^* - \bar{\lambda} I)$, whence $\sum_{i=1}^{n} c_i \psi_i = \sum_{i=1}^{n} c_i v_i = 0$. Since $\{\psi_1, \ldots, \psi_n\}$ is linearly independent, $c_1 = c_2 = \ldots = c_n = 0$ and $\phi = 0$. The sum in (3.4) is therefore a direct sum. To complete the proof we need to show that $\{u_1, \ldots, u_n\}$ is linearly independent. If $\sum_{i=1}^{n} c_i u_i = 0$, then $\sum_{i=1}^{n} c_i (S-\lambda I) u_i = 0$ and it follows as before that $c_1 = c_2 = \ldots = c_n = 0$. The theorem is therefore proved since $\dim \mathrm{U} = \dim[\mathcal{N}(T^* - \bar{\lambda} I) \ominus \mathcal{N}(S^* - \bar{\lambda} I] =$ def $(T-\lambda I) -$ def $(S-\lambda I)$. □

Theorem 3.1 will subsequently be applied in situations in which there are two densely defined operators A and B which are *adjoint to each other* or form an *adjoint pair* in the sense that

$$(Ax, y) = (x, By) \quad \text{for all } x \in \mathcal{D}(A) \text{ and all } y \in \mathcal{D}(B), \tag{3.5}$$

or equivalently $A \subset B^*$ and $B \subset A^*$. On setting $T = A$ and $S = B^*$ in Theorem 3.1 we have the following corollary.

Corollary 3.2. Let A and B be closed, densely defined operators in H which form an adjoint pair. Suppose there exists $\lambda \in \Pi(A)$ such that $\bar{\lambda} \in \Pi(B)$. Then

$$\mathcal{D}(B^*) = \mathcal{D}(A) \dot{+} \mathcal{N}((A^* - \bar{\lambda} I)(B^* - \lambda I)), \tag{3.6}$$

$$\mathcal{D}(A^*) = \mathcal{D}(B) \dot{+} \mathcal{N}((B^* - \lambda I)(A^* - \bar{\lambda} I)), \tag{3.7}$$

and, if def$(A-\lambda I)$ and def$(B-\bar{\lambda} I)$ are finite,

$$\begin{aligned} \dim[\mathcal{D}(B^*)/\mathcal{D}(A)] &= \dim[\mathcal{D}(A^*)/\mathcal{D}(B)] \\ &= \operatorname{def}(A-\lambda I) + \operatorname{def}(B - \bar{\lambda} I). \end{aligned} \tag{3.8}$$

Thus def$(A-\lambda I)+$def$(B-\bar{\lambda} I)$ is constant for all $\lambda \in \Pi(A)$ such that $\bar{\lambda} \in \Pi(B)$. ■

Proof. If $\bar{\lambda} \in \Pi(B)$, the sets $\mathcal{R}(B-\bar{\lambda} I)$ and hence $\mathcal{R}(B^* - \lambda I)$ are closed and def $(B^* - \lambda I) =$ nul $(B - \bar{\lambda} I) = 0$. Thus (3.6) follows from (3.1) on setting $T = A$ and $S = B^*$ and (3.2) gives (3.8). Similarly (3.7) is a consequence of (3.1) with $T = B$ and $S = A^*$. □

If S is an extension of A, that is, $A \subset S$, then $\Pi(S) \subset \Pi(A)$ where the inclusion may be proper. In general $\Pi(A) \neq \varnothing$ does not imply $\Pi(S) \neq \varnothing$.

However, for an adjoint pair A, B satisfying the hypothesis of Corollary 3.2 there exists an extension S of A with $A \subset S \subset B^*$ for which the implication in the previous sentence is valid. In fact we have more, namely, the following result of Višik[1] which is an extension of a result of Calkin[1] for symmetric operators.

Theorem 3.3. Let A and B be closed, densely defined operators in H which form an adjoint pair and suppose there exists a $\lambda \in \Pi(A)$ such that $\bar{\lambda} \in \Pi(B)$. Then there exists a closed operator S such that $A \subset S \subset B^*$ and $\lambda \in \rho(S)$. ■

Proof. Let us denote $\mathcal{N}((A^* - \bar{\lambda}I)(B^* - \lambda I))$ *by* M, and $\mathcal{N}(B^* - \lambda I)$ by N_0, and set $N = M \ominus N_0$. Define S to be the restriction of B^* to

$$\mathcal{D}(S) = \mathcal{D}(A) \dot{+} N. \tag{3.9}$$

To verify that the sum is a direct sum, let $\phi \in \mathcal{D}(A) \cap N$. Then $(A - \lambda I)\phi = (B^* - \lambda I)\phi \in \mathcal{N}(A^* - \bar{\lambda}I) = \mathcal{R}(A - \lambda I)^{\perp}$, whence $(A - \lambda I)\phi = 0$ and $\phi = 0$ since $\lambda \in \Pi(A)$. The sum is therefore direct. We shall now prove that the above-defined S has the asserted properties.

From (3.9), we have the orthogonal sum

$$\begin{aligned}
\mathcal{R}(S - \lambda I) &= \mathcal{R}(A - \lambda I) \oplus (B^* - \lambda I)N \\
&= \mathcal{R}(A - \lambda I) \oplus (B^* - \lambda I)M \\
&= \mathcal{R}(A - \lambda I) \oplus \mathcal{N}(A^* - \bar{\lambda}I) \quad (\text{since } \mathcal{R}(B^* - \lambda I) = H) \\
&= H.
\end{aligned}$$

To prove that S is closed let $(u_n) \subset \mathcal{D}(S)$ be such that $u_n \to u$ and $(S - \lambda I)u_n \to f$. In accordance with (3.9),

$$u_n = v_n + w_n, \qquad v_n \in \mathcal{D}(A), \qquad w_n \in N,$$

$$(S - \lambda I)u_n = (A - \lambda I)v_n + (B^* - \lambda I)w_n$$

and, since $(B^* - \lambda I)w_n \in \mathcal{N}(A^* - \bar{\lambda}I) = \mathcal{R}(A - \lambda I)^{\perp}$,

$$\|(S - \lambda I)u_n\|^2 = \|(A - \lambda I)v_n\|^2 + \|(B^* - \lambda I)w_n\|^2.$$

Consequently $((A - \lambda I)v_n)$ and $((B^* - \lambda I)w_n)$ are Cauchy sequences and, since $\lambda \in \Pi(A)$, the sequence (v_n) converges to a limit $v \in \mathcal{D}(A)$, where $(A - \lambda I)v_n \to (A - \lambda I)v$. It follows that (w_n) is convergent and since B^* is closed, the convergence $w_n \to w$ and the afore-mentioned convergence of $((B^* - \lambda I)w_n)$ imply that $(B^* - \lambda I)w_n \to (B^* - \lambda I)w$, with w in N since N is closed. We conclude that $f = (A - \lambda I)v + (B^* - \lambda I)w = (S - \lambda I)u$ and hence S is closed.

Finally, we prove that $\text{nul}\,(S - \lambda I) = 0$. Suppose to the contrary that $(S - \lambda I)u = 0$ for some $u \in \mathcal{D}(S)$, with $u \neq 0$. From (3.9), $u = v + w$ with $v \in \mathcal{D}(A)$ and $w \in N$, and so

$$(A - \lambda I)v = -(B^* - \lambda I)w.$$

But $(B^* - \lambda I)w \in \mathcal{N}(A^* - \bar{\lambda}I) = \mathcal{R}(A - \lambda I)^\perp$ and hence $(A - \lambda I)v = (B^* - \lambda I)w = 0$. Now $\lambda \in \Pi(A)$ implies that $v = 0$, while $w \in N$ implies that $w \perp \mathcal{N}(B^* - \lambda I)$ as well as $w \in \mathcal{N}(B^* - \lambda I)$. We must therefore have $w = 0$, proving that $\text{nul}\,(S - \lambda I) = 0$. $(S - \lambda I)^{-1}$ is therefore a closed operator defined on H and the Closed-Graph Theorem yields the asserted conclusion, that $(S - \lambda I)^{-1} \in \mathcal{B}(H)$. □

Definition 3.4. An adjoint pair of closed, densely defined operators A, B is said to be *compatible* if $\Pi(A, B) := \{\lambda : \lambda \in \Pi(A), \bar{\lambda} \in \Pi(B), \text{def}(A - \lambda I) < \infty, \text{def}(B - \bar{\lambda}I) < \infty\} \neq \emptyset$. We shall call the set $\Pi(A, B)$ the *joint field of regularity* of A, B. A closed operator S is said to be *regularly solvable with respect to a compatible pair* A, B if $A \subset S \subset B^*$ and $\Pi(A, B) \cap \Delta_4(S) \neq \emptyset$, where $\Delta_4(S) = \{\lambda : (S - \lambda I) \text{ is Fredholm with } \text{ind}\,(S - \lambda I) = 0\}$. If $A \subset S \subset B^*$ and $\rho(S) \neq \emptyset$, S is said to be *well-posed with respect to A, B.* ■

Theorem 3.3 establishes the existence of an operator S which is well-posed with respect to any given compatible pair A, B. Note that if $A \subset S \subset B^*$ and $\lambda \in \rho(S)$ then $\lambda \in \Pi(A)$ and $\bar{\lambda} \in \rho(S^*) \subset \Pi(B)$ so that if $A - \lambda I$ and $B - \bar{\lambda}I$ have finite deficiency, A and B are compatible. The terminology 'regularly solvable' comes from Višik's paper [1], while the notion of being 'well-posed' was introduced by Zhikhar in his work on J-self-adjoint operators [1] which will be studied in §5 below.

Theorem 3.5. Let S be regularly solvable with respect to the compatible pair A, B. Then, if $\lambda \in \pi(A, B) \cap \Delta_4(S)$,

$$\mathcal{D}(S) = \mathcal{D}(A) \dotplus \mathcal{N}((A^* - \bar{\lambda}I)(S - \lambda I)), \tag{3.10}$$

$$\mathcal{D}(S^*) = \mathcal{D}(B) \dotplus \mathcal{N}((B^* - \lambda I)(S^* - \bar{\lambda}I)), \tag{3.11}$$

and

$$\dim \mathcal{D}(S)/\mathcal{D}(A) = \text{def}\,(A - \lambda I), \tag{3.12}$$

$$\dim \mathcal{D}(S^*)/\mathcal{D}(B) = \text{def}\,(B - \bar{\lambda}I). \tag{3.13}$$ ■

Proof. This is an immediate consequence of Theorem 3.1. □

It is of interest to note that if S is well posed in Theorem 3.5 and $\lambda \in \rho(S)$, the direct sum in (3.10) is orthogonal with respect to the inner product $((S - \lambda I)u, (S - \lambda I)v)$. For if $u \in \mathcal{D}(A)$ and $v \in \mathcal{N}((A^* - \bar{\lambda}I)(S - \lambda I))$,

$$\begin{aligned}((S - \lambda I)u, (S - \lambda I)v) &= ((A - \lambda I)u, (S - \lambda I)v) \\ &= (u, (A^* - \bar{\lambda}I)(S - \lambda I)v) = 0.\end{aligned}$$

Furthermore, since $\lambda \in \rho(S)$, we have $\|(S - \lambda I)u\| \geqslant k_\lambda \|u\|$ for $u \in \mathcal{D}(S)$ and so

$$\begin{aligned}\|(S - \lambda I)u\| &\leqslant \|Su\| + |\lambda|\,\|u\| \\ &\leqslant (1 + 2|\lambda| k_\lambda^{-1})\,\|(S - \lambda I)u\|.\end{aligned}$$

Thus $\|(S-\lambda I)\bullet\|$ is a norm on $\mathscr{D}(S)$ which is equivalent to the graph norm $\|\bullet\|_S$. Similarly the direct sum in (3.11) is orthogonal with respect to the inner product $((S^*-\lambda I)u, (S^*-\bar{\lambda}I)v)$ and $\|(S^*-\bar{\lambda}I)\bullet\|$ is equivalent to $\|\bullet\|_{S^*}$ on $\mathscr{D}(S^*)$. By the same reasoning the direct sums in (3.6) and (3.7) are orthogonal with respect to $((B^*-\lambda I)u, (B^*-\lambda I)v)$ and $((A^*-\bar{\lambda}I)u, (A^*-\bar{\lambda}I)v)$ respectively but these are not inner products since they vanish on $\mathscr{N}(B^*-\lambda I)$ and $\mathscr{N}(A^*-\bar{\lambda}I)$.

On using Theorem 3.5 we can characterize all operators S which are regularly solvable with respect to a given compatible pair A, B in terms of the sesquilinear form

$$\beta[\![u, v]\!] = (B^*u, v) - (u, A^*v) \qquad (u\in\mathscr{D}(B^*), v\in\mathscr{D}(A^*)), \tag{3.14}$$

the domains $\mathscr{D}(S)$ and $\mathscr{D}(S^*)$ being defined by "boundary conditions" on $\mathscr{D}(B^*)$ and $\mathscr{D}(A^*)$ respectively, determined by $\beta[\![\bullet,\bullet]\!]$.

Theorem 3.6. If S is regularly solvable with respect to the compatible pair A, B and $\lambda\in\Pi(A, B)\cap\Delta_4(S)$,

$$\mathscr{D}(S) = \{u : u\in D(B^*) \text{ and } \beta[\![u, \phi]\!] = 0 \text{ for all } \phi\in\mathscr{N}((B^*-\lambda I)(S^*-\bar{\lambda}I))\}, \tag{3.15}$$

$$\mathscr{D}(S^*) = \{v : v\in D(A^*) \text{ and } \beta[\![\psi, v]\!] = 0 \text{ for all } \psi\in\mathscr{N}((A^*-\bar{\lambda}I)(S-\lambda I))\}. \tag{3.16}$$ ■

Proof. If $u\in\mathscr{D}(S)\subset\mathscr{D}(B^*)$ and $\phi\in\mathscr{N}((B^*-\lambda I)(S^*-\bar{\lambda}I))$,

$$\begin{aligned}\beta[\![u, \phi]\!] &= (Su, \phi) - (u, A^*\phi)\\ &= ((S-\lambda I)u, \phi) - (u, (S^*-\bar{\lambda}I)\phi)\\ &= 0\end{aligned}$$

so that $\mathscr{D}(S)$ is contained in the set in (3.15). Conversely if u belongs to the set in (3.15) then $u\in\mathscr{D}(B^*)$ and for all $v\in\mathscr{D}(S^*)$,

$$\begin{aligned}(B^*u, v) - (u, S^*v) &= (B^*u, v) - (u, A^*v)\\ &= \beta[\![u, v]\!]\\ &= \beta[\![u, w]\!] + \beta[\![u, \phi]\!]\end{aligned}$$

(where $v = w+\phi$, $w\in\mathscr{D}(B)$ and $\phi\in\mathscr{N}((B^*-\lambda I)(S^*-\bar{\lambda}I))$ from (3.11))

$$\begin{aligned}&= \beta[\![u, w]\!]\\ &= (B^*u, w) - (u, A^*w)\\ &= (B^*u, w) - (u, Bw)\\ &= 0.\end{aligned}$$

Consequently $u \in \mathscr{D}(S)$ and $B^*u = Su$. The proof of (3.15) is complete and (3.16) is established in a similar way. □

A characterization of all operators S which are regularly solvable with respect to a compatible pair A, B is also given by Visik [1]. Visik makes a closer scrutiny of $\mathscr{D}(S)/\mathscr{D}(A)$ than we have, and instead of (3.10) he obtains a direct sum involving an operator which is well-posed with respect to A, B. However, we shall see in §10 that the straightforward results of this section are ideally suited for application to ordinary differential operators.

4. Symmetric and self-adjoint operators

4.1. *General remarks*

A linear operator S with domain and range in a Hilbert space H is said to be symmetric if its domain $\mathscr{D}(S)$ is dense in H and

$$(Sx, y) = (x, Sy) \quad \text{for all } x, y \in \mathscr{D}(S). \tag{4.1}$$

Since $\mathscr{D}(S)$ is dense in H, the operator S^* exists and (4.1) is equivalent to

$$S \subset S^*. \tag{4.2}$$

Consequently, as the adjoint is always closed, a symmetric operator is closable and

$$S \subset \bar{S} \subset S^* = \bar{S}^*.$$

If $S = S^*$ then S is said to be self-adjoint. A self-adjoint operator is therefore closed and has no proper symmetric extensions. This is because if $S = S^*$ and $S \subset T$ for a symmetric operator T then $S \subset T \subset T^* \subset S^* = S$ and hence $S = T$. The notions of symmetry and self-adjointness are distinct, in general, as long as $\mathscr{D}(S) \neq H$; there are even *maximal* symmetric operators, i.e. symmetric operators with no proper symmetric extensions, which are not self-adjoint. On the other hand, if $S \in \mathscr{B}(H)$ is symmetric it is necessarily self-adjoint, while if S is self-adjoint and $\mathscr{D}(S) = H$ then S, being closed, is also bounded by virtue of the Closed-Graph Theorem.

It is usually a straightforward matter to determine whether or not an operator is symmetric, but self-adjointness is a much more delicate property to establish. A class of operators which frequently occurs, especially in the study of differential equations, is the class of operators which are *essentially self-adjoint*. A symmetric operator S is *essentially self-adjoint* if its closure $\bar{S}$ is self-adjoint. The following result is readily established.

Theorem 4.1. The following statements are equivalent:

(i) S is essentially self-adjoint,

(ii) $\bar{S} = S^*$ is the unique self-adjoint extension of S,

(iii) S^* is symmetric,
(iv) S^* is self-adjoint. ∎

There is a very rich general theory for self-adjoint operators due largely to the Spectral Theorem, which establishes that a self-adjoint operator can be represented as a multiplication operator on some integration space $L^2(-\infty, \infty; \sigma)$. This theorem is not given in this book as it is extensively treated by other authors, e.g. Akhiezer and Glazman [1, Vol.I §6], and also because we strive to use methods which are applicable to operators other than self-adjoint ones.

If T is symmetric, (Tu, u) is real for all $u \in \mathscr{D}(T)$ and hence $\Theta(T)$ is a subset of the real line, an interval in fact since it is convex. For a closed symmetric operator T, the set $\tilde{\Delta}(T) = \mathbb{C} \setminus \bar{\Theta}(T)$ discussed in §2 therefore includes the upper and lower half-planes $\mathbb{C}_{\pm} = \{\lambda \in \mathbb{C}: \operatorname{im}\lambda \gtrless 0\}$. The set $\tilde{\Delta}(T)$ is not connected if, and only if, $\Theta(T) = \mathbb{R}$ and in this case the connected components $\tilde{\Delta}_1(T)$ and $\tilde{\Delta}_2(T)$ coincide with $\mathbb{C}_-$ and $\mathbb{C}_+$ respectively. The deficiency indices of a closed symmetric operator T are therefore given by

$$m_+(T) \equiv m_1(T) = \operatorname{def}(T - \lambda I), \qquad \lambda \in \mathbb{C}_-,$$

$$m_-(T) \equiv m_2(T) = \operatorname{def}(T - \lambda I), \qquad \lambda \in \mathbb{C}_+,$$

these being constants in view of Theorem 2.3. Since $\mathbb{C}_{\pm}$ lie in $\tilde{\Delta}(T)$ it follows that $\operatorname{nul}(T - \lambda I) = 0$ and $\mathscr{R}(T - \lambda I)$ is closed for $\lambda \in \mathbb{C}_{\pm}$ (by Theorem 2.3), and we have

$$\mathscr{N}(T^* - \bar{\lambda} I) = \mathscr{R}(T - \lambda I)^{\perp}, \mathscr{N}(T^* - \bar{\lambda} I)^{\perp} = \mathscr{R}(T - \lambda I), \qquad \lambda \in \mathbb{C}_{\pm}, \tag{4.3}$$

and as $\mathscr{R}(T^* - \bar{\lambda} I)$ is also closed on account of Theorem I.3.7,

$$\mathscr{N}(T - \lambda I) = \mathscr{R}(T^* - \bar{\lambda} I)^{\perp}, \mathscr{N}(T - \lambda I)^{\perp} = \mathscr{R}(T^* - \bar{\lambda} I), \qquad \lambda \in \mathbb{C}_{\pm}. \tag{4.4}$$

The closed subspaces

$$\mathscr{N}_{\pm} \equiv \mathscr{N}_{\pm}(T) := \mathscr{N}(T^* \mp iI) = \mathscr{R}(T \pm iI)^{\perp} \tag{4.5}$$

are called the *deficiency subspaces* of T, their dimension being $m_{\pm}(T)$.

Theorem 4.2. (i) A symmetric operator T is self-adjoint if, and only if, $\mathscr{R}(T - \lambda I) = \mathscr{R}(T - \bar{\lambda} I) = H$ for some, and hence all, $\lambda \notin \mathbb{R}$.

(ii) A symmetric operator T is essentially self-adjoint if, and only if, $\mathscr{R}(T - \lambda I)$ and $\mathscr{R}(T - \bar{\lambda} I)$ are dense subspaces of H for some, and hence all, $\lambda \notin \mathbb{R}$.

(iii) If T is symmetric and $\mathscr{R}(T - \lambda I) = H$ for some $\lambda \in \mathbb{R}$, then T is self-adjoint. ∎

Proof. (i) If T is self-adjoint then for any non-real λ, the space $\mathscr{R}(T - \lambda I)$ is closed and $\operatorname{def}(T - \lambda I) = \operatorname{nul}(T - \bar{\lambda} I) = 0$, whence $\mathscr{R}(T - \lambda I) = H$. Conversely, suppose that T is symmetric and $\mathscr{R}(T - \lambda I) = \mathscr{R}(T - \bar{\lambda} I) = H$ for some $\lambda \notin \mathbb{R}$. Since $T \subset \bar{T} \subset T^*$, it follows that $\mathscr{R}(\bar{T}-\lambda I) = \mathscr{R}(T^* - \lambda I) = H$ and similarly for $\bar{\lambda}$. Hence $\operatorname{nul}(T^* - \lambda I) = \operatorname{def}(\bar{T} - \bar{\lambda} I) = 0$. For any $u \in \mathscr{D}(T^*)$ there exists some $v \in \mathscr{D}(T)$ such that $(T^* - \lambda I)u = (T - \lambda I)v$ and thus $(T^* - \lambda I)(u - v) = 0$. This yields $u = v \in \mathscr{D}(T)$ and $T = T^*$.

(ii) Since $\mathscr{R}(\bar{T} - \lambda I)$ is the closure of $\mathscr{R}(T - \lambda I)$ for $\lambda \notin \mathbb{R}$ this part is an immediate consequence of (i).

(iii) For any $u \in \mathscr{D}(T^*)$ there exists some $v \in \mathscr{D}(T)$ such that $(T^* - \lambda I)(u - v) = 0$. The result follows since $\mathscr{N}(T^* - \lambda I) = \mathscr{R}(T - \lambda I)^{\perp} = \{0\}$. □

Lemma 4.3. Let T be self-adjoint and $\lambda \in \rho(T)$. Then

$$\|(T - \lambda I)^{-1}\| = 1/\operatorname{dist}[\lambda, \sigma(T)], \qquad \|T(T - \lambda I)^{-1}\| = \sup_{\mu \in \sigma(T)} |\mu(\mu - \lambda)^{-1}|. \blacksquare$$

Proof. For any $\lambda \in \rho(T)$, the map $T_\lambda := (T - \lambda I)^{-1}$ is a bounded normal operator in H and hence $\|T_\lambda\| = \sup\{|\mu| : \mu \in \sigma(T_\lambda)\}$, the spectral radius of T_λ, from Kato [1, *V*-(2.4)]. The identity

$$T - \xi I = (\lambda - \xi)[T_\lambda - I/(\xi - \lambda)](T - \lambda I)$$

gives that for $\xi \neq \lambda$, $\xi \in \sigma(T)$ if, and only if, $1/(\xi - \lambda) \in \sigma(T_\lambda)$, whence

$$\|T_\lambda\| = \sup[1/|\xi - \lambda| : \xi \in \sigma(T)] = 1/\operatorname{dist}[\lambda, \sigma(T)].$$

Similarly $\xi \in \sigma(T)$ if, and only if, $1 + \lambda/(\xi - \lambda) \in \sigma(I + \lambda T_\lambda) = \sigma(T T_\lambda)$ and

$$\begin{aligned} \|T T_\lambda\| &= \sup\{|1 + \lambda/(\xi - \lambda)| : \xi \in \sigma(T)\} \\ &= \sup\{|\xi||\xi - \lambda|^{-1} : \xi \in \sigma(T)\}. \end{aligned}$$ □

Theorem 4.4. Let T be a self-adjoint operator in H. Then $\Theta(T)$ is bounded below if, and only if, $\sigma(T)$ is bounded below and the lower bounds are equal:

$$\inf\{\lambda : \lambda \in \Theta(T)\} = \inf\{\lambda : \lambda \in \sigma(T)\}. \blacksquare$$

Proof. Let l_T and l_σ denote the lower bounds of $\Theta(T)$ and $\sigma(T)$ respectively. If $\Theta(T)$ is bounded below the set $\tilde{\Delta}(T)$ in §2 is connected and includes all real numbers $\lambda < l_T$. Furthermore, by Theorem 4.2(i), $\operatorname{def}(T - \lambda I) = 0$ for all $\lambda \in \tilde{\Delta}(T)$ and hence $\tilde{\Delta}(T) \subset \rho(T)$ by Theorem 2.3. We have therefore shown that $l_\sigma \geqslant l_T$. To prove the converse we make use of Lemma 4.3. Suppose $\sigma(T)$ is bounded below by l and set $T' = T - lI$. Then T' is self-adjoint and $\sigma(T') \subset [0, \infty)$. Thus for any $\alpha > 0$ we see from Lemma 4.3 that $\|(T' + \alpha I)^{-1}\| = \alpha^{-1}$. It follows that for any $u \in \mathscr{D}(T)$,

$$\begin{aligned} \|u\|^2 &\leqslant \alpha^{-2}\|(T' + \alpha I)u\|^2 \\ &= \alpha^{-2}\|T'u\|^2 + 2\alpha^{-1}(T'u, u) + \|u\|^2 \end{aligned}$$

and

$$0 \leqslant \alpha^{-1}\|T'u\|^2 + 2(T'u, u).$$

On allowing $\alpha \to \infty$ we get $(T'u, u) \geqslant 0$ so that $\Theta(T) \subset [l, \infty)$. Therefore $l_T \geqslant l_\sigma$ and the theorem is proved. □

4.2. *Symmetric and self-adjoint extensions of symmetric operators*

If in Corollary 3.2, A is a closed symmetric operator T and $B = T$, the decomposition (3.6) of $\mathscr{D}(T^*)$ takes the following form due to von Neumann.

Theorem 4.5. Let T be a closed symmetric operator in H. Then $\mathscr{D}(T)$ and the deficiency subspaces $\mathscr{N}_\pm$ of T are closed subspaces of $\mathrm{H}(T^*)$ and

$$\mathscr{D}(T^*) = \mathscr{D}(T) \oplus_* \mathscr{N}_+ \oplus_* \mathscr{N}_- \tag{4.6}$$

where $\oplus_*$ indicates the orthogonal sum with respect to the graph inner product $(\bullet, \bullet)_{T^*}$ of T^*. Thus

$$\dim \mathscr{D}(T^*)/\mathscr{D}(T) = \mathrm{m}_+(T) + \mathrm{m}_-(T). \tag{4.7}$$

If $\lambda \in \Pi(T)$ is real,

$$\mathscr{D}(T^*) = \mathscr{D}(T) \dotplus \mathscr{N}((T^* - \lambda I)^2) \tag{4.8}$$

and

$$\dim \mathscr{D}(T^*)/\mathscr{D}(T) = 2 \operatorname{def}(T - \lambda I). \tag{4.9}$$ ■

Proof. From (3.6) with $A = B = T$ and $\lambda = \mathrm{i}$ we have that

$$\mathscr{D}(T^*) = \mathscr{D}(T) \dotplus \mathscr{N}((T^*)^2 + I).$$

Since T and T^* are closed and $T \subset T^*$, the spaces $\mathscr{D}(T)$ and $\mathscr{N}_\pm$ are closed subspaces of $\mathrm{H}(T^*)$. Also $\mathscr{N}_+$ and $\mathscr{N}_-$ are disjoint subspaces of $\mathscr{N}((T^*)^2 + I)$. But any $\phi \in \mathscr{N}((T^*)^2 + I)$ can be written

$$\begin{aligned} \phi &= (1/2\mathrm{i})[(T^* + \mathrm{i}I)\phi - (T^* - \mathrm{i}I)\phi] \\ &= \phi_+ + \phi_- \end{aligned}$$

say, where $(T^* - \mathrm{i}I)\phi_+ = 0$ and $(T^* + \mathrm{i}I)\phi_- = 0$. Hence $\mathscr{N}(T^{*2} + I) = \mathscr{N}_+ \dotplus \mathscr{N}_-$ and

$$\mathscr{D}(T^*) = \mathscr{D}(T) \dotplus \mathscr{N}_+ \dotplus \mathscr{N}_-.$$

For any $u \in \mathscr{D}(T)$ and $\phi_\pm \in \mathscr{N}_\pm$,

$$\begin{aligned} (u, \phi_\pm)_{T^*} &= (Tu, T^*\phi_\pm) + (u, \phi_\pm) \\ &= \mp \mathrm{i}[(Tu, \phi_\pm) - (u, T^*\phi_\pm)] \\ &= 0 \end{aligned}$$

and

$$\begin{aligned}(\phi_+, \phi_-)_{T^*} &= (\mathrm{i}\phi_+, -\mathrm{i}\phi_-) + (\phi_+, \phi_-)\\ &= 0.\end{aligned}$$

The orthogonal sum decomposition (4.6) is therefore established. The rest of the theorem follows directly from Corollary 3.2. □

From the result proved in §2 that for any closed operator T, $\operatorname{def}(T-\lambda I)$ is constant in any connected component of $\Pi(T)$, it follows that if a closed symmetric operator T has a real number λ in $\Pi(T)$, then T must have equal deficiency indices $m_{\pm}(T)$ so that (4.9) is a consequence of (4.7).

The analogue of Theorem 3.3 is now the following result of Calkin [1].

Theorem 4.6. Let T be a closed symmetric operator. Then given any $\lambda \notin \mathbb{R}$, there exists a closed operator S such that $T \subset S \subset T^*$ and $\lambda \in \rho(S)$. If there exists a real $\lambda \in \Pi(T)$ then there exists a self-adjoint extension S of T with $\lambda \in \rho(S)$. ■

Proof. The first part of the theorem is an immediate consequence of Theorem 3.3 with $A = B = T$. To prove the second part we show that the operator S constructed in the proof of Theorem 3.3 is symmetric. The fact that S is then self-adjoint will follow from Theorem 4.2 (iii). If $u_1, u_2 \in \mathscr{D}(S)$, we have from (3.9), with $A = B = T$ and λ real, that for $i = 1, 2$,

$$u_i = v_i + w_i, \quad v_i \in \mathscr{D}(T), \quad w_i \in \mathscr{N}((T^*-\lambda I)^2) \ominus \mathscr{N}(T^*-\lambda I).$$

Thus

$$\begin{aligned}((S-\lambda I)u_1, u_2) &= ((T-\lambda I)v_1, v_2+w_2) + ((T^*-\lambda I)w_1, v_2+w_2)\\ &= (v_1, (T-\lambda I)v_2) + (v_1, (T^*-\lambda I)w_2)\\ &\quad + (w_1, (T-\lambda I)v_2) + ((T^*-\lambda I)w_1, w_2)\\ &= (v_1, (T-\lambda I)v_2) + (v_1, (T^*-\lambda I)w_2) + (w_1, (T-\lambda I)v_2)\end{aligned}$$

(since $(T^*-\lambda I)w_1 \in \mathscr{N}((T^*-\lambda I))$ and $w_2 \perp \mathscr{N}(T^*-\lambda I))$

$$= (u_1, (S-\lambda I)u_2).$$

S is therefore symmetric and the theorem is proved. □

Definition 4.7. A *partial isometry* from a Hilbert space H_1 into a Hilbert space H_2 is a linear map $V: H_1 \to H_2$ which is such that $\|Vx\|_{H_2} = \|x\|_{H_1}$ for all x in some closed subspace $\mathscr{I}(V)$ of H_1 and $Vx = 0$ for $x \in \mathscr{I}(V)^{\perp}$. $\mathscr{I}(V)$ is called the *initial set* of V and $V\mathscr{I}(V)$ the *final set* of V. ■

Theorem 4.8. There is a one–one correspondence between the set of all closed symmetric extensions of a closed symmetric operator T and the set of all partial isometries V from $\mathcal{N}_+$ into $\mathcal{N}_-$, where $\mathcal{N}_+$ and $\mathcal{N}_-$ are the deficiency subspaces of T. If V is one such partial isometry with initial set $\mathcal{I}(V) \subset \mathcal{N}_+$, the corresponding closed symmetric extension T_V of T is the restriction of T^* to $\mathcal{D}(T_V)$ where

$$\mathcal{D}(T_V) := \{\phi + \phi_+ + V\phi_+ : \phi \in \mathcal{D}(T),\ \phi_+ \in \mathcal{I}(V)\}. \tag{4.10}$$

T_V is self-adjoint if, and only if, V is a unitary map of $\mathcal{N}_+$ onto $\mathcal{N}_-$.

If $\dim \mathcal{I}(V) < \infty$,

$$\mathrm{m}_\pm(T_V) = \mathrm{m}_\pm(T) - \dim \mathcal{I}(V). \tag{4.11}$$ ■

Proof. If S is a closed symmetric extension of T, $T \subset S \subset S^* \subset T^*$ and from (4.6) any $u \in \mathcal{D}(S)$ can be written as $u = \phi + \phi_+ + \phi_-$, where $\phi \in \mathcal{D}(T)$ and $\phi_\pm \in \mathcal{N}_\pm$ with $\phi_+ + \phi_- \in \mathcal{D}(S)$. Also $Su = T\phi + T^*\phi_+ + T^*\phi_- = T\phi + \mathrm{i}\phi_+ - \mathrm{i}\phi_-$ and since S is symmetric,

$$\begin{aligned} 0 &= (Su, u) - (u, Su) \\ &= (T\phi, \phi) - (\phi, T\phi) + (T\phi, \phi_+ + \phi_-) - (\phi, T^*(\phi_+ + \phi_-)) \\ &\quad + (T^*(\phi_+ + \phi_-), \phi) - (\phi_+ + \phi_-, T\phi) + (T^*(\phi_+ + \phi_-), \phi_+ + \phi_-) \\ &\quad - (\phi_+ + \phi_-, T^*(\phi_+ + \phi_-)) \\ &= 2\mathrm{i}\,\mathrm{im}\,(\mathrm{i}\phi_+ - \mathrm{i}\phi_-, \phi_+ + \phi_-) \\ &= 2\mathrm{i}(\|\phi_+\|^2 - \|\phi_-\|^2). \end{aligned} \tag{4.12}$$

Let V be the linear isometry defined on the subspace

$$\mathcal{I}(V) = \{\phi_+ \in \mathcal{N}_+ : \phi_+ + \phi_- \in \mathcal{D}(S) \text{ for some } \phi_- \in \mathcal{N}_-\} \tag{4.13}$$

by $V\phi_+ = \phi_-$; it is an isometry in view of (4.12). $\mathcal{I}(V)$ is a closed subspace of H. To see this, let $\phi_+^{(n)} \in \mathcal{I}(V)$, $\phi_+^{(n)} + \phi_-^{(n)} \in \mathcal{D}(S)$ and $\phi_+^{(n)} \to \phi_+$ in H. Then $\phi_+ \in \mathcal{N}_+$ since $\mathcal{N}_+$ is closed in H and also, since $(\phi_-^{(n)})$ is a Cauchy sequence on account of (4.12), $\phi_-^{(n)} \to \phi_-$ for some $\phi_- \in \mathcal{N}_-$. Furthermore, $S(\phi_+^{(n)} + \phi_-^{(n)}) = \mathrm{i}\phi_+^{(n)} - \mathrm{i}\phi_-^{(n)} \to \mathrm{i}\phi_+ - \mathrm{i}\phi_-$ and since S is closed, we must therefore have $\phi_+ + \phi_- \in \mathcal{D}(S)$ and $\phi_+ \in \mathcal{I}(V)$, which is consequently closed. Defining V to be the zero operator on $\mathcal{I}(V)^\perp$, we see that it is a partial isometry of $\mathcal{N}_+$ into $\mathcal{N}_-$ and S has the domain (4.10).

Conversely, let T_V be the restriction of T^* to the domain given by (4.10), where V is a partial isometry from $\mathcal{N}_+$ into $\mathcal{N}_-$. Then $T \subset T_V$ and it is an easy matter to check that T_V is symmetric. Since $\mathcal{I}(V)$, and hence $V\mathcal{I}(V)$, are closed subspaces of H, they are also closed in $\mathrm{H}(T^*)$, as is $\mathcal{D}(T)$. From

$$\mathcal{D}(T_V) = \mathcal{D}(T) \oplus_* \mathcal{I}(V) \oplus_* V\mathcal{I}(V)$$

it therefore follows that $\mathcal{D}(T_V)$ is closed in $\mathrm{H}(T^*)$ and T_V is a closed operator.

For $u = \phi + \phi_+ + V\phi_+ \in \mathscr{D}(T_V)$,

$$(T_V + iI)u = (T + iI)\phi + 2i\,\phi_+$$

and so, since $\mathscr{I}(V) \subset \mathscr{N}_+ = \mathscr{R}(T+iI)^{\perp}$,

$$\mathscr{R}(T_V + iI) = \mathscr{R}(T+iI) \oplus \mathscr{I}(V), \tag{4.14}$$

where $\oplus$ is the orthogonal sum in H. Similarly

$$\mathscr{R}(T_V - iI) = \mathscr{R}(T - iI) \oplus V\mathscr{I}(V). \tag{4.15}$$

We conclude from Theorem 4.2(i) that T_V is self-adjoint if, and only if, $\mathscr{I}(V) = \mathscr{N}_+$ and $V\mathscr{I}(V) = \mathscr{N}_-$, that is, V is unitary from $\mathscr{N}_+$ onto $\mathscr{N}_-$.

From (4.14) and (4.15),

$$\begin{aligned}\mathscr{R}(T_V + iI)^{\perp} &= \mathscr{R}(T+iI)^{\perp} \cap \mathscr{I}(V)^{\perp} \\ &= \mathscr{R}(T+iI)^{\perp} \ominus \mathscr{I}(V)\end{aligned}$$

and

$$\mathscr{R}(T_V - iI)^{\perp} = \mathscr{R}(T - iI)^{\perp} \ominus V\mathscr{I}(V).$$

These identities give (4.11) since $\mathscr{I}(V)$ and $V\mathscr{I}(V)$ have the same dimension. □

Theorem 4.9. Let T_V be a maximal symmetric or self-adjoint extension of a closed symmetric operator T which corresponds to an isometric map V between $\mathscr{N}_+$ and $\mathscr{N}_-$ as in Theorem 4.8. Then

$$\mathscr{D}(T_V) = \{u \in \mathscr{D}(T^*) : \beta[\![u, \phi_+ + V\phi_+]\!] = 0 \text{ for all } \phi_+ \in \mathscr{N}_+\}, \tag{4.16}$$

where $\beta[\![\bullet,\bullet]\!]$ is the sesquilinear form

$$\beta[\![u,v]\!] = (T^*u, v) - (u, T^*v) \qquad (u, v \in \mathscr{D}(T^*)). \tag{4.17}$$

■

Proof. The proof is similar to that of Theorem 3.6, on using (4.10). □

In comparing Theorems 3.6 and 4.9 with $A = B = T$, one should note that if S is a closed symmetric extension of T then $\mathbb{C}_{\pm} \subset \Pi(T) \cap \Pi(S)$ and $\Delta_4(S)$ is an open set in $\mathbb{C}$ as was shown in Theorem I.3.18. Hence, if $\Delta_4(S) \neq \varnothing$ there exists a non-real $\lambda \in \Pi(T) \cap \Delta_4(S)$. Hence, in this case $\operatorname{nul}(S - \lambda I) = \operatorname{def}(S - \lambda I) = 0$ and S is either self-adjoint or maximal symmetric with $\lambda \in \rho(S)$. Conversely, if S is either self-adjoint or maximal symmetric, $\rho(S) \neq \varnothing$ and so $\Pi(T) \cap \Delta_4(S) \supset \Pi(S) \cap \Delta_4(S) \supset \rho(S) \neq \varnothing$. Therefore the notions of regularly solvable and well-posed operators S merge into one when T and S are symmetric, the operators S being either self-adjoint or maximal symmetric extensions of T.

Corollary 4.10. Let T be a closed symmetric operator in H. Then
(i) T is self-adjoint if, and only if, $\mathrm{m}_+(T) = \mathrm{m}_-(T) = 0$,
(ii) T has a self-adjoint extension if, and only if, $\mathrm{m}_+(T) = \mathrm{m}_-(T)$,
(iii) T is maximal symmetric if, and only if, either $\mathrm{m}_+(T) = 0$ or $\mathrm{m}_-(T) = 0$, but not both. ■

Proof. (i) is already given in Theorem 4.2(i). The rest follows easily from Theorem 4.8. □

Finally in this section we examine the various subdivisions of the spectrum of a closed symmetric operator T. The point, continuous and residual spectra of T are denoted by $\sigma_p(T)$, $\sigma_c(T)$, $\sigma_r(T)$, these sets being defined in §I.1. The following results are readily established.

Theorem 4.11. Let T be a closed symmetric operator. Then:
(i) $\sigma_p(T)$ and $\sigma_c(T)$ are subsets of $\mathbb{R}$;
(ii) $\mathbb{C}_+$(or $\mathbb{C}_-$) is either in $\sigma_r(T)$ or $\rho(T)$; T is self-adjoint if, and only if, $\mathbb{C}_+ \cup \mathbb{C}_- \subset \rho(T)$.
If T is self-adjoint:
(iii) $\sigma_r(T) = \varnothing$;
(iv) $\lambda \in \sigma_p(T)$ if, and only if, $\mathscr{R}(T - \lambda I)$ is not dense in H;
(v) $\lambda \in \sigma_c(T)$ if, and only if, $\mathscr{R}(T - \lambda I) \subsetneq \overline{\mathscr{R}(T - \lambda I)} = H$. ■

5. J-symmetric and J-self-adjoint operators

5.1. *General remarks*

Definition 5.1. An operator J defined on a Hilbert space H is a *conjugation* operator if, for all $x, y \in H$,

$$(Jx, Jy) = (y, x), \qquad J^2 x = x. \tag{5.1}$$ ■

A simple example in any L^2 space is the complex conjugation $x \mapsto \bar{x}$.

The definition implies that a conjugation J is a conjugate linear, norm-preserving bijection on H and that

$$(Jx, y) = (Jy, x) \quad \text{for all } x, y \in H. \tag{5.2}$$

Furthermore, if T is a densely defined linear operator in H, it is easily shown that

$$(JTJ)^* = JT^*J. \tag{5.3}$$

Definition 5.2. A densely defined linear operator T in H is said to be *J-symmetric*, for a conjugation operator J, if

$$JTJ \subset T^*.$$

T is said to be *J-self-adjoint* if $JTJ = T^*$. If $JTJ = T$, the operator T is said to be *real with respect to J*. ∎

From the definition, if T is J-symmetric then $T \subset JT^*J$ and this implies that T is closable, the closure $\bar{T}$ being also J-symmetric. Note also that a real J-symmetric operator is symmetric.

Theorem 5.3. If T is a closed symmetric operator which is real with respect to some conjugation operator J then $m_+(T) = m_-(T)$ and hence T has a self-adjoint extension. ∎

Proof. From (5.3), $T^* = (JTJ)^* = JT^*T$ and so

$$\begin{aligned} m_+(T) = \text{nul}(T^* - iI) &= \text{nul}(JT^*J - iI) \\ &= \text{nul}[J(T^* + iI)J] \\ &= m_-(T). \end{aligned}$$

The result then follows from Corollary 4.10 (ii). □

Lemma 5.4. Let T be a closed J-symmetric operator in H. Then, for any $\lambda \in \mathbb{C}$,

$$\text{nul}(T - \lambda I) \leqslant \text{nul}(T^* - \bar{\lambda} I). \tag{5.4}$$

If T is J-self-adjoint equality holds in (5.4). ∎

Proof. From $T \subset JT^*J$ we have that $\mathcal{N}(T - \lambda I) \subset \mathcal{N}(JT^*J - \lambda I) = \mathcal{N}\big(J(T^* - \bar{\lambda} I)J\big)$, with equality when T is J-self-adjoint. The lemma follows since J is an injection. □

Theorem 5.5. A closed J-symmetric operator is J-self-adjoint if, and only if, $\text{def}(T - \lambda I) = 0$ for some, and hence all, $\lambda \in \Pi(T)$. ∎

Proof. If T is J-self-adjoint and $\lambda \in \Pi(T)$, then $\mathcal{R}(T - \lambda I)$ is closed and, from Lemma 5.4, $\text{def}(T - \lambda I) = \text{nul}(T^* - \bar{\lambda} I) = \text{nul}(T - \lambda I) = 0$.

Conversely, suppose $\text{def}(T - \lambda I) = 0$ for some $\lambda \in \Pi(T)$. Then $\mathcal{R}(T - \lambda I) = H$ and

$$\mathcal{R}(JTJ - \bar{\lambda} I) = \mathcal{R}\big(J(T - \lambda I)J\big) = H. \tag{5.5}$$

If $JTJ \neq T^*$, there exists a non-zero $u \in \mathcal{D}(T^*) \backslash \mathcal{D}(JTJ)$. From (5.5), $(T^* - \bar{\lambda} I)u = (JTJ - \bar{\lambda} I)v$ for some $v \in \mathcal{D}(JTJ)$ and since $JTJ \subset T^*$ we have $(T^* - \bar{\lambda} I)\ (u - v) = 0$ or $u - v \in \mathcal{N}(T^* - \bar{\lambda} I)$. But $\mathcal{N}(T^* - \bar{\lambda} I) = \mathcal{R}(T - \lambda I)^{\perp} = \{0\}$ and consequently $u = v \in \mathcal{D}(JTJ)$, contradicting the assumption that $JTJ \neq T^*$. The theorem is therefore proved. □

5.2 *J-self-adjoint and regularly solvable extensions of J-symmetric operators*

If T is J-symmetric, $A = T$ and $B = JTJ$, then we see from (5.3) that $A \subset B^*$ and $B \subset A^*$. In other words A and B are an adjoint pair in the sense of §3. Also if $\lambda \in \Pi(A)$ and $x \in \mathscr{D}(JTJ)$,

$$\begin{aligned}\|(B-\bar{\lambda}I)x\| &= \|J(T-\lambda I)Jx\| = \|(T-\lambda I)Jx\| \\ &\geqslant k_\lambda \|Jx\| \\ &= k_\lambda \|x\|\end{aligned}$$

so that $\bar{\lambda} \in \Pi(B)$. Moreover, if $\lambda \in \Pi(T)$ then $\operatorname{def}(T-\lambda I) = \operatorname{nul}(T^* - \bar{\lambda}I) = \operatorname{nul}(JT^*J - \lambda I) = \operatorname{def}(JTJ - \bar{\lambda}I)$. We therefore have from Corollary 3.2

Theorem 5.6. Let T be a closed J-symmetric operator in H with $\Pi(T) \neq \varnothing$. Then for $\lambda \in \Pi(T)$,

$$\mathscr{D}(JT^*J) = \mathscr{D}(T) \dot{+} \mathscr{N}((T^* - \bar{\lambda}I)(JT^*J - \lambda I)) \tag{5.6}$$

and, if $\operatorname{def}(T-\lambda I) < \infty$,

$$\dim \mathscr{D}(JT^*J)/\mathscr{D}(T) = 2\operatorname{def}(T-\lambda I). \tag{5.7}$$ ■

An immediate consequence of Theorem 5.6 is the following result, a version of which was proved by Zhikhar [1].

Theorem 5.7. If T is a closed J-symmetric operator, $\operatorname{def}(T-\lambda I)$ is constant for all $\lambda \in \Pi(T)$. ■

Therefore, with reference to Theorem 2.3, $\operatorname{def}(T-\lambda I)$ is constant in $\Pi(T)$ for a closed J-symmetric operator T even if the external domain of regularity $\tilde{\Delta}(T)$ is not connected. One naturally compares the situation with that which prevails when T is symmetric; in which case the deficiency indices $m_+(T)$ and $m_-(T)$ are, in general, distinct and are equal if, and only if, T has a self-adjoint extension. It is not surprising therefore to learn of the following result of Galindo[1]. We shall give the proof of this result given by Knowles in [1].

Theorem 5.8. Any J-symmetric operator has a J-self-adjoint extension. ■

Proof. The proof uses Zorn's Lemma. Let $\mathscr{P}$ be the set of all J-symmetric extensions of the given closed J-symmetric operator T and let $\mathscr{P}$ be ordered by extension, i.e. $A < B$ if and only if $A \subset B$. Let $\mathscr{M} = \{T_\alpha\}$ be any totally ordered subset of $\mathscr{P}$ and define the operator $\hat{T}$ by

$$\mathscr{D}(\hat{T}) = \bigcup_\alpha \mathscr{D}(T_\alpha), \qquad \hat{T}x = T_\alpha x \text{ if } x \in \mathscr{D}(T_\alpha).$$

Since $\mathcal{M}$ is totally ordered $\hat{T}$ is well-defined, $\mathcal{D}(\hat{T})$ is a linear subspace of H and $\hat{T}$ is an upper bound of $\mathcal{M}$ in $\mathcal{P}$. Therefore $\mathcal{P}$ has a maximal element by Zorn's Lemma. This means that T has a maximal J-symmetric extension, which we denote by S. We shall prove that S is J-self-adjoint: suppose it is not; then $S \underset{\neq}{\subset} JS^*J$ and there exists a non-zero $y \in \mathcal{D}(JS^*J) \setminus \mathcal{D}(S)$. Define the operator S_1 by

$$\begin{aligned} &\mathcal{D}(S_1) = \mathcal{D}(S) + \mathrm{sp}\{y\}, \\ &S_1 x = Sz + JS^*J(\lambda y), \qquad x = z + \lambda y, \quad z \in \mathcal{D}(S). \end{aligned} \tag{5.8}$$

Then S_1 is linear and $S \subset S_1$. Also for $x_i = z_i + \lambda_i y$ $(i = 1, 2)$ in $\mathcal{D}(S_1)$,

$$\begin{aligned} (S_1 x_1, J x_2) &= (Sz_1 + JS^*J(\lambda_1 y), J(z_2 + \lambda_2 y)) \\ &= (Sz_1, Jz_2) + \lambda_2 (Sz_1, Jy) + (z_2, S^*J(\lambda_1 y)) \\ &\qquad + \lambda_2 (y, S^*J(\lambda_1 y)) \end{aligned}$$

(from (5.1))

$$= (z_1, JSz_2) + (z_1, S^*J(\lambda_2 y)) + \lambda_1 (Sz_2, Jy) + \lambda_1 \lambda_2 (y, S^*Jy)$$

(since $JS \subset S^*J$)

$$= (z_1, JSz_2) + (z_1, S^*J(\lambda_2 y)) + (\lambda_1 y, JSz_2) + (\lambda_1 y, S^*J(\lambda_2 y))$$

(from (5.2))

$$= (x_1, JS_1 x_2).$$

Consequently, S_1 is J-symmetric. Since S is a maximal J-symmetric operator we have a contradiction and hence S is J-self-adjoint. □

Theorem 5.9. Let T be closed and J-symmetric and with $\Pi(T) \neq \emptyset$. Then, for all $\lambda \in \Pi(T)$ there exists a J-self-adjoint extension S of T with $\lambda \in \rho(S)$. ■

Proof. If S is a J-self-adjoint extension of T, then $T \subset S = JS^*J \subset JT^*J$ and the theorem will follow from Theorem 3.3 if we show that the operator S constructed there is J-symmetric. The fact that S is then J-self-adjoint is a consequence of Theorem 5.5.

Let $u_1, u_2 \in \mathcal{D}(S)$. From (3.9) with $A = T, B = JTJ$, and $\lambda \in \Pi(T)$, we have $u_i = v_i + w_i$, $v_i \in \mathcal{D}(T)$, $w_i \in \mathcal{N}((T^* - \bar{\lambda} I)\ (JT^*J - \lambda I)) \ominus \mathcal{N}(JT^*J - \lambda I)$ for $i = 1, 2$. Thus

$$\begin{aligned} ((S - \lambda I)u_1, Ju_2) &= ((T - \lambda I)v_1, Jv_2) + ((T - \lambda I)v_1, Jw_2) \\ &\quad + ((JT^*J - \lambda I)w_1, Jv_2) + ((JT^*J - \lambda I)w_1, Jw_2) \\ &= (v_1, J(T - \lambda I)v_2) + (v_1, (T^* - \bar{\lambda} I)Jw_2) + (v_2, (T^* - \bar{\lambda} I)Jw_1) \end{aligned}$$

from the hypotheses, (5.2) and since $(JT^*J - \lambda I)w_1$ is orthogonal to Jw_2. The final assertion follows since $(JT^*J-\lambda I)w_1 \in \mathcal{N}(T^*-\bar{\lambda}I)$ and $w_2 \perp \mathcal{N}(JT^*J-\lambda I)$ implies that $w_2 \in \mathcal{R}(JTJ-\bar{\lambda}I) = \mathcal{R}(J(T-\lambda I)J)$, whence $Jw_2 \in \mathcal{R}(T-\lambda I) = \mathcal{N}(T^*-\bar{\lambda}I)^{\perp}$. Similarly

$$\begin{aligned}(u_1, J(S-\lambda I)u_2) &= (v_1, J(T-\lambda I)v_2) + (v_1, (T^*-\bar{\lambda}I)Jw_2) \\ &\quad + (w_1, J(T-\lambda I)v_2) + (w_1, (T^*-\bar{\lambda}I)Jw_2) \\ &= (v_1, J(T-\lambda I)v_2) + (v_1, (T^*-\bar{\lambda}I)Jw_2) + ((T-\lambda I)v_2, Jw_1) \\ &= ((S-\lambda I)u_1, Ju_2).\end{aligned}$$

It follows that S is J-symmetric and the theorem is proved. □

Let T and S be closed J-symmetric operators with $T \subset S$, so that $T \subset S \subset JS^*J \subset JT^*J$. In accordance with Definition 3.4 we shall say that S is a *regularly solvable* extension of T if $\Pi(T) \cap \Delta_4(S) \neq \varnothing$ and *well-posed* if $\rho(S) \neq \varnothing$. This definition of 'well-posed' differs from that of Zhikhar [1] in that he only requires $\Pi(S) \neq \varnothing$. However, the two notions are identical for J-self-adjoint operators S for then $\Pi(S) = \rho(S)$, by Theorem 5.5. Note that by Theorem 5.5, if S is well-posed in our sense then it is J-self-adjoint.

All the regularly solvable J-symmetric extensions of a given J-symmetric operator T can be characterized in terms of the sesquilinear form

$$\beta[\![u, v]\!] = (JT^*Ju, v) - (u, T^*v), \tag{5.9}$$

as shown by the following special case of Theorem 3.6.

Theorem 5.10. If S is a regularly solvable J-symmetric extension of a closed J-symmetric operator T and $\lambda \in \Pi(T) \cap \Delta_4(S)$, then S is the restriction of JT^*J to

$$\mathcal{D}(S) = \{u : u \in \mathcal{D}(JT^*J) \text{ and } \beta[\![u, J\phi]\!] = 0 \text{ for all } \phi \in \mathcal{N}((T^*-\bar{\lambda}I)(JS^*J-\lambda I))\}. \tag{5.10}$$ ■

For a given closed J-symmetric operator T with $\Pi(T) \neq \varnothing$, Zhikhar [1] characterizes all the J-symmetric extensions S of T with $\Pi(S) \neq \varnothing$. Zhikhar's work is extended in Knowles [2] where a description is given of all the J-self-adjoint extensions of T, again assuming that $\Pi(T) \neq \varnothing$. This latter assumption has been removed by Race [1]. When $\Pi(T) = \varnothing$ the type of direct-sum composition used here and also in the cited works of Knowles and Zhikhar is not available. Race's method rests on the result that a closed J-symmetric extension S of T is J-self-adjoint if, and only if,

$$2 \dim \mathcal{D}(S)/\mathcal{D}(T) = \dim \mathcal{D}(JT^*J)/\mathcal{D}(T).$$

Also

$$\mathcal{D}(JT^*J) = \mathcal{D}(T) \oplus \mathcal{N}((T^*J)^2 + I)$$

where $\oplus$ denotes the orthogonal sum with respect to the inner product

$$(T^*Jx, T^*Jy) + (Jx, Jy)$$

on $\mathscr{D}(JT^*J)$.

6. Accretive and sectorial operators

Definition 6.1. A linear operator T acting in a Hilbert space H is said to be *accretive* if $\mathrm{re}\,(Tu, u) \geqslant 0$ for all $u \in \mathscr{D}(T)$, and *quasi-accretive* if $T + \alpha I$ is accretive for some $\alpha > 0$. The operator T is said to be *dissipative* (*quasi-dissipative*) if $-T$ is accretive (quasi-accretive). ■

If T is symmetric, (Tu, u) is real for all $u \in \mathscr{D}(T)$ and T is quasi-accretive if, and only if, $\Theta(T)$ is bounded below. In this case we say that T is *bounded below* or *lower semi-bounded*; we write $T \geqslant \gamma$ if $(Tu, u) \geqslant \gamma \|u\|^2$ for all $u \in \mathscr{D}(T)$, and the supremum of all the numbers γ with this property is called the *lower bound* of T. *Boundedness above* and the *upper bound* are defined similarly.

From Theorem 2.6 we obtain the following result.

Proposition 6.2. A densely defined accretive operator is closable and its closure is accretive. ■

In view of this, we shall usually consider closed accretive operators.

If $T \in \mathscr{C}(H)$ is accretive, the half plane $\{\lambda : \mathrm{re}\,\lambda < 0\}$ lies in the exterior domain of regularity $\tilde{\Delta}(T)$ of T and by Theorem 2.3, $T - \lambda I$ is semi-Fredholm with zero nullity and constant deficiency for $\mathrm{re}\,\lambda < 0$. If $\mathrm{def}\,(T - \lambda I) = 0$ for $\mathrm{re}\,\lambda < 0$ then

$$\{\lambda : \mathrm{re}\,\lambda < 0\} \subset \rho(T),\ \|(T - \lambda I)^{-1}\| \leqslant |\mathrm{re}\,\lambda|^{-1} \text{ for } \mathrm{re}\,\lambda < 0. \tag{6.1}$$

Definition 6.3. A linear operator $T \in \mathscr{C}(H)$ is said to be *m-accretive* if it satisfies (6.1). If $T + \alpha I$ is m-accretive for some $\alpha > 0$, then T is said to be *quasi-m-accretive*. Similarly for *m-dissipative* and *quasi-m-dissipative* operators. ■

Theorem 6.4. An m-accretive operator is densely defined, accretive, and has no proper accretive extensions. ■

Proof. Let T be m-accretive and $\lambda = -\alpha < 0$. Then, from (6.1), for all $u \in \mathscr{D}(T)$,

$$\begin{aligned} \|u\|^2 &\leqslant \alpha^{-2} \|(T + \alpha I)u\|^2 \\ &= \alpha^{-2}[\|Tu\|^2 + 2\alpha\,\mathrm{re}\,(Tu, u) + \alpha^2 \|u\|^2] \end{aligned}$$

and

$$0 \leqslant \alpha^{-1} \|Tu\|^2 + 2\,\mathrm{re}\,(Tu, u).$$

On allowing $\alpha \to \infty$, we get re $(Tu, u) \geqslant 0$ and so T is accretive. To prove that $\mathscr{D}(T)$ is dense in H, assume to the contrary that there exists a non-zero $v \perp \mathscr{D}(T)$. Then, for re $\lambda < 0$,

$$0 = ((T-\lambda I)^{-1}v, v)$$
$$= (w, (T-\lambda I)w)$$

where $w = (T-\lambda I)^{-1}v$. Since T is accretive,

$$0 = \text{re}\,([T-\lambda I]w, w)$$
$$\geqslant |\text{re}\,\lambda|\,\|w\|^2,$$

whence $w = 0$ and $v = 0$. This contradiction implies that $\mathscr{D}(T)$ is dense in H.

If S is an accretive extension of T, which we may assume is closed, without loss of generality, then, by Theorem 2.3, $(S-\lambda I)^{-1}$ exists for re $\lambda < 0$ and is an extension of $(T-\lambda I)^{-1}$. But by (6.1), $\lambda \in \rho(T)$ and hence $(T-\lambda I)^{-1}$ is defined on H. It follows that $S = T$ and the proof is complete. □

Theorem 6.5. A densely defined accretive operator T is m-accretive if, and only if, $\mathscr{R}(T-\lambda I) = H$ for some, and hence all, $\lambda \in \{\mu: \text{re}\,\mu < 0\}$. ■

Proof. If T is m-accretive, that $\mathscr{R}(T-\lambda I) = H$ if re $\lambda < 0$ follows from the definition. Conversely, given that $\mathscr{R}(T-\lambda I) = H$ for some λ such that re $\lambda < 0$, we conclude that $\mathscr{R}(\bar{T}-\lambda I) = H = \mathscr{R}(T-\lambda I)$, the existence of the closure $\bar{T}$ being guaranteed by Proposition 6.2. If $u \in \mathscr{D}(\bar{T})$, there exists some $v \in \mathscr{D}(T)$ such that $(\bar{T}-\lambda I)u = (T-\lambda I)v$ and hence $u - v \in \mathscr{N}(\bar{T}-\lambda I)$. But nul $(\bar{T}-\lambda I) = 0$ for re $\lambda < 0$ since $\bar{T}$ is accretive, and therefore we have proved that $T = \bar{T}$, i.e. T is closed. The theorem follows from Theorem 2.3. □

Theorem 6.6. If T is m-accretive, so is its adjoint T^*. ■

Proof. Note that the existence of T^* is ensured by Theorem 6.4. Under the hypothesis, $\lambda \in \rho(T)$ whenever re $\lambda < 0$ and so $\bar{\lambda} \in \rho(T^*)$ for all λ in the left half-plane, by Theorem 1.1. In particular, $\mathscr{R}(T^*-\lambda I) = H$ whenever re $\lambda < 0$ and so, in view of Theorem 6.5, it suffices to show that T^* is accretive; note that $\mathscr{D}(T^*)$ is dense since T is closable. Let $u \in \mathscr{D}(T^*)$ and $\varepsilon > 0$. Since $\mathscr{R}(T+\varepsilon I) = H$ there exists $v \in \mathscr{D}(T)$ such that $(T^*+\varepsilon I)u = (T+\varepsilon I)v$ and hence

$$((T^*+\varepsilon I)u, u) = ((T+\varepsilon I)v, u) = (v, (T^*+\varepsilon I)u)$$
$$= (v, (T+\varepsilon I)v).$$

Consequently

$$\text{re}\,(T^*u, u) + \varepsilon\|u\|^2 > 0$$

and the accretiveness of T^* follows on allowing $\varepsilon \to 0+$. The theorem is proved. □

Theorem 6.7. (i) A closed symmetric operator T is m-accretive if, and only if, T is self-adjoint and $T \geqslant 0$.

(ii) A closed J-symmetric operator T is m-accretive if, and only if, T is J-self-adjoint and accretive. ■

Proof. (i) If T is self-adjoint and $T \geqslant 0$, the set $\tilde{\Delta}(T)$ contains $\{\lambda: \operatorname{re}\lambda < 0\}$. Hence $\operatorname{def}(T-\lambda I) = 0$ when $\operatorname{re}\lambda < 0$ and T is m-accretive on account of Theorem 2.3. Conversely, if T is symmetric and m-accretive, $T \geqslant 0$ and $\operatorname{def}(T-\lambda I) = 0$ whenever $\operatorname{re}\lambda < 0$. From Theorem 4.2(ii) the self-adjointness of T follows.

(ii) If T is J-self-adjoint and accretive, $\{\lambda: \operatorname{re}\lambda < 0\} \subset \Pi(T)$ and hence, by Theorem 5.5, $\operatorname{def}(T-\lambda I) = 0$ whenever $\operatorname{re}\lambda < 0$. By Theorem 2.3, T is therefore m-accretive. If T is J-symmetric and m-accretive, it is accretive and $\operatorname{def}(T-\lambda I) = 0$ when $\operatorname{re}\lambda < 0$. The rest follows from Theorem 5.5. □

Theorem 6.8. If $T = \mathrm{i}S^*$, where S is a maximal symmetric operator, then T is m-accretive. ■

Proof. Since S is maximal symmetric it is closed and one of its deficiency indices, $\mathrm{m}_+(S)$ say, is zero. From (4.6),

$$\mathscr{D}(T) = \mathscr{D}(S^*) = \mathscr{D}(S) \dot{+} \mathscr{N}_-(S)$$

so that for $u \in \mathscr{D}(T)$, we have

$$u = v + w, \qquad v \in \mathscr{D}(S), \qquad w \in \mathscr{N}_-(S),$$

$$Tu = \mathrm{i}Sv + \mathrm{i}S^*w = \mathrm{i}Sv + w.$$

Hence

$$\begin{aligned}(Tu, u) &= (\mathrm{i}Sv + w, v + w) = \mathrm{i}(Sv, v) + \mathrm{i}(Sv, w) + (w, v) + \|w\|^2 \\ &= \mathrm{i}(Sv, v) + \mathrm{i}(v, S^*w) + (w, v) + \|w\|^2 \\ &= \mathrm{i}(Sv, v) + 2\mathrm{i}\,\operatorname{im}(w, v) + \|w\|^2\end{aligned}$$

and

$$\operatorname{re}(Tu, u) = \|w\|^2 \geqslant 0.$$

T is therefore accretive. Also, for $\operatorname{re}\lambda < 0$, we have $\operatorname{def}(T-\lambda I) = \operatorname{def}[\mathrm{i}(S^* + \mathrm{i}\lambda I)] = \operatorname{nul}(S - \mathrm{i}\bar{\lambda}I) = 0$ since $\operatorname{im}(\mathrm{i}\bar{\lambda}) = \operatorname{re}\lambda < 0$, and hence $\mathrm{i}\bar{\lambda} \in \Pi(S)$. The theorem follows from Theorem 2.3.

Definition 6.9. A linear operator T in a Hilbert space H is said to be *sectorial* if its numerical range lies in a sector $\{z \in \mathbb{C}: \operatorname{re} z \geqslant \gamma, |\arg(z-\gamma)| \leqslant \theta < \frac{1}{2}\pi\}$ for some $\gamma \in \mathbb{R}$ and $\theta \in [0, \frac{1}{2}\pi)$; γ is called a *vertex* of T and θ a *semi-angle*. We say that T is *m-sectorial* if it is sectorial and quasi-m-accretive. If for some $\beta \in (-\pi, \pi)$ and $\delta \in \mathbb{C}$, $e^{i\beta}(T + \delta I)$ is m-sectorial, we say that T is *quasi-m-sectorial.* ■

We shall return to sectorial operators in Chapter IV when we shall be examining sesquilinear forms. An important property of m-accretive operators which is neither used nor proved in this book, is that such an operator possesses a square root $T^{\frac{1}{2}}$ which is m-sectorial and whose numerical range lies in the sector $\{z \in \mathbb{C} : \operatorname{re} z \geqslant 0, |\arg z| \leqslant \frac{1}{4}\pi\}$. A discussion of this and related results may be found in Kato [1, V-§3.11].

7. Relative boundedness and relative compactness

Let T and P be linear operators with domains in a Banach space X and ranges in a Banach space Y.

Definition 7.1. P is said to be *T-bounded* if $\mathscr{D}(T) \subset \mathscr{D}(P)$ and there exist non-negative constants a and b, such that

$$\|Pu\|_Y \leqslant a\|u\|_X + b\|Tu\|_Y \quad \text{for all } u \in \mathscr{D}(T). \tag{7.1}$$

The infimum of the constants b which satisfy (7.1) for some $a \geqslant 0$ is called the *T-bound* of P. ■

If $\mathscr{D}(T)$ is equipped with the graph norm $\|\bullet\|_T$, the operator P is T-bounded if and only if the restriction of P to $\mathscr{D}(T)$ is bounded as a map from the normed vector space $\mathrm{X}(T)$ determined by $\mathscr{D}(T)$ and $\|\bullet\|_T$ into Y.

Proposition 7.2. Let $T \in \mathscr{C}(X, Y)$ and let P be closable with $\mathscr{D}(T) \subset \mathscr{D}(P)$. Then P is T-bounded and the inclusion map of $\mathrm{X}(T)$ into $\mathrm{X}(\bar{P})$ is continuous. If $\mathscr{D}(T) = \mathscr{D}(P)$ and P is closed then $\mathrm{X}(T)$ and $\mathrm{X}(P)$ are topologically isomorphic. ■

Proof. Let P_T be the restriction of P to $\mathscr{D}(T)$. Then P is T-bounded if and only if $P_T \in \mathscr{B}(\mathrm{X}(T), Y)$, and in view of the Closed-Graph Theorem it therefore suffices to prove that P_T is closed as a map from $\mathrm{X}(T)$ into Y. Let $u_n \in \mathrm{X}(T)$ be such that $u_n \underset{\mathrm{X}(T)}{\to} u$ and $P_T u_n \underset{Y}{\to} v$. Then $u \in \mathrm{X}(T)$, and since P is closable, $P_T u_n = Pu_n \to Pu = P_T u$. The operator P_T is therefore closed and P is T-bounded. The rest of the proposition is immediate. □

Definition 7.3. P is said to be *T-compact* if $\mathscr{D}(T) \subset \mathscr{D}(P)$ and for any sequence (u_n) in $\mathscr{D}(T)$ which is such that $(\|u_n\|_T)$ is bounded, (Pu_n) contains a subsequence which is convergent in Y. ■

A T-compact operator P is necessarily T-bounded, because otherwise there would exist a sequence (u_n) in $\mathscr{D}(T)$ which is such that $\|u_n\|_T = 1$ and

$\|Pu_n\|_Y \to \infty$, in contradiction to the T-compactness of P. In fact, we shall see in Corollary 7.7 below that much more can be said in certain circumstances.

If T is closable, $\mathscr{D}(T)$ is dense in the Banach space $\mathrm{X}(\bar{T})$ and any T-bounded operator P can be extended uniquely by continuity to an operator $\tilde{P} \in \mathscr{B}(\mathrm{X}(\bar{T}), Y)$. In this case P is T-compact if, and only if, $\tilde{P} \in \mathscr{K}(\mathrm{X}(\bar{T}), Y)$. If P is closable then $\tilde{P}$ is the restriction of $\bar{P}$ to $\mathscr{D}(\bar{T})$.

The notion of *weak convergence* is now needed. A sequence (u_n) in a Banach space X is said to *converge weakly* to a limit u in X if $((f, u_n))$ converges to (f, u) for every $f \in X^*$. This is written $u_n \underset{X}{\rightharpoonup} u$, and $u_n \rightharpoonup u$ if there is no ambiguity about X. We need to know that every weakly convergent sequence is bounded and also that in a reflexive space X every bounded sequence has a weakly convergent subsequence. The proof of these results may be found in Taylor [1, Chapter 4].

Proposition 7.4. Let T be closable and let P be T-bounded. Then

(i) P is T-compact if, and only if, $\tilde{P} \in \mathscr{K}(\mathrm{X}(\bar{T}), Y)$;

(ii) if P is closable, P is T-compact if, and only if, $\bar{P}$ is $\bar{T}$-compact;

(iii) if P is T-compact and $(u_n) \subset \mathscr{D}(\bar{T})$ is such that $u_n \rightharpoonup 0$ in $\mathrm{X}(\bar{T})$ then $\tilde{P}u_n \underset{Y}{\to} 0$;

(iv) if X and Y are reflexive and $\tilde{P}u_n \underset{Y}{\to} 0$ whenever $(u_n) \subset \mathscr{D}(\bar{T})$ is such that $u_n \rightharpoonup 0$ in $\mathrm{X}(\bar{T})$ then P is T-compact. ■

Proof. Parts (i) and (ii) follow from the remarks preceding the proposition.

(iii) Any weakly convergent sequence (u_n) in $\mathrm{X}(\bar{T})$ is bounded in $\mathrm{X}(\bar{T})$, and if P is T-compact we see from (i) that $(\tilde{P}u_n)$ contains a subsequence $(\tilde{P}u_{n(k)})$ which converges in Y, say $\tilde{P}u_{n(k)} \underset{Y}{\to} v$. But $\tilde{P} \in \mathscr{B}(\mathrm{X}(\bar{T}), Y)$ and $u_n \underset{\mathrm{X}(T)}{\rightharpoonup} 0$ imply that $\tilde{P}u_n \underset{Y}{\rightharpoonup} 0$ and hence $\tilde{P}u_{n(k)} \underset{Y}{\to} 0$. Moreover, any convergent subsequence of $(\tilde{P}u_n)$ converges to 0 in Y and so $\tilde{P}u_n \underset{Y}{\to} 0$.

(iv) Since X and Y are reflexive, so are $X \times Y$ and its closed subspace $\mathscr{G}(\bar{T})$, the graph of $\bar{T}$. The correspondence $x \leftrightarrow \{x, \bar{T}x\}$ defines a natural congruence (i.e. an isometric isomorphism) between $\mathrm{X}(\bar{T})$ and $\mathscr{G}(\bar{T})$ and from this it follows that $\mathrm{X}(\bar{T})$ is also reflexive. Any bounded sequence (u_n) in $\mathrm{X}(\bar{T})$ therefore contains a subsequence $(u_{n(k)})$ which converges weakly in $\mathrm{X}(\bar{T})$, $u_{n(k)} \underset{\mathrm{X}(\bar{T})}{\rightharpoonup} v$ say. From the hypothesis $\tilde{P}u_{n(k)} = \tilde{P}(u_{n(k)} - v) + \tilde{P}v \underset{Y}{\to} \tilde{P}v$, whence $\tilde{P} \in \mathscr{K}(\mathrm{X}(\bar{T}), Y)$. By (i), P is therefore T-compact. □

Corollary 7.5. Let T be a closable operator and P a T-bounded operator from a Hilbert space X to a Hilbert space Y. Then P is T-compact if, and only if, $\tilde{P}u_n \underset{Y}{\to} 0$ whenever $(u_n) \subset \mathscr{D}(\bar{T})$ is such that $u_n \underset{X}{\rightharpoonup} 0$ and $\bar{T}u_n \underset{Y}{\rightharpoonup} 0$. ■

Proof. Let P be T-compact and let $u_n \underset{X}{\rightharpoonup} 0$ and $\bar{T}u_n \underset{Y}{\rightharpoonup} 0$. Then, for any $\phi \in \mathscr{D}(\bar{T})$,

$$(u_n, \phi)_{\bar{T}} = (u_n, \phi)_X + (\bar{T}u_n, \bar{T}\phi)_Y \to 0$$

and hence $u_n \rightharpoonup 0$ in $X(\bar{T})$. From Proposition 7.4 (iii) it follows that $\tilde{P}u_n \underset{Y}{\to} 0$. To prove the reverse implication, it is sufficient to prove that if $u_n \rightharpoonup 0$ in $X(\bar{T})$ then $u_n \underset{X}{\rightharpoonup} 0$ and $\bar{T}u_n \underset{Y}{\rightharpoonup} 0$, since we may then apply Proposition 7.4 (iv). We therefore suppose that $u_n \rightharpoonup 0$ in $X(\bar{T})$, i.e.

$$(\{u_n, \bar{T}u_n\}, \{\phi, \psi\})_{X \times Y} = (u_n, \phi)_X + (\bar{T}u_n, \psi)_Y \to 0$$

for all $\{\phi, \psi\} \in \mathscr{G}(\bar{T})$. But $(u_n, \phi)_X + (\bar{T}u_n, \psi)_Y = 0$ for all $\{\phi, \psi\} \perp \mathscr{G}(\bar{T})$ and hence

$$(u_n, \phi)_X + (\bar{T}u_n, \psi)_Y \to 0 \quad \text{for all } \{\phi, \psi\} \in X \times Y.$$

It follows that $u_n \underset{X}{\rightharpoonup} 0$ and $\bar{T}u_n \underset{Y}{\rightharpoonup} 0$. □

Theorem 7.6. Let X and Y be Banach spaces, let S be a T-bounded operator and suppose there exists a T-compact operator P which is such that $S - P$ has T-bound $k \in [0, \alpha)$ for some given $\alpha > 0$. Then

(i) if P is closable, S has T-bound $< \alpha$;

(ii) if X and Y are reflexive and T is closable, S has T-bound $< \alpha$. ■

Proof. We are required to prove that there exists a constant $b \in [0, \alpha)$ such that $\|Su\|_Y \leqslant a\|u\|_X + b\|Tu\|_Y$ for some $a \geqslant 0$ and all $u \in \mathscr{D}(T)$. Suppose not; then there is a sequence (u_n) in $\mathscr{D}(T)$ such that

$$\|Tu_n\|_Y = 1, \qquad \|Su_n\|_Y > n\|u_n\|_X + \alpha - 1/n. \tag{7.2}$$

From (7.2), $u_n \underset{X}{\to} 0$ and $(\|u_n\|_T)$ is bounded.

(i) Since P is T-compact, (Pu_n) contains a convergent subsequence $(Pu_{n(m)})$. But $u_n \underset{X}{\to} 0$ and P closable imply that $Pu_{n(m)} \underset{Y}{\to} 0$. From the hypothesis,

$$\|(S - P)u\|_Y \leqslant c\|u\|_X + d\|Tu\|_Y \quad \text{for all } u \in \mathscr{D}(T), \tag{7.3}$$

where $c \geqslant 0$ and $d \in (k, \alpha)$. Hence

$$\|Su_{n(m)}\|_Y \leqslant c\|u_{n(m)}\|_X + d\|Tu_{n(m)}\|_Y + \|Pu_{n(m)}\|_Y$$

and

$$\limsup_{m \to \infty} \|Su_{n(m)}\|_Y \leqslant d < \alpha.$$

This contradicts (7.2) and we infer that S has T-bound $< \alpha$.

(ii) As we observed in the proof of Proposition 7.4 (iv), $\mathrm{X}(\bar{T})$ is reflexive if X and Y are reflexive. Therefore the sequence (u_n) in (7.2), being bounded in $\mathrm{X}(\bar{T})$, must contain a weakly convergent subsequence in $\mathrm{X}(\bar{T})$, say $(u_{n(k)})$, with $u_{n(k)} \rightharpoonup v$ in $\mathrm{X}(\bar{T})$. Since $\mathrm{X}(\bar{T})$ is continuously embedded in X, it follows that $u_{n(k)} \underset{X}{\rightharpoonup} v$ and consequently $v = 0$, since $u_n \underset{X}{\to} 0$. But from $u_{n(k)} \rightharpoonup 0$ in $\mathrm{X}(\bar{T})$ and the assumption that P is T-compact we conclude that $Pu_{n(k)} = \tilde{P}u_{n(k)} \underset{Y}{\to} 0$ on account of Proposition 7.4 (iii). The rest of the proof follows that of (i) without change. □

Corollary 7.7. Let P be a T-compact operator from X to Y, X and Y being reflexive Banach spaces. If P or T is closable, then P has T-bound zero. ■

Corollary 7.8. Let X and Y be reflexive and P closable. Then P is T-compact if, and only if, P is $(T+P)$-compact. ■

Proof. Let P be T-compact. Then, by Corollary 7.7, P has T-bound zero and so there exists $a \geqslant 0$ such that $\|Pu\|_Y \leqslant a\|u\|_X + \frac{1}{2}\|Tu\|_Y$ for all $u \in \mathscr{D}(T)$. Also since $\mathscr{D}(P) \supset \mathscr{D}(T)$ we have $\mathscr{D}(T+P) = \mathscr{D}(T)$ and so for all $u \in \mathscr{D}(T)$,

$$\|Tu\|_Y \leqslant 2(\|Tu\|_Y - \|Pu\|_Y + a\|u\|_X) \leqslant 2[\|(T+P)u\|_Y + a\|u\|_X].$$

Thus T is $(T+P)$-bounded and the $(T+P)$-compactness of P follows. Similarly $T+P$ is T-bounded if P is $(T+P)$-compact and the reverse implication follows. □

If in Theorem 7.6(ii), $\|\tilde{S} - \tilde{P}\| < \alpha$, then the ball measure of non-compactness $\beta(\tilde{S})$ of $\tilde{S}$ is $< \alpha$. It is therefore natural to inquire if there is a connection between the T-bound of S and $\beta(\tilde{S})$. This is not clear in general but the next theorem establishes a connection in special circumstances. In order to understand it, one must first be familiar with the terms and ideas introduced in §V.5 below.

Theorem 7.9. Let T be a closable linear operator with domain and range in the space $\mathrm{L}^s(\Omega)$, where $1 \leqslant s < \infty$ and Ω is a domain in $\mathbb{R}^n$, with $n \geqslant 1$, and let A be a closed non-empty subset of $\bar{\Omega}$ such that $A \cap \Omega$ has zero Lebesgue measure. Let S be T-bounded and suppose there exists an $\tilde{S}$-compact A-admissible family of domains $\mathscr{F}_A$. Then if $\beta(\tilde{S}) < \alpha$ there exist real numbers $a \geqslant 0$ and $b \in [0, \alpha)$ and an open subset V of $\Omega \setminus A$ with $\bar{V} \subset \bar{\Omega} \setminus A$ such that

$$\|Su\|_{s,\Omega} \leqslant a\|u\|_{s,V} + b\|u\|_T \quad \text{for all } u \in \mathscr{D}(T). \tag{7.4}$$

Therefore S has T-bound $< \alpha$ and in addition, if $(\operatorname{supp} u) \cap V = \varnothing$,

$$\|Su\|_{s,\Omega} \leqslant b\|u\|_T. \tag{7.5}$$

Proof. Suppose that (7.4) does not hold. There is therefore a sequence (u_n) in $\mathscr{D}(T)$ such that

$$\| u_n \|_T = 1, \quad \| Su_n \|_{s,\Omega} > n \| u_n \|_{s,\Omega\setminus A(1/n)} + \alpha - 1/n, \tag{7.6}$$

where $A(1/n) = \{x \in \Omega\colon \text{dist}\,(x, A) < 1/n\}$—see §V.5. Since S is T-bounded (7.6) implies that for any fixed $m \in \mathbb{N}$,

$$\| u_n \|_{s,\Omega\setminus A(1/m)} \to 0 \tag{7.7}$$

as $n \to \infty$. Also $\| u_n \|_{s,\Omega} \leqslant \| u_n \|_T = 1$ and for $\phi \in L^{s'}(\Omega)$, with $1/s' + 1/s = 1$,

$$\begin{aligned} \left| \int_\Omega u_n \bar{\phi} \right| &= \left| \left(\int_{\Omega\setminus A(1/m)} + \int_{A(1/m)} \right) u_n \bar{\phi} \right| \\ &\leqslant \| u_n \|_{s,\Omega\setminus A(1/m)} \| \phi \|_{s',\Omega} + \| \phi \|_{s',A(1/m)} \\ &\to 0 \end{aligned}$$

on allowing n and then m to tend to infinity. This means that $u_n \rightharpoonup 0$ in $L^s(\Omega)$ and in fact $u_n \rightharpoonup 0$ in $X(\bar{T})$. For if I is the natural embedding $X(\bar{T}) \to L^s(\Omega)$ its adjoint I^* has dense range in $X(\bar{T})^*$ and for $\phi \in L^{s'}(\Omega)$,

$$\begin{aligned} (I^* \phi, u_n) &= \int_\Omega \phi \bar{u}_n \\ &\to 0. \end{aligned}$$

Thus $u_n \rightharpoonup 0$ in $X(\bar{T})$ since $\| u_n \|_T = 1$ and $\mathscr{R}(I^*)$ is dense. This in turn implies that $\| Su_n \|_{s,U} \to 0$ for all U in the A-admissible family $\mathscr{F}_A$. We now make use of the result $\tilde{\beta}(\tilde{S}) = \Gamma_{\tilde{S}}^{1/s}(0) = \Gamma_S^{1/s}(0)$ established in Theorem V.5.7 under hypotheses which cover those here. Since $\tilde{\beta}(\tilde{S}) < \alpha$ it is possible to choose $b \in [0, \alpha)$ and $\varepsilon > 0$ such that $\| Su_n \|_{s,A(\varepsilon)} < b$. Thus for $U \in \mathscr{F}_A$ satisfying $\Omega \setminus A(\varepsilon) \subset U$,

$$\begin{aligned} \| Su_n \|_{s,\Omega} &\leqslant \| Su_n \|_{s,A(\varepsilon)} + \| Su_n \|_{s,\Omega\setminus A(\varepsilon)} \\ &< b + \| Su_n \|_{s,U} \end{aligned}$$

and

$$\limsup_{n\to\infty} \| Su_n \|_{s,\Omega} \leqslant b < \alpha$$

in contradiction to (7.6). The theorem is therefore proved. □

Corollary 7.10. If S is T-compact in Theorem 7.9, then S has T-bound zero and given any $\varepsilon > 0$ there exists an open subset V of $\Omega \setminus A$ with $\bar{V} \subset \bar{\Omega} \setminus A$ such that

$$\| Su \|_{s,\Omega} \leqslant \varepsilon \| u \|_T \text{ for all } u \in \mathscr{D}(T) \text{ with } (\operatorname{supp} u) \cap V = \varnothing. \tag{7.8}$$ ■

Definition 7.11. An operator S which satisfies (7.8) in the sense defined in Corollary 7.10 is said to be *T-small at A*. ■

This concept was introduced by Jörgens and Weidmann [1].

8. Stability results

In this section our prime concern will be in determining properties of an operator T which remain unchanged when T is subjected to a *T-bounded perturbation P*, i.e. properties of T which are inherited by $S = T+P$. X and Y are Banach spaces and the operators T, P, S considered have domains in X and ranges in Y.

Lemma 8.1. If T and P are closable and $\mathscr{D}(T) \subset \mathscr{D}(P)$ then P is T-bounded with T-bound α if, and only if, $\bar{P}$ is $\bar{T}$-bounded with $\bar{T}$-bound α. ■

Proof. If T has T-bound α, then for any $b > \alpha$,

$$\| Pu \|_Y \leqslant a \|u\|_X + b \| Tu \|_Y \tag{8.1}$$

for some $a \geqslant 0$ and all $u \in \mathscr{D}(T) \subset \mathscr{D}(P)$. From (8.1) it readily follows that $\mathscr{D}(\bar{T}) \subset \mathscr{D}(\bar{P})$ and $\|\bar{P}u\|_Y \leqslant a \ \|u\|_X + b\|\bar{T}u\|_Y$ for $u \in \mathscr{D}(\bar{T})$. Thus $\bar{P}$ is $\bar{T}$-bounded and, since $P \subset \bar{P}$, it follows that $\bar{P}$ has $\bar{T}$-bound α. Conversely if $\bar{P}$ has $\bar{T}$-bound α then P has T-bound $\leqslant \alpha$. The first part of the proof now implies that P has T-bound α. □

Theorem 8.2. Let P be T-bounded with T-bound < 1. Then $S = T+P$ is closable if, and only if, T is closable and $\mathscr{D}(\bar{S}) = \mathscr{D}(\bar{T})$; in this case $\bar{S} = \bar{T} + \bar{P}$ if P is closable. In particular S is closed if, and only if, T is closed. ■

Proof. Since P has T-bound < 1, (8.1) is satisfied for some $b < 1$ and so for all $u \in \mathscr{D}(T)$,

$$-a \| u \|_X + (1-b) \| Tu \|_Y \leqslant \| Su \|_Y \leqslant a \| u \|_X + (1+b) \| Tu \|_Y. \tag{8.2}$$

For $c > a$, the functional $u \mapsto \| Su \|_Y + c \| u \|_X$ therefore defines a norm on $\mathscr{D}(T)$ which is equivalent to the graph norm $\| \bullet \|_T$ of T. The norms $\|\bullet\|_S$ and $\|\bullet\|_T$ are thus equivalent on $\mathscr{D}(T)$ and since $\mathscr{D}(S) = \mathscr{D}(T)$, the operators T and S are closable together and $\mathscr{D}(\bar{T}) = \mathscr{D}(\bar{S})$. Whenever $u \in \mathscr{D}(\bar{T})$ and $u_n \in \mathscr{D}(T)$ $(n \in \mathbb{N})$ are such that $u_n \to u$ in $X(\bar{T})$, we see from (8.1) that (Pu_n) converges in Y and if P is closable, $\mathscr{D}(\bar{T}) \subset \mathscr{D}(\bar{P})$, $Pu_n \underset{Y}{\to} \bar{P}u$ and $(T+P)u_n \underset{Y}{\to} (\bar{T}+\bar{P})u$. Consequently $\bar{S} \subset \bar{T}+\bar{P}$, and since we have already proved that $\mathscr{D}(\bar{S}) = \mathscr{D}(\bar{T}) = \mathscr{D}(\bar{T}+\bar{P})$ we conclude that $\bar{S} = \bar{T}+\bar{P}$. □

Proposition 8.3. Let P be T-compact and suppose that either P is closable or T is closable and X and Y are reflexive. Then P is S-compact ($S = T+P$). ■

Proof. The hypothesis and Corollary 7.7 imply that P has T-bound zero. Thus (8.2) holds for arbitrary $b > 0$ and any sequence (u_n) in $\mathscr{D}(S) = \mathscr{D}(T)$ is either bounded with respect to both norms $\|\bullet\|_S$ and $\|\bullet\|_T$ simultaneously or unbounded with respect to both. Thus as P is T-compact it is also S-compact. □

Theorem 8.4. Let $X = Y = H$, a Hilbert space, and let $T \in \mathscr{C}(H)$. Suppose also that for some $\alpha \in [0, 2\pi)$ and $\gamma \in \mathbb{C}$ the set

$$\Delta_{\alpha,\gamma} := \{z \in \mathbb{C}: \alpha \leqslant \arg(z-\gamma) < \pi + \alpha\} \tag{8.3}$$

lies in $\tilde{\Delta}(T) \cap \rho(T)$, where $\tilde{\Delta}(T) = \mathbb{C} \setminus \bar{\Theta}(T)$. Then if P is T-bounded with T-bound < 1, the map $S = T+P$ is closed and there exists an $R > 0$ such that $\lambda = re^{i(\alpha+\pi/2)} \in \rho(S)$ for $r \geqslant R$. ■

Proof. We first observe that we can assume, without loss of generality, that $\gamma = 0$ in (8.3). This is because $\Delta_{\alpha,\gamma} \subset \tilde{\Delta}(T) \cap \rho(T)$ is equivalent to $\Delta_{\alpha,0} \subset \tilde{\Delta}(T-\gamma I) \cap \rho(T-\gamma I)$, $T-\gamma I$ is closed (by Theorem 8.2), and $P+\gamma I$ has $(T-\gamma I)$-bound < 1. Also, the fact that $S = T+P$ is closed follows from Theorem 8.2.

Since $\Theta(T) \subset \mathbb{C} \setminus \Delta_{\alpha,0}$ we see that $\arg(Tu, u) \in [\pi+\alpha, 2\pi+\alpha]$ for any $u \in \mathscr{D}(T)$. If $\lambda = re^{i(\alpha+\pi/2)}$ it therefore follows that $\arg[\bar{\lambda}(Tu, u)] \in [\frac{1}{2}\pi, \frac{3}{2}\pi]$ for all $u \in \mathscr{D}(T)$, and so $\mathrm{re}[\bar{\lambda}(Tu, u)] \leqslant 0$. Hence

$$\begin{aligned}\|Tu\|^2 + |\lambda|^2 \|u\|^2 &= \|(T-\lambda I)u\|^2 + 2\,\mathrm{re}[\bar{\lambda}(Tu,u)] \\ &\leqslant \|(T-\lambda I)u\|^2.\end{aligned} \tag{8.4}$$

If P has T-bound < 1, then

$$\|Pu\| \leqslant a\|u\| + b\|Tu\| \qquad (u \in \mathscr{D}(T)),$$

for some $a \geqslant 0$ and $b \in (0, 1)$. This gives, for any $\varepsilon > 0$,

$$\begin{aligned}\|Pu\|^2 &\leqslant a^2\|u\|^2 + b^2\|Tu\|^2 + 2ab\|u\|\,\|Tu\| \\ &\leqslant a^2(1+1/\varepsilon)\|u\|^2 + b^2(1+\varepsilon)\|Tu\|^2 \\ &= a'^2\|u\|^2 + b'^2\|Tu\|^2\end{aligned} \tag{8.5}$$

where $0 < b' < 1$, on choosing ε sufficiently small. On choosing $r = |\lambda| \geqslant a'/b'$, (8.4) and (8.5) yield

$$\begin{aligned}\|Pu\|^2 &\leqslant b'^2(\|Tu\|^2 + |\lambda|^2\|u\|^2) \\ &\leqslant b'^2\|(T-\lambda I)u\|^2.\end{aligned} \tag{8.6}$$

Since $\lambda = re^{i(\alpha+\pi/2)} \in \rho(T)$ by hypothesis, (8.6) gives $\| P(T-\lambda I)^{-1} \| \leqslant b' < 1$. From the identity

$$S - \lambda I = [I + P(T-\lambda I)^{-1}](T - \lambda I)$$

it follows that $\lambda \in \rho(S)$. □

The above theorem is mainly of interest for its consequences.

Corollary 8.5. Let P be T-bounded with T-bound < 1, $T \in \mathscr{C}(H)$ and $S = T + P$. Then:

(i) if T is maximal symmetric and P is symmetric, S is maximal symmetric or self-adjoint;

(ii) if T is self-adjoint and P is symmetric, S is self-adjoint;

(iii) if T is J-self-adjoint for some conjugation J, $\Delta_{\alpha,\gamma} \subset \tilde{\Delta}(T)$ and P is J-symmetric, then S is J-self-adjoint and $\rho(S) \neq \varnothing$;

(iv) if T is m-accretive and P is accretive, S is m-accretive. ■

Proof. (i) If T is maximal symmetric, $\operatorname{def}(T - \lambda I) = 0$ for λ in either the upper or lower half-plane, by Corollary 4.10 (iii); take the upper half-plane $\mathbb{C}_+$. Since T is symmetric, $\Theta(T)$ lies in $\mathbb{R}$ and hence $\mathbb{C}_+ \subset \rho(T) \cap \tilde{\Delta}(T)$. The hypothesis of Theorem 8.4 is therefore satisfied with $\Delta_{0,0} = \mathbb{C}_+$ and so we conclude that $\lambda = ik \in \rho(S)$ for k positive and large enough. Since S is a closed symmetric operator, $\Theta(S) \subset \mathbb{R}$ and $\operatorname{def}(S - \lambda I) = 0$ for all $\lambda \in \mathbb{C}_+$, on account of Theorem 2.3. S is therefore maximal symmetric or self-adjoint by Corollary 4.10.

(ii) The proof of this is similar to that of (i).

(iii) For a J-self-adjoint operator T we have $\operatorname{def}(T - \lambda I) = \operatorname{nul}(T - \lambda I) = 0$ for all $\lambda \in \tilde{\Delta}(T)$; whence $\tilde{\Delta}(T) \subset \rho(T)$. The hypothesis of Theorem 8.4 is therefore satisfied. S is thus a closed J-symmetric operator with non-empty resolvent set $\rho(S)$. But $\rho(S) \subset \Pi(S)$ and $\operatorname{def}(S - \lambda I) = 0$ for $\lambda \in \rho(S)$. The result follows from Theorem 5.5.

(iv) The choice $\Delta_{\pi/2,0}$ is now valid in Theorem 8.4. S is obviously closed and accretive and by Theorem 8.4, $\operatorname{def}(S - \lambda I) = 0$ for all large negative λ. The whole left-half plane in $\mathbb{C}$ therefore lies in $\rho(S)$, on account of Theorem 2.3, and S is m-accretive. □

Corollary 8.6. Let T be a closable operator in the Hilbert space H and let P be closable and T-bounded with T-bound < 1. Then:

(i) if $\bar{T}$ is maximal symmetric and P is symmetric then $\bar{T} + \bar{P}$ is maximal symmetric or self-adjoint;

(ii) if T is essentially self-adjoint and P is symmetric then $T + P$ is essentially self-adjoint;

(iii) if $\bar{T}$ is J-self-adjoint, $\Delta_{\alpha,\gamma} \subset \tilde{\Delta}(T)$, and P is J-symmetric, then $\bar{T} + \bar{P}$ is J-self-adjoint;

(iv) if $\bar{T}$ is m-accretive and P is accretive then $\bar{T} + \bar{P}$ is m-accretive. ■

Proof. $\overline{(T+P)} = \bar{T} + \bar{P}$ by Lemma 8.1 and Theorem 8.2. The corollary is then an immediate consequence of Corollary 8.5. □

The requirement in parts (i), (ii), and (iv) of Corollary 8.5 that P has T-bound < 1 can be weakened using an idea due to Wüst [2].

Theorem 8.7. Corollary 8.5(i), (ii) and (iv) continue to hold if we drop the assumption that P has T-bound < 1 and assume instead that $T + \mu P$ is closed for all $\mu \in [0, 1]$. ■

Proof. Note that the present hypothesis is weaker than that in Corollary 8.5 on account of Theorem 8.2. From the relations $T + \mu P \in \mathscr{C}(H)$ and $\mathscr{D}(T + \mu P) = \mathscr{D}(T)$, and from Proposition 7.2, we infer that the Hilbert spaces $\mathrm{H}(T)$ and $\mathrm{H}(T + \mu P)$ are topologically isomorphic and hence that P is $(T + \mu P)$-bounded. Therefore there exist non-negative constants a_μ and b_μ such that

$$\|Pf\| \leqslant a_\mu \|f\| + b_\mu \|(T + \mu P)f\| \quad \text{for all } f \in \mathscr{D}(T).$$

If $|\mu' - \mu| < 1/(2b_\mu)$,

$$\|Pf\| \leqslant a_\mu \|f\| + b_\mu \|(T + \mu' P)f\| + b_\mu |\mu' - \mu| \, \|Pf\|$$

and

$$\|Pf\| \leqslant 2a_\mu \|f\| + 2b_\mu \|(T + \mu' P)f\|. \tag{8.7}$$

There are finitely many numbers in $[0, 1]$, say $\mu_1, \mu_2, \ldots, \mu_m$, such that $[0, 1]$ is covered by the open intervals $\{\mu' : |\mu' - \mu_j| < (2b_{\mu_j})^{-1}\}$ $(j = 1, 2, \ldots, m)$. If $b = \max\{2b_{\mu_j} : j = 1, 2, \ldots, m\}$ and $k \in \mathbb{N}$ is chosen such that $b/k < 1$, then we find from (8.7) that $k^{-1}P$ has $(T + \mu P)$-bound < 1 for all $\mu \in [0, 1]$. Consequently, on applying Corollary 8.5(i), (ii), (iv) with the perturbation $k^{-1}P$ and the operators $T, T + k^{-1}P, T + 2k^{-1}P, \ldots$ in turn, the theorem follows. □

Corollary 8.5 does not hold in general if we merely assume that P has T-bound 1. For if $P = -T$ it has T-bound 1 but S is the proper restriction of the zero operator to $\mathscr{D}(T)$. However, Wüst [1] has proved the following extension of Corollary 8.6(ii).

Theorem 8.8. Let T be essentially self-adjoint and let P be a symmetric operator with $\mathscr{D}(T) \subset \mathscr{D}(P)$ and $\|Pf\| \leqslant a\|f\| + \|Tf\|$ for some $a \geqslant 0$ and all $f \in \mathscr{D}(T)$. Then $S = T + P$ is essentially self-adjoint. ■

Proof. From Theorem 4.2(ii) it suffices to prove that $\mathscr{R}(S - \lambda I)$ is dense in H for $\lambda = \pm \mathrm{i}$. Let $t_n \in (0, 1)$ for $n \in \mathbb{N}$ and $t_n \to 1$. The operators $S_n = T + t_n P$ are essentially self-adjoint, by Corollary 8.6(ii), and also, by hypothesis,

$$\begin{aligned}\|(S-S_n)f\| = (1-t_n)\|Pf\| &\leqslant a\|f\| + \|Tf\| - t_n\|Pf\| \\ &\leqslant a\|f\| + \|(T+t_nP)f\| \\ &= a\|f\| + \|S_nf\|. \qquad (8.8)\end{aligned}$$

Let $h \perp \mathscr{R}(S-\lambda I)$. Since S_n is essentially self-adjoint there exists an $f_n \in \mathscr{D}(S_n) = \mathscr{D}(T)$ such that

$$\|h-(S_n-\lambda I)f_n\| \leqslant 1/n, \qquad n \in \mathbb{N},$$

and consequently

$$h = \lim_{n\to\infty} (S_n - \lambda I)f_n. \qquad (8.9)$$

Furthermore, since S_n is symmetric and $\lambda = \pm \mathrm{i}$ we have $\|(S_n-\lambda I)^{-1}\| \leqslant 1$ and hence $\|f_n\| \leqslant \|(S_n-\lambda I)f_n\|$. It follows from this and (8.9) that

$$\limsup \|f_n\| \leqslant \|h\|, \qquad \limsup \|Sf_n\| \leqslant 2\|h\|. \qquad (8.10)$$

Thus, on using (8.8),

$$\limsup \|(S-S_n)f_n\| \leqslant (a+2)\|h\|.$$

Since $\mathscr{D}(T)$ is dense in H then, for any $\varepsilon > 0$, there exists an $h_\varepsilon \in \mathscr{D}(T)$ such that $\|h-h_\varepsilon\| < \varepsilon$. We therefore have

$$\begin{aligned}\|h\|^2 &= \lim (h, (S_n-\lambda I)f_n) \\ &= \lim (h, (S_n - S)f_n) \qquad (\text{since } h \perp \mathscr{R}(S-\lambda I)) \\ &= \lim [(h-h_\varepsilon, (S_n-S)f_n) + (h_\varepsilon, (S_n-S)f_n)] \\ &\leqslant \|h-h_\varepsilon\| \limsup \|(S_n-S)f_n\| + \limsup [\|(S_n-S)h_\varepsilon\| \, \|f_n\|] \\ &\leqslant (a+2)\varepsilon\|h\| + \limsup (1-t_n)\|Ph_\varepsilon\| \, \|f_n\| \\ &= (a+2)\varepsilon\|h\|.\end{aligned}$$

As this is true for all $\varepsilon > 0$ we conclude that $h = 0$ and the theorem is proved. □

Theorem 8.9. Let T be m-accretive in H and let P be T-bounded with T-bound < 1. Then $S = T + P$ is closed and $\lambda \in \rho(S)$ for all large negative λ. ■

Proof. The fact that S is closed follows from Theorem 8.2. For $\lambda < 0$ and $u \in \mathscr{D}(T)$,

$$\begin{aligned}\|(T-\lambda I)u\|^2 &= \|Tu\|^2 - 2\lambda \operatorname{re}(Tu, u) + |\lambda|^2\|u\|^2 \\ &\geqslant \|Tu\|^2 + |\lambda|^2\|u\|^2, \qquad (8.11)\end{aligned}$$

since T is accretive. As in (8.5), there exist numbers $a \geqslant 0$ and $b \in [0,1)$ such that for all $u \in \mathscr{D}(T)$ and all $\lambda < 0$,

$$\begin{aligned}\|Pu\|^2 &\leqslant a^2\|u\|^2 + b^2\|Tu\|^2 \\ &\leqslant a^2\|u\|^2 + b^2\|(T-\lambda I)u\|^2\end{aligned}$$

from (8.11). Since $\lambda \in \rho(T)$, we deduce that for all $\phi \in H$,

$$\begin{aligned}\|P(T-\lambda I)^{-1}\phi\|^2 &\leqslant a^2\|(T-\lambda I)^{-1}\phi\|^2 + b^2\|\phi\|^2 \\ &\leqslant (a^2/\lambda^2 + b^2)\|\phi\|^2,\end{aligned}$$

on using (2.1). Hence $\|P(T-\lambda I)^{-1}\| < 1$ for λ negative and large enough, and the result follows from the identity $S - \lambda I = [I + P(T-\lambda I)^{-1}](T-\lambda I)$. □

When T is self-adjoint and P symmetric in Theorem 8.9 we can be more precise.

Theorem 8.10. Let T be self-adjoint and bounded below in H, with lower bound γ_T. Let P be symmetric and T-bounded with T-bound < 1 and, for all $u \in \mathscr{D}(T)$, let

$$\|Pu\| \leqslant a\|u\| + b\|Tu\|, \qquad a \geqslant 0, \qquad b \in [0,1). \tag{8.12}$$

Then $T+P$ is self-adjoint and bounded below by

$$\gamma = \gamma_T - \max\{a/(1-b),\, a + b|\gamma_T|\}. \tag{8.13}$$ ■

Proof. $T+P$ is self-adjoint by virtue of Corollary 8.5. The theorem will follow if we can show that any $\lambda < \gamma$ lies in $\rho(T+P)$, since, by Theorem 4.4, the lower bound of $T+P$ coincides with that of its spectrum. If $\lambda < \gamma < \gamma_T$ then $\lambda \in \rho(T)$ and, on using Lemma 4.3 and

$$\|P(T-\lambda I)^{-1}\phi\| \leqslant a\|(T-\lambda I)^{-1}\phi\| + b\|T(T-\lambda I)^{-1}\phi\| \qquad (\phi \in H),$$

we obtain

$$\begin{aligned}\|P(T-\lambda I)^{-1}\| &\leqslant a(\gamma_T-\lambda)^{-1} + b\sup\{|\xi|(\xi-\lambda)^{-1}\colon \xi \geqslant \gamma_T\} \\ &= a(\gamma_T-\lambda)^{-1} + b\max\{1, |\gamma_T|(\gamma_T-\lambda)^{-1}\} \\ &= \max\{a(\gamma_T-\lambda)^{-1} + b,\, (a+b|\gamma_T|)(\gamma_T-\lambda)^{-1}\} \\ &< 1.\end{aligned}$$

From this and the identity $(T+P-\lambda I) = [I + P(T-\lambda I)^{-1}](T-\lambda I)$ we conclude that $\lambda \in \rho(T+P)$. □

Two interesting results ensue from Theorem 8.10; see Weidmann [1, p. 271].

Proposition 8.11. Let T be self-adjoint and non-negative and let S be symmetric with $\mathscr{D}(T) \subset \mathscr{D}(S)$ and $\|Sf\| \leqslant \|Tf\|$ for all $f \in \mathscr{D}(T)$. Then

$$|(Sf, f)| \leqslant (Tf, f) \quad \text{for all } f \in \mathscr{D}(T). \qquad \blacksquare$$

Proof. We apply Theorem 8.10 to T and $P = \xi S$, for $\xi \in (-1, 1)$, so that $a = 0$, $b = |\xi|$, and $\gamma = \gamma_T = 0$. $T + \xi S$ is therefore self-adjoint and non-negative for every $\xi \in (-1, 1)$. On allowing $\xi \to \pm 1$ in $((T + \xi S)f, f) \geqslant 0$ we obtain

$$((T+S)f, f) \geqslant 0, \quad ((T-S)f, f) \geqslant 0 \qquad \text{for all } f \in \mathscr{D}(T),$$

and hence the asserted result follows. □

Proposition 8.12. Let T and S be self-adjoint and non-negative.

(i) If $\mathscr{D}(T) \subset \mathscr{D}(S)$ and $\|Sf\| \leqslant \|Tf\|$ for all $f \in \mathscr{D}(T)$, then $\mathscr{D}(T^{\frac{1}{2}}) \subset \mathscr{D}(S^{\frac{1}{2}})$ and $\|S^{\frac{1}{2}}f\| \leqslant \|T^{\frac{1}{2}}f\|$ for all $f \in \mathscr{D}(T^{\frac{1}{2}})$.

(ii) If $\mathscr{D}(T) \subset \mathscr{D}(S)$ then $\mathscr{D}(T^{\frac{1}{2}}) \subset \mathscr{D}(S^{\frac{1}{2}})$. In particular $\mathscr{D}(T) = \mathscr{D}(S)$ implies that $\mathscr{D}(T^{\frac{1}{2}}) = \mathscr{D}(S^{\frac{1}{2}})$. ■

Proof. (i) From Proposition 8.11, for $f \in \mathscr{D}(T) \subset \mathscr{D}(S) \subset \mathscr{D}(S^{\frac{1}{2}})$,

$$\|S^{\frac{1}{2}}f\|^2 = (Sf, f) \leqslant (Tf, f) = \|T^{\frac{1}{2}}f\|^2.$$

Since $\mathscr{D}(T)$ is a core of $T^{\frac{1}{2}}$, $\mathscr{D}(T^{\frac{1}{2}}) \subset \mathscr{D}(S^{\frac{1}{2}})$ and $\|S^{\frac{1}{2}}f\| \leqslant \|T^{\frac{1}{2}}f\|$ for $f \in \mathscr{D}(T^{\frac{1}{2}})$.

(ii) Proposition 7.2 and the fact that $\mathscr{D}(T) \subset \mathscr{D}(S)$ imply that S is T-bounded,

$$\|Sf\| \leqslant \alpha(\|Tf\| + \|f\|) \quad \text{for all } f \in \mathscr{D}(T),$$

say, for some $\alpha \geqslant 0$. Thus

$$\begin{aligned} \|Sf\|^2 &\leqslant 2\alpha^2(\|Tf\|^2 + \|f\|^2) \\ &\leqslant 2\alpha^2(\|Tf\|^2 + 2(Tf, f) + \|f\|^2) \\ &= 2\alpha^2\|(T+I)f\|^2. \end{aligned}$$

Part (i) now gives $\mathscr{D}(S^{\frac{1}{2}}) \supset \mathscr{D}([T+I]^{\frac{1}{2}}) = \mathscr{D}(T^{\frac{1}{2}})$. □

9. Multiplication operators

Let Ω be a subset of $\mathbb{R}^n$ which is measurable with respect to some positive measure $\mathrm{d}\sigma$. Then $L^2(\Omega; \mathrm{d}\sigma)$ is the set of all (equivalence classes) of functions f which are measurable on Ω and satisfy $\int_\Omega |f|^2\, \mathrm{d}\sigma < \infty$, functions being equivalent if they are equal almost everywhere in Ω with respect to $\mathrm{d}\sigma$; the set $L^2(\Omega; \mathrm{d}\sigma)$ is a Hilbert space with inner product

$$(f, g) = \int_\Omega f\bar{g}\, \mathrm{d}\sigma.$$

Definition 9.1. If q is a measurable function on Ω, the *maximal operator of multiplication by* q is denoted by $\mathrm{M}(q)$ and defined by:

$$\mathrm{M}(q)f = qf, \qquad f \in \mathscr{D}(\mathrm{M}(q)) := \{f : f, qf \in \mathrm{L}^2(\Omega; \mathrm{d}\sigma)\}. \qquad \blacksquare$$

Theorem 9.2. Let q be measurable on Ω. Then
(i) $\mathscr{D}(\mathrm{M}(q))$ is dense in $\mathrm{L}^2(\Omega; \mathrm{d}\sigma)$;
(ii) $\mathrm{M}(q)^* = \mathrm{M}(\bar{q})$;
(iii) $\mathrm{M}(q)$ is closed.
(iv) The following are equivalent:
(a) $\mathscr{R}(\mathrm{M}(q))$ is dense in $\mathrm{L}^2(\Omega; \mathrm{d}\sigma)$,
(b) $q(x) \neq 0$ a.e. in Ω,
(c) $\mathrm{M}(q)$ is injective.
If one (and hence all) of (a), (b), (c) is satisfied then $\mathrm{M}(q)^{-1}$ exists and $\mathrm{M}(q)^{-1} = \mathrm{M}(q_1)$ where

$$q_1(x) = \begin{cases} q(x)^{-1} & \text{when } q(x) \neq 0, \\ 0 & \text{otherwise.} \end{cases} \qquad \blacksquare$$

Proof. (i) For each $m \in \mathbb{N}$, let χ_{Ω_m} be the characteristic function of $\Omega_m = \{x \in \Omega : |q(x)| \leqslant m\}$. Then $\Omega_m \subset \Omega_{m+1}$ and $\Omega = \bigcup_{m=1}^{\infty} \Omega_m$. Given $f \in \mathrm{L}^2(\Omega, \mathrm{d}\sigma)$, we have $f_m = \chi_{\Omega_m} f \in \mathscr{D}(\mathrm{M}(q))$ and

$$\|f - f_m\|^2 = \int_{\bigcup_{j>m} \Omega_j} |f|^2 \,\mathrm{d}\sigma \to 0$$

as $m \to \infty$. $\mathscr{D}(\mathrm{M}(q))$ is therefore dense.

(ii) Clearly $\mathscr{D}(\mathrm{M}(\bar{q})) = \mathscr{D}(\mathrm{M}(q))$ and for $u, v \in \mathscr{D}(\mathrm{M}(q))$,

$$(\mathrm{M}(q)u, v) = \int_\Omega qu\bar{v}\,\mathrm{d}\sigma = \int_\Omega u(\overline{\bar{q}v})\,\mathrm{d}\sigma = (u, \mathrm{M}(\bar{q})v),$$

whence $\mathrm{M}(\bar{q}) \subset \mathrm{M}(q)^*$. Let $u \in \mathscr{D}(\mathrm{M}(q)^*)$. Then for any $\phi \in \mathscr{D}(\mathrm{M}(q))$,

$$(\phi, \mathrm{M}(q)^*u) = (\mathrm{M}(q)\phi, u) = \int_\Omega q\phi\bar{u}\,\mathrm{d}\sigma$$

and

$$\int_\Omega \phi[\overline{\mathrm{M}(q)^*u - \bar{q}\,u}]\,\mathrm{d}\sigma = 0.$$

If Ω_m $(m = 1, 2, \ldots)$ are the sets in (i) then $\chi_{\Omega_m} f \in \mathscr{D}(\mathrm{M}(q))$ for any $f \in \mathrm{L}^2(\Omega; \mathrm{d}\sigma)$ and so, on putting $\phi = \chi_{\Omega_m}[\mathrm{M}(q)^*u - \bar{q}u]$, we find that $\mathrm{M}(q)^*u(x) = \bar{q}(x)u(x)$ a.e. on Ω_m. Since this is true for all m and $\Omega = \bigcup_{m=1}^{\infty} \Omega_m$, we have $M(q)^*u = \bar{q}u$ and thus $M(q)^* = M(\bar{q})$.

(iii) From (ii), $M(q) = M(\bar{q})^*$ and therefore $M(q)$ is closed.

(iv) $(a) \Rightarrow (b)$: Every $f \in L^2(\Omega; d\sigma)$ which vanishes outside the set $M = \{x \in \Omega: q(x) = 0\}$ is orthogonal to $\mathscr{R}(M(q))$. Hence if $\mathscr{R}(M(q))$ is dense in $L^2(\Omega; d\sigma)$, $M(q)^\perp = \{0\}$ and M is of zero measure.

$(b) \Rightarrow (c)$: If $M(q)f = 0$ then $q(x)f(x) = 0$ a.e. in Ω and (b) implies that $f(x) = 0$ a.e. in Ω and $f = 0$ in $L^2(\Omega; d\sigma)$.

$(c) \Rightarrow (a)$: Let $h \in \mathscr{R}(M(q))^\perp = \mathscr{N}(M(q)^*) = \mathscr{N}(M(\bar{q}))$. Then $\bar{h} \in \mathscr{D}(M(q))$ and $M(q)\bar{h} = 0$, whence $h = 0$ from (c); and (a) is proved.

Under the assumptions in (iv), $M(q)^{-1}$ exists and its domain is

$$\begin{aligned}\mathscr{D}(M(q)^{-1}) &= \mathscr{R}(M(q)) \\ &= \{g \in L^2(\Omega; d\sigma)\text{: there exists an } f \in L^2(\Omega; d\sigma) \text{ such that } g = qf\} \\ &= \{g \in L^2(\Omega; d\sigma): q_1 g \in L^2(\Omega; d\sigma)\}\end{aligned}$$

and $M(q)^{-1}g = f = q_1 g$. □

It is clear from Theorem 9.2 that $M(q)$ is J-self-adjoint, where J denotes complex conjugation; and $M(q)$ is self-adjoint if, and only if, q is real-valued.

10. Second-order linear differential operators

10.1. *Quasi-differential equations*

The following notation will be used in this section; see Chapter V for more details. If I is an interval we shall denote by $L^1_{loc}(I)$ the set of functions which are measurable on I and are integrable on all compact subintervals of I; thus, functions in $L^1_{loc}[a, b)$, for instance, are such that $\int_a^X |f(t)|\,dt < \infty$ for all $X \in [a, b)$. The spaces $L^p_{loc}(I)$ $(1 < p \leqslant \infty)$ are defined similarly. The set of functions which are absolutely continuous on I will be denoted by $AC(I)$ while $AC_{loc}(I)$ will stand for the set of functions which are absolutely continuous on all compact subintervals of I. We also use the notation $C^\alpha(I)$ $(\alpha \in \mathbb{N})$ and $C_0^\infty(I)$.

We shall consider second-order equations of the form

$$\tau\phi := -(p\phi')' + r\phi' + q\phi = \lambda w\phi, \qquad \lambda \in \mathbb{C}, \tag{10.1}$$

on an interval (a, b), with $-\infty \leqslant a < b \leqslant \infty$, where p and w are real-valued and r and q are complex-valued functions on (a, b). In order for $\tau\phi$ to be defined almost everywhere with respect to Lebesgue measure in (a, b) it is sufficient that ϕ and $p\phi'$ belong to $AC_{loc}(a, b)$. Accordingly, we define ϕ to be a solution of the equation $\tau\sigma = f$ if ϕ and $p\phi' \in AC_{loc}(a, b)$ and $\tau\phi = f$ a.e. in (a, b). With this interpretation τ is called a *quasi-differential expression* and $\tau\phi = f$ a *quasi-differential equation*. Also

$$\phi^{[1]} := p\phi' \tag{10.2}$$

is called the *quasi-derivative* of ϕ. The equation (10.1) is no longer a second-order differential equation in the generally accepted sense since ϕ'' need not exist even a.e. in (a, b). The main benefit of defining τ to be a quasi-differential expression is that it enables us to study (10.1) under the following weak assumptions on the coefficients:

$$\begin{aligned}&\text{(i) } p(x) \neq 0 \text{ a.e. on } (a,b) \text{ and } 1/p \in \mathrm{L}^1_{\mathrm{loc}}(a,b),\\ &\text{(ii) } r/p, q \in \mathrm{L}^1_{\mathrm{loc}}(a,b),\\ &\text{(iii) } w(x) > 0 \text{ a.e. on } (a,b) \text{ and } w \in \mathrm{L}^1_{\mathrm{loc}}(a,b).\end{aligned} \tag{10.3}$$

These conditions will always be assumed.

The usual existence theorem (see Naimark [1, §16.2]) continues to be valid for the expression τ in (10.1) and we shall quote the result below without proof. The fact that r and q are complex-valued does not affect the proof, and the usual consequences continue to hold.

Theorem 10.1. Let $f \in \mathrm{L}^1_{\mathrm{loc}}(a,b)$ and suppose that (10.3) (i) and (ii) are satisfied. Then given any complex numbers c_0 and c_1 and any $x_0 \in (a,b)$ there exists a unique solution of $\tau\phi = f$ in (a,b) which satisfies $\phi(x_0) = c_0$, $\phi^{[1]}(x_0) = c_1$. ∎

A simple consequence of Theorem 10.1 is that the solutions of (10.1) form a 2-dimensional vector space over $\mathbb{C}$. If (α_0, α_1) and (β_0, β_1) are linearly independent vectors in $\mathbb{C}^2$ then the solutions $\phi_1(\bullet, \lambda)$, $\phi_2(\bullet, \lambda)$ of (10.1) which satisfy $\phi_1(x_0, \lambda) = \alpha_0$, $\phi_1^{[1]}(x_0, \lambda) = \alpha_1$, $\phi_2(x_0, \lambda) = \beta_0$, $\phi_2^{[1]}(x, \lambda) = \beta_1$ for some $x_0 \in (a,b)$, form a basis for the space of solutions of (10.1).

Definition 10.2. The equation (10.1) is said to be *regular* at a if

$$a \in \mathbb{R};\ \frac{1}{p}, \frac{r}{p}, q, w \in \mathrm{L}^1_{\mathrm{loc}}[a,b). \tag{10.4}$$

Otherwise (10.1) is said to be *singular* at a. Similarly we define (10.1) to be regular or singular at b. If (10.1) is regular at a and b $(> a)$ we say that it is *regular* on $[a,b]$; in this case

$$a, b \in \mathbb{R};\ \frac{1}{p}, \frac{r}{p}, q, w \in \mathrm{L}^1(a,b). \tag{10.5}$$ ∎

If (10.1) is regular at a, say, then Theorem 10.1 continues to hold even when $x_0 = a$.

Let ϕ and ψ be such that ϕ, $\phi^{[1]}$, ψ, $\psi^{[1]}$ are absolutely continuous on a compact interval $[\alpha, \beta] \subset (a,b)$ and suppose that $r \in \mathrm{AC}[\alpha, \beta]$. Then, on integration by parts,

$$\int_\alpha^\beta [\bar{\psi}\tau\phi - \phi(\overline{\tau^+\psi})] = [\phi, \psi](\beta) - [\phi, \psi](\alpha) \tag{10.6}$$

where

$$\tau^+\psi = -(p\psi')' - (\bar{r}\psi)' + \bar{q}\psi \tag{10.7}$$

and

$$[\phi,\psi] := \phi\bar{\psi}^{[1]} - \bar{\psi}\phi^{[1]} + r\phi\bar{\psi}. \tag{10.8}$$

The expression τ^+ is called the *Lagrange* or *formal adjoint* of τ and (10.6) is called *Green's formula* on $[\alpha,\beta]$; τ^+ is similar in form to τ as long as we assume, as we shall always do, that

$$\text{(iv)}\quad r\in \mathrm{AC}_{\mathrm{loc}}(a,b), \qquad r'\in \mathrm{L}^1_{\mathrm{loc}}(a,b); \tag{10.9}$$

τ is said to be *formally symmetric* if $\tau^+ = \tau$ and *formally J-symmetric* if $\tau^+ = \bar{\tau}$. It is an easy matter to verify that

$$\tau^{++} := (\tau^+)^+ = \tau. \tag{10.10}$$

Also note that $\tau^+\psi = \lambda w\psi$ is regular on $[a,b]$ if $r'\in \mathrm{L}^1(a,b)$ in addition to (10.5).

Our ultimate objective in this section is to characterize various linear operators generated by $(1/w)\tau$ in the weighted Hilbert space $\mathrm{L}^2(a,b;w) \equiv \mathrm{L}^2\big((a,b);w\mathrm{d}x\big)$ (*cf.* §9). We shall use the notation

$$(f,g) := \int_a^b f\bar{g}w, \qquad \|f\| := \left(\int_a^b |f|^2 w\right)^{\frac{1}{2}}, \tag{10.11}$$

for the $\mathrm{L}^2(a,b;w)$ inner product and norm throughout.

10.2. *The regular problem on* $[a,b]$

Let

$$\mathfrak{D}(\tau) := \{\phi : \phi \text{ and } \phi^{[1]}\in \mathrm{AC}[a,b], (1/w)\tau\phi\in \mathrm{L}^2(a,b;w)\}$$

and let $T(\tau)$ be the operator in L^2 $(a,b\colon w)$ defined by $T(\tau)\phi = (1/w)\tau\phi$ for $\phi\in\mathfrak{D}(\tau)$. Then $T(\tau)$ is clearly the operator with the largest domain defined by $(1/w)\tau$ in $\mathrm{L}^2(a,b;w)$ and it is called the *maximal operator* generated by $(1/w)\tau$ in $\mathrm{L}^2(a,b;w)$; we define $\mathfrak{D}(\tau^+)$ and $T(\tau^+)$ similarly.

Lemma 10.3. Let (10.1) be regular on $[a,b]$, $r\in\mathrm{AC}[a,b]$, $r'\in \mathrm{L}^1(a,b)$, and $f\in \mathrm{L}^2(a,b;w)$. Then the equation $\tau\phi = wf$ has a solution $\phi\in\mathfrak{D}(\tau)$ which satisfies $\phi(a) = \phi^{[1]}(a) = \phi(b) = \phi^{[1]}(b) = 0$ if, and only if, f is orthogonal to the solutions of $\tau^+\psi = 0$ in $\mathrm{L}^2(a,b;w)$. ■

Proof. Since $f\in \mathrm{L}^2(a,b;w)$ and $w\in \mathrm{L}^1(a,b)$ we have that $wf\in \mathrm{L}^1(a,b)$ and so, by Theorem 10.1, there exists a solution ϕ of $\tau\phi = wf$ which satisfies $\phi(a) = \phi^{[1]}(a) = 0$, and this solution is unique. Also, $r'\in \mathrm{L}^1(a,b)$ implies that the equation $\tau^+\psi = 0$ is regular on $[a, b]$ and hence it has unique solutions ψ_1 and ψ_2 satisfying

$$\psi_1(b) = 1, \quad \psi_1^{[1]}(b) = 0, \qquad \psi_2(b) = 0, \quad \psi_2^{[1]}(b) = 1, \tag{10.12}$$

and $\{\psi_1, \psi_2\}$ is a basis for the solution space of $\tau^+\psi = 0$. From (10.6),

$$\begin{aligned} (f, \psi_j) &= \int_a^b (\tau\phi)\bar{\psi}_j \\ &= \int_a^b \phi\overline{(\tau^+\psi_j)} + [\phi, \psi_j](b) - [\phi, \psi_j](a) \\ &= [\phi, \psi_j](b) \\ &= \begin{cases} -\phi^{[1]}(b) + r(b)\phi(b) & \text{when } j = 1, \\ \phi(b) \quad \text{when } j = 2, \end{cases} \end{aligned}$$

and the lemma follows. □

In the remainder of this subsection we shall assume that $\tau\phi = \lambda w\phi$ and $\tau^+\psi = \lambda w\psi$ are regular on $[a, b]$ and also that $r \in \mathrm{AC}[a, b]$.

Lemma 10.4. Given arbitrary complex numbers $\lambda, \alpha_0, \alpha_1, \beta_0, \beta_1$, there exists a function $\phi \in \mathcal{N}([T(\tau^+) - \bar{\lambda}I][T(\tau) - \lambda I]) \subset \mathfrak{D}(\tau)$ such that

$$\phi(a) = \alpha_0, \quad \phi^{[1]}(a) = \alpha_1, \qquad \phi(b) = \beta_0, \quad \phi^{[1]}(b) = \beta_1. \qquad ■$$

Proof. Let $\{\psi_1, \psi_2\}$ be a basis of the space of solutions of $[(1/w)\tau^+ - \bar{\lambda}]\psi = 0$, determined by (10.12). The Gram determinant $\det_{i,j=1,2} [(\psi_i, \psi_j)]$ is non-zero. Otherwise the system of equations

$$\gamma_1(\psi_1, \psi_1) + \gamma_2(\psi_2, \psi_1) = 0,$$

$$\gamma_1(\psi_1, \psi_2) + \gamma_2(\psi_2, \psi_2) = 0$$

has a non-zero solution $\{\gamma_1, \gamma_2\}$ and so $\gamma_1\psi_1 + \gamma_2\psi_2$, being orthogonal to ψ_1 and ψ_2 in $\mathrm{L}^2(a, b; w)$, is zero. This contradicts the fact that ψ_1 and ψ_2 are linearly independent. It follows that there exist $\mu_1, \mu_2 \in \mathbb{C}$ such that

$$\mu_1(\psi_1, \psi_1) + \mu_2(\psi_2, \psi_1) = -\beta_1 + r(b)\beta_0$$

$$\mu_1(\psi_1, \psi_2) + \mu_2(\psi_2, \psi_2) = \beta_0.$$

Hence $f = \mu_1\psi_1 + \mu_2\psi_2$ satisfies

$$(f, \psi_j) = \begin{cases} -\beta_1 + r(b)\beta_0 & \text{when } j = 1, \\ \beta_0 & \text{when } j = 2. \end{cases}$$

Let u be the solution of: $[(1/w)\tau-\lambda]u=f$, with $u(a)=u^{[1]}(a)=0$. Then $u\in\mathcal{N}([T(\tau^+)-\bar{\lambda}I][T(\tau)-\lambda I])$ by construction and

$$\begin{aligned}(f,\psi_j) &= \int_a^b [(\tau-\lambda w)u]\bar{\psi}_j \\ &= \int_a^b u\overline{[(\tau^+-\bar{\lambda}w)\psi_j]}+[u,\psi_j](b)-[u,\psi_j](a) \\ &= [u,\psi_j](b) \\ &= \begin{cases} -u^{[1]}(b)+r(b)u(b) & \text{when } j=1, \\ u(b) & \text{when } j=2. \end{cases}\end{aligned}$$

Thus $u(b)=\beta_0$ and $u^{[1]}(b)=\beta_1$. Similarly we can construct some $v\in\mathcal{N}([T(\tau^+)-\bar{\lambda}I][T(\tau)-\lambda I])$ such that $v(a)=\alpha_0, v^{[1]}(a)=\alpha_1, v(b)=0, v^{[1]}(b)=0$. The function $\phi=u+v$ is then seen to satisfy the lemma. □

Let $T_0(\tau)$ denote the restriction of $T(\tau)$ to the subspace

$$\mathfrak{D}_0(\tau):=\{\phi:\phi\in\mathfrak{D}(\tau) \text{ and } \phi(a)=\phi^{[1]}(a)=\phi(b)=\phi^{[1]}(b)=0\}.$$

Theorem 10.5. $\mathfrak{D}_0(\tau)$ is a dense subspace of $L^2(a,b;w)$, $T_0(\tau)$ is closed, and

$$T_0(\tau)^*=T(\tau^+), \qquad T_0(\tau^+)=T(\tau)^*. \tag{10.13}$$ ■

Proof. Let h be orthogonal to $\mathfrak{D}_0(\tau)$ in $L^2(a,b;w)$ and let ψ satisfy $(1/w)\tau^+\psi=h$. Then for all $\phi\in\mathfrak{D}_0(\tau)$,

$$\begin{aligned}(T_0(\tau)\phi,\psi) &= \int_a^b (\tau\phi)\bar{\psi} \\ &= \int_a^b \phi\overline{\tau^+\psi}\end{aligned}$$

(from (10.6) and since $\phi\in\mathfrak{D}_0(\tau)$)

$$\begin{aligned}&=(\phi,h)\\ &=0.\end{aligned}$$

Thus $\psi\in\mathcal{R}[T_0(\tau)]^\perp$. From Lemma 10.3, $\mathcal{R}[T_0(\tau)]=\mathcal{N}[T(\tau^+)]^\perp$, where $\mathcal{N}[T(\tau^+)]$ is the finite-dimensional, and hence closed, null-space of $T(\tau^+)$. Consequently $\mathcal{R}[T_0(\tau)]^\perp=\mathcal{N}[T(\tau^+)]$ and $h=T(\tau^+)\psi=0$. $\mathfrak{D}_0(\tau)$ is therefore dense in $L^2(a,b;w)$.

From (10.6), for $\phi\in\mathfrak{D}_0(\tau)$ and $\psi\in\mathfrak{D}(\tau^+)$,

$$(T_0(\tau)\phi,\psi)-(\phi,T(\tau^+)\psi)=\int_a^b(\bar{\psi}\tau\phi-\phi\overline{\tau^+\psi})=0$$

and so $T(\tau^+) \subset T_0(\tau)^*$. To prove the reverse, take any $u \in \mathscr{D}[T_0(\tau)^*]$ and set $h = T_0(\tau)^* u$. By Theorem 10.1, there exists a solution ψ of $(1/w)\tau^+\psi = h$, and $\psi \in \mathfrak{D}(\tau^+)$ since $h \in L^2(a, b; w)$. For all $\phi \in \mathfrak{D}_0(\tau)$,

$$\begin{aligned}(T_0(\tau)\phi, u) &= (\phi, h)\\ &= \int_a^b \phi\overline{\tau^+\psi}\\ &= \int_a^b \bar{\psi}\tau\phi = (T_0(\tau)\phi, \psi).\end{aligned}$$

Thus $u - \psi \in \mathscr{R}[T_0(\tau)]^\perp = \mathscr{N}[T(\tau^+)]$. In particular, $u \in \mathfrak{D}(\tau^+)$ and $T_0(\tau)^* u = h = T(\tau^+)\psi = T(\tau^+)u$. This completes the proof of $T_0(\tau)^* = T(\tau^+)$. Moreover, since $\mathfrak{D}_0(\tau^+)$ is dense in $L^2(a, b; w)$, so is the larger set $\mathfrak{D}(\tau^+)$ and therefore by Theorem 1.5, $T_0(\tau)$ is closable. Also $T_0(\tau) \subset T_0(\tau)^{**} = T(\tau^+)^*$ and the theorem will follow once we have proved that $T(\tau^+)^* \subset T_0(\tau)$, since then (10.13) and the fact that $T_0(\tau)$ is closed will be established.

Let $z \in \mathscr{D}[T(\tau^+)^*]$. Since $T_0(\tau^+) \subset T(\tau^+)$ we have $T(\tau^+)^* \subset T_0(\tau^+)^* = T(\tau)$ from above and so $z \in \mathfrak{D}(\tau)$ with $T(\tau^+)^* z = (1/w)\tau z$. For any $u \in \mathscr{D}[T(\tau^+)]$ $(= \mathfrak{D}(\tau^+))$,

$$\begin{aligned}0 &= (T(\tau^+)u, z) - (u, T(\tau^+)^* z)\\ &= \int_a^b (\bar{z}\tau^+ u - u\overline{\tau z})\\ &= \overline{[z, u]}(a) - \overline{[z, u]}(b).\end{aligned}$$

Lemma 10.4 asserts that $u \in \mathfrak{D}(\tau^+)$ can be chosen to have any prescribed values at a and b. It therefore follows that $z(a) = z^{[1]}(a) = z(b) = z^{[1]}(b) = 0$ and so $z \in \mathfrak{D}_0(\tau)$. Thus $T(\tau^+)^* \subset T_0(\tau)$ and the theorem is proved. Notice that we are free to interchange τ and τ^+ in results proved for τ on account of (10.10) and the fact that τ and τ^+ are of the same form and give rise to regular differential equations on $[a, b]$. □

An immediate consequence of Theorem 10.5 is that $T_0(\tau)$ and $T_0(\tau^+)$ form an adjoint pair of closed, densely defined operators and so do $T(\tau)$ and $T(\tau^+)$. Our next step will be to use Theorem 3.6 to characterize all the operators which are regularly solvable with respect to $T_0(\tau)$ and $T_0(\tau^+)$ when $T_0(\tau)$ and $T_0(\tau^+)$ are compatible. We must therefore assume that there exists some $\lambda \in \Pi[T_0(\tau), T_0(\tau^+)]$. This requirement is not unduly stringent. It is satisfied if $\tau^+ = \tau$, since $T_0(\tau)$ is then symmetric, and also when $T_0(\tau)$ and $T_0(\tau^+)$ are quasi-accretive. If $r = 0$ then $\tau^+ = \bar{\tau}$ and $T(\tau^+) = JT(\tau)J$ where J is complex conjugation. Thus, in this case, $T_0(\tau)$ is J-symmetric and the aforementioned condition becomes $\Pi(T_0(\tau)) \neq \varnothing$. This is satisfied if im q is semi-bounded above or below since, taking im $q \leqslant k$ to illustrate, for all $\phi \in \mathfrak{D}_0(\tau)$,

$$\operatorname{im}(T_0(\tau)\phi, \phi) = \operatorname{im} \int_a^b (p|\phi'|^2 + q|\phi|^2)$$
$$\leqslant k\|\phi\|^2,$$

and hence $\Theta(T_0(\tau)) \subset \{\lambda : \operatorname{im} \lambda \leqslant k\}$.

Theorem 10.6. Let S be regularly solvable with respect to the compatible pair $T_0(\tau), T_0(\tau^+)$. Then S is the restriction of $T(\tau)$ to the set of functions $u \in \mathfrak{D}(\tau)$ which satisfy linearly independent boundary conditions

$$\alpha_{j1} u(a) + \alpha_{j2} u^{[1]}(a) + \beta_{j1} u(b) + \beta_{j2} u^{[1]}(b) = 0 \qquad (j = 1, 2). \quad (10.14)$$

S^* is the restriction of $T(\tau^+)$ to the set of functions $v \in \mathfrak{D}(\tau^+)$ which satisfy linearly independent boundary conditions

$$\gamma_{j1} \bar{v}(a) + \gamma_{j2} \bar{v}^{[1]}(a) + \delta_{j1} \bar{v}(b) + \delta_{j2} \bar{v}^{[1]}(b) = 0 \qquad (j = 1, 2). \quad (10.15)$$

Here the α_{jk}, β_{jk}, γ_{jk}, δ_{jk} are complex numbers satisfying

$$\alpha_{j1}\gamma_{k2} - \alpha_{j2}\gamma_{k1} + r(a)\alpha_{j2}\gamma_{k2} = \beta_{j1}\delta_{k2} - \beta_{j2}\delta_{k1} + r(b)\beta_{j2}\delta_{k2} \quad (10.16)$$

for $j, k = 1, 2$. Conversely, if $\Pi[T_0(\tau), T_0(\tau^+)] \neq \varnothing$ and if D_1 (D_2) is the set of functions in $\mathfrak{D}(\tau)$ ($\mathfrak{D}(\tau^+)$) which satisfy the linearly independent boundary conditions (10.14) ((10.15)) and (10.16) is satisfied, then $S := T(\tau) \restriction D_1$ is regularly solvable with respect to $T_0(\tau)$ and $T_0(\tau^+)$ and $S^* = T(\tau^+) \restriction D_2$. S is self-adjoint (J-self-adjoint) if, and only if, $\tau^+ = \tau$ ($\tau^+ = \bar{\tau}$) and

$$\gamma_{jk} = \bar{\alpha}_{jk} \text{ and } \delta_{jk} = \bar{\beta}_{jk} \ (\gamma_{jk} = \alpha_{jk} \text{ and } \delta_{jk} = \beta_{jk}) \ (j, k = 1, 2). \qquad \blacksquare$$

Proof. In Theorem 3.6 we set $A = T_0(\tau)$ and $B = T_0(\tau^+)$, so that $A^* = T(\tau^+)$ and $B^* = T(\tau)$, by Theorem 10.5. We therefore have, in the notation (3.14),

$$\beta[\![u, v]\!] = (T(\tau)u, v) - (u, T(\tau^+)v) \qquad (u \in \mathfrak{D}(\tau), v \in \mathfrak{D}(\tau^+))$$
$$= \int_a^b (\bar{v}\tau u - u\overline{\tau^+ v})$$
$$= [u, v](b) - [u, v](a).$$

Furthermore, from Theorem 3.5, for $\lambda \in \Pi[T_0(\tau), T_0(\tau^+)] \cap \Delta_4(S)$,

$$\dim \mathcal{N}([T(\tau) - \lambda I](S^* - \bar{\lambda} I)) = \dim \mathscr{D}(S^*)/\mathscr{D}(T_0(\tau))$$
$$= \operatorname{def}[T_0(\tau^+) - \bar{\lambda} I]$$
$$= \operatorname{nul}[T(\tau) - \lambda I]$$
$$= 2,$$

and similarly $\mathcal{N}([T(\tau^+) - \bar{\lambda} I](S - \lambda I))$ is of dimension 2 for the above λ.

Let $\{\phi_1, \phi_2\}$ and $\{\psi_1, \psi_2\}$ be bases for $\mathcal{N}([T(\tau)-\lambda I](S^*-\bar{\lambda}I))$ and $\mathcal{N}([T(\tau^+)-\bar{\lambda}I](S-\lambda I))$ respectively. Then, by Theorem 3.6,

$$\mathscr{D}(S) = \{u : u \in \mathfrak{D}(\tau), [u, \phi_j](b) - [u, \phi_j](a) = 0 \ (j = 1, 2)\} \tag{10.17}$$

and

$$\mathscr{D}(S^*) = \{v : v \in \mathfrak{D}(\tau^+), [\psi_j, v](b) - [\psi_j, v](a) = 0 \ (j = 1, 2)\}. \tag{10.18}$$

Since $\phi_k \in \mathscr{D}(S^*)$ we must have

$$[\psi_j, \phi_k](b) - [\psi_j, \phi_k](a) = 0 \qquad (j, k = 1, 2). \tag{10.19}$$

If we put

$$\left.\begin{aligned} -\alpha_{j1} &= r(a)\bar{\phi}_j(a) + \bar{\phi}_j^{[1]}(a), & \alpha_{j2} &= \bar{\phi}_j(a), \\ \beta_{j1} &= r(b)\bar{\phi}_j(b) + \bar{\phi}_j^{[1]}(b), & \beta_{j2} &= -\bar{\phi}_j(b), \end{aligned}\right\} \tag{10.20}$$

$$\left.\begin{aligned} \gamma_{j1} &= -r(a)\psi_j(a) + \psi_j^{[1]}(a) & \gamma_{j2} &= -\psi_j(a), \\ \delta_{j1} &= r(b)\psi_j(b) - \psi_j^{[1]}(b), & \delta_{j2} &= \psi_j(b), \end{aligned}\right\} \tag{10.21}$$

we see that (10.17)–(10.19) can be rewritten in the form (10.14)–(10.16). The boundary conditions (10.14) and (10.15), or equivalently (10.17) and (10.18), are linearly independent, for if there exist $c_1, c_2 \in \mathbb{C}$ such that

$$\sum_{j=1}^{2} c_j\{[u, \phi_j](b) - [u, \phi_j](a)\} = 0 \qquad (u \in \mathfrak{D}(\tau)),$$

then $z = \bar{c}_1\phi_1 + \bar{c}_2\phi_2 \in \mathcal{N}([T(\tau)-\lambda I](S^*-\bar{\lambda}I)) \subset \mathfrak{D}(\tau^+)$ and

$$[u, z](b) - [u, z](a) = 0 \qquad (u \in \mathfrak{D}(\tau)).$$

This gives

$$(T(\tau)u, z) - (u, \tau^+ z) = 0$$

and hence $z \in \mathscr{D}(T(\tau)^*) = \mathfrak{D}_0(\tau^+)$. Consequently $[T_0(\tau^+)-\bar{\lambda}I]z = (S^*-\bar{\lambda}I)z \in \mathcal{N}[T(\tau)-\lambda I] = \mathscr{R}[T_0(\tau^+)-\bar{\lambda}I]^\perp$ and so $[T_0(\tau^+)-\bar{\lambda}I]z = 0$. But $\bar{\lambda} \in \Pi[T_0(\tau^+)]$ and hence $z = 0$. Since ϕ_1 and ϕ_2 are linearly independent we must have $c_1 = c_2 = 0$ and the linear independence of (10.17), and hence (10.14), is established. Similarly for (10.15).

Suppose now that we are given the linearly independent boundary conditions (10.14) and (10.15) subject to (10.16), and let us select some $\lambda \in \Pi(T_0(\tau), T_0(\tau^+))$. By Lemma 10.4 there exist functions ϕ_1 and ϕ_2 in $\mathcal{N}([T(\tau)-\lambda I][T(\tau^+)-\bar{\lambda}I])$ which satisfy (10.20) and functions ψ_1 and ψ_2 in $\mathcal{N}([T(\tau^+)-\bar{\lambda}I][T(\tau)-\lambda I])$ which satisfy (10.21). The sets D_1 and D_2 are then given by (10.17) and (10.18), and (10.16) assumes the form (10.19). We first show that the linear independence of (10.14) and (10.15) implies that ϕ_1 and ϕ_2 are linearly independent modulo $\mathfrak{D}_0(\tau^+)$ (i.e. $c_1\phi_1 + c_2\phi_2 \in \mathfrak{D}_0(\tau^+)$ implies c_1

$= c_2 = 0$) and ψ_1 and ψ_2 are linearly independent modulo $\mathfrak{D}_0(\tau)$. If $\phi = c_1\phi_1 + c_2\phi_2 \in \mathfrak{D}_0(\tau^+)$ then, for all $u \in \mathfrak{D}(\tau)$, since $T(\tau)^* = T_0(\tau^+)$,

$$\begin{aligned} 0 &= (T(\tau)u, \phi) - (u, T_0(\tau^+)\phi) \\ &= \int_a^b (\bar{\phi}\tau u - u\overline{\tau^+\phi}) \\ &= [u, \phi](b) - [u, \phi](a) \\ &= \sum_{j=1}^{2} \bar{c}_j\{[u, \phi_j](b) - [u, \phi_j](a)\}. \end{aligned}$$

Since (10.14) are linearly independent, we conclude that $c_1 = c_2 = 0$ and hence that ϕ_1 and ϕ_2 are linearly independent modulo $\mathfrak{D}_0(\tau^+)$. Similarly ψ_1 and ψ_2 are linearly independent modulo $\mathfrak{D}_0(\tau)$. Next, we define the sets

$$D_1' = \{u : u = u_0 + \sum_{j=1}^{2} c_j\psi_j \text{ for some } u_0 \in \mathfrak{D}_0(\tau) \text{ and some } c_1, c_2 \in \mathbb{C}\}, \tag{10.22}$$

$$D_2' = \{v : v = v_0 + \sum_{j=1}^{2} c_j\phi_j \text{ for some } v_0 \in \mathfrak{D}_0(\tau^+) \text{ and some } c_1, c_2 \in \mathbb{C}\}. \tag{10.23}$$

If $u \in D_1'$ then $u \in \mathfrak{D}(\tau)$ and

$$[u, \phi_j](b) - [u, \phi_j](a) = 0$$

since $u_0 \in \mathfrak{D}_0(\tau)$ and (10.19) is satisfied. Therefore $D_1' \subset D_1$ and similarly $D_2' \subset D_2$. From Corollary 3.2 with $A = T_0(\tau)$ and $B = T_0(\tau^+)$ we see that $\dim \mathfrak{D}(\tau)/\mathfrak{D}_0(\tau) = \operatorname{def}[T_0(\tau) - \lambda I] + \operatorname{def}[T_0(\tau^+) - \bar{\lambda} I] = \operatorname{nul}[T(\tau^+) - \bar{\lambda} I] + \operatorname{nul}[T(\tau) - \lambda I] = 4$. Since D_1 is determined by the two linearly independent boundary conditions in (10.17) it follows that $D_1/\mathfrak{D}_0(\tau)$ has dimension 2. But clearly $\dim D_1'/\mathfrak{D}_0(\tau) = 2$. Therefore $D_1 = D_1'$ and similarly $D_2 = D_2'$.

Let $S = T(\tau)\restriction D_1$ and $S_2 = T(\tau^+)\restriction D_2$. We shall prove that $S_2 = S^*$. For $u \in D_1$ and $v \in D_2$,

$$\begin{aligned} (Su, v) - (u, S_2 v) &= \int_a^b (\bar{v}\tau u - u\overline{\tau^+ v}) \\ &= [u, v](b) - [u, v](a) \\ &= 0 \end{aligned}$$

on using (10.22), (10.23), and (10.19). Consequently $S_2 \subset S^*$ and $S \subset S_2^*$. If $v \in \mathscr{D}(S^*)$, then $v \in \mathfrak{D}(\tau^+)$ since $T_0(\tau) \subset S \subset T(\tau)$ gives $T_0(\tau^+) \subset S^* \subset T(\tau^+)$,

and

$$[\psi_j, v](b) - [\psi_j, v](a) = \int_a^b (\bar{v}\tau\psi_j - \psi_j\overline{\tau^+ v})$$

$$= (S\psi_j, v) - (\psi_j, S^*v) = 0.$$

This proves that $v \in D_2' = D_2$ and $S^* \subset S_2$. Hence $S_2 = S^*$.

Finally we prove that $\lambda \in \Delta_4(S)$ and so $\bar{\lambda} \in \Delta_4(S^*)$. By (10.22),

$$\mathscr{R}(S - \lambda I) = \mathscr{R}[T_0(\tau) - \lambda I] \oplus [T(\tau) - \lambda I]\Psi,$$

where Ψ is the linear span of $\{\psi_1, \psi_2\}$, the sum being orthogonal since $[T(\tau) - \lambda I]\Psi \subset \mathscr{N}[T(\tau^+) - \bar{\lambda}I] = \mathscr{R}[T_0(\tau) - \lambda I]^{\perp}$. Since $\mathscr{R}[T_0(\tau) - \lambda I]$ is closed and $[T(\tau) - \lambda I]\Psi$ is finite-dimensional and hence closed, $\mathscr{R}(S - \lambda I)$ is closed. Also, since $\operatorname{nul}(S - \lambda I) \leqslant \operatorname{nul}[T(\tau) - \lambda I] = 2$ and $\operatorname{def}(S - \lambda I) \leqslant \operatorname{def}[T_0(\tau) - \lambda I] = \operatorname{nul}[T(\tau^+) - \bar{\lambda}I] = 2$ we have $\lambda \in \Delta_3(S)$. But, from (3.2),

$$\operatorname{ind}(S - \lambda I) + 2 = \dim[D_1'/\mathfrak{D}_0(\tau)] = 2,$$

whence $\operatorname{ind}(S - \lambda I) = 0$ and $\lambda \in \Delta_4(S)$. The rest of the theorem is obvious. □

10.3. *The case of one singular end-point*

We shall suppose that $r \in \mathrm{AC}_{\mathrm{loc}}[a, b)$ and also that (10.1) and $\tau^+\phi = \lambda w\phi$ are regular at a and singular at b, the argument being similar when a is the singular end-point and b the regular one. The *maximal operator* $T(\tau)$ is now defined by $T(\tau)\phi = (1/w)\tau\phi$ on the domain

$$\mathfrak{D}(\tau) := \{\phi : \phi, \phi^{[1]} \in \mathrm{AC}_{\mathrm{loc}}[a, b);\ \phi, (1/w)\tau\phi \in \mathrm{L}^2(a, b; w)\}. \tag{10.24}$$

Let $T_0'(\tau)$ denote the restriction of $T(\tau)$ to the subspace

$$\mathfrak{D}_0'(\tau) := \{\phi : \phi \in \mathfrak{D}(\tau), \phi(a) = \phi^{[1]}(a) = 0,$$
$$\text{and } \phi = 0 \text{ outside a compact subset of } [a, b)\}. \tag{10.25}$$

Theorem 10.7. $T_0'(\tau)$ is densely defined and closable in $\mathrm{L}^2(a, b; w)$ and if $T_0(\tau)$ denotes its closure, we have

$$T_0(\tau)^* = T(\tau^+), \qquad T_0(\tau^+) = T(\tau)^*. \tag{10.26}$$

For any $\lambda \in \Pi[T_0(\tau), T_0(\tau^+)]$, the sum $\operatorname{def}[T_0(\tau) - \lambda I] + \operatorname{def}[T_0(\tau^+) - \bar{\lambda}I]$ is constant and

$$2 \leqslant \operatorname{def}[T_0(\tau) - \lambda I] + \operatorname{def}[T_0(\tau^+) - \bar{\lambda}I] \leqslant 4. \tag{10.27}$$

If $\tau^+ = \bar{\tau}$ then $T_0(\tau)$ is J-symmetric with respect to complex conjugation J and $\operatorname{def}[T_0(\tau) - \lambda I]$ is either 1 or 2 for all $\lambda \in \Pi(T_0(\tau))$. If $\tau^+ = \tau$ then $T_0(\tau)$ is symmetric, its deficiency indices are equal, and $\mathrm{m}_{\pm}(T_0(\tau))$ is 1 or 2. ■

Proof. For $c \in (a,b)$, we shall denote the operators and domains defined in $[a,c]$ as in §10.2 by $\mathfrak{D}_{0,c}(\tau)$, $T_{0,c}(\tau)$, etc. Suppose that there is an $h \in L^2(a,b;w)$ which is orthogonal to $\mathfrak{D}'_0(\tau)$. Defining any $\phi \in \mathfrak{D}_{0,c}(\tau)$ to be zero outside $[a,c]$ we have $\mathfrak{D}_{0,c}(\tau) \subset \mathfrak{D}'_0(\tau)$ and $h \perp \mathfrak{D}_{0,c}(\tau)$ in $L^2(a,c;w)$. But by Theorem 10.5, $\mathfrak{D}_{0,c}(\tau)$ is dense in $L^2(a,c;w)$ and hence $h = 0$ a.e. in $[a,c]$. Since c is arbitrary, $h = 0$ in $L^2(a,b;w)$ and $\mathfrak{D}'_0(\tau)$ is dense in $L^2(a,b;w)$.

Functions in $\mathfrak{D}'_0(\tau)$ vanish in a left neighbourhood of b and the Green formula (10.6) therefore gives

$$(T'_0(\tau)\phi, \psi) = (\phi, T(\tau^+)\psi) \qquad (\phi \in \mathfrak{D}'_0(\tau), \psi \in \mathfrak{D}(\tau^+)),$$

and so $T(\tau^+) \subset [T'_0(\tau)]^*$. Since $\mathfrak{D}'_0(\tau)$, and similarly $\mathfrak{D}'_0(\tau^+)$, is dense in $L^2(a,b;w)$, the subspaces $\mathfrak{D}(\tau^+)$ and $\mathscr{D}([T'_0(\tau)]^*)$, being larger than $\mathfrak{D}'_0(\tau^+)$, are also dense and T'_0 is therefore closable, by Theorem 1.5. Moreover, $T(\tau^+) \subset [T_0(\tau)]^*$ and so we must show that $[T_0(\tau)]^* \subset T(\tau^+)$ in order to establish (10.26). Let $u \in \mathscr{D}([T_0(\tau)]^*)$. For all $\phi \in \mathfrak{D}_{0,c}(\tau) \subset \mathfrak{D}'_0(\tau)$,

$$\begin{aligned}([T_0(\tau)]^*u, \phi)_{L^2(a,c;w)} &= ([T_0(\tau)]^*u, \phi)\\ &= (u, T_0(\tau)\phi)\\ &= (u, T_{0,c}(\tau)\phi)_{L^2(a,c;w)}\end{aligned}$$

and so, on account of Theorem 10.5, $u \in \mathscr{D}([T_{0,c}(\tau)]^*) = \mathscr{D}(T_c(\tau^+))$ and $[T_0(\tau)]^*u(x) = T_c(\tau^+)u(x) = (1/w)\tau^+u(x)$ for a.e. $x \in [a,c]$. Since c is arbitrary in (a,b), the functions u and $u^{[1]}$ are in $AC_{\mathrm{loc}}[a,b)$ and $(1/w)\tau^+u = [T_0(\tau)]^*u$, as well as u, lies in $L^2(a,b;w)$. Consequently $[T_0(\tau)]^* \subset T(\tau^+)$ and so $[T_0(\tau)]^* = T(\tau^+)$. Furthermore, since $T_0(\tau^+)$ is closed, $T(\tau)^* = T_0(\tau^+)^{**} = T_0(\tau^+)$ and (10.26) is proved.

If $\lambda \in \Pi[T_0(\tau), T_0(\tau^+)]$ then

$$\begin{aligned}\operatorname{def}[T_0(\tau)-\lambda I] + \operatorname{def}[T_0(\tau^+)-\bar{\lambda} I] &= \operatorname{nul}[T(\tau^+)-\bar{\lambda} I] + \operatorname{nul}[T(\tau)-\lambda I]\\ &\leqslant 4.\end{aligned}$$

From Corollary 3.2, with $\mathfrak{D}_0(\tau) = \mathscr{D}[T_0(\tau)]$,

$$\operatorname{def}[T_0(\tau)-\lambda I] + \operatorname{def}[T_0(\tau^+)-\bar{\lambda} I] = \dim \mathfrak{D}(\tau)/\mathfrak{D}_0(\tau)$$

and hence (10.27) will follow if we can exhibit two functions ϕ_1 and ϕ_2 in $\mathfrak{D}(\tau)$ which are linearly independent modulo $\mathfrak{D}_0(\tau)$. Let $\phi_1, \phi_2 \in \mathfrak{D}(\tau)$ be such that $\det(\phi_j^{[k-1]}(a)) \neq 0$, $\phi_j(c) = \phi_j^{[1]}(c) = 0$, and $\phi_j = 0$ outside $[a,c]$ $(j = 1,2)$. The existence of such functions is guaranteed by Lemma 10.4. The system of linear equations

$$\begin{aligned}c_1\phi_1(a) + c_2\phi_2(a) &= 0,\\ c_1\phi_1^{[1]}(a) + c_2\phi_2^{[1]}(a) &= 0\end{aligned}$$

has only the trivial solution $c_1 = c_2 = 0$; this means that ϕ_1 and ϕ_2 are linearly independent modulo $\mathfrak{D}_0(\tau)$ and (10.27) is proved.

If $\tau^+ = \bar{\tau}$ then $T_0(\tau)$ is J-symmetric and, because $\operatorname{def}[T_0(\tau^+)-\bar{\lambda}I] = \operatorname{def}[T_0(\tau)-\lambda I]$, the deficiency index of $T_0(\tau)$ is either 1 or 2. $T_0(\tau)$ is symmetric when $\tau^+ = \tau$ and the assertion in the theorem that $T_0(\tau)$ has equal deficiency indices implies that the pair of values (2,0) or (0,2) for $(\mathrm{m}_+(T_0(\tau)), \mathrm{m}_-(T_0(\tau)))$ which are consistent with (10.27) are in fact excluded. The reason for this is that if $\operatorname{def}[T_0(\tau)-\lambda I] = 2$ for some $\lambda \in \mathbb{C}$ then it has the same value for all $\lambda \in \mathbb{C}$. The proof of this will be deferred until Corollary 10.11. □

In the case in which $r = 0$ and q is real the result that the deficiency indices of $T_0(\tau)$ are equal and are at least 1 was first proved by H.Weyl [2] in a celebrated paper using complex-variable techiniques. The cases (i) $\mathrm{m}_\pm(T_0(\tau)) = 1$ and (ii) $\mathrm{m}_\pm(T_0(\tau)) = 2$ are both possible, as we shall see in due course, and these are termed (i) the limit-point and (ii) the limit-circle case respectively (see Definition 10.12 below), the terminology being a reflection of the geometric nature of Weyl's proof. Weyl's method was applied by Sims [1] in the case when $p = w = 1$, $r = 0$, and q is a complex function whose imaginary part is semi-bounded. We shall return to Sims' result in §10.5 below. The operator-theoretic technique used here was first applied to determining the range of values assumed by the deficiency indices of real even-order symmetric differential operators by Glazman (see Akhiezer and Glazman, [1, volume II, Appendix 2]) and the necessary modifications to deal with J-symmetric differential operators may be found in the work of Knowles [2] and Zhikar [1].

Lemma 10.8 **(Liouville).** If ϕ_1 and ϕ_2 are solutions of $\tau\phi = \lambda w\phi$ in $[a,b]$ then for any $c \in [a,b)$,

$$(\phi_1\phi_2^{[1]} - \phi_1^{[1]}\phi_2)(x) = (\phi_1\phi_2^{[1]} - \phi_1^{[1]}\phi_2)(c)\exp\int_c^x \frac{r}{p}\,\mathrm{d}t. \tag{10.28}$$

The solutions ϕ_1 and ϕ_2 are linearly independent if, and only if, $(\phi_1\phi_2^{[1]} - \phi_1^{[1]}\phi_2)(c) \neq 0$ for some, and hence all, $c \in [a,b)$. ■

Proof. Setting $F = (\phi_1\phi_2^{[1]} - \phi_1^{[1]}\phi_2)$ we have $F' = (r/p)F$, whence (10.28). If ϕ_1 and ϕ_2 are linearly dependent, $c_1\phi_1 + c_2\phi_2 = 0$ for constants c_1 and c_2 which are not both zero. The system of linear equations

$$c_1\phi_1(c) + c_2\phi_2(c) = 0,$$
$$c_2\phi_1^{[1]}(c) + c_2\phi_2^{[1]}(c) = 0$$

therefore has a non-trivial solution and so $F(c) = 0$. Conversely, if $F(c) = 0$ for some $c \in [a,b)$ the above system of linear equations has a non-trivial solution c_1, c_2 and hence $\phi = c_1\phi_1 + c_2\phi_2$ satisfies $\phi(c) = 0$ and $\phi^{[1]}(c) = 0$. It follows from the existence theorem, Theorem 10.1, that $\phi = 0$ and hence that ϕ_1 and ϕ_2 are linearly dependent. □

Lemma 10.9 (Variation-of-parameters formula). Let ϕ_1 and ϕ_2 be linearly independent solutions of $\tau\phi = \lambda w\phi$ in $[a, b)$ and suppose $(\phi_1\phi_2^{[1]} - \phi_1^{[1]}\phi_2)(c) = 1$ for some $c \in [a, b)$. Then u satisfies $\tau u = \lambda wu + fw$, with $f \in L^1_{\text{loc}}[a, b)$ if, and only if,

$$u(x) = c_1\phi_1(x) + c_2\phi_2(x) + \int_c^x [\phi_1(x)\phi_2(t) - \phi_1(t)\phi_2(x)]\exp\left(-\int_c^t \frac{r}{p}\,\mathrm{d}s\right) f(t)w(t)\,\mathrm{d}t \tag{10.29}$$

for some complex constants c_1 and c_2. ■

Proof. Notice that the supposition $(\phi_1\phi_2^{[1]} - \phi_1^{[1]}\phi_2)(c) = 1$ involves no loss of generality in view of Lemma 10.8, as it can be achieved by multiplying ϕ_1 and ϕ_2 by suitable constants. We have

$$\frac{1}{w}\tau\left[u(x) - \int_c^x [\phi_1(x)\phi_2(t) - \phi_1(t)\phi_2(x)]\exp\left(-\int_c^t \frac{r}{p}\,\mathrm{d}s\right) f(t)w(t)\,\mathrm{d}t\right]$$

$$= \frac{1}{w}\tau u(x) - \left[(\phi_1\phi_2^{[1]} - \phi_2\phi_1^{[1]})(x)\exp\left(-\int_c^x \frac{r}{p}\,\mathrm{d}s\right) f(x)\right.$$

$$\left. + \lambda\int_c^x [\phi_1(x)\phi_2(t) - \phi_1(t)\phi_2(x)]\exp\left(-\int_c^t \frac{r}{p}\,\mathrm{d}s\right) f(t)w(t)\,\mathrm{d}t\right]$$

$$= \left(\frac{1}{w}\tau u - \lambda u - f\right)(x)$$

$$+ \lambda\left[u(x) - \int_c^x [\phi_1(x)\phi_2(t) - \phi_1(t)\phi_2(x)]\exp\left(-\int_c^t \frac{r}{p}\,\mathrm{d}s\right) f(t)w(t)\,\mathrm{d}t\right].$$

The lemma follows from this since $\{\phi_1, \phi_2\}$ is a basis for the 2-dimensional space of solutions of $(1/w)\tau\phi = \lambda\phi$. □

Theorem 10.10. Suppose that for some constant K_0 and $c \in [a, b)$,

$$\left|\exp\left(-\int_c^x \frac{r}{p}\,\mathrm{d}t\right)\right| \leqslant K_0 \quad \text{for all } x \in [a, b), \tag{10.30}$$

and that for some $\lambda_0 \in \mathbb{C}$ all solutions of $\tau\phi = \lambda_0 w\phi$ are in $L^2(a, b; w)$. Then all solutions of $\tau\phi = \lambda w\phi$ are in $L^2(a, b; w)$ for every $\lambda \in \mathscr{C}$. ■

Proof. Let ϕ_1 and ϕ_2 be linearly independent solutions of $\tau\phi = \lambda_0 w\phi$ which satisfy $(\phi_1\phi_2^{[1]} - \phi_1^{[1]}\phi_2)(c) = 1$. From Lemma 10.9, any solution of $\tau\phi = \lambda w\phi$ satisfies

$$\phi(x) = c_1\phi_1(x) + c_2\phi_2(x) + (\lambda - \lambda_0)\int_c^x [\phi_1(x)\phi_2(t) - \phi_1(t)\phi_2(x)]\exp\left(-\int_c^t \frac{r}{p}\,\mathrm{d}s\right)\phi(t)w(t)\,\mathrm{d}t$$

for some c_1 and c_2 in $\mathbb{C}$. Setting $\|\phi_j\|_{L^2(c,b;w)} \leqslant K_1$ and using the Cauchy–Schwarz inequality yields, for $x \geqslant c$,

$$|\phi(x)| \leqslant |c_1|\,|\phi_1(x)| + |c_2|\,|\phi_2(x)| + |\lambda - \lambda_0| K_0 K_1 \left(|\phi_1(x)| + |\phi_2(x)|\right)\left(\int_c^x |\phi|^2 w\right)^{\frac{1}{2}}$$

and

$$\left(\int_c^z |\phi|^2 w\right)^{\frac{1}{2}} \leqslant (|c_1| + |c_2|)K_1 + 2|\lambda - \lambda_0| K_0 K_1^2 \left(\int_c^z |\phi|^2 w\right)^{\frac{1}{2}}.$$

Hence, if $|\lambda - \lambda_0| \leqslant 1/(4K_0 K_1^2)$, we obtain

$$\int_c^z |\phi|^2 w \leqslant 4(|c_1| + |c_2|)^2 K_1^2$$

and so $\phi \in L^2(c, b; w)$. Since the bound K_1 can be made arbitrarily small by choosing c close enough to b, and since also $\phi \in L^2(a, c; w)$ for any $c < b$, the theorem is proved. □

Corollary 10.11. If $\tau^+ = \tau$ then $T_0(\tau)$ has equal deficiency indices and the only possible values are 1 and 2. ■

Proof. If $\tau^+ = \tau$ we must have that $\bar{r} = -r$ and hence r is wholly imaginary-valued. (10.30) is therefore satisfied and thus, from Theorem 10.10, $m_+(T_0(\tau)) = 2$ if and only if $m_-(T_0(\tau)) = 2$. The corollary thus follows from Theorem 10.7. □

Corollary 10.11 was first proved by Weyl in [2] and in his terminology the two cases are distinguished as follows.

Definition 10.12. If $\tau^+ = \tau$, the equation $\tau\phi = \lambda w \phi$ is said to be in the *limit-point case at b* if, for $\operatorname{im} \lambda \neq 0$, there is precisely one solution in $L^2(a, b; w)$. Otherwise $\tau\phi = \lambda w \phi$ is said to be in the *limit-circle case at b*, when all solutions are in $L^2(a, b; w)$ for all $\lambda \in \mathbb{C}$. ■

Note that the definition does not preclude the possibility of $\tau\phi = \lambda w \phi$, with λ real, having no $L^2(a, b; w)$ solution in the limit-point case.

By Lemma 10.4 there exist $u_1, u_2 \in \mathfrak{D}(\tau^+)$ such that, for some $c \in (a, b)$,

$$u_1(a) = 0,\ u_1^{[1]}(a) = 1,\ u_2(a) = 1,\ u_2^{[1]}(a) = 0,$$

and

$$u_j(x) = u_j^{[1]}(x) = 0 \quad \text{for } x \geqslant c \qquad (j = 1, 2).$$

Hence, for all $\phi \in \mathfrak{D}(\tau)$ and $j = 1, 2$,

$$(\phi, T(\tau^+)u_j) - (T(\tau)\phi, u_j) = [\phi, u_j](a) = (\phi \bar{u}_j^{[1]} - \phi^{[1]}\bar{u}_j + r\phi\bar{u}_j)(a).$$

It follows that there exists a positive constant K such that, for all $\phi \in \mathfrak{D}(\tau)$,

$$|\phi(a)| + |\phi^{[1]}(a)| \leqslant K[\|\phi\| + \|T(\tau)\phi\|]. \tag{10.31}$$

In particular,

$$\phi(a) = \phi^{[1]}(a) = 0 \quad \text{if } \phi \in \mathfrak{D}_0(\tau). \tag{10.32}$$

Theorem 10.13. The sum $\operatorname{def}[T_0(\tau) - \lambda I] + \operatorname{def}[T_0(\tau^+) - \bar{\lambda} I]$ equals 2 for $\lambda \in \Pi[T_0(\tau), T_0(\tau^+)]$ if, and only if,

$$[\phi, \psi](b) := \lim_{x \to b-} [\phi, \psi](x) = 0 \qquad (\phi \in \mathfrak{D}(\tau), \psi \in \mathfrak{D}(\tau^+)). \tag{10.33}$$

In this case

$$\mathfrak{D}_0(\tau) \equiv \mathscr{D}[T_0(\tau)] = \{\phi : \phi \in \mathfrak{D}(\tau), \phi(a) = \phi^{[1]}(a) = 0\}. \tag{10.34}$$ ■

Proof. If $\operatorname{def}[T_0(\tau) - \lambda I] + \operatorname{def}[T_0(\tau^+) - \bar{\lambda} I] = 2$, it follows from Corollary 3.2 with $A = T_0(\tau)$ and $B = T_0(\tau^+)$ that the quotient space $\mathfrak{D}(\tau)/\mathfrak{D}_0(\tau)$ has dimension 2 and

$$\mathfrak{D}(\tau) = \mathfrak{D}_0(\tau) \dotplus \mathcal{N}([T(\tau^+) - \bar{\lambda} I][T(\tau) - \lambda I]).$$

Let ϕ_1 and ϕ_2 be functions in $\mathcal{N}([T(\tau^+) - \bar{\lambda} I][T(\tau) - \lambda I])$ which, for $c \in (a, b)$, satisfy

$$\phi_j^{[k-1]}(a) = \begin{cases} 1 & (j = k), \\ 0 & (j \neq k), \end{cases}$$
$$\phi_j^{[k-1]}(c) = 0 \qquad (j, k = 1, 2),$$
$$\phi_j(x) = 0 \qquad (x > c, j = 1, 2).$$

The existence of such functions is guaranteed by Lemma 10.4. It is readily seen that ϕ_1 and ϕ_2 are linearly independent modulo $\mathfrak{D}_0(\tau)$ and hence that $\{\phi_1, \phi_2\}$ is a basis of $\mathfrak{D}(\tau)/\mathfrak{D}_0(\tau)$. Any $\phi \in \mathfrak{D}(\tau)$ can therefore be written as $\phi = \phi_0 + c_1\phi_1 + c_2\phi_2$, for some $\phi_0 \in \mathfrak{D}_0(\tau)$ and constants c_1 and c_2. If $\psi \in \mathfrak{D}(\tau^+)$ we see from Green's formula that

$$[\phi_0, \psi](x) - [\phi_0, \psi](a) = \int_a^x (\bar{\psi}\tau\phi_0 - \phi_0\overline{\tau^+\psi})$$

and since the integrand on the right-hand side is in $L^1(a, b)$, the limit $[\phi_0, \psi](b) := \lim_{x \to b-} [\phi_0, \psi](x)$ exists and

$$\begin{aligned}[\phi_0, \psi](b) &= [\phi_0, \psi](a) + (T_0(\tau)\phi_0, \psi) - (\phi_0, T(\tau^+)\psi) \\ &= 0\end{aligned}$$

by (10.26) and since $\phi_0(a) = \phi_0^{[1]}(a) = 0$. Furthermore $[\phi_j, \psi](b) = 0$ for $j = 1, 2$ and (10.33) follows.

Conversely, suppose that (10.33) is satisfied. We shall first establish (10.34). To this end, let T_1 denote the restriction of $T(\tau)$ to the subspace in (10.34). From Green's formula and (10.33),

$$(T_1\phi, \psi) = (\phi, T(\tau^+)\psi) \qquad (\phi \in \mathscr{D}(T_1), \psi \in \mathscr{D}[T(\tau^+)]),$$

whence $T_1 \subset T(\tau^+)^* = T_0(\tau)$. Since $T_0(\tau) \subset T_1$ from (10.32), we conclude that $T_1 = T_0(\tau)$.

If ϕ_1 and ϕ_2 are the functions defined in the first part of the proof, it follows from (10.34) that $u \in \mathfrak{D}(\tau)$ if and only if

$$u - [u(a)\phi_1 + u^{[1]}(a)\phi_2] \in \mathfrak{D}_0(\tau).$$

Consequently $\mathfrak{D}(\tau)/\mathfrak{D}_0(\tau)$ has dimension 2 and by Corollary 3.2, $\operatorname{def}[T_0(\tau) - \lambda I] + \operatorname{def}[T_0(\tau^+) - \bar{\lambda} I] = 2$ for $\lambda \in \Pi[T_0(\tau), T_0(\tau^+)]$. □

Theorem 10.14. Let $T_0(\tau)$ and $T_0(\tau^+)$ be compatible and, for $\lambda \in \Pi[T_0(\tau), T_0(\tau^+)]$, $\operatorname{def}[T_0(\tau) - \lambda I] = \operatorname{def}[T_0(\tau^+) - \bar{\lambda} I] = 1$. Every closed operator S which is regularly solvable with respect to $T_0(\tau)$ and $T_0(\tau^+)$ is the restriction of $T(\tau)$ to the set of functions $u \in \mathfrak{D}(\tau)$ which satisfy a boundary condition

$$\alpha_2 u(a) + \alpha_2 u^{[1]}(a) = 0, \tag{10.35}$$

and S^* is the restriction of $T(\tau^+)$ to the set of functions $v \in \mathfrak{D}(\tau^+)$ which satisfy a boundary condition

$$\gamma_1 \bar{v}(a) + \gamma_2 \bar{v}^{[1]}(a) = 0, \tag{10.36}$$

where $\alpha_1, \alpha_2, \gamma_1, \gamma_2$ are complex numbers satisfying

$$\alpha_1\gamma_2 - \alpha_2\gamma_1 + r(a)\alpha_2\gamma_2 = 0. \tag{10.37}$$

Conversely, if D_1 (D_2) is the set of functions in $\mathfrak{D}(\tau)$ ($\mathfrak{D}(\tau^+)$) which satisfy (10.35) ((10.36)) and (10.37) is satisfied, then $S = T(\tau)\restriction D_1$ is regularly solvable with respect to $T_0(\tau)$ and $T_0(\tau^+)$ and $S^* = T(\tau^+)\restriction D_2$. S is self-adjoint (J-self-adjoint) if, and only if, $\tau^+ = \tau$ ($\tau^+ = \bar{\tau}$) and $\gamma_j = \bar{\alpha}_j$ ($\gamma_j = \alpha_j$) for $j = 1, 2$. ■

Proof. The proof is virtually identical to that of Theorem 10.6 and we give only a brief sketch. In the first part there exist functions $\phi \in \mathscr{N}([T(\tau) - \lambda I]\ (S^* - \bar{\lambda} I))$ and $\psi \in \mathscr{N}([T(\tau^+) - \bar{\lambda} I]\ (S - \lambda I))$ such that, on account of Theorem 10.13,

$$\mathscr{D}(S) = \{u : u \in \mathfrak{D}(\tau), [u, \phi](a) = 0\},$$

$$\mathscr{D}(S^*) = \{v : v \in \mathfrak{D}(\tau^+), [\psi, v](a) = 0\},$$

and $[\psi, \phi](a) = 0$; note that $\mathscr{N}([T(\tau) - \lambda I]\ (S^* - \bar{\lambda} I))$ and $\mathscr{N}([T(\tau^+) - \bar{\lambda} I]\ (S - \lambda I))$ are one-dimensional in view of Theorem 3.5. $\mathscr{D}(S)$ and $\mathscr{D}(S^*)$ can be described as in (10.35) and (10.36) by choice of ϕ and ψ.

To prove the converse we first show that $D_1 = \{u : u = u_0 + c\psi, u_0 \in \mathfrak{D}_0(\tau), c \in \mathbb{C}\}$ and $D_2 = \{v : v = v_0 + c\phi, v_0 \in \mathfrak{D}_0(\tau^+), c \in \mathbb{C}\}$. The rest of the proof follows that of Theorem 10.6. □

If in Theorem 10.14 we had assumed only that $\operatorname{def}[T_0(\tau) - \lambda I] + \operatorname{def}[T_0(\tau^+) - \bar{\lambda} I] = 2$ there would be the additional possibilities of either $\operatorname{def}[T_0(\tau) - \lambda I] = 0$ or $\operatorname{def}[T_0(\tau^+) - \bar{\lambda} I] = 0$ to investigate. In the first instance we have $\lambda \in \rho[T_0(\tau)]$ and in the second $\bar{\lambda} \in \rho[T_0(\tau^+)]$. If $\lambda \in \rho[T_0(\tau)]$, then $T_0(\tau)$ has no proper closed extension S with $\lambda \in \Delta_4(S)$ since $\operatorname{def}(S - \lambda I) \leqslant \operatorname{def}[T_0(\tau) - \lambda I] = 0$ implies $\lambda \in \rho(S)$ and $S = T_0(\tau)$. Similarly, if $\bar{\lambda} \in \rho[T_0(\tau^+)]$ then $T(\tau)$ has no proper closed restriction S with $\lambda \in \Delta_4(S)$.

Finally we have the following immediate consequences of Theorem 3.6.

Theorem 10.15. Let $T_0(\tau)$ and $T_0(\tau^+)$ be compatible and suppose that $\operatorname{def}[T_0(\tau) - \lambda I] + \operatorname{def}[T_0(\tau^+) - \bar{\lambda} I] = 4$ for $\lambda \in \Pi[T_0(\tau), T_0(\tau^+)]$. Then every closed operator S which is regularly solvable with respect to $T_0(\tau)$ and $T_0(\tau^+)$ is the restriction of $T(\tau)$ to the set of functions $u \in \mathfrak{D}(\tau)$ which satisfy linearly independent boundary conditions

$$[u, \phi_j](b) - [u, \phi_j](a) = 0 \qquad (j = 1, 2), \tag{10.38}$$

where $\{\phi_1, \phi_2\}$ is a basis for $\mathscr{D}(S^*)/\mathfrak{D}_0(\tau^+)$, and S^* is the restriction of $T(\tau^+)$ to the set of functions $v \in \mathfrak{D}(\tau^+)$ which satisfy linearly independent boundary conditions

$$[\psi_j, v](b) - [\psi_j, v](a) = 0 \qquad (j = 1, 2) \tag{10.39}$$

where $\{\psi_1, \psi_2\}$ is a basis for $\mathscr{D}(S)/\mathfrak{D}_0(\tau)$ and

$$[\psi_j, \phi_k](b) - [\psi_j, \phi_k](a) = 0 \qquad (j, k = 1, 2). \tag{10.40}$$

Conversely, for arbitrary functions ϕ_1 and ϕ_2 (ψ_1 and ψ_2) in $\mathfrak{D}(\tau^+)$ ($\mathfrak{D}(\tau)$) which are linearly independent modulo $\mathfrak{D}_0(\tau^+)$ ($\mathfrak{D}_0(\tau)$), if D_1 (D_2) is the set of functions in $\mathfrak{D}(\tau)$ ($\mathfrak{D}(\tau^+)$) which satisfy (10.38) ((10.39)) and (10.40) is satisfied, then $S = T(\tau) \upharpoonright D_1$ is regularly solvable with respect to $T_0(\tau)$ and $T_0(\tau^+)$, and $S^* = T(\tau^+) \upharpoonright D_2$. S is self-adjoint (J-self adjoint) if, and only if, $\tau^+ = \tau$ ($\tau^+ = \bar{\tau}$) and $\psi_j = \phi_j$ ($\psi_j = \bar{\phi}_j$) for $j = 1, 2$. ■

Theorem 10.16. Let $T_0(\tau)$ and $T_0(\tau^+)$ be compatible with $\operatorname{def}[T_0(\tau) - \lambda I] = 1$ and $\operatorname{def}[T_0(\tau^+) - \bar{\lambda} I] = 2$ for some $\lambda \in \Pi[T_0(\tau), T_0(\tau^+)]$. Then Theorem 10.15 holds with $\phi_1 = \phi_2$ for every closed operator S which is regularly solvable with respect to $T_0(\tau)$ and $T_0(\tau^+)$ with $\lambda \in \Delta_4(S)$. If $\operatorname{def}[T_0(\tau) - \lambda I] = 2$ and $\operatorname{def}[T_0(\tau^+) - \bar{\lambda} I] = 1$ for some $\lambda \in \Pi[T_0(\tau), T_0(\tau^+)]$, Theorem 10.15 holds with $\psi_1 = \psi_2$ for every S with $\lambda \in \Delta_4(S)$. ■

Finally in this section we establish local smoothness properties of members of $\mathfrak{D}(\tau)$.

Theorem 10.17. Let (10.3) and (10.9) be satisfied, suppose that $1/|p|$ and $1/w \in L^{\infty}_{\mathrm{loc}}[a, b)$, and let a be a regular end point of (10.1). Then, for any $\beta \in (a, \gamma)$ and $\gamma \in (a, b)$, we have for all $\phi \in \mathfrak{D}(\tau)$,

$$\sup_{x \in [a, \beta]} [|\phi(x)| + |\phi^{[1]}(x)|] \leqslant K_{\gamma}(\|\phi\| + \|\tau\phi\|) \exp \int_a^{\gamma} \left(\frac{1}{|p|} + \frac{|r|}{|p|} + |q| \right) \tag{10.41}$$

where K_{γ} depends on γ. ■

Proof. By Lemma 10.4 with $r = q = 0$, there exists a function θ such that $\theta, \theta^{[1]} \in AC[\beta, \gamma]$ with $(\theta^{[1]})' \in L^2(\beta, \gamma; w)$ and

$$\theta(\beta) = 1, \qquad \theta^{[1]}(\beta) = 0, \qquad \theta(\gamma) = \theta^{[1]}(\gamma) = 0.$$

We extend this function to $[a, b)$ by setting $\theta(x) = 1$ for $x \in [a, \beta)$ and $\theta(x) = 0$ for $x \in (\gamma, b)$. The extended function θ therefore has the properties: $\theta, \theta^{[1]} \in AC[a, b)$ and $(\theta^{[1]})' \in L^2(a, b; w)$.

For $\phi \in \mathfrak{D}(\tau)$ and $x \in [a, \gamma]$,

$$\begin{aligned}(\theta\phi)(x) &= -\int_x^{\gamma} (\theta\phi)' \\ &= -\int_x^{\gamma} \frac{1}{p}(\theta\phi^{[1]} + \theta^{[1]}\phi)\end{aligned}$$

and hence

$$\begin{aligned}|(\theta\phi)(x)| &\leqslant \int_x^{\gamma} \frac{1}{|p|}|\theta\phi^{[1]} - \theta^{[1]}\phi| + 2\int_a^{\gamma} \frac{1}{|p|}|\theta^{[1]}\phi| \\ &\leqslant \int_x^{\gamma} \frac{1}{|p|}|\theta\phi^{[1]} - \theta^{[1]}\phi| + K_{\gamma}\|\phi\|.\end{aligned} \tag{10.42}$$

Also, we have, since $\theta^{[1]}(\gamma) = 0$,

$$\begin{aligned}(\theta\phi^{[1]})(x) &= -\int_x^{\gamma} (\theta\phi^{[1]})' \\ &= -\int_x^{\gamma} (\theta\phi^{[1]\prime} + \theta'\phi^{[1]}) \\ &= \int_x^{\gamma} \theta\left(\tau\phi - \frac{r}{p}\phi^{[1]} - q\phi\right) + (\theta^{[1]}\phi)(x) + \int_x^{\gamma} \phi\theta^{[1]\prime}\end{aligned}$$

and

$$(\theta\phi^{[1]} - \theta^{[1]}\phi)(x) = \int_x^{\gamma}\left[\theta\left(\tau\phi - \frac{r}{p}\phi^{[1]} - q\phi\right) + \phi\theta^{[1]\prime}\right].$$

Hence

$$
\begin{aligned}
|(\theta\phi^{[1]}-\theta^{[1]}\phi)(x)| &\leqslant K_\gamma(\|\tau\phi\|+\|\phi\|)+\int_x^\gamma\left(\frac{|r|}{|p|}|\theta\phi^{[1]}|+|q|\,|\theta\phi|\right)\\
&\leqslant K_\gamma(\|\tau\phi\|+\|\phi\|)+\\
&\quad+\int_x^\gamma\left(\frac{|r|}{|p|}|\theta\phi^{[1]}-\theta^{[1]}\phi|+\frac{|r|}{|p|}|\theta^{[1]}\phi|+|q|\,|\theta\phi|\right)\\
&\leqslant K_\gamma(\|\tau\phi\|+\|\phi\|)\\
&\quad+\int_x^\gamma\left(\frac{|r|}{|p|}|\theta\phi^{[1]}-\theta^{[1]}\phi|+|q||\theta\phi|\right). \qquad (10.43)
\end{aligned}
$$

Let $\Psi(x)=|(\theta\phi)(x)|+|(\theta\phi^{[1]}-\theta^{[1]}\phi)(x)|$. Then, from (10.42) and (10.43), for $x\in[a,\gamma]$,

$$\Psi(x)\leqslant K_\gamma(\|\phi\|+\|\tau\phi\|)+\int_x^\gamma\left(\frac{1}{|p|}+\frac{|r|}{|p|}+|q|\right)\Psi. \qquad (10.44)$$

Let $\alpha=K_\gamma(\|\phi\|+\|\tau\phi\|)$ and $R(x)=\int_x^\gamma\left(\frac{1}{|p|}+\frac{|r|}{|p|}+|q|\right)\Psi$ in (10.44). Then from (10.44),

$$
\begin{aligned}
R'(x) &= -\left(\frac{1}{|p|}+\frac{|r|}{|p|}+|q|\right)(x)\Psi(x)\\
&\geqslant -\left(\frac{1}{|p|}+\frac{|r|}{|p|}+|q|\right)(x)[\alpha+R(x)]
\end{aligned}
$$

and

$$
\begin{aligned}
&\frac{\mathrm{d}}{\mathrm{d}x}\left\{R(x)\exp\left[-\int_x^\gamma\left(\frac{1}{|p|}+\frac{|r|}{|p|}+|q|\right)\right]\right\}\\
&\qquad\geqslant -\alpha\left(\frac{1}{|p|}+\frac{|r|}{|p|}+|q|\right)(x)\exp\left[-\int_x^\gamma\left(\frac{1}{|p|}+\frac{|r|}{|p|}+|q|\right)\mathrm{d}t\right].
\end{aligned}
$$

On integrating over $[x,\gamma]$ we obtain

$$R(x)\leqslant\alpha\exp\left[\int_x^\gamma\left(\frac{1}{|p|}+\frac{|r|}{|p|}+|q|\right)-1\right]$$

and from (10.44),

$$\Psi(x)\leqslant\alpha\exp\int_x^\gamma\left(\frac{1}{|p|}+\frac{|r|}{|p|}+|q|\right).$$

What we have just proved is the Gronwall inequality for (10.44). Since $\Psi(x)=|\phi(x)|+|\phi^{[1]}(x)|$ for $x\in[a,\beta]$, (10.41) is proved. □

An immediate consequence of (10.41) is that if a sequence (ϕ_n) in $\mathfrak{D}(\tau)$ is such that (ϕ_n) and $(\tau\phi_n)$ are convergent in $L^2(a, b; w)$, then $(\phi_n(x))$ and $(\phi_n^{[1]}(x))$ are uniformly convergent on compact subsets of $[a, b)$.

If we assume that $p \in C^1[a, b)$, τ is a differential expression and the restriction of τ to $C_0^\infty(a, b)$ is a densely defined operator in $L^2(a, b; w)$ as long as $q \in L^2_{\text{loc}}(a, b)$. This operator has a smaller domain than $T_0(\tau)$ but in fact we have the following important results. This is the natural place for them although they involve properties of the Sobolev spaces $W^{2,2}(a, b)$ and $W_0^{2,2}(a, b)$ which are not derived until Chapters V and IX. The space $W^{2,2}(a, b)$ consists of functions ϕ which are such that $\phi' \in AC_{\text{loc}}[a, b)$ and $\phi, \phi', \phi'' \in L^2(a, b)$. It is a Hilbert space with inner product

$$(\phi, \psi)_{2,2,(a,b)} := \int_a^b (\phi''\bar{\psi}'' + \phi'\bar{\psi}' + \phi\bar{\psi});$$

$W_0^{2,2}(a, b)$ is the closure of $C_0^\infty(a, b)$ in $W^{2,2}(a, b)$. In general, Sobolev spaces are defined in terms of distributions but the above description is established in Corollary V.3.9.

Theorem 10.18. Suppose that a is a regular end-point of (10.1) and that

$$\left.\begin{array}{ll}\text{(i) } p > 0, & p \in C^1[a, b),\\ \text{(ii) } r \in C^1[a, b), & \\ \text{(iii) } w > 0, & 1/w \in L^\infty_{\text{loc}}[a, b),\\ \text{(iv) } q \in L^2_{\text{loc}}(a, b). & \end{array}\right\} \tag{10.45}$$

Then $C_0^\infty(a, b)$ is a core of $T_0(\tau)$. ∎

Proof. It is sufficient to show that, for $\phi \in \mathfrak{D}_0'(\tau)$ (see (10.25)), there exists a sequence (ϕ_n) in $C_0^\infty(a, b)$ such that $\phi_n \to \phi$ and $\tau\phi_n \to \tau\phi$ in $L^2(a, b; w)$. If $\operatorname{supp}\phi \equiv I \subset [a, b)$, we have from Theorem 10.17 and (10.45) that ϕ and ϕ' are bounded and also

$$\begin{aligned}\phi'' &= -(1/p)(\tau\phi + p'\phi - r\phi' - q\phi)\\ &\in L^2(a, b).\end{aligned}$$

Thus $\phi \in W^{2,2}(a, b)$ and since $\phi(a) = \phi'(a) = 0$ while $\operatorname{supp}\phi \subset [a, b)$, we conclude from Lemma IX.7.1 that $\phi \in W_0^{2,2}(a, b)$. Consequently there exist $C_0^\infty(a, b)$ functions ϕ_n $(n \in \mathbb{N})$ such that $\phi_n \to \phi$ in $W^{2,2}(a, b)$. Also given any closed subinterval I_0 of $[a, b)$ containing I, we may suppose that $\operatorname{supp}\phi_n \subset I_0$ for each n. The proof follows easily from this. □

Theorem 10.19. Let a be a regular point of (10.1) and suppose that in addition to (10.45),

$$p, \frac{1}{p}, p', r, q, w, \frac{1}{w} \in L^\infty(a, b). \tag{10.46}$$

Then $\mathfrak{D}_0(\tau) = W_0^{2,2}(a,b)$ and $\mathfrak{D}(\tau) = W^{2,2}(a,b)$. ■

Proof. In view of Theorem 10.18 it is sufficient to prove that $H(T(\tau)) = W^{2,2}(a,b)$; recall that $H(T(\tau))$ is the space determined by $\mathfrak{D}(\tau)$ and the graph norm of $T(\tau)$. If $\phi \in W^{2,2}(a,b)$ then $\phi' \in AC_{loc}[a,b)$ and $\tau\phi \in L^2(a,b;w)$ on account of (10.46). Thus $\phi \in \mathfrak{D}(\tau)$, and since it may easily be shown that $\|\tau\phi\| \leqslant K\|\phi\|_{2,2,(a,b)}$ we see that $W^{2,2}(a,b) \subset H(T(\tau))$ both algebraically and topologically, that is, the identification map is continuous. Conversely, let $\phi \in \mathfrak{D}(\tau)$ and let $[a,b) = \bigcup_{i=1}^{\infty} I_i$ where the I_i are disjoint intervals of length not exceeding 1. From (10.46) and the equation

$$\phi'' = -(1/p)(\tau\phi + p'\phi' - r\phi' - q\phi)$$

it follows that there exists a constant $K > 0$ such that

$$|\phi''| \leqslant K(|\tau\phi| + |\phi'| + |\phi|)$$

and hence, for each $i \in \mathbb{N}$,

$$\int_{I_i} |\phi''|^2 \leqslant K\left(\int_{I_i} |\tau\phi|^2 w + \int_{I_i} |\phi'|^2 + \int_{I_i} |\phi|^2 w\right).$$

Since each I_i has length $\leqslant 1$, we have from Theorem V.4.14 and a similarity transformation that, given any $\varepsilon > 0$ there exists a constant K_ε, which is independent of I_i, such that

$$\int_{I_i} |\phi'|^2 \leqslant \varepsilon \int_{I_i} |\phi''|^2 + K_\varepsilon \int_{I_i} |\phi|^2.$$

We therefore conclude that

$$\int_{I_i} |\phi''|^2 \leqslant K\left(\int_{I_i} |\tau\phi|^2 w + \int_{I_i} |\phi|^2 w\right),$$

and this in turn gives $\phi'' \in L^2(a,b)$, $\phi' \in L^2(a,b)$, and

$$\|\phi''\|^2_{L^2(a,b)} \leqslant K(\|\tau\phi\|^2 + \|\phi\|^2), \qquad \|\phi'\|^2_{L^2(a,b)} \leqslant \varepsilon\|\tau\phi\|^2 + K_\varepsilon\|\phi\|^2.$$

Consequently $\phi \in W^{2,2}(a,b)$ and the proof is complete. □

10.4. *The case of two singular end-points*

We now define $T(\tau)$ and $T_0'(\tau)$ to be the restrictions of $(1/w)\tau$ to the subspaces

$$\mathfrak{D}(\tau) := \{\phi : \phi, \phi^{[1]} \in AC_{loc}(a,b) \text{ and } \phi, (1/w)\tau\phi \in L^2(a,b;w)\},$$

$$\mathfrak{D}_0'(\tau) := \{\phi : \phi \in \mathfrak{D}(\tau),\ \phi = 0 \text{ outside a compact subset of } (a,b)\}.$$

It follows by a proof similar to that of Theorem 10.7 that $\mathfrak{D}_0'(\tau)$ is dense in

$L^2(a,b;w)$ and $T_0'(\tau)$ is closable. Also if $T_0(\tau)$ denotes the closure of $T_0'(\tau)$,

$$T_0(\tau)^* = T(\tau^+), \qquad T_0(\tau^+) = T(\tau)^*. \tag{10.47}$$

The problem on (a,b) is effectively reduced to the problems with one singular end-point on the intervals $(a,c]$ and $[c,b)$, where $c \in (a,b)$, in the following way. First we define $T(\tau;a)$ and $T(\tau;b)$ to be the operators determined by $(1/w)\tau$ on the domains

$$\mathfrak{D}(\tau,a) := \{\phi : \phi, \phi^{[1]} \in AC_{loc}(a,c];\quad \phi, (1/w)\tau\phi \in L^2(a,c;w)\},$$
$$\mathfrak{D}(\tau,b) := \{\phi : \phi, \phi^{[1]} \in AC_{loc}[c,b);\quad \phi, (1/w)\tau\phi \in L^2(c,b;w)\},$$

and denote by $T_0(\tau;a)$ and $T_0(\tau;b)$ the closures of the operators $T_0'(\tau;a)$ and $T_0'(\tau;b)$ defined in §10.3 on the intervals $(a,c]$ and $[c,b)$ respectively. Let $\tilde{T}_0'(\tau)$ be the orthogonal sum

$$\tilde{T}_0'(\tau) = T_0'(\tau;a) \oplus T_0'(\tau;b)$$

in

$$L^2(a,b;w) = L^2(a,c;w) \oplus L^2(c,b;w).$$

It is easy to verify that $\tilde{T}_0'(\tau)$ is densely defined and closable in $L^2(a,b;w)$ and its closure $\tilde{T}_0(\tau)$ is given by

$$\tilde{T}_0(\tau) = T_0(\tau;a) \oplus T_0(\tau;b).$$

Also

$$\operatorname{nul}[\tilde{T}_0(\tau) - \lambda I] = \operatorname{nul}[T_0(\tau;a) - \lambda I] + \operatorname{nul}[T_0(\tau;b) - \lambda I],$$
$$\operatorname{def}[\tilde{T}_0(\tau) - \lambda I] = \operatorname{def}[T_0(\tau;a) - \lambda I] + \operatorname{def}[T_0(\tau;b) - \lambda I],$$

and $\mathscr{R}[\tilde{T}_0(\tau) - \lambda I]$ is closed if, and only if, $\mathscr{R}[T_0(\tau;a) - \lambda I]$ and $\mathscr{R}[T_0(\tau;b) - \lambda I]$ are both closed. These results imply in particular that

$$\Pi[\tilde{T}_0(\tau)] = \Pi[T_0(\tau;a)] \cap \Pi[T_0(\tau;b)].$$

Theorem 10.20. $\tilde{T}_0(\tau) \subset T_0(\tau)$, $T(\tau) \subset T(\tau,a) \oplus T(\tau,b)$ and

$$\dim \mathscr{D}[T_0(\tau)]/\mathscr{D}[\tilde{T}_0(\tau)] = 2.$$

If $\lambda \in \Pi[\tilde{T}_0(\tau)] \cap \Delta_3[T_0(\tau) - \lambda I]$,

$$\operatorname{ind}[T_0(\tau) - \lambda I] = 2 - \operatorname{def}[T_0(\tau;a) - \lambda I] - \operatorname{def}[T_0(\tau;b) - \lambda I]$$

and in particular, if $\lambda \in \Pi[T_0(\tau)]$,

$$\operatorname{def}[T_0(\tau) - \lambda I] = \operatorname{def}[T_0(\tau;a) - \lambda I] + \operatorname{def}[T_0(\tau;b) - \lambda I] - 2. \qquad \blacksquare$$

Proof. It is clear that $\tilde{T}_0'(\tau) \subset T_0'(\tau)$ and hence, on taking closures, $\tilde{T}_0(\tau) \subset T_0(\tau)$. That $T(\tau) \subset T(\tau;a) \oplus T(\tau;b)$ is obvious. The rest of the proof will follow from Theorem 3.1 once we have shown that

$$\mathscr{D}[\tilde{T}_0(\tau)] = \{\phi : \phi \in \mathscr{D}[T_0(\tau)] \quad \text{and} \quad \phi(c) = \phi^{[1]}(c) = 0\}. \tag{10.48}$$

To simplify the notation, we shall denote the intervals (a, b), $(a, c]$ and $[c, b)$ by I, I_1 and I_2 respectively and write T_{0i} $(i = 1, 2)$ for $T_0(\tau; a)$ and $T_0(\tau; b)$ respectively. Also H_i $(i = 1, 2)$ will stand for $L^2(a, c; w)$ and $L^2(c, b; w)$.

If $\phi \in \mathscr{D}[\tilde{T}_0(\tau)]$ then $\phi = \phi_1 + \phi_2$ where $\phi_i \in \mathscr{D}(T_{0i})$ $(i = 1, 2)$ and so, by definition, there exist sequences $(\phi_{n,i})$ in $\mathscr{D}(T'_{0i})$ such that $\phi_{n,i} \to \phi_i$ and $(1/w)\tau\phi_{n,i} \to (1/w)\tau\phi_i$ in H_i. Since $\phi_{n,i}(c) = \phi_{n,i}^{[1]}(c) = 0$ we have that $\phi^{(n)} := \phi_{n,1} + \phi_{n,2} \in \mathscr{D}[T'_0(\tau)]$ and as $\phi^{(n)} \to \phi$ and $(1/w)\tau\phi^{(n)} \to (1/w)\tau\phi$ in $L^2(a, b; w)$ we conclude that $\phi \in \mathscr{D}[T_0(\tau)]$. Furthermore $\phi(c) = \phi^{[1]}(c) = 0$, by (10.32).

Conversely, let $\phi \in \mathscr{D}[T_0(\tau)]$ and suppose that $\phi(c) = \phi^{[1]}(c) = 0$. There exists a sequence $(\phi^{(n)})$ in $\mathscr{D}[T'_0(\tau)]$ such that $\phi^{(n)} \to \phi$ and $(1/w)\tau\phi^{(n)} \to (1/w)\tau\phi$ in $L^2(a, b; w)$. Let $\phi_{n,i}$ denote the restriction of $\phi^{(n)}$ to I_i and similarly define ϕ_i. Clearly, $\phi_{n,1} \in \mathfrak{D}(\tau, a)$, $\phi_{n,2} \in \mathfrak{D}(\tau, b)$, and $\phi_{n,i} \to \phi_i$ and $(1/w)\tau\phi_{n,i} \to (1/w)\tau\phi_i$ in H_i. Since $\phi(c) = \phi^{[1]}(c) = 0$ we conclude from (10.31) that $\phi_{n,i}(c) \to 0$ and $\phi_{n,i}^{[1]}(c) \to 0$. Now let $\theta_1, \theta_2 \in \mathfrak{D}(\tau, a)$ be such that

$$\theta_1(c) = 1, \quad \theta_1^{[1]}(c) = 0, \qquad \theta_2(c) = 0, \quad \theta_2^{[1]}(c) = 1$$

and $\theta_j(x) = \theta_j^{[1]}(x) = 0$ for $x \in (a, \gamma]$, where $\gamma \in (a, c)$; the existence of such functions is guaranteed by Lemma 10.4. Define

$$\psi_n(x) = \phi_{n,1}(x) - [\phi_{n,1}(c)\theta_1(x) + \phi_{n,1}^{[1]}(c)\theta_2(x)] \qquad (x \in (a, c]).$$

Then $\psi_n(c) = \psi_n^{[1]}(c) = 0$, and $\psi_n \to \phi_1$ and $(1/w)\tau\psi_n \to (1/w)\tau\phi_1$ in H_1. Consequently $\phi_1 \in \mathscr{D}(T_{01})$. A similar argument gives $\phi_2 \in \mathscr{D}(T_{02})$ and hence $\phi = \phi_1 + \phi_2 \in \mathscr{D}[\tilde{T}_0(\tau)]$. This proves (10.48) and the remainder of the theorem follows on appeal to Theorem 3.1. □

Corollary 10.21. Let $\Pi[T_0(\tau), T_0(\tau^+)] \neq \varnothing$. Then $T_0(\tau) = T(\tau)$ and $T_0(\tau^+) = T(\tau^+)$ if, and only if, $\operatorname{def}[T_0(\tau; a) - \lambda I] + \operatorname{def}[T_0(\tau^+; a) - \bar{\lambda} I] = \operatorname{def}[T_0(\tau; b) - \lambda I] + \operatorname{def}[T_0(\tau^+; b) - \bar{\lambda} I] = 2$ for some, and hence all $\lambda \in \Pi[T_0(\tau), T_0(\tau^+)]$.

If $\tau^+ = \tau$ then $T_0(\tau)$ is self-adjoint if, and only if, $\tau\phi = \lambda w\phi$ is in the limit point case at a and b. If $\tau^+ = \bar{\tau}$ then $T_0(\tau)$ is J-self-adjoint if, and only if, $\operatorname{def}[T_0(\tau; a) - \lambda I] = \operatorname{def}[T_0(\tau; b) - \lambda I] = 1$ for all $\lambda \in \Pi[T_0(\tau)]$. ■

Proof. We first observe that since $T_0(\tau) \supset \tilde{T}_0(\tau)$,

$$\Pi := \Pi[T_0(\tau), T_0(\tau^+)] \subset \Pi[T_0(\tau; a), T_0(\tau^+; a)] \cap \Pi[T_0(\tau; b), T_0(\tau^+; b)]$$
$$=: \Pi_1 \cap \Pi_2$$

say. Let $m(\lambda) = \operatorname{def}[T_0(\tau) - \lambda I] + \operatorname{def}[T_0(\tau^+) - \bar{\lambda} I]$ and define $m_1(\lambda)$ and $m_2(\lambda)$ similarly for $T_0(\tau; a)$ and $T_0(\tau; b)$ respectively. From Corollary 3.2, $m(\lambda)$, $m_1(\lambda)$ and $m_2(\lambda)$ are constant for λ in Π, Π_1 and Π_2 respectively, and from

Theorem 10.7, $2 \leqslant m_i(\lambda) \leqslant 4$ for $\lambda \in \Pi_i$ $(i = 1, 2)$. Moreover, Theorem 10.20 yields

$$m(\lambda) = m_1(\lambda) + m_2(\lambda) - 4 \qquad (\lambda \in \Pi),$$

and we conclude that $m(\lambda) = 0$ for some, and hence all, $\lambda \in \Pi$ if, and only if, $m_1(\lambda) = m_2(\lambda) = 2$. But if $m(\lambda) = 0$ we must have $\operatorname{def}[T_0(\tau) - \lambda I] = \operatorname{def}[T_0(\tau^+) - \bar{\lambda} I] = 0$, whence $T_0(\tau) = T(\tau)$ and $T_0(\tau^+) = T(\tau^+)$. Conversely, if $T_0(\tau) = T(\tau)$ and $T_0(\tau^+) = T(\tau)$, then for $\lambda \in \Pi$ we have $m(\lambda) = 0$. In the remainder of the theorem, $m_1(\lambda) = 2\operatorname{def}[T_0(\tau, a) - \lambda I]$ and $m_2(\lambda) = 2\operatorname{def}[T_0(\tau; b) - \lambda I]$, by Theorem 10.7. □

10.5. *The limit-point, limit-circle results of Sims*

Let $\tau\phi = -(p\phi')' + q\phi$ and suppose (10.4) is satisfied on $[a, b)$ with $w = 1$; a is therefore a regular end-point of (10.1). Furthermore, assume that $\operatorname{im} q$ is semibounded on (a, b); for definiteness we suppose that

$$\operatorname{im} q(x) \leqslant M \quad \text{for a.e. } x \in (a, b), \tag{10.49}$$

the case when $\operatorname{im} q$ is bounded below being similar. Sims applied the complex-variable method of Weyl to obtain an analogue for τ of the limit-point, limit-circle theory. We shall now deduce Sims' basic results from the results of this chapter. The key result is the following.

Theorem 10.22. For any $\lambda \in \mathbb{C}$ with $\operatorname{im} \lambda > M$ there exists a solution ϕ of $(\tau - \lambda)\phi = 0$ in (a, b) such that

$$\int_a^b \operatorname{im}(\lambda - q)|\phi|^2 < \infty. \tag{10.50}$$ ■

Proof. Let $\tau_\lambda = \tau - \lambda$ and $\operatorname{im} \lambda > M$. We shall apply Theorem 10.7 to τ_λ and with weight $w = \operatorname{im}(\lambda - q)$. For $\phi \in \mathscr{D}[T_0(\tau_\lambda)]$,

$$\operatorname{im}(T_0(\tau_\lambda)\phi, \phi) = \operatorname{im} \int_a^b [p|\phi'|^2 + (q - \lambda)|\phi|^2]$$

$$= \int_a^b \operatorname{im}(q - \lambda)|\phi|^2 = -\|\phi\|^2.$$

Hence $0 \in \mathbb{C} \setminus \bar{\Theta}[T_0(\tau_\lambda)] \subset \Pi[T_0(\tau_\lambda)]$ and we conclude from Theorem 10.7 that $\operatorname{def} T_0(\tau_\lambda) = \operatorname{nul} T(\bar{\tau}_\lambda) \geqslant 1$. There is therefore at least one solution of $\tau_\lambda \phi = 0$, or $\tau\phi = \lambda\phi$, in $L^2(a, b; w)$ and the theorem is proved. □

Next, we rule out the possibility of there being a different number of solutions of $\tau\phi = \lambda\phi$ satisfying (10.50) for different values of λ with $\operatorname{im} \lambda > M$. We do this by means of an analogue of Theorem 10.10.

Theorem 10.23. If, for some $\lambda_0 \in \mathbb{C}$, all solutions of $\tau\phi = \lambda_0 \phi$ are in $L^2(a, b) \cap L^2(a, b; |\text{im}\, q|)$ then, for all $\lambda \in \mathbb{C}$, all solutions of $\tau\phi = \lambda\phi$ are in $L^2(a, b) \cap L^2(a, b; |\text{im}\, q|)$. ■

Proof. Let ϕ_1 and ϕ_2 be linearly independent solutions of $\tau\phi = \lambda_0 \phi$ which are in $L^2(a, b) \cap L^2(a, b; |\text{im}\, q|)$ and also satisfy, for some $c \in [a, b)$, the equation $(\phi_1 \phi_2^{[1]} - \phi_1^{[1]} \phi_2)(c) = 1$. By Lemma 10.9 any solution of $\tau\phi = \lambda\phi$ satisfies

$$\phi(x) = c_1 \phi_1(x) + c_2 \phi_2(x) + (\lambda - \lambda_0) \int_c^x [\phi_1(x)\phi_2(t) - \phi_1(t)\phi_2(x)]\phi(t)\mathrm{d}t$$

for some c_1, c_2 in $\mathbb{C}$. Since ϕ_1 and ϕ_2 are in $L^2(a, b)$, Theorem 10.10 yields $\phi \in L^2(a, b)$. By the same argument as in the proof of Theorem 10.10 we get

$$|\phi(x)| \leqslant |c_1|\, |\phi_1(x)| + |c_2|\, |\phi_2(x)| + |\lambda - \lambda_0| K_1 \|\phi\| (|\phi_1(x)| + |\phi_2(x)|)$$

where $K_1 \geqslant \|\phi_j\|_{L^2(c, b)}$, and hence, with $w = |\text{im}\, q|$,

$$w^{\frac{1}{2}}(x)|\phi(x)| \leqslant |c_1| w^{\frac{1}{2}}(x)|\phi_1(x)| + |c_2| w^{\frac{1}{2}}(x)|\phi_2(x)|$$
$$+ |\lambda - \lambda_0| K_1 \|\phi\| w^{\frac{1}{2}}(x)(|\phi_1(x)| + |\phi_2(x)|)$$

The rest of the proof follows that of Theorem 10.10. □

Theorems 10.22 and 10.23 imply that the number of solutions of $\tau\phi = \lambda\phi$ in $L^2(a, b; \text{im}(\lambda - q))$ is constant for all λ with $\text{im}\, \lambda > M$, this number being either 1 or 2. Since any $\phi \in L^2(a, b; \text{im}(\lambda - q))$ lies in $L^2(a, b)$ but not conversely there are three possible cases:

Definition 10.24. The equation $\tau\phi = \lambda\phi$ is said to be in *case I, II*, or *III* at b according to the following circumstances:
Case *I* There is precisely one $L^2(a, b)$ solution for $\text{im}\, \lambda > M$.
Case *II* There is precisely one $L^2(a, b; \text{im}\, [\lambda - q])$ solution for $\text{im}\, \lambda > M$ but all solutions are in $L^2(a, b)$ (for all $\lambda \in \mathbb{C}$).
Case *III* All solutions are in $L^2(a, b) \cap L^2(a, b; |\text{im}\, q|)$ for all $\lambda \in \mathbb{C}$. ■

If q is real, Case *II* is vacuous and Cases *I* and *III* become Weyl's limit-point and limit-circle cases respectively. We shall prove in the next section that each of the three cases in Definition 10.24 does exist. Case *I* prevails if, and only if, $\text{def}[T_0(\tau) - \lambda I] = 1$ for $\lambda \in \Pi[T_0(\tau)]$. Hence, when both a and b are singular end-points, we see from Corollary 10.21 that *$T_0(\tau)$ is J-self-adjoint in $L^2(a, b)$ if, and only if, $\tau\phi = \lambda\phi$ is in Case I at a and b.*

While Theorems 10.14 and 10.15 characterize all regularly solvable extensions of $T_0(\tau)$, more can be said under the special circumstances of this section. For $\text{im}\, \lambda > M$, some J-self-adjoint extensions of $T_0(\tau)$ can be described in terms of linearly independent solutions θ and ψ of $\tau u = \lambda u$ in $[a, b)$ which satisfy the following conditions:

$$[\psi, \bar{\theta}] = 1, \qquad \psi \in L^2(a, b) \cap L^2(a, b; |\text{im}\, q|). \tag{10.51}$$

Since θ and ψ are linearly independent solutions of $\tau u = \lambda u$, it follows that $[\theta, \bar{\psi}](\bullet)$ is a non-zero constant on $[a, b)$ and so the first requirement in (10.15) involves no loss of generality. Furthermore, the existence of a ψ satisfying the second condition in (10.51) is guaranteed by Theorem 10.22.

We need the following preliminary result about the operator R_λ defined on $L^2(a,b)$ by

$$(R_\lambda f)(x) = \psi(x,\lambda) \int_a^x \theta(t,\lambda) f(t)\mathrm{d}t + \theta(x,\lambda) \int_x^b \psi(t,\lambda) f(t)\mathrm{d}t, \quad \operatorname{im}\lambda > M. \tag{10.52}$$

Let $\|(1+|\operatorname{im} q|)^{\frac{1}{2}} \bullet\|$ be the norm on $L^2(a,b) \cap L^2(a,b;|\operatorname{im} q|)$.

Lemma 10.25. Suppose τ is in case *II* or *III* at b. Then R_λ is a bounded linear map of $L^2(a,b)$ into $L^2(a,b) \cap L^2(a,b;|\operatorname{im} q|)$. The range of R_λ is

$$\Lambda = \{u : u \in \mathfrak{D}(\tau), \quad [u, \bar{\theta}](a) = [u, \bar{\psi}](b) = 0\}$$

and $(\tau - \lambda)R_\lambda f = f$ for all $f \in L^2(a,b)$. ■

Proof. Let $\Phi = R_\lambda f$ and $f \in L^2(a,b)$. Then $\Phi \in AC_{\mathrm{loc}}[a,b)$ and

$$\Phi^{[1]}(x) = \psi^{[1]}(x,\lambda) \int_a^x \theta(t,\lambda) f(t)\mathrm{d}t + \theta^{[1]}(x,\lambda) \int_x^b \psi(t,\lambda) f(t)\mathrm{d}t.$$

Hence $\Phi^{[1]} \in AC_{\mathrm{loc}}[a,b)$,

$$\Phi^{[1]\prime}(x) = \psi^{[1]\prime}(x,\lambda) \int_a^x \theta f + \theta^{[1]\prime}(x,\lambda) \int_x^b \psi f + [\theta, \bar{\psi}](x) f(x)$$

and, from (10.51),

$$(\tau - \lambda)\Phi = f.$$

Furthermore $[\Phi, \bar{\theta}](a) = [\Phi, \bar{\psi}](b) = 0$.

To prove that R_λ is a bounded map of $L^2(a,b)$ into $L^2(a,b) \cap L^2(a,b;|\operatorname{im} q|)$ we first consider functions $f \in L^2(a,b)$ which vanish on $[\beta, b)$ for some $\beta \in (a,b)$. Since $(\tau - \lambda)\Phi = f$, we have

$$\begin{aligned} 2\mathrm{i} \int_a^\beta \operatorname{im}[f\bar{\Phi}] &= \int_a^\beta [\bar{\Phi}(\tau - \lambda)\Phi - \Phi(\bar{\tau} - \bar{\lambda})\bar{\Phi}] \\ &= 2\mathrm{i} \int_a^\beta \operatorname{im}(q - \lambda)|\Phi|^2 + [\Phi, \Phi](\beta) - [\Phi, \Phi](a). \end{aligned} \tag{10.53}$$

We also have

$$[\Phi, \Phi](\beta) = [\psi, \psi](\beta) \left| \int_a^\beta \theta f \right|^2, \qquad [\Phi, \Phi](a) = [\theta, \theta](a) \left| \int_a^\beta \psi f \right|^2$$

and

$$\begin{aligned}[\psi,\psi](\beta)-[\psi,\psi](a) &= \int_a^\beta [\bar{\psi}(-\psi^{[1]})'-\psi(-\bar{\psi}^{[1]})'] \\ &= \int_a^\beta [\bar{\psi}(\tau-q)\psi-\psi(\bar{\tau}-\bar{q})\bar{\psi}] \\ &= 2\mathrm{i}\int_a^\beta \operatorname{im}(\lambda-q)|\psi|^2.\end{aligned}$$

Since $\theta \in \mathrm{L}^2(a,b)$ and $\psi \in \mathrm{L}^2(a,b) \cap \mathrm{L}^2(a,b;|\operatorname{im} q|)$, there exists a positive constant K such that

$$|[\Phi,\Phi](\beta)| \leqslant K\|f\|^2, \qquad |[\Phi,\Phi](a)| \leqslant K\|f\|^2,$$

where $\|f\|^2 = \int_a^b |f|^2$. On substituting these inequalities in (10.53) we obtain

$$\int_a^\beta \operatorname{im}(\lambda-q)|\Phi|^2 \leqslant \frac{1}{\gamma-M}\|f\|\left(\int_a^\beta \operatorname{im}(\lambda-q)|\Phi|^2\right)^{\frac{1}{2}} + K\|f\|^2$$

and this yields

$$\int_a^\beta \operatorname{im}(\lambda-q)|\Phi|^2 \leqslant K\|f\|^2. \tag{10.54}$$

Suppose now that f is arbitrary in $\mathrm{L}^2(a,b)$ and let $f_\beta(x) = f(x)$ for $x \in [a,\beta]$ and $f_\beta(x) = 0$ in (β,b). If $\Phi_\beta := R_\lambda f_\beta$ it is readily seen that $\Phi_\beta(x) \to \Phi(x)$ as $\beta \to \infty$, uniformly on compact sub-intervals of $[a,b)$. Hence if β' is fixed, we see from (10.54) that

$$\begin{aligned}\int_a^{\beta'} \operatorname{im}(\lambda-q)|\Phi|^2 &= \lim_{\beta\to b-}\int_a^{\beta'} \operatorname{im}(\lambda-q)|\Phi_\beta|^2 \\ &\leqslant K\|f\|^2.\end{aligned}$$

On allowing $\beta' \to b-$, it follows that R_λ is bounded as a map from $\mathrm{L}^2(a,b)$ into $\mathrm{L}^2(a,b) \cap \mathrm{L}^2(a,b;|\operatorname{im} q|)$. We have also shown that $\mathscr{R}(R_\lambda) \subset \Lambda$.

Finally we prove that Λ is the range of R_λ. If $u \in \Lambda$ then $(\tau-\lambda)u \in \mathrm{L}^2(a,b)$ and

$$\begin{aligned}&(R_\lambda[(\tau-\lambda)u])(x) \\ &= \psi(x)\int_a^x \theta[-(u^{[1]})'+(q-\lambda)u] + \theta(x)\int_x^b \psi[-(u^{[1]})'+(q-\lambda)u] \\ &= \psi(x)([u,\bar{\theta}](x)-[u,\bar{\theta}](a)) + \theta(x)([u,\bar{\psi}](b)-[u,\bar{\psi}](x))\end{aligned}$$

(on integrating by parts and using $(\tau-\lambda)\theta = (\tau-\lambda)\psi = 0$)

$$= u(x)[\psi,\bar{\theta}] = u(x),$$

by (10.51). Consequently Λ is the range of R_λ and the proof is concluded. □

Retaining the notation of Lemma 10.25 we have

Theorem 10.26. Suppose τ is in case *II* or *III* at b and let $S(\tau)$ be the restriction of $T(\tau)$ to the subspace Λ in Lemma 10.25. Then $S(\tau)$ is J-self-adjoint, $\{\lambda : \text{im}\,\lambda > M\} \subset \rho(S(\tau))$, and $R_\lambda = [S(\tau) - \lambda I]^{-1}$. If τ is in case *II* at b,

$$\Lambda = \{u : u \in \mathfrak{D}(\tau),\ [u, \bar{\theta}](a) = 0,\ u \in L^2(a, b; |\text{im}\, q|)\}. \qquad (10.55)$$ ■

Proof. By Lemma 10.25, R_λ is the resolvent of $S(\tau)$ for $\text{im}\,\lambda > M$ and $S(\tau)$ is a closed operator with $\text{def}[S(\tau) - \lambda I] = 0$. If $u \in \mathcal{N}[S(\tau) - \lambda I]$ we must have that $u = c_1\theta + c_2\psi$ and use of $[u, \bar{\theta}](a) = [u, \bar{\psi}](b) = 0$ gives $c_1 = c_2 = 0$. Hence $\text{nul}[S(\tau) - \lambda I] = 0$ and $\lambda \in \rho[S(\tau)]$. To establish the J-self-adjointness of $S(\tau)$ we must prove that $S(\tau)$ is J-symmetric. First we note that if $u, v \in \Lambda$ then $u = R_\lambda f$ and $v = R_\lambda g$ for some $f, g \in L^2(a, b)$ and hence from

$$[u, \bar{v}](x) = [(R_\lambda f)(R_\lambda g)^{[1]} - (R_\lambda f)^{[1]}(R_\lambda g)](x)$$

we get $[u, \bar{v}](a) = [u, \bar{v}](b) = 0$. This implies that $S(\tau)$ is J-symmetric and consequently, by Theorem 5.5, it is J-self-adjoint.

If u lies in the subspace Λ in (10.55), we see from Lemma 10.25 that $u - R_\lambda(\tau - \lambda)u$ lies in $\mathcal{N}[T(\tau) - \lambda I] \cap L^2(a, b; |\text{im}\, q|)$. Therefore in case II, $u - R_\lambda(\tau - \lambda)u = K\psi$ for some constant K. Since $[u, \bar{\theta}](a) = 0$, this gives $K = 0$ and hence $u = R_\lambda(\tau - \lambda)u$ lies in $\mathcal{D}[S(\tau)]$. It follows that $\mathcal{D}[S(\tau)]$ is the set in (10.55). □

If τ is in case *I* at b, Theorem 10.14 characterizes all the closed regularly solvable extensions $S(\tau)$ of $T_0(\tau)$. The boundary condition (10.35) and (10.37) can be equivalently written as

$$[u, \bar{\theta}](a) = 0, \qquad [\theta, \theta](a) = 0,$$

where $(\tau - \lambda)\theta = 0$. If we construct R_λ as in (10.52) with this θ and some $\psi \in \mathcal{N}(T(\tau) - \lambda I) \cap L^2(a, b;\ |\text{im}\ q|)$, we see from (10.53) that Lemma 10.25 continues to hold if $\text{im}\,[\Phi, \Phi](\beta) \geqslant 0$, that is,

$$2\int_a^\beta \text{im}\,(\lambda - q)|\psi|^2 + \text{im}[\psi, \psi](a) \geqslant 0.$$

In this case R_λ is the resolvent of $S(\tau)$ for $\text{im}\,\lambda > M$ and $S(\tau)$ is J-self-adjoint with domain in $L^2(a, b;\ |\text{im}\, q|)$.

10.6. *Examples*

We make no attempt at generality in this section but content ourselves with the case when $\tau\phi = -\phi'' + q\phi$ and the real and imaginary parts of $q(x)$ are powers of x. Our objective is to demonstrate the existence of each of the three cases of Sims discussed in §10.5. General criteria for case I will be obtained in §VII.3.2.

Lemma 10.27. Let $k(x, \bullet)$ be measurable on $[a,b)$ for each $x \in [a,b)$ and suppose that for all $x, t \in [a, b)$, $|k(x, t)| \leqslant R(t)$ where $R \in L^1(a,b)$. Then, if K_1 and K_2 are any constants, there exist unique solutions of the integral equations

$$\eta_1(x) = K_1 + \int_a^x k(x, t)\eta_1(\tau)\,dt, \tag{10.56}$$

$$\eta_2(x) = K_2 - \int_x^b k(x, t)\eta_2(t)\,dt, \tag{10.57}$$

and for $x \in [a, b)$,

$$|\eta_i(x)| \leqslant |K_i| \exp \int_a^b R(t)\,dt \qquad (i = 1, 2.) \tag{10.58}$$ ■

Proof. The proof of existence of solutions is by the method of successive approximations. In the case of (10.56) we define the sequence

$$\phi_1(x) = K_1, \qquad \phi_{n+1}(x) = K_1 + \int_a^x k(x, t)\phi_n(t)\,dt \quad (n = 1, 2, \ldots).$$

Then it follows by induction that

$$\begin{aligned} |\phi_{n+1}(x) - \phi_n(x)| &\leqslant K_1 r^n(x)/n! \qquad (n = 1, 2, \ldots) \\ &\leqslant K_1 r^n(b)/n! \end{aligned}$$

where $r(x) = \int_a^x R(t)\,dt$. The sequence $(\phi_n(x))$ therefore converges uniformly to a limit $\phi(x)$ on $[a, b)$ and ϕ is readily seen to satisfy (10.56). If there are two solutions ϕ and ψ, then

$$\phi(x) - \psi(x) = \int_a^x k(x, t)[\phi(t) - \psi(t)]\,dt$$

and from this it follows that if $\sup_{t \in [a,x]} |\phi(t) - \psi(t)| =: M$ then

$$|\phi(x) - \psi(x)| \leqslant Mr^n(x)/n! \leqslant Mr^n(b)/n!$$

Hence $M = 0$ and $\psi = \phi$ on $[a, b)$. The equation (10.57) is treated in the same way. The inequalities (10.58) are instances of Gronwall's inequality and their proof is similar to that in the proof of Theorem 10.17. □

Theorem 10.28. Let $q(t) = -a^2 t^\alpha - ib^2 t^\beta$ on $[1, \infty)$ where a and b are non-zero real constants, and let $\tau\phi = -\phi'' + q\phi$. If $\beta > 0$ we have the following results:

(i) if $\alpha < 2\beta+2$ then τ is in case I at ∞;
(ii) if $\alpha = 2\beta+2$ then τ is in case I at ∞ when $b^2/a \geqslant \beta$ and in case II when $b^2/a < \beta$;
(iii) if $\alpha > 2\beta+2$ then τ is in case III at ∞.

If $\beta \leqslant 0$ then cases II and III coincide and τ is in case I at ∞ if, and only if, $\alpha < 2$. ■

Proof. Let $\operatorname{im}\lambda > 0$ and

$$y(x) = [\lambda - q(x)]^{-\frac{1}{4}}\eta(x), \tag{10.59}$$

$$\xi(x) = \int_1^x [\lambda - q(t)]^{\frac{1}{2}}\mathrm{d}t, \tag{10.60}$$

taking $\operatorname{im}[\lambda - q(t)]^{\frac{1}{2}} > 0$. A calculation yields that if η is a solution of

$$\eta(x) = A\exp[-\mathrm{i}\xi(x)] + \int_1^x \{\sin[\xi(x)-\xi(t)]\}\, R(t)\eta(t)\,\mathrm{d}t, \tag{10.61}$$

where A is a non-zero constant and

$$R(t) = -\frac{q''(t)}{4[\lambda - q(t)]^{\frac{3}{2}}} - \frac{5q'^2(t)}{16[\lambda - q(t)]^{\frac{5}{2}}},$$

then y in (10.59) satisfies $\tau y = \lambda y$. From (10.61), $\eta_1(x) = \exp[\mathrm{i}\xi(x)]\eta(x)$ satisfies

$$\eta_1(x) = A + \int_1^x \{\exp \mathrm{i}[\xi(x)-\xi(t)]\}\{\sin[\xi(x)-\xi(t)]\}R(t)\eta_1(t)\,\mathrm{d}t. \tag{10.62}$$

If $\beta > 0$ we have as $t \to \infty$,

$$R(t) = O(t^{-\frac{1}{2}\beta - 2}),$$

$$[\lambda - q(t)]^{\frac{1}{2}} = \begin{cases} at^{\frac{1}{2}\alpha} + (\mathrm{i}b^2/2a)\,t^{\beta - \frac{1}{2}\alpha} + O(t^{2\beta - \frac{3}{2}\alpha}) & \text{if } \alpha > \beta, \\ \mathrm{e}^{\pi\mathrm{i}/4}\,b\,t^{\frac{1}{2}\beta}[1 + O(t^{\alpha-\beta})] & \text{if } \alpha < \beta, \\ (a^2 + \mathrm{i}b^2)^{\frac{1}{2}}t^{\frac{1}{2}\alpha}[1 + O(t^{-\beta})] & \text{if } \alpha = \beta. \end{cases} \tag{10.63}$$

Hence $R \in \mathrm{L}^1(1, \infty)$, and Lemma 10.27 ensures the existence of a unique non-trivial solution of (10.62) since

$$|\{\exp \mathrm{i}[\xi(x)-\xi(t)]\}\sin[\xi(x)-\xi(t)]| \leqslant 1.$$

Also, by (10.58), $\eta_1(x) = O(1)$ and on substituting in (10.62) we obtain

$$\eta_1(x) = A + \frac{\mathrm{i}}{2}\int_1^\infty R(t)\eta_1(t)\,\mathrm{d}t + O\left(\int_1^x \exp(-2\operatorname{im}[\xi(x)-\xi(t)])|R(t)|\mathrm{d}t\right)$$
$$+ O\left(\int_x^\infty |R(t)|\mathrm{d}t\right). \tag{10.64}$$

Let $\alpha < 2\beta + 2$. We see from (10.63) that as $x \to \infty$,

$$|[\lambda - q(t)]^{-\frac{1}{4}}| \asymp x^{-\frac{1}{4}\max\{\alpha, \beta\}},$$

where $\asymp$ means that the ratio of the two sides lies between two positive constants, and

$$\operatorname{im} \xi(x) \begin{cases} = \dfrac{b^2}{2a} \dfrac{x^{\beta - \frac{1}{2}\alpha + 1}}{(\beta - \frac{1}{2}\alpha + 1)} [1 + o(1)] & \text{if } \beta < \alpha, \\ \asymp x^{\frac{1}{2}\beta + 1} & \text{if } \beta \geqslant \alpha. \end{cases}$$

On writing the second integral in (10.64) as $\int_1^{\frac{1}{2}x} + \int_{\frac{1}{2}x}^{x}$ and using the established facts that $R \in L^1(1, \infty)$ and $\operatorname{im} \xi(x) \to \infty$, we see from (10.64) that

$$\eta_1(x) \sim A + \frac{\mathrm{i}}{2} \int_1^{\infty} R(t)\eta_1(t)\mathrm{d}t.$$

If this constant on the right-hand side is zero, we have from (10.62) that

$$\eta_1(x) = -\frac{\mathrm{i}}{2} \int_x^{\infty} R(t)\eta_1(t)\,\mathrm{d}t + \frac{1}{2\mathrm{i}} \int_1^{x} \{\exp 2\mathrm{i}[\xi(x) - \xi(t)]\}\, R(t)\eta_1(t)\,\mathrm{d}t$$

and hence

$$|\eta_1(x)| \leqslant \tfrac{1}{2} \int_x^{\infty} |R(t)|\,|\eta_1(t)|\,\mathrm{d}t + \tfrac{1}{2} \int_1^{x} |R(t)|\,|\eta_1(t)|\,\mathrm{d}t$$

$$= \tfrac{1}{2} \int_1^{\infty} |R(t)|\,|\eta_1(t)|\,\mathrm{d}t.$$

This implies that $\int_1^{\infty} |R(t)|\,\mathrm{d}t \geqslant 2$. Let $\lambda = \mathrm{i}\nu$, $\nu > 0$, and $\gamma = \max\{\alpha, \beta\}$. Then, for a sufficiently large constant k, depending on a and b, we have

$$R(t) = \begin{cases} O(t^{\gamma - 2}/\nu^{\frac{3}{2}}) & \text{if } t \leqslant k\nu^{1/\gamma}, \\ O(t^{-\frac{1}{2}\gamma - 2}) & \text{if } t > k\nu^{1/\gamma}. \end{cases}$$

From these estimates it follows that $\int_1^{\infty} |R(t)|\,\mathrm{d}t \to 0$ as $\nu \to \infty$. Hence for sufficiently large ν,

$$K := A + \frac{\mathrm{i}}{2} \int_1^{\infty} R(t)\eta_1(t)\,\mathrm{d}t \neq 0$$

and $\eta(x) \sim K \exp[-\mathrm{i}\xi(x)]$. From (10.59) we conclude that

$$|y(x)| \asymp |[\lambda - q(x)]^{-\frac{1}{4}}| \exp[\operatorname{im} \xi(x)]$$

is exponentially large for large x and hence $y \notin L^2(1, \infty)$. We are therefore in case *I* at ∞.

If $\alpha = 2\beta + 2$ we see from (10.63) that

$$\operatorname{im} \xi(x) = (b^2/2a) \log x + O(1)$$

and

$$|y(x)| \asymp x^{-\frac{1}{4}\alpha + b^2/2a}.$$

Hence $y \notin L^2(1, \infty)$ if $b^2/a \geqslant \beta$ and τ is in case *I* at ∞. If $b^2/a < \beta$ then $y \in L^2(1, \infty)$ and

$$|\operatorname{im} q(x)|\,|y(x)|^2 \asymp x^{-1+b^2/a}$$

so that $y \notin L^2(1, \infty; |\operatorname{im} q|)$. A second solution of $\tau\phi = \lambda\phi$ is given by (10.59) where η now satisfies the integral equation

$$\eta(x) = A \exp[\mathrm{i}\xi(x)] - \int_x^\infty \{\sin[\xi(x) - \xi(t)]\} R(t)\eta(t)\,\mathrm{d}t. \qquad (10.65)$$

We now obtain $\eta(x) = O[\exp(-\operatorname{im} \xi(x))]$ and

$$y(x) = O(x^{-\frac{1}{4}\alpha - b^2/2a})$$

and

$$|\operatorname{im} q(x)|\,|y(x)|^2 = O(x^{-1-b^2/a}).$$

Thus $y \in L^2(1, \infty) \cap L^2(1, \infty; |\operatorname{im} q|)$ and we conclude that τ is in case *II* at ∞ when $\alpha = 2\beta + 2$ and $b^2/a < \beta$.

If $\alpha > 2\beta + 2$ then $\operatorname{im} \xi(x)$ is bounded on $[1, \infty)$ and hence the solutions of (10.61) and (10.65) are both bounded. The corresponding solutions of $\tau\phi = \lambda\phi$ determined by (10.59) are linearly independent and satisfy

$$y(x) = O(x^{-\alpha/4}), \ |\operatorname{im} q|\,|y(x)|^2 = O(x^{\beta - \alpha/2})$$

and hence all solutions lie in $L^2(1, \infty) \cap L^2(1, \infty; |\operatorname{im} q|)$; τ is therefore in case *III* at ∞.

Finally we consider the case $\beta \leqslant 0$. Cases *II* and *III* coincide since $\operatorname{im} q$ is now bounded. If $\alpha > 2$ then $\operatorname{im} \xi(x) = O(1)$ and the two solutions determined by (10.61) and (10.65) in (10.59) satisfy

$$y(x) = O(x^{-\alpha/4}).$$

They therefore lie in $L^2(1, \infty)$ and τ is in case *III*. If $\alpha \leqslant 2$ then

$$\operatorname{im} \xi(x) \asymp \begin{cases} x^{1-\alpha/2} & \text{for } 0 < \alpha < 2, \\ \log x & \text{for } \alpha = 2, \\ x & \text{for } \alpha \leqslant 0. \end{cases}$$

Thus η in (10.61), and with it the solution y in (10.59), is large at infinity so that $y \notin L^2(1, \infty)$. τ is therefore in Case *I* at ∞ when $\alpha \leqslant 2$. The theorem is therefore proved. □

The last part of Theorem 10.28 is valid when $b = 0$ and hence q is real. The expression τ is then formally symmetric and the aforementioned result has the interpretation, when $q(t) = -a^2 t^\alpha$, that τ is in Weyl's limit-point case at ∞ if $\alpha \leqslant 2$, and is otherwise in the limit-circle case at infinity.

IV

Sesquilinear forms in Hilbert spaces

The vital tool in the Hilbert-space approach to elliptic boundary-value problems is the celebrated Lax–Milgram Theorem. The essence of the method is the interpretation of the problem in a weak or variational sense involving a sesquilinear form defined in a natural way by the problem and acting on some Sobolev space. Under appropriate conditions on the sesqulinear form the existence of a weak solution is ensured by the Lax–Milgram Theorem. This technique will be motivated and discussed in detail in Chapter VI. An important consequence of the Lax–Milgram Theorem is Kato's First Representation Theorem which associates a unique m-sectorial operator with a closed, densely defined sectorial sesquilinear form. This generalizes a well-known result of Friedrichs concerning the so-called Friedrichs extension of a lower semi-bounded symmetric operator. The results in §§4,5 on the perturbation of sesquilinear forms will be our principal tools in Chapter X for locating the essential spectra of differential operators. Finally, in §6, we give Stampacchia's generalization of the Lax–Milgram Theorem to variational inequalites in readiness for our discussion of Stampacchia's weak maximum principle and capacity in Chapter VI.

1. Bounded coercive forms and the Lax-Milgram Theorem

A *sesquilinear form* (or *form*) with domain H, a complex Hilbert space, is a complex-valued function a defined on $H \times H$ which is such that $a[\![u, v]\!]$ is linear in u and conjugate-linear in v. We shall denote $a[\![u, u]\!]$ by $a[\![u]\!]$ and call $a[\![\bullet]\!]$ the *quadratic form* associated with $a[\![\bullet,\bullet]\!]$. The inner product $(\bullet,\bullet)_H$ on H is clearly a sesquilinear form; we shall denote it by $1[\![\bullet,\bullet]\!]$. If a_1 and a_2 are sesquilinear forms on H and $\lambda \in \mathbb{C}$ we define the forms $a_1 + a_2$ and λa_1 by

$$(a_1 + a_2)[\![u,v]\!] = a_1[\![u,v]\!] + a_2[\![u,v]\!],$$

$$(\lambda a_1)[\![u,v]\!] = \lambda(a_1[\![u,v]\!]).$$

The form $a+\lambda 1$ will always be denoted by $a+\lambda$:

$$(a+\lambda)[\![u,v]\!] = a[\![u,v]\!] + \lambda(u,v)_H.$$

The *adjoint form* a^* of a is defined by

$$a^*[\![u,v]\!] := \overline{a[\![v,u]\!]}$$

and a is said to be *symmetric* if $a^* \equiv a$, i.e. for all $u, v \in H$

$$a^*[\![u,v]\!] = \bar{a}[\![v,u]\!] = a[\![u,v]\!].$$

The *real* and *imaginary parts* of a form a are respectively

$$a_1 = \frac{1}{2}(a+a^*), \qquad a_2 = \frac{1}{2i}(a-a^*).$$

They are both symmetric and $a = a_1 + ia_2$. Moreover

$$a_1[\![u]\!] = \operatorname{re} a[\![u]\!], \quad a_2[\![u]\!] = \operatorname{im} a[\![u]\!].$$

A form a is said to be *bounded* on $H \times H$ if there exists a constant $M > 0$ such that

$$|a[\![x,y]\!]| \leqslant M \,\|x\|_H \,\|y\|_H \quad \text{for all } x, y \in H. \tag{1.1}$$

A form a is said to be *coercive* on H if there exists a constant $m > 0$ such that

$$|a[\![x,x]\!]| \geqslant m \,\|x\|_H^2 \quad \text{for all } x \in H. \tag{1.2}$$

We now give the celebrated *Lax–Milgram Theorem* for bounded coercive forms.

Theorem 1.1. Let a be a bounded coercive form on a Hilbert space H_0 with bounds m and M as in (1.1) and (1.2). Then, for any $F \in H_0^*$, the adjoint of H_0, there exists an $f \in H_0$ such that

$$a[\![f,\phi]\!] = (F,\phi) \equiv F(\phi) \quad \text{for all } \phi \in H_0. \tag{1.3}$$

The map $\hat{A} : f \mapsto F$ defined by (1.3) is a linear bijection of H_0 onto H_0^* and

$$m \leqslant \|\hat{A}\| \leqslant M, \qquad M^{-1} \leqslant \|\hat{A}^{-1}\| \leqslant m^{-1}. \tag{1.4}$$ ■

Proof. For $f \in H_0$, the map $F : \phi \mapsto a[\![f,\phi]\!]$ is conjugate-linear and from (1.1)

$$|a[\![f,\phi]\!]| \leqslant (M \,\|f\|_{H_0}) \,\|\phi\|_{H_0}. \tag{1.5}$$

The map F therefore belongs to H_0^*. Also, $\hat{A} : f \mapsto F$ is a well-defined linear map of H_0 into H_0^* and from (1.5),

$$\|\hat{A}f\|_{H_0^*} \leqslant M \,\|f\|_{H_0} \quad \text{for all } f \in H_0. \tag{1.6}$$

The coercivity condition (1.2) gives, for $f \in H_0$,

$$\begin{aligned} m\,\|f\|_{H_0}^2 &\leqslant |a[\![f,f]\!]| \\ &= |(\hat{A}f, f)| \leqslant \|\hat{A}f\|_{H_0^*} \,\|f\|_{H_0} \end{aligned}$$

and hence

$$\| \hat{A}f \|_{H_0^*} \geqslant m \| f \|_{H_0}. \tag{1.7}$$

Thus $\hat{A}$ has an inverse which is bounded on its domain in H_0^*. It only remains to prove that the latter domain, in other words $\mathscr{R}(\hat{A})$, is all of H_0^*. From (1.7), if $(\hat{A}f_n)$ is a convergent sequence in $\mathscr{R}(\hat{A})$ then (f_n) is a Cauchy sequence in H_0 and hence converges to $f \in H_0$. Therefore $\hat{A}f$ is the limit of $(\hat{A}f_n)$, in view of (1.6), and so $\mathscr{R}(\hat{A})$ is a closed subspace of H_0^*. To prove that $\mathscr{R}(\hat{A})$ is dense in H_0^* (and hence $\mathscr{R}(\hat{A}) = H_0^*$) assume to the contrary that there exists $x_0 \in {}^0\mathscr{R}(\hat{A})$ with $x_0 \neq 0$. Then we have the contradiction

$$0 = (\hat{A}x_0, x_0) = a[\![x_0, x_0]\!] \geqslant m \| x_0 \|_{H_0}^2.$$

The theorem is therefore proved. □

In the next important result we apply the Lax–Milgram Theorem to the situation where the Hilbert space H_0 is continuously embedded in another Hilbert space H, the embedding map $E: H_0 \to H$ having dense range in H. For $\phi \in H_0$ and $x \in H$,

$$\begin{aligned} |(x, E\phi)_H| &\leqslant \| x \|_H \, \| E\phi \|_H \\ &\leqslant (\| E \| \, \| x \|_H) \, \| \phi \|_{H_0}. \end{aligned}$$

Hence $E^*: x \mapsto (x, E\bullet)_H$ is a linear map of H into H_0^* and $\| E^*x \|_{H_0} \leqslant \| E \| \, \| x \|_H$. Since E has dense range in H, the map E^* is an injection and E^* has dense range in H^* since E is injective. We therefore have the triplet of spaces

$$H_0 \overset{E}{\to} H \overset{E^*}{\to} H_0^* \tag{1.8}$$

with continuous embedding maps E and E^* having dense ranges. As the notation suggests, E^* is the adjoint of E, since on identifying H with its adjoint,

$$(E^*x, \phi) = (x, E\phi)_H \qquad (x \in H, \phi \in H_0). \tag{1.9}$$

Corollary 1.2. Let a satisfy (1.1) and (1.2) on H_0 and let $\hat{A}$ be the linear bijection of H_0 onto H_0^* given in Theorem 1.1. The operator $A = (E^*)^{-1} \hat{A} E^{-1}$ has the following properties:

(i) $$\mathscr{D}(A) = \{ z \in EH_0 : a[\![E^{-1}z, x]\!] = (f, Ex)_H \text{ for some } f \in H \text{ and all } x \in H_0 \}; \quad f := Az; \tag{1.10}$$

(ii) $\mathscr{D}(A)$ is a dense subspace of H and $E^{-1}\mathscr{D}(A)$ is dense in H_0;

(iii) $A: \mathscr{D}(A) \to H$ is closed and has range H;

(iv) $A^{-1} \in \mathscr{B}(H)$ and $\| A^{-1} \| \leqslant m^{-1} \| E \|^2$;

(v) A^* is the operator determined by (1.10) with a replaced by its adjoint a^*; also, A is self-adjoint if, and only if, a is symmetric. ■

Proof. $A = E^{*-1} \hat{A} E^{-1}$ is clearly the operator obtained by restricting $\hat{A}$ to act only in H. We have that $z\in\mathscr{D}(A)$ if, and only if, $z = Ew$ with $w\in H_0$ and $\hat{A}w = E^*Az\in\mathscr{R}(E^*)$; also

$$\begin{aligned} a[\![w,x]\!] &= (\hat{A}w, x) \qquad (x\in H_0) \\ &= (E^*Az, x) \\ &= (Az, Ex)_H. \end{aligned}$$

Furthermore, since E and E^* have dense ranges and $\hat{A}$ is an isomorphism of H_0 onto H_0^*, the subspace $\mathscr{D}(A)$ is dense in H and $E^{-1}\,\mathscr{D}(A)$ is dense in H_0. Also, $\mathscr{R}(A) = \mathscr{R}((E^*)^{-1}\hat{A}E^{-1}) = H$ and from (1.2), for $z = Ew\in\mathscr{D}(A)$,

$$\begin{aligned} (Az, z)_H &= a[\![w,w]\!] \\ &\geqslant m\|w\|_{H_0}^2 \\ &\geqslant m\|E\|^{-2}\|z\|_H^2. \end{aligned}$$

Hence A^{-1} exists and (iii) and (iv) are proved.

Since a^* also satisfies (1.1) and (1.2) there is an operator B, say, associated with a^* with properties analogous to those of A above. Moreover, for $Ew\in\mathscr{D}(A)$ and $Ex\in\mathscr{D}(A^*)$,

$$\begin{aligned} (A^*Ex, Ew)_H &= (Ex, AEw)_H \\ &= \overline{a[\![w,x]\!]} \\ &= a^*[\![x,w]\!]. \end{aligned}$$

By (ii) the last identity continues to hold for all $w\in H_0$ and hence from (i) applied to a^* and B it follows that $A^*\subset B$. Similarly, if $Ew\in\mathscr{D}(A)$ and $Ex\in\mathscr{D}(B)$,

$$\begin{aligned} (Ew, BEx)_H &= \overline{a^*[\![x,w]\!]} \\ &= a[\![w,x]\!] \\ &= (AEw, Ex)_H. \end{aligned}$$

Consequently $B\subset A^*$ and (v) is established. □

Another important consequence of Theorem 1.1 is the following characterization of the adjoint of the space $W_0^{1,2}(\Omega)$ (see §V.3.1). This adjoint space is usually denoted by $W^{-1,2}(\Omega)$; more generally the adjoint of $W_0^{m,p}(\Omega)$ for $1 < p < \infty$ is denoted by $W^{-m,p'}(\Omega)$, where $1/p + 1/p' = 1$.

Corollary 1.3. Let $F\in W^{-1,2}(\Omega)$, the adjoint of $W_0^{1,2}(\Omega)$. Then there exist $g_i\in L^2(\Omega)$ $(i = 0, 1, \ldots, n)$ such that $F = g_0 - \sum_{i=1}^{n} D_i g_i$ in the sense of distributions, i.e.

$$(F,\phi) = (g_0,\phi)_{2,\Omega} + \sum_{i=1}^{n}(g_i, D_i\phi)_{2,\Omega} \quad \text{for all } \phi\in C_0^\infty(\Omega).$$

Conversely, if $F = g_0 - \sum_{i=1}^{n} D_i g_i$ in the above sense for some $g_i \in L^2(\Omega)$ $(i = 0, 1, \ldots, n)$ then $F \in W^{-1,2}(\Omega)$. ■

Proof. We apply Theorem 1.1 with $a[\![u,v]\!] = (u,v)_{1,2,\Omega}$ (see §V.3.1) and $H_0 = W_0^{1,2}(\Omega)$. For any $F \in W^{-1,2}(\Omega)$ there exists $f \in W_0^{1,2}(\Omega)$ such that $(F, \phi) = (f, \phi)_{1,2,\Omega}$ for all $\phi \in C_0^\infty(\Omega)$. Consequently $F = f - \Delta f$, which is of the asserted form.

If $F = g_0 - \sum_{i=1}^{n} D_i g_i$, with $g_i \in L^2(\Omega)$ $(i = 0, 1, \ldots, n)$, then for all $\phi \in C_0^\infty(\Omega)$,

$$(F, \phi) = (g_0, \phi)_{2,\Omega} + \sum_{i=1}^{n} (g_i, D_i \phi)_{2,\Omega}$$

$$\leqslant \left(\| g_0 \|_{2,\Omega}^2 + \sum_{i=1}^{n} \| g_i \|_{2,\Omega}^2 \right)^{\frac{1}{2}} \| \phi \|_{1,2,\Omega}.$$

Since $C^\infty(\Omega)$ is dense in $W_0^{1,2}(\Omega)$, it follows that $\phi \mapsto (F, \phi)$ defines a bounded, conjugate-linear functional on $W_0^{1,2}(\Omega)$ and hence $F \in W^{-1,2}(\Omega)$. □

We shall have more to say about $W^{-1,2}(\Omega)$ in Remark VI.1.7 below. For $W^{-m,p'}(\Omega)$ in the general case see Adams [1, Theorem 3.10].

2. Sectorial forms

We now turn to forms a which are such that $a[\![u,v]\!]$ is only defined for u and v in a subspace $\mathscr{D}(a)$ of H. The subspace $\mathscr{D}(a)$ is called the *domain* of a. Our aim is to determine properties of a which allow a to be extended in a natural way to a bounded coercive form on a Hilbert space H_0 which is continuously embedded and is dense in H. Corollary 1.2 will then apply.

The *numerical range* of the form a is the set

$$\Theta(a) := \{ a[\![u]\!] : u \in \mathscr{D}(a), \| u \|_H = 1 \}$$

in $\mathbb{C}$. The form a is said to be *sectorial* if $\Theta(a)$ is a subset of a sector

$$\{ z \in \mathbb{C} : |\arg(z - \gamma)| \leqslant \theta, 0 \leqslant \theta < \tfrac{1}{2}\pi, \gamma \in \mathbb{R} \}; \tag{2.1}$$

γ will be called a *vertex* of a and θ a corresponding *semi-angle* of a; γ and θ are clearly not unique. If a is also symmetric, that is, $a[\![u,v]\!] = \overline{a[\![v,u]\!]}$ for all $u, v \in \mathscr{D}(a)$, then $a[\![u]\!] \geqslant \gamma$ for all $u \in \mathscr{D}(a)$ with $\| u \|_H = 1$, and we say that a is *lower semi-bounded* or *bounded below* by γ; we write $a \geqslant \gamma$ to indicate this. If $\gamma \geqslant 0$, the form a is said to be *non-negative*. Similarly the notions of boundedness above and non-positiveness may be defined.

If a is sectorial with a vertex γ and semi-angle θ then

$$\text{re } a[\![u]\!] \geqslant \gamma \|u\|_H^2,$$
$$|\text{im } a[\![u]\!]| \leqslant (\tan\theta)\,(\text{re } a[\![u]\!] - \gamma\|u\|_H^2),$$

for all $u \in \mathscr{D}(a)$. Hence, if a_1 is the real part of a, then $a_1 - \gamma + \alpha$ is an inner product on $\mathscr{D}(a)$ for all $\alpha > 0$, and the associated norms, for different $\alpha > 0$, are equivalent. Also, for $u \in \mathscr{D}(a)$,

$$|(a-\gamma)[\![u]\!]|^2 \leqslant (\sec^2\theta)\,\{(a_1-\gamma)[\![u]\!]\}^2$$

and hence

$$(a_1-\gamma)[\![u]\!] \leqslant |(a-\gamma)[\![u]\!]| \leqslant (\sec\theta)\,(a_1-\gamma)[\![u]\!]. \tag{2.2}$$

Lemma 2.1. Let a be a sesquilinear form with domain $\mathscr{D}(a)$ in H and suppose, for some constant K, that $|a[\![x]\!]| \leqslant K t[\![x]\!]$ for all $x \in \mathscr{D}(a)$, where t is a non-negative form on $\mathscr{D}(a)$. Then

$$|a[\![x,y]\!]| \leqslant \varepsilon K\,(t[\![x]\!]\,t[\![y]\!])^{\frac{1}{2}} \quad \text{for all } x,y \in \mathscr{D}(a), \tag{2.3}$$

where $\varepsilon = 1$ if a is symmetric and $\varepsilon = 2$ otherwise. ■

Proof. Let a be symmetric and suppose that $a[\![x,y]\!]$ is real. The latter assumption involves no loss of generality as (2.3) is unchanged if we replace x by $e^{-i\theta}x$ with $\theta = \arg a[\![x,y]\!]$. From the polarization identity

$$a[\![x,y]\!] = \tfrac{1}{4}(a[\![x+y]\!] - a[\![x-y]\!])$$

we get

$$|a[\![x,y]\!]| \leqslant \tfrac{1}{4}K\,(t[\![x+y]\!] + t[\![x-y]\!])$$
$$= \tfrac{1}{2}K\,(t[\![x]\!] + t[\![y]\!]).$$

Hence, with $c^4 = t[\![y]\!]/t[\![x]\!]$ $(t[\![x]\!] \neq 0)$,

$$|a[\![x,y]\!]| = |a[\![cx,(1/c)y]\!]| \leqslant \tfrac{1}{2}K\left(c^2 t[\![x]\!] + \frac{1}{c^2}t[\![y]\!]\right)$$
$$= K\,(t[\![x]\!]\,t[\![y]\!])^{\frac{1}{2}}.$$

The non-symmetric case follows from the symmetric on putting $a = a_1 + ia_2$. □

It follows from Lemma 2.1 that for all $\alpha \geqslant 0$,

$$|(a-\gamma+\alpha)[\![u,v]\!]| \leqslant (1+\tan\theta)\,(a_1-\gamma+\alpha)^{\frac{1}{2}}[\![u]\!]\,(a_1-\gamma+\alpha)^{\frac{1}{2}}[\![v]\!]. \tag{2.4}$$

The sectorial form a is said to be *closed* if $\mathscr{D}(a)$ is complete with respect to the inner product $a_1 - \gamma + 1$. The Hilbert space determined by $\mathscr{D}(a)$ and the inner product $a_1 - \gamma + 1$ in this case will be denoted by $\mathrm{H}(a)$ and we shall write

$$\|u\|_{\mathrm{H}(a)} = (a_1 - \gamma + 1)^{\frac{1}{2}}[\![u]\!].$$

The inclusion map of $\mathscr{D}(a)$ into H gives a natural injection of $\mathrm{H}(a)$ into H which is continuous since

$$\|u\|_H \leqslant \|u\|_{\mathrm{H}(a)}. \tag{2.5}$$

If a is not closed we define $\mathrm{H}(a)$ to be the completion of $\mathscr{D}(a)$ with respect to $\|\bullet\|_{\mathrm{H}(a)}$. By (2.5), Cauchy sequences in $\mathrm{H}(a)$ are also Cauchy sequences in H and hence to each $x \in \mathrm{H}(a)$ there corresponds a unique $y \in H$. The map $E: \mathrm{H}(a) \to H$ determined by $x \mapsto y$ is therefore well-defined, its restriction to $\mathscr{D}(a)$ being the inclusion $\mathscr{D}(a) \subset H$. If E is an injection, and hence a continuous embedding of $\mathrm{H}(a)$ into H, then a is said to be *closable*. Thus a is closable if and only if any Cauchy sequence (x_n) in $\mathrm{H}(a)$ which converges to zero in H must also converge to zero in $\mathrm{H}(a)$ and, from (2.4),

$$\lim_{n\to\infty} a[\![x_n]\!] = 0.$$

It also follows from (2.4) that if $(x_n), (y_n) \in \mathscr{D}(a)$ are Cauchy sequences in $\mathrm{H}(a)$ then $(a[\![x_n, y_n]\!])$ converges in $\mathbb{C}$. If $x_n \to x$ and $y_n \to y$ in $\mathrm{H}(a)$, and a is closable, we define the *closure* of a to be the form $\bar{a}$ defined on $\mathrm{H}(a)$ by

$$\bar{a}[\![x,y]\!] = \lim_{n\to\infty} a[\![x_n, y_n]\!].$$

It is easy to see that a is closable if, and only if, it has a closed extension and $\bar{a}$ is the minimal closed extension of a.

Definition 2.2. If a is a closed sectorial form, a subspace D of H is said to be a *core of a* if D is a dense subspace of $\mathrm{H}(a)$. Equivalently, a is the closure of the restriction of a to D. ■

A typical example of a sectorial form which is not closable is given by

$$a[\![x,y]\!] = x(0)\bar{y}(0)$$

with $\mathscr{D}(a) = \mathrm{C}[0,1]$ in $H = \mathrm{L}^2(0,1)$. If (x_n) is a Cauchy sequence in $\mathrm{H}(a)$ and $x_n \underset{H}{\to} 0$, then $(x_n(0))$ is a Cauchy sequence in $\mathbb{C}$ and hence tends to a limit α, say. However, even though $x_n \underset{H}{\to} 0$ it does not follow that $\alpha = 0$, and hence $a[\![x_n]\!] \not\to 0$.

An important class of examples of closable sectorial forms is provided by the following theorem.

Theorem 2.3. Let T be a sectorial operator in H and define

$$t[\![u,v]\!] := (Tu, v)_H, \qquad \mathscr{D}(t) = \mathscr{D}(T).$$

Then t is sectorial and closable. ■

Proof. It is clear that t is sectorial with the same vertex γ and semi-angle θ as T. Let $(x_n) \subset \mathscr{D}(t)$ be a Cauchy sequence in $\mathrm{H}(t)$ and suppose that $x_n \to 0$ in H. We must show that $x_n \to 0$ in $\mathrm{H}(t)$. If $t_1 = \mathrm{re}\, t$,

$$\begin{aligned}\|x_n\|^2_{\mathrm{H}(t)} &= (t_1 - \gamma + 1) [\![x_n]\!] \\ &\leqslant |(t - \gamma + 1) [\![x_n]\!]| \\ &\leqslant |(t-\gamma+1)[\![x_n, x_n - x_m]\!]| + |(t-\gamma+1)[\![x_n, x_m]\!]| \\ &\leqslant (1 + \tan\theta) \|x_n\|_{\mathrm{H}(t)} \|x_n - x_m\|_{\mathrm{H}(t)} + |(Tx_n, x_m)_H| + (1 + |\gamma|) \|x_n\|_H \|x_m\|_H,\end{aligned}$$

by (2.4). Given any $\varepsilon > 0$, there exists an $N \in \mathbb{N}$ such that for $n, m > N$,

$$\|x_n\|^2_{\mathrm{H}(t)} < \varepsilon + |(Tx_n, x_m)_H|.$$

On allowing $m \to \infty$ we get $\|x_n\|^2_{\mathrm{H}(t)} < \varepsilon$ for $n > N$ and the result follows. □

If a is a closed sectorial form with vertex γ and semi-angle θ, then a^* is closed and sectorial with vertex γ and semi-angle θ since $\mathrm{H}(a^*) = \mathrm{H}(a)$ and $\lambda \in \Theta(a^*)$ if, and only if, $\bar{\lambda} \in \Theta(a)$.

Theorem 2.4 (The First Representation Theorem). Let a be a closed, densely defined sectorial form in H and let E denote the natural embedding $\mathrm{H}(a) \to H$. Then there exists an m-sectorial operator A with domain and range in H which has the following properties:

(i) $\mathscr{D}(A) = \{z \in E\mathrm{H}(a)\colon a[\![E^{-1}z, x]\!] = (f, Ex)_H$

$$\text{for some } f \in H \text{ and all } x \in \mathrm{H}(a)\}; \qquad f := Az; \tag{2.6}$$

(ii) $\mathscr{D}(A)$ is dense in H and $E^{-1}\mathscr{D}(A)$ is dense in $\mathrm{H}(a)$;

(iii) A^* is m-sectorial and is the operator determined by (2.6) with a replaced by a^*;

(iv) if $u \in \mathrm{H}(a)$ and $v \in H$, and

$$a[\![u, x]\!] = (v, Ex)_H \tag{2.7}$$

for all x in a core of a, then $Eu \in \mathscr{D}(A)$ and $v = AEu$. A is uniquely determined by (i) and (ii);

(v) $\Theta(A)$ is a dense subset of $\Theta(a)$. ■

Proof. By (2.4),

$$|(a - \gamma + 1) [\![x, y]\!]| \leqslant (1 + \tan\theta) \|x\|_{\mathrm{H}(a)} \|y\|_{\mathrm{H}(a)}$$

and we also have

$$\begin{aligned}|(a - \gamma + 1) [\![x]\!]| &\geqslant (a_1 - \gamma + 1) [\![x]\!] \\ &= \|x\|^2_{\mathrm{H}(a)}.\end{aligned}$$

Hence $a-\gamma+1$ is bounded on $\mathrm{H}(a)\times\mathrm{H}(a)$ and coercive on $\mathrm{H}(a)$. It therefore follows from Corollary 1.2 that there exists an operator $A_\gamma\in\mathscr{C}(H)$ such that

$$(a-\gamma+1)[\![w,x]\!] = (A_\gamma Ew, Ex)_H.$$

The operator $A = A_\gamma+\gamma-1$ satisfies (i) and (ii), is in $\mathscr{C}(H)$ and $\gamma-1\in\rho(A)$. Since $\Theta(A)$ is a subset of $\Theta(a)$, A is therefore m-sectorial, by Theorem III-2.3. Similarly (iii) follows in view of the remark preceding this theorem.

By (iii) $\mathscr{D}(A^*)\subset\mathscr{R}(E)$ and since $\mathscr{R}(E)$ is dense in H the core of a featured in (iv) is dense in H. Consequently (2.7) holds for all $Ex\in\mathscr{D}(A^*)$, whence

$$\begin{aligned}(v,Ex)_H &= a[\![u,x]\!]\\ &= \overline{a^*[\![x,u]\!]}\\ &= (Eu, A^*Ex)_H\end{aligned}$$

and (iv) is proved.

For (v) we use the fact that $E^{-1}\mathscr{D}(A)$ is dense in $\mathrm{H}(a)$ and $a[\![x,x]\!] = (AEx,Ex)_H$ for all $Ex\in\mathscr{D}(A)$. □

Definition 2.5. We shall call the unique m-sectorial operator A determined by Theorem 2.4 in terms of a closed, densely defined sectorial form a the *operator associated with a.* We also call a the *form of A* and $\mathrm{H}(a)$ the *form domain of A*; as is the custom we use $\mathscr{Q}(A)$ to denote the form domain of A. ■

If A_0 is a given, densely defined sectorial operator in H, the form

$$a_0[\![x,y]\!] = (A_0x,y)_H \qquad (x,y\in\mathscr{D}(A_0)) \tag{2.8}$$

is closable and sectorial by Theorem 2.3. If we denote the closure of a_0 by a, this a satisfies Theorem 2.4 and the m-sectorial operator A associated with a is an extension of A_0 in view of Theorem 2.4(iv). Hence, if A_0 is already m-sectorial, we have $A = A_0$. There is therefore a one–one correspondence between the set of m-sectorial operators and the set of closed, densely defined sectorial forms, the correspondence being determined by the identity

$$a[\![x,y]\!] = (Ax,y)_H \qquad (x\in\mathscr{D}(A),\ y\in\mathscr{D}(a)). \tag{2.9}$$

Note that in (2.9) we have omitted the embedding map E of (2.6); we shall always do this when it is clear from the context whether $\mathscr{D}(a)$ is to be regarded as a subspace of H or the Hilbert space $\mathrm{H}(a)$.

A particular case of the correspondence determined by (2.9) is that between the lower semi-bounded self-adjoint operators A and the set of closed, densely defined, lower semi-bounded forms a. If a is the closure of a form a_0 defined by (2.8) for a lower semi-bounded symmetric operator A_0, the lower semi-bounded, self-adjoint operator A associated with a is called the *Friedrichs extension* of A_0, after Friedrichs who first investigated such operators. We shall continue to call A the Friedrichs extension of A_0 when A_0 is any densely defined sectorial operator in H.

It is readily shown that the sum $a = a_1 + a_2$, with domain $\mathscr{D}(a_1) \cap \mathscr{D}(a_2)$, of closed sectorial forms a_1 and a_2 is also closed and sectorial. Thus if $\mathscr{D}(a_1) \cap \mathscr{D}(a_2)$ is dense in H, the m-sectorial operators A, A_1, A_2 associated with a, a_1, a_2 respectively are defined. A is called the *form sum* of A_1 and A_2 and written $A = A_1 \dot{+} A_2$. It follows from part (iv) of Theorem 2.4 that $A_1 \dot{+} A_2$ is an extension of the operator sum $A_1 + A_2$ defined on $\mathscr{D}(A_1) \cap \mathscr{D}(A_2)$. The operator sum may not be densely defined and even if it is, it may neither be m-sectorial nor have an m-sectorial closure. If $\mathscr{D}(A_1) \cap \mathscr{D}(A_2)$ is dense in H, $A_1 + A_2$ is a densely defined sectorial operator and so its Friedrichs extension, A_F say, is defined. It is not even true in general that $A_F = A_1 \dot{+} A_2$ (see Kato [1, Example VI-2.19]).

Theorem 2.6. Let A be the Friedrichs extension of the densely defined sectorial operator A_0 and let S be any sectorial extension of A_0 with $\mathscr{D}(S) \subset \mathscr{D}(a)$, where a is the form of A. Then $S \subset A$ so that A is the only m-sectorial extension of A_0 with domain in $\mathscr{D}(a)$. ■

Proof. We omit the natural embedding $E\colon \mathrm{H}(a) \to H$ throughout and identify $\mathrm{H}(a)$ with the subspace $\mathscr{D}(a)$ of H. Define the sectorial form

$$s[\![x, y]\!] = (Sx, y)_H \qquad (x, y \in \mathscr{D}(S)).$$

Since $\mathscr{D}(s) = \mathscr{D}(S) \supset \mathscr{D}(A_0) = \mathscr{D}(a_0)$ and $A_0 \subset S$, the closure $\bar{s}$ of s is an extension of a_0 and hence of a. Thus for $x \in \mathscr{D}(S)$ and $y \in \mathscr{D}(a)$ we have

$$a[\![x, y]\!] = \bar{s}[\![x, y]\!] = (Sx, y)_H.$$

Consequently, from Theorem 2.4(iv), $S \subset A$ and the theorem is proved. □

Corollary 2.7. Let A be the Friedrichs extension of the lower semi-bounded symmetric operator A_0. Then A is the restriction of A_0^* to $\mathscr{D}(A_0^*) \cap \mathscr{D}(a)$, where a is the form of A; A and A_0 have the same lower bound. ■

Proof. Let S denote the restriction of A_0^* to $\mathscr{D}(A_0^*) \cap \mathscr{D}(a)$. As $\mathscr{D}(A_0)$ is a core of a, for any $\phi \in \mathscr{D}(a)$ there exists a sequence (ϕ_n) in $\mathscr{D}(A_0)$ which converges to ϕ in the form domain $\mathscr{Q}(A) = \mathrm{H}(a)$ of A and hence in H. Thus for any $\phi \in \mathscr{D}(S) = \mathscr{D}(A_0^*) \cap \mathscr{D}(a)$,

$$\begin{aligned}(S\phi, \phi)_H &= \lim_{n\to\infty} (A_0^*\phi, \phi_n)_H \\ &= \lim_{n\to\infty} (\phi, A_0\phi_n)_H \\ &= \lim_{n\to\infty} a[\![\phi, \phi_n]\!] \\ &= a[\![\phi, \phi]\!].\end{aligned}$$

Consequently S is sectorial and $\Theta(S) \subset \Theta(a)$. It follows from Theorem 2.6 that $S \subset A$. But $A_0 \subset A$ implies $A \subset A_0^*$, whence $A \subset S$ and so $A = S$ as asserted. The fact that A and A_0 have the same lower bound comes from Theorem 2.4(v) and the denseness of $\mathscr{D}(A_0)$ in $\mathscr{Q}(A)$. □

Before proceeding we need to recall some facts about self-adjoint operators; these may be found in Kato [1, Theorem V-3.35]. A non-negative self-adjoint operator S has a unique square root $S^{\frac{1}{2}}$, satisfying $(S^{\frac{1}{2}})^2 = S$, which is non-negative and self-adjoint. Furthermore, $\mathscr{D}(S)$ is a core of $S^{\frac{1}{2}}$, i.e. $\mathscr{D}(S)$ is dense in $\mathrm{H}(S^{\frac{1}{2}})$, the Hilbert space determined by $\mathscr{D}(S^{\frac{1}{2}})$ and the graph inner product.

Theorem 2.8. (The Second Representation Theorem). Let s be a closed, densely defined, non-negative symmetric form and let S be the associated non-negative self-adjoint operator. Then $\mathscr{D}(s) = \mathscr{D}(S^{\frac{1}{2}})$ and

$$s[\![x, y]\!] = (S^{\frac{1}{2}}x, S^{\frac{1}{2}}y)_H \qquad (x, y \in \mathscr{D}(S^{\frac{1}{2}})). \tag{2.10}$$

Thus $\mathscr{Q}(S) = \mathrm{H}(S^{\frac{1}{2}})$. ■

Proof. For $x, y \in \mathscr{D}(S) \subset \mathscr{D}(S^{\frac{1}{2}})$,

$$s[\![x, y]\!] = (Sx, y)_H = (S^{\frac{1}{2}}x, S^{\frac{1}{2}}y)_H.$$

Since $\mathscr{D}(S)$ is dense in both $\mathrm{H}(S^{\frac{1}{2}})$ and $\mathscr{Q}(S)$, the theorem follows by continuity. □

Theorem 2.9. Let s and S be as in Theorem 2.8. Then the following are equivalent:

(i) S has compact resolvent (i.e. $(S - \lambda I)^{-1}$ is compact for some $\lambda \in \rho(S)$),
(ii) $S^{\frac{1}{2}}$ has compact resolvent,
(iii) $\{x : x \in \mathscr{D}(S),\ s[\![x]\!] + \|x\|_H^2 \leqslant 1\}$ is relatively compact in H. ■

Proof. Parts (ii) and (iii) are equivalent, from Theorem 2.8, since $(S^{\frac{1}{2}} + \mathrm{i}I)^{-1}$ is compact if, and only if, $\mathrm{H}(S^{\frac{1}{2}})$ is compactly embedded in H and (iii) merely expresses the latter property.

To prove (i) implies (ii), let (u_n) be a sequence in H with $\|u_n\|_H \leqslant 1$ and set $v_n = (S^{\frac{1}{2}} + \mathrm{i}I)^{-1}u_n$ and $w_n = (S^{\frac{1}{2}} - \mathrm{i}I)^{-1}v_n = (S + I)^{-1}u_n$. Then (w_n) is pre-compact in H on account of (i), and since $w_n \in \mathscr{D}(S)$ we have from (2.10) and Theorem 2.4,

$$\begin{aligned}\|v_n\|_H^2 &= \|(S^{\frac{1}{2}} - \mathrm{i}I)w_n\|_H^2\\ &= (Sw_n, w_n)_H + \|w_n\|^2\\ &= (u_n, w_n)_H.\end{aligned}$$

Hence $\|v_n - v_m\|_H^2 = (u_n - u_m, w_n - w_m)_H \leqslant 2\|w_n - w_m\|_H$; this implies that (v_n)

is precompact in H and (ii) follows. Conversely, if (ii) is satisfied, $(S^{\frac{1}{2}} \pm iI)^{-1}$ are compact and so $(S+I)^{-1} = (S^{\frac{1}{2}} + iI)^{-1}(S^{\frac{1}{2}} - iI)^{-1} \in \mathscr{K}(H)$, whence (i). □

If A is m-sectorial with form a, we call the lower semi-bounded, self-adjoint operator T associated with the closed, densely defined, and lower-semibounded symmetric form $t = \text{re } a = \frac{1}{2}(a + a^*)$ the *real part* of A and write $T = \text{re } A$. While $T = \frac{1}{2}(A + A^*)$ when $A \in \mathscr{B}(H)$, this is not true in general; indeed $\mathscr{D}(A) \cap \mathscr{D}(A^*)$ may not be dense in H. We do, however, have the following result of Kato [1, Theorem VI-3.2].

Theorem 2.10. Let A be an m-sectorial operator with vertex O and semi-angle θ. Then $T = \text{re } A$ is non-negative and there exists a symmetric operator $B \in \mathscr{B}(H)$ such that $\|B\| \leqslant \tan \theta$ and

$$A = T^{\frac{1}{2}}(I + iB)T^{\frac{1}{2}}, \qquad A^* = T^{\frac{1}{2}}(I - iB)T^{\frac{1}{2}}.$$ ■

Proof. A^* is m-sectorial with form a^* and so $T = \text{re } A = \text{re } A^*$. If $k = \text{im } a = (1/2i)\,(a - a^*)$ we see from the hypothesis that $|k[\![u]\!]| \leqslant (\tan \theta) t[\![u]\!]$ and hence, by Lemma 2.1 and Theorem 2.8, for all $u, v \in \mathscr{D}(a) = \mathscr{D}(t) = \mathscr{D}(T^{\frac{1}{2}})$,

$$\begin{aligned} |k[\![u, v]\!]| &\leqslant (\varepsilon \tan \theta) t^{\frac{1}{2}}[\![u]\!] t^{\frac{1}{2}}[\![v]\!] \\ &\leqslant (\varepsilon \tan \theta) \|T^{\frac{1}{2}} u\|_H \|T^{\frac{1}{2}} v\|_H. \end{aligned} \tag{2.11}$$

Furthermore, if $T^{\frac{1}{2}}u = T^{\frac{1}{2}}u'$ and $T^{\frac{1}{2}}v = T^{\frac{1}{2}}v'$, (2.11) implies that

$$k[\![u, v]\!] - k[\![u', v']\!] = k[\![u - u', v]\!] + k[\![u', v - v']\!] = 0.$$

It follows that $k[\![u, v]\!]$ is determined by $T^{\frac{1}{2}}u$ and $T^{\frac{1}{2}}v$ and is a bounded symmetric form in $x = T^{\frac{1}{2}}u$ and $y = T^{\frac{1}{2}}v$ on the range M of $T^{\frac{1}{2}}$. By continuity, $k[\![\bullet, \bullet]\!]$ can be extended to be bounded on $\bar{M} \times \bar{M}$, where $\bar{M}$ is the closure of $M = \mathscr{R}(T^{\frac{1}{2}})$ in H. Hence there exists a symmetric operator B which is bounded on $\bar{M}$ and is such that

$$k[\![u, v]\!] = (BT^{\frac{1}{2}}u, T^{\frac{1}{2}}v)_H \qquad (u, v \in \mathscr{D}(a)).$$

On setting $Bx = 0$ for $x \in M^{\perp}$, B can be regarded as an operator in $\mathscr{B}(H)$ and this does not increase the value of $\|B\|$. On applying Theorem 2.8 we obtain

$$\begin{aligned} a[\![u, v]\!] &= (t + ik)[\![u, v]\!] \\ &= ((I + iB)T^{\frac{1}{2}}u, T^{\frac{1}{2}}v)_H \qquad (u, v \in \mathscr{D}(a)). \end{aligned}$$

If $u \in \mathscr{D}(A)$ and $v \in \mathscr{D}(a) = \mathscr{D}(T^{\frac{1}{2}})$,

$$\begin{aligned} (Au, v) &= a[\![u, v]\!] \\ &= ((I + iB)T^{\frac{1}{2}}u, T^{\frac{1}{2}}v)_H. \end{aligned}$$

Thus $(I+\mathrm{i}B)T^{\frac{1}{2}}u \in \mathscr{D}(T^{\frac{1}{2}})$ and $A \subset T^{\frac{1}{2}}(I+\mathrm{i}B)T^{\frac{1}{2}}$. But, for $u \in \mathscr{D}(T^{\frac{1}{2}}(I+\mathrm{i}B)T^{\frac{1}{2}})$,

$$(T^{\frac{1}{2}}(I+\mathrm{i}B)T^{\frac{1}{2}}u, u)_H = \|T^{\frac{1}{2}}u\|_H^2 + \mathrm{i}(BT^{\frac{1}{2}}u, T^{\frac{1}{2}}u)_H$$

from which it follows that $T^{\frac{1}{2}}(I+\mathrm{i}B)T^{\frac{1}{2}}$ is sectorial. Thus $T^{\frac{1}{2}}(I+\mathrm{i}B)T^{\frac{1}{2}}$ cannot be a proper extension of the m-sectorial operator A and so the identity is established. The result for A^* follows similarly since $\operatorname{im} a^* = -\operatorname{im} a$. □

3. The polar decomposition of closed operators

We first need the following important result. In this section H_1 and H_2 are complex Hilbert spaces.

Theorem 3.1. Let $T \in \mathscr{C}(H_1, H_2)$ be densely defined in H_1. Then T^*T is a non-negative self-adjoint operator with domain and range in H_1. Moreover $\mathscr{D}(T^*T)$ is a core of T. ■

Proof. The sesquilinear form

$$t[\![x, y]\!] = (Tx, Ty)_{H_2}, \qquad \mathscr{D}(t) = \mathscr{D}(T)$$

is densely defined and non-negative in H_1, and since

$$(t+1)[\![x]\!] = \|Tx\|_{H_2}^2 + \|x\|_{H_1}^2 =: \|x\|_T^2$$

and $\mathscr{D}(T)$ is complete with respect to $\|\bullet\|_T$, t is closed. Hence from Theorem 2.4 there exists a non-negative self-adjoint operator S in H_1 whose domain is a core of T and which is such that

$$(Sx, y)_{H_1} = (Tx, Ty)_{H_2} \qquad (x \in \mathscr{D}(S),\ y \in \mathscr{D}(T)). \tag{3.1}$$

This implies that $S \subset T^*T$. Hence $\mathscr{D}(T^*T)$ is dense in H_1 and as T^*T is obviously symmetric and S is self-adjoint we must have $S = T^*T$.

From Theorem 2.8 and (3.1), if we write $G = S^{\frac{1}{2}}$ for the positive square root of S, we obtain

$$(Gx, Gy)_{H_1} = (Tx, Ty)_{H_2} \qquad (x \in \mathscr{D}(G) = \mathscr{D}(T)). \tag{3.2}$$

The map $Gx \mapsto Tx$ of $\mathscr{R}(G)$ onto $\mathscr{R}(T)$ is therefore an isometry and can be extended by continuity to an isometry U of $\overline{\mathscr{R}(G)}$ onto $\overline{\mathscr{R}(T)}$. This map U can be further extended to an operator in $\mathscr{B}(H_1, H_2)$ by setting $Ux = 0$ for $x \in \mathscr{R}(G)^\perp = \mathscr{N}(G)$. We have the orthogonal sum decompositions

$$H_1 = \overline{\mathscr{R}(G)} \oplus \mathscr{R}(G)^\perp, \qquad H_2 = \overline{\mathscr{R}(T)} \oplus \mathscr{R}(T)^\perp \tag{3.3}$$

and U is a unitary map between the closed subspaces $\overline{\mathscr{R}(G)}$ and $\overline{\mathscr{R}(T)}$ of H_1 and H_2 respectively with $Ux = 0$ for $x \in \mathscr{R}(G)^\perp$. U is therefore a *partial* *isometry* with *initial set* $\overline{\mathscr{R}(G)}$ and *final set* $\overline{\mathscr{R}(T)}$. Its adjoint $U^* \in \mathscr{B}(H_2, H_1)$ is easily seen to be a partial isometry with initial set $\overline{\mathscr{R}(T)}$ and final set $\overline{\mathscr{R}(G)}$, and

$$U^*Ux = x \quad \text{for } x \in \overline{\mathscr{R}(G)}, \tag{3.4}$$

$$UU^*y = y \quad \text{for } y \in \overline{\mathscr{R}(T)}. \tag{3.5}$$

The formula

$$T = UG, \qquad \mathscr{D}(G) = \mathscr{D}(T) \tag{3.6}$$

is called the *polar decomposition* of T; G is called the *absolute value* of T and written $|T|$. The polar decomposition (3.6) is unique in the sense that if (3.6) is satisfied for a non-negative self-adjoint operator G in H_1 and U is a partial isometry with initial set $\overline{\mathscr{R}(G)}$ and final set $\overline{\mathscr{R}(T)}$, then $G = |T|$ and U is uniquely determined. To see this we first establish easily that

$$T^* = GU^*, \tag{3.7}$$

so that $T^*T = GU^*UG$. Since (3.4) remains valid, it follows that $G^2 = T^*T$ and hence $G = |T|$ by the uniqueness of the square root. This determines U on $\overline{\mathscr{R}(G)}$ and as U is a partial isometry with initial set $\overline{\mathscr{R}(G)}$, it is uniquely determined. □

Theorem 3.2. The polar decomposition of T^* is

$$T^* = U^*|T^*|, \qquad |T^*|^{\frac{1}{2}} = (TT^*)^{\frac{1}{2}}. \tag{3.8}$$

Also

$$\mathscr{N}(T) = \mathscr{N}(|T|), \qquad \mathscr{R}(T) = \mathscr{R}(|T^*|). \tag{3.9}$$

Proof. We first note that $|T^*|$ is a non-negative self-adjoint operator in H_2 and claim that

$$|T^*| = U|T|U^*. \tag{3.10}$$

To prove this let $G' = UGU^*$ with $G = |T|$. The assertion (3.10) will be established if we can show that G' is a non-negative self-adjoint operator and $G'^2 = TT^*$. Since U^* has initial set $\overline{\mathscr{R}(T)}$ it follows that $G'x = 0$ for $x \in \mathscr{R}(T)^\perp$, and $\mathscr{R}(G') \subset \overline{\mathscr{R}(T)}$. In $\overline{\mathscr{R}(T)}$, G' is unitarily equivalent to the restriction of G to $\overline{\mathscr{R}(G)}$ and so G' is non-negative and self-adjoint. Also, by (3.4), $G'^2 = UGU^*UGU^* = UG^2U^*$ and hence, by (3.6) and (3.7), $G'^2 = TT^*$ as required.

By (3.4), (3.7), and (3.10), $T^* = U^*|T^*|$, and as we have already shown that the polar decomposition is unique, (3.8) follows. That $\mathscr{N}(T) = \mathscr{N}(|T|)$ is obvious, while $\mathscr{R}(T) = \mathscr{R}(|T^*|)$ is a consequence of $T = |T^*|U$ and $|T^*| = TU^*$, these identities being obtained from (3.10). □

In the case when $H_1 = H_2 = H$ and T is self-adjoint, the polar decomposition of T can be used to decompose H into the orthogonal sum of subspaces on which the quadratic form (Tx, x) is positive definite, negative definite, or zero.

Theorem 3.3. Let T be a self-adjoint operator on H and let $T = U|T|$ be its polar decomposition. Then $U^* = U$, $TU = UT$, $|T|U = U|T|$, and H has the orthogonal sum decomposition

$$H = M_+ \oplus M_0 \oplus M_- \tag{3.11}$$

where $M_\pm = \{x \in H : Ux = \pm x\}$ and $M_0 = \mathcal{N}(T)$. The decomposition (3.11) reduces T to its positive, zero, and negative parts:

$$TM_\pm \subset M_\pm, \qquad TM_0 \subset M_0, \qquad Tx = \begin{cases} \pm|T|x & (x \in M_\pm), \\ 0 & (x \in M_0). \end{cases} \tag{3.12}$$ ■

Proof. By Theorem 3.2 and the self-adjointness of T,

$$T = U|T| = T^* = U^*|T^*| = U^*|T|.$$

Since the polar decomposition is unique, this gives $U^* = U$ and hence $U^2x = x$ for $x \in \overline{\mathcal{R}(|T|)}$. Also, from (3.7), $U|T| = T = T^* = |T|U$ and $TU = U|T|U = |T| = UT$.

If $x \in \overline{\mathcal{R}(|T|)}$ and $x_\pm = \frac{1}{2}(I \pm U)x$, then $Ux_\pm = \pm x_\pm$ so that $x_\pm \in M_\pm$. By (3.3), and since $\mathcal{R}(|T|)^\perp = \mathcal{N}(|T|) = \mathcal{N}(T)$ by (3.9), we conclude that any $x \in H$ can be written as

$$x = x_+ + x_0 + x_-, \qquad x_0 \in M_0.$$

For any $x_+ \in M_+$ and $y_- \in M_-$,

$$(x_+, y_-) = -(Ux_+, Uy_-) = -(x_+, y_-)$$

and so $M_+ \perp M_-$. Since M_0 is orthogonal to both M_+ and M_- the orthogonal decomposition (3.11) is established.

If $x_+ \in M_+$ then $UTx_+ = TUx_+ = Tx_+$, so that $TM_+ \subset M_+$; similarly, $TM_- \subset M_-$; furthermore $TM_0 \subset M_0$ since $M_0 = \mathcal{N}(T)$. The operator T, and similarly $|T|$, is therefore reduced by (3.11). Also, for $x_\pm \in M_\pm$ we have $Tx_\pm = \pm TUx_\pm = \pm|T|x_\pm$ and (3.12) follows. □

From Theorem 3.3, it follows that

$$|T|Tx = |T|^2x_+ - |T|^2x_- = T|T|x \tag{3.13}$$

and for $\lambda \in \rho(T)\backslash\{0\}$,

$$(T - \lambda I)^{-1}x = (|T| - \lambda I)^{-1}x_+ - (|T| - \lambda I)^{-1}x_- - \lambda^{-1}x_0$$

whence, for $\mu \in \rho(|T|)$,

$$(T - \lambda I)^{-1}(|T| - \mu I)^{-1} = (|T| - \mu I)^{-1}(T - \lambda I)^{-1}. \tag{3.14}$$

T and $|T|$ therefore have commuting resolvents. □

Theorem 3.4. Let T be self-adjoint in H and $\lambda \in \mathbb{C}$. Then for all $u \in \mathcal{D}(T) = \mathcal{D}(|T|)$,

$$\|Tu\| = \|\,|T|u\|, \quad |(Tu, u)| \leqslant (|T|u, u),$$

$$\|(T+\lambda I)u\| \leqslant \|(|T|+|\lambda|I)u\|,$$

$$(|T+\lambda I|u, u) \leqslant ((|T|+|\lambda|I)u, u). \qquad \blacksquare$$

Proof. The key to this theorem is Proposition III-8.11 and the inequality

$$\begin{aligned}\|(T+\lambda)u\|^2 &= \|Tu\|^2 + 2(\operatorname{re}\lambda)(Tu, u) + |\lambda|^2\,\|u\|^2\\ &\leqslant \|\,|T|u\|^2 + 2|\lambda|(|T|u, u) + |\lambda|^2\,\|u\|^2\\ &= \|(|T|+|\lambda|I)u\|^2,\end{aligned}$$

the inequality $|(Tu, u)| \leqslant (|T|u, u)$ being obtained from Proposition III-8.11. This proposition applies to $|T+\lambda I|$ and $|T|+|\lambda|I$ since

$$\mathcal{D}(|T|+|\lambda|I) = \mathcal{D}(|T|) = \mathcal{D}(T) = \mathcal{D}(|T+\lambda I|). \qquad \square$$

4. Forms of general self-adjoint operators and their perturbations.

In §2 we established a one-one correspondence between the set of all closed, lower semi-bounded, sesquilinear forms a which are densely defined in a Hilbert space H and the set of all lower semi-bounded, self-adjoint operators A, the correspondence being determined by the First Representation Theorem 2.4. If $A \geqslant 0$ is the self-adjoint operator associated with the non-negative form a, the Second Representation Theorem 2.8 asserts that $\mathcal{D}(a) = \mathcal{D}(A^{\frac{1}{2}})$ and $a[\![x, y]\!] = (A^{\frac{1}{2}}x, A^{\frac{1}{2}}y)_H$. In this case (see Definition 2.5) a is the form of A and the form domain of A is $\mathcal{Q}(A) = \mathrm{H}(a) = \mathrm{H}(A^{\frac{1}{2}})$, a Hilbert space with inner product

$$(x, y)_{\mathcal{Q}(A)} = (A^{\frac{1}{2}}x, A^{\frac{1}{2}}y)_H + (x, y)_H. \tag{4.1}$$

For an *arbitrary* self-adjoint operator A we now define the *form domain* $\mathcal{Q}(A)$ to be $\mathcal{Q}(|A|)$, so that

$$(x, y)_{\mathcal{Q}(A)} := (|A|^{\frac{1}{2}}x, |A|^{\frac{1}{2}}y)_H + (x, y)_H, \tag{4.2}$$

and furthermore, we define the *form* a of A to be

$$a[\![x, y]\!] := (U|A|^{\frac{1}{2}}x, |A|^{\frac{1}{2}}y) \qquad (x, y \in \mathcal{Q}(A)), \tag{4.3}$$

where U is the partial isometry in the polar decomposition $A = U|A|$ of A. Since $U \in \mathcal{B}(H)$ and, by Theorem 3.3, U commutes with $|A|$, then U also commutes with $|A|^{\frac{1}{2}}$ (see Kato [1, Theorem V-3.35]). Hence

$$a[\![x, y]\!] = (Ax, y)_H \qquad (x \in \mathcal{D}(A),\ y \in \mathcal{Q}(A)), \tag{4.4}$$

so that our definition is consistent with Theorem 2.4. If A is lower semi-bounded, $A+\alpha I \geqslant 0$ for some $\alpha > 0$, say, then

$$\mathscr{Q}(A) = \mathrm{H}(|A|^{\frac{1}{2}}) = \mathrm{H}\big((A+\alpha I)^{\frac{1}{2}}\big) \tag{4.5}$$

and

$$\|\bullet\|_{\mathscr{Q}(A)} \sim [\|(A+\alpha I)^{\frac{1}{2}}\bullet\|_H^2 + \|\bullet\|_H^2]^{\frac{1}{2}} \tag{4.6}$$

where $\sim$ indicates that the norms are equivalent. To see this we first recall that $\mathscr{D}(A)$ is dense in $\mathscr{Q}(A)$ and $\mathrm{H}\big((A+\alpha I)^{\frac{1}{2}}\big)$. It is therefore enough to show that (4.6) holds on $\mathscr{D}(A) = \mathscr{D}(A+\alpha I)$. For $x \in \mathscr{D}(A)$, we get from Theorem 2.8,

$$\begin{aligned}
\|(A+\alpha I)^{\frac{1}{2}}x\|^2 &= ((A+\alpha I)x, x)_H \\
&= (Ax, x)_H + \alpha\|x\|_H^2 \\
&= (|A|Ux, x)_H + \alpha\|x\|_H^2 \\
&= (|A|^{\frac{1}{2}}Ux, |A|^{\frac{1}{2}}x)_H + \alpha\|x\|_H^2 \\
&= (U|A|^{\frac{1}{2}}x, |A|^{\frac{1}{2}}x)_H + \alpha\|x\|_H^2 \\
&\leqslant \||A|^{\frac{1}{2}}x\|_H^2 + \alpha\|x\|_H^2.
\end{aligned} \tag{4.7}$$

Also, since U commutes with $A+\alpha I$ and hence with $(A+\alpha I)^{\frac{1}{2}}$,

$$\begin{aligned}
\||A|^{\frac{1}{2}}x\|_H^2 &= (|A|x, x)_H \\
&= ((A+\alpha I)Ux, x)_H - \alpha(Ux, x)_H \\
&= (U(A+\alpha I)^{\frac{1}{2}}x, (A+\alpha I)^{\frac{1}{2}}x)_H - \alpha(Ux, x)_H \\
&\leqslant \|(A+\alpha I)^{\frac{1}{2}}x\|_H^2 + \alpha\|x\|_H^2.
\end{aligned} \tag{4.8}$$

The equivalence in (4.6) is therefore established. Also, in this case of $A+\alpha I \geqslant 0$, the form of a can be written

$$a[\![x, y]\!] = ((A+\alpha I)^{\frac{1}{2}}x, (A+\alpha I)^{\frac{1}{2}}y)_H - \alpha(x, y)_H \tag{4.9}$$

since this is clearly true for $x \in \mathscr{D}(A)$ and hence on $\mathscr{Q}(A)$ by continuity.

The norm on $\mathscr{Q}(A)$ given in (4.2) can be written $\|x\|_{\mathscr{Q}(A)}^2 = \|(|A|^{\frac{1}{2}} + \mathrm{i}I)x\|_H^2$ since $|A|^{\frac{1}{2}}$ is self-adjoint. In practice we shall often work with $\|(|A|^{\frac{1}{2}} + \mathrm{i}bI)\bullet\|_H$, where $b > 0$. All these norms are equivalent to $\|\bullet\|_{\mathscr{Q}(A)}$ since, if $0 < b_1 < b_2$,

$$\begin{aligned}
\|(|A|^{\frac{1}{2}} + \mathrm{i}b_1 I)x\|_H^2 &= \||A|^{\frac{1}{2}}x\|_H^2 + b_1^2\|x\|_H^2 \\
&\leqslant \||A|^{\frac{1}{2}}x\|_H^2 + b_2^2\|x\|_H^2 \\
&= \|(|A|^{\frac{1}{2}} + \mathrm{i}b_2 I)x\|_H^2 \\
&\leqslant (b_2^2/b_1^2)\|(|A|^{\frac{1}{2}} + \mathrm{i}b_1 I)x\|_H^2.
\end{aligned} \tag{4.10}$$

Let $\mathfrak{Q} \equiv \mathscr{Q}(A)$, where A is an *arbitrary* self-adjoint operator and let $\mathfrak{Q}^*$ denote the adjoint of $\mathfrak{Q}$. The inclusion map $E: \mathfrak{Q} \to H$ is a continuous embedding with norm $\leqslant 1$ and as $\mathscr{D}(|A|^{\frac{1}{2}})$ is a dense subspace of H, the map E has dense range in H. The map $E^*: x \mapsto (x, E\,\bullet)_H$ is a linear injection of H into $\mathfrak{Q}^*$ with $\|E^*\| \leqslant 1$. Since E is injective, E^* has dense range in $\mathfrak{Q}^*$ and we have the triplet

$$\mathfrak{Q} \xrightarrow{E} H \xrightarrow{E^*} \mathfrak{Q}^*$$

(cf. (1.8)). Hereafter in this section we shall omit the inclusion map E when it is clear from the context that $\mathfrak{Q}$ is to be identified with the subspace $\mathscr{D}(|A|^{\frac{1}{2}})$ of H.

By (4.3), for all $x, y \in \mathfrak{Q}$,

$$\begin{aligned} |a[\![x, y]\!]| &\leqslant \| \, |A|^{\frac{1}{2}} x \|_H \| \, |A|^{\frac{1}{2}} y \|_H \\ &\leqslant \|x\|_{\mathfrak{Q}} \|y\|_{\mathfrak{Q}}. \end{aligned}$$

This implies that for every $x \in \mathfrak{Q}$ we have $a[\![x, \bullet]\!] \in \mathfrak{Q}^*$ and hence there exists a bounded linear map $\hat{A}: \mathfrak{Q} \to \mathfrak{Q}^*$ such that for all $x, y \in \mathfrak{Q}$,

$$a[\![x, y]\!] = (\hat{A}x, y) \equiv (\hat{A}x)(y). \tag{4.11}$$

We have the following relationship between A and $\hat{A}$:

$$\begin{aligned} &\mathscr{D}(A) = \{\phi \in \mathfrak{Q}: \hat{A}\phi \in \mathscr{R}(E^*)\}, \\ &A\phi = E^{*-1}\hat{A}\phi \quad \text{for } \phi \in \mathscr{D}(A). \end{aligned} \tag{4.12}$$

This is because if ϕ lies in $\mathscr{D}(E^{*-1}\hat{A})$, $\hat{A}\phi = E^*\hat{\phi}$ say, where $\hat{\phi} \in H$ and for all $y \in \mathfrak{Q}$,

$$\begin{aligned} (|A|^{\frac{1}{2}} U\phi, |A|^{\frac{1}{2}} y)_H &= a[\![\phi, y]\!] \\ &= (E^*\hat{\phi}, y) \\ &= (\hat{\phi}, y)_H. \end{aligned}$$

Thus $\phi \in \mathscr{D}(A)$ and $A\phi = |A| U\phi = \hat{\phi} = E^{*-1}\hat{A}\phi$; in other words, $E^{*-1}\hat{A} \subset A$. Conversely, if $\phi \in \mathscr{D}(A)$ and $y \in \mathfrak{Q}$,

$$\begin{aligned} (\hat{A}\phi, y) = a[\![\phi, y]\!] &= (A\phi, y)_H \\ &= (E^* A\phi, y) \end{aligned}$$

which gives the reverse inclusion $A \subset E^{*-1}\hat{A}$; (4.12) is therefore proved. Note that by (4.12) A is a map between $\mathfrak{Q}$ and H and is a map in H when $\mathfrak{Q}$ is identified with $\mathscr{D}(|A|^{\frac{1}{2}})$. More precisely, we should have written $A = E^{*-1}\hat{A}E^{-1}$ but this dual role of A in (4.12) will not cause confusion as the precise meaning will always be clear from the context.

We denote the form of $|A|$ by $|a|$:

$$|a|[\![x, y]\!] := (|A|^{\frac{1}{2}} x, |A|^{\frac{1}{2}} y)_H \qquad (x, y \in \mathfrak{Q}), \tag{4.13}$$

and define $|\hat{A}|: \mathfrak{Q} \to \mathfrak{Q}^*$ by

$$(|\hat{A}| x, y) = |a|[\![x, y]\!] \qquad (x, y \in \mathfrak{Q}). \tag{4.14}$$

Lemma 4.1. Let $\|x\|_{\mathfrak{Q}} = \|(|A|^{\frac{1}{2}} + \mathrm{i}bI)x\|_H$, with $b > 0$, be the norm on $\mathfrak{Q}$ (see (4.10)). Then

(i) $|\hat{A}| + b^2 E^*\colon \mathfrak{Q} \to \mathfrak{Q}^*$ is unitary and $\mathfrak{Q}^*$ is a Hilbert space with respect to the inner product

$$(\phi^*, \psi^*)_{\mathfrak{Q}^*} = (\phi, \psi)_{\mathfrak{Q}}, \qquad (|\hat{A}| + b^2 E^*) : \phi \mapsto \phi^* \text{ and } \psi \mapsto \psi^*;$$

(ii) if $z \in \mathbb{R}\setminus\{0\}$ then $(A - \mathrm{i}zI)^{-1} \in \mathscr{B}(H, \mathfrak{Q}) \subset \mathscr{B}(H)$, $(\hat{A} - \mathrm{i}zE^*)^{-1} \in \mathscr{B}(\mathfrak{Q}^*, \mathfrak{Q})$, and when $|z| \geqslant b^2$,

$$\|(A - \mathrm{i}zI)^{-1}\|_{\mathscr{B}(H,\mathfrak{Q})}, \qquad \|(\hat{A} - \mathrm{i}zE^*)^{-1}\|_{\mathscr{B}(\mathfrak{Q}^*,\mathfrak{Q})} \leqslant 1 + b^2/|z|. \tag{4.15}$$ ∎

Proof. (i) By the Riesz Representation Theorem, if $\phi^* \in \mathfrak{Q}^*$, there is a $\phi \in \mathfrak{Q}$ such that $(\phi^*, \psi) = (\phi, \psi)_{\mathfrak{Q}}$ for all $\psi \in \mathfrak{Q}$. Hence

$$\begin{aligned}(\phi^*, \psi) &= (|A|^{\frac{1}{2}}\phi, |A|^{\frac{1}{2}}\psi)_H + b^2(\phi, \psi)_H \\ &= |a|[\![\phi, \psi]\!] + b^2(\phi, \psi)_H \\ &= ((|\hat{A}| + b^2 E^*)\phi, \psi)\end{aligned}$$

and consequently $|\hat{A}| + b^2 E^*$ is surjective. Also

$$\begin{aligned}\|(|\hat{A}| + b^2 E^*)\phi\|_{\mathfrak{Q}^*} &= \sup_{\|\psi\|_{\mathfrak{Q}} \leqslant 1} |((|\hat{A}| + b^2 E^*)\phi, \psi)| \\ &= \sup_{\|\psi\|_{\mathfrak{Q}} \leqslant 1} |(\phi, \psi)_{\mathfrak{Q}}| = \|\phi\|_{\mathfrak{Q}}\end{aligned}$$

which proves that $|\hat{A}| + b^2 E^*\colon \mathfrak{Q} \to \mathfrak{Q}^*$ is unitary. For any $\phi^* \in \mathfrak{Q}^*$, $\phi^* = (|\hat{A}| + b^2 E^*)\phi$ for some $\phi \in \mathfrak{Q}$ and $\|\phi^*\|_{\mathfrak{Q}^*} = \|\phi\|_{\mathfrak{Q}}$. Thus $\mathfrak{Q}^*$ is a Hilbert space with the inner product indicated.

(ii) Since A is self-adjoint in H, it follows that $\mathscr{R}(A - \mathrm{i}zI) = H$ and so $\mathscr{R}(\hat{A} - \mathrm{i}zE^*)$ is dense in $\mathfrak{Q}^*$. Let $\phi \in \mathscr{R}(A - \mathrm{i}zI) = H$, $\theta = (|A|^{\frac{1}{2}} - \mathrm{i}bI)^{-1}\phi$ and $|z| \geqslant b^2$. Then, since A and $|A|^{\frac{1}{2}}$ commute, we have

$$\begin{aligned}&\|(A - \mathrm{i}zI)^{-1}\phi\|_{\mathfrak{Q}}^2 \\ &= \|(|A|^{\frac{1}{2}} + \mathrm{i}bI)(A - \mathrm{i}zI)^{-1}\phi\|_H^2 \\ &= \|(|A| + b^2 I)(|A|^{\frac{1}{2}} - \mathrm{i}bI)^{-1}(A - \mathrm{i}zI)^{-1}\phi\|_H^2 \\ &= \|(|A| + b^2 I)(A - \mathrm{i}zI)^{-1}\theta\|_H^2 \\ &= \|A(A - \mathrm{i}zI)^{-1}\theta\|_H^2 + 2b^2(|A|(A - \mathrm{i}z)^{-1}\theta, (A - \mathrm{i}z)^{-1}\theta)_H \\ &\qquad + b^4\|(A - \mathrm{i}zI)^{-1}\theta\|_H^2 \\ &\leqslant \|(A - \mathrm{i}zI)(A - \mathrm{i}zI)^{-1}\theta\|_H^2 + 2b^2\|A(A - \mathrm{i}zI)^{-1}\theta\|_H\|(A - \mathrm{i}zI)^{-1}\theta\|_H\end{aligned}$$

$$\leqslant \|\theta\|_H^2 + 2b^2 \|\theta\|_H \|(A - \mathrm{i}zI)^{-1}\theta\|_H$$
$$\leqslant (1 + 2b^2/|z|)\|\theta\|_H^2$$
$$\leqslant (1 + b^2/|z|)^2 \|\theta\|_H^2.$$

In the preceding calculation we have used the inequalities $\|A\psi\|_H$, $|z|\,\|\psi\|_H \leqslant \|(A - \mathrm{i}zI)\psi\|_H$ which follow from the identity

$$\|(A - \mathrm{i}zI)\psi\|_H^2 = \|A\psi\|_H^2 + z^2\|\psi\|_H^2.$$

Furthermore, if $\phi^* = E^*\phi$,

$$\begin{aligned}\|\phi^*\|_{\mathfrak{Q}^*} &= \sup_{\|\psi\|_{\mathfrak{Q}} \leqslant 1} |(\phi^*, \psi)| \\ &= \sup_{\|\psi\|_{\mathfrak{Q}} \leqslant 1} |(\phi, \psi)_H| \\ &= \sup_{\|\psi\|_{\mathfrak{Q}} \leqslant 1} |((|A|^{\frac{1}{2}} - \mathrm{i}bI)^{-1}\phi, (|A|^{\frac{1}{2}} + \mathrm{i}bI)\psi)_H| \\ &= \|(|A|^{\frac{1}{2}} - \mathrm{i}bI)^{-1}\phi\|_H \\ &= \|\theta\|_H.\end{aligned}$$

Hence $\|\theta\|_H \leqslant \|\phi\|_H$ and $(A - \mathrm{i}zI)^{-1} \in \mathscr{B}(H, \mathfrak{Q})$ with norm satisfying (4.15). If $0 < |z| < b^2$, the same argument applies but the norm has a different bound.

Since $\mathscr{N}(\hat{A} - \mathrm{i}zE^*) = \mathscr{N}(A - \mathrm{i}zI) = \{0\}$, the map $(\hat{A} - \mathrm{i}zE^*)^{-1}$ exists and for $\phi^* = E^*\phi \in \mathscr{R}(\hat{A} - \mathrm{i}zE^*)$ and $\psi \in \mathfrak{Q}$,

$$\begin{aligned}|((\hat{A} - \mathrm{i}zE^*)^{-1}\phi^*, \psi)_{\mathfrak{Q}}| &= |((A - \mathrm{i}zI)^{-1}\phi, \psi)_{\mathfrak{Q}}| \\ &\leqslant \|(A - \mathrm{i}zI)^{-1}\phi\|_{\mathfrak{Q}}\|\psi\|_{\mathfrak{Q}}.\end{aligned}$$

Consequently, from the above, when $|z| \geqslant b^2$,

$$\begin{aligned}\|(\hat{A} - \mathrm{i}zE^*)^{-1}\phi^*\|_{\mathfrak{Q}} &\leqslant \|(A - \mathrm{i}zI)^{-1}\phi\|_{\mathfrak{Q}} \\ &\leqslant (1 + b^2/|z|)\|\theta\|_H \\ &= (1 + b^2/|z|)\|\phi^*\|_{\mathfrak{Q}^*}.\end{aligned}$$

Thus $(\hat{A} - \mathrm{i}zE^*)^{-1}$ is bounded on $\mathscr{R}(\hat{A} - \mathrm{i}zE^*)$ when $|z| \geqslant b^2$ and similarly for $0 < |z| < b^2$. But $(\hat{A} - \mathrm{i}zE^*)^{-1} \in \mathscr{C}(\mathfrak{Q}^*, \mathfrak{Q})$ since $(\hat{A} - \mathrm{i}zE^*) \in \mathscr{B}(\mathfrak{Q}, \mathfrak{Q}^*)$. Also $\mathscr{R}(\hat{A} - \mathrm{i}zE^*)$ is dense in $\mathfrak{Q}^*$. Therefore $(\hat{A} - \mathrm{i}zE^*)^{-1} \in \mathscr{B}(\mathfrak{Q}^*, \mathfrak{Q})$ and (4.15) is satisfied. □

Theorem 4.2. Let p be a bounded sesquilinear form on $\mathfrak{Q} \times \mathfrak{Q}$, where $\mathfrak{Q} = \mathscr{Q}(A)$ is the form domain of the self-adjoint operator A in H, and define

$$\hat{P}: \mathfrak{Q} \to \mathfrak{Q}^*, \qquad (\hat{P}x, y) = p[\![x, y]\!] \qquad (x, y \in \mathfrak{Q}).$$

Suppose that for some $\gamma > 0$,

$$|p[\![x]\!]| \leqslant \beta \| \, |A|^{\frac{1}{2}} x \|_H^2 + \gamma \|x\|_H^2 \qquad (x \in \mathfrak{Q}), \tag{4.16}$$

where $0 \leqslant \beta < 1$ if p is symmetric and $0 \leqslant \beta < \frac{1}{2}$ otherwise. Then, if a is the form of A, the following results hold:

(i) $s = a + p$ is bounded on $\mathfrak{Q} \times \mathfrak{Q}$ and

$$\hat{S}\colon \mathfrak{Q} \to \mathfrak{Q}^*, \qquad (\hat{S}x, y) = s[\![x, y]\!] \quad (x, y \in \mathfrak{Q})$$

is defined;

(ii) $(\hat{S} - \mathrm{i}zE^*)^{-1} \in \mathscr{B}(\mathfrak{Q}^*, \mathfrak{Q})$ for z real and $|z|$ large enough; if $S = E^{*-1}\hat{S}$ then $(S - \mathrm{i}zI)^{-1} \in \mathscr{B}(H, \mathfrak{Q}) \subset \mathscr{B}(H)$ and S is closed as an operator in H;

(iii) for $|z|$ large enough,

$$(A - \mathrm{i}zI)^{-1} - (S - \mathrm{i}zI)^{-1} = (\hat{S} - \mathrm{i}zE^*)^{-1} \, \hat{P}(A - \mathrm{i}zI)^{-1} \tag{4.17}$$

where $(A - \mathrm{i}zI)^{-1}$ on the right-hand side is understood as a map of H into $\mathfrak{Q}$;

(iv) if A is lower semi-bounded S is m-sectorial;

(v) if p is symmetric S is self-adjoint;

(vi) if $\hat{P}(A - \mathrm{i}zI)^{-1} \in \mathscr{K}(H, \mathfrak{Q}^*)$ then for $|z|$ large enough,

$$(A - \mathrm{i}zI)^{-1} - (S - \mathrm{i}zI)^{-1} \in \mathscr{K}(H, \mathfrak{Q}) \subset \mathscr{K}(H). \qquad \blacksquare$$

Remarks. The bound (4.16) is clearly unnecessary for (i) to hold. In (ii), if p is symmetric, $(S - \mathrm{i}zI)^{-1} \in \mathscr{B}(H)$ for all real $z \neq 0$ on account of (v). We shall see in Chapter IX, Theorem 2.4 that (vi) implies that A and S have the same essential spectra σ_{ek} $(k = 1, 2, 3, 4)$.

Proof of Theorem 4.2. (i) Let $b^2 = \gamma/\beta$, with $\beta > 0$, and note that

$$\|x\|_{\mathfrak{Q}}^2 = \| (|A|^{\frac{1}{2}} + \mathrm{i}bI) x \|_H^2 = \| \, |A|^{\frac{1}{2}} x \|_H^2 + b^2 \|x\|_H^2$$

(see (4.10)). Since a and p are bounded on $\mathfrak{Q} \times \mathfrak{Q}$ so is s, and $\hat{S}$ is defined.

(ii) From the equation $\hat{S} = \hat{A} + \hat{P}$ and Lemma 4.1 (ii) we see that for real $z \neq 0$,

$$\begin{aligned} \hat{S} - \mathrm{i}zE^* &= \hat{A} - \mathrm{i}zE^* + \hat{P} \\ &= [I_{Q^*} + \hat{P}(\hat{A} - \mathrm{i}zE^*)^{-1}](\hat{A} - \mathrm{i}zE^*) \end{aligned} \tag{4.18}$$

where $I_{\mathfrak{Q}^*}$ is the identity on $\mathfrak{Q}^*$. From (4.16) we obtain

$$|p[\![x]\!]| \leqslant \beta \|x\|_{\mathfrak{Q}}^2$$

and so in view of Lemma 2.1,

$$|p[\![x, y]\!]| \leqslant \varepsilon \beta \|x\|_{\mathfrak{Q}} \|y\|_{\mathfrak{Q}} \qquad (x, y \in \mathfrak{Q}),$$

with $\varepsilon = 1$ if p is symmetric and $\varepsilon = 2$ otherwise. Hence

$$\begin{aligned}\|\hat{P}x\|_{\mathfrak{Q}^*} &= \sup_{\|y\|_{\mathfrak{Q}} \leqslant 1} |p[\![x, y]\!]| \\ &\leqslant \varepsilon\beta \|x\|_{\mathfrak{Q}}.\end{aligned}$$

For z large enough it now follows from (4.15) that if $x \in \mathfrak{Q}^*$

$$\begin{aligned}\|\hat{P}(\hat{A} - izE^*)^{-1}x\|_{Q^*} &\leqslant \varepsilon\beta \|(\hat{A} - izE^*)^{-1}x\|_{\mathfrak{Q}} \\ &\leqslant \varepsilon\beta(1 + b^2/|z|)\|x\|_{\mathfrak{Q}^*}\end{aligned}$$

and this, together with the fact that $\varepsilon\beta(1+b^2/|z|) < 1$ for $|z|$ large enough, proves that $\|\hat{P}(\hat{A} - izE^*)^{-1}\| < 1$. Therefore, in (4.18),

$[I_{\mathfrak{Q}^*} + \hat{P}(\hat{A} - izE^*)^{-1}]^{-1} \in \mathscr{B}(\mathfrak{Q}^*)$ and consequently $(\hat{S} - izE^*)^{-1} \in \mathscr{B}(\mathfrak{Q}^*, \mathfrak{Q})$. Furthermore, $(S - izI)^{-1} = (\hat{S} - izE^*)^{-1}E^* \in \mathscr{B}(H, \mathfrak{Q}) \subset \mathscr{B}(H)$, and in particular, $S \in \mathscr{C}(H)$.

(iii) This is an immediate consequence of (4.18).

(iv) Let $A + \alpha > 0$ and $\alpha \geqslant 0$. Then $a + \alpha > 0$ and from (4.16), (4.8) and (4.9),

$$|p[\![x]\!]| \leqslant \beta[(a + 2\alpha)[\![x]\!] + b^2\|x\|_H^2] \quad \text{for all } x \in \mathfrak{Q}.$$

This yields

$$\begin{aligned}\operatorname{re} s[\![x]\!] &\geqslant a[\![x]\!] - \beta[(a+\alpha)[\![x]\!] + (b^2 + \alpha)\|x\|_H^2] \\ &= (1-\beta)(a+\alpha)[\![x]\!] - (\gamma + \alpha\beta + \alpha)\|x\|_H^2.\end{aligned}$$

Consequently, for a sufficiently large positive δ,

$$\operatorname{re}(s+\delta)[\![x]\!] \geqslant (1-\beta)(a+\alpha)[\![x]\!] + \|x\|_H^2 \qquad (x \in \mathfrak{Q}). \tag{4.19}$$

Moreover,

$$\begin{aligned}\operatorname{im} s[\![x]\!] &\leqslant \beta[(a+\alpha)[\![x]\!] + (b^2 + \alpha)\|x\|_H^2] \\ &\leqslant K \operatorname{re}(S + \delta)[\![x]\!] \end{aligned} \tag{4.20}$$

from (4.19), for some $K > 0$. The inequalities (4.19) and (4.20) together imply that S is sectorial. From (ii) we therefore conclude that def $(S - \lambda I) = 0$ for λ outside $\Theta(S)$, and hence S is m-sectorial.

(v) If p is symmetric so is s; and by (ii) and Theorem III.6.7(i), S is self-adjoint.

(vi) This follows from (ii) and (4.17). □

Corollary 4.3. Let A and S be self-adjoint operators with forms a and s and form domains $\mathscr{Q}(A) = \mathscr{Q}(S) = \mathfrak{Q}$. Then (iii) and (vi) in Theorem 4.2 hold with $p = s - a$. ■

Proof. Since $\mathscr{Q}(A) = \mathscr{Q}(S)$, it follows from Proposition III.7.2 that $\mathscr{Q}(A)$ and $\mathscr{Q}(S)$ are topologically isomorphic and hence, for all $x \in \mathfrak{Q}$,

$$|p[\![x]\!]| \leqslant |s[\![x]\!]| + |a[\![x]\!]| \leqslant K\|x\|_{\mathfrak{Q}}^2.$$

$\hat{P}: \mathfrak{Q} \to \mathfrak{Q}^*$ is therefore defined and the identity (4.18) holds. Since $(\hat{A} - \mathrm{i}zE^*)^{-1}$ and $(\hat{S} - \mathrm{i}zE^*)^{-1}$ both exist and lie in $\mathscr{B}(\mathfrak{Q}^*, \mathfrak{Q})$ for any $z \neq 0$, the corollary is a consequence of (4.18). □

Let P be an operator with domain $\mathscr{D}(P) \subset \mathscr{D}(A)$ and

$$|(Px, x)_H| \leqslant \beta(|A|x, x)_H + \gamma\|x\|_H^2 \qquad (x \in \mathscr{D}(P)),$$

where A is self-adjoint and the constants β and γ are as in (4.16). Suppose that $\mathscr{D}(P)$ is a core of a. Then, by Lemma 2.1, $(Px, y)_H$ can be extended by continuity to a form $p[\![x, y]\!]$ on $\mathscr{Q}(A)$ which satisfies (4.16). The operator S in Theorem 4.2 is then a closed extension of $A + P$. In the case when P is symmetric S is called the *pseudo-Friedrichs extension* of $A + P$ (see Kato [1, Theorem VI-3.11]).

Finally in this section we prove an equivalent form of the requirement in (vi) of Theorem 4.2.

Theorem 4.4. Let $\mathfrak{Q} = \mathscr{Q}(A)$ be the form domain of a self-adjoint operator A in H and let $p = p_1 + \mathrm{i}p_2$, where p_1 and p_2 are the bounded symmetric forms on $\mathfrak{Q} \times \mathfrak{Q}$ of self-adjoint operators P_1 and P_2, i.e.,

$$p_j[\![x, y]\!] = (U_j|P_j|^{\frac{1}{2}}x, |P_j|^{\frac{1}{2}}y)_H \qquad (x, y \in \mathfrak{Q}),$$

where U_j is the partial isometry in the polar decomposition $P_j = U_j|P_j|$ $(j = 1, 2)$. Then $\hat{P}(A - \mathrm{i}zI)^{-1} \in \mathscr{K}(H, \mathfrak{Q}^*)$ if

$$|P_j|^{\frac{1}{2}}(A - \mathrm{i}zI)^{-1} \in \mathscr{K}(H) \qquad (j = 1, 2).$$ ■

Proof. Let

$$B_j := |P_j|^{\frac{1}{2}}(|A|^{\frac{1}{2}} + \mathrm{i}bI)^{-1} \qquad (j = 1, 2).$$

Since $\mathfrak{Q} \subset \mathscr{D}(|P_j|^{\frac{1}{2}})$ by hypothesis, it follows from Proposition III.7.2 that $\mathfrak{Q}$ is continuously embedded in $\mathscr{Q}(P_j)$ and, for some $K > 0$,

$$\|\,|P_j|^{\frac{1}{2}}x\|_H \leqslant K\|x\|_{\mathfrak{Q}} \qquad (x \in \mathfrak{Q}, j = 1, 2),$$

and

$$\begin{aligned}\|B_jx\|_H &\leqslant K\|(|A|^{\frac{1}{2}} + \mathrm{i}bI)^{-1}x\|_{\mathfrak{Q}} \\ &= K\|x\|_H \qquad (x \in \mathfrak{Q},\ j = 1, 2).\end{aligned}$$

Thus $B_j \in \mathscr{B}(H)$ $(j = 1, 2)$. Furthermore, for $x, y \in \mathfrak{Q}$,

$$\begin{aligned}(\hat{P}_jx, y) &= (U_j|P_j|^{\frac{1}{2}}x, |P_j|^{\frac{1}{2}}y)_H \\ &= (U_jB_j(|A|^{\frac{1}{2}} + \mathrm{i}bI)x, B_j(|A|^{\frac{1}{2}} + \mathrm{i}bI)y)_H \\ &= (B_j^*U_jB_j(|A|^{\frac{1}{2}} + \mathrm{i}bI)x, (|A|^{\frac{1}{2}} + \mathrm{i}bI)y)_H\end{aligned}$$

and consequently for $x \in Q$,

$$\begin{aligned} \| \hat{P}_j x \|_{\mathfrak{Q}^*} &\leqslant \| B_j^* U_j B_j (|A|^{\frac{1}{2}} + ibI) x \|_H \\ &= \| B_j^* U_j | P_j |^{\frac{1}{2}} x \|_H \qquad (j = 1, 2). \end{aligned}$$

Thus, for $x \in H$,

$$\| \hat{P}_j (A - izI)^{-1} x \|_{\mathfrak{Q}^*} \leqslant \| B_j^* U_j | P_j |^{\frac{1}{2}} (A - izI)^{-1} x \|_H$$

and

$$\| \hat{P} (A - izI)^{-1} x \|_{\mathfrak{Q}^*} \leqslant \sum_{j=1}^{2} \| B_j^* U_j | P_j |^{\frac{1}{2}} (A - izI)^{-1} x \|_H .$$

Since $|P_j|^{\frac{1}{2}}(A - izI)^{-1} \in \mathscr{K}(H)$, $U_j \in \mathscr{B}(H)$ and $B_j^* \in \mathscr{B}(H)$, it follows that $\hat{P}(A - izI)^{-1} \in \mathscr{K}(H, \mathfrak{Q}^*)$. □

5. Perturbation of sectorial forms.

In this section a is a closed, densely defined sectorial form and A is the m-sectorial operator associated with a in H as in Theorem 2.4. Let $a_1 = \mathrm{re}\, a$ and $a_2 = \mathrm{im}\, a$ and suppose that, for some $\alpha \geqslant 0$ and $M > 0$,

$$\Theta(a) \subset \{z : \mathrm{re}\, z + \alpha > 0 \text{ and } |\mathrm{im}\, z| \leqslant M(\mathrm{re}\, z + \alpha)\}. \tag{5.1}$$

The symmetric form a_1 is closed and bounded below by $-\alpha$ in H. We recall that the semi-bounded self-adjoint operator A_1 associated with a_1 is called the real part of A, written $A_1 = \mathrm{re} A$.

The *form domain* $\mathfrak{Q} = \mathscr{Q}(A)$ of A is defined to be the form domain $\mathscr{Q}(A_1)$ of A_1. In view of (4.5) and (4.9), $\mathfrak{Q} = \mathrm{H}((A_1 + \alpha I)^{\frac{1}{2}})$ with norm (4.6) and

$$a_1[\![x, y]\!] = ((A_1 + \alpha I)^{\frac{1}{2}} x, (A_1 + \alpha I)^{\frac{1}{2}} y)_H - \alpha (x, y)_H. \tag{5.2}$$

It is convenient to take for the norm on $\mathfrak{Q}$ the norm equivalent to (4.6) defined by

$$\| x \|_{\mathfrak{Q}}^2 = \| (A_1 + \alpha I)^{\frac{1}{2}} x \|_H^2 + b^2 \| x \|_H^2, \qquad b \neq 0. \tag{5.3}$$

From (5.1), for $x \in \mathfrak{Q}$,

$$\begin{aligned} |(a + \alpha)[\![x]\!]| &\leqslant (a_1 + \alpha)[\![x]\!] + |a_2[\![x]\!]| \\ &\leqslant (1 + M)(a_1 + \alpha)[\![x]\!] \end{aligned}$$

and from Lemma 2.1

$$\begin{aligned} |(a + \alpha)[\![x, y]\!]| &\leqslant \varepsilon (1 + M) \{ (a_1 + \alpha)[\![x]\!] (a_1 + \alpha)[\![y]\!] \}^{\frac{1}{2}} \\ &\leqslant \varepsilon (1 + M) \| x \|_{\mathfrak{Q}} \| y \|_{\mathfrak{Q}} \qquad (x, y \in \mathfrak{Q}), \end{aligned} \tag{5.4}$$

where $\varepsilon = 1$ if a is symmetric and $\varepsilon = 2$ otherwise. Thus a is bounded on $\mathfrak{Q} \times \mathfrak{Q}$ and we may define

$$\hat{A} : \mathfrak{Q} \to \mathfrak{Q}^*, \qquad (\hat{A} x, y) = a[\![x, y]\!] \quad (x, y \in \mathfrak{Q}).$$

We claim that

$$A = E^{*-1}\hat{A} \tag{5.5}$$

where E^*: $H \to \mathfrak{Q}^*$ is the embedding defined in §4. To prove this let $A_0 = E^{*-1}\hat{A}$; then for $x \in \mathscr{D}(A_0)$,

$$(A_0 x, x)_H = (E^* A_0 x, x) = a[\![x]\!]$$

which implies that A_0 is sectorial. If $x \in \mathscr{D}(A)$ and $y \in \mathfrak{Q}$ we have from Theorem 2.4 that

$$(\hat{A}x, y) = a[\![x, y]\!] = (Ax, y)_H = (E^* Ax, y)$$

and hence $A \subset A_0$. Since A is m-sectorial it cannot have a proper sectorial extension and so $A = A_0$, thus establishing (5.5).

For any real $z \neq 0$, the point $-(\alpha+z^2)$ lies outside the set in (5.1) and hence is in the resolvent set of A. Consequently $[A+(\alpha+z^2)I]^{-1} \in \mathscr{B}(H)$ and $\hat{A}+(\alpha+z^2)E^*$ is a bounded operator from $\mathfrak{Q}$ to $\mathfrak{Q}^*$ with dense range, namely $\mathscr{R}(E^*)$. Moreover for $x \in \mathfrak{Q}$,

$$\begin{aligned}
\|[\hat{A}+(\alpha+z^2)E^*]x\|_{\mathfrak{Q}^*}\|x\|_{\mathfrak{Q}} &\geqslant |([\hat{A}+(\alpha+z^2)E^*](x), x)| \\
&= |(a+\alpha+z^2)[\![x]\!]| \\
&\geqslant (a_1+\alpha+z^2)[\![x]\!] \\
&= \|(A_1+\alpha I)^{\frac{1}{2}}x\|_H^2 + z^2\|x\|_H^2 \\
&\geqslant [z^2/(b^2+z^2)]\|x\|_{\mathfrak{Q}}^2.
\end{aligned}$$

Hence $[\hat{A}+(\alpha+z^2)E^*]^{-1}$ is closed and bounded on a dense subspace of $\mathfrak{Q}^*$, whence

$$[\hat{A}+(\alpha+z^2)E^*]^{-1} \in \mathscr{B}(\mathfrak{Q}^*, \mathfrak{Q}) \tag{5.6}$$

and

$$\|[\hat{A}+(\alpha+z^2)E^*]^{-1}\|_{\mathscr{B}(\mathfrak{Q}^*, \mathfrak{Q})} \leqslant 1 + b^2/z^2. \tag{5.7}$$

Since $[A+(\alpha+z^2)I]^{-1} = [\hat{A}+(\alpha+z^2)E^*]^{-1}E^*$, it follows that $[A+(\alpha+z^2)I]^{-1} \in \mathscr{B}(H, \mathfrak{Q})$ with

$$\|[A+(\alpha+z^2)I]^{-1}\|_{\mathscr{B}(H, \mathfrak{Q})} \leqslant 1 + b^2/z^2. \tag{5.8}$$

Theorem 5.1. Let $\mathfrak{Q}$ be the form domain of the m-sectorial operator A in H associated with the closed, densely defined sectorial form a and let $A_1 := \operatorname{re} A > -\alpha$. Let p be a sesquilinear form defined on $\mathfrak{Q}$ and such that for some $\gamma > 0$,

$$|p[\![x]\!]| \leqslant \beta\|(A_1+\alpha I)^{\frac{1}{2}}x\|_H^2 + \gamma\|x\|_H^2 \qquad (x \in \mathfrak{Q}), \tag{5.9}$$

where $0 \leqslant \beta < 1$ if p is symmetric and $0 \leqslant \beta < \frac{1}{2}$ otherwise. Then

(i) $s = a + p$ is closed and sectorial for all p with $0 \leqslant \beta < 1$ in (5.9);
(ii) $\hat{S}$: $\mathfrak{Q} \to \mathfrak{Q}^*$: $(\hat{S}x, y) = s[\![x, y]\!]$ is defined;

(iii) $S := E^{*-1}\hat{S}$ is the m-sectorial operator associated with s and for large enough real z,

$$(\hat{S}+z^2E^*)^{-1} \in \mathscr{B}(\mathfrak{Q}^*, \mathfrak{Q}) \quad \text{and} \quad (S+z^2I)^{-1} \in \mathscr{B}(H, \mathfrak{Q}) \subset \mathscr{B}(H);$$

(iv) $\hat{P}: \mathfrak{Q} \to \mathfrak{Q}^*: (\hat{P}x, y) = p[\![x, y]\!]$ is defined and for large enough real z,

$$(A+z^2I)^{-1} - (S+z^2I)^{-1} = (\hat{S}+z^2E^*)^{-1}\hat{P}(A+z^2I)^{-1}, \tag{5.10}$$

where $(A+z^2I)^{-1}$ is understood as a map of H into $\mathfrak{Q}$ on the right-hand side;
(v) if $\hat{P}(A+z^2I)^{-1} \in \mathscr{K}(H, \mathfrak{Q}^*)$ then for large enough real z,

$$(A+z^2I)^{-1} - (S+z^2I)^{-1} \in \mathscr{K}(H, \mathfrak{Q}) \subset \mathscr{K}(H). \qquad \blacksquare$$

Proof. There is no loss of generality if we take $\alpha = 0$, because in general we can replace A_1 by $A_1 + \alpha I$ throughout. In (5.3) we take $b^2 = \gamma/\beta$ ($\beta > 0$) so that

$$\|x\|_{\mathfrak{Q}}^2 = \|A_1^{\frac{1}{2}}x\|_H^2 + \gamma\beta^{-1}\|x\|_H^2.$$

(i) By (5.8), for $x \in \mathfrak{Q}$,

$$\begin{aligned} s_1[\![x]\!] &= \operatorname{re} s[\![x]\!] = a_1[\![x]\!] + p_1[\![x]\!] \\ &\geq \|A_1^{\frac{1}{2}}x\|_H^2 - (\beta\|A_1^{\frac{1}{2}}x\|_H^2 + \gamma\|x\|_H^2) \end{aligned}$$

so that, for large enough real z,

$$(s_1+z^2)[\![x]\!] \geq (1-\beta)\|x\|_{\mathfrak{Q}}^2. \tag{5.11}$$

Also, for some $K > 0$,

$$(s_1+z^2)[\![x]\!] \leq K\|x\|_{\mathfrak{Q}}^2. \tag{5.12}$$

Hence $(s_1+z^2)^{\frac{1}{2}}[\![\bullet]\!]$ is a norm on $\mathfrak{Q}$ which is equivalent to $\|\bullet\|_{\mathfrak{Q}}$. Since $\mathfrak{Q}$ is complete it is sufficient to prove that s is sectorial in order to establish that s is closed. This is so since

$$\begin{aligned} |s_2[\![x]\!]| &\leq |a_2[\![x]\!]| + |p_2[\![x]\!]| \\ &\leq Ma_1[\![x]\!] + \beta\|A_1^{\frac{1}{2}}x\|_H^2 + \gamma\|x\|_H^2 \quad \text{(by (5.1))} \\ &\leq M'\|x\|_{\mathfrak{Q}}^2 \\ &\leq M''(s_1+z^2)[\![x]\!] \end{aligned}$$

by (5.12), for large enough z.

(ii) This is a consequence of Lemma 2.1 and the fact that

$$\begin{aligned} |s[\![x]\!]| &\leq |a[\![x]\!]| + |p[\![x]\!]| \\ &\leq (1+M)a_1[\![x]\!] + \beta\|x\|_{\mathfrak{Q}}^2 \\ &\leq K\|x\|_{\mathfrak{Q}}^2. \end{aligned}$$

$s[\![\bullet, \bullet]\!]$ is therefore bounded on $\mathfrak{Q} \times \mathfrak{Q}$ and $\hat{S}$ is defined.

(iii) By (5.9) and Lemma 2.1,

$$|p[\![x, y]\!]| \leqslant \varepsilon\beta \|x\|_{\mathfrak{Q}} \|y\|_{\mathfrak{Q}} \qquad (x, y \in \mathfrak{Q}), \tag{5.13}$$

where $\varepsilon = 1$ if p is symmetric and $\varepsilon = 2$ otherwise. Consequently, from (5.7) with $\alpha = 0$ and $z \neq 0$, for $x \in \mathfrak{Q}^*$,

$$\begin{aligned}\|\hat{P}(\hat{A}+z^2E^*)^{-1}x\|_{\mathfrak{Q}^*} &= \sup_{\|y\|_{\mathfrak{Q}} \leqslant 1} |p[\![(\hat{A}+z^2E^*)^{-1}x, y]\!]| \\ &\leqslant \varepsilon\beta \|(\hat{A}+z^2E^*)^{-1}x\|_{\mathfrak{Q}} \\ &\leqslant \varepsilon\beta(1+b^2/z^2)\|x\|_{\mathfrak{Q}^*}.\end{aligned}$$

Thus, since $\varepsilon\beta < 1$, for large enough z,

$$\|\hat{P}(A+z^2E^*)^{-1}\|_{\mathscr{B}(\mathfrak{Q}^*)} < 1. \tag{5.14}$$

On writing

$$\begin{aligned}\hat{S}+z^2E^* &= \hat{A}+z^2E^*+\hat{P} \\ &= [I_{\mathfrak{Q}^*}+\hat{P}(\hat{A}+z^2E^*)^{-1}](\hat{A}+z^2E^*)\end{aligned} \tag{5.15}$$

where $I_{\mathfrak{Q}^*}$ is the identity on $\mathfrak{Q}^*$, we see that $(\hat{S}+z^2E^*)^{-1} \in \mathscr{B}(\mathfrak{Q}^*, \mathfrak{Q})$ for z large enough. The rest follows by the argument used to establish (5.5).

(iv) This follows from (iii).

(v) Since $(\hat{S}+z^2E^*)^{-1} \in \mathscr{B}(\mathfrak{Q}^*, \mathfrak{Q})$, (v) is an immediate consequence of (5.10). □

We shall see in Theorem IX.2.4 that the conclusion of (v) above has the important implication that the essential spectra σ_{ek} of S and A are coincident for $k=1, 2, 3, 4$.

Corollary 5.2. Let s and a be closed sectorial forms on $\mathfrak{Q}$ and let $p=s-a$. Then (iv) and (v) of Theorem 5.1 hold. ■

Proof. As in Corollary 4.3, $\mathscr{Q}(S)$ and $\mathscr{Q}(A)$ are topologically isomorphic and $\hat{P}$ is defined. From the identity (5.15), and the knowledge that $(\hat{A}+z^2E^*)^{-1}$ and $(\hat{S}+z^2E^*)^{-1}$ now both exist and lie in $\mathscr{B}(\mathfrak{Q}^*, \mathfrak{Q})$ for z large enough, the corollary follows. □

There is also the following analogue of Theorem 4.4.

Theorem 5.3. Let $\mathfrak{Q}$ be the form domain of an m-sectorial operator A in H and let $p = p_1 + ip_2$ where p_1 and p_2 are the bounded symmetric forms on $\mathfrak{Q} \times \mathfrak{Q}$ of self-adjoint operators P_1 and P_2 respectively. Then $\hat{P}(A+z^2I)^{-1} \in \mathscr{K}(H, \mathfrak{Q}^*)$ if

$$|P_j|^{\frac{1}{2}}(A+z^2I)^{-1} \in \mathscr{K}(H) \qquad (j = 1, 2). \tag{5.16}$$ ■

The proof is similar to that of Theorem 4.4.

Theorem 5.4. Let S be the m-sectorial operator associated with a closed, densely defined sectorial form s and let $T = \operatorname{re} S$. Then S has a compact resolvent if and only if T has. ∎

Proof. We shall use Corollary 5.2, with $a = \operatorname{re} s = t$ and $p = \mathrm{i} \operatorname{im} s$, to prove that $(S+z^2I)^{-1}$ and $(T+z^2I)^{-1}$ are compact together when z is large enough.

Since s is sectorial,

$$|p[\![x]\!]| \leqslant M(t+\alpha)[\![x]\!] \qquad (x \in \mathfrak{Q})$$

if $\Theta(s)$ lies in the sector (5.1), and hence by Lemma 2.1,

$$|p[\![x, y]\!]| \leqslant \varepsilon M\{(t+\alpha)[\![x]\!](t+\alpha)[\![y]\!]\}^{\frac{1}{2}}. \tag{5.17}$$

Let $(T+z^2I)^{-1} \in \mathscr{K}(H)$. Then if (u_n) is a bounded sequence in H and $v_n = (T+z^2I)^{-1} u_n$, the sequence (v_n) is precompact in H. Also, by (5.17) and since $v_n \in \mathscr{D}(T)$,

$$\begin{aligned} \|\hat{P}v_n\|^2_{\mathfrak{Q}^*} &= \sup_{\|\phi\|_{\mathfrak{Q}} \leqslant 1} |p[\![v_n, \phi]\!]|^2 \\ &\leqslant \varepsilon^2 M^2 (t+\alpha)[\![v_n]\!] \\ &= \varepsilon^2 M^2 ((T+\alpha I)v_n, v_n)_H \\ &\leqslant K(|(u_n, v_n)_H| + \|v_n\|^2_H) \\ &\leqslant K(\|v_n\|_H + \|v_n\|^2_H). \end{aligned}$$

It follows that $(\hat{P}v_n)$ is precompact in $\mathfrak{Q}^*$ and that $\hat{P}(T+z^2I)^{-1}$ is therefore in $\mathscr{K}(H, \mathfrak{Q}^*)$. This and Corollary 5.2(v) yield $(S+z^2I)^{-1} \in \mathscr{K}(H)$.

To prove the converse we reverse the roles of S and T and now write $\hat{T} = \hat{S} - \hat{P}$, so that (5.10) becomes

$$(S+z^2I)^{-1} - (T+z^2I)^{-1} = -(\hat{T}+z^2E^*)^{-1}\hat{P}(S+z^2I)^{-1}.$$

Let $(S+z^2I)^{-1} \in \mathscr{K}(H)$, let (u_n) be bounded in H and set $v_n = (S+z^2I)^{-1} u_n$. From (5.17),

$$\begin{aligned} \|\hat{P}v_n\|^2_{\mathfrak{Q}^*} &\leqslant \varepsilon^2 M^2 (t+\alpha)[\![v_n]\!] \\ &\leqslant \varepsilon^2 M^2 |(s+\alpha)[\![v_n]\!]| \\ &= \varepsilon^2 M^2 |((S+\alpha I)v_n, v_n)_H| \\ &\leqslant K(|(u_n, v_n)_H| + \|v_n\|^2_H) \\ &\leqslant K(\|v_n\|_H + \|v_n\|^2_H). \end{aligned}$$

As before, $(T+z^2I)^{-1} \in \mathscr{K}(H)$. □

6. Variational inequalities.

The Lax–Milgram Theorem was generalized by Stampacchia [1] to a result concerning the existence of a solution of an inequality satisfied on a closed convex subset of a real Hilbert space H by a bounded coercive form. Such inequalities are referred to as *variational inequalities.* They occur naturally in questions involving the minimizing of quadratic functionals on closed convex sets. More explicitly, consider the well known result that if $\mathfrak{K}$ is a closed, convex, non-empty subset of a real Hilbert space H, then for any $f \in H$ there exists a unique $u_0 \in \mathfrak{K}$ such that

$$\| u_0 - f \| = \inf \{ \| u - f \| : u \in \mathfrak{K} \}, \tag{6.1}$$

i.e. u_0 is the projection of f on $\mathfrak{K}$. Let $v \in \mathfrak{K}$ and $\lambda \in [0, 1]$ and define $F \in C^1[0, 1]$ by

$$F(\lambda) = \| \lambda u_0 + (1 - \lambda) v - f \|^2.$$

Since $\mathfrak{K}$ is convex, we have $\lambda u_0 + (1 - \lambda) v \in \mathfrak{K}$ and hence from the definition of u_0, $F(\lambda) \geqslant F(1)$ for $\lambda \in [0, 1]$. Therefore F has a minimum at 1 and $F'(1) \leqslant 0$. We conclude that

$$F'(1) = 2[(u_0, u_0 - v) - (f, u_0 - v)] \leqslant 0$$

and so, for any $f \in H$,

$$(u_0, v - u_0) \geqslant (f, v - u_0) \qquad (v \in \mathfrak{K}). \tag{6.2}$$

This is an example of a variational inequality. Stampacchia's result establishes the existence of a unique $u_0 \in \mathfrak{K}$ satisfying (6.2) for all $f \in H$. Note that (6.2) also implies (6.1). For we have

$$0 \leqslant (u_0 - f, v - u_0) = (u_0 - f, v - f) - \| u_0 - f \|^2,$$

and consequently

$$\| u_0 - f \| \leqslant \| v - f \|$$

for all $v \in \mathfrak{K}$ and $f \in H$. Thus the solution of the variational inequality (6.2) is equivalent to the existence of the minimum in (6.1).

Let H be a real Hilbert space and let $a[\![\bullet, \bullet]\!]$ be a sesquilinear form on $H \times H$; since $a[\![\bullet, \bullet]\!]$ is now linear in both variables as H is real, it is more natural to call $a[\![\bullet, \bullet]\!]$ a *bilinear form.* We shall do this and reserve the term sesquilinear form for the complex case. We make the following assumptions on $a[\![\bullet, \bullet]\!]$:

(i) $a[\![\bullet, \bullet]\!]$ is bounded on $H \times H$, i.e. there exists a positive constant M such that for all $x, y \in H$,

$$|a[\![x, y]\!]| \leqslant M \| x \|_H \| y \|_H, \tag{6.3}$$

(ii) $a[\![\bullet, \bullet]\!]$ is coercive and positive on H, i.e. there exists a positive constant m such that for all $x \in H$,

$$a[\![x]\!] \equiv a[\![x, x]\!] \geqslant m \| x \|_H^2. \tag{6.4}$$

Let $\mathfrak{K}$ be a closed convex subset of H and for $u \in \mathfrak{K}$ define the convex cone

$$\mathfrak{K}_u := \{v \in H : u + \varepsilon v \in \mathfrak{K} \text{ for some } \varepsilon > 0\}. \tag{6.5}$$

Clearly $\mathfrak{K}_u = H$ if, and only if, u is an interior point of $\mathfrak{K}$. Also, if $\mathfrak{K}$ is an affine subvariety, i.e., $\mathfrak{K} = u_0 + G$ for some subspace G of H, then $\mathfrak{K}_u = \mathfrak{K} - u = G$.

Theorem 6.1. Let a be a bilinear form on $H \times H$ which satisfies (6.3) and (6.4) and let $\mathfrak{K}$ be a closed, convex, non-empty subset of H. Then, given $F \in H^*$ there exists a unique $u \in \mathfrak{K}$ such that for all $\phi \in \mathfrak{K}_u$,

$$a[\![u, \phi]\!] \geq (F, \phi). \tag{6.6}$$ ■

If $\mathfrak{K}_u$ is a subspace of H, as is the case when u is an interior point of $\mathfrak{K}$ or when $\mathfrak{K}$ is an affine subvariety, then $-\phi \in \mathfrak{K}_u$ whenever $\phi \in \mathfrak{K}_u$. Hence (6.6) implies $a[\![u, \phi]\!] \leq (F, \phi)$ and consequently $a[\![u, \phi]\!] = (F, \phi)$ for $\phi \in \mathfrak{K}_u$. The case $\mathfrak{K}_u = H$ is of course the Lax–Milgram Theorem for real H and positive bilinear forms.

Before proving Theorem 6.1 we need some preparatory lemmas. The first establishes the uniqueness of u in (6.6).

Lemma 6.2. Let a and $\mathfrak{K}$ be as in Theorem 6.1 and suppose that for $j = 1, 2$,

$$a[\![u_j, \phi_j]\!] \geq (F_j, \phi_j)$$

for all $\phi_j \in \mathfrak{K}_{u_j}$, where $u_j \in \mathfrak{K}$ and $F_j \in H^*$. Then

$$\|u_1 - u_2\|_H \leq m^{-1} \|F_1 - F_2\|_{H^*} \tag{6.7}$$

where m is the constant in (6.4). ■

Proof. Since $u_2 - u_1 \in \mathfrak{K} - u_1 \subset \mathfrak{K}_{u_1}$ and $u_1 - u_2 \in \mathfrak{K} - u_2 \subset \mathfrak{K}_{u_2}$ we have, by hypothesis,

$$a[\![u_1, u_2 - u_1]\!] \geq (F_1, u_2 - u_1), \qquad a[\![u_2, u_1 - u_2]\!] \geq (F_2, u_1 - u_2).$$

On adding and using (6.4) we obtain

$$\begin{aligned} m\|u_1 - u_2\|_H^2 &\leq a[\![u_1 - u_2, u_1 - u_2]\!] \\ &\leq (F_1 - F_2, u_1 - u_2) \\ &\leq \|F_1 - F_2\|_{H^*} \|u_1 - u_2\|_H, \end{aligned}$$

whence the result. □

Next we prove Theorem 6.1 in the special case when a is symmetric, i.e. $a[\![u, v]\!] = a[\![v, u]\!]$. This result will be of some importance in Chapters VI and VIII.

Proposition 6.3. Let a and $\mathfrak{K}$ satisfy the hypothesis of Theorem 6.1 and let a be symmetric. Then, given $F \in H^*$ there exists a unique $u \in \mathfrak{K}$ such that

$$a[\![u, \phi]\!] \geqslant (F, \phi) \quad \text{for all } \phi \in \mathfrak{K}_u.$$

If $F = 0$, the unique solution u of $a[\![u, \phi]\!] \geqslant 0$, $\phi \in \mathfrak{K}_u$, satisfies

$$a[\![u]\!] = \inf\{a[\![v]\!]: v \in \mathfrak{K}\}. \qquad \blacksquare$$

Proof. Let

$$I(v) = a[\![v]\!] - 2(F, v) \qquad (v \in \mathfrak{K}),$$

and let $d = \inf\{I(v): v \in \mathfrak{K}\}$. We shall first prove that $d > -\infty$ and then show that $d = I(u)$ for some unique $u \in \mathfrak{K}$. From (6.4),

$$\begin{aligned} I(v) &\geqslant m\|v\|_H^2 - 2\|F\|_{H^*}\|v\|_H \\ &\geqslant m\|v\|_H^2 - (m\|v\|_H^2 + m^{-1}\|F\|_{H^*}^2) \\ &= -m^{-1}\|F\|_{H^*}^2 > -\infty \end{aligned}$$

and hence $d > -\infty$. Let (u_k) be a sequence in $\mathfrak{K}$ which satisfies $I(u_k) < d + 1/k$. Then, since $\mathfrak{K}$ is convex, $\frac{1}{2}(u_k + u_j) \in \mathfrak{K}$ and

$$\begin{aligned} a[\![u_k - u_j]\!] &= 2a[\![u_k]\!] + 2a[\![u_j]\!] - 4a[\![\tfrac{1}{2}(u_k + u_j)]\!] \\ &= 2I(u_k) + 2I(u_j) - 4I(\tfrac{1}{2}(u_k + u_j)) \\ &\leqslant 4d + 2(1/k + 1/j) - 4d \\ &= 2(1/k + 1/j). \end{aligned}$$

On using (6.4) we see that

$$\|u_k - u_j\|_H^2 \leqslant 2m^{-1}(1/k + 1/j)$$

and therefore (u_k) is a Cauchy sequence in $\mathfrak{K}$. Since $\mathfrak{K}$ is closed the sequence converges to a limit u, say, in $\mathfrak{K}$. Also, from (6.3), $a[\![u_k]\!] \to a[\![u]\!]$ and hence $I(u) = \lim_{k \to \infty} I(u_k) = d$.

Let $\phi \in \mathfrak{K}_u$, i.e. $u + \varepsilon\phi \in \mathfrak{K}$ for some $\varepsilon > 0$. Then, since $\mathfrak{K}$ is convex, $u + \delta\phi = (\delta/\varepsilon)(u + \varepsilon\phi) + (1 - \delta/\varepsilon)u \in \mathfrak{K}$ for $\delta \in (0, \varepsilon)$ and so, from above,

$$\begin{aligned} 0 &\leqslant I(u + \delta\phi) - I(u) \\ &= \delta^2 a[\![\phi]\!] + 2\delta(a[\![u, \phi]\!] - (F, \phi)). \end{aligned}$$

On dividing by δ and then allowing $\delta \to 0$ we get $a[\![u, \phi]\!] \geqslant (F, \phi)$ as required. The uniqueness has already been proved in Lemma 6.2. □

In order to prove Theorem 6.1 we decompose a into $a = \alpha + \beta$ where

$$\alpha[\![u, v]\!] = \tfrac{1}{2}(a[\![u, v]\!] + a[\![v, u]\!]), \qquad \beta[\![u, v]\!] = \tfrac{1}{2}(a[\![u, v]\!] - a[\![v, u]\!]).$$

α is symmetric and β is skew-symmetric, i.e. $\beta[\![u, v]\!] = -\beta[\![v, u]\!]$. Furthermore, both the linear forms α and β satisfy (6.3) with the same constant M; also α satisfies (6.4) and $\beta[\![u]\!] = 0$. If we set $a_t = \alpha + t\beta$ for some non-negative number t we obtain

$$|a_t[\![x, y]\!]| \leqslant (M + tN)\|x\|_H \|y\|_H, \tag{6.8}$$

where

$$N = \sup\{|\beta[\![x, y]\!]| : \|x\|_H = \|y\|_H = 1\} \leqslant M,$$

and also, by (6.4),

$$a_t[\![x]\!] = \alpha[\![x]\!] \geqslant m\|x\|_H^2. \tag{6.9}$$

Lemma 6.4. Let τ be a fixed positive number and suppose that for any $F \in H^*$ there exists a unique $u \in \mathfrak{K}$ such that for all $\phi \in \mathfrak{K}_u$,

$$a_\tau[\![u, \phi]\!] \geqslant (F, \phi). \tag{6.10}$$

Then there exists a unique $u \in \mathfrak{K}$ such that for all $\phi \in \mathfrak{K}_u$,

$$a_t[\![u, \phi]\!] \geqslant (F, \phi) \tag{6.11}$$

if $\tau \leqslant t \leqslant \tau + t_0$, where $0 < t_0 < m/N$ with $N \neq 0$. ■

Proof. We write

$$a_t = a_\tau + (t - \tau)\beta$$

and, for $v \in H$,

$$G[\![v, \phi]\!] = (F, \phi) - (t - \tau)\beta[\![v, \phi]\!].$$

The map $\phi \mapsto G[\![v, \phi]\!]$ lies in H^* for each $v \in H$, and so, in view of hypothesis (6.10), for each $v \in H$ there exists a unique $u \in \mathfrak{K}$ such that for all $\phi \in \mathfrak{K}_u$,

$$a_\tau[\![u, \phi]\!] \geqslant G[\![v, \phi]\!].$$

We shall now prove that the map $T: v \mapsto u$ so determined is a contraction map of H into $\mathfrak{K}$ if $\tau \leqslant t \leqslant \tau + t_0$. Let $u_j = Tv_j$ $(j = 1, 2)$. Then, by Lemma 6.2,

$$\begin{aligned}\|u_1 - u_2\|_H &\leqslant m^{-1} \sup\{|G[\![v_1, \phi]\!] - G[\![v_2, \phi]\!]| : \|\phi\|_H = 1\} \\ &= m^{-1}|t - \tau| \sup\{|\beta[\![v_1 - v_2, \phi]\!]| : \|\phi\|_H = 1\} \\ &\leqslant Nm^{-1}|t - \tau|\, \|v_1 - v_2\|_H \leqslant (Nm^{-1}t_0)\|v_1 - v_2\|_H;\end{aligned}$$

T is therefore a contraction map of H into $\mathfrak{K}$ and has a unique fixed point (see Gilbarg and Trudinger [1, Theorem 5.5.1]) $u \in \mathfrak{K}$ which satisfies

$$a_\tau[\![u, \phi]\!] \geqslant G[\![u, \phi]\!] \quad \text{for all } \phi \in \mathfrak{K}_u.$$

This is equivalent to (6.11) and the lemma is proved. □

Proof of Theorem 6.1. In view of Proposition 6.3 the hypotheses of Lemma 6.4 are satisfied when $\tau = 0$ and hence we may conclude that there is a unique $u \in \mathfrak{K}$

which satisfies (6.11) for $0 \leqslant t \leqslant t_0$. We now apply Lemma 6.4 again, this time with $\tau = t_0$, to show that there is a unique solution of (6.11) whenever $0 < t < 2t_0$. Continuing in this way we reach $t = 1$ after a finite number of steps. Theorem 6.1 is therefore proved. □

Corollary 6.5. Let a be a bilinear form which is bounded on $H \times H$ and satisfies (6.4) on a closed subspace H_0 of H. Let $\mathfrak{K}$ be a closed, convex, non-empty subset of H which is such that $\mathfrak{K} - h_0 \subset H_0$ for some $h_0 \in H$. Then there exists a unique $u \in \mathfrak{K}$ such that for all $\phi \in \mathfrak{K}_u$,

$$a[\![u, \phi]\!] \geqslant 0.$$ ■

Proof. $\mathfrak{K}^0 := \mathfrak{K} - h_0$ is a closed convex subset of H_0, and since the hypothesis of Theorem 6.1 is satisfied on H_0 we conclude that there exists a unique $u_0 \in \mathfrak{K}^0$ such that for all $\phi \in \mathfrak{K}^0_{u_0}$,

$$a[\![u_0, \phi]\!] \geqslant -a[\![h_0, \phi]\!];$$

note that the map $\phi \mapsto -a[\![h_0, \phi]\!]$ lies in H^* on account of the boundedness of a on $H \times H$. But $u \in \mathfrak{K}$ if, and only if, $u - h_0 \in \mathfrak{K}^0$. Thus with $u = u_0 + h_0$, we have $\phi \in \mathfrak{K}^0_{u_0}$ if, and only if, $\phi \in \mathfrak{K}_u$ and u is the unique solution of

$$a[\![u, \phi]\!] \geqslant 0 \qquad (\phi \in \mathfrak{K}_u).$$ □

V
Sobolev spaces

Sobolev spaces are function spaces which provide a natural framework for much of the modern theory of partial differential equations, a theory which is greatly aided by the possibility of embedding one Sobolev space in a variety of other such spaces, in L^p spaces, or even in spaces of continuous functions, the corresponding embedding maps being continuous and often compact. The existence of compact embeddings is important, since it makes it possible to reduce elliptic boundary-value problems to questions involving the Fredholm–Riesz–Schauder theory of compact linear operators, and is at the heart of much work on the asymptotic distribution of eigenvalues of elliptic operators.

In this chapter we aim to provide a reasonable selection of the most important results in the theory of Sobolev spaces, an encyclopaedic treatment being impossible owing to limitations of space and our knowledge. Special emphasis is placed upon the embeddings mentioned above and on the question as to whether they have any properties, such as compactness, better than mere continuity. Some results concerning the k-set contractive nature of certain embedding maps are given, both for bounded and unbounded space domains: we also estimate the approximation numbers of embedding maps and use these estimates to classify these embeddings.

We begin, however, with some preparatory material.

1. Function spaces, approximation, and regularization

Throughout we shall reserve the symbol Ω for a non-empty open subset of $\mathbb{R}^n$, with closure $\bar{\Omega}$ and boundary $\partial\Omega$. Points of $\mathbb{R}^n$ will be denoted by $x = (x_1, x_2, \ldots, x_n) = (x_i)$, and we shall write $|x| = (x_1^2 + x_2^2 + \ldots + x_n^2)^{\frac{1}{2}}$; given any $r > 0$, we put $B(x,r) = \{y \in \mathbb{R}^n : |x - y| < r\}$. Given any $\alpha \in \mathbb{N}_0^n$, say $\alpha = (\alpha_1, \alpha_2, \ldots, \alpha_n)$, we write

$$|\alpha| = \sum_{j=1}^{n} \alpha_j, \qquad \alpha! = \prod_{j=1}^{n} \alpha_j!, \qquad x^\alpha = \prod_{j=1}^{n} x_j^{\alpha_j}, \qquad D^\alpha = \prod_{j=1}^{n} D_j^{\alpha_j},$$

where $D_j = \partial/\partial x_j$.

1.1. *Spaces of continuous functions*

Given any $k \in \mathbb{N}_0$, we shall denote by $C^k(\Omega)$ the vector space of all continuous, real- (or complex-) valued functions u on Ω such that for all $\alpha \in \mathbb{N}_0^n$ with $|\alpha| \leqslant k$, the function $D^\alpha u$ exists and is continuous on Ω. We define $C^k(\bar{\Omega})$ to be the vector space of all bounded functions u in $C^k(\Omega)$ such that u and all its derivatives $D^\alpha u$ with $|\alpha| \leqslant k$ can be extended so as to be bounded and continuous on $\bar{\Omega}$. When $k = 0$ the superscript 0 will usually be omitted and we shall simply write $C(\Omega)$ and $C(\bar{\Omega})$. We put

$$C^\infty(\Omega) = \bigcap_{k=1}^{\infty} C^k(\Omega), \qquad C^\infty(\bar{\Omega}) = \bigcap_{k=1}^{\infty} C^k(\bar{\Omega}).$$

Given any $\lambda \in (0, 1]$, we denote by $C^{0,\lambda}(\Omega)$ the vector space of all continuous functions which satisfy a local Hölder condition on Ω; that is, $u \in C^{0,\lambda}(\Omega)$ if, and only if, given any compact set $K \subset \Omega$, there is a constant $C > 0$ such that $|u(x) - u(y)| \leqslant C|x - y|^\lambda$ for all $x, y \in K$. If $k \in \mathbb{N}$, we set

$$C^{k,\lambda}(\Omega) = \{u \in C^k(\Omega) \colon D^\alpha u \in C^{0,\lambda}(\Omega) \text{ for all } \alpha \in \mathbb{N}_0^n \text{ with } |\alpha| = k\}.$$

We shall also need spaces of Hölder-continuous functions on $\bar{\Omega}$: with $k \in \mathbb{N}_0$ and $\lambda \in (0, 1]$ we write

$$C^{k,\lambda}(\bar{\Omega}) = \{u \in C^k(\bar{\Omega}) \colon \text{given any } \alpha \in \mathbb{N}_0^n \text{ with } |\alpha| = k, \text{ there exists } C > 0 \text{ such that for all } x, y \in \Omega,\ |D^\alpha u(x) - D^\alpha u(y)| \leqslant C|x - y|^\lambda\}.$$

It is routine to verify that $C^k(\bar{\Omega})$ and $C^{k,\lambda}(\bar{\Omega})$ ($\lambda \in (0, 1], k \in \mathbb{N}_0$) become Banach spaces when furnished with the norms

$$|||u|||_{k,\Omega} = \max_{|\alpha| \leqslant k} \sup_{x \in \Omega} |D^\alpha u(x)|,$$

$$|||u|||_{k,\lambda,\Omega} = |||u|||_{k,\Omega} + \max_{|\alpha| = k} \sup_{x,y \in \Omega, x \neq y} |D^\alpha u(x) - D^\alpha u(y)|/|x - y|^\lambda,$$

respectively.

The various relations which exist between these spaces may be summarized as follows:

Theorem 1.1. Let $k \in \mathbb{N}_0$ and let $0 < \nu < \lambda \leqslant 1$. Then:

(i) $C^{k+1}(\bar{\Omega}) \subset C^k(\bar{\Omega})$;

(ii) $C^{k,\lambda}(\bar{\Omega}) \subset C^{k,\nu}(\bar{\Omega}) \subset C^k(\bar{\Omega})$.

If Ω is bounded, both the natural embeddings corresponding to (ii) are compact. If Ω is convex, then

(iii) $C^{k+1}(\bar{\Omega}) \subset C^{k,1}(\bar{\Omega})$;

(iv) $C^{k+1}(\bar{\Omega}) \subset C^{k,\nu}(\bar{\Omega})$.

If Ω is bounded and convex, the embeddings in (i) and (iv) are compact. ∎

The proof is left as an exercise: it uses the Arzelà-Ascoli compactness theorem, while convexity is simply an *ad hoc* condition imposed to enable the Mean-Value Theorem to be invoked: sufficient conditions weaker than convexity are known. For example, if Ω has the property that there is a constant M such that given any $x, y \in \Omega$, there exist points $x_0 (= x), x_1, \ldots, x_n$ $(= y)$ such that the segment with endpoints x_i and x_{i+1} lies in Ω $(i = 0, 1, \ldots, n-1)$ and

$$\sum_{i=0}^{n-1} |x_i - x_{i+1}| \leqslant M|x-y|,$$

then it is a routine matter to show that, for all $k \in \mathbb{N}_0$ and all $\lambda \in (0, 1]$, we have $C^{k+1}(\bar{\Omega}) \subset C^{k,\lambda}(\bar{\Omega})$. That some condition on Ω is necessary is shown by taking $\Omega = \{(x, y) \in \mathbb{R}^2 : y < |x|^{\frac{1}{2}}, x^2 + y^2 < 1\}$, $\beta \in (1, 2)$ and $\alpha \in (\frac{1}{2}\beta, 1)$. Consideration of the function u defined by $u(x, y) = (\operatorname{sgn} x)\, y^\beta$ if $y > 0$, with $u(x, y) = 0$ if $y \leqslant 0$, shows that $C^1(\bar{\Omega}) \not\subset C^{0,\alpha}(\bar{\Omega})$. However, Fraenkel [4] has shown that if $\partial\Omega$ is uniformly of class $C^{0,\alpha}$ (see §4 below) then $C^1(\bar{\Omega}) \subset C^{0,\alpha}(\bar{\Omega})$ for all $\alpha \in (0, 1]$; the same paper also gives various examples in which embeddings of the type discussed above fail to exist.

The *support* of a function u on Ω is the set, supp u, defined to be the closure of $\{x \in \Omega : u(x) \neq 0\}$. In other words, it is the complement of the largest open set on which u is zero. For $k \in \mathbb{N}_0 \cup \{\infty\}$ we shall write $C_0^k(\Omega)$ to stand for the linear subspace of $C^k(\Omega)$ which consists of all those functions with compact support contained in Ω; $C_0^{k,\gamma}(\Omega)$ has the corresponding meaning, when $\gamma \in (0, 1]$ and $k \in \mathbb{N}_0$. These functions vanish near the boundary of Ω. Since the space $C_0^\infty(\Omega)$ will often be used, it is important to know that it really does contain functions other than the zero function. To establish this it is enough to consider the function $\phi : \mathbb{R}^n \to \mathbb{R}$ defined by

$$\phi(x) = \begin{cases} \exp[-1/(1-|x|^2)] & \text{if } |x| < 1, \\ 0 \quad \text{if } |x| \geqslant 1. \end{cases} \tag{1.1}$$

A straightforward computation shows that ϕ is a non-negative, infinitely differentiable function with support the closed unit ball in $\mathbb{R}^n$: it is thus in $C_0^\infty(\mathbb{R}^n)$. Moreover, since $\int_{\mathbb{R}^n} \phi(x)dx > 0$, the function $\psi := \phi / \int_{\mathbb{R}^n} \phi(x)dx$ belongs to $C_0^\infty(\mathbb{R}^n)$ and has unit integral over $\mathbb{R}^n$.

A function $u \in C^\infty(\Omega)$ is said to be *analytic* in Ω if it can be expanded in a convergent power series about every point of Ω: that is, if given any $x \in \Omega$, there exists $r > 0$ such that for all y with $|y - x| < r$,

$$u(y) = \sum_{|\alpha|=0}^{\infty} \frac{(y-x)^\alpha}{\alpha!} D^\alpha u(x),$$

the series being absolutely and uniformly convergent on $B(x, r)$. The function ϕ given above is not analytic on $\mathbb{R}^n$ because all its derivatives vanish on the whole of the unit sphere.

A particular space of smooth functions which we shall need is the *Schwartz space* $\mathfrak{S} \equiv \mathscr{S}(\mathbb{R}^n)$, which is the linear subspace of $C^\infty(\mathbb{R}^n)$ consisting of all those functions which, together with all their derivatives, go to zero at infinity faster than any power of x. Thus $u \in \mathfrak{S}$ if, and only if, $u \in C^\infty(\mathbb{R}^n)$ and for all $\alpha, \beta \in \mathbb{N}_0^n$ we have $\sup\{|x^\alpha D^\beta u(x)| : x \in \mathbb{R}^n\} < \infty$.

In Schwartz's Theory of Distributions the space $C_0^\infty(\Omega)$ is given a topology which makes it a locally convex topological vector space. The dual of this space is denoted by $\mathscr{D}'(\Omega)$ and is called the *space of distributions* on Ω. We shall have very little need for this theory but for details we refer to Hörmander [1].

1.2. *Functions in Lebesgue spaces*

Given any measurable subset S of $\mathbb{R}^n$ and any $p \in [1, \infty]$, the norm in the Lebesgue space $L^p(S)$ is given by

$$\|u\|_{p,S} := \left(\int_S |u|^p\right)^{1/p} \quad \text{if } 1 \leqslant p < \infty, \qquad \|u\|_{\infty,S} := \operatorname*{ess\,sup}_{s \in S} |u(s)|.$$

If there is no ambiguity, we shall omit the subscript S and simply write $\|u\|_p$ for such norms. The set of functions which are in $L^p(K)$ for every compact subset K of S is denoted by $L^p_{\text{loc}}(S)$. Note that if S is closed, a compact subset K of S may include part of the boundary of S. If the Lebesgue measure dx in $L^p(S)$ is replaced by a measure of the form $w(x)dx$ we obtain the *weighted space* $L^p(S; w)$ which has norm

$$\|u\|_{L^p(S;w)} := \left(\int_S |u|^p w\right)^{1/p}.$$

If Ω is an open set, any $f \in L^1_{\text{loc}}(\Omega)$ may be identified with an $F \in \mathscr{D}'(\Omega)$ by

$$(F, \phi) = \int_\Omega f\phi \qquad (\phi \in C_0^\infty(\Omega));$$

see Hörmander [1]. By p' we shall mean the number defined by

$$1/p + 1/p' = 1,$$

with the understanding that $p' = \infty$ if $p = 1$, and $p' = 1$ if $p = \infty$.

Let f and g be real- (or complex-) valued functions defined on $\mathbb{R}^n$. The *convolution* of f and g, written $f * g$, is the function defined by

$$(f * g)(x) = \int_{\mathbb{R}^n} f(x - y)\, g(y)\, dy \qquad (x \in \mathbb{R}^n).$$

Of course, $f * g$ will not exist unless suitable restrictions are imposed upon f and g; if it does exist, then $f * g = g * f$. We now give a basic result concerning convolutions, due to W. H. Young.

Theorem 1.2. Let $p, q, r \in [1, \infty]$, let $p^{-1} + q^{-1} = 1 + r^{-1}$, and suppose that $f \in L^p(\mathbb{R}^n)$ and $g \in L^q(\mathbb{R}^n)$. Then $f * g \in L^r(\mathbb{R}^n)$ and

$$\|f * g\|_r \leqslant \|f\|_p \, \|g\|_q. \tag{1.2}$$ ■

Proof. When $p = q'$ and $r = \infty$ the result follows immediately from Hölder's inequality. If either $p = 1$ and $q = r < \infty$ or $q = 1$ and $p = r < \infty$, the triangle inequality is enough to give the result. Thus suppose, for example, that $q = 1, p = r < \infty$, and that f and g are continuous functions with compact support. Then by the L^p triangle inequality applied to Riemann sums approximating the integral we have

$$\begin{aligned}\|f * g\|_p &= \left\| \int_{\mathbb{R}^n} f(\bullet - y) g(y) \, dy \right\|_p \leqslant \int_{\mathbb{R}^n} \|f\|_p |g(y)| \, dy \\ &= \|f\|_p \|g\|_1,\end{aligned}$$

as required. The case of general functions $f \in L^p(\mathbb{R}^n)$ and $g \in L^1(\mathbb{R}^n)$ follows from the density of continuous functions with compact support in the Lebesgue spaces.

It remains to establish (1.2) when $1 < p < q'$, $\max\{p, q\} < r < \infty$, and $q < \infty$. Put $s = pr/(r-p)$ and $t = qr/(r-q)$, and notice that

$$s^{-1} + t^{-1} + r^{-1} = 1.$$

By Hölder's inequality,

$$\begin{aligned}|(f * g)(x)| &\leqslant \int_{\mathbb{R}^n} |f(y)|^{p/r} |f(y)|^{1-p/r} |g(x-y)|^{1-q/r} |g(x-y)|^{q/r} \, dy \\ &\leqslant \left(\int_{\mathbb{R}^n} |g(x-y)|^{t(1-q/r)} \, dy \right)^{1/t} \left(\int_{\mathbb{R}^n} |g(x-y)|^q |f(y)|^p \, dy \right)^{1/r} \\ &\qquad \times \left(\int_{\mathbb{R}^n} |f(y)|^{s(1-p/r)} \, dy \right)^{1/s},\end{aligned}$$

and hence

$$\|f * g\|_r^r \leqslant \|g\|_q^{rq/t} \|g\|_q^q \|f\|_p^p \|f\|_p^{rp/s} = \|f\|_p^r \|g\|_q^r.$$

The proof is complete. □

We now proceed to show the utility of convolutions in the approximation of various kinds of functions by smooth functions. First we give a useful technical lemma.

Lemma 1.3. Let $1 \leqslant p < \infty$; let $f \in L^p(\mathbb{R}^n)$ and given any $x \in \mathbb{R}^n$ define a function f_x on $\mathbb{R}^n$ by $f_x(y) = f(x+y)$ $(y \in \mathbb{R}^n)$. Then $\|f_x - f\|_p \to 0$ as $|x| \to 0$. ■

Proof. Suppose first that $g \in C_0(\mathbb{R}^n)$. Then g is uniformly continuous on $\mathbb{R}^n$, and consequently $g_x \to g$ uniformly on $\mathbb{R}^n$ as $x \to 0$. Since supp g and supp g_x are contained in a common compact set for all small enough $|x|$, it follows directly that $\|g_x - g\|_p \to 0$ as $x \to 0$. Now given any $f \in L^p(\mathbb{R}^n)$ and any $\varepsilon > 0$, there exists $g \in C_0(\mathbb{R}^n)$ such that $\|f - g\|_p < \frac{1}{3}\varepsilon$. Hence

$$\|f_x - g_x\|_p = \|f - g\|_p < \tfrac{1}{3}\varepsilon,$$

and

$$\|f_x - f\|_p \leqslant \|f_x - g_x\|_p + \|g_x - g\|_p + \|g - f\|_p < \|g_x - g\|_p + \tfrac{2}{3}\varepsilon.$$

For small enough $|x|$, say $|x| < \delta$, we have $\|g_x - g\|_p < \frac{1}{3}\varepsilon$. Thus $\|f_x - f\|_p < \varepsilon$ if $|x| < \delta$. The proof is complete. □

Note that this result is false if $p = \infty$, since if $\|f_x - f\|_\infty \to 0$ as $|x| \to 0$, then f must be uniformly continuous on $\mathbb{R}^n$. However, the proof does show that the lemma holds for $p = \infty$ if $f \in C_0(\mathbb{R}^n)$.

Now let $\rho \in C_0^\infty(\mathbb{R}^n)$ be non-negative, with $\rho(x) = 0$ if $|x| \geqslant 1$, and $\int_{\mathbb{R}^n} \rho(x)\,dx = 1$. Such a function is often called a *mollifier*; an example is obtained from the function defined in (1.1). Given any $u \in L^1_{loc}(\Omega)$ and any $\varepsilon > 0$, put

$$u_\varepsilon(x) = \varepsilon^{-n} \int_\Omega \rho[(x-y)/\varepsilon]\,u(y)\,dy$$

provided that $x \in \Omega$ and $\varepsilon < \text{dist}(x, \partial\Omega)$: the function u_ε is called a *regularization* of u. For subsequent reference we note that $u_\varepsilon(x) = \int_{B(0,1)} \rho(z) u(x - \varepsilon z)\,dz$ if $x \in \Omega$ and $\varepsilon < \text{dist}(x, \partial\Omega)$. It is clear that if Ω' is an open set with compact closure contained in Ω, then $u_\varepsilon \in C^\infty(\Omega')$ if $0 < \varepsilon < \text{dist}(\Omega', \partial\Omega)$.

Lemma 1.4. Let Ω be an open subset of $\mathbb{R}^n$, let $u \in C(\Omega)$, and let Ω' be any bounded, open subset of Ω with $\bar{\Omega}' \subset \Omega$. Then u_ε converges uniformly on Ω' to u as $\varepsilon \to 0+$. ■

Proof. If $0 < \varepsilon < \frac{1}{2}\text{dist}(\Omega', \partial\Omega)$, then

$$\sup_{x \in \Omega'} |u(x) - u_\varepsilon(x)| \leqslant \sup_{x \in \Omega'} \int_{B(0,1)} \rho(y)|u(x) - u(x - \varepsilon y)|\,dy$$

$$\leqslant \sup_{x \in \Omega'} \sup_{|y| \leqslant 1} |u(x) - u(x - \varepsilon y)|.$$

Since u is uniformly continuous on $\{x \in \mathbb{R}^n : \text{dist}(x, \Omega') < \varepsilon\}$, the result follows. □

Theorem 1.5. Let Ω be an open subset of $\mathbb{R}^n$, let $p \in [1, \infty)$, let $u \in L^p(\Omega)$ and suppose that u is extended to all of $\mathbb{R}^n$ by setting it equal to 0 on $\mathbb{R}^n \setminus \Omega$, this extension still being denoted by u. Then $\|u_\varepsilon - u\|_{p,\Omega} \to 0$ as $\varepsilon \to 0+$. ■

Proof. First suppose that $1 < p < \infty$. For all $x \in \Omega$ and all $\varepsilon > 0$,

$$|u_\varepsilon(x)| \leqslant \left(\int_{\mathbb{R}^n} \varepsilon^{-n}\rho[(x-y)/\varepsilon]\,\mathrm{d}y\right)^{1/p'}\left(\int_{\mathbb{R}^n} \varepsilon^{-n}\rho[(x-y)/\varepsilon]|u(y)|^p\,\mathrm{d}y\right)^{1/p}$$
$$= \left(\int_{\mathbb{R}^n} \varepsilon^{-n}\rho[(x-y)/\varepsilon]|u(y)|^p\,\mathrm{d}y\right)^{1/p}.$$

Thus by Fubini's Theorem,

$$\|u_\varepsilon\|_{p,\Omega}^p \leqslant \int_\Omega |u(y)|^p\,\mathrm{d}y \int_{\mathbb{R}^n} \varepsilon^{-n}\rho[(x-y)/\varepsilon]\,\mathrm{d}x = \|u\|_{p,\Omega}^p;$$

that is,

$$\|u_\varepsilon\|_{p,\Omega} \leqslant \|u\|_{p,\Omega}. \tag{1.3}$$

Let $\delta > 0$. Since $C_0(\Omega)$ is dense in $L^p(\Omega)$, there exists $v \in C_0(\Omega)$ such that $\|u - v\|_{p,\Omega} < \frac{1}{3}\delta$, and so for all $\varepsilon > 0$ we have $\|u_\varepsilon - v_\varepsilon\|_{p,\Omega} < \frac{1}{3}\delta$, by (1.3). Since v is uniformly continuous on Ω and has compact support in Ω, the proof of Lemma 1.4 shows that $v_\varepsilon \to v$ uniformly on Ω as $\varepsilon \to 0+$; thus $\|v_\varepsilon - v\|_{p,\Omega} < \frac{1}{3}\delta$ for small enough ε, say $\varepsilon \in (0, \varepsilon_0)$. Then for $0 < \varepsilon < \varepsilon_0$,

$$\|u_\varepsilon - u\|_{p,\Omega} \leqslant \|u_\varepsilon - v_\varepsilon\|_{p,\Omega} + \|v_\varepsilon - v\|_{p,\Omega} + \|v - u\|_{p,\Omega} < \delta,$$

and the result is established, save for the case $p = 1$. However, when $p = 1$, an application of Fubini's Theorem shows that

$$\|u_\varepsilon - u\|_{1,\Omega} \leqslant \int_{B(0,1)} \rho(y)\|u - u_{-\varepsilon y}\|_{1,\mathbb{R}^n}\,\mathrm{d}y,$$

so that in view of Lemma 1.3,

$$\lim_{\varepsilon \to 0+} \|u_\varepsilon - u\|_{1,\Omega} = 0.$$ □

Theorem 1.6. Let $p \in [1, \infty]$, let $u \in L^p(\mathbb{R}^n)$, and suppose that $\phi \in \mathfrak{S}$. Then $u * \phi \in C^\infty(\mathbb{R}^n)$ and $D^\alpha(u * \phi) = u * D^\alpha\phi$ for all $\alpha \in \mathbb{N}_0^n$. ■

Proof. Since $\phi \in \mathfrak{S}$ we have $D^\alpha\phi \in L^{p'}(\mathbb{R}^n)$ for all $\alpha \in \mathbb{N}_0^n$. Use of Hölder's inequality shows that

$$\int_{\mathbb{R}^n} u(y)\,D^\alpha\phi(x-y)\,\mathrm{d}y$$

($= (u * D^\alpha\phi)(x)$) converges absolutely and uniformly on $\mathbb{R}^n$. Differentiation and integration may thus be interchanged, and we have $D^\alpha(u * \phi) = u * D^\alpha\phi$. □

In particular, $u * \phi \in C^\infty(\mathbb{R}^n)$ if $\phi \in C_0^\infty(\mathbb{R}^n)$; and in fact the same proof shows that the theorem holds if $\phi \in C_0^\infty(\mathbb{R}^n)$ and the condition on u is relaxed to the requirement that $u \in L^1_{\mathrm{loc}}(\mathbb{R}^n)$.

Lemma 1.7. Given any two functions u and v on $\mathbb{R}^n$ such that $u * v$ exists, $\operatorname{supp}(u * v) \subset \{x+y : x \in \operatorname{supp} u, y \in \operatorname{supp} v\}$. ■

Proof. Since $(u * v)(x) = \int_{\operatorname{supp} v} u(x-y)v(y)\,dy$, it follows that if $(u * v)(x) \neq 0$, then $x - y \in \operatorname{supp} u$ for some $y \in \operatorname{supp} v$; that is, $x \in \operatorname{supp} u + \operatorname{supp} v$. □

Proposition 1.8. Let Ω be an open subset of $\mathbb{R}^n$ and let $p \in [1, \infty)$. Then $C_0^\infty(\Omega)$ is dense in $L^p(\Omega)$. ■

Proof. We use a mollifier ρ. Let $u \in L^p(\Omega)$, extend u to $\mathbb{R}^n$ by setting it equal to 0 in $\mathbb{R}^n \setminus \Omega$ and suppose that u has compact support in Ω. Then by Lemma 1.7 together with Theorems 1.5 and 1.6, $u_\varepsilon \in C_0^\infty(\Omega)$ if $\varepsilon (> 0)$ is small enough, and $u_\varepsilon \to u$ in $L^p(\Omega)$ as $\varepsilon \to 0+$. Since such functions u form a dense subset of $L^p(\Omega)$, the proof is complete. □

1.3. *Partitions of unity.*

The next lemma is useful in the construction of infinitely differentiable functions of compact support.

Lemma 1.9. Let K be a compact subset of an open subset Ω of $\mathbb{R}^n$. Then there exists $\psi \in C_0^\infty(\Omega)$ such that $0 \leqslant \psi \leqslant 1$ and $\psi(x) = 1$ for all x in a neighbourhood of K. ■

Proof. Let $0 < \varepsilon < \varepsilon_1 < \varepsilon + \varepsilon_1 < \operatorname{dist}(K, \mathbb{R}^n \setminus \Omega)$, and for any $\eta > 0$ set $K_\eta = \{x \in \mathbb{R}^n \colon \operatorname{dist}(x, K) \leqslant \eta\}$: K_η is evidently compact. Define a function $u: \mathbb{R}^n \to \mathbb{R}$ by $u(x) = 1$ $(x \in K_{\varepsilon_1})$, $u(x) = 0$ $(x \in \mathbb{R}^n \setminus K_{\varepsilon_1})$. Then $u \in L^1(\mathbb{R}^n)$ and $\operatorname{supp} u = K_{\varepsilon_1}$; hence $\operatorname{supp} u_\varepsilon \subset K_{\varepsilon+\varepsilon_1} \subset \Omega$ and $u_\varepsilon(x) = 1$ for all $x \in K_{\varepsilon_1 - \varepsilon}$. It follows that $\psi := u_\varepsilon$ has all the required properties. □

With this lemma the existence of 'partitions of unity' is easy to establish.

Theorem 1.10. Let Γ be a collection of open subsets of $\mathbb{R}^n$ with union Ω. Then there is a sequence (ϕ_i) in $C_0^\infty(\Omega)$, with each $\phi_i \geqslant 0$, such that:

(a) each ϕ_i has its support in some member of Γ ;

(b) $\sum_{i=1}^{\infty} \phi_i(x) = 1$ for each $x \in \Omega$;

(c) given any compact set $K \subset \Omega$, there exist $m \in \mathbb{N}$ and an open set $V \supset K$ such that $\sum_{i=1}^{m} \phi_i(x) = 1$ for all $x \in V$. ■

Proof. Let S be a countable dense subset of Ω, and let $(B_i)_{i \in \mathbb{N}}$ be a sequence that contains every closed ball B_j with centre $p_j \in S$, with rational radius r_j and which is contained in some member of Γ ; let $V_j = B(p_j, \frac{1}{2}r_j)$. Evidently $\Omega = \cup_{i=1}^{\infty} V_i$. By Lemma 1.9, given any $i \in \mathbb{N}$ there is a function $\psi_i \in C_0^{\infty}(B_i)$ such that $0 \leqslant \psi_i \leqslant 1$, $\psi_i = 1$ in V_i and $\operatorname{supp} \psi_i \subset B_i$. Put $\phi_1 = \psi_1$ and $\phi_{j+1} = \psi_{j+1} \Pi_{i=1}^{j} (1 - \psi_i)$ $(j \geqslant 1)$. Clearly $\phi_i = 0$ outside B_i; (a) follows. By induction,

$$\sum_{i=1}^{j} \phi_i = 1 - \prod_{i=1}^{j} (1 - \psi_i)$$

for all $j \in N$. Since $\psi_i = 1$ in V_i, we see that $\Sigma_{i=1}^{m} \phi_i(x) = 1$ if $x \in \cup_{i=1}^{m} V_i$, and (b) follows. Finally, if K is compact, then $K \subset \cup_{i=1}^{m} V_i$ for some m, and (c) is now immediate, □

The family $\{\phi_i : i \in \mathbb{N}\}$ is said to be a *locally finite partition of unity in Ω, subordinate to the covering Γ*. The use of the words 'locally finite' is justified because, in view of (b) and (c), every point of Ω has a neighbourhood which intersects the supports of only a finite number of the ϕ_i.

1.4. *The maximal function*

We now introduce the *maximal function*, in view of its importance in L^p theory and because of an application to be made to Sobolev spaces later in the chapter. Given any measurable subset A of $\mathbb{R}^n$, we denote by $|A|$ its Lebesgue n-measure.

Definition 1.11. Let $f \in L^1_{\text{loc}}(\mathbb{R}^n)$. The *maximal function* Mf is defined by

$$(Mf)(x) = \sup_{r>0} |B(x,r)|^{-1} \int_{B(x,r)} |f(y)| \, dy. \qquad \blacksquare$$

Of course, Mf may take infinite values; it is measurable since it is the supremum of a family of measurable functions. Our concern here is to show that if $p \in (1, \infty]$, then M acts as a bounded map of $L^p(\mathbb{R}^n)$ to itself. To do this we need the following 'covering lemma'.

Lemma 1.12. Let E be a measurable subset of $\mathbb{R}^n$ and let $\mathfrak{F}$ be a family of open balls in $\mathbb{R}^n$ with $E \subset \cup_{B \in \mathfrak{F}} B$ and $\sup_{B \in \mathfrak{F}} \operatorname{diam} B < \infty$. Then there is a positive constant C, depending only upon n, and balls $B_1, B_2, \ldots$ in $\mathfrak{F}$ (finite or countably infinite in number), with $B_i \cap B_j = \varnothing$ if $i \neq j$, such that

$$\sum_k |B_k| \geqslant C|E|. \qquad \blacksquare$$

Proof. Choose $B_1 \in \mathfrak{F}$ so that $\operatorname{diam} B_1 \geqslant \frac{1}{2} \sup\{\operatorname{diam} B : B \in \mathfrak{F}\}$. Now suppose that $B_1, \ldots, B_k$ have been chosen. If there is no ball in $\mathfrak{F}$ disjoint from $B_1, \ldots, B_k$ our sequence (B_i) terminates at B_k; otherwise, choose $B_{k+1} \in \mathfrak{F}$ disjoint from $B_1, \ldots, B_k$ and with $\operatorname{diam} B_{k+1} \geqslant \frac{1}{2} \sup\{\operatorname{diam} B : B \in \mathfrak{F}, B \cap B_i = \varnothing \text{ for } i = 1, \ldots, k\}$.

If $\Sigma_k |B_k| = \infty$, the lemma follows. Suppose there are infinitely many B_k and $\Sigma_k |B_k| < \infty$. For each k let S_k be the open ball with the same centre as B_k and with diam $S_k = 5$ diam B_k. We claim that $E \subset \cup_k S_k$. To do this it is enough to prove that for all $B \in \mathfrak{F}$ we have $B \subset \cup_k S_k$; and evidently we need only consider those $B \in \mathfrak{F}$ with B distinct from the B_k. Given such a B, let k be the smallest natural number m such that diam $B_{m+1} < \frac{1}{2}$ diam B: note that diam $B_m \to 0$ as $m \to \infty$ since $\Sigma_m |B_m| < \infty$. Since diam $B > 2$ diam B_{k+1}, it follows that $B \cap B_j \neq \varnothing$ for some j, $1 \leqslant j \leqslant k$, for otherwise the selection procedure for (B_i) would be contradicted. Since, moreover, $\frac{1}{2}$ diam $B \leqslant$ diam B_j, it is clear that $B \subset S_j$, and our claim that $E \subset \cup_k S_k$ is justified. Hence

$$|E| \leqslant \sum_k |S_k| = 5^n \sum_k |B_k|,$$

as required. If there are only finitely many B_i's, say $B_1, \ldots, B_k$, then $B \cap B_j \neq \varnothing$ for some $j \in \{1, \ldots, k\}$ and the proof follows as before. □

Theorem 1.13. Let $p \in (1, \infty]$ and suppose that $f \in L^p(\mathbb{R}^n)$. Then $Mf \in L^p(\mathbb{R}^n)$ and there is a constant C, which depends only on p and n, such that

$$\| Mf \|_p \leqslant C \| f \|_p.$$ ■

Proof. If $p = \infty$ the result is obvious, and $C = 1$. Suppose that $1 < p < \infty$; and let $t > 0$. Put $g(x) = f(x)$ if $|f(x)| \geqslant \frac{1}{2}t$, with $g(x) = 0$ otherwise. Then $|f(x)| \leqslant |g(x)| + \frac{1}{2}t$, $Mf(x) \leqslant Mg(x) + \frac{1}{2}t$, and

$$E_t := \{x \in \mathbb{R}^n : Mf(x) > t\} \subset \{x \in \mathbb{R}^n : Mg(x) > \tfrac{1}{2}t\} := F_t. \tag{1.4}$$

Note that E_t and F_t are measurable because Mf is measurable. For each $x \in F_t$, there is a ball $B(x)$ with centre x such that

$$\int_{B(x)} |g(y)| \, dy > \tfrac{1}{2} t |B(x)|. \tag{1.5}$$

Note that since $f \in L^p(\mathbb{R}^n)$ we have $g \in L^1(\mathbb{R}^n)$, because $\int_{\mathbb{R}^n} |g(x)| \, dx = \int_{|f(x)| \geqslant \frac{1}{2}t} |f(x)| \, dx \leqslant \left(\frac{2}{t}\right)^{p-1} \int_{\mathbb{R}^n} |f(x)|^p \, dx$. Thus by (1.5),

$$|B(x)| < 2t^{-1} \|g\|_1$$

for all $x \in F_t$. Since $F_t \subset \cup_{x \in F_t} B(x)$ we may invoke Lemma 1.12 to conclude that there is a sequence of pairwise disjoint balls B_i such that

$$\sum_i |B_i| \geqslant C |F_t|. \tag{1.6}$$

With the aid of (1.5) and (1.6) we now find that

$$\int_{\cup B_i} |g(y)|\,\mathrm{d}y > \tfrac{1}{2}t \sum_i |B_i| \geqslant \tfrac{1}{2}tC|F_t|.$$

Thus, with (1.4), we see that

$$|E_t| \leqslant \frac{2}{Ct}\|g\|_1 = \frac{2}{Ct}\int_{|f(x)|\geqslant\frac{1}{2}t} |f(x)|\,\mathrm{d}x. \tag{1.7}$$

Put $\lambda(t) = |E_t|$; then as we shall see in a moment,

$$\int_{\mathbb{R}^n} [Mf(x)]^p\,\mathrm{d}x = p\int_0^\infty t^{p-1}\lambda(t)\,\mathrm{d}t. \tag{1.8}$$

Hence, in view of (1.7),

$$\begin{aligned}\|Mf\|_p^p &= p\int_0^\infty t^{p-1}|E_t|\,\mathrm{d}t \leqslant 2pC^{-1}\int_0^\infty t^{p-2}\left(\int_{|f(x)|\geqslant\frac{1}{2}t}|f(x)|\,\mathrm{d}x\right)dt\\ &= 2pC^{-1}\int_{\mathbb{R}^n}|f(x)|\left(\int_0^{2|f(x)|} t^{p-2}\,\mathrm{d}t\right)\mathrm{d}x\\ &= 2^pC^{-1}p(p-1)^{-1}\int_{\mathbb{R}^n}|f(x)|^p\,\mathrm{d}x,\end{aligned}$$

which is what we need.

All that remains is to establish (1.8). This follows immediately from the general result that if $h\in L^1(\mathbb{R}^n)$ and $h\geqslant 0$, then

$$\int_{\mathbb{R}^n} h(x)\,\mathrm{d}x = \int_0^\infty \mu(t)\,\mathrm{d}t, \tag{1.9}$$

where $\mu(t)$ is the Lebesgue measure of $\{x\in\mathbb{R}^n: h(x) > t\}$. To prove (1.9) first note that $\{(x,t)\in\mathbb{R}^{n+1}: h(x) > t\}$ is a measurable subset of $\mathbb{R}^{n+1}$, since it can be represented as

$$\bigcup_{w\in\mathbb{Q}} \left(\{x\in\mathbb{R}^n: h(x) > w\}\times\{t\in\mathbb{R}: t < w\}\right)$$

Now put $S = \{(x,t)\in\mathbb{R}^{n+1}: 0 < t < h(x)\}$. Since

$$S = \{(x,t)\in\mathbb{R}^{n+1}: h(x) > t\}\cap\{(x,t)\in\mathbb{R}^{n+1}: t > 0\}$$

it is clear that S is measurable; and thus its characteristic function χ_S is also measurable. For a.a. $x\in\mathbb{R}^n$ we have $\chi_S(x,\bullet) = \chi_{(0,h(x))}\in L^1(\mathbb{R})$; also the function $x\mapsto \int\chi_{(0,h(x))}(t)\,\mathrm{d}t = h(x)$ is in $L^1(\mathbb{R}^n)$. Hence the repeated integral

$$\int_{\mathbb{R}^n}\left(\int_{\mathbb{R}}\chi_S(x,t)\,\mathrm{d}t\right)\mathrm{d}x$$

exists and equals $\int_{\mathbb{R}^n} h(x)\,\mathrm{d}x$. By Tonelli's theorem, $\chi_S \in L^1(\mathbb{R}^{n+1})$ and

$$\int_{\mathbb{R}^{n+1}} \chi_S = \int_{\mathbb{R}^n} h,$$

$$\int_{\mathbb{R}^{n+1}} \chi_S = \int_{\mathbb{R}} \left(\int_{\mathbb{R}^n} \chi_S(x,t)\,\mathrm{d}x \right) \mathrm{d}t = \int_0^\infty \mu(t)\,\mathrm{d}t.$$

Thus (1.9) follows. □

1.5. *The Fourier transform*

To conclude this section we give some basic information about Fourier transforms. The *Fourier transform* $\mathbf{F}(f)$, or $\hat{f}$, of an element f of $L^1(\mathbb{R}^n)$ is defined by

$$\hat{f}(\xi) = \int_{\mathbb{R}^n} \mathrm{e}^{-2\pi \mathrm{i} x \cdot \xi} f(x)\,\mathrm{d}x \qquad (\xi \in \mathbb{R}^n),$$

where $x \cdot \xi = \sum_{j=1}^{n} x_j \xi_j$. Evidently $\hat{f}(\xi)$ is well-defined for all $\xi \in \mathbb{R}^n$, and $\|\hat{f}\|_\infty \leqslant \|f\|_1$.

Theorem 1.14. Suppose that f and g belong to $L^1(\mathbb{R}^n)$. Then $\mathbf{F}(f * g) = \mathbf{F}(f)\mathbf{F}(g)$. ■

Proof. By Fubini's theorem we have, for all $\xi \in \mathbb{R}^n$,

$$\begin{aligned} \mathbf{F}(f*g)(\xi) &= \int_{\mathbb{R}^n}\int_{\mathbb{R}^n} \mathrm{e}^{-2\pi \mathrm{i} x\cdot\xi} f(x-y) g(y)\,\mathrm{d}y\,\mathrm{d}x \\ &= \int_{\mathbb{R}^n}\int_{\mathbb{R}^n} \mathrm{e}^{-2\pi \mathrm{i}(x-y)\cdot\xi} f(x-y)\,\mathrm{e}^{-2\pi \mathrm{i} y\cdot\xi} g(y)\,\mathrm{d}y\,\mathrm{d}x \\ &= \mathbf{F}(f)(\xi) \int_{\mathbb{R}^n} \mathrm{e}^{-2\pi \mathrm{i} y\cdot\xi} g(y)\,\mathrm{d}y \\ &= \mathbf{F}(f)(\xi)\mathbf{F}(g)(\xi). \end{aligned}$$ □

Lemma 1.15. Let $f \in \mathfrak{S}$. Then $D^\alpha \hat{f} = \hat{g}$ for all $\alpha \in \mathbb{N}_0^n$, where $g(x) = (-2\pi \mathrm{i} x)^\alpha f(x)$; $\mathbf{F}(D^\beta f)(\xi) = (2\pi i \xi)^\beta \mathbf{F}(f)(\xi)$ for all $\beta \in \mathbb{N}_0^n$; and $\hat{f} \in \mathfrak{S}$. ■

Proof. Differentiation of

$$\hat{f}(\xi) = \int_{\mathbb{R}^n} \mathrm{e}^{-2\pi \mathrm{i} x\cdot\xi} f(x)\,\mathrm{d}x$$

under the integral sign gives

$$D^\alpha \hat{f}(\xi) = \int_{\mathbb{R}^n} e^{-2\pi i x\cdot\xi}(-2\pi i x)^\alpha f(x)\,dx,$$

and is justified since the integral obtained is uniformly convergent on $\mathbb{R}^n$. Hence $\hat{f} \in C^\infty(\mathbb{R}^n)$ and $D^\alpha \hat{f} = \hat{g}$. Integration by parts shows that

$$(2\pi i\xi)^\beta \hat{f}(\xi) = \int_{\mathbb{R}^n} e^{-2\pi i x\cdot\xi} D^\beta f(x)\,dx.$$

It remains to prove that $\hat{f} \in \mathfrak{S}$. By the results just established, $D^\beta[\xi^\alpha \hat{f}(\xi)] = \hat{h}(\xi)$, where

$$h(x) = (-1)^{|\beta|}(2\pi i)^{|\beta|-|\alpha|} x^\beta D^\alpha f(x) \qquad (x \in \mathbb{R}^n).$$

Hence $D^\beta[\xi^\alpha \hat{f}(\xi)]$ is bounded on $\mathbb{R}^n$, for all $\alpha, \beta \in \mathbb{N}_0^n$. Use of the product rule for derivatives and induction on $|\beta|$ now shows that $\xi^\alpha D^\beta \hat{f}(\xi)$ is bounded, for all $\alpha, \beta \in \mathbb{N}_0^n$. This implies that $\hat{f} \in \mathfrak{S}$. □

Theorem 1.16 (The Riemann–Lebesgue Lemma). Let $f \in L^1(\mathbb{R}^n)$. Then $\hat{f}$ is continuous on $\mathbb{R}^n$ and $\hat{f}(\xi) \to 0$ as $|\xi| \to \infty$. ■

Proof. If $f \in \mathfrak{S}$, then as we have just seen, $\hat{f} \in \mathfrak{S}$ and the result holds. If f merely belongs to $L^1(\mathbb{R}^n)$, then since $\mathfrak{S}$ is dense in $L^1(\mathbb{R}^n)$ (cf. Prop. 1.8), there is a sequence (f_j) in $\mathfrak{S}$ such that $\|f_j - f\|_1 \to 0$ as $j \to \infty$. But $\|\hat{f}_j - \hat{f}\|_\infty \leqslant \|f_j - f\|_1$, and so $\hat{f}_j \to \hat{f}$ uniformly on $\mathbb{R}^n$. The result is now clear. □

Lemma 1.17. Let $a \in \mathbb{R}$ and $a > 0$, and define a function f by $f(x) = e^{-\pi a|x|^2}$ $(x \in \mathbb{R}^n)$. Then $\hat{f}(\xi) = a^{-n/2} e^{-\pi|\xi|^2/a}$ $(\xi \in \mathbb{R}^n)$. ■

Proof. It is enough to deal with the case $a = 1$ as the general situation may be reduced to this by a change of variable. By Fubini's Theorem, for all $\xi \in \mathbb{R}^n$,

$$\hat{f}(\xi) = \int_{\mathbb{R}^n} \exp(-2\pi i x\cdot\xi - \pi|x|^2)\,dx = \prod_{j=1}^n \int_{\mathbb{R}} \exp(-2\pi i x_j\xi_j - \pi x_j^2)\,dx_j.$$

But

$$\int_{\mathbb{R}} \exp(-2\pi i x_j\xi_j - \pi x_j^2)\,dx_j = e^{-\pi\xi_j^2} \int_{\mathbb{R}} \exp(-\pi(x_j + i\xi_j)^2\,dx_j,$$

and application of Cauchy's theorem enables us to see that

$$\int_{\mathbb{R}} \exp(-\pi(x_j + i\xi_j)^2)\,dx_j = \int_{\mathbb{R}} \exp(-\pi x_j^2)\,dx_j = 1.$$

The result follows. □

Lemma 1.18. If $f, g \in \mathfrak{S}$, then $\displaystyle\int_{\mathbb{R}^n} f(x)\hat{g}(x)\,dx = \int_{\mathbb{R}^n} \hat{f}(x)g(x)\,dx.$ ■

Proof. By Fubini's theorem,

$$\int_{\mathbb{R}^n} f(x)\hat{g}(x)\,dx = \int_{\mathbb{R}^n}\int_{\mathbb{R}^n} f(x)g(y)e^{-2\pi i x\cdot y}\,dy\,dx = \int_{\mathbb{R}^n} \hat{f}(x)g(x)\,dx. \qquad \square$$

We now set about the Fourier inversion theorem. To do this we define a function $\check{f}$ by

$$\check{f}(x) = \int_{\mathbb{R}^n} e^{2\pi i x\cdot\xi} f(\xi)\,d\xi = \hat{f}(-x) \qquad (x \in \mathbb{R}^n).$$

Theorem 1.19 (The Fourier Inversion Theorem). If $f \in \mathfrak{S}$, then $(\hat{f})\check{} = f$. ■

Proof. Given any $\varepsilon > 0$, put $\phi(\xi) = \exp(2\pi i x\cdot\xi - \pi\varepsilon^2|\xi|^2)$, with $x, \xi \in \mathbb{R}^n$. Then by Lemma 1.17, for all $y \in \mathbb{R}^n$,

$$\begin{aligned}\hat{\phi}(y) &= \int_{\mathbb{R}^n} \exp[-2\pi i(y-x)\cdot\xi]\exp(-\pi\varepsilon^2|\xi|^2)\,d\xi \\ &= \varepsilon^{-n}\exp(-\pi|x-y|^2/\varepsilon^2).\end{aligned}$$

Hence $\hat{\phi}(y) = \varepsilon^{-n}g[(x-y)/\varepsilon]$, where $g(x) = \exp(-\pi|x|^2)$. Thus

$$\begin{aligned}\int_{\mathbb{R}^n} \exp(-\pi\varepsilon^2|\xi|^2 + 2\pi i x\cdot\xi)\hat{f}(\xi)\,d\xi &= \int_{\mathbb{R}^n} \hat{f}(\xi)\phi(\xi)\,d\xi = \int_{\mathbb{R}^n} f(y)\hat{\phi}(y)\,dy \\ &= \int_{\mathbb{R}^n} f(y)\varepsilon^{-n}g[(x-y)/\varepsilon]\,dy = \int_{\mathbb{R}^n} f(x-\varepsilon z)g(z)\,dz \to f(x)\end{aligned}$$

as $\varepsilon \to 0+$, uniformly on $\mathbb{R}^n$ since f is uniformly continuous. However, for each $x \in \mathbb{R}^n$,

$$\begin{aligned}\int_{\mathbb{R}^n} \exp(-\pi\varepsilon^2|\xi|^2 + 2\pi i x\cdot\xi)\hat{f}(\xi)\,d\xi &\to \int_{\mathbb{R}^n} \exp(2\pi i x\cdot\xi)\hat{f}(\xi)\,d\xi \\ &= (\hat{f})\check{}(x).\end{aligned}$$

The proof is complete. □

Corollary 1.20. The Fourier transform is an isomorphism of $\mathfrak{S}$ onto itself. ■

Theorem 1.21. The Fourier transform can be extended to a unitary isomorphism of $L^2(\mathbb{R}^n)$ onto itself. ■

Proof. Since $\mathfrak{S}$ is dense in $L^2(\mathbb{R}^n)$ and the Fourier transform is an isomorphism of $\mathfrak{S}$ onto itself, it is enough to prove that $\|\hat{f}\|_2 = \|f\|_2$ for all $f \in \mathfrak{S}$. Let $f \in \mathfrak{S}$ and put $g(x) = \overline{f(-x)}$ $(x \in \mathbb{R}^n)$. Then $\hat{g} = \overline{\hat{f}}$ and with the help of Theorem 1.19 we have

$$\|f\|_2^2 = \int_{\mathbb{R}^n} f(x)\overline{f(x)}\,dx = \int_{\mathbb{R}^n} f(x)g(-x)\,dx = (f * g)(0)$$

$$= \int_{\mathbb{R}^n} \mathsf{F}(f * g)(\xi)\,d\xi = \int_{\mathbb{R}^n} \hat{f}(\xi)\overline{\hat{f}(\xi)}\,d\xi = \|\hat{f}\|_2^2. \qquad \square$$

2. Weak derivatives

As before, Ω will stand for a non-empty, open subset of $\mathbb{R}^n$. Let $f \in L^1_{\text{loc}}(\Omega)$, and suppose that there is a function $g \in L^1_{\text{loc}}(\Omega)$ such that for all $\phi \in C_0^\infty(\Omega)$,

$$\int_\Omega f D_i \phi = -\int_\Omega g\phi.$$

Then g is said to be a *weak derivative* of f in Ω with respect to x_i, and we write $g = D_i f$ (weakly). More generally, if $\alpha \in \mathbb{N}_0^n$ and $h \in L^1_{\text{loc}}(\Omega)$ is such that

$$\int_\Omega f D^\alpha \phi = (-1)^{|\alpha|} \int_\Omega h\phi$$

for all $\phi \in C_0^\infty(\Omega)$, then we say that h is an αth weak derivative of f in Ω and write $h = D^\alpha f$ (weakly).

Lemma 2.1. Let $f \in L^1_{\text{loc}}(\Omega)$, let $\alpha \in \mathbb{N}_0^n$, and let h_1 and h_2 be αth weak derivatives of f. Then $h_1(x) = h_2(x)$ for a.e. $x \in \Omega$. ■

Proof. Let ρ be a mollifier, and note that given any fixed $x \in \Omega$, the function $\xi \mapsto \varepsilon^{-n}\rho[(\xi - x)/\varepsilon]$ is in $C_0^\infty(\Omega)$ provided that ε (> 0) is small enough; thus

$$\int_\Omega [h_1(\xi) - h_2(\xi)]\varepsilon^{-n}\rho[(\xi - x)/\varepsilon]\,d\xi = 0;$$

that is, the regularization $(h_1 - h_2)_\varepsilon$ of $h_1 - h_2$ is zero at x, if ε is small enough. By Theorem 1.5, $(h_1 - h_2)_\varepsilon \to h_1 - h_2$ in $L^1(B(y, \delta))$ as $\varepsilon \to 0$, for all $y \in \Omega$, if δ (> 0) is sufficiently small. Since $(h_1 - h_2)_\varepsilon(x) = 0$ for all $x \in \Omega$, if ε is small enough, the result follows. $\square$

In view of this result we shall refer to *the* αth weak derivative of a function, functions equal almost everywhere being identified, as usual.

It is clear that if $f \in C^{|\alpha|}(\Omega)$, then the classical and weak derivatives of f, up to those of order $|\alpha|$, coincide modulo sets of measure zero; thus the notion of weak differentiation extends the classical idea. This extension is a proper one:

there are functions which have a weak derivative but no corresponding classical derivative. Perhaps the simplest example of such a function is the function f defined on $\mathbb{R}^2$ by

$$f(x_1, x_2) = f_1(x_1) + f_2(x_2) \qquad ((x_1, x_2) \in \mathbb{R}^2),$$

where f_1 and f_2 are continuous, nowhere (classically) differentiable functions on R. Suppose that $\phi \in C_0^\infty(\mathbb{R}^2)$; then

$$\iint_{\mathbb{R}^2} f(x_1, x_2) D_1 D_2 \phi(x_1, x_2)\, dx_1\, dx_2$$

$$= \int_{\mathbb{R}} f_1(x_1) \int_{\mathbb{R}} D_2 D_1 \phi(x_1, x_2)\, dx_2\, dx_1 + \int_{\mathbb{R}} f_2(x_2) \int_{\mathbb{R}} D_1 D_2 \phi(x_1, x_2)\, dx_1\, dx_2$$

$$= 0,$$

and hence the weak derivative $D_1 D_2 f$ exists and is the zero function on $\mathbb{R}^2$; the corresponding classical derivative evidently does not exist. Again, the function f given by $f(x) = |x|^{2-n}$ $(x \neq 0;\ n > 2)$ is not in $C^1(\mathbb{R}^n)$, no matter how it is defined at 0, but all of its weak derivatives $D_i f$ of order 1 exist, with $(D_i f)(x) = (2-n)x_i/|x|^n$.

It is also possible that a given function may not have a weak derivative: the Heaviside step function $H: \mathbb{R} \to \mathbb{R}$ defined by $H(x) = 1$ if $x \geqslant 0$, with $H(x) = 0$ otherwise, is such a function. This is because if H had a weak derivative DH in $\mathbb{R}$, then for all $\phi \in C_0^\infty(\mathbb{R})$,

$$\int_{\mathbb{R}} \phi DH\, dx = -\int_{\mathbb{R}} H D\phi\, dx = -\int_0^\infty D\phi\, dx = \phi(0). \tag{2.1}$$

Let $0 < a < b < \infty$ and put $\psi = \chi_{(a,b)} \operatorname{sgn} DH$; there is a sequence (ψ_m) in $C_0^\infty((a, b))$ such that $\psi_m \to \psi$ in $L^1((a, b))$, with $|\psi_m(x)| \leqslant 2$ for all $m \in \mathbb{N}$ and all $x \in (a, b)$. There is a subsequence of (ψ_m), again denoted by (ψ_m) for simplicity, such that $\psi_m(x) \to \psi(x)$ a.e. in (a, b). Thus by the Dominated Convergence Theorem,

$$0 = \int_{\mathbb{R}} \psi_m DH\, dx \to \int_a^b |DH|\, dx,$$

and hence $DH(x) = 0$ a.e. in (a, b). Since a and b are arbitrary positive numbers, and since the same argument can be used for negative numbers, we conclude that DH must be the zero function. Hence

$$\int_{\mathbb{R}} \phi DH\, dx = 0$$

for all $\phi \in C_0^\infty(R)$, and (2.1) is contradicted.

This last example shows that our extension of the notion of differentiation is by no means as wide as in the theory of distributions, where a derivative need not be a function: the derivative of H in the sense of distributions is the Dirac measure.

We now turn to the fundamental properties of weak derivatives. We shall use a mollifier ρ and the corresponding regularization f_ε of a suitable function f on Ω, where

$$f_\varepsilon(x) = \varepsilon^{-n} \int_\Omega \rho[(x-y)/\varepsilon] f(y)\,dy$$

provided that $0 < \varepsilon < \operatorname{dist}(x, \partial\Omega)$ and $x \in \Omega$.

Lemma 2.2. Let $u \in L^1_{loc}(\Omega)$, let $\alpha \in \mathbb{N}_0^n$ and suppose that the weak derivative $D^\alpha u$ of u in Ω exists. Then $(D^\alpha u_\varepsilon)(x) = (D^\alpha u)_\varepsilon(x)$ if $x \in \Omega$ and $0 < \varepsilon < \operatorname{dist}(x, \partial\Omega)$. ■

Proof. Suppose that $0 < \varepsilon < \operatorname{dist}(x, \partial\Omega)$. Then since the interchange of integration and differentiation is evidently justified,

$$\begin{aligned}(D^\alpha u_\varepsilon)(x) &= \int_\Omega \varepsilon^{-n} D_x^\alpha \rho[(x-y)/\varepsilon] u(y)\,dy \\ &= (-1)^{|\alpha|}\varepsilon^{-n}\int_\Omega u(y) D_y^\alpha\, \rho[(x-y)/\varepsilon]\,dy \\ &= \varepsilon^{-n}\int_\Omega \rho[(x-y)/\varepsilon] D^\alpha u(y)\,dy \\ &= (D^\alpha u)_\varepsilon(x).\end{aligned}$$ □

Lemma 2.3. Let $u, v \in L^1_{loc}(\Omega)$ and let $\alpha \in \mathbb{N}_0^n$. Then $v = D^\alpha u$ (weakly) in Ω if, and only if, there is a sequence $(u^{(m)})$ in $C^{|\alpha|}(\Omega)$ such that for every compact subset K of Ω we have $u^{(m)} \to u$ in $L^1(K)$ and $D^\alpha u^{(m)} \to v$ in $L^1(K)$ as $m \to \infty$. If $v = D^\alpha u$ (weakly) in Ω the $u^{(m)}$ may be taken to be in $C^\infty(\Omega)$; if in addition u has compact support in Ω, the $u^{(m)}$ may be taken to be in $C_0^\infty(\Omega)$. ■

Proof. First suppose that $v = D^\alpha u$ (weakly) in Ω, and let K be any compact subset of Ω. Then by Theorem 1.5, the sequence $(u_{1/k})_{k\in\mathbb{N}}$ of regularized functions is in $C^\infty(\mathbb{R}^n)$ and converges to u in $L^1(K)$; and by Lemma 2.2, as $k \to \infty$,

$$D^\alpha u_{1/k} = (D^\alpha u)_{1/k} \to D^\alpha u \quad \text{in } L^1(K).$$

If supp u is compact, then evidently $u_{1/k}$ is in $C_0^\infty(\Omega)$ for large k.

Conversely, suppose that a sequence $(u^{(m)})$ of the kind described in the lemma exists, and let $\phi \in C_0^\infty(\Omega)$ with $K = \text{supp}\,\phi$. Then for each $m \in \mathbb{N}$,

$$\int_\Omega u^{(m)} D^\alpha \phi \, dx = (-1)^{|\alpha|} \int_\Omega \phi D^\alpha u^{(m)} \, dx,$$

and since these integrals are effectively over K, the fact that $u^{(m)} \to u$ and $D^\alpha u^{(m)} \to v$ in $L^1(K)$ shows that

$$\int_\Omega u D^\alpha \phi \, dx = (-1)^{|\alpha|} \int_\Omega \phi v \, dx.$$

Hence $v = D^\alpha u$ (weakly). □

Many of the results familiar for derivatives taken in the classical sense can be established with little difficulty for weak derivatives. Thus if $u \in L^1_{\text{loc}}(\Omega)$ and $\alpha, \beta \in \mathbb{N}_0^n$, a simple integration-by-parts argument shows that if any one of the weak derivatives $D^{\alpha+\beta} u$, $D^\alpha(D^\beta u)$, $D^\beta(D^\alpha u)$ exists, then they all do, and coincide a.e. in Ω. The product rule for differentiation holds: if $u \in L^1_{\text{loc}}(\Omega)$ has a weak derivative $D_j u$ in Ω, and $\psi \in C^\infty(\Omega)$, then $D_j(u\psi) = \psi D_j u + u D_j \psi$ weakly in Ω. To prove this, simply note that for all $\phi \in C_0^\infty(\Omega)$,

$$\begin{aligned}\int_\Omega u\psi D_j\phi &= \int_\Omega u[D_j(\psi\phi) - \phi D_j\psi] \\ &= -\int_\Omega \phi(\psi D_j u + u D_j \psi),\end{aligned}$$

which establishes the result. Moreover, a function u has a weak derivative $D_j u$ on Ω if, and only if, it has a corresponding weak derivative on a neighbourhood of each point in Ω. To see this, suppose that u has a weak derivative $(D_j u)_{V(x)}$ on a neighbourhood $V(x)$ of each $x \in \Omega$, and let $\phi \in C_0^\infty(\Omega)$, with $K = \text{supp}\,\phi$. Since K is compact, it may be covered by finitely many of the neighbourhoods $V(x)$, say $V(x_1), \ldots, V(x_k)$. Let (ϕ_i) be a partition of unity subordinate to the $V(x)$, so that $\phi = \sum_{i=1}^k \phi\phi_i$ in a neighbourhood of K. Then

$$\begin{aligned}\int_\Omega u D_j \phi &= \int_\Omega u \sum_{i=1}^k D_j(\phi\phi_i) = \sum_{i=1}^k \int_{V(x_i)} u D_j(\phi\phi_i) \\ &= -\sum_{i=1}^k \int_{V(x_i)} (D_j u)_{V(x_i)} \phi\phi_i \\ &= -\int_\Omega \phi \sum_{i=1}^k \phi_i (D_j u)_{V(x_i)},\end{aligned}$$

which shows that the weak derivative $D_j u$ exists on Ω. The converse is obvious.

The next result gives an important connection between weak and classical differentiation.

Theorem 2.4. Let $u \in \mathrm{C}(\Omega)$, and suppose that for all $\alpha \in \mathbb{N}_0^n$ with $|\alpha| \leqslant k \in \mathbb{N}$, the weak derivative $\mathrm{D}^\alpha u$ of u on Ω exists and is in $\mathrm{C}(\Omega)$. Then $u \in \mathrm{C}^k(\Omega)$. ∎

Proof. It is enough to establish the result when $k = 1$, since then induction will do the rest. If $\phi \in \mathrm{C}_0^\infty(\Omega)$, then ϕu and $\mathrm{D}_j(\phi u) = (\mathrm{D}_j\phi)u + \phi \mathrm{D}_j u$ are continuous functions with compact support in Ω. It is thus enough to deal with the case when u has compact support K in Ω. By Lemma 2.2, there exist an open set Ω', with $K \subset \Omega' \subset \overline{\Omega'} \subset \Omega$, and a positive number ε_0, such that $(\mathrm{D}_j u_\varepsilon)(x) = (\mathrm{D}_j u)_\varepsilon(x)$ for all $x \in \Omega'$ and all $\varepsilon \in (0, \varepsilon_0)$. Moreover, $u_\varepsilon \in \mathrm{C}^\infty(\Omega')$ if $0 < \varepsilon < \varepsilon_0$; and $u_\varepsilon \to u$ and $\mathrm{D}_j u_\varepsilon \to \mathrm{D}_j u$ uniformly on K as $\varepsilon \to 0$. Hence $\mathrm{D}_j u$ is the classical derivative of u with respect to x_j, and the proof is complete. □

Finally we deal with the chain rule, and the positive and negative parts of a function.

Lemma 2.5. Let $f \in \mathrm{C}^1(\mathbb{R})$ and suppose that the derivative f' of f is in $\mathrm{L}^\infty(\mathbb{R})$; suppose also that u is real-valued and locally integrable over Ω, and that all its first-order weak derivatives on Ω exist. Then all the first-order weak derivatives on Ω of $f \circ u$ exist, and $\mathrm{D}_j(f \circ u) = (f' \circ u)\mathrm{D}_j u$ for $j = 1, 2, \ldots, n$. ∎

Proof. Let $(u^{(m)})$ be a sequence of functions in $\mathrm{C}^\infty(\Omega)$ related to u as in Lemma 2.3. Then given any compact subset K of Ω,

$$\int_K |f \circ u^{(m)} - f \circ u| \leqslant \|f'\|_{\infty,\mathbb{R}} \int_K |u^{(m)} - u| \to 0 \text{ as } m \to \infty,$$

and

$$\int_K |(f' \circ u^{(m)})\mathrm{D}_j u^{(m)} - (f' \circ u)\mathrm{D}_j u|$$

$$\leqslant \|f'\|_{\infty,\mathbb{R}} \int_K |\mathrm{D}_j u^{(m)} - \mathrm{D}_j u| + \int_K |f' \circ u^{(m)} - f' \circ u|\, |\mathrm{D}_j u|.$$

Since $u^{(m)} \to u$ in $\mathrm{L}^1(K)$, there is a subsequence of $(u^{(m)})$, again denoted by $(u^{(m)})$ for convenience, which converges a.e. on K to u; and because f' is continuous, $(f' \circ u^{(m)})$ converges a.e. on K to $f' \circ u$. Thus by the Dominated Convergence Theorem,

$$\int_K |f' \circ u^{(m)} - f' \circ u|\, |D_j u| \to 0$$

as $m \to \infty$. The result now follows from Lemma 2.3. □

Proposition 2.6. Let u be a real-valued function in $L^1_{loc}(\Omega)$, and suppose that all its first-order weak derivatives on Ω exist. Then the first-order weak derivatives on Ω of $u^+ = \max(u, 0)$, $u^- = -\min(u, 0)$ and $|u|$ all exist, and for $j = 1, 2, \ldots, n$ and $x \in \Omega$ we have

$$D_j u^+(x) = \begin{cases} D_j u(x) & \text{if } u(x) > 0, \\ 0 & \text{if } u(x) \leqslant 0, \end{cases} \qquad D_j u^-(x) = \begin{cases} 0 & \text{if } u(x) \geqslant 0, \\ -D_j u(x) & \text{if } u(x) < 0, \end{cases}$$

and

$$D_j|u(x)| = \begin{cases} D_j u(x) & \text{if } u(x) > 0, \\ 0 & \text{if } u(x) = 0, \\ -D_j u(x) & \text{if } u(x) < 0. \end{cases} \quad ■$$

Proof. Given any $\varepsilon > 0$, put

$$f(\varepsilon, u(x)) = \begin{cases} [u^2(x) + \varepsilon^2]^{\frac{1}{2}} - \varepsilon & \text{if } u(x) > 0, \\ 0 & \text{if } u(x) \leqslant 0. \end{cases}$$

By Lemma 2.5, it follows that for all $\phi \in C_0^\infty(\Omega)$,

$$\int_\Omega f(\varepsilon, u(x)) D_j\phi(x)\,dx = -\int_{\Omega^+} \phi(x) u(x) [u^2(x) + \varepsilon^2]^{-\frac{1}{2}} D_j u(x)\,dx,$$

where $\Omega^+ = \{x \in \Omega : u(x) > 0\}$. Then the Dominated Convergence Theorem shows that if we let $\varepsilon \to 0+$, we obtain

$$\int_\Omega u^+ D_j\phi = -\int_{\Omega^+} \phi D_j u.$$

This establishes the result for u^+; those for u^- and $|u|$ are derived by use of the relations $u^- = (-u)^+$ and $|u| = u^+ + u^-$. □

3. Sobolev spaces: definitions and basic properties

3.1. *Foundations*

Let $k \in \mathbb{N}$ and suppose that $p \in [1, \infty]$. The *Sobolev space* $W^{k,p}(\Omega)$ is defined to be the vector space of all elements u of $L^p(\Omega)$ such that, for all $\alpha \in \mathbb{N}_0^n$ with $|\alpha| \leqslant k$, we have $D^\alpha u \in L^p(\Omega)$, the derivatives being taken in the weak sense. It is made into a normed vector space by endowing it with the norm $\|\bullet\|_{k,p,\Omega}$, where

$$\|u\|_{k,p,\Omega} := \left(\sum_{|\alpha| \leqslant k} \|D^\alpha u\|_{p,\Omega}^p \right)^{1/p} \quad \text{if } 1 \leqslant p < \infty,$$

and

$$\|u\|_{k,\infty,\Omega} := \sum_{|\alpha| \leqslant k} \|D^\alpha u\|_{\infty,\Omega}.$$

When $p = 2$, the spaces may be regarded as inner-product spaces, with inner product given by

$$(u, v)_{k,2,\Omega} = \int_\Omega \sum_{|\alpha| \leqslant k} (D^\alpha u)\overline{(D^\alpha v)},$$

the complex conjugate being necessary, of course, only when the spaces are complex.

Theorem 3.1. For any $k \in \mathbb{N}$ and any $p \in [1, \infty)$, the Sobolev space $W^{k,p}(\Omega)$ is a separable Banach space. It is reflexive if $p \in (1, \infty)$. ■

Proof. Let (u_m) be a Cauchy sequence in $W^{k,p}(\Omega)$. Then (u_m) is a Cauchy sequence in $L^p(\Omega)$, and hence it converges, to $u \in L^p(\Omega)$, say. Similarly, for each $\alpha \in \mathbb{N}_0^n$ with $0 < |\alpha| \leqslant k$, the sequence $(D^\alpha u_m)$ converges in $L^p(\Omega)$, to $u^{(\alpha)}$, say. Then for all $\phi \in C_0^\infty(\Omega)$, use of Hölder's inequality shows that

$$\left| \int_\Omega \phi (D^\alpha u_m - u^{(\alpha)}) \right| \leqslant \| D^\alpha u_m - u^{(\alpha)} \|_{p,\Omega} \| \phi \|_{p',\Omega} \to 0$$

as $m \to \infty$. Hence

$$\int_\Omega \phi u^{(\alpha)} = \lim_{m \to \infty} \int_\Omega \phi D^\alpha u_m = \lim_{m \to \infty} (-1)^{|\alpha|} \int_\Omega u_m D^\alpha \phi$$
$$= (-1)^{|\alpha|} \int_\Omega u D^\alpha \phi.$$

Thus $D^\alpha u = u^{(\alpha)}$, and $u_m \to u$ in $W^{k,p}(\Omega)$. This proves that $W^{k,p}(\Omega)$ is complete. As for the remainder of the theorem, let $N = \# \{\alpha \in \mathbb{N}_0^n : |\alpha| \leqslant k\}$, and observe that $W^{k,p}(\Omega)$ is isometrically isomorphic to a closed linear subspace V of the product of N copies of $L^p(\Omega)$, the isomorphism being established by means of the map $u \mapsto (D^\alpha u)_{|\alpha| \leqslant k}$. Since $L^p(\Omega)$ is separable, and reflexive if $1 < p < \infty$, it follows that V and hence $W^{k,p}(\Omega)$ are also separable and reflexive for these values of p. □

Theorem 3.2. Let $k \in \mathbb{N}$ and $p \in [1, \infty)$. The linear subspace $C^\infty(\Omega) \cap W^{k,p}(\Omega)$ of $W^{k,p}(\Omega)$ is dense in $W^{k,p}(\Omega)$. ■

Proof. Let $(\Omega_j)_{j \in \mathbb{N}}$ be a sequence of open subsets of Ω with union Ω and such that $\bar{\Omega}_j$ is a compact subset of Ω_{j+1} for each $j \in \mathbb{N}$, with $\Omega_1 = \varnothing$, and suppose that the sets $(\Omega_{j+2} \setminus \bar{\Omega}_j)$ form an open covering of Ω. We could, for example, take $\Omega_j = \{x \in \Omega : |x| < j, \operatorname{dist}(x, \partial\Omega) > j^{-1}\}$ $(j \geqslant 2)$. Let (ϕ_j) be a partition of unity subordinate to this covering, so that $\operatorname{supp} \phi_j \subset \Omega_{j+2} \setminus \bar{\Omega}_j$ for each j.

Now let $u \in W^{k,p}(\Omega)$ and let $\varepsilon > 0$. For each $j \in \mathbb{N}$, the function $u\phi_j$ has weak derivatives in Ω of all orders not greater than k; and $\operatorname{supp}(u\phi_j) \subset \Omega_{j+2} \setminus \bar{\Omega}_j$, so that for small enough $\varepsilon_j > 0$, the regularization $v_j := (u\phi_j)_{\varepsilon_j}$ has support in $\Omega_{j+3}\backslash\bar{\Omega}_{j-1}$ (with the convention that $\Omega_0 = \varnothing$) and $\|v_j - u\phi_j\|_{k,p,\Omega} < \varepsilon/2^j$. Hence given any compact subset K of Ω, only a finite number of the v_j are non-zero on K, and $v := \sum_{j=1}^{\infty} v_j$ is in $C^\infty(\Omega)$. Moreover,

$$D^\alpha v(x) - D^\alpha u(x) = \sum_{j=1}^{\infty} D^\alpha(v_j - u\phi_j)(x),$$

and since for each $x \in \Omega$ there are at most four non-zero terms in this series we see that there is a positive constant C such that for each $x \in \Omega$,

$$|D^\alpha v(x) - D^\alpha u(x)|^p \leqslant C \sum_{j=1}^{\infty} |D^\alpha(v_j - u\phi_j)(x)|^p.$$

Hence

$$\|v - u\|_{k,p,\Omega} \leqslant C^{1/p} \sum_{j=1}^{\infty} \varepsilon/2^j = C^{1/p}\varepsilon,$$

and the proof is complete. □

This result, due to Meyers and Serrin [1], shows that $W^{k,p}(\Omega)$ may be regarded as the completion of the set of all those functions in $C^\infty(\Omega)$ which have finite norm.

An important closed linear subspace of $W^{k,p}(\Omega)$ is obtained by taking the closure of $C_0^\infty(\Omega)$ in $W^{k,p}(\Omega)$. This subspace, denoted by $W_0^{k,p}(\Omega)$, is in general a *proper* subspace and is of particular interest in connection with the Dirichlet problem for elliptic equations. Moreover, it is possible to establish useful relationships involving $W_0^{k,p}(\Omega)$ without the need for special restrictions on the boundary of Ω. Note that $W_0^{k,p}(\mathbb{R}^n) = W^{k,p}(\mathbb{R}^n)$ if $k \in \mathbb{N}$ and $p \in [1, \infty)$. To prove this, observe that by Theorem 3.2, given any $u \in W^{k,p}(\mathbb{R}^n)$, there is a sequence (u_m) in $C^\infty(\mathbb{R}^n) \cap W^{k,p}(\mathbb{R}^n)$ such that $\|u_m - u\|_{k,p,\mathbb{R}^n} \to 0$ as $m \to \infty$. Let $\phi \in C_0^\infty(\mathbb{R}^n)$ be such that $0 \leqslant \phi \leqslant 1$, with $\phi(x)$ equal to 1 if $|x| < 1$ and equal to 0 if $|x| > 2$; put $\phi_m(x) = \phi(x/m)$ $(m \in \mathbb{N})$. Then it is routine to check that $\|\phi_m u - u\|_{k,p,\mathbb{R}^n} \to 0$ as $m \to \infty$; and since, for each $\alpha \in \mathbb{N}_0^n$ with $|\alpha| \leqslant k$, the function $D^\alpha\phi_m$ is in $L^\infty(\mathbb{R}^n)$, with bound independent of m, it follows that

$$\begin{aligned}
\|\phi_m u_m - u\|_{k,p,\mathbb{R}^n} &\leqslant \|\phi_m(u_m - u)\|_{k,p,\mathbb{R}^n} + \|\phi_m u - u\|_{k,p,\mathbb{R}^n} \\
&\leqslant C\|u_m - u\|_{k,p,\mathbb{R}^n} + \|\phi_m u - u\|_{k,p,\mathbb{R}^n} \\
&\to 0 \quad \text{as } m \to \infty.
\end{aligned}$$

Since $\phi_m u_m \in C_0^\infty(\mathbb{R}^n)$, u must belong to $W_0^{k,p}(\mathbb{R}^n)$ and the proof is complete.

If Ω is a *bounded* open subset of $\mathbb{R}^n$, however, then $W_0^{k,p}(\Omega) \neq W^{k,p}(\Omega)$, with the same conditions on k and p as above. To see this, let u be the function equal to 1 at all points of Ω. Then $u \in W^{k,p}(\Omega)$. But given any $\phi \in C_0^\infty(\Omega)$,

$$\|u-\phi\|_{1,1,\Omega} = \int_\Omega |u(x)-\phi(x)|\,dx + \int_\Omega \sum_{i=1}^n |D_i\phi(x)|\,dx.$$

Moreover, representation of $\phi(x)$ as an indefinite integral of $D_1\phi$ shows that there is a constant c (> 1) depending only on Ω, such that

$$\int_\Omega |\phi(x)|\,dx \leqslant c \int_\Omega |D_1\phi(x)|\,dx;$$

the proof of Theorem 3.19 gives full details of this procedure. Hence

$$\begin{aligned}\|u-\phi\|_{1,1,\Omega} &\geqslant \int_\Omega |u(x)|\,dx - \int_\Omega |\phi(x)|\,dx + \int_\Omega \sum_{i=1}^n |D_i\phi(x)|\,dx \\ &\geqslant |\Omega| - (c-1)\int_\Omega \sum_{i=1}^n |D_i\phi(x)|\,dx \\ &\geqslant |\Omega| - (c-1)\|u-\phi\|_{1,1,\Omega},\end{aligned}$$

and thus

$$\|u-\phi\|_{1,1,\Omega} \geqslant |\Omega|/c.$$

Since $W^{k,p}(\Omega)$ is continuously embedded in $W^{1,1}(\Omega)$, it follows that there is a positive constant λ such that for all $\phi \in C_0^\infty(\Omega)$,

$$\|u-\phi\|_{k,p,\Omega} \geqslant \lambda,$$

which shows that $u \notin W_0^{k,p}(\Omega)$.

For other results in this direction, relating to *unbounded* sets Ω, we refer to Lions [1].

Corollary 3.3. Let $k \in \mathbb{N}$ and $p \in [1, \infty)$, and suppose that Ω is unbounded. Then $\{u : u \in C^\infty(\Omega) \cap W^{k,p}(\Omega),\ \text{supp}\, u \text{ bounded}\}$ is dense in $W^{k,p}(\Omega)$. ■

Proof. Let $u \in W^{k,p}(\Omega)$, let ϕ_m be as in the discussion following the proof of Theorem 3.2 and let (u_m) be a sequence in $C^\infty(\Omega) \cap W^{k,p}(\Omega)$ with $\|u_m - u\|_{k,p,\Omega} \to 0$ as $m \to \infty$. Then just as in the earlier discussion, $\phi_m u_m \in C^\infty(\Omega)$, $\|\phi_m u_m - u\|_{k,p,\Omega} \to 0$ as $m \to \infty$, and supp $(\phi_m u_m)$ is bounded. □

In view of the importance of $W_0^{m,p}(\Omega)$, criteria adequate to detect elements of this subspace are of great utility. One such criterion is the following.

Theorem 3.4. Let Ω be a non-empty open subset of $\mathbb{R}^n$ with $\Omega \neq \mathbb{R}^n$, and for each $x \in \mathbb{R}^n$ put $d(x) = \text{dist}\,(x, \mathbb{R}^n \backslash \Omega)$; let $p \in (1, \infty)$ and $m \in \mathbb{N}$. Then if $u \in W^{m,p}(\Omega)$ and $u/d^m \in L^p(\Omega)$, it follows that $u \in W_0^{m,p}(\Omega)$. ■

Proof. First suppose that Ω is bounded. Let $0 < \varepsilon < \delta$; define $u^{(\delta)}$ by $u^{(\delta)}(x) = u(x)$ if $d(x) \geqslant \delta$, with $u^{(\delta)}(x) = 0$ otherwise; let ρ be a mollifier and put $\rho^{(\varepsilon)}(x) = \varepsilon^{-n}\rho(x/\varepsilon)$; given any function g defined on Ω we continue to denote by g the extension of g which is zero in $\mathbb{R}^n \setminus \Omega$. Note that $\rho^{(\varepsilon)} * u^{(\delta)} \in C_0^\infty(\Omega)$.

Now suppose that $d(x) > \delta + \varepsilon$, with $x \in \Omega$, and $\alpha \in \mathbb{N}_0^n$, with $|\alpha| \leqslant m$. Then since $B(x, \varepsilon) \subset B(x, \delta + \varepsilon) \subset \Omega$, and $d(y) > \delta$ if $y \in B(x, \varepsilon)$,

$$\begin{aligned} D^\alpha(\rho^{(\varepsilon)} * u^{(\delta)})(x) &= D^\alpha \int_{B(x,\varepsilon)} \rho^{(\varepsilon)}(x-y)\, u^{(\delta)}(y)\,dy \\ &= D^\alpha \int_{\mathbb{R}^n} \rho^{(\varepsilon)}(x-y)\, u(y)\,dy \\ &= \int_{\mathbb{R}^n} \rho^{(\varepsilon)}(x-y)\, D^\alpha u(y)\,dy; \end{aligned}$$

that is,

$$D^\alpha(\rho^{(\varepsilon)} * u^{(\delta)})(x) = (\rho^{(\varepsilon)} * D^\alpha u)(x) \quad \text{if} \quad d(x) > \delta + \varepsilon \text{ and } x \in \Omega. \tag{3.1}$$

It follows that, with $M(g)$ as the maximal function defined by

$$M(g)(x) = \sup_{r>0} \left(\int_{B(x,r)} |g(y)|\,dy / |B(0,r)| \right),$$

$$\begin{aligned} |D^\alpha(\rho^{(\varepsilon)} * u^{(\delta)})(x)| &\leqslant M(D^\alpha u)(x)\omega_n \sup_{B(x,\varepsilon)} \rho[(x-y)/\varepsilon] \\ &= M(D^\alpha u)(x)\omega_n \sup_{z \in B(0,1)} \rho(z) \\ &= M(D^\alpha u)(x) b(\rho), \text{ say.} \end{aligned}$$

Next, suppose that $d(x) \leqslant k\varepsilon$ for some $k \in \mathbb{N}$, and that $|\alpha| \leqslant m$. Then

$$\begin{aligned} |D^\alpha(\rho^{(\varepsilon)} * u^{(\delta)})(x)| &= \left| \int_{\mathbb{R}^n} \varepsilon^{-n-|\alpha|} (D^\alpha \rho)[(x-y)/\varepsilon] u^{(\delta)}(y)\,dy \right| \\ &\leqslant \varepsilon^{-n-|\alpha|} \int_{B(x,\varepsilon)} |(D^\alpha \rho)[(x-y)/\varepsilon] u^{(\delta)}(y)/d^m(y)| \varepsilon^m (k+1)^m\,dy \\ &\leqslant \omega_n \varepsilon^{m-|\alpha|}(k+1)^m M(u/d^m)(x) \sup_{z \in B(0,1)} |(D^\alpha \rho)(z)|. \end{aligned}$$

Thus if we take $\delta = 2\varepsilon = 2/j \ (j \in \mathbb{N})$ and $k = 3$, we see that for all $x \in \Omega$, for all $\alpha \in \mathbb{N}_0^n$ with $|\alpha| \leqslant m$, and for all $j \in \mathbb{N}$,

$$\begin{aligned} |D^\alpha(\rho^{(1/j)} * u^{(2/j)})(x)| &\leqslant \max\{b(\rho)M(D^\alpha u)(x),\ 4^m c(\alpha,\rho) M(u/d^m)(x)\} \\ &= G_{\alpha,\rho}(x), \text{ say,} \end{aligned} \tag{3.2}$$

where

$$c(\alpha, \rho) = \omega_n \sup_{z \in B(0,1)} |(D^\alpha \rho)(z)|.$$

Since, by Theorem 1.13, M maps $L^p(\mathbb{R}^n)$ boundedly into itself, it follows that $G_{\alpha,\rho} \in L^p(\mathbb{R}^n)$. Moreover, for each $\alpha \in \mathbb{N}_0^n$ with $|\alpha| \leqslant m$, $\rho^{(1/j)} * D^\alpha u \to D^\alpha u$ in $L^p(\Omega)$ as $j \to \infty$, and hence there is a subsequence $(j(k))$ of the sequence of all positive integers such that for all $\alpha \in \mathbb{N}_0^n$ with $|\alpha| \leqslant m$, $(\rho^{(1/j(k))} * D^\alpha u)(x) \to D^\alpha u(x)$ a.e. in Ω. Thus by (3.1),

$$D^\alpha(\rho^{(1/j(k))} * u^{(2/j(k))})(x) \to D^\alpha u(x) \tag{3.3}$$

a.e. in Ω. Together with Lebesgue's Dominated Convergence Theorem, (3.2) and (3.3) show that $\rho^{(1/j(k))} * u^{(2/j(k))} \to u$ in $W^{m,p}(\Omega)$; that is, $u \in W_0^{m,p}(\Omega)$. If Ω is unbounded, we simply apply the above argument to $u\phi_k$, where $\phi_k \in C_0^\infty(\mathbb{R}^n)$, $\phi_k(x) = 1$ if $|x| < k$, 0 if $|x| > 2k$, $0 \leqslant \phi_k \leqslant 1$, and note that $u\phi_k \to u$ in $W^{m,p}(\Omega)$. □

Remark 3.5. The proof of Theorem 3.4, due to D. J. Harris, uses the fact that the maximal function M maps $L^p(\mathbb{R}^n)$ boundedly into itself, and so the hypothesis that $p > 1$ is essential, since M does not map $L^1(\mathbb{R}^n)$ into itself. However, a related result holds even when $p = 1$, since then a Whitney decomposition of Ω into cubes may be used to establish it. This line of argument is due to C. Kenig, and we indicate the main steps in it for the case $k = 1$, the proof for $k > 1$ being similar.

By the Whitney Decomposition Theorem (Stein [1, pp. 167–170]), $\Omega = \bigcup_{j=1}^{\infty} Q_j$, where each Q_j is a closed cube with sides parallel to the coordinate axes, $Q_j^\circ \cap Q_k^\circ = \varnothing$ if $j \neq k$, and for each $j \in \mathbb{N}$,

$$\operatorname{diam} Q_j \leqslant \operatorname{dist}(Q_j, \mathbb{R}^n \backslash \Omega) \leqslant 4 \operatorname{diam} Q_j.$$

Let $\varepsilon_1 \in (0, \frac{1}{4})$, let $x^{(j)}$ be the centre of Q_j, let l_j be the length of each side of Q_j, and put $Q_j^* = (1+\varepsilon_1)(Q_j - x^{(j)}) + x^{(j)}$; then $Q_j \subset Q_j^*$ and the Q_j^* need not be pairwise disjoint. Let $\phi \in C_0^\infty(\mathbb{R}^n)$ be such that $0 \leqslant \phi \leqslant 1$, $\phi(x) = 1$ for all $x \in Q_0 := [-\frac{1}{2}, \frac{1}{2}]^n$, and $\phi(x) = 0$ for all $x \notin (1+\varepsilon_1)Q_0$; for each $j \in \mathbb{N}$ put $\phi_j(x) = \phi[(x - x^{(j)})/l_j]$ $(x \in \mathbb{R}^n)$. Then $\phi_j(x) = 1$ for all $x \in Q_j$, $\phi_j(x) = 0$ for all $x \notin Q_j^*$, and there is a constant A such that for all $x \in \mathbb{R}^n$, all $i \in \{1, 2, \ldots, n\}$ and all $j \in \mathbb{N}$,

$$|D_i \phi_j(x)| \leqslant A (\operatorname{diam} Q_j)^{-1}.$$

Put $\phi_j^*(x) = \phi_j(x)/\Phi(x)$, where $\Phi(x) = \Sigma_{i=1}^\infty \phi_i(x)$ $(x \in \Omega)$; then $\Sigma_{j=1}^\infty \phi_j^*(x) = 1$ for all $x \in \Omega$. It is also shown in Stein [1] that each point of Ω is contained in at most 12^n of the Q_j^*. Finally, put $d_j = \operatorname{dist}(Q_j, \mathbb{R}^n \backslash \Omega)$ and note that $\sqrt{n} l_j \leqslant d_j \leqslant 4\sqrt{n} l_j$ $(j \in \mathbb{N})$.

Now let $u \in W^{1,p}(\Omega)$ and suppose that $u/d \in L^p(\Omega)$, where $p \in [1, \infty)$. For each $x \in \Omega$ we have $u(x) = \sum_{j=1}^{\infty} u_j(x)$, where $u_j(x) = u(x)\phi_j^*(x)$; then given any $\varepsilon > 0$, we may write $u(x) = \sum_{d_j > \varepsilon} u_j(x) + \sum_{d_j \leqslant \varepsilon} u_j(x) := v(x) + w(x)$. Since there are points $y^{(j)} \in Q_j^*$ and $z^{(j)} \in Q_j$ such that $\operatorname{dist}(y^{(j)}, \mathbb{R}^n \setminus \Omega) = \operatorname{dist}(Q_j^*, \mathbb{R}^n \setminus \Omega)$ and $|z^{(j)} - y^{(j)}| \leqslant \frac{1}{2}\varepsilon_1 \sqrt{n} l_j$, it follows that $d_j \leqslant \operatorname{dist}(Q_j^*, \mathbb{R}^n \setminus \Omega) + \frac{1}{2}\varepsilon_1 \sqrt{n}\, l_j$, and so

$$\operatorname{dist}(Q_j^*, \mathbb{R}^n \setminus \Omega) \geqslant d_j - \tfrac{1}{2}\varepsilon_1 \sqrt{n}\, l_j \geqslant (1 - \tfrac{1}{2}\varepsilon_1) d_j \geqslant \tfrac{1}{2} d_j,$$

for all $j \in \mathbb{N}$. Since the series for v has a finite number of terms only, each term being a function with compact support in Ω, we have $v \in W_0^{1,p}(\Omega)$. To show that $u \in W_0^{1,p}(\Omega)$, it is thus sufficient to prove that $\|u - v\|_{1,p,\Omega} = \|w\|_{1,p,\Omega} \leqslant C(\varepsilon)$, where $C(\varepsilon) \to 0$ as $\varepsilon \to 0$.

Let $x \in \operatorname{supp} u_j$ for some $j \in \mathbb{N}$ with $d_j \leqslant \varepsilon$. Then $x \in Q_j^*$; thus

$$d(x) \leqslant d_j + (1 + \varepsilon_1)\sqrt{n}\, l_j \leqslant d_j(2 + \varepsilon_1) \leqslant \varepsilon(2 + \varepsilon_1) < 3\varepsilon.$$

Put $\Omega(\varepsilon) = \{x \in \Omega : d(x) < 3\varepsilon\}$. Then

$$\|w\|_{p,\Omega}^p = \int_{\Omega(\varepsilon)} \left| \sum_{d_j \leqslant \varepsilon} u_j \right|^p \leqslant \int_{\Omega(\varepsilon)} \left| \sum_{j=1}^{\infty} |u| \phi_j^* \right|^p = \int_{\Omega(\varepsilon)} |u|^p \to 0$$

as $\varepsilon \to 0$. To estimate $\|D_i w\|_{p,\Omega}$ note that

$$D_i w = (D_i u)\left(\sum_{d_j \leqslant \varepsilon} \phi_j^* \right) + u \sum_{d_j \leqslant \varepsilon} D_i \phi_j^*$$

and that

$$D_i \phi_j^* = (1/\Phi) D_i \phi_j - \phi_j (D_i \Phi)/\Phi^2.$$

If $x \in \operatorname{supp} \phi_j$ then $d(x) \leqslant d_j(2 + \varepsilon_1) \leqslant 3d_j$ and $|D_i \phi_j(x)| \leqslant A(\sqrt{n}\, l_j)^{-1} \leqslant 12A/d(x)$; also $\Phi(x)$ is a finite sum, with $D_i \Phi(x) = \sum_{m=1}^{\infty} D_i \phi_m(x)$; further, $d(x) \leqslant 3d_m$ and $|D_i \phi_m(x)| \leqslant 12A/d(x)$ if in addition $x \in \operatorname{supp} \phi_m$. Hence for all $x \in \Omega$,

$$|D_i w(x)| \leqslant$$

$$|D_i u(x)| \left| \sum_{d_j \leqslant \varepsilon} \phi_j^*(x) \right| + 12A|u(x)| \sum_{d_j \leqslant \varepsilon} \chi_{Q_j^*}(x) \left(1 + \sum_{m=1}^{\infty} \chi_{Q_m^*}(x) \right) \Big/ d(x),$$

from which it follows that

$$\|D_i w\|_{p,\Omega} \leqslant \left(\int_{\Omega(\varepsilon)} |D_i u|^p \right)^{1/p} + 12A(1 + 12^n) \left(\int_{\Omega(\varepsilon)} (|u|/d)^p \right)^{1/p}$$

$$\to 0 \quad \text{as } \varepsilon \to 0.$$

It is now clear that $u \in W_0^{1,p}(\Omega)$, as required.

If Ω is bounded and has a suitably smooth boundary, a converse of Theorem 3.4 has been established by Kadlec and Kufner [1], who show that if

$u \in W_0^{m,p}(\Omega)$, then $d^{-(m-|\alpha|)} D^\alpha u \in L^p(\Omega)$ for all $\alpha \in \mathbb{N}_0^n$ with $|\alpha| \leqslant m$, provided that $m \in \mathbb{N}$ and $p \in (1, \infty)$. We shall return to this point in Chapter X, §6.2.

3.2. *Embeddings and inequalities for* $W_0^{k,p}(\Omega)$

We now turn to embedding results involving the spaces $W_0^{k,p}(\Omega)$, results of great importance in the development of the theory.

Theorem 3.6. Suppose that $1 \leqslant p < n$, and put $p^* = np/(n-p)$ (the so-called *Sobolev conjugate* of p). Then $W_0^{1,p}(\Omega)$ is continuously embedded in $L^{p^*}(\Omega)$, and there is a constant C, depending only upon p and n, such that for all $u \in W_0^{k,p}(\Omega)$,

$$\|u\|_{p^*,\Omega} \leqslant C\|\nabla u\|_{p,\Omega},$$

where $\|\nabla u\|_{p,\Omega} = \left\| \left(\sum_{j=1}^n |D_j u|^2 \right)^{\frac{1}{2}} \right\|_{p,\Omega}$. If $q \in [1, p^*)$ and Ω is bounded, $W_0^{1,p}(\Omega)$ is compactly embedded in $L^q(\Omega)$. ■

Proof. First suppose that $u \in C_0^1(\Omega)$: we may clearly suppose that u is extended to be 0 outside Ω. Then for any $x = (x_1, \ldots, x_n) \in \Omega$,

$$u(x) = \int_{-\infty}^{x_1} D_1 u(\xi_1, x_2, \ldots, x_n)\, d\xi_1 = -\int_{x_1}^{\infty} D_1 u(\xi_1, x_2, \ldots, x_n)\, d\xi_1,$$

and hence

$$|u(x)| \leqslant \tfrac{1}{2} \int_{-\infty}^{\infty} |D_1 u(\xi_1, x_2, \ldots, x_n)|\, d\xi_1.$$

By use of similar estimates which involve integration over the other co-ordinates we find that

$$|2u(x)|^{n/(n-1)} \leqslant \left[\left(\int_{-\infty}^{\infty} |D_1 u(\xi_1, x_2, \ldots, x_n)|\, d\xi_1 \right) \cdots \left(\int_{-\infty}^{\infty} |D_n u(x_1, \ldots, x_{n-1}, \xi_n)|\, d\xi_n \right) \right]^{1/(n-1)}.$$

Integration of this with respect to x_1 and use of Hölder's inequality in the form

$$\int (v_1 v_2 \cdots v_{n-1})^{1/(n-1)} \leqslant \left(\int v_1\right)^{1/(n-1)} \cdots \left(\int v_{n-1}\right)^{1/(n-1)}$$

gives

$$\int_{-\infty}^{\infty} |2u(x_1, x_2, \ldots, x_n)|^{n/(n-1)}\, dx_1$$

$$\leqslant \left(\int_{-\infty}^{\infty} |D_1 u(x)|\, dx_1 \right)^{1/(n-1)} \prod_{j=2}^{n} \left(\int_{-\infty}^{\infty} \int_{-\infty}^{\infty} |D_j u|\, dx_j\, dx_1 \right)^{1/(n-1)}.$$

We now integrate successively with respect to $x_2, \ldots, x_n$, and note that each time precisely $n-1$ factors are specifically involved in the integration. Thus

$$\int_\Omega |2u(x)|^{n/(n-1)}\,\mathrm{d}x \leqslant \prod_{j=1}^{n} \left(\int_\Omega |\mathrm{D}_j u(x)|\,\mathrm{d}x \right)^{n/(n-1)};$$

that is,

$$\|u\|_{n/(n-1),\Omega} \leqslant \tfrac{1}{2} \prod_{j=1}^{n} \|\mathrm{D}_j u\|_{1,\Omega}^{1/n}. \tag{3.4}$$

Use of the arithmetic–geometric inequality now shows that

$$\begin{aligned} \|u\|_{n/(n-1),\Omega} &\leqslant (1/2n) \int_\Omega (|\mathrm{D}_1 u| + \cdots + |\mathrm{D}_n u|) \\ &\leqslant (1/2\sqrt{n}) \int_\Omega (|\mathrm{D}_1 u|^2 + \cdots + |\mathrm{D}_n u|^2)^{\frac{1}{2}} \\ &= (1/2\sqrt{n}) \|\nabla u\|_{1,\Omega}, \end{aligned} \tag{3.5}$$

the second inequality following from use of Schwarz's inequality in the form $\sum_1^n a_r \leqslant n^{\frac{1}{2}} \left(\sum_1^n a_r^2 \right)^{\frac{1}{2}}$. If $p = 1$, (3.5) is the inequality we want. We now concentrate on the case $p > 1$.

The next idea is to apply (3.5) to $v := |u|^{(n-1)p/(n-p)}$. Since $(n-1)p/(n-p) > 1$, the function v is in $\mathrm{C}_0^1(\Omega)$; and so this application is legitimate, giving

$$\begin{aligned} \left(\int_\Omega |u|^{p^*} \right)^{(n-1)/n} &\leqslant (1/2\sqrt{n}) \int_\Omega [(n-1)p/(n-p)] |u|^{n(p-1)/(n-p)} |\nabla u| \\ &\leqslant [(n-1)p/2(n-p)\sqrt{n}] \|\nabla u\|_{p,\Omega} \left(\int_\Omega |u|^{p^*} \right)^{1/p'}, \end{aligned}$$

the final step being a consequence of Hölder's inequality. Division of both sides by $\left(\int_\Omega |u|^{p^*} \right)^{1/p'}$ (assumed, without loss of generality, to be non-zero) now gives

$$\|u\|_{p^*,\Omega} \leqslant C \|\nabla u\|_{p,\Omega},$$

with

$$C = (n-1)p/2(n-p)\sqrt{n}.$$

To remove the condition that $u \in \mathrm{C}_0^1(\Omega)$, suppose that u merely belongs to $\mathrm{W}_0^{1,p}(\Omega)$. Then there is a sequence (u_m) in $\mathrm{C}_0^\infty(\Omega)$ which converges to u in $\mathrm{W}^{1,p}(\Omega)$. By what has been proved already, (u_m) is a Cauchy sequence in $\mathrm{L}^{p^*}(\Omega)$, and so converges, to U, say. There is a subsequence of (u_m) which converges pointwise a.e. to U and to u; hence $U = u \in \mathrm{L}^{p^*}(\Omega) \cap \mathrm{L}^p(\Omega)$. Now let $m \to \infty$ in $\|u_m\|_{p^*,\Omega} \leqslant C\|Du_m\|_{p,\Omega}$ to obtain the required inequality. This completes the proof of the first part of the theorem.

For the second part, suppose that Ω is bounded and that $1 \leqslant q < p^*$. Let $B = \{u \in C_0^1(\Omega): \|u\|_{1,p,\Omega} \leqslant 1\}$, let $\varepsilon > 0$, let ρ be a mollifier and let u_ε be the corresponding regularization of u; put $B_\varepsilon = \{u_\varepsilon : u \in B\}$. Then given any $u \in B$ and any $x \in \Omega$,

$$|u_\varepsilon(x)| \leqslant \int_{|y| \leqslant 1} \rho(y)|u(x-\varepsilon y)|\mathrm{d}y \leqslant \|u\|_{1,\Omega} \sup_{|y| \leqslant 1} \rho(y);$$

and for $j = 1, 2, \ldots, n$ and small enough $\varepsilon > 0$ we have, by Lemma 2.2,

$$|\mathrm{D}_j u_\varepsilon(x)| \leqslant \int_{|y| \leqslant 1} |\mathrm{D}_j u(x-\varepsilon y)| \; |\rho(y)|\mathrm{d}y \leqslant \|\mathrm{D}_j u\|_{1,\Omega} \sup_{|y| \leqslant 1} |\rho(y)|.$$

Thus B_ε is a bounded, equicontinuous subset of $C(\bar{\Omega})$; by the Ascoli–Arzelà theorem it is therefore relatively compact in $C(\bar{\Omega})$, and hence relatively compact in $L^1(\Omega)$. However, if $u \in B$,

$$\begin{aligned} |u(x) - u_\varepsilon(x)| &\leqslant \int_{|y| \leqslant 1} \rho(y)|u(x) - u(x-\varepsilon y)|\mathrm{d}y \\ &\leqslant \int_{|y| \leqslant 1} \rho(y) \int_0^{\varepsilon|y|} \left|\frac{\partial}{\partial r} u(x - r\omega)\right| \mathrm{d}r \, \mathrm{d}y, \end{aligned}$$

where $\omega = y/|y|$. Hence for all $u \in B$,

$$\begin{aligned} \|u - u_\varepsilon\|_{1,\Omega} &\leqslant \varepsilon \int_\Omega |\nabla u| \leqslant \varepsilon \|\nabla u\|_{p,\Omega} |\Omega|^{1/p'} \\ &\leqslant c\varepsilon |\Omega|^{1/p'}, \end{aligned}$$

where c $(\leqslant 1)$ is a positive constant independent of u and ε. Since B_ε is totally bounded in $L^1(\Omega)$, this last inequality shows that B is also totally bounded in $L^1(\Omega)$. This proves that the dense linear subspace $C_0^1(\Omega)$ of $W_0^{1,p}(\Omega)$ is compactly embedded in $L^1(\Omega)$, and hence that the final part of the theorem holds when $q = 1$.

To deal with $q > 1$ we use Hölder's inequality and the first part of the theorem to show that for all $u \in W_0^{1,p}(\Omega)$ we have, with $\lambda + (1-\lambda)(p^*)^{-1} = q^{-1}$, $0 < \lambda < 1$,

$$\|u\|_{q,\Omega} = \left(\int_\Omega |u|^{q\lambda + (1-\lambda)q}\right)^{1/q} \leqslant \|u\|_{1,\Omega}^\lambda \|u\|_{p^*}^{1-\lambda} \leqslant \|u\|_{1,\Omega}^\lambda (C\|\nabla u\|_{p,\Omega})^{1-\lambda}.$$

It follows immediately that any bounded subset of $W_0^{k,p}(\Omega)$ is relatively compact when viewed as a subset of $L^q(\Omega)$, and the proof is complete. □

Extensions of this theorem may be made to higher-order Sobolev spaces. Thus if $1 \leqslant p < \frac{1}{2}n$, then by Theorem 3.6, if $u \in W_0^{2,p}(\Omega)$ then $u \in W_0^{1,p^*}(\Omega)$, which in turn implies that $u \in L^{np/(n-2p)}(\Omega)$; moreover, since $u \in L^p(\Omega) \cap L^{np/(n-2p)}(\Omega)$ an application of Hölder's inequality shows that $u \in L^s(\Omega)$ for all $s \in [p, np/(n-2p)]$. We summarize the results which can be obtained by this straightforward process of iteration as follows:

Theorem 3.7. Suppose that $p \geqslant 1$, $k \in \mathbb{N}$, $kp < n$ and $s \in [p, np/(n-kp)]$. Then $W_0^{k,p}(\Omega)$ is continuously embedded in $L^s(\Omega)$. If, in addition, Ω is bounded and $s < np/(n-kp)$, then the embedding is compact. ■

Note that Theorem 3.6 cannot be extended to allow $p = n$ (a similar remark applies to Theorem 3.7). For if this were the case it would imply that every element of $W_0^{1,n}(\Omega)$ is in $L^\infty(\Omega)$. However, the function defined by

$$u(x) = \begin{cases} \log\log(1+|x|^{-1}) - \log\log 2 & \text{if } 0 < |x| < 1, \\ 0 & \text{if } |x| \geqslant 1, \end{cases}$$

is in $W_0^{1,n}(\mathbb{R}^n)$: in fact

$$\begin{aligned} \int_{|x|<1} |\nabla u|^n &= \omega_n \int_0^1 [(r+r^2)\log(1+r^{-1})]^{-n} r^{n-1}\, dr \\ &\leqslant \omega_n \int_0^1 (r^2+r)^{-1}[\log(1+r^{-1})]^{-n}\, dr \\ &= \omega_n (n-1)^{-1} \int_0^1 \frac{d}{dr}[\log(1+r^{-1})]^{1-n}\, dr \\ &< \infty. \end{aligned}$$

However, $\operatorname{ess\,sup}|u(x)| = \infty$, and so the theorem does not hold when $p = n > 1$.

Another very useful inequality is that provided by the following theorem: in it, and subsequently, we shall employ the notation

$$\|\nabla^m u\|_{p,\Omega} = \left(\sum_{|\alpha|=m} \|D^\alpha u\|_{p,\Omega}^p \right)^{1/p}.$$

Theorem 3.8. Let $m, j \in \mathbb{N}_0$, with $0 \leqslant j < m$, let $q, r \in [1, \infty]$ and let p be defined by

$$1/p = 1/q + (j/m)(1/r - 1/q).$$

Then there is a constant C, depending only on j, m, n, q, and r, such that for all $u \in C_0^m(\mathbb{R}^n)$,

$$\|\nabla^j u\|_{p,\mathbb{R}^n} \leqslant C \|\nabla^m u\|_{r,\mathbb{R}^n}^{j/m} \|u\|_{q,\mathbb{R}^n}^{(m-j)/m}. \tag{3.6}$$ ■

Proof. We first suppose that $j = 1$, $m = 2$, and $n = 1$, so that $2/p = 1/q + 1/r$. We shall assume momentarily that $1 \leqslant q < \infty$ and $1 < r < \infty$. Let $a, b \in \mathbb{R}$, with $a < b$, put $l = \frac{1}{4}(b-a)$, and let $x_1 \in (a, a+l)$ and $x_2 \in (a+3l, b)$. By the Mean-Value Theorem, there exists $\xi \in (x_1, x_2)$ such that

$$u(x_2) - u(x_1) = (x_2 - x_1)u'(\xi).$$

Moreover, for all $x \in (a, b)$,

$$u'(x) = u'(\xi) + \int_\xi^x u''(y)\,\mathrm{d}y;$$

these two equalities give

$$|u'(x)| \leqslant \frac{|u(x_1)| + |u(x_2)|}{2l} + \int_a^b |u''(y)|\,\mathrm{d}y.$$

Now integrate with respect to x_1 over $(a, a+l)$, and with respect to x_2 over $(a+3l, b)$: we obtain

$$l^2|u'(x)| \leqslant \tfrac{1}{2}\int_a^b |u(y)|\,\mathrm{d}y + l^2\int_a^b |u''(y)|\,\mathrm{d}y.$$

To this inequality is applied Hölder's inequality in the form $A + B \leqslant (A^p + B^p)^{1/p}2^{1/p'}$, that is, $(A+B)^p \leqslant 2^{p-1}\,(A^p + B^p)$; thus

$$l^{2p}|u'(x)|^p \leqslant 2^{p-1}\left[2^{-p}\left(\int_a^b |u(y)|\,\mathrm{d}y\right)^p + l^{2p}\left(\int_a^b |u''(y)|\,\mathrm{d}y\right)^p\right].$$

Hence

$$2|u'(x)|^p \leqslant l^{-2p}\left(\int_a^b |u(y)|\,\mathrm{d}y\right)^p + 2^p\left(\int_a^b |u''(y)|\,\mathrm{d}y\right)^p$$

and

$$\int_a^b |u'(x)|^p\,\mathrm{d}x$$
$$\leqslant c^p l^{1+p-p/r}\left(\int_a^b |u''(y)|^r\,\mathrm{d}y\right)^{p/r} + c^p l^{-(1+p-p/r)}\left(\int_a^b |u(y)|^q\,\mathrm{d}y\right)^{p/q}, \tag{3.7}$$

the final step following from Hölder's inequality. Here c is a constant: it can be taken to be 64, although this is not the best possible value.

To establish (3.6) for our special case it will evidently be enough to prove that for any $L > 0$,

$$\int_0^L |u'(x)|^p\,\mathrm{d}x \leqslant 2c^p\left(\int_0^\infty |u''(y)|^r\,\mathrm{d}y\right)^{p/2r}\left(\int_0^\infty |u(y)|^q\,\mathrm{d}y\right)^{p/2q}. \tag{3.8}$$

To do this, first observe that if u'' is identically zero on $(0, \infty)$ then so is u', since u has compact support; and hence (3.8) holds. We may therefore suppose that

u'' is not identically zero on $(0, \infty)$. Let $k \in \mathbb{N}$, with $k > 1$, and apply (3.7) with $a = 0$ and $b = L/k$. If the first term on the right-hand side of (3.7) is greater than the second term, we write $I_1 = [0, L/k]$; if not, we consider intervals of the form $[0, \delta + L/k], \delta > 0$, and increase δ until the two terms are equal, say when $\delta = \delta_1$, and set $I_1 = [0, \delta_1 + L/k]$. We thus have

$$\int_{I_1} |u'(x)|^p \, dx$$
$$\leqslant \begin{cases} 2c^p (L/k)^{1+p-p/r} \left(\int_0^L |u''|^r \, dy \right)^{p/r} & \text{if } I_1 = [0, L/k], \\ 2c^p \left(\int_{I_1} |u''(y)|^r \, dy \right)^{p/2r} \left(\int_{I_1} |u(y)|^q \, dy \right)^{p/2q} & \text{otherwise.} \end{cases}$$

If the length $l(I_1)$ of I_1 is such that $l(I_1) \geqslant L$, then (3.8) follows immediately. If $l(I_1) < L$, we repeat this process, starting at the right-hand end-point of I_1, and construct an interval I_2. In this way we form at most k intervals $I_1, I_2, \ldots,$ until $[0, L]$ is covered; the last I_i is such that $l(I_1) + \cdots + l(I_i) \geqslant L > l(I_1) + \cdots + l(I_{i-1})$. It follows that

$$\begin{aligned} \int_0^L |u'(x)|^p \, dx &\leqslant \sum_j \int_{I_j} |u'(x)|^p \, dx \\ &\leqslant 2kc^p (L/k)^{1+p-p/r} \left(\int_0^L |u''(y)|^r \, dy \right)^{p/r} \\ &\quad + 2c^p \sum_j \left(\int_{I_j} |u''(y)|^r \, dy \right)^{p/2r} \left(\int_{I_j} |u(y)|^q \, dy \right)^{p/2q} \\ &\leqslant 2kc^p (L/k)^{1+p-p/r} \left(\int_0^L |u''(y)|^r \, dy \right)^{p/r} \\ &\quad + 2c^p \left(\int_0^\infty |u''(y)|^r \, dy \right)^{p/2r} \left(\int_0^\infty |u(y)|^q \, dy \right)^{p/2q}, \end{aligned} \tag{3.9}$$

the final step being a consequence of Hölder's inequality and the fact that $p/2r + p/2q = 1$. Since $r > 1$ we may let $k \to \infty$ in (3.9) to obtain (3.8), and hence (3.6).

To establish (3.6) when n is arbitrary, we apply the result just proved and obtain, for $i = 1, 2, \ldots, n$,

$$\int |D_i u(x)|^p \, dx_i \leqslant C^p \left(\int |D_i^2 u(x)|^r \, dx_i \right)^{p/2r} \left(\int |u(x)|^q \, dx_i \right)^{p/2q}.$$

Note that the constant C which appears here is independent of q and r, and hence of p.

Integration of this with respect to the other variables, together with use of

Hölder's inequality, gives

$$\int_{\mathbb{R}^n} |D_i u(x)|^p \, dx \leqslant C^p \left(\int_{\mathbb{R}^n} |D_i^2 u(x)|^r \, dx \right)^{p/2r} \left(\int_{\mathbb{R}^n} |u(x)|^q \, dx \right)^{p/2q}, \tag{3.10}$$

which is (3.6), in this case.

This completes the proof of (3.6) when $j = 1$ and $m = 2$, provided that $1 \leqslant q < \infty$ and $1 < r < \infty$. To deal with the exceptional cases $q = \infty$ and $r = 1$ or ∞, again with $j = 1$ and $m = 2$, it is enough to let $q \to \infty$ and r tend to 1 or ∞ in (3.10). All that now remains is to lift the restrictions that $j = 1$ and $m = 2$, and merely to require that $0 \leqslant j < m$. If $j = 0$ the result is obvious; to deal with the general case we use induction on m and on j as follows, denoting all constants by C for simplicity. First we establish (3.6) when $j = m - 1$. There is nothing to prove when $m = 1$; suppose that (3.6) holds when $m = k$ and $j = k - 1$. Let $u \in C_0^{k+1}(\mathbb{R}^n)$, let $\alpha \in \mathbb{N}_0^n$ be such that $|\alpha| = k - 1$, and apply (3.6) with $j = 1$ and $m = 2$ to $D^\alpha u$. With $p(k+1)$ and $\tilde{q}$ defined by

$$1/p(k+1) = 1/q + k(1/r - 1/q)/(k+1),$$
$$1/\tilde{q} = 2q^{-1}/(k+1) + r^{-1}(k+1)/(k+1)$$

this gives

$$\|\nabla D^\alpha u\|_{p(k+1), \mathbb{R}^n} \leqslant C \|\nabla^2 D^\alpha u\|_{r, \mathbb{R}^n}^{\frac{1}{2}} \|D^\alpha u\|_{\tilde{q}, \mathbb{R}^n}^{\frac{1}{2}},$$

so that

$$\|\nabla^k u\|_{p(k+1), \mathbb{R}^n} \leqslant C \|\nabla^{k+1} u\|_{r, \mathbb{R}^n}^{\frac{1}{2}} \|\nabla^{k-1} u\|_{\tilde{q}, \mathbb{R}^n}^{\frac{1}{2}}.$$

Use of the inductive hypothesis now shows that

$$\|\nabla^k u\|_{p(k+1), \mathbb{R}^n} \leqslant C \|\nabla^{k+1} u\|_{r, \mathbb{R}^n}^{\frac{1}{2}} \|\nabla^k u\|_{p(k+1), \mathbb{R}^n}^{1/2k'} \|u\|_{q, \mathbb{R}^n}^{1/2k},$$

from which we have (3.6) with $m = k + 1$ and $j = k$. Thus (3.6) holds with $j = m - 1$, for all $m \in \mathbb{N}$.

To complete the proof we use backward induction on j, for fixed m. Assume that (3.6) holds for some j with $2 \leqslant j \leqslant m - 1$, define $\tilde{p}(j)$ by $1/\tilde{p}(j) = 1/q + (j-1)(1/r - 1/q)m$ and let $u \in C_0^m(\mathbb{R}^n)$. Then by what has just been proved,

$$\|\nabla^{j-1} u\|_{\tilde{p}(j-1), \mathbb{R}^n} \leqslant C \|\nabla^j u\|_{\tilde{p}(j), \mathbb{R}^n}^{(j-1)/j} \|u\|_{q, \mathbb{R}^n}^{1/j},$$

so that with the aid of the inductive hypothesis we have

$$\|\nabla^{j-1} u\|_{\tilde{p}(j-1), \mathbb{R}^n} \leqslant C \|\nabla^m u\|_{r, \mathbb{R}^n}^{(j-1)/m} \|u\|_{q, \mathbb{R}^n}^{(m-j+1)/m},$$

which is (3.6) with j replaced by $j - 1$. The proof is complete. □

Remark 3.9. The inequality (3.6) also holds for any $u \in W_0^{m,r}(\Omega) \cap L^q(\Omega)$, where Ω is any open subset of $\mathbb{R}^n$ and $q, r < \infty$. This is because, given any such u, there is a sequence $(u^{(k)})$ in $C_0^\infty(\Omega)$ (formed by regularization) which converges to u in $W^{m,r}(\Omega)$ and in $L^q(\Omega)$. Application of (3.6) to $u^{(k)} - u^{(l)}$, with $l > k$, shows that $\|\nabla^j u^{(k)} - \nabla^j u^{(l)}\|_{p,\Omega} \leqslant \varepsilon_k \to 0$ as $k \to \infty$; and since there is a subsequence of $(\nabla^j u^{(l)})$ which converges a.e. on Ω to $\nabla^j u$, Fatou's lemma shows that $\|\nabla^j u^{(k)} - \nabla^j u\|_{p,\Omega} \leqslant \varepsilon_k$, and hence $\|\nabla^j u^{(k)}\|_{p,\Omega} \to \|\nabla^j u\|_{p,\Omega}$ as $k \to \infty$. We may now pass to the limit, as $k \to \infty$, in (3.6) applied to $u^{(k)}$, and the result follows.

3.3. *More embeddings: continuity properties.*

We now direct our attention to circumstances in which it is possible to say that certain Sobolev spaces may be embedded in spaces other than Lebesgue spaces, such as spaces of continuous functions. To help to set the scene for this we first point out certain absolute continuity properties which elements of Sobolev spaces possess, and to do this a little notation will be useful.

Let $a, b \in \mathbb{R}^n$, $a \neq b$, and let L be the line through a and b: $L = \{ta + (1-t)b : t \in \mathbb{R}\}$. Suppose that $L \cap \Omega \neq \varnothing$. Then there is a (finite or infinite) sequence of open, pairwise-disjoint intervals (J_i) in $\mathbb{R}$ such that

$$L \cap \Omega = \bigcup_i \{ta + (1-t)b : t \in J_i\}.$$

A (real- or complex-valued) function u on Ω is said to be *absolutely continuous on L* if, and only if, the function $t \mapsto u(ta + (1-t)b)$ is absolutely continuous on each compact subinterval of J_i, for each i. Given any $x \in \mathbb{R}^n$ let $L(x^{(i)})$ be the line $\{(x_1, \ldots, x_{i-1}, t, x_{i+1}, \ldots, x_n) : t \in \mathbb{R}\}$, and set $x_i' = (x_1, \ldots, x_{i-1}, x_{i+1}, \ldots, x_n) \in \mathbb{R}^{n-1}$. For $i = 1, 2, \ldots, n$ we define $AC_i(\Omega)$ to be the family of all (real- or complex-valued) functions u on Ω such that $M := \{x_i' : L(x^{(i)}) \cap \Omega \neq \varnothing$ and u is not absolutely continuous on $L(x^{(i)})\}$ has $(n-1)$-dimensional Lebesgue measure zero. We denote $\cap_{i=1}^n AC_i(\Omega)$ by $AC(\Omega)$ and $AC_{loc}(\Omega)$ stands for the set of functions in $AC(K)$ for every compact subset K of Ω. Of course, if $u \in AC_i(\Omega)$ then it has a classical derivative with respect to x_i almost everywhere in Ω, since u is absolutely continuous for almost all lines $L(x^{(i)})$; we shall denote this classical derivative by $\partial u/\partial x_i$ to distinguish it from any weak derivative $D_i u$ which it may possess in Ω. These two derivatives are related, under mild hypotheses, as we see from the following result.

Proposition 3.10. Let $u \in L^1_{loc}(\Omega) \cap AC_i(\Omega)$ and suppose that $\partial u/\partial x_i \in L^1_{loc}(\Omega)$. Then $\partial u/\partial x_i$ coincides with the weak derivative $D_i u$ on Ω. ■

Proof. Given any $\phi \in C_0^\infty(\Omega)$ we have, by Fubini's theorem,

$$\int_\Omega u(x) D_i\phi(x)\,dx = \int_{\mathbb{R}^{n-1}} \left(\int_{L(x^{(i)})} u(x) D_i\phi(x)\,dx_i \right) dx_1 \ldots dx_{i-1}\,dx_{i+1} \ldots dx_n$$

$$= -\int_{\mathbb{R}^{n-1}} \left(\int_{L(x^{(i)})} \frac{\partial u}{\partial x_i}(x)\,\phi(x)\,dx_i \right) dx_1 \ldots dx_{i-1}\,dx_{i+1} \ldots dx_n$$

$$= -\int_\Omega \frac{\partial u}{\partial x_i}(x)\phi(x)\,dx,$$

and the result follows. □

Theorem 3.11. Let $u \in L^1_{loc}(\Omega)$ and suppose that u has a weak derivative $D_i u$ on Ω (which, by definition, is in $L^1_{loc}(\Omega)$). Then there is a function $\tilde{u} \in AC_i(\Omega)$ which is equal a.e. on Ω to u, and has classical derivative $\partial\tilde{u}/\partial x_i$ equal a.e. on Ω to $D_i u$. ■

Proof. We extend u to the whole of $\mathbb{R}^n$ by setting it equal to zero in $\mathbb{R}^n \setminus \Omega$. Let $(K_m)_{m\in\mathbb{N}}$ be a sequence of compact subsets of Ω, with union Ω and with $K_m \subset K_{m+1}$ for all $m \in \mathbb{N}$. For each $m \in \mathbb{N}$ let $\phi_m \in C_0^\infty(\Omega)$, with $\phi_m(x) = 1$ for all $x \in K_m$; and write $u_m = u\phi_m$. Clearly $u_m \in L^1(\Omega)$, $D_i u_m \in L^1(\Omega)$, and for all $x \in K_m$ we have $u_m(x) = u(x)$ and $D_i u_m(x) = D_i u(x)$. Now define a function U_m by

$$U_m(x) = \int_{-\infty}^{x_i} D_i u_m(x_1, \ldots, x_{i-1}, t, x_{i+1}, \ldots, x_n)\,dt$$

for all x such that

$$\int_{-\infty}^{\infty} |D_i u_m(x_1, \ldots, x_{i-1}, t, x_{i+1}, \ldots, x_n)|\,dt < \infty;$$

in other words, for all $x_i' \in \mathbb{R}^{n-1}$ save possibly those in a set of $(n-1)$-dimensional Lebesgue measure zero. It is clear that $U_m \in AC_i(\Omega)$ for each $m \in \mathbb{N}$; moreover, we claim that $U_m(x) = u_m(x)$ for almost all $x \in \Omega$. To justify this claim, observe that by Lemma 2.3, there is a sequence $(u_m^{(k)})_{k\in\mathbb{N}}$ in $C_0^\infty(\Omega)$ such that as $k \to \infty$,

$$\|u_m^{(k)} - u_m\|_{1,\Omega} \to 0 \quad \text{and} \quad \|D_i u_m^{(k)} - D_i u_m\|_{1,\Omega} \to 0.$$

Hence

$$\|u_m - U_m\|_{1,\Omega} \leqslant \lim_{k\to\infty} \|u_m - u_m^{(k)}\|_{1,\Omega} + \limsup_{k\to\infty} \|u_m^{(k)} - U_m\|_{1,\Omega}$$

$$= \limsup_{k\to\infty} \|u_m^{(k)} - U_m\|_{1,\Omega}.$$

But

$$|u_m^{(k)}(x) - U_m(x)| = \left| \int_{-\infty}^{x_i} [D_i u_m^{(k)}(x_1, \ldots, x_{i-1}, t, x_{i+1}, \ldots, x_n) - D_i u_m(x_1, \ldots, x_{i-1}, t, x_{i+1}, \ldots, x_n)] \, dt \right|$$

and thus

$$\limsup_{k \to \infty} \|u_m^{(k)} - U_m\|_{1,\Omega} \leqslant 2(\operatorname{diam} K_m) \lim_{k \to \infty} \|D_i u_m^{(k)} - D_i u_m\|_{1,\Omega} = 0.$$

It follows that $\|u_m - U_m\|_{1,\Omega} = 0$, and our claim is established.

To conclude the proof, define $\tilde{u}$ by $\tilde{u}(x) = U_m(x)$ if $x \in K_m$ $(m \in \mathbb{N})$. Then $\tilde{u} \in AC_i(\Omega)$, $\tilde{u}$ is equal a.e. on Ω to u; and by Proposition 3.10, $\partial\tilde{u}/\partial x_i$ is equal a.e. on Ω to $D_i u$. □

Corollary 3.12. Let $a, b \in \mathbb{R}$, $a < b$, $I = (a, b)$ and $p \in [1, \infty]$, and suppose that $u \in W^{1,p}(I)$. Then u is absolutely continuous on I; that is, there is a representative in the equivalence class u which is absolutely continuous on I. More generally, if $m \in \mathbb{N}$ and $u \in W^{m,p}(I)$, then $u^{(m-1)}$ is absolutely continuous on I. ■

We now continue our study of the Sobolev space $W_0^{1,p}(\Omega)$, and since the case $p < n$ has already been dealt with by Theorem 3.6, we concentrate on the situation when $p \geqslant n$. The consideration of integrals of potential type will be of great assistance to us, and we turn to this.

Given any $\mu \in (0, 1)$ and any suitable function f on a *bounded*, open subset Ω of $\mathbb{R}^n$, define a function $V_\mu f$ by

$$(V_\mu f)(x) = \int_\Omega |x - y|^{-(1-\mu)n} f(y) \, dy \qquad (x \in \Omega).$$

Lemma 3.13. Let $p, q \in [1, \infty]$ and suppose that $0 \leqslant \delta := p^{-1} - q^{-1} < \mu$. Then $V_\mu \in \mathscr{B}[L^p(\Omega), L^q(\Omega)]$ and

$$\|V_\mu\| \leqslant [(1-\delta)/(\mu-\delta)]^{1-\delta} \omega_n^{1-\mu} |\Omega|^{\mu-\delta}. \qquad ■$$

Proof. Define $r(\geqslant 1)$ by

$$1/r = 1 + 1/q - 1/p = 1 - \delta,$$

and write $h(x) = |x|^{-(1-\mu)n}$. Then given any $f \in L^p(\Omega)$, Hölder's inequality shows that

$$|(V_\mu f)(x)| \leqslant \int_\Omega [h(x-y)]^{r/q} [h(x-y)]^{r(1-p^{-1})} |f(y)|^{p/q} |f(y)|^{p\delta} \, dy$$

$$\leqslant \left(\int_\Omega [h(x-y)]^r |f(y)|^p \, dy \right)^{1/q} \left(\int_\Omega [h(x-y)]^r \, dy \right)^{1-p^{-1}} \left(\int_\Omega |f(y)|^p \, dy \right)^{\delta}.$$

Hence

$$\|V_\mu f\|_{q,\Omega} \leqslant \|f\|_{p,\Omega} \sup_{x\in\Omega}\left(\int_\Omega [h(x-y)]^r \, dy\right)^{1/r}.$$

To estimate the integral on the right-hand side let $R > 0$ be so chosen that $|\Omega| = |B(x,R)| = \omega_n R^n$. Then for all $x \in \Omega$ (see Lemma 5.14),

$$\int_\Omega [h(x-y)]^r \, dy \leqslant \int_{B(x,R)} |x-y|^{-(1-\mu)nr} \, dy$$
$$= n^{-1}[(1-\delta)/(\mu-\delta)]\,\omega_n{}^{(1-\mu)/(1-\delta)}|\Omega|^{(\mu-\delta)/(1-\delta)}.$$

The result follows. □

The next two lemmas enable us to show that $W_0^{1,n}(\Omega)$ is embedded in an Orlicz space.

Lemma 3.14. Let $p \in (1,\infty)$, let Ω be bounded, let $f \in L^p(\Omega)$, and put $g = V_{1/p} f$. Then there are positive constants c_i $(i = 1, 2)$, depending only upon n and p, such that

$$\int_\Omega \exp(|g|/c_1 \|f\|_{p,\Omega})^{p'} \, dx \leqslant c_2 |\Omega|.$$ ■

Proof. Lemma 3.13 shows that if $p \leqslant q < \infty$, then

$$\int_\Omega |g(x)|^q \, dx \leqslant q^{1+q/p'} \omega_n^{q/p'} |\Omega| \, \|f\|_{p,\Omega}^q.$$

Then if $q \geqslant p-1$, so that $p'q \geqslant p$, we have

$$\int_\Omega |g(x)|^{p'q} \, dx \leqslant p'q\,(\omega_n p'q \|f\|_{p,\Omega}^{p'})^q |\Omega|.$$

It follows that if we let n_0 be the integer part of p, then for any $m \in \mathbb{N}$, and any $c_1 > 0$,

$$\int_\Omega \sum_{k=n_0}^{m} \frac{1}{k!}\left(\frac{|g(x)|}{c_1\|f\|_{p,\Omega}}\right)^{p'k} dx \leqslant p'|\Omega| \sum_{k=n_0}^{\infty}\left(\frac{p'\omega_n}{c_1^{p'}}\right)^k \frac{k^k}{(k-1)!}.$$

The series on the right-hand side converges if $c_1^{p'} > e\omega_n p'$. We make a choice of c_1 so that this inequality holds, and use the Monotone Convergence Theorem to show that m may be replaced by infinity on the left-hand side. This gives the desired result, save that the first few terms in the series expansion of the exponential are missing. However, these terms can be estimated by means of

Hölder's inequality in terms of $|\Omega|$: thus if $k < p-1$, then

$$\int_\Omega |g|^{p'k} \leqslant \left(\int_\Omega |g|^{p'(p-1)}\right)^{k/(p-1)} |\Omega|^{1-k/(p-1)} \leqslant \text{const.}\,|\Omega|.$$

The proof is complete. □

Lemma 3.15. Let Ω be bounded and let $u \in W_0^{1,1}(\Omega)$. Then for almost all x in Ω,

$$u(x) = \frac{1}{n\omega_n}\int_\Omega |x-y|^{-n} \sum_{i=1}^{n} (x_i - y_i) D_i u(y)\, dy \tag{3.11}$$

and

$$|u(x)| \leqslant (V_{1/n}|\nabla u|)(x)/n\omega_n. \tag{3.12}$$ ■

Proof. First suppose that $u \in C_0^1(\Omega)$, so that we may and shall assume that u is defined on the whole of $\mathbb{R}^n$ and is zero outside Ω. Then for all $x \in \Omega$, and any $\omega \in \mathbb{R}^n$ with $|\omega| = 1$

$$u(x) = -\int_0^\infty \frac{\partial}{\partial r} u(x + r\omega)\, dr.$$

Thus

$$n\omega_n u(x) = -\int_0^\infty \int_{|\omega|=1} \frac{\partial}{\partial r} u(x + r\omega)\, d\omega\, dr$$

$$= \int_\Omega |x-y|^{-n} \sum_{i=1}^{n} (x_i - y_i) D_i u(y)\, dy,$$

as required. To deal with a general $u \in W_0^{1,1}(\Omega)$, let (u_m) be a sequence in $C_0^\infty(\Omega)$ which converges to u in $W_0^{1,1}(\Omega)$, and observe that by Lemma 3.13, $V_{1/n} \in \mathscr{B}(L^1(\Omega))$. This implies that since $D_i u_m \to D_i u$ in $L^1(\Omega)$ as $m \to \infty$ $(i = 1, \ldots, n)$, the sequence of functions

$$\left(x \mapsto \int_\Omega |x-y|^{-n} \sum_{i=1}^{n} (x_i - y_i) D_i u_m(y)\, dy\right)_{m \in \mathbb{N}}$$

converges to

$$x \mapsto \int_\Omega |x-y|^{-n} \sum_{i=1}^{n} (x_i - y_i) D_i u(y)\, dy$$

in $L^1(\Omega)$ as $m \to \infty$. Hence there is a subsequence which converges pointwise a.e. in Ω to this limiting function, and the proof of (3.11) is complete. That (3.12) holds is now clear. □

Now let $u \in W_0^{1,n}(\Omega)$. Then by Lemma 3.14 and (3.12) we see that there are positive constants c_1 and c_2, which depend only upon n, such that

$$\int_\Omega \exp\left[|u|/(c_1 \|\nabla u\|_{n,\Omega})\right]^{n'} \leqslant c_2 |\Omega|. \tag{3.13}$$

This significant result can be interpreted by means of Orlicz spaces, which may be viewed as generalizations of the L^p spaces. An *Orlicz function* is a non-negative, continuous, convex function ϕ on $[0, \infty)$, with $\lim_{t \to 0+} \phi(t)/t = 0$ (so that $\phi(0) = 0$) and $\lim_{t \to \infty} \phi(t)/t = \infty$. The *Orlicz class* $L_\phi(\Omega)$ is the set of all Lebesgue-measurable functions (or, more precisely, equivalence classes of functions, functions equal a.e. on Ω being identified) $u: \Omega \to \mathbb{R}$ such that

$$\int_\Omega \phi(|u(x)|)\,dx < \infty.$$

The *Orlicz space* $L^\phi(\Omega)$ is the linear span of $L_\phi(\Omega)$, endowed with the so-called Luxembourg norm given by

$$\|u\|_{\phi,\Omega} := \inf\{\lambda > 0 : \int_\Omega \phi(|u(x)|/\lambda)\,dx \leqslant 1\}.$$

It can be shown that $L^\phi(\Omega)$ is a Banach space and that if Ω is bounded, $L^\infty(\Omega) \subset L^\phi(\Omega) \subset L^1(\Omega)$; in general, it is neither reflexive nor separable. The next theorem, due to Trudinger [1], follows from (3.13) immediately.

Theorem 3.16. Let $n > 1$, suppose that Ω is bounded, and define ϕ by $\phi(t) = \exp(t^{n/(n-1)} - 1)$ for all $t \geqslant 0$. Then $W_0^{1,n}(\Omega)$ is continuously embedded in $L^\phi(\Omega)$, and thus also in $L^q(\Omega)$ for every $q \in [1, \infty)$. ■

An extension of this result to higher-order Sobolev spaces may easily be obtained. Thus if $u \in W_0^{k,1}(\Omega)$, (3.12) may be generalized without difficulty to give

$$(k-1)!\, n\omega_n |u(x)| \leqslant V_{k/n} |D^k u(x)| \tag{3.14}$$

if $1 \leqslant k < n$. This, together with Lemma 3.14, can be used to show that there are positive constants c_1, c_2 (depending only upon n and p) such that if $u \in W_0^{k,p}(\Omega)$ and $n = kp$, then

$$\int_\Omega \exp\left(\frac{|u(x)|}{c_1 \|\nabla^k u\|_{p,\Omega}}\right)^{p'} dx \leqslant c_2 |\Omega|.$$

This leads immediately to

Theorem 3.17. Let Ω be bounded, suppose $k \in \mathbb{N}$ is such that $k < n$, and define ϕ by $\phi(t) = \exp(t^{n/(n-k)} - 1)$ for all $t \geqslant 0$. Then $W_0^{k,n/k}(\Omega)$ is continuously embedded in $L^\phi(\Omega)$. ■

Conditions on an Orlicz function ϕ sufficient to ensure that $W_0^{k,n/k}(\Omega)$ is compactly embedded in $L^\phi(\Omega)$ can be given, but we postpone any further discussion of this matter to §6 which deals with embeddings and approximation numbers.

Having dealt with $W_0^{1,n}(\Omega)$ we next turn to $W_0^{1,p}(\Omega)$ when $p > n$. In this case the elements of the space can be identified with Hölder-continuous functions.

Theorem 3.18. Let Ω be bounded and suppose that $n < p < \infty$. Then $W_0^{1,p}(\Omega)$ is continuously embedded in $C^{0,\gamma}(\bar{\Omega})$, where $\gamma = 1 - n/p$. ■

Proof. Lemma 3.13 with $q = \infty$ and $\mu = n^{-1}$ shows that if $u \in L^p(\Omega)$, then

$$\| V_{1/n} u \|_{\infty,\Omega} \leqslant [(1-p^{-1})/(n^{-1}-p^{-1})]^{1/p'} \omega_n^{1/n'} |\Omega|^{n^{-1}-p^{-1}} \|u\|_{p,\Omega};$$

while Lemma 3.15 tells us that if $u \in W_0^{1,1}(\Omega)$, then for almost all x in Ω,

$$|u(x)| \leqslant n^{-1} \omega_n^{-1} (V_{1/n} |\nabla u|)(x).$$

Thus if $u \in W_0^{1,p}(\Omega)$, then

$$\|u\|_{\infty,\Omega} \leqslant C |\Omega|^{n^{-1}-p^{-1}} \|\nabla u\|_{p,\Omega}, \tag{3.15}$$

where

$$C = n^{-1} [(1-p^{-1})/(n^{-1}-p^{-1})]^{1/p'} \omega_n^{-1/n}.$$

Given any $u \in W_0^{1,p}(\Omega)$, there is a sequence (u_m) in $C_0^\infty(\mathbb{R}^n)$ which converges to u in $W_0^{1,p}(\Omega)$. By (3.15), (u_m) is a Cauchy sequence in $C(\bar{\Omega})$ and hence converges to an element of $C(\bar{\Omega})$. This shows that u may be identified, after a possible change on a set of measure zero, with a continuous function in $C(\bar{\Omega})$; and (3.15) then proves that $W_0^{1,p}(\Omega)$ is continuously embedded in $C(\bar{\Omega})$.

To proceed further, let $u \in C_0^1(\Omega)$, extend u by zero outside Ω, let B be any ball in $\mathbb{R}^n$, with radius R, say, and put

$$u_B = |B|^{-1} \int_B u(x)\,dx.$$

For any $x, y \in B$,

$$u(x) - u(y) = -\int_0^{|x-y|} \frac{\partial}{\partial r} u(x + r\omega)\,dr \qquad (\omega = (y-x)/|y-x|).$$

Thus

$$u(x) - u_B = -|B|^{-1} \int_B \int_0^{|x-y|} \frac{\partial}{\partial r} u(x + r\omega)\,dr\,dy.$$

Put $y - x = \rho\omega$, with $|\omega| = 1$ and $0 \leqslant \rho \leqslant \rho_0 \leqslant 2R$. Then

$$\begin{aligned} |u(x) - u_B| &\leqslant |B|^{-1} \int_0^{\rho_0} \int_{|\omega|=1} \int_0^{\rho_0} \left| \frac{\partial}{\partial r} u(x + r\omega) \right| \rho^{n-1}\,dr\,d\omega\,d\rho \\ &= \frac{(2R)^n}{n|B|} \int_0^{\rho_0} \int_{|\omega|=1} \left| \frac{\partial}{\partial r} u(x + r\omega) \right| d\omega\,dr \\ &\leqslant \frac{(2R)^n}{n|B|} \int_B |x-y|^{1-n} |\nabla u(y)|\,dy. \end{aligned}$$

Together with Lemma 3.13, with $q=\infty$ and $\mu=n^{-1}$, this shows that for all $x\in B$,

$$|u(x)-u_B| \leqslant CR^{\gamma}\,\|\nabla u\|_{p,B}, \tag{3.16}$$

where C depends only upon n and p. Thus for all x, y in B,

$$\begin{aligned}|u(x)-u(y)| &\leqslant |u(x)-u_B|+|u_B-u(y)|\\ &\leqslant 2CR^{\gamma}\,\|\nabla u\|_{p,B}.\end{aligned} \tag{3.17}$$

We obtain from (3.15) and (3.17) the inequality,

$$\|\!|u|\!\|_{0,\gamma,\Omega} \leqslant C_1[1+(\operatorname{diam}\Omega)^{\gamma}]\,\|\nabla u\|_{p,\Omega}, \tag{3.18}$$

for any $u\in C_0^1(\Omega)$. The same holds for any $u\in W_0^{1,p}(\Omega)$ as we may proceed as before to take a sequence of functions in $C_0^{\infty}(\Omega)$ which converges to u in $W_0^{1,p}(\Omega)$, use (3.18) to show that this sequence is a Cauchy sequence in $C^{0,\gamma}(\bar{\Omega})$ and then pass to the limit, using (3.18) again. This completes the proof. □

A compactness result follows easily.

Theorem 3.19. Let Ω be bounded, let $n<p<\infty$, and put $\gamma=1-n/p$. Then $W_0^{1,p}(\Omega)$ is compactly embedded in $C^{0,\lambda}(\bar{\Omega})$ for any $\lambda\in(0,\gamma)$. ■

Proof. The result is immediate in view of Theorems 3.18 and 1.1. □

Higher-order versions of our last two theorems follow by a routine process of iteration. We summarize as follows:

Theorem 3.20. Let Ω be bounded, let $k\in\mathbb{N}$, $l\in\mathbb{N}_0$, $\gamma\in(0,1]$, and suppose that $p\in[1,\infty)$ is such that $(k-l-\gamma)p\geqslant n$. Then $W_0^{k,p}(\Omega)$ is continuously embedded in $C^{l,\gamma}(\bar{\Omega})$; the embedding is compact if $(k-l-\gamma)p>n$. ■

Corollary 3.21. Let $a,b\in\mathbb{R}$ with $a<b$, $I=(a,b)$, $p>1$, and $m\in\mathbb{N}$, and suppose that $u\in W_0^{m,p}(I)$. Then $u^{(m-1)}$ is absolutely continuous on $[a,b]$ and $u^{(j)}(a)=u^{(j)}(b)=0$ for $j=0,1,\ldots,m-1$. ■

Proof. The absolute continuity follows directly from Corollary 3.12 and Theorem 3.20. As for the rest of the corollary, let $j\in\{0,1,\ldots,m-1\}$. By Theorem 3.20, $u^{(j)}$ is the limit, in $C(\bar{I})$, of a sequence of functions in $C_0^{\infty}(I)$, and so must be zero at the end-points of I. □

3.4. *Poincaré inequalities*

We conclude Section 3 by giving various inequalities often referred to as being of *Poincaré type*. The first one below has the advantage that it applies to certain unbounded sets Ω.

Theorem 3.22. Let Ω lie between two parallel hyperplanes at a distance l apart, and suppose that $p \in [1, \infty)$. Then for all $u \in W_0^{1,p}(\Omega)$,

$$\|u\|_{p,\Omega} \leqslant l \|\nabla u\|_{p,\Omega}. \tag{3.19}$$ ■

Proof. It is clearly enough to prove the result when $u \in C_0^1(\Omega)$, and to assume that u has been extended by 0 on $\mathbb{R}^n \backslash \Omega$. But then, assuming without loss of generality that Ω lies between the hyperplanes $x_1 = 0$ and $x_1 = l$, we have for all $x \in \Omega$,

$$|u(x)| = \left| \int_{-\infty}^{x_1} D_1 u(t, x_2, \ldots, x_n)\, dt \right| \leqslant \left(\int_0^l |D_1 u(t, x_2, \ldots, x_n)|^p\, dt \right)^{1/p} l^{1/p'}.$$

Hence

$$\int_0^l |u(x_1, x_2, \ldots, x_n)|^p\, dx_1 \leqslant l^p \int_0^l |D_1 u(t, x_2, \ldots, x_n)|^p\, dt,$$

and thus

$$\int_\Omega |u(x)|^p\, dx \leqslant l^p \int_\Omega |\nabla u(x)|^p\, dx,$$

as required. □

It is an entirely routine matter to check that Theorem 3.22 can be extended to give the result that for all $u \in W_0^{k,p}(\Omega)$,

$$\|\nabla^j u\|_{p,\Omega} \leqslant l^{k-j} \|\nabla^k u\|_{p,\Omega}$$

whenever $j \in \mathbb{N}$ with $j < k$. This shows that $W_0^{k,p}(\Omega)$ may also be normed by $\|\nabla^k u\|_{p,\Omega}$, and that this new norm is equivalent to the old one.

If Ω is *bounded*, a similar inequality can be derived by combining Lemmas 3.13 and 3.15, namely,

$$\|u\|_{p,\Omega} \leqslant (|\Omega|/\omega_n)^{1/n} \|\nabla u\|_{p,\Omega}, \tag{3.20}$$

whenever $u \in W_0^{1,p}(\Omega)$ and $1 \leqslant p \leqslant \infty$. Moreover, again if Ω is bounded, Theorem 3.6, together with Hölder's inequality, yields the inequality

$$\|u\|_{q,\Omega} \leqslant C |\Omega|^{1/n + 1/q - 1/p} \|\nabla u\|_{p,\Omega} \tag{3.21}$$

for all $u \in W_0^{1,p}(\Omega)$, provided that $1 \leqslant p < n$ and $q \in [1, np/(n-p)]$.

We can even give a result with a similar flavour for elements of $W^{1,p}(\Omega)$.

Theorem 3.23. Let Ω be convex and bounded, and let $p \in [1, \infty]$. Then for all $u \in W^{1,p}(\Omega)$ we have, with $u_\Omega = |\Omega|^{-1} \int_\Omega u(x)\, dx$ and $d = \operatorname{diam} \Omega$,

$$\|u - u_\Omega\|_{p,\Omega} \leqslant (\omega_n/|\Omega|)^{1-1/n} d^n \|\nabla u\|_{p,\Omega}. \tag{3.22}$$ ■

Proof. Just as in the proof of (3.16) we have that for almost all x in Ω,

$$|u(x) - u_\Omega| \leqslant \frac{d^n}{n|\Omega|} \int_\Omega |x-y|^{1-n} |\nabla u(y)| \, dy.$$

The proof is now completed by use of Lemma 3.13. □

Note that (3.22) gives rise to the inequality

$$\|u\|_{p,\Omega} \leqslant \left(\frac{\omega_n}{|\Omega|}\right)^{1-1/n} d^n \|\nabla u\|_{p,\Omega} + |\Omega|^{-1+1/p} |\textstyle\int_\Omega u(x)\,dx|, \tag{3.23}$$

which is occasionally a more useful form of inequality. Of course, (3.22) may be recovered from (3.23) by applying it to $v := u - u_\Omega$, so that $v_\Omega = 0$; thus (3.22) and (3.23) are equivalent; both inequalities will also be said to be of Poincaré type.

If we anticipate some of the results to be proved in §4, Theorem 3.23 can be sharpened as follows.

Theorem 3.24. Let Ω be convex and bounded, let $p \in [1, n)$ and let $q \in [p, p^*]$, where p^* is the Sobolev conjugate of p, given by $p^* = np/(n-p)$. Then there is a constant C such that for all $u \in W^{1,p}(\Omega)$,

$$\|u - u_\Omega\|_{q,\Omega} \leqslant C \|\nabla u\|_{p,\Omega}.$$ ■

Proof. By Theorem 4.2 and the remarks before it, $\partial\Omega$ is uniformly of class $C^{0,1}$; by the remarks following the definition of minimal smoothness, $\partial\Omega$ is minimally smooth; by Theorem 4.13, there is a constant C_1 such that for all $u \in W^{1,p}(\Omega)$,

$$\begin{aligned} \|u - u_\Omega\|_{q,\Omega} &\leqslant C_1 \|u - u_\Omega\|_{1,p,\Omega} \\ &\leqslant C_2(\|u - u_\Omega\|_{p,\Omega} + \|\nabla u\|_{p,\Omega}), \end{aligned}$$

where C_2 is another positive constant independent of u. Since by Theorem 3.23,

$$\|u - u_\Omega\|_{p,\Omega} \leqslant C_3 \|\nabla u\|_{p,\Omega},$$

the result follows. □

We conclude this section by remarking that for the space $W_0^{1,p}(\Omega)$, a necessary and sufficient condition for an inequality of Poincaré type to hold on an unbounded Ω will be given in Chapter VIII.

4. The boundary of Ω

To deal with the spaces $W^{k,p}(\Omega)$, as distinct from their subspaces $W_0^{k,p}(\Omega)$, some conditions on the boundary of the open set Ω are needed. There are various ways in which the counterparts of the embedding theorems and inequalities established for $W_0^{k,p}(\Omega)$ in §3 may be proved for $W^{k,p}(\Omega)$; perhaps the most natural method of procedure is to try to extend functions defined on Ω to the whole of $\mathbb{R}^n$, with *preservation of their main properties*; and then adapt the techniques used in §3. To perform the extension procedure, conditions on $\partial\Omega$ are required, and we devote this section to a discussion of the most common conditions to be imposed. Our treatment owes much to the paper of Fraenkel [3].

4.1. *Boundaries of class* $C^{k,\gamma}$

Definition 4.1. Let $n \in \mathbb{N}\backslash\{1\}$, $k \in \mathbb{N}_0$ and $\gamma \in [0, 1]$. The boundary $\partial\Omega$ of an open set $\Omega \subset \mathbb{R}^n$ is said to be *of class* $C^{k,\gamma}$ if:

(a) $\partial\Omega = \partial\bar{\Omega}$;

(b) given any point $a \in \partial\Omega$, there exist a neighbourhood $U(a)$ of a (an open subset of $\mathbb{R}^n$), local Cartesian coordinates $y = (y_1, \ldots, y_n) = (y', y_n)$ (where $y' = (y_1, \ldots, y_{n-1})$), with $y = 0$ at $x = a$, a convex, open subset G of $\mathbb{R}^{n-1}$ with $0 \in G$, and a function $h \in C^{k,\gamma}(\bar{G})$ such that $\partial\Omega \cap U(a)$ has a representation

$$y_n = h(y'), \qquad y' \in G. \tag{4.1}$$

If $\gamma = 0$ we shall write C^k in place of $C^{k,0}$ above; if for each $a \in \partial\Omega$, the function $h \in C^\infty(\bar{G})$ we shall say that $\partial\Omega$ is of *class* C^∞. If $\gamma = 0$ and $k = 0$ we shall write C in place of $C^{0,0}$ above. Finally, if for each $a \in \partial\Omega$ the corresponding function h is analytic on G we say that $\partial\Omega$ is *analytic*. ■

Some remarks on this definition may be helpful at this stage. The condition $\partial\Omega = \partial\bar{\Omega}$ is there simply to prevent Ω from being on both sides of $\partial\Omega$, and rules out such sets as $\mathbb{R}^n\backslash\{(x_1, 0, \ldots, 0) \in \mathbb{R}^n : x_1 \geqslant 0\}$ and the open unit ball in $\mathbb{R}^n$ with the line of points $\{(x_1, 0, \ldots, 0) \in \mathbb{R}^n : 0 \leqslant x < 1\}$ removed. In fact (see Fraenkel [3], p. 396), if Ω is bounded and connected, condition (a) is redundant, since it is implied by (b).

The local coordinates y and the underlying coordinates x, in terms of which Ω is defined, are of course related by an affine mapping A_a; that is, $y = A_a x = L_a(x - a)$, where $L_a \in \mathscr{B}(\mathbb{R}^n)$ is orthogonal and so is simply a rotation. Neighbourhoods of a certain type will be useful in the subsequent work: with $Q_m(a, \rho)$ standing for the open cube in $\mathbb{R}^m$ with centre a and of side 2ρ, we set

$$V_{\lambda,\mu}(a) = \{y \in \mathbb{R}^n : y' \in Q_{n-1}(0, \lambda), \ |y_n - h(y')| < \mu\}, \tag{4.2}$$

where

$$y_n > h(y') \text{ if, and only if, } y \in V_{\lambda,\mu}(a) \cap L_a(\Omega); \tag{4.3}$$

and put

$$U_{\lambda,\mu}(a) = L_a^{-1} V_{\lambda,\mu}(a). \tag{4.4}$$

Here the positive numbers λ and μ, and of course the function h, may depend upon a. It is clear that any neighbourhood $U(a)$ of a as in Definition 4.1 contains a neighbourhood $U_{\lambda,\mu}(a)$ provided that the sign of y_n is appropriately chosen. These particular neighbourhoods lead naturally to the idea of a boundary $\partial\Omega$ which is *uniformly of class* $C^{k,\gamma}$, by which is meant that there are positive numbers λ, μ and M such that given any $a \in \partial\Omega$, there is a neighbourhood $U_{\lambda,\mu}(a)$ of a with the property that $\| h_a \| \leqslant M$; here h_a is as in (4.1), the dependence upon a being recorded by the subscript, and the norm that of $C^{k,\gamma}(\bar{Q}_{n-1}(0,\lambda))$. It is not difficult to check that if $\partial\Omega$ is bounded and of class $C^{k,\gamma}$, then it is uniformly of class $C^{k,\gamma}$. This is not true, in general, if $\partial\Omega$ is unbounded: see Fraenkel [3, p. 406].

As for examples of open sets which have boundaries of the kind we have introduced in Definition 4.1, it is easy to check that any open cube in $\mathbb{R}^n$ has boundary of class $C^{0,1}$. This is, however, but a very special case of the next result.

Theorem 4.2. Let Ω be a convex, open subset of $\mathbb{R}^n$. Then $\partial\Omega$ is of class $C^{0,1}$. ∎

Proof. Given any $a \in \partial\Omega$, let local coordinates y be chosen at $x = a$ so that $y_n = 0$ is a supporting hyperplane of $A_a(\Omega)$ with $y_n > 0$ if $y \in A_a(\Omega)$, and so that there is a point $q = (0, \ldots, 0, k)$ in $A_a(\Omega)$. Since $A_a(\Omega)$ is open, there exists $\rho > 0$ such that $B(q, \rho) \subset A_a(\Omega)$; put $z = y - q$, introduce a Minkowski functional p by

$$p(z) = \inf\{\lambda > 0 : \lambda^{-1}z + q \in A_a(\bar{\Omega})\},$$

and note that for all $z, z_0 \in \mathbb{R}^n$,

$$|p(z) - p(z_0)| \leqslant |z - z_0|/\rho.$$

Given any $z \in \mathbb{R}^n \backslash \{0\}$, write $r = |z|$ and $\eta = z'/|z|$, and restrict attention to those z with $z_n < 0$ and $|\eta| = (\eta_1^2 + \cdots + \eta_{n-1}^2)^{\frac{1}{2}} \leqslant \frac{1}{2}$. Then, with $z = y - q$, we have

$$z' = y' = r\eta, \qquad -z_n = k - y_n = r(1 - |\eta|^2)^{\frac{1}{2}}. \tag{4.5}$$

Thus $z = (z', z_n) = \big(r\eta, r(1 - |\eta|^2)^{\frac{1}{2}}\big) := Z(r, \eta)$. Put $R(\eta) = 1/p\big(Z(1, \eta)\big)$. Hence

$$p(z) = p\big(Z(r, \eta)\big) = p\big(rZ(1, \eta)\big) = rp\big(Z(1, \eta)\big) = r/R(\eta).$$

Since a point $y = z + q$ belongs to $A_a(\partial\Omega)$ if, and only if, $p(z) = 1$, the equation $r = R(\eta)$ provides a local representation of $\partial\Omega$; and if $|\eta|, |\eta_0| \leqslant \frac{1}{2}$, we obtain, from the inequalities

$$R(\eta)(1 - |\eta|^2)^{\frac{1}{2}} \leqslant k, \qquad R(\eta_0)(1 - |\eta_0|^2)^{\frac{1}{2}} \leqslant k, \qquad \rho \leqslant R(\eta) \leqslant 2k/\sqrt{3},$$

$$|R(\eta) - R(\eta_0)| \leqslant \rho^{-1}|Z(1, \eta) - Z(1, \eta_0)| \leqslant \rho^{-1}(2k/\sqrt{3})^2\{(\eta - \eta_0)^2 + [(1 - \eta^2)^{\frac{1}{2}} - (1 - \eta_0^2)^{\frac{1}{2}}]^2\}^{\frac{1}{2}},$$

the estimate

$$|R(\eta) - R(\eta_0)| \leqslant (2/\sqrt{3})^3 k^2 |\eta - \eta_0|/\rho. \tag{4.6}$$

In view of (4.5), when $r = R(\eta)$ we have

$$y_n = (2k)^{-1}[k^2 + |y|^2 - R(\eta(y))^2] := f(y', y_n), \text{ say.}$$

The inequalities (4.6) now enable us to show that there is a positive number ε_0 such that given any $\varepsilon \in (0, \varepsilon_0)$, there exists $\delta > 0$ such that for each fixed $y' \in \mathbb{R}^{n-1}$ with $|y'| < \delta$, the map $f(y', \bullet)$ is a contraction mapping of the metric space $[-\varepsilon, \varepsilon]$ into itself and so has a unique fixed point $h(y')$. That h is Lipschitz-continuous follows immediately from (4.6). The proof is complete. □

4.2. *Other conditions on the boundary*

It is now time to relate our boundaries of class $C^{k,\gamma}$ with those which turn up with other commonly imposed conditions on Ω. We begin with the *segment property*.

Definition 4.3. An open subset Ω of $\mathbb{R}^n$ is said to have the segment property if, and only if, given any $a \in \partial\Omega$, there exist an open subset W of $\mathbb{R}^n$, with $a \in W$, and an element b of $\mathbb{R}^n \setminus \{0\}$ such that if $x \in \bar{\Omega} \cap W$ and $t \in (0, 1)$, then $x + tb \in \Omega$. ■

Theorem 4.4. An open subset Ω of $\mathbb{R}^n$ has the segment property if, and only if, $\partial\Omega$ is of class C. ■

Proof. Suppose first that $\partial\Omega$ is of class C. Given any $a \in \partial\Omega$, there will be a neighbourhood $U_{\lambda,2\mu}(a)$ of a (cf. (4.4)) for small enough λ and μ. Set $W = U_{\lambda,\mu}(a)$ and $b = L_a^{-1}((0, 0, \ldots, v))$ for small enough $v > 0$: it is easy to check that the segment property holds.

Conversely, suppose that Ω has the segment property, let $a \in \partial\Omega$ and let W and b be as in Definition 4.3. Choose local coordinates y, with $y = 0$ at $x = a$ and $\nabla y_n = b/|b|$; and set

$$\mathfrak{A}_y := A_a(\{x + tb : 0 < t < 1\}) = \{(y', y_n + t|b|) : 0 < t < 1\}.$$

Consider $\ell := \{y : y' = 0, |y_n| \leqslant \beta\}$, where $0 < \beta < |b|$ and β is chosen so small that $\ell \subset A_a(W)$. The points y of ℓ with $y_n > 0$ are in $A_a(\Omega)$ since they belong to $\mathfrak{A}_0$, while those with $y_n < 0$ are not in $A_a(\bar{\Omega})$ because $0 \in \mathfrak{A}_y$ and $0 \notin \Omega$. Put

$$\ell_u = \{(y', y_n) : y' = 0, \tfrac{1}{2}\beta \leqslant y_n \leqslant \beta\}, \quad \ell_d = \{(y', y_n) : y' = 0, -\beta \leqslant y_n \leqslant -\tfrac{1}{2}\beta\}.$$

Since ℓ_u, ℓ_d and $\overline{(\ell \setminus \ell_u) \setminus \ell_d}$ are compact in $\mathbb{R}^n$, they have finite, open covers of cubes with centres in ℓ and lying in the open sets $\Omega \cap W$, $(\mathbb{R}^n \setminus \bar{\Omega}) \cap W$ and W, respectively: strictly speaking we should write $A_a(\Omega \cap W)$, etc. here, but no ambiguity will arise from our omission of A_a, here and later, for ease of reading. It follows that there exists $\lambda > 0$ such that these covers together contain an open box V in $\mathbb{R}^n$, with

$$V = Q \times (-\beta, \beta) \subset W, \quad \text{where } Q = Q_{n-1}(0, \lambda),$$

such that

$$V_u := Q \times (\tfrac{1}{2}\beta, \beta) \subset \Omega \quad \text{and} \quad V_d := Q \times (-\beta, -\tfrac{1}{2}\beta) \subset \mathbb{R}^n \setminus \bar{\Omega}.$$

Now define $h: Q \to \mathbb{R}$ by

$$h(y') = \inf\{y_n : (y', y_n) \in \bar{\Omega} \cap V\}.$$

Note that for each y', there exists y_n such that $(y', y_n) \in \bar{\Omega} \cap V$ since $V_u \subset \Omega$. Moreover, since $V_u \subset \Omega$ and $V_d \subset \mathbb{R}^n \setminus \bar{\Omega}$ we have $\frac{1}{2}\beta \geqslant h(y') \geqslant -\frac{1}{2}\beta$; and as $\bar{\Omega}$ is closed, the infimum is attained. Let $y \in V$. Then

$$y \in \Omega \text{ if } y_n > h(y'); \quad \text{and} \quad y \in \mathbb{R}^n \setminus \bar{\Omega} \text{ if } y_n < h(y') \tag{4.7}$$

since, respectively, $y \in \mathfrak{A}_{(y', h(y'))}$ and $(y', h(y')) \in \bar{\Omega}$; and because of the definition of $h(y')$ as an infimum. This shows that given any $y' \in Q$, the point $(y', h(y')) \in \partial\Omega$ and is the only point (y', z) in V to lie on $\partial\Omega$.

To complete the proof all that is left is to show that h is continuous. Given any $c' \in Q$ and any $\varepsilon \in (0, \frac{1}{2}\beta)$, put $c_\pm = (c', h(c') \pm \varepsilon)$. Since $|h(y')| \leqslant \frac{1}{2}\beta$ we see, in view of (4.7), that $c_+ \in \Omega \cap V$ and $c_- \in (\mathbb{R}^n \setminus \bar{\Omega}) \cap V$. Evidently there exists $\delta > 0$ such that $Q_n(c_+, \delta) \subset \Omega \cap V$ and $Q_n(c_-, \delta) \subset (\mathbb{R}^n \setminus \bar{\Omega}) \cap V$. It now follows that $|h(y') - h(c')| < \varepsilon$ if $y' \in Q_{n-1}(c', \delta)$. To see this, suppose otherwise: then either $h(y') \geqslant h(c') + \varepsilon$, which contradicts the minimum property of $h(y')$, since $Q_n(c_+, \delta) \subset \Omega$; or $h(y') \leqslant h(c') - \varepsilon$, which means that $\mathfrak{A}_{(y', h(y'))}$ contains points of $Q_n(c_-, \delta) \subset \mathbb{R}^n \setminus \bar{\Omega}$, and again gives a contradiction. □

Another frequently used condition on Ω is the *uniform* C^k*-regularity* property which we give below and then relate to our conditions on $\partial\Omega$.

Definiton 4.5. Let $k \in \mathbb{N}_0$ and let Ω be an open subset of $\mathbb{R}^n$. We say that Ω has the *uniform* C^k*-regularity* property if there exist a locally finite open cover (U_j) of $\partial\Omega$ and corresponding C^k homeomorphisms Φ_j of U_j onto $B(0, 1)$ such that:

(i) for some $\delta > 0$,

$$\cup_j \Phi_j^{-1} B(0, \tfrac{1}{2}) \supset \Omega_\delta := \{x \in \Omega : \operatorname{dist}(x, \partial\Omega) < \delta\};$$

(ii) for each j,

$$\Phi_j(\Omega \cap U_j) = \{y \in B(0, 1) : y_n > 0\};$$

(iii) there is an $N \in \mathbb{N}$ such that any $N+1$ distinct sets U_j have empty intersection;

(iv) there is a sequence of points $c^j \in \mathbb{R}^n$ such that $\|\Phi_j\|$ and $\|\Phi_j^{-1} - c^j\|$ are bounded independently of j, the norms being those of $C^k(\bar{U}_j, \mathbb{R}^n)$ and $C^k(\bar{B}(0, 1), \mathbb{R}^n)$, respectively. ■

Theorem 4.6. (i) Let $\partial\Omega$ be bounded and of class C^k for some $k \in \mathbb{N}_0$. Then Ω has the uniform C^k-regularity property.

(ii) Let Ω have the uniform C^k-regularity property for some $k \in \mathbb{N}$. Then $\partial\Omega$ is of class C^k. ■

Proof. (i) Since $\partial\Omega$ is of class C^k, given any $a \in \partial\Omega$ there is a C^k homeomorphism Z_a given by

$$z = Z_a(y) := (y', y_n - h(y', a)), \qquad y = Z_a^{-1}(z) := (z', z_n + h(z', a));$$

it maps the neighbourhood $V_{\lambda,\mu}(a)$ onto the open cube $W := Q_{n-1}(0, \lambda) \times (-\mu, \mu)$, and here, since $\partial\Omega$ is uniformly of class C^k, we may suppose that λ and μ can be chosen independently of a. Put $\rho = \min\{\lambda, \mu\}$; then clearly $B(0, \rho) \subset W$. Now define, for certain points $a^j \in \partial\Omega$ to be specified in a moment,

$$U_j = A_{a^j}^{-1} Z_{a^j}^{-1} B(0, \rho), \qquad \Phi_j = \rho^{-1} Z_{a^j} A_{a^j}.$$

Plainly $\Phi_j(U_j) = B(0, 1)$, and condition (ii) of Definition 4.5 is satisfied.

Next, observe that $\{A_a^{-1} Z_a^{-1} B(0, \frac{1}{2}\rho) : a \in \partial\Omega\}$ is an open covering of the compact set $\partial\Omega$; the a^j above are given by taking any finite subcover. The corresponding finite cover (U_j) and finite sequence (Φ_j) evidently satisfy conditions (iii) and (iv), and all that remains is to show that condition (i) of Definition 4.5 is satisfied. However, $B := \cup_j \Phi_j^{-1} B(0, \frac{1}{2})$ is an open cover of $\partial\Omega$; if for all $\delta > 0$, $\Omega_\delta \not\subset B$ then for all $\delta > 0$, $(\mathbb{R}^n \backslash B) \cap \Omega_\delta \neq \varnothing$ which shows that $\operatorname{dist}(\mathbb{R}^n \backslash B, \partial\Omega) = 0$. This contradicts the fact that $\mathbb{R}^n \backslash B$ and $\partial\Omega$ are closed disjoint sets, with $\partial\Omega$ compact.

(ii) Let Φ_{ji} be the ith coordinate map of Φ_j, and put $z_i = \Phi_{ji}(x)$, $x_i = (\Phi_j^{-1})_i(z)$ $(i = 1, 2, \ldots, n)$. Given any $a \in \partial\Omega$, there exists j such that $\partial\Omega$ has the implicit equation $\Phi_{jn}(x) = 0$ in a neighbourhood U_j of a, and $(D_i\Phi_{jn})(a) \neq 0$ for some i, say $i = l$, since the Jacobian $\det(D_i\Phi_{jk}) \neq 0$. Choose the y_n-axis of local coordinates at $x = a$ parallel to the x_l-axis: the Implicit-Function Theorem now gives the local representation $y_n = h(y')$ of $\partial\Omega$. That $\partial\Omega = \partial\bar{\Omega}$ follows from condition (ii) of Definition 4.5. The proof is complete. □

We shall see later on that the restriction $k \geqslant 1$ in (ii) of the theorem is essential; for the moment, however, we shall concentrate on showing how various properties of Sobolev spaces may be established under appropriate restrictions on Ω.

4.3. *A density property*

Theorem 4.7. Let $p \in [1, \infty)$, let $k \in \mathbb{N}$, and suppose that Ω has boundary of class C. Then the set of restrictions to Ω of all functions in $C_0^\infty(\mathbb{R}^n)$ is dense in $W^{k,p}(\Omega)$. ■

Proof. Let $\phi \in C_0^\infty(\mathbb{R}^n)$ be such that there is a constant $M > 0$ with

$$\phi(x) = \begin{cases} 1 & \text{if } |x| < 1, \\ 2 & \text{if } |x| > 2, \end{cases}$$

and such that $|D^\alpha \phi(x)| \leqslant M$ for all $x \in \mathbb{R}^n$ and for all $\alpha \in \mathbb{N}_0^n$ with $|\alpha| \leqslant k$. Given any $\varepsilon \in (0, 1)$, set $\phi_\varepsilon(x) = \phi(\varepsilon x)$ $(x \in \mathbb{R}^n)$. Let $u \in W^{k,p}(\Omega)$ and put $U_\varepsilon = \phi_\varepsilon u$: then $U_\varepsilon \in W^{k,p}(\Omega)$ and supp U_ε is compact; for all $\alpha \in \mathbb{N}_0^n$ with $|\alpha| \leqslant k$,

$$|D^\alpha U_\varepsilon(x)| = \left| \sum_{\beta \leqslant \alpha} \binom{\alpha}{\beta} D^\beta u(x) \cdot D^{\alpha-\beta} \phi_\varepsilon(x) \right| \leqslant M \sum_{\beta \leqslant \alpha} \binom{\alpha}{\beta} |D^\beta u(x)|$$

for all $x \in \Omega$. Now put $\Omega^{(\varepsilon)} = \{x \in \Omega: |x| > 1/\varepsilon\}$. Then

$$\begin{aligned} \|u - U_\varepsilon\|_{k,p,\Omega} &= \|u - U_\varepsilon\|_{k,p,\Omega^{(\varepsilon)}} \leqslant \|u\|_{k,p,\Omega^{(\varepsilon)}} + \|U_\varepsilon\|_{k,p,\Omega^{(\varepsilon)}} \\ &\leqslant C \|u\|_{k,p,\Omega^{(\varepsilon)}} \end{aligned}$$

for some constant C, independent of ε. Since $\|u\|_{k,p,\Omega^{(\varepsilon)}} \to 0$ as $\varepsilon \to 0$, it follows that u can be approximated arbitrarily closely in $W^{k,p}(\Omega)$ by functions with compact support in $\mathbb{R}^n$. This was, in fact, indicated in the discussion following Theorem 3.2.

In view of this we shall assume that $u \in W^{k,p}(\Omega)$ has compact support K, and shall content ourselves with proving that u may be approximated arbitrarily closely by elements of $C_0^\infty(\mathbb{R}^n)$. If $K \subset \Omega$, this is clear, as we need only consider the regularization of u to obtain the desired approximations. We shall therefore suppose that $K \cap \partial\Omega \neq \varnothing$. By Theorem 4.4, Ω has the segment property; and so to each $x \in \partial\Omega$ there corresponds a neighbourhood $W(x)$, as in Definition 4.3. Put $F = K \cap [\bar{\Omega} \setminus \bigcup_{x \in \partial\Omega} W(x)]$: then F is compact and $F \subset \Omega$. Let W_0 be an open set such that $F \subset W_0 \subset\subset \Omega$. Since $K \cap \partial\Omega$ is compact, it may be covered by finitely many of the $W(x)$, say $W_1, W_2, \ldots, W_l$, and thus the sets $W_0, W_1, \ldots, W_l$ form an open covering of K. Let $W_0', W_1', \ldots, W_l'$ be open sets which form an open covering of K, with $W_j' \subset\subset W_j$ for each j. To construct these sets, for any $x \in W_i$ let $r(x) > 0$ be so small that $B(x, 2r(x)) \subset W_i$; observe that since K is compact and covered by the sets $B(x, r(x))$ when x runs through K, it is covered by finitely many balls $B(x_k, r(x_k))$ $(k = 1, 2, \ldots, m)$; and set $W_j' = \cup_{x_k \in W_j} B(x_k, r(x_k))$.

Now let $\psi_0, \psi_1, \ldots, \psi_l$ form a partition of unity subordinate to $W_0', W_1', \ldots, W_l'$, with supp $\psi_i \subset W_i'$ for each i; and put $u_i = \psi_i u$ $(i = 1, \ldots, l)$. Then $u = \sum_{j=0}^{l} u_j$, with supp $u_i \subset W_i'$ and each u_j belonging to $W^{k,p}(\Omega)$. It is accordingly enough to prove the theorem for each u_j. Since $W_0' \subset\subset \Omega$, our discussion of the case $K \subset \Omega$ above shows that the theorem holds for u_0. For $j \geqslant 1$, extend u_j to all of $\mathbb{R}^n$ by setting it equal to 0 on $\mathbb{R}^n \setminus \Omega$, and put $\Gamma_j = \bar{W}_j' \cap \partial\Omega$; then $u_j \in W^{k,p}(\mathbb{R}^n \setminus \Gamma_j)$. To see this, let $\phi \in C_0^\infty(\mathbb{R}^n \setminus \Gamma_j)$. Then

$K' := \operatorname{supp} \phi \cap \operatorname{supp} u_j \subset \Omega$; and if $\zeta \in C_0^\infty(\Omega)$ takes the value 1 throughout K', then $\zeta\phi \in C_0^\infty(\Omega)$ and

$$\begin{aligned}\int_{\mathbb{R}^n\setminus\Gamma_j} u_j\, D^\alpha\phi\, dx &= \int_{K'} u_j\, D^\alpha(\zeta\phi)\, dx = \int_\Omega u_j\, D^\alpha(\zeta\phi)\, dx \\ &= (-1)^{|\alpha|}\int_\Omega D^\alpha u_j\, \zeta\phi\, dx = (-1)^{|\alpha|}\int_K \phi\, D^\alpha u_j\, dx \\ &= (-1)^{|\alpha|}\int_{\mathbb{R}^n\setminus\Gamma_j} \phi\, D^\alpha u_j\, dx\end{aligned}$$

for all $\alpha \in \mathbb{N}_0^n$ with $|\alpha| \leqslant k$. In the last integral we have defined $D^\alpha u_j$ to be zero outside Ω. This establishes our claim that $u_j \in W^{k,p}(\mathbb{R}^n\setminus\Gamma_j)$. Let b be the element associated with W_j in Definition 4.3 and put $\Gamma^{(t)} = \Gamma_j - tb$, where

$$0 < t < \min\{1, |b|^{-1} \operatorname{dist}(W_j', \mathbb{R}^n\setminus W_j)\}.$$

Then $\Gamma^{(t)} \subset W_j$, since if $x \in \Gamma^{(t)}$ then $\operatorname{dist}(x, W_j') \leqslant \operatorname{dist}(x, \Gamma_j) \leqslant |tb| < \operatorname{dist}(W_j', \mathbb{R}^n\setminus W_j)$. Moreover, $\Gamma^{(t)} \cap \bar{\Omega} = \emptyset$, since if $x \in \Gamma^{(t)} \cap \bar{\Omega}$, then $x \in \bar{\Omega} \cap W_j$ and $x + tb \in \Gamma_j$, so that $x + tb \notin \Omega$, which contradicts the segment property. Since $\Gamma^{(t)}$ is compact, it follows that $\operatorname{dist}(\Gamma^{(t)}, \Omega) > 0$. Define an element $u_{j,t}$ of $W^{k,p}(\mathbb{R}^n\setminus\Gamma^{(t)})$ by $u_{j,t}(x) = u_j(x + tb)$. Since $\Omega \subset \mathbb{R}^n\setminus\Gamma^{(t)}$ we have $u_{j,t} \in W^{k,p}(\Omega)$. Evidently, for all $\alpha \in \mathbb{N}_0^n$ with $|\alpha| \leqslant k$, we have $D^\alpha u_{j,t} \to D^\alpha u_j$ in $L^p(\Omega)$ as $t \to 0$: thus $\lim_{t\to 0} \|u_{j,t} - u_j\|_{k,p,\Omega} = 0$. It is thus enough to approximate $u_{j,t}$ by functions in $C_0^\infty(\mathbb{R}^n)$, and to do this, since $\operatorname{dist}(\Gamma^{(t)}, \Omega) > 0$, we simply consider regularizations of $u_{j,t}$. The proof is complete. □

4.4. *Extension properties*

As indicated in our opening remarks in this section, one of the main reasons for the imposition of restrictions on the boundary of an open set Ω is to enable us to extend functions originally defined only on Ω. We can now illustrate this procedure, beginning with Hölder-continuous functions.

Proposition 4.8. Let $k \in \mathbb{N}$, let $\gamma \in [0, 1]$, suppose that Ω is a bounded open subset of $\mathbb{R}^n$ with boundary of class $C^{k,\gamma}$, and let Ω_0 be an open set which contains $\bar{\Omega}$. Then given any $u \in C^{k,\gamma}(\bar{\Omega})$, there is a function $U \in C_0^{k,\gamma}(\Omega_0)$ such that $U = u$ in Ω and

$$|||U|||_{k,\gamma,\Omega_0} \leqslant C|||u|||_{k,\gamma,\Omega}, \tag{4.8}$$

where C depends only upon k, Ω, and Ω_0. ■

Proof. Let $a \in \partial\Omega$. Exactly as in the proof of Theorem 4.6 (i), we see that there is a neighbourhood $V = V(a)$ of a and a $C^{k,\gamma}$ homeomorphism $\Phi = \Phi_a$ of V

onto B(0, 1) such that $\Phi(V \cap \Omega) = \{y \in B(0,1): y_n > 0\}$ and $\Phi(V \cap \partial\Omega) = \{y \in B(0,1): y_n = 0\}$. Define $\tilde{u}$ by $\tilde{u}(y) = (u \circ \Phi^{-1})(y)$ for all $y \in B(0,1)$ with $y_n \geqslant 0$; and extend $\tilde{u}$ to the whole of B (0, 1) by writing $\tilde{u}(y) = \Sigma_{i=1}^{k+1} c_i \tilde{u}(y', -y_n/i)$ if $y \in B(0,1)$ and $y_n < 0$, where the c_i are constants determined by the set of equations

$$\sum_{i=1}^{k+1} c_i \left(\frac{-1}{i}\right)^m = 1 \qquad (m = 0, 1, \ldots, k).$$

Since the determinant of this set of equations is of Vandermonde type, it can easily be shown to be non-zero; thus the c_i are well-defined constants. It is a simple matter to check that $\tilde{u} \in C^{k,\gamma}(B(0,1))$, and consequently $w := \tilde{u} \circ \Phi \in C^{k,\gamma}(\bar{B})$ for some ball B centred at a: the function w provides a $C^{k,\gamma}$ extension of u to $\Omega \cup \bar{B}$, and direct calculation using Fraenkel [2, Formula A] shows that for this extension, (4.8) holds with Ω_0 replaced by $\Omega \cup \bar{B}$.

Now cover $\partial\Omega$ by finitely many balls B_i $(i = 1, 2, \ldots, N)$ chosen just as B was, and to each B_i let w_i be the corresponding extension; we arrange for the B_i to be so small that $\Omega \cup (\cup_{i=1}^N B_i) \subset \Omega_0$. Let Ω^0 be an open set with $\Omega^0 \subset\subset \Omega$, and such that Ω^0 and the B_i form a finite open covering of $\bar{\Omega}$. Let $\phi_0, \phi_1, \ldots, \phi_N$ be a partition of unity subordinate to the covering Ω^0, $B_1, \ldots, B_N$, and put

$$U = u\phi_0 + \sum_{i=1}^{N} w_i \phi_i,$$

with the understanding that $w_i\phi_i$ is 0 if $\phi_i = 0$. It is now straightforward to check that U has all the required properties, and the proof is complete. □

Functions defined on $\partial\Omega$ can also be extended, with preservation of their properties under appropriate circumstances. Then let $\partial\Omega$ be of class $C^{k,\gamma}$, and let ϕ be a real- (or complex-) valued function defined on $\partial\Omega$. We say that $\phi \in C^{k,\gamma}(\partial\Omega)$ if, in the notation of the proof of Proposition 4.8, $\phi \circ \Phi_a^{-1} \in C^{k,\gamma}(\{y \in B(0,1): y_n = 0\})$ for all $a \in \partial\Omega$.

Proposition 4.9. Let $k \in \mathbb{N}$, let $\gamma \in [0, 1]$, suppose that Ω is a bounded, open set in $\mathbb{R}^n$ with boundary of class $C^{k,\gamma}$ and let Ω' be an open set which contains $\bar{\Omega}$. Then given any $\phi \in C^{k,\gamma}(\partial\Omega)$, there is a function $\phi' \in C_0^{k,\gamma}(\Omega')$ such that $\phi' = \phi$ on $\partial\Omega$. ■

Proof. Given any $a \in \partial\Omega$, $\tilde{\phi} := \phi \circ \Phi_a^{-1} \in C^{k,\gamma}(\{y \in B(0,1): y_n = 0\})$. Define $\tilde{\psi}(y) = \tilde{\phi}(y')$ for all $y \in B(0,1)$ and put $\psi(x) = (\tilde{\psi} \circ \Phi)(x)$ for all $x \in V(a)$. Evidently $\psi \in C^{k,\gamma}(\bar{B})$ for some ball $B = B(a)$, and $\psi = \phi$ on $B \cap \partial\Omega$. Cover $\partial\Omega$ by finitely many balls $B_1, \ldots, B_N$ such as B, and let ψ_i be the corresponding $C^{k,\gamma}(\bar{B}_i)$ function. The proof is now completed just as in Proposition 4.8.

□

Next we turn to Sobolev spaces. For elements of $W_0^{k,p}(\Omega)$ there is no problem in extending them to $\mathbb{R}^n$. We simply take any $u \in W_0^{k,p}(\Omega)$, let (ϕ_m) be a sequence in $C_0^\infty(\Omega)$ which converges to u in $W^{k,p}(\Omega)$, and observe that by setting each ϕ_m equal to zero in $\mathbb{R}^n \backslash \Omega$ the sequence (ϕ_m) becomes a Cauchy sequence in $W_0^{k,p}(\mathbb{R}^n)$, and hence converges, to $v \in W_0^{k,p}(\mathbb{R}^n)$, say. Letting R stand for the restriction operator which associates to each element of $W^{k,p}(\mathbb{R}^n)$ its restriction to Ω, we see that $Rv = u$. Evidently v is independent of the particular choice of sequence (ϕ_m) approximating u; and since v is thus uniquely determined by u we may write $v = Eu$, where E is the *extension operator* from $W_0^{k,p}(\Omega)$ to $W^{k,p}(\mathbb{R}^n)$; in this case E has norm 1.

For the spaces $W^{k,p}(\Omega)$ the procedure is more difficult, at least if we are to obtain the best results known. We begin with a lemma concerning the approximation of Lipschitz-continuous functions by C^∞ functions and due to Fraenkel [1] in the form given.

Lemma 4.10. Let $f: \mathbb{R}^n \to \mathbb{R}$ be such that for some constant M,

$$|f(x) - f(y)| \leqslant M|x - y|$$

for all $x, y \in \mathbb{R}^n$, and suppose that there is an open set $G \subset \mathbb{R}^n$ with $f(x) > 0$ for all $x \in G$. Then given any $\varepsilon \in (0, 1)$, there is a function $g \in C^\infty(G)$ such that for all $x \in G$,

$$(1+\varepsilon)^{-2} \leqslant g(x)/f(x) \leqslant (1-\varepsilon)^{-2} \tag{4.9}$$

and

$$|(D^\alpha g)(x)| \leqslant B_\alpha M^{|\alpha|} [\varepsilon f(x)]^{1-|\alpha|} \tag{4.10}$$

for all $\alpha \in \mathbb{N}_0^n$ with $|\alpha| \geqslant 1$. Here B_α is a constant which depends only on α. ■

Proof. Let $\rho \in C_0^\infty(\mathbb{R}^n)$ be the particular mollifier given by

$$\rho(x) = \begin{cases} c\exp[-1/(1-|x|^2)] & \text{if } |x| < 1, \\ 0 & \text{if } |x| \geqslant 1, \end{cases}$$

where c is so chosen that $\int_{\mathbb{R}^n} \rho(x)\, dx = 1$. For any $\delta > 0$ write $K(x, \delta) = \delta^{-n}\rho(x/\delta)$; thus $\int_{\mathbb{R}^n} K(x, \delta)\, dx = 1$.

Let $\varepsilon \in (0, 1)$ and put $\varepsilon f/M = h$. We claim that the function $S_\varepsilon f$ defined by

$$(S_\varepsilon f)(x^0) = \int_{\mathbb{R}^n} K\big(x - x^0, h(x)\big) f(x)\, dx \qquad (x^0 \in G) \tag{4.11}$$

will do as the function g. To prove this, fix $x^0 \in G$, put $K\big(x - x^0, h(x)\big) = k(x)$ and represent $x - x^0$ by polar coordinates (r, θ), where $r = |x - x^0|$ and $\theta = (x - x^0)/r$. Note that supp $k = \{x \in \mathbb{R}^n : r \leqslant h(x)\}$. We assert that supp k is a starlike set with

$$B\big(x^0, h(x^0)(1+\varepsilon)^{-1}\big) \subset \operatorname{supp} k \subset B\big(x^0, h(x^0)(1-\varepsilon)^{-1}\big). \tag{4.12}$$

For, with θ fixed, $h(x)-r$ is a strictly decreasing function of r since h is Lipschitz-continuous with Lipschitz constant ε (<1), and is bounded above and below by $h(x^0)-(1\mp\varepsilon)r$ respectively. It thus has a unique zero in the interval $[h(x^0)(1+\varepsilon)^{-1}, h(x^0)(1-\varepsilon)^{-1}]$, and our assertion is established. Note also that if $|h(x)-h(x^0)|\leqslant \varepsilon r$, then

$$1-\varepsilon r/h(x)\leqslant h(x^0)/h(x)\leqslant 1+\varepsilon r/h(x); \tag{4.13}$$

and, of course, $r/h(x)\leqslant 1$ on supp k.

Now let $\eta>0$ be so small that $B:=\bar{\mathrm{B}}(x^0,\eta)\subset G$ and write $k_b(x)=K(x-b,h(x))$, $D=\cup_{b\in B}\operatorname{supp} k_b$, $c=\min_{b\in B}f(b)(1+\varepsilon)^{-1}>0$,

$$f_\delta(x)=\int_{\mathbb{R}^n}K(z-x,\delta)f(z)\,\mathrm{d}z \qquad (x\in D,\ 0<\delta<c/M).$$

Then $f_\delta\in \mathrm{C}^\infty(D)$; and on D we have $f_\delta(x)>0$, since if $|z-x|\leqslant\delta$ then $f(z)\geqslant f(x)-M\delta\geqslant c-M\delta>0$, the fact that $f(x)\geqslant c$ following from (4.13). Moreover, $|\nabla f_\delta|\leqslant M$ on D since $f_\delta\in\mathrm{C}^\infty(D)$ and

$$|f_\delta(x+\xi)-f_\delta(x)|/|\xi|=\left|\int_{\mathbb{R}^n}K(z-x,\delta)\,[f(z+\xi)-f(z)]\,|\xi|^{-1}\,\mathrm{d}z\right|\leqslant M.$$

Hence f is Lipschitz-continuous on D, with Lipschitz constant M. As $\delta\to 0$ so $f_\delta(x)\to f(x)$ uniformly on D, since

$$|f(x)-f_\delta(x)|=\left|\int_{\mathbb{R}^n}K(z-x,\delta)\,[f(x)-f(z)]\,\mathrm{d}z\right|\leqslant M\delta;$$

and hence $(\mathrm{D}^\alpha S_\varepsilon f_\delta)(x)\to(\mathrm{D}^\alpha S_\varepsilon f)(x)$ uniformly on B for all $\alpha\in\mathbb{N}_0^n$. We now claim that

$$\int_{\mathbb{R}^n}\rho(y)/(1+\varepsilon|y|)^2\,\mathrm{d}y\leqslant(S_\varepsilon f_\delta)(x^0)/f(x^0)$$

$$\leqslant\int_{\mathbb{R}^n}\rho(y)\,(1-\varepsilon|y|)^2\,\mathrm{d}y \tag{4.14}$$

and

$$|(\mathrm{D}^\alpha S_\varepsilon f_\delta)(x^0)|\leqslant M^{|\alpha|}[\varepsilon f_\delta(x^0)]^{1-|\alpha|}B_\alpha \tag{4.15}$$

$(\alpha\in\mathbb{N}_0^n,\ |\alpha|\geqslant 1)$. Granted (4.14) and (4.15), the bounds (4.9) and (4.10) follow immediately in view of the above remarks about uniform convergence as $\delta\to 0$. It is therefore enough to establish (4.14) and (4.15) with f_δ replaced by f, and under the assumption that f is of class C^∞.

To do this, map supp k onto $\bar{\mathrm{B}}(0,1)$ by the transformation

$$(x-x^0)/h(x)=y=(s,\theta), \qquad s=|y|=r/h(x).$$

Then

$$\frac{\partial s}{\partial r}=\frac{1}{h(x)}\left(1-\frac{r}{h(x)}\frac{\partial h}{\partial r}\right), \qquad \left|\frac{\partial h}{\partial r}\right|\leqslant\varepsilon, \tag{4.16}$$

and use of (4.13) gives

$$(1-\varepsilon s)^2 \leqslant h(x^0)\frac{\partial s}{\partial r} \leqslant (1+\varepsilon s)^2 \qquad (x\in \operatorname{supp} k), \tag{4.17}$$

recalling that $\operatorname{supp} k = \{x\in \mathbb{R}^n : r \leqslant h(x)\}$ and so $s = r/h(x) \leqslant 1$ for $x\in \operatorname{supp} k$. Let $\varepsilon S_\varepsilon f/M = \tilde{h}$. Then, with standard notation,

$$\begin{aligned}\tilde{h}(x^0) &= \int_{\mathbb{R}^n} K(x-x^0, h(x))\, h(x)\,\mathrm{d}x \\ &= \int_{S^{n-1}}\int_0^\infty \rho\left(\frac{x-x^0}{h(x)}\right)\left(\frac{r}{h(x)}\right)^{n-1} \mathrm{d}r\,\mathrm{d}\omega_\theta \\ &= \int_{S^{n-1}}\int_0^1 \rho(y)\, s^{n-1}\left(\frac{\partial s}{\partial r}\right)^{-1} \mathrm{d}s\,\mathrm{d}w_\theta = \int_{\mathbb{R}^n} \rho(y)\left(\frac{\partial s}{\partial r}\right)^{-1} \mathrm{d}y.\end{aligned}$$

This, together with (4.17), immediately gives (4.14).

As for the derivatives of $\tilde{h}$, we proceed as follows. Put $\rho(y) = \rho_*(|y|)$ and note that

$$\begin{aligned}\frac{\partial}{\partial x_j^0}\rho\left(\frac{x-x^0}{h(x)}\right) &= -\rho_*'\left(\frac{r}{h(x)}\right)\left(\frac{1}{h(x)}\right)\frac{\partial r}{\partial x_j} \\ &= -\frac{\partial}{\partial x_j}\rho_*\left(\frac{r}{h(x)}\right) + \rho_*'\left(\frac{r}{h(x)}\right) r \frac{\partial}{\partial x_j}\left(\frac{1}{h(x)}\right).\end{aligned}$$

Integrating by parts the term with $-\partial \rho_*/\partial x_j$, we have

$$(\mathrm{D}_j \tilde{h})(x^0) = -\int_{\mathbb{R}^n} \rho_1\left(\frac{x-x^0}{h(x)}\right)[h(x)]^{-n}\,\mathrm{D}_j h(x)\,\mathrm{d}x,$$

where

$$\rho_1(y) = (n-1)\,\rho_*(|y|) + |y|\,\rho_*'(|y|).$$

Now differentiate with respect to x^0 and transform to y as before: for $|\gamma| = 1$ we have, in view of (4.13), (4.16), (4.17) and the bound $|\mathrm{D}^\gamma h(x)| \leqslant \varepsilon$,

$$\begin{aligned}|(\mathrm{D}^{\beta+\gamma}\tilde{h})(x^0)| &= \left|\int_{\mathbb{R}^n} (\mathrm{D}^\beta \rho_1)(y)\,[h(x)]^{-|\beta|-1}\left(\frac{\partial s}{\partial r}\right)^{-1}(\mathrm{D}^\gamma h)(x)\,\mathrm{d}y\right| \\ &\leqslant \varepsilon[h(x^0)]^{-|\beta|}\int_{\mathbb{R}^n} |(\mathrm{D}^\beta \rho_1)(y)|\,(1+\varepsilon|y|)^{|\beta|}(1-\varepsilon|y|)^{-1}\,\mathrm{d}y.\end{aligned}$$

This gives (4.10), with

$$B_{\beta+\gamma} = \int_{\mathbb{R}^n} |(\mathrm{D}^\beta \rho_1)(y)|\,(1+|y|)^{|\beta|}(1-|y|)^{-1}\,\mathrm{d}y \qquad (|\gamma| = 1).$$

The proof is complete. □

Our application of this result will be with the Lipschitz-continuous function f as the distance function from a closed subset of $\mathbb{R}^n$.

We now introduce the class of open sets Ω for which extension theorems will be proved. Our treatment follows that of Stein [1], who calls an open subset Ω of $\mathbb{R}^n$ a *special Lipschitz domain* if it is a rotation of a set of the form

$$\{x \in \mathbb{R}^n : x_n > \phi(x')\},$$

where $\phi : \mathbb{R}^{n-1} \to \mathbb{R}$ is Lipschitz-continuous;

$$M := \sup\{|\phi(x) - \phi(y)|/|x-y| : x, y \in \mathbb{R}^{n-1}, x \neq y\} \quad (< \infty)$$

is called the *bound* of Ω. An open set $\Omega \subset \mathbb{R}^n$ is said to have *minimally smooth boundary* $\partial\Omega$ if there exist $\varepsilon > 0$, $N \in \mathbb{N}$, $M > 0$, and a sequence $(U_i)_{i \in \mathbb{N}}$ of open subsets of $\mathbb{R}^n$ such that:

(i) given any $x \in \partial\Omega$, then $B(x, \varepsilon) \subset U_i$ for some $i \in \mathbb{N}$;

(ii) no point in $\mathbb{R}^n$ belongs to more than N of the U_i;

(iii) given any $i \in \mathbb{N}$, there is a special Lipschitz domain Ω_i, with bound $\leqslant M$, such that $U_i \cap \Omega = U_i \cap \Omega_i$.

The family of all open sets $\Omega \subset \mathbb{R}^n$ with minimally smooth boundary is a wide one: it obviously includes all bounded, open sets with boundary of class $C^{0,1}$; and it can be shown (cf. Fraenkel [3]) that it contains every open set with boundary uniformly of class $C^{0,1}$.

The strategy now is to prove the promised extension theorem first of all for special Lipschitz domains, and then to extend it to sets with minimally smooth boundary. One piece of notation will be convenient: for any open set $\Omega \subset \mathbb{R}^n$ we write

$$W(\Omega) = \bigcup_{k \in \mathbb{N}_0, p \in [1,\infty)} W^{k,p}(\Omega). \tag{4.18}$$

Theorem 4.11. Let Ω be a special Lipschitz domain in $\mathbb{R}^n$. Then there is a map $E : W(\Omega) \to W(\mathbb{R}^n)$ such that given any $u \in W(\Omega)$, the restriction REu of Eu to Ω coincides with u; and given any $k \in \mathbb{N}_0$ and any $p \in [1, \infty)$, the restriction of E to $W^{k,p}(\Omega)$ belongs to $\mathscr{B}(W^{k,p}(\Omega), W^{k,p}(\mathbb{R}^n))$ and has norm depending only upon n, k and the bound of Ω. ■

Proof. First we claim that there is a function $\psi \in C([1, \infty))$ such that $\psi(\lambda) = O(\lambda^{-N})$ as $\lambda \to \infty$, for all $N \in \mathbb{N}$; and with the extra properties that

$$\int_1^\infty \psi(\lambda)\,d\lambda = 1, \qquad \int_1^\infty \lambda^k \psi(\lambda)\,d\lambda = 0 \quad \text{for all } k \in \mathbb{N}. \tag{4.19}$$

In fact, the function ψ defined by

$$\psi(\lambda) = e(\pi\lambda)^{-1} \operatorname{im} \exp[-\omega(\lambda-1)^{\frac{1}{4}}] \qquad (\lambda \in [1, \infty)),$$

where $\omega = \exp(-\frac{1}{4}\pi i)$, has all these properties. To verify this, take the contour γ illustrated below, consider an appropriate branch of $\exp[-\omega(z-1)^{\frac{1}{4}}]$,

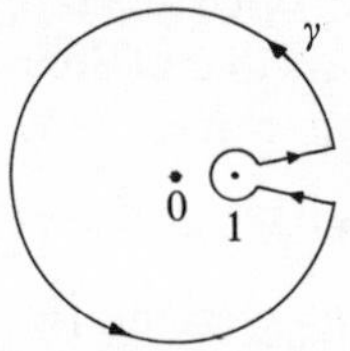

observe that by Cauchy's Residue Theorem,

$$\int_\gamma z^{-1}\exp[-\omega(z-1)^{\frac{1}{4}}]dz = 2\pi i e^{-1},$$

$$\int_\gamma z^{k-1}\exp[-\omega(z-1)^{\frac{1}{4}}]\,dz = 0 \quad (k \in \mathbb{N}),$$

and use the customary limiting procedure.

Next, let f be defined by $f(x) = \operatorname{dist}(x, \bar{\Omega})$ $(x \in \mathbb{R}^n)$; clearly f is Lipschitz-continuous on $\mathbb{R}^n$ and positive on $\mathbb{R}^n \backslash \bar{\Omega} := G$. Let g be the function in $C^\infty(G)$, corresponding to f, which appears in Lemma 4.10, with ε fixed, say $\varepsilon = \frac{1}{2}$. We now claim that there is a constant c, which depends only on the bound of Ω, such that if $x \in G$, then

$$cg(x) \geqslant \phi(x') - x_n. \tag{4.20}$$

To establish this, let Γ_- be the cone with vertex at the origin given by $\Gamma_- = \{x \in \mathbb{R}^n : M|x'| < |x_n|, x_n < 0\}$, and for any $p \in \mathbb{R}^n$ let $\Gamma_-(p) = p + \Gamma_-$. Since ϕ is Lipschitz-continuous, if $p \in \partial\Omega$ so that $p_n = \phi(p')$, then $\Gamma_-(p) \subset G$. Now let $x \in G$ and let $p := (x', \phi(x')) \in \partial\Omega$. Then $x \in \Gamma_-(p)$, and no point of $\bar{\Omega}$ is closer to x than the boundary of $\Gamma_-(p)$. It follows that

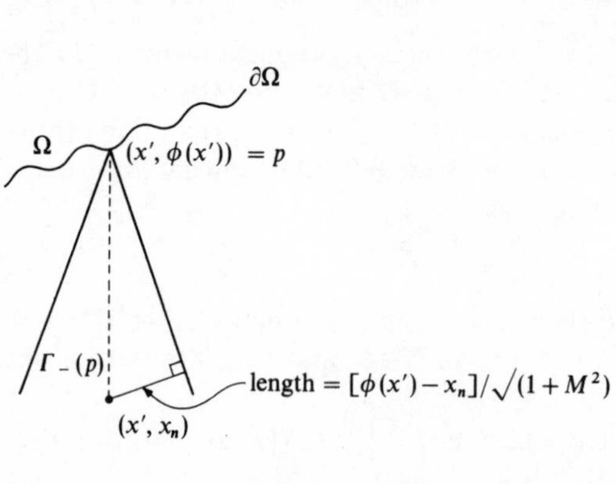

$f(x) \geqslant [\phi(x') - x_n]/\sqrt{(1+M^2)}$; and hence by Lemma 4.10,

$$g(x) \geqslant \tfrac{4}{9} f(x) \geqslant \tfrac{4}{9}[\phi(x') - x_n]/\sqrt{(1+M^2)}.$$

Our claim follows, and we have (4.20) with $c = \frac{9}{4}\sqrt{(1+M^2)}$. We shall now write $g^* = 2cg$, so that $g^*(x) \geqslant 2[\phi(x') - x_n]$.

Suppose that $u \in W^{k,p}(\Omega)$ for some $k \in \mathbb{N}_0$ and $p \in [1, \infty)$. For the moment we shall assume that $u \in C^\infty(\Omega)$ and that for all $\alpha \in \mathbb{N}_0^n$, the function $D^\alpha u$ is bounded on Ω and may be extended by continuity so as to be continuous and bounded on $\bar{\Omega}$. Define Eu by

$$(Eu)(x) = \begin{cases} u(x) & \text{if } x \in \bar{\Omega}, \\ \displaystyle\int_1^\infty \psi(\lambda)\, u(x', x_n + \lambda g^*(x))\, d\lambda & \text{if } x \in \mathbb{R}^n \backslash \bar{\Omega} = G. \end{cases}$$

Since $g^*(x) \geqslant 2[\phi(x') - x_n]$ we have, if $\lambda \geqslant 1$ and $x \in G$,

$$x_n + \lambda g^*(x) \geqslant x_n + 2[\phi(x') - x_n] > \phi(x'),$$

so that $(x', x_n + \lambda g^*(x)) \in \Omega$. This, together with the assumed boundedness of u and the properties of ψ, ensures that the above integral is well-defined.

Put $\Omega_- = \{x \in \mathbb{R}^n : \phi(x') > x_n\}$ with $\bar{\Omega}_- \cup \bar{\Omega} = \mathbb{R}^n$ and $\bar{\Omega}_- \cap \bar{\Omega} \neq \emptyset$. Evidently $Eu \in C^\infty(\bar{\Omega})$; we claim that $Eu \in C^\infty(\bar{\Omega}_-)$ and that all the derivatives of Eu are bounded on $\bar{\Omega}_-$ (that is, they are bounded on Ω_- and may be extended by continuity to $\bar{\Omega}_-$). That $Eu \in C^\infty(\Omega_-)$ follows from the facts that the integral is in $C^\infty(\Omega_-)$ and that for each $\alpha \in \mathbb{N}_0^n$, the integral $\int_1^\infty \psi(\lambda)\, D^\alpha u(x', x_n + \lambda g^*(x))\, d\lambda$ is very well-behaved, as we see on applying Formula A of Fraenkel [2], which gives a formula for general derivatives of composite functions, and making use of (4.19). Now let $x \in \Omega_-$, and suppose that $x \to x^0 \in \partial\Omega_-$, so that $x^0 \in \bar{\Omega}_- \cap \bar{\Omega}$. We claim that for all $\alpha \in \mathbb{N}_0^n$,

$$(D^\alpha Eu)(x) \to \lim_{y \to x^0,\, y \in \Omega} (D^\alpha u)(y). \tag{4.21}$$

To illustrate the procedure used to establish (4.21) we take the particular case where $D^\alpha = D_j^2$ $(1 \leqslant j \leqslant n-1)$. We have, for $x \in \Omega_-$,

$$\begin{aligned}(D_j^2 Eu)(x) = \int_1^\infty \big(&D_j^2 u(z_\lambda) + 2\lambda (D_j g^*)(x)\,(D_j D_n u)(z_\lambda) \\ &+ [\lambda (D_j g^*)(x)]^2\, (D_n^2 u)(z_\lambda) \\ &+ \lambda (D_j^2 g^*)(x)\,(D_n u)(z_\lambda)\big)\, \psi(\lambda)\, d\lambda,\end{aligned} \tag{4.22}$$

where $z_\lambda = (x', x_n + \lambda g^*(x))$. Let $x \to x^0$. Since by Lemma 4.10, $g^*(x) \to 0$ and $D_j g^*(x)$ remains bounded, we have by virtue of (4.19) that the first three terms on the right-hand side of (4.22) converge to

$$\lim_{y \to x^0} (D_j^2 u)(y).$$

The remaining term needs more delicate treatment, since $D_j^2 g^*$ may be unbounded. But

$$(\mathrm{D}_n u)(z_\lambda) = \mathrm{D}_n u\left(x', x_n + g^*(x)\right) + (\lambda - 1) g^*(x)\, \mathrm{D}_n^2 u\left(x', x_n + g^*(x)\right) + O\left([(\lambda - 1) g^*(x)]^2\right),$$

and if we use this in the final term, together with (4.10) and (4.19), (4.21) follows, when $\mathrm{D}^\alpha = \mathrm{D}_j^2$. The proof of (4.21) for the general case is achieved by similar means, using Fraenkel [2], Formula A, and we leave it to the industrious reader to supply the necessary details. Thus Eu is continuous and bounded in $\bar{\Omega}_-$, and u and Eu coincide on $\bar{\Omega} \cap \bar{\Omega}_-$, together with all their derivatives.

We now prove that $Eu \in \mathrm{C}^1(\mathbb{R}^n)$, by showing that given any $x \in \mathbb{R}^n$, $(Eu)(y) - (Eu)(x) = (y - x)\cdot(\nabla Eu)(y) + o(|y - x|)$ as $y \to x$. This is clear if $x \in \Omega \cup \Omega_-$, and so we assume that $x \in \partial\Omega$ $(= \partial\Omega_-)$. Suppose that $y \in \bar{\Omega}_-$; the argument if $y \in \bar{\Omega}$ is similar. We claim that y and x can be joined by a polygonal path which, apart from y and x, lies entirely in Ω_-, and has total length $\leqslant c|x - y|$, for some constant c independent of x and y; and that this path can be taken to consist of two line segments, one joining y to a point $z \in \Omega_-$, the other joining z to x. To justify this, observe that either $y \in \bar{\Gamma}_-(x)$ (in which case we may choose $z = y$) or $\bar{\Gamma}_-(x) \cap \bar{\Gamma}_-(y) \neq \varnothing$ (and we choose z to be the nearest point of this intersection to y). Then

$$(Eu)(y) - (Eu)(z) = (y - z)\cdot(\nabla Eu)(y) + o(|y - z|),$$

$$(Eu)(z) - (Eu)(x) = (z - x)\cdot(\nabla Eu)(z) + o(|z - x|).$$

Thus

$$(Eu)(y) - (Eu)(x) = (y - z)\cdot(\nabla Eu)(y) + (z - x)\cdot(\nabla Eu)(z) + o(|y - z|) + o(|z - x|);$$

and since $(\nabla Eu)(z) - (\nabla Eu)(y) = o(1)$ as $|x - y| \to 0$, it follows that $Eu \in \mathrm{C}^1(\mathbb{R}^n)$. A similar argument shows that $Eu \in \mathrm{C}^k(\mathbb{R}^n)$ for all $k \in \mathbb{N}$; thus $Eu \in C^\infty(\mathbb{R}^n)$.

Now we prove that

$$\|Eu\|_{k,p,\mathbb{R}^n} \leqslant C_{k,n,M} \|u\|_{k,p,\Omega}. \tag{4.23}$$

First suppose that $k = 0$. Fix $x' \in \mathbb{R}^{n-1}$ and suppose that $\phi(x') = 0$. Since $\psi(\lambda) = O(\lambda^{-2})$, there is a constant A such that $|\psi(\lambda)| \leqslant A\lambda^{-2}$, and hence if $y < 0$,

$$|(Eu)(x', y)| \leqslant A \int_1^\infty |u(x', y + \lambda g^*(x', y))| \lambda^{-2}\, \mathrm{d}\lambda. \tag{4.24}$$

Since $g^*(x', y) \geqslant 2(\phi(x') - y)$ it follows that $g^*(x', y) \geqslant -2y = 2|y|$; also, $\phi(x') - y \geqslant \operatorname{dist}((x', y), \bar{\Omega})$, so that $g^*(x', y) \leqslant a|y|$ for some constant a.

Put $s = y+\lambda g^*(x', y)$; then (4.24) gives

$$|(Eu)(x',y)| \leqslant A \int_{|y|}^{\infty} |u(x',s)|(s-y)^{-2} g^*(x',y)\, ds$$
$$\leqslant Aa|y| \int_{|y|}^{\infty} |u(x',s)| s^{-2}\, ds \qquad (y<0). \qquad (4.25)$$

We may now use Hardy's inequality (cf. Hardy, Littlewood, and Polyá [1]) to obtain

$$\left(\int_{-\infty}^{0} |(Eu)(x',y)|^p\, dy\right)^{1/p} \leqslant A'\left(\int_{0}^{\infty} |u(x',y)|^p\, dy\right)^{1/p},$$

under the assumption that $\phi(x') = 0$. If this condition is removed, then an appropriate translation in y shows that

$$\left(\int_{-\infty}^{\infty} |(Eu)(x',y)|^p\, dy\right)^{1/p} \leqslant A'\left(\int_{\phi(x')}^{\infty} |u(x',y)|^p\, dy\right)^{1/p},$$

and from this (4.23) follows immediately, when $k = 0$.

When $k > 0$ the procedure is similar. Suppose, for example, that $k = 2$, and consider $D_j^2\, Eu$ as a typical term. We use the representation (4.22) of this term, and handle the first three terms on the right-hand side in the same way as above, using this time the estimate $|\psi(\lambda)| \leqslant A\lambda^{-4}$ $(\lambda \geqslant 1)$. As for the last term, we write

$$D_n u(x', x_n+\lambda g^*(x)) = D_n u(x', x_n+g^*(x)) + \int_{x_n+g^*(x)}^{x_n+\lambda g^*(x)} D_n^2\, u(x',t)\, dt$$

and use this in (4.22). Since $\int_1^{\infty} \lambda\psi(\lambda)\, d\lambda = 0$, the contribution of $D_n u(x', x_n+g^*(x))$ is zero, and we simply have to estimate

$$|x_n|^{-1} \int_1^{\infty} \left(\int_{x_n+g^*(x)}^{x_n+\lambda g^*(x)} |D_n^2\, u(x',t)|\, dt\right) \lambda^{-3}\, d\lambda.$$

However, interchange of the order of integration reduces this to a term similar to (4.25), and we may then proceed as before to complete the proof when $k = 2$. For general $k \in \mathbb{N}$, differentiation under the integral sign in (4.22) produces derivatives $D^{\alpha}u$ with $1 \leqslant |\alpha| \leqslant k$. For these derivatives under the integral sign with $|\alpha| < k$ we write the Taylor expansion of the derivatives about $(x', x_n+g^*(x))$ up to order k with integral remainder, and then proceed as above. Thus if $|\alpha| = k_0 < k$ and $v = \mathrm{D}^\alpha u$, then

$$v(x', x_n + \lambda g^*(x)) = \sum_{j=0}^{l+1} \frac{[(\lambda-1)g^*(x)]^j}{j!} (\mathrm{D}_n^j v)(x', x_n + g^*(x))$$

$$+ \frac{1}{l!} \int_{g^*(x)}^{\lambda g^*(x)} [\lambda g^*(x) - t]^{l-1} (\mathrm{D}_n^l v)(x', x_n + t)\, \mathrm{d}t \quad (k_0 + l = k).$$

Of these terms, only the integral gives a non-zero contribution, in view of (4.19), but it can be majorized by

$$A[\lambda g^*(x)]^{l-1} \int_{g^*(x)}^{\lambda g^*(x)} |(\mathrm{D}_n^l v)(x', x_n + t)|\, \mathrm{d}t$$

and the argument is now as before.

To conclude the proof of the theorem it is merely necessary to remove the smoothness hypotheses on u. Suppose that $u \in \mathrm{W}^{k,p}(\Omega)$, and let $\eta \in \mathrm{C}_0^\infty(\Gamma_-)$ be such that $\eta \geqslant 0$ and $\int_{\mathbb{R}^n} \eta(x)\, \mathrm{d}x = 1$. The function η acts like a mollifier and for $\varepsilon > 0$ we may form a function u_ε, given by

$$u_\varepsilon(x) = \varepsilon^{-n} \int_\Omega \eta[(x-y)/\varepsilon]\, u(y)\, \mathrm{d}y;$$

this is defined for all $x \in \Omega$ since, for $x \in \Omega$ and $(x-y)/\varepsilon \in \Gamma_-$ we have $y_n > x_n > \phi(x)$ and so $y \in \Omega$. It is even defined for all x in a neighbourhood of $\bar{\Omega}$, if ε is small enough; u_ε behaves like a regularization of u, is of class C^∞ in a neighbourhood of $\bar{\Omega}$, and in view of the analogues of (1.3), Theorem 1.5, and Lemma 2.2, it follows that

$$\|u_\varepsilon\|_{k,p,\Omega} \leqslant \|u\|_{k,p,\Omega}, \qquad \|u_\varepsilon - u\|_{k,p,\Omega} \to 0 \quad \text{as } \varepsilon \to 0. \tag{4.26}$$

By (4.26) and (4.23), we have for all $m \in \mathbb{N}$,

$$\|Eu_{1/m}\|_{k,p,\mathbb{R}^n} \leqslant C_{k,n,M} \|u\|_{k,p,\Omega}; \tag{4.27}$$

thus since $\|u_{1/m} - u_{1/l}\|_{k,p,\Omega} \to 0$ as $l, m \to \infty$, the sequence $(Eu_{1/m})_{m \in \mathbb{N}}$ is a Cauchy sequence in $\mathrm{W}^{k,p}(\Omega)$ and consequently converges, to v, say, where by (4.27),

$$\|v\|_{k,p,\mathbb{R}^n} \leqslant C_{k,n,M} \|u\|_{k,p,\Omega}.$$

This element v is the required extension of u, and can thus be taken as Eu. That $E \in \mathscr{B}(\mathrm{W}^{k,p}(\Omega), \mathrm{W}^{k,p}(\mathbb{R}^n))$ follows immediately; and the proof is complete. □

We can now give the main extension theorem.

Theorem 4.12. Let Ω be an open subset of $\mathbb{R}^n$ with minimally smooth boundary. Then there is a map $E: \mathrm{W}(\Omega) \to \mathrm{W}(\mathbb{R}^n)$ such that given any $u \in \mathrm{W}(\Omega)$, the restriction REu of Eu to Ω coincides with u; and given any $k \in \mathbb{N}_0$

and any $p \in [1, \infty)$, the restriction of E to $W^{k,p}(\Omega)$ is in $\mathscr{B}(W^{k,p}(\Omega), W^{k,p}(\mathbb{R}^n))$. ■

Proof. Given any $U \subset \mathbb{R}^n$ and any $\varepsilon > 0$, let $U^{\varepsilon} = \{x \in \mathbb{R}^n : B(x, \varepsilon) \subset U\}$. Let $(U_i)_{i \in \mathbb{N}}$ be as in the definition of a set with minimally smooth boundary, let ρ be a mollifier, put $\rho_{\varepsilon}(x) = \varepsilon^{-n}\rho(x/\varepsilon)$, let χ_i be the characteristic function of $U_i^{\frac{3}{4}\varepsilon}$, and set $\lambda_i(x) = (\chi_i * \rho_{\frac{1}{4}\varepsilon})(x)$. Then for each $i \in \mathbb{N}$ we have: $\operatorname{supp} \lambda_i \subset U_i$, $\lambda_i(x) = 1$ if $x \in U_i^{\frac{1}{2}\varepsilon}$ (and thus if $x \in U_i^{\varepsilon}$), $\lambda_i \in C_0^{\infty}(U_i)$, and bounds for the derivatives of λ_i can be taken to be independent of i and depending only upon the L^1 norm of the corresponding derivatives of $\rho_{\frac{1}{4}\varepsilon}$.

Now put

$$U_0 = \{x \in \mathbb{R}^n : \operatorname{dist}(x, \Omega) < \tfrac{1}{4}\varepsilon\}, \; U_+ = \{x \in \mathbb{R}^n : \operatorname{dist}(x, \partial\Omega) < \tfrac{1}{4}\varepsilon\},$$

$$U_- = \{x \in \Omega : \operatorname{dist}(x, \partial\Omega) > \tfrac{1}{4}\varepsilon\}.$$

Let χ_0, χ_+, χ_- be the corresponding characteristic functions, with regularizations $\lambda_0 = \chi_0 * \rho_{\frac{1}{4}\varepsilon}$, $\lambda_{\pm} = \chi_{\pm} * \rho_{\frac{1}{4}\varepsilon}$, and observe that $\lambda_0(x) = 1$ $(x \in \bar{\Omega})$, $\lambda_+(x) = 1$ if $\operatorname{dist}(x, \partial\Omega) \leqslant \frac{1}{2}\varepsilon$, and $\lambda_-(x) = 1$ if $x \in \Omega$ and $\operatorname{dist}(x, \partial\Omega) \geqslant \frac{1}{2}\varepsilon$; while the supports of λ_0, λ_+, and λ_- are contained respectively in the $\frac{1}{2}\varepsilon$-neighbourhood of Ω, the ε-neighbourhood of $\partial\Omega$, and in Ω. These functions λ_0 and $\lambda_{\pm}$ are also bounded in $\mathbb{R}^n$, as are all their partial derivatives. Put

$$\Lambda_{\pm} = \lambda_0 \lambda_{\pm}/(\lambda_+ + \lambda_-),$$

and note that since $\operatorname{supp} \lambda_0 \subset \{x \in \mathbb{R}^n : \lambda_+(x) + \lambda_-(x) \geqslant 1\}$, all the derivatives of $\Lambda_{\pm}$ are bounded on $\mathbb{R}^n$, while $\Lambda_+ + \Lambda_-$ is 1 on $\bar{\Omega}$ and 0 outside the $\frac{1}{2}\varepsilon$-neighbourhood of Ω.

To each U_i there corresponds a special Lipschitz domain Ω_i; let E^i be the extension operator for $W^{k,p}(\Omega_i)$, the existence of which follows from Theorem 4.11. Given any $u \in W^{k,p}(\Omega)$ define Eu by

$$(Eu)(x) = \Lambda_+(x)\left(\sum_{i=1}^{\infty} \lambda_i(x) E^i(\lambda_i u)(x)\right)\left(\sum_{i=1}^{\infty} \lambda_i^2(x)\right)^{-1} + \Lambda_-(x)u(x) \, (x \in \mathbb{R}^n). \tag{4.28}$$

Then:

(i) If $x \in \operatorname{supp} \Lambda_+$ (or if $\operatorname{dist}(x, \partial\Omega) \leqslant \frac{1}{2}\varepsilon$), then $x \in U_i^{\frac{1}{2}\varepsilon}$ for some $i \in \mathbb{N}$, and hence $\Sigma_{i=1}^{\infty} \lambda_i^2(x) \geqslant 1$.

(ii) For each $x \in \mathbb{R}^n$, the sums in (4.28) involve at most $N + 1$ non-zero terms, in view of condition (ii) in the definition of minimally smooth boundaries.

(iii) Since $\operatorname{supp} \Lambda_- \subset \Omega$, the term $\Lambda_-(x)u(x)$ is well-defined.

(iv) Since the $\lambda_i u$ are defined on the Ω_i, the terms $E^i(\lambda_i u)$ are well-defined.

(v) If $x \in \Omega$ then $(Eu)(x) = u(x)$.

Our object is to prove that given any $u \in W^{k,p}(\Omega)$,

$$\|Eu\|_{k,p,\mathbb{R}^n} \leqslant A_{k,n}(\Omega)\, \|u\|_{k,p,\Omega}. \tag{4.29}$$

To do this we use the result that if $A(x) = \Sigma_{i=1}^{\infty} a_i(x)$, where for each x at most N of the $a_i(x)$ are non-zero, then

$$\|A\|_p \leqslant N^{1-1/p}\left(\sum_{i=1}^{\infty} \|a_i\|_p^p\right)^{1/p}. \tag{4.30}$$

This result follows immediately from the inequality

$$|A(x)|^p \leqslant N^{p-1} \sum_{i=1}^{\infty} |a_i(x)|^p,$$

which results from Hölder's inequality. To prove (4.29) when $k = 0$, we use the properties of the λ_i, (i)–(v) above, (4.30) and Theorem 4.11 to obtain

$$\begin{aligned}\|Eu\|_{p,\mathbb{R}^n} &\leqslant N^{1-1/p}\left(\sum_{i=1}^{\infty}\int_{U_i} |E^i(\lambda_i u)|^p \,\mathrm{d}x\right)^{1/p} + \left(\int_{\Omega} |u(x)|^p \,\mathrm{d}x\right)^{1/p} \\ &\leqslant AN^{1-1/p}\left(\sum_{i=1}^{\infty}\int_{\Omega} |\lambda_i u|^p \,\mathrm{d}x\right)^{1/p} + \left(\int_{\Omega} |u(x)|^p \,\mathrm{d}x\right)^{1/p} \\ &\leqslant 2AN\left(\int_{\Omega} |u(x)|^p \,\mathrm{d}x\right)^{1/p} + \left(\int_{\Omega} |u(x)|^p \,\mathrm{d}x\right)^{1/p},\end{aligned}$$

the final step following from the inequality $(\Sigma_{i=1}^{\infty} \lambda_i)^{1/p} \leqslant (N+1)^{1/p}$. Thus (4.29) is established when $k = 0$. When $k > 0$, the same kind of argument may be carried through since for all $\alpha \in \mathbb{N}_0^n$ with $|\alpha| \leqslant k$, we have that $\mathrm{D}^\alpha \lambda_i$ $(i \in \mathbb{N})$ and $\mathrm{D}^\alpha \Lambda_\pm$ are uniformly bounded. The proof is complete. □

A remarkable feature of this theorem, which is given in Stein [1], is that the amount of smoothness required of the boundary of Ω is independent of k and p: earlier results in this direction (cf. Nécas [1]) did not have this attractive and valuable feature.

4.5. *Embedding theorems for* $\mathrm{W}^{k,p}(\Omega)$

Turning now to embedding theorems for the spaces $\mathrm{W}^{k,p}(\Omega)$, the results of §3 and our extension theorems make them very simple to handle. Thus let Ω be a bounded open subset of $\mathbb{R}^n$ with minimally smooth boundary, and let Ω_0 be a bounded open set such that $\bar{\Omega} \subset \Omega_0$. By Lemma 1.9 we know that there is a function $\phi \in \mathrm{C}_0^\infty(\Omega_0)$ such that $0 \leqslant \phi \leqslant 1$ and $\phi(x) = 1$ for all x in a neighbourhood of $\bar{\Omega}$. Hence by Theorem 4.12, if $u \in \mathrm{W}^{k,p}(\Omega)$, with $k \in \mathbb{N}$ and $p \in [1, \infty)$, then $Eu \in \mathrm{W}^{k,p}(\mathbb{R}^n)$ and $\phi Eu \in \mathrm{W}_0^{k,p}(\Omega_0)$, by Theorem 3.2. It follows that

$$\begin{aligned}\|u\|_{k,p,\Omega} &\leqslant \|\phi Eu\|_{k,p,\Omega_0} \leqslant C(\phi)\|Eu\|_{k,p,\Omega_0} \leqslant C(\phi)\|Eu\|_{k,p,\mathbb{R}^n} \\ &\leqslant C(\phi)C_1(k,p,\Omega)\|u\|_{k,p,\Omega}.\end{aligned} \tag{4.31}$$

These upper and lower bounds for $\|\phi Eu\|_{k,p,\Omega_0}$ in terms of $\|u\|_{k,p,\Omega}$ mean that inequalities known to hold for the element ϕEu of $W_0^{k,p}(\Omega_0)$ may be translated into similar inequalities relating to $u \in W^{k,p}(\Omega)$. For example, if $kp < n$ and $s \in [p, np/(n-kp)]$, then by Theorem 3.7, $W_0^{k,p}(\Omega_0)$ is continuously embedded in $L^s(\Omega_0)$, and hence for all $u \in W^{k,p}(\Omega)$,

$$\|u\|_{s,\Omega} \leqslant \|\phi Eu\|_{s,\Omega_0} \leqslant C_2 \|\phi Eu\|_{k,p,\Omega_0} \leqslant C_2 C(\phi) C_1(k,p,\Omega)\|u\|_{k,p,\Omega};$$

that is, $W^{k,p}(\Omega)$ is continuously embedded in $L^s(\Omega)$. Moreover, if in addition, $s < np/(n-kp)$ and B is a bounded subset of $W^{k,p}(\Omega)$, then $\phi E(B)$ is bounded in $W_0^{k,p}(\Omega_0)$; and hence is relatively compact when viewed as a subset of $L^s(\Omega_0)$; and thus B is relatively compact as a subset of $L^s(\Omega)$. This shows that $W^{k,p}(\Omega)$ is compactly embedded in $L^s(\Omega)$. We may, in a similar manner, take over the other inequalities and embedding theorems of §3 to the spaces $W^{k,p}(\Omega)$. The position may be partially summarized by the following Theorem.

Theorem 4.13. Let Ω be a bounded, open subset of $\mathbb{R}^n$ with minimally smooth boundary. Then Theorems 3.7, 3.17 and 3.20 hold with the Sobolev space W_0 replaced by the corresponding space W. This information is summarized diagrammatically as follows:

$$W^{k,p}(\Omega) \to \begin{cases} L^s(\Omega) & \text{for } kp < n,\ s \in [p, np/(n-kp)] \\ L^{\phi}(\Omega) & \text{for } kp = n,\ p > 1 \\ C^{l,\gamma}(\bar{\Omega}) & \text{for } l \in \mathbb{N}_0,\ \gamma \in (0,1],\ p \in [1,\infty),\ (k-l-\gamma)p \geqslant n \end{cases}$$

and

$$W^{k,p}(\Omega) \rightrightarrows \begin{cases} L^s(\Omega) & \text{for } kp < n,\ s \in [p, np/(n-kp)) \\ C^{l,\gamma}(\bar{\Omega}) & \text{for } l \in \mathbb{N}_0,\ \gamma \in (0,1],\ p \in [1,\infty),\ (k-l-\gamma)p > n \end{cases}$$

where '$\to$' denotes continuous embedding, '$\rightrightarrows$' stands for compact embedding and $\phi(t) = \exp(t^{n/(n-k)} - 1)$. ■

Compactness of the embedding when $kp = n$ is investigated in Theorem 6.6. The analogue of Theorem 3.8 is also worth recording.

Theorem 4.14. Let $m, j \in \mathbb{N}_0$, with $0 \leqslant j < m$, let $q, r \in [1, \infty)$, and let p be defined by

$$\frac{1}{p} = \frac{1}{q} + \frac{j}{m}\left(\frac{1}{r} - \frac{1}{q}\right).$$

Let Ω be a bounded open subset of $\mathbb{R}^n$ with minimally smooth boundary. Then there is a constant C such that for all $u \in W^{m,r}(\Omega) \cap L^q(\Omega)$,

$$\|\nabla^j u\|_{p,\Omega} \leqslant C \|u\|_{m,r,\Omega}^{j/m} \|u\|_{q,\Omega}^{(m-j)/m}. \tag{4.32}$$ ■

Proof. Choose Ω_0 and ϕ as in the remarks immediately before Theorem 4.13; observe that $\phi Eu \in W_0^{m,r}(\Omega_0) \cap L^q(\Omega_0)$ if $u \in W^{m,r}(\Omega) \cap L^q(\Omega)$; use Remark 3.9 and (4.31). □

Another, and more elementary, way of establishing embedding theorems for the spaces $W^{k,p}(\Omega)$ is by way of the so-called *cone property.*

Definition 4.15. An open set $\Omega \subset \mathbb{R}^n$ is said to have the *cone property* if there is a cone $V := \{x \in \mathbb{R}^n : x_n \geqslant 0, |x'| \leqslant bx_n, |x| \leqslant a\}$, with $a > 0$ and $b > 0$, such that every point x in $\bar{\Omega}$ is the vertex of a cone $V_x \subset \bar{\Omega}$ congruent to V (that is, $V_x = x + L_x(V)$, where L_x is a rotation operator). The set Ω is said to have the *strong cone property* if there are positive constants c and d such that given any $x, y \in \bar{\Omega}$ with $|x-y| := s < d$, there are cones

$$V_{x,s} := V_x \cap B(x,s) \text{ and } V_{y,s} := V_y \cap B(y,s) \subset \bar{\Omega} \text{ with } |V_{x,s} \cap V_{y,s}| \geqslant cs^n.$$

■

If Ω has bounded boundary of class C^1, then it can be shown (cf. Fraenkel [3], p. 409) that Ω has the strong cone property. The set

$$\Theta := \{(x_1, x_2) \in \mathbb{R}^2 : 0 < x_1 < 1, |x_2| < x_1^4\}$$

has a cusp at the origin and so does not even have the cone property. However, the function u defined by $u(x) = \log x_1 \quad (x \in \Theta)$ is in $W^{2,2}(\Theta)$, since

$$\|u\|_{2,2,\Theta}^2 = 2\int_0^1 s^4[(\log s)^2 + s^{-2} + s^{-4}]\,ds < \infty;$$

but evidently $u \notin C(\Theta)$, and so the analogue of Theorem 3.20 does not hold for $W^{2,2}(\Theta)$. This emphasizes the need for some restrictions on the boundary if results like Theorem 3.20 are to hold. We shall not go into this matter very deeply here, and shall be content to give some idea of how the cone condition can be used to establish embedding theorems.

Suppose, for example, that Ω has the cone property and that $p \in (n, \infty)$. Let $u \in C^\infty(\Omega) \cap W^{1,p}(\Omega)$, choose any point in Ω and take it as the origin of coordinates; let $\phi \in C^\infty([0,\infty))$ be such that $0 \leqslant \phi \leqslant 1$, $\phi(r) = 1$ for $0 \leqslant r \leqslant \frac{1}{2}a$ (a as in Definition 4.15), and $\phi(r) = 0$ for $r \geqslant a$. Then

$$u(0) = -\int_0^a \frac{\partial}{\partial r}(\phi u)\,dr,$$

and so

$$\begin{aligned}
|u(0)| &= \text{const.}\left|\int_{V_0} \frac{\partial}{\partial r}(\phi u)\,|x|^{1-n}\,dx\right| \\
&\leqslant \text{const.}\|u\|_{1,p,\Omega}\left(\int_{V_0} |x|^{(1-n)p'}\,dx\right)^{1/p'} \\
&\leqslant \text{const.}\|u\|_{1,p,\Omega}\int_0^a r^{-1+(p-n)(p-1)^{-1}}\,dr \\
&\leqslant \text{const.}\,\|u\|_{1,p,\Omega}
\end{aligned}$$

since $p > n$. Since, by Theorem 3.2, $C^\infty(\Omega) \cap W^{1,p}(\Omega)$ is dense in $W^{1,p}(\Omega)$, it follows that given any $v \in W^{1,p}(\Omega)$, there is a sequence (v_m) in $C^\infty(\Omega) \cap W^{1,p}(\Omega)$ which converges to v in $W^{1,p}(\Omega)$, and that for all $k, l \in \mathbb{N}$ and all $x \in \Omega$,

$$|v_k(x) - v_l(x)| \leqslant \text{const.} \|v_k - v_l\|_{1,p,\Omega},$$

$$|v_l(x)| \leqslant \text{const.} \|v_l\|_{1,p,\Omega}.$$

Hence $v \in C_B(\Omega)$, the space of all continuous, bounded functions on Ω; the same methods can be used to prove that, in general, if Ω has the cone property then

$$W^{m,p}(\Omega) \to C_B(\Omega) \quad \textit{if} \quad n < mp.$$

We cannot, however, go as far as we would like and claim that $v \in C(\bar{\Omega})$: to do this, more constraints have to be imposed. The difficulty is that we are unable, with merely the cone condition available, to prove that v is *uniformly continuous* on Ω. To do this, the strong cone condition can be used. Thus suppose that Ω has the strong cone property and again assume that $n < p < \infty$. Let $u \in C^\infty(\Omega) \cap W^{1,p}(\Omega)$, and let $y, z \in \Omega$, with $s := |y - z| \leqslant d$ (d as in Definition 4.15); put $\Omega_1 = V_{y,s} \cap V_{z,s}$. Then

$$|u(y) - u(z)|cs^n \leqslant \left| \int_{\Omega_1} [u(y) - u(z)]\,dx \right|$$

$$\leqslant \left| \int_{\Omega_1} [u(y) - u(x)]\,dx \right| + \left| \int_{\Omega_1} [u(x) - u(z)]\,dx \right|.$$

Now take the origin to be at y. Thus

$$\left| \int_{\Omega_1} [u(y) - u(x)]\,dx \right| \leqslant \left| \int \int_0^s t^{n-1} \int_0^t \frac{\partial u}{\partial r}\,dr\,dt\,d\omega \right|$$

$$= \left| \int \int_0^s \frac{\partial u}{\partial r} \int_r^s t^{n-1}\,dt\,dr\,d\omega \right|$$

$$\leqslant \frac{s^n}{n} \int_{V_{y,s}} \left| \frac{\partial u}{\partial r} \right| |x|^{1-n}\,dx$$

$$\leqslant \frac{s^n}{n} \|\nabla u\|_{p,\Omega} \left(\int_{V_{y,s}} |x|^{(1-n)p'}\,dx \right)^{1/p'}$$

$$\leqslant \text{const.} \|\nabla u\|_{p,\Omega}\, s^{\gamma + n},$$

where $\gamma = 1 - n/p$. Then for all $y, z \in \Omega$, with $|y - z| = s$,

$$|u(y) - u(z)| \leqslant \text{const.} \|\nabla u\|_{p,\Omega} s^\gamma.$$

Hence u is not only bounded on Ω, but is uniformly continuous on Ω, and may thus be extended by continuity to the whole of $\bar{\Omega}$, the extension being in $C^{\gamma}(\bar{\Omega})$. The usual approximation procedure now shows that any element of $W^{1,p}(\Omega)$ may be identified with an element of $C^{\gamma}(\bar{\Omega})$.

The cases $p \leqslant n$ may be handled similarly; the details of this, and of more complicated embedding theorems involving $W^{k,p}(\Omega)$ and analogous to those of §3, we leave to the reader. This task is a worthwhile one, even bearing in mind the very general results of this type already available in Theorem 4.13 under the hypothesis of minimal smoothness, because of the simplicity of the proofs and the apparent lack of a direct connection between the cone conditions and the minimal smoothness assumption (see Adams [1], pp. 67–8 and Fraenkel [3], p. 409 for some information about connections).

4.6. *More about compact embeddings*

So far we have been forced to impose the condition that $\partial\Omega$ should be at least minimally smooth in order to obtain a compact embedding of one Sobolev space in a Lebesgue space, or in another Soboleve space. It is, on occasion, possible to weaken even this mild smoothness requirement, and we give two examples of this phenomenon, beginning with a very simple result in which no condition whatever is imposed on $\partial\Omega$.

Theorem 4.16 Let Ω be a bounded open subset of $\mathbb{R}^n$, and let $k \in \mathbb{N}$ and $p \in (1, \infty)$. Then for any $q \in [1, p)$, the space $W^{k,p}(\Omega)$ is compactly embedded in $W^{k-1,q}(\Omega)$. ■

Proof. By induction on k, it is enough to prove that $W^{1,p}(\Omega)$ is compactly embedded in $L^q(\Omega)$, with $1 \leqslant q < p$. Moreover, it is enough to prove that $W^{1,p}(\Omega)$ is compactly embedded in $L^1(\Omega)$, since we may define $\theta \in [0, 1)$ by $1/q = 1 - \theta + \theta/p$, use Hölder's inequality to obtain the inequality

$$\|u\|_{q,\Omega} \leqslant \|u\|_{1,\Omega}^{1-\theta} \|u\|_{p,\Omega}^{\theta} \tag{4.33}$$

and apply (4.33), with $u = u_m - u_l$, to a sequence (u_m) which is convergent in $L^1(\Omega)$ and bounded in $L^p(\Omega)$.

We therefore take a bounded set $B \subset W^{1,p}(\Omega)$ and seek to prove that it is relatively compact in $L^1(\Omega)$. Hölder's inequality shows immediately that B is bounded in $L^1(\Omega)$, and we simply have to prove that B is equicontinuous in the mean to obtain the result (cf. Fučik, John, and Kufner [1, Theorem 2.13.1]). That is, we must show that given any $\varepsilon > 0$, there exists $\delta > 0$ such that

$$\sup_{u \in B} \int_{\Omega} |u(x+h) - u(x)| \, dx < \varepsilon \quad \text{if } |h| < \delta.$$

Here the understanding is that u is extended by 0 outside Ω. Let $\varepsilon > 0$ and, for any $\eta > 0$, put $\Omega_\eta = \{x \in \Omega : \text{dist}(x, \partial\Omega) > \eta\}$. Then

$$\sup_{u\in B} \int_{\Omega\setminus\Omega_{2\delta}} |u(x+h) - u(x)|\,dx \leqslant 2\left(\sup_{u\in B} \|u\|_{p,\Omega}\right)|\Omega\setminus\Omega_{2\delta}|^{1/p'} < \varepsilon$$

if δ is small enough. Moreover, if $u \in C^\infty(\Omega) \cap W^{1,p}(\Omega)$,

$$\int_{\Omega_{2\delta}} |u(x+h) - u(x)|\,dx = \int_{\Omega_{2\delta}} \left| \int_0^1 \frac{d}{dt} u(x+th)\,dt \right| dx$$

$$\leqslant |h| \int_0^1 \int_{\Omega_\delta} |\nabla u(y)|\,dy\,dt.$$

This completes the proof, in view of the density of $C^\infty(\Omega) \cap W^{1,p}(\Omega)$ in $W^{1,p}(\Omega)$, by Theorem 3.2. □

Our second example is usually associated with Rellich, who proved it when $p = 2$. Note that $\partial\Omega$ merely has to be of class C.

Theorem 4.17. Let Ω be a bounded, open subset of $\mathbb{R}^n$ with boundary of class C, and let $p \in [1, \infty)$. Then $W^{1,p}(\Omega)$ is compactly embedded in $L^p(\Omega)$. ■

Proof. Let $\varepsilon > 0$. We claim that there is an open set $\Omega' \subset\subset \Omega$ such that for all $u \in W^{1,p}(\Omega)$ with $\|u\|_{1,p,\Omega} \leqslant 1$,

$$\int_{\Omega\setminus\Omega'} |u(x)|^p\,dx \leqslant \left(\frac{\varepsilon}{6}\right)^p. \tag{4.34}$$

To prove this, let $v \in C^\infty(\bar{\Omega})$ and let, with the usual notation (cf. (4.2)),

$$V_{\lambda,\beta} = \{y \in \mathbb{R}^n : y' \in Q_{n-1}(0, \lambda), |y_n - h(y')| < \beta\},$$

where h is the continuous function which gives the local representation of $\partial\Omega$; the points y with $y' \in Q_{n-1}(0, \lambda)$ and $h(y') < y_n < h(y') + \beta$ correspond to points of Ω. For points (y', z_n) with $h(y') < z_n < h(y') + \frac{1}{2}\beta$ we have, if $h(y') < \tau < h(y') + \beta$,

$$v(y', z_n) = -\int_{z_n}^{\tau} D_n v(y', \xi_n)\,d\xi_n + v(y', \tau).$$

Hence

$$|v(y', z_n)|^p \leqslant 2^{p-1} \left| \int_{z_n}^{\tau} D_n v(y', \xi_n) d\xi_n \right|^p + 2^{p-1} |v(y', \tau)|^p$$

$$\leqslant 2^{p-1} \beta^{p/p'} \int_{h(y')}^{h(y')+\beta} |D_n v(y', \xi_n)|^p d\xi_n + 2^{p-1} |v(y', \tau)|^p.$$

Integration of this with respect to τ throughout the interval $[h(y'), h(y') + \beta]$

gives

$$\beta|v(y', z_n)|^p \leqslant 2^p\beta^p \int_{h(y')}^{h(y')+\beta} |D_n v(y', \xi_n)|^p d\xi_n + 2^p \int_{h(y')}^{h(y')+\beta} |v(y', \tau)|^p d\tau.$$

Now integrate this with respect to y' throughout $Q_{n-1}(0, \lambda) := Q$, and with respect to z_n throughout $(h(y'), h(y')+\delta)$, where $0 < \delta \leqslant \beta$. Then we obtain

$$\int_Q \int_{h(y')}^{h(y')+\delta} |v(y', z_n)|^p dz_n\, dy' \leqslant 2^p\beta^{p-1}\delta \int_Q \int_{h(y')}^{h(y')+\beta} |D_n(y', \xi_n)|^p d\xi_n\, dy' + 2^p\delta\beta^{-1} \int_Q \int_{h(y')}^{h(y')+\beta} |v(y', \tau)|^p d\tau\, dy'. \tag{4.35}$$

In view of Theorem 4.7, (4.35) holds for all $v \in W^{1,p}(\Omega)$. All that now remains is to choose δ sufficiently small, and to apply (4.35), with v replaced by u, in connection with each of the finitely many sets $V_{\lambda,\beta}$ which form a covering of $\partial\Omega$: (4.34) then follows, with Ω' of the form $\{x \in \Omega: \text{dist}(x, \partial\Omega) > \rho\}$ for some small enough $\rho > 0$ (cf. Theorem 4.6(i)).

Now cover Ω' by open cubes $Q^{(1)}, \ldots, Q^{(m)}$, each with sides of length $d > 0$. From (3.23) it follows that

$$\|u\|_{p,\Omega'}^p \leqslant \|u\|_{p,\cup_1^m Q^{(j)}}^p \leqslant \sum_{j=1}^m \int_{Q^{(j)}} |u(x)|^p dx$$
$$\leqslant 2^p \omega_n^{p(1-1/n)} d^p \int_{\cup_1^m Q^{(j)}} |\nabla u(x)|^p dx + 2^p d^{n(1-p)} \sum_{j=1}^m \left| \int_{Q^{(j)}} u(x)\, dx \right|^p.$$

Thus if we choose $d = \delta\varepsilon$, for some suitably small δ (independent of ε), it follows that

$$\|u\|_{p,\Omega'}^p \leqslant \tfrac{1}{2}(\varepsilon/6)^p \int_{\cup_1^m Q^{(j)}} |\nabla u(x)|^p dx + 2^p(\delta\varepsilon)^{n(1-p)} \sum_{j=1}^m \left| \int_{Q^{(j)}} u(x)\, dx \right|^p. \tag{4.36}$$

Now define a map $T: W^{1,p}(\Omega) \to \mathbb{C}^m$ by $Tu = (\int_{Q^{(j)}} u(x)\, dx)$. This map is bounded and linear; and since $\dim \mathbb{C}^m < \infty$, the map T is compact. Hence there are elements $u_1, u_2, \ldots, u_s$ of the unit ball of $W^{1,p}(\Omega)$ such that for any $u \in W^{1,p}(\Omega)$ with $\|u\|_{1,p,\Omega} \leqslant 1$, there exists $t \in \{1, 2, \ldots, s\}$ such that

$$2^p(\delta\varepsilon)^{n(1-p)} \sum_{j=1}^m \left| \int_{Q^{(j)}} [u(x) - u_t(x)]\, dx \right|^p < \tfrac{1}{2}(\varepsilon/3)^p. \tag{4.37}$$

Thus we see, from (4.36) and (4.37), that

$$\|u - u_t\|_{p,\Omega'} < \tfrac{1}{3}\varepsilon;$$

and hence, with the aid of (4.34), we have

$$\|u - u_t\|_{p,\Omega} \leqslant \|u - u_t\|_{p,\Omega'} + \|u - u_t\|_{p,\Omega\setminus\Omega'} < \varepsilon,$$

for all $u \in W^{1,p}(\Omega)$ with $\|u\|_{1,p,\Omega} \leqslant 1$. The unit ball of $W^{1,p}(\Omega)$ is therefore totally bounded as a subset of $L^p(\Omega)$, and the proof is complete. □

Slight variants of this important result can easily be given.

Theorem 4.18. Let Ω be a bounded, open subset of $\mathbb{R}^n$, let $p \in [1, \infty)$ and let k, $l \in \mathbb{N}_0$, with $l > k$. Then $W_0^{l,p}(\Omega)$ is compactly embedded in $W_0^{k,p}(\Omega)$. If, in addition, the boundary of Ω is of class C, then $W^{l,p}(\Omega)$ is compactly embedded in $W^{k,p}(\Omega)$. ■

Proof. First we show that $W_0^{1,p}(\Omega)$ is compactly embedded in $L^p(\Omega)$, with no conditions imposed on $\partial\Omega$. Let B be an open ball in $\mathbb{R}^n$ with $\Omega \subset\subset B$. As explained in the discussion following Proposition 4.9, elements of $W_0^{1,p}(\Omega)$ may be extended by 0 outside Ω, and so may, when extended, be regarded as elements of $W^{1,p}(B)$. Let E be the corresponding extension operator. Since ∂B is of class C, $W^{1,p}(B)$ is compactly embedded in $L^p(B)$. Thus if (u_m) is a bounded sequence in $W_0^{1,p}(\Omega)$ then $\{Eu_m\}$ is a bounded sequence in $W^{1,p}(B)$ and hence has a subsequence $\{Eu_{m(r)}\}$ which converges in $L^p(B)$: thus $\{u_{m(r)}\}$ is convergent in $L^p(\Omega)$, and the proof is complete.

To deal with the rest of the theorem it is enough to show that for each $l \in \mathbb{N}$, the space $W^{l,p}(\Omega)$ is compactly embedded in $W^{l-1,p}(\Omega)$ if $\partial\Omega$ is of class C, because $W^{l-1,p}(\Omega)$ is continuously embedded in $W^{k,p}(\Omega)$ for all $k \in \mathbb{N}$ with $k < l$. An induction on l will do this. □

Of course, special cases of the assertions of this theorem about $W_0^{l,p}(\Omega)$ have been given earlier: cf. Theorem 3.7, for example. However, the present proof has its own interest.

4.7. *The Poincaré inequality*

We have already seen inequalities of Poincaré type valid for $W_0^{k,p}(\Omega)$ and $W^{1,p}(\Omega)$ (cf. the discussion at the end of §3). The inequality obtained for elements of $W^{1,p}(\Omega)$, however, required that Ω be convex and bounded. Here we weaken the convexity condition, replace it by the requirement that $\partial\Omega$ be of class C, and give an inequality valid for elements of $W^{k,p}(\Omega)$.

Theorem 4.19. Let Ω be a bounded domain in $\mathbb{R}^n$ with boundary of class C, let $k \in \mathbb{N}$ and suppose that $p \in [1, \infty)$. Then there is a positive constant c such that for all $u \in W^{k,p}(\Omega)$,

$$\|u\|_{k,p,\Omega} \leqslant c\left(\sum_{|\alpha|=k} \int_\Omega |D^\alpha u(x)|^p \, dx + \sum_{|\alpha|<k} \left|\int_\Omega D^\alpha u(x) \, dx\right|^p\right)^{1/p}. \tag{4.38}$$ ■

Proof. Suppose the result is false. Then there is a sequence (u_m) in $W^{k,p}(\Omega)$

such that for all $m \in \mathbb{N}$,

$$1 = \|u_m\|_{k,p,\Omega}^p > m\left(\sum_{|\alpha| = k} \int_\Omega |D^\alpha u_m(x)|^p \, dx + \sum_{|\alpha| < k} \left| \int_\Omega D^\alpha u_m(x) \, dx \right|^p \right). \tag{4.39}$$

Thus for all α with $|\alpha| = k$, we have $D^\alpha u_m \to 0$ in $L^p(\Omega)$ as $m \to \infty$. Also, by Theorem 4.18, there is a subsequence $(u_{m(l)})$ of (u_m) which converges in $W^{k-1,p}(\Omega)$, to u, say. Hence $u_{m(l)} \to u$ in $W^{k,p}(\Omega)$; and as $D^\alpha u = 0$ for all α with $|\alpha| = k$, the function u is a polynomial of degree at most $k-1$ (cf. Theorem 3.11). Since by (4.39),

$$\lim_{l \to \infty} \int_\Omega D^\alpha u_{m(l)}(x) \, dx = \int_\Omega D^\alpha u(x) \, dx = 0$$

whenever $|\alpha| < k$, it follows that $u = 0$, which contradicts the equality

$$1 = \lim_{l \to \infty} \|u_{m(l)}\|_{k,p,\Omega} = \|u\|_{k,p,\Omega}.$$ □

Inequality (4.38) is sometimes called the (general) *Poincaré inequality*. It requires only mild smoothness conditions on $\partial\Omega$ to be satisfied, but as we shall presently see, it is false in the absence of any conditions on $\partial\Omega$.

4.8. *The approximation of bad boundaries by very smooth ones*

Our object here is to show that given a bounded domain (an open connected set), with possibly unpleasant boundary, there is a domain with smooth boundary, contained in the original domain, and with boundary arbitrarily close to that of the original domain. The treatment is due to Amick (unpublished), who based it upon work of Kellogg [1].

Theorem 4.20. Let Ω be a bounded domain in $\mathbb{R}^n$ and let U be an open set with $\bar{U} \subset \Omega$. Then there is a domain V with analytic boundary such that $\bar{U} \subset V \subset \bar{V} \subset \Omega$. ■

Proof. Let $x_0 \in \Omega$, let $0 < L < \operatorname{dist}(x_0, \partial\Omega)$, and let $Q^{(0)}$ be a closed cube with centre x_0 and sides of length L. Decompose $Q^{(0)}$ into 3^n congruent closed cubes with sides of length $\frac{1}{3}L$, and denote by Q_0 the unique $\frac{1}{3}L$-sided cube that contains x_0. By extension outward from $Q^{(0)}$ form a covering of $\mathbb{R}^n$ by closed cubes Q_i $(i \in \mathbb{N}_0)$ of side $\frac{1}{3}L$. Given any $i \in \mathbb{N}_0$, let $Q^{(i)}$ be the closed cube of side L, with centre the same as the centre of Q_i, and formed by the union of 3^n of the Q_j. Next, set $\Gamma_1 = \{Q_i : Q^{(i)} \subset \Omega\}$, and let $\tilde{\Gamma}_1 = \{\mathfrak{U} : \mathfrak{U}$ is a union of elements of Γ_1, $\mathfrak{U}$ is connected, $Q_0 \subset \mathfrak{U}\}$. Since $Q_0 \in \tilde{\Gamma}_1$ we have $\tilde{\Gamma}_1 \neq \varnothing$ and thus $\tilde{\Gamma}_1$ has an element, $\mathfrak{U}_1$ say, which is maximal in the sense of inclusion: $\mathfrak{U}_1 = \cup_{\mathfrak{U} \in \tilde{\Gamma}_1} \mathfrak{U}$, $\mathfrak{U}_1 = \bar{\mathfrak{U}}_1 \subset \Omega$, $\mathfrak{U}_1$ is connected, and $\mathfrak{U}_1$ is a finite union of the Q_i.

For each $i \in \mathbb{N}_0$, decompose Q_i into 2^n closed cubes with sides of length $\frac{1}{6}L$. This gives a covering of $\mathbb{R}^n$ by cubes $Q_{i,1}$ $(i \in \mathbb{N}_0)$, each of side $\frac{1}{6}L$: the enumeration is so arranged that $Q_{0,1}$ is one of the 2^n cubes which contain x_0. We now proceed as before to form Γ_2 and $\tilde{\Gamma}_2$ just as Γ_1 and $\tilde{\Gamma}_1$ were formed, but with $Q_{i,1}$ in place of Q_i for Γ_2, and $Q_{0,1}$ in place of Q_0 for $\tilde{\Gamma}_2$. There will be a maximal element $\mathfrak{U}_2$ of $\tilde{\Gamma}_2$ such that $Q_{0,1} \subset \mathfrak{U}_2$, $\mathfrak{U}_2$ is connected, $\mathfrak{U}_2$ is a finite union of $Q_{i,1}$'s, and $\mathfrak{U}_1 \subset \mathfrak{U}_2^\circ \subset \mathfrak{U}_2 = \overline{\mathfrak{U}}_2 \subset \Omega$.

We now continue this procedure inductively, halving the side of the cubes at each stage. In this way we obtain a sequence, $(\mathfrak{U}_i)_{i \in \mathbb{N}}$, of closed connected sets such that for each $i \in \mathbb{N}$, we have $x_0 \in \mathfrak{U}_i \subset \mathfrak{U}_{i+1}^\circ \subset \mathfrak{U}_{i+1} \subset \Omega$ and $\mathfrak{U}_i$ is a finite union of cubes of side $2^{-i+1}L/3$. We claim that $\cup_{i=1}^\infty \mathfrak{U}_i = \Omega$. To prove this, let $x \in \Omega$: since Ω is path-connected, there is a piecewise-smooth path γ: $[0, 1] \to \Omega$ which joins x_0 to x. Let $d = \operatorname{dist}(\gamma([0, 1]), \partial\Omega)$ (> 0), and choose $i \in \mathbb{N}$ so large that $L\sqrt{n}/2^{i-1} < d$. Let $z \in \gamma([0, 1])$; then z belongs to some closed cube $Q(z)$ with sides of length $2^{-i+1}L/3$; and if we let B be the union of all the cubes of side $2^{-i+1}L/3$ which have points in common with $Q(z)$, then $z \in B$ and $\operatorname{diam} B = 2^{-i+1}\sqrt{n}L < d$. Hence $B \subset \Omega$; thus $z \in Q(z) \in \Gamma_i$, and in fact, $z \in \mathfrak{U}_i$ since $z \in \cup_{z_1 \in \gamma([0,1])} Q(z_1) \in \tilde{\Gamma}_i$. Since z is an arbitrary point of $\gamma([0, 1])$ it follows that $x \in \mathfrak{U}_i$, which establishes our claim.

Now consider the given open set $U \subset\subset \Omega$. Since $\Omega = \cup_{i=1}^\infty \mathfrak{U}_i$, and the $\mathfrak{U}_i$ are nested, there exists $i_0 \in \mathbb{N}$ such that $U \subset \mathfrak{U}_{i_0}^\circ$. Put $T = \mathbb{R}^n \backslash \partial\mathfrak{U}_{i_0}$ and define $f: T \to \mathbb{R}$ by

$$f(x) = \int_{\partial\mathfrak{U}_{i_0}} |x - z|^{1-n}\, dz \qquad (x \in T).$$

Since the integrand is analytic with respect to x, the function f is analytic in T; also $f(x) \to \infty$ as $x \to \partial\mathfrak{U}_{i_0}$. Put $M = \max_{x \in \bar{U}} f(x)$, and observe that if $K > M$, then

$$U \subset W_K := \{x \in T : f(x) < K\} \cap \mathfrak{U}_{i_0}^\circ.$$

Let $S = \{x \in \mathfrak{U}_{i_0}^\circ : \nabla f(x) = 0\}$. By Sard's Theorem (cf. Lloyd [1]), the Lebesgue measure of $f(S)$ in $\mathbb{R}$ is zero, and thus there exists $K' > M$ such that $K' \notin f(S)$. Since $\partial W_{K'} = \{x \in \mathfrak{U}_{i_0}^\circ : f(x) = K'\}$, we see that for all $x \in \partial W_{K'}$ $\nabla f(x) \neq 0$. Thus by the Implicit-Function Theorem (cf. Dieudonné [1, p. 272]), $\partial W_{K'}$ is analytic. We then have $U \subset \bar{U} \subset W_{K'}$ and $\partial W_{K'}$ is analytic. The difficulty with this open set $W_{K'}$ is that we do not know that it is connected, since $\mathfrak{U}_{i_0}^\circ$ may be disconnected: it has all the other properties demanded of the set V. However, if U is connected, then some open connected component of $W_{K'}$, say W, must be such that $U \subset\subset W \subset\subset \Omega$; and as $\partial W \subset \partial W_{K'}$, ∂W is analytic.

To complete the proof, let $N_i = \{A : A \text{ is a domain in } \mathbb{R}^n, x_0 \in A \subset \mathfrak{U}_i\}$ and put $M_i = \cup_{A \in N_i} A$ $(i \in \mathbb{N})$. Just as we showed above that $\Omega = \cup_{i=1}^\infty \mathfrak{U}_i$, it follows that $\Omega = \cup_{i=1}^\infty M_i$; moreover, each M_i is a domain and $x_0 \in M_i \subset \mathfrak{U}_i$.

We may thus choose i_1 so large that $\bar{U} \subset M_{i_1}$. Application of the first part of our proof to M_{i_1} then shows that there is a domain V such that $M_{i_1} \subset\subset V \subset\subset \Omega$, and ∂V is analytic. The proof is complete. $\square$

4.9. *A counterexample*

Here we exhibit a bounded domain in $\mathbb{R}^2$ which illustrates the need for restrictions on the boundary if embedding theorems and inequalities of the kind we have been discussing (with the exception of Theorem 4.16) are to hold. The example is due to Fraenkel [3], although somewhat similar sets had been considered earlier by Courant and Hilbert [1, p. 521] and Mazja [1]. It consists of a subset S of $\mathbb{R}^2$ referred to as 'rooms and passages' and made up of an infinite sequence of square boxes ('rooms') of decreasing sizes joined together by pipes ('passages'). We give the details below, and illustrate by means of Fig. 1.

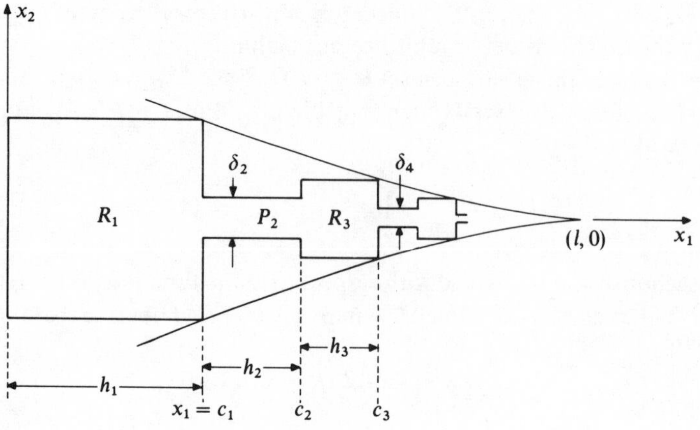

Fig. 1.

Let $(h_m)_{m\in\mathbb{N}}$ and $(\delta_{2m})_{m\in\mathbb{N}}$ be sequences of positive real numbers such that

$$\sum_1^{\infty} h_m := l < \infty; 0 < C \leqslant h_{m+1}/h_m \leqslant 1 \quad \text{and} \quad 0 < \delta_{2m} \leqslant h_{2m+1} \tag{4.40}$$

for all $m \in \mathbb{N}$, where C is independent of m. For each $m \in \mathbb{N}$ set

$$c_m = \sum_{i=1}^{m} h_i, \tag{4.41}$$

and for each positive odd integer j set

$$R_j = (c_j - h_j, c_j) \times (-\tfrac{1}{2}h_j, \tfrac{1}{2}h_j),\ P_{j+1} = [c_j, c_j + h_{j+1}] \times (-\tfrac{1}{2}\delta_{j+1}, \tfrac{1}{2}\delta_{j+1}). \tag{4.42}$$

Finally define $S \subset \mathbb{R}^2$ by

$$S = \bigcup_{j \in \mathbb{N},\, j\,\text{odd}} (R_j \cup P_{j+1}).$$

The numbers h_m and δ_{2m} may be chosen to our convenience, subject to the constraints above. Note that the real reason for the success of this kind of set S as a source of counterexamples is that the limit point $(l,0) \in \partial S$ prevents ∂S from being of class C: it is not possible to find local coordinates (y_1, y_2) such that ∂S has a local representation of the form $y_2 = h(y_1)$, with h continuous, in any neighbourhood of $(l,0)$. However, S is not an especially pathological set, since ∂S is a rectifiable Jordan curve, and in the typical cases with which we shall be concerned, it can be shown (cf. Fraenkel [3, Appendix I]) that given any bounded neighbourhood V of $\bar{S}$, there is a $C^{0,\lambda}$ homeomorphism of V that maps S onto the open unit ball in $\mathbb{R}^2$, and that S has the uniform C-regularity property.

Theorem 4.21. (i) Let $h_m = m^{-\frac{3}{2}}$ and $\delta_{2m} = (2m)^{-\frac{5}{2}}$ $(m \in \mathbb{N})$. Then given any $q \in (2, \infty)$, $W^{1,2}(S)$ is not embedded in $L^q(S)$.

(ii) Let $a \geqslant 3$ and suppose that $\delta_{2m} = h_{2m}^a$ for each $m \in \mathbb{N}$. Then $W^{1,2}(S)$ is not compactly embedded in $L^2(S)$.

(iii) Let $a > 3$ and suppose that $\delta_{2m} = h_{2m}^a$ for all $m \in \mathbb{N}$; let Ω be the interior of the union of $\bar{S}$ and the reflection of S in the x_2-axis. Then the Poincaré inequality (4.38) does not hold for $W^{1,2}(\Omega)$. ∎

Proof. (i) Define a function u on S by

$$u(x) = \begin{cases} j/\log 2j := K_j & \text{if } x \in R_j, \\ K_j + (K_{j+2} - K_j)(x_1 - c_j)/h_{j+1} & \text{if } x \in P_{j+1}, \end{cases}$$

for each positive odd integer j. We claim that $u \in W^{1,2}(S)$. This is because

$$\int_S |u(x)|^2\, dx = \int_{\cup R_j} |u(x)|^2\, dx + \int_{\cup P_{j+1}} |u(x)|^2\, dx$$

$$\leqslant \sum_j \left((j/\log 2j)^2 j^{-3} + [(j+2)/\log 2j]^2 (j+1)^{-4} \right) < \infty$$

and

$$\int_S |\nabla u(x)|^2\, dx = \sum_j (K_{j+2} - K_j)^2 (j+1)^{-1} < \infty$$

since $|K_{j+2}-K_j| < 2/\log(2j+4)$. However, if $q\in(2,\infty)$, then since $\sum_j j^{q-3}/(\log 2j)^q$ is divergent, $u\notin L^q(S)$. This proves (i).

(ii) Consider the sequence $u_1, u_3, \ldots$, where for each odd $k\in\mathbb{N}$,

$$u_k(x) = \begin{cases} 1/h_k & \text{if } x\in R_k, \\ 0 & \text{if } x\in S\backslash(P_{k-1}\cup R_k\cup P_{k+1}), \end{cases}$$

and $\nabla u(x) = \pm\left(1/(h_k h_{k\mp 1}), 0\right)$, in $P_{k\mp 1}$. Then by (4.40),

$$\int_S |u_k(x)|^2\, dx = 1+\tfrac{1}{3}h_k^{-2}(h_{k-1}\delta_{k-1}+h_{k+1}\delta_{k+1}) \leqslant 1+\text{const. } h_k^2,$$

and

$$\int_S |\nabla u_k(x)|^2\, dx = (h_k h_{k-1})^{-2}\, h_{k-1}\,\delta_{k-1} + (h_k h_{k+1})^{-2}\, h_{k+1}\delta_{k+1} \leqslant \text{const.}$$

Hence (u_k) is bounded in $W^{1,2}(S)$. But if $j\neq k$,

$$\int_S |u_j(x)-u_k(x)|^2\, dx > \int_{R_j\cup R_k} |u_j(x)-u_k(x)|^2\, dx = 2,$$

and so (u_k) has no subsequence which converges in $L^2(S)$. Part (ii) follows.

(iii) Extend the functions u_k $(k=3,5,\ldots)$ of (ii) to Ω as odd functions of x_1. Then

$$\int_\Omega |\nabla u_k(x)|^2\, dx \to 0 \quad \text{as } k\to\infty;$$

and

$$\int_\Omega u_k(x)\, dx = 0, \qquad \int_\Omega |u_k(x)|^2\, dx > 2$$

for $k = 3,5,\ldots$. It is now clear that no inequality of the form

$$\int_\Omega |u(x)|^2\, dx \leqslant \text{const.}\left(\int_\Omega |\nabla u(x)|^2\, dx + \left|\int_\Omega u(x)\, dx\right|^2\right)$$

can hold for arbitrary elements of $W^{1,2}(S)$. □

This theorem shows the usefulness of S as an 'all-purpose' counter-example. We shall return in §5 to discuss the Poincaré inequality in more detail in the light of this particular example.

5. Measures of non-compactness

Here we link up the ball measure of non-compactness $\tilde{\beta}(I)$ of the embedding I: $W^{1,p}(\Omega)\to L^p(\Omega)$ (with $1\leqslant p<\infty$), the quantities $\|I\|_{\mathscr{K}}$ and $\alpha(I)$ defined in (II.2.2) and (II.2.1) and the Poincaré inequality of §4.7.

5.1. *Bounded and precompact sets in* $L^p(\Omega)$, with $1 \leqslant p < \infty$

We denote by $\mathscr{F}_r(X,Y)$ the set of linear operators from X to Y which are bounded and have finite rank; we write $\mathscr{F}_r(X, X)$ as $\mathscr{F}_r(X)$. Recall that $\tilde{\psi}(K)$ is the ball measure of non-compactness of the bounded set K.

Theorem 5.1. For any bounded set $K \subset L^p(\Omega)$, with $1 \leqslant p < \infty$,

$$2\tilde{\psi}(K) \geqslant \inf_{P \in \mathscr{F}_r(L^p(\Omega))} \sup\{\|f - Pf\|_{p,\Omega} : f \in K, \|f\|_{p,\Omega} = 1\} \geqslant \tilde{\psi}(K), \tag{5.1}$$

that is,

$$2\tilde{\beta}(I_K) \geqslant \alpha(I_K) \geqslant \tilde{\beta}(I_K),$$

where I_K is the identity map on $\bar{K}$, the closure of K. ■

Proof. Let $\varepsilon > \tilde{\psi}(K)$. Then there exist $g_1, g_2, \ldots, g_N \in L^p(\Omega)$ such that for any $f \in K$ and some $i \in \{1, 2, \ldots, N\}$,

$$\|f - g_i\|_{p,\Omega} < \varepsilon. \tag{5.2}$$

Set any function in $L^p(\Omega)$ to be zero outside Ω and choose Ω' to be a cube which is such that, for a given $\varepsilon > 0$,

$$\left(\int_{\Omega \setminus \Omega'} |g_i|^p \, dx\right)^{1/p} < \tfrac{1}{2}\delta \qquad (i = 1, 2, \ldots, N). \tag{5.3}$$

Let $\Omega' = \Omega \cap \cup_{j=1}^m Q_j$, where the Q_j are congruent cubes of diameter h, and define

$$(Pf)(x) = \sum_{j=1}^m f_{Q_j} \chi_{Q_j}(x), \qquad f_{Q_j} = |Q_j|^{-1} \int_{Q_j} f(y) \, dy.$$

Then

$$\begin{aligned} \|g_i - Pg_i\|_{p,\Omega'}^p &= \sum_{j=1}^m \int_{Q_j} \left| |Q_j|^{-1} \int_{Q_j} [g_i(x) - g_i(y)] \, dy \right|^p dx \\ &\leqslant \sum_{j=1}^m \int_{Q_j} |Q_j|^{-1} \int_{Q_j} |g_i(x) - g_i(y)|^p \, dy \, dx \\ &\leqslant \sup_{|z| < h} \int_{\Omega} |g_i(x) - g_i(x+z)|^p \, dx \to 0 \end{aligned}$$

as $h \to 0$, by Lemma 1.3. From the latter result and (5.3) it follows that we may choose h such that

$$\|g_i - Pg_i\|_{p,\Omega} < \delta \qquad (i = 1, 2, \ldots, N). \tag{5.4}$$

Also

$$\|Pf\|_{p,\Omega}^p = \sum_{j=1}^{m} \int_{Q_j} \left| |Q_j|^{-1} \int_{Q_j} f(y)\,dy \right|^p dx$$
$$\leqslant \sum_{j=1}^{m} \int_{Q_j} |Q_j|^{-1} \int_{Q_j} |f(y)|^p\,dy\,dx \leqslant \|f\|_{p,\Omega}^p, \tag{5.5}$$

and so $P \in \mathscr{F}_r(L^p(\Omega))$, with rank $\leqslant m$. From (5.2) – (5.5) we therefore obtain

$$\|f - Pf\|_{p,\Omega} \leqslant \|f - g_i\|_{p,\Omega} + \|g_i - Pg_i\|_{p,\Omega} + \|P(g_i - f)\|_{p,\Omega}$$
$$< \varepsilon + \delta + \|g_i - f\|_{p,\Omega}$$
$$< 2\varepsilon + \delta,$$

and hence $\alpha(I_{\tilde{K}}) \leqslant 2\tilde{\psi}(K)$. The second inequality in (5.1) is proved in Proposition II.2.7 since $\tilde{\psi}(K) = \beta(I_K)$ □

Corollary 5.2. A set $K \subset L^p(\Omega)$ (with $1 \leqslant p < \infty$) is precompact if, and only if, it is bounded and given any $\varepsilon > 0$ there exists $P_\varepsilon \in \mathscr{F}_r(L^p(\Omega))$ such that for all $f \in K$,

$$\|f - P_\varepsilon f\|_{p,\Omega} \leqslant \varepsilon \|f\|_{p,\Omega}.$$ ■

Corollary 5.3. Let $T: X \to L^p(\Omega)$ (with $1 \leqslant p < \infty$) be linear and compact, where X is any Banach space. Then $\alpha(T) = 0$. This implies in particular that $L^p(\Omega)$ has the approximation property; that is, any compact operator in $L^p(\Omega)$ is the uniform limit of $\mathscr{F}_r(L^p(\Omega))$ maps. ■

Proof. Since T is compact its range is precompact in $L^p(\Omega)$. Hence from Corollary 5.2, $\|Tf - P_\varepsilon Tf\|_{p,\Omega} \leqslant \varepsilon \|Tf\|_{p,\Omega} \leqslant \varepsilon \|T\|\,\|f\|_X$ and T is therefore the uniform limit of the operators $P_\varepsilon T$ which clearly lie in $\mathscr{F}_r(X, L^p(\Omega))$. Thus $\alpha(T) = 0$. □

Corollary 5.4. If $T \in \mathscr{B}(X, L^p(\Omega))$ (with $1 \leqslant p < \infty$) then

$$\|T\|_{\mathscr{K}} = \alpha(T) = \operatorname{dist}[T, \mathscr{F}_r(X, L^p(\Omega))].$$ ■

Proof. Given $\delta > 0$ there exists a $K \in \mathscr{K}(X, L^p(\Omega))$ such that $\|T - K\| < \|T\|_{\mathscr{K}} + \delta$. From Corollary 5.3 there is a $P \in \mathscr{F}_r(X, L^p(\Omega))$ such that $\|K - P\| < \delta$, and this gives $\|T - P\| < \|T\|_{\mathscr{K}} + 2\delta$; whence $\alpha(T) \leqslant \|T\|_{\mathscr{K}}$. The reverse inequality is trivial. □

In application the following version of Corollary 5.2 is very useful; it is related to the criterion of M. Riesz for precompactness via the Ascoli–Arzelà Theorem used in the proof of Theorem 4.16 and proved in Fučik, John and Kufner [1, Theorem 2.13.1].

Corollary 5.5. A set $K \subset L^p(\Omega)$ is precompact if, and only if, it is bounded and, given $\varepsilon > 0$, there exists a bounded subset Ω' of Ω such that K is precompact in $L^p(\Omega')$ and

$$\int_{\Omega \setminus \Omega'} |f|^p \, dx \leqslant \varepsilon^p \|f\|_{p,\Omega}^p \qquad (f \in K). \tag{5.6} \blacksquare$$

Proof. Let K be precompact and $\varepsilon > 0$. Then K is bounded and from the proof of Theorem 5.1 there exists a bounded subset Ω' of Ω and $P \in \mathscr{F}_r(L^p(\Omega))$ such that

$$\|f - Pf\|_{p,\Omega} \leqslant \varepsilon \|f\|_{p,\Omega} \qquad (f \in K)$$

with $Pf(x) = 0$ in $\Omega \setminus \Omega'$. Hence (5.6) is satisfied. Also, arguing as in the proof of Theorem 5.1 but now in $L^p(\Omega')$, and, for a given $\delta > 0$, choosing the g_i such that $\|f - g_i\|_{p,\Omega'} < \delta$, we obtain a $P_\delta \in \mathscr{F}_r(L^p(\Omega))$ such that

$$\|f - P_\delta f\|_{p,\Omega'} \leqslant \delta \|f\|_{p,\Omega} \qquad (f \in K).$$

This implies that I_K is compact as a map from $L^p(\Omega)$ into $L^p(\Omega')$ and hence K is precompact in $L^p(\Omega')$.

Conversely, let K be precompact in $L^p(\Omega')$ and suppose (5.6) is satisfied. From Theorem 5.1, given $\varepsilon > 0$, there exists a $P \in \mathscr{F}_r(L^p(\Omega'))$ such that

$$\|f - Pf\|_{p,\Omega'} \leqslant \varepsilon \|f\|_{p,\Omega'} \qquad (f \in K).$$

Set $Rf(x) = \chi_{\Omega'}(x)\, Pf(x)$. Then $R \in \mathscr{F}_r(L^p(\Omega))$ and

$$\begin{aligned} \|f - Rf\|_{p,\Omega}^p &= \|f\|_{p,\Omega \setminus \Omega'}^p + \|f - Pf\|_{p,\Omega'}^p \\ &< 2\varepsilon^p \|f\|_{p,\Omega}^p. \end{aligned}$$

Corollary 5.2 now implies that K is precompact. □

5.2. *A formula for* $\alpha(I)$ *and* $\tilde{\beta}(I)$ *where* $I: W^{1,p}(\Omega) \to L^p(\Omega)$ *with* $1 \leqslant p < \infty$

We know from Theorem 4.17 that if Ω is bounded and has boundary of class C then I is compact. However, even if there are no conditions imposed on $\partial\Omega$, then given any open set $\Omega_0 \subset\subset \Omega$, Theorem 4.20 ensures that if Ω is connected then there is a domain V with analytic boundary, such that $\Omega_0 \subset\subset V \subset\subset \Omega$. Hence the natural embedding of $W^{1,p}(\Omega)$ into $L^p(\Omega_0)$ is compact, since it may be represented as a composition of the natural maps

$$W^{1,p}(\Omega) \to W^{1,p}(V) \to L^p(V) \to L^p(\Omega_0),$$

in which the outer two maps are continuous and the middle one is compact. This clarifies the role of the boundary when I is not compact and suggests that we should look at L^p integrals over boundary strips. However, it need not be the whole of $\partial\Omega$ which is responsible for any lack of compactness: Theorem 4.21 (ii) provides an example of a bounded domain Ω for which I is not compact and in this example the difficulty arises because of one particular point in $\partial\Omega$.

These considerations lead us to consider any bounded domain Ω in $\mathbb{R}^n$ and to associate with any closed, non-empty subset A of $\bar{\Omega}$ and any $\varepsilon > 0$ the set

$$A(\varepsilon) := \{x \in \Omega : \text{dist}(x, A) < \varepsilon\}.$$

Given any such A a family $\mathfrak{U}_A$ of domains will be called *A-admissible* if:

(i) each $U \in \mathfrak{U}_A$ is contained in $\Omega \backslash A$;

(ii) given any $\varepsilon > 0$ there exists $U \in \mathfrak{U}_A$ such that $\Omega \backslash A(\varepsilon) \subset U \subset \bar{U} \subset \bar{\Omega} \backslash A$.

If Ω is unbounded we shall use the same terminology as above but with a slight change: A will now be permitted to be empty, in which case we shall set

$$A(\varepsilon) = \{x \in \Omega : |x| > 1/\varepsilon\};$$

while if $A \neq \varnothing$ we shall write

$$A(\varepsilon) = \{x \in \Omega : \text{dist}(x, A) < \varepsilon \text{ or } |x| > 1/\varepsilon\}.$$

Next let X be a Banach space with norm $\|\bullet\|_X$, let Ω be a domain in $\mathbb{R}^n$ and $T \in \mathscr{B}(X, \mathrm{L}^p(\Omega))$, with $1 \leqslant p < \infty$. Given any set A as above we write, for each $\varepsilon > 0$,

$$\Gamma_T(\varepsilon, A) = \sup\{\|Tu\|^p_{p, A(\varepsilon)} : \|u\|_X = 1\}.$$

If there is no ambiguity about which set A is meant we shall write this simply as $\Gamma_T(\varepsilon)$. Since $\Gamma_T(\varepsilon)$ is monotonic decreasing as $\varepsilon \to 0$, the limit $\lim_{\varepsilon \to 0} \Gamma_T(\varepsilon)$ exists and will be denoted by $\Gamma_T(0, A)$ or $\Gamma_T(0)$.

An A-admissible family of domains $\mathfrak{U}_A$ will be called a *T-compact family* if, for all $U \in \mathfrak{U}_A$, the operator T is compact when viewed as a map from X to $\mathrm{L}^p(U)$. On account of Corollary 5.3, $\mathfrak{U}_A$ is a T-compact family if, and only if, for any $U \in \mathfrak{U}_A$ and $\varepsilon > 0$ there exists a $P \in \mathscr{F}_r(X, \mathrm{L}^p(U))$ such that

$$\|Tf - Pf\|_{p,U} \leqslant \varepsilon \|f\|_X \qquad (f \in X). \tag{5.7}$$

We need the following simple result.

Lemma 5.6. Given $P \in \mathscr{F}_r(X, \mathrm{L}^p(\Omega))$ and $\varepsilon > 0$, there exist an $R \in \mathscr{F}_r(X, \mathrm{L}^p(\Omega))$ and $\Omega' \subset\subset \Omega$ such that $\|P - R\| < \varepsilon$ and the range of R is a subset of $\mathrm{C}_0^\infty(\Omega')$. If $\Omega_0 \subset \Omega$, $P \in \mathscr{F}_r(X, \mathrm{L}^p(\Omega))$ and $\varepsilon > 0$ are given there exists $R \in \mathscr{F}_r(X, \mathrm{L}^p(\Omega))$ such that $\|(P - R)f\|_{p, \Omega_0} \leqslant \varepsilon \|f\|_X$ and the range of R is a subset of $\mathrm{C}_0^\infty(\Omega_0)$. ■

Proof. There exist linearly independent functions u_i $(i = 1, 2, \ldots, N)$ in $\mathrm{L}^p(\Omega)$ such that $\|u_i\|_{p,\Omega} = 1$ and

$$Pf = \sum_{i=1}^{N} c_i(f)\, u_i \qquad (f \in X).$$

On the finite-dimensional range of P all norms are equivalent and hence there exists a positive constant K such that

$$\sum_{i=1}^{N} |c_i(f)| \leqslant K\|Pf\|_{p,\Omega} \leqslant K\|P\|\,\|f\|_X.$$

We choose $\phi_i \in C_0^\infty(\Omega)$ such that $\|u_i - \phi_i\| < \varepsilon/(K\|P\|)$ $(i = 1, 2, \ldots, N)$ and set $Rf = \Sigma_{i=1}^N c_i(f)\phi_i$. Then $R \in \mathscr{F}_r(X, L^p(\Omega))$ and

$$\|Pf - Rf\|_{p,\Omega} \leqslant \sum_{i=1}^{N} |c_i(f)|\,\|u_i - \phi_i\|_{p,\Omega} \leqslant \varepsilon\|f\|_X.$$

Furthermore, supp $Rf \subset \cup_{i=1}^N$ supp $\phi_i \subset\subset \Omega$ and the first part of the lemma is proved. The second part is proved analogously on observing that $f \mapsto \chi_{\Omega_0} Pf \in \mathscr{F}_r(X, L^p(\Omega_0))$ and then choosing $\phi_i \in C_0^\infty(\Omega_0)$.

Theorem 5.7. Let $T \in \mathscr{B}(X, L^p(\Omega))$ with $1 \leqslant p < \infty$, let A be a closed subset of $\partial\Omega$ ($A \neq \varnothing$ if Ω is bounded), and suppose there exists an A-admissible family $\mathfrak{U}_A$ which is T-compact. Then $\Gamma_T(0) = [\tilde{\beta}(T)]^p = [\alpha(T)]^p$, where $\tilde{\beta}(T)$ is the ball measure of non-compactness of T and $\alpha(T) = \lim_{n\to\infty} a_n(T) = \inf\{\|T - P\| : P \in \mathscr{F}_r(X, L^p(\Omega))\}$. ■

Proof. We prove first that $\Gamma_T(0) \leqslant [\tilde{\beta}(T)]^p$. Suppose that this inequality is false, in which case there exists a $\delta > 0$ such that

$$\zeta := [\Gamma_T(0) - 2\delta]^{1/p} - \tilde{\beta}(T) - \delta > 0.$$

There are functions $g_1, g_2, \ldots, g_l \in L^p(\Omega)$ such that, given any $u \in X$ with $\|u\|_X < 1$, the bound $\|Tu - g_i\|_{p,\Omega} \leqslant \tilde{\beta}(T) + \delta$ holds for some $i \in \{1, 2, \ldots, l\}$. Choose $\varepsilon_1 > 0$ so small that $\Gamma_T(\varepsilon_1) \leqslant \Gamma_T(0) + \delta$; there exists $u_1 \in X$, with $\|u_1\|_X < 1$, such that $\|Tu_1\|_{p,A(\varepsilon_1)}^p \geqslant \Gamma_T(0) - \delta$. Let $\varepsilon_2 \in (0, \varepsilon_1)$ be such that $\|Tu_1\|_{p,A(\varepsilon_2)}^p \leqslant \delta$, so that

$$\|Tu_1\|_{p,A(\varepsilon_1)\backslash A(\varepsilon_2)}^p \geqslant \Gamma_T(0) - 2\delta.$$

Hence, for some $l_1 \in \{1, 2, \ldots, l\}$,

$$\begin{aligned}[\Gamma_T(0) - 2\delta]^{1/p} &\leqslant \|Tu_1\|_{p,A(\varepsilon_1)\backslash A(\varepsilon_2)} \\ &\leqslant \|g_{l_1}\|_{p,A(\varepsilon_1)\backslash A(\varepsilon_2)} + \tilde{\beta}(T) + \delta,\end{aligned}$$

which shows that

$$\zeta \leqslant \|g_{l_1}\|_{p,A(\varepsilon_1)\backslash A(\varepsilon_2)}.$$

Now let $u_2 \in X$, with $\|u_2\|_X < 1$, be such that $\|Tu_2\|_{p,A(\varepsilon_2)}^p > \Gamma_T(0) - \delta$; there exists $\varepsilon_3 \in (0, \varepsilon_2)$ and $l_2 \in \{1, 2, \ldots, l\}$ such that

$$\zeta \leqslant \|g_{l_2}\|_{p,A(\varepsilon_2)\backslash A(\varepsilon_3)}.$$

Proceeding in this way we obtain a strictly decreasing sequence (ε_i) of positive numbers and a sequence (l_i) of integers, with $l_i \in \{1, 2, \ldots, l\}$ for all i, such that for $i = 2, 3, \ldots,$

$$\zeta \leqslant \|g_{l_i}\|_{p, A(\varepsilon_i)\setminus A(\varepsilon_{i+1})}.$$

Infinitely many of the l_i must be equal, to l_0 say; but this implies that $\|g_{l_0}\|_{p,\Omega} = \infty$ and we have a contradiction. Thus $\Gamma_T(0) \leqslant [\tilde{\beta}(T)]^p$.

Since $\tilde{\beta}(T) \leqslant \alpha(T)$ by Proposition II. 2.7, we need to prove that $[\alpha(T)]^p \leqslant \Gamma_T(0)$. Let $\delta > 0$ and choose $\varepsilon > 0$ such that $\Gamma_T(\varepsilon) < \Gamma_T(0) + \delta$. Let $U \in \mathfrak{U}_A$ be such that $\Omega\setminus A(\varepsilon) \subset U \subset \bar{U} \subset \bar{\Omega}\setminus A$. Then

$$\|Tf\|^p_{p,\Omega\setminus U} \leqslant [\Gamma_T(0) + \delta]\,\|f\|^p_X. \tag{5.8}$$

By (5.7) there exists a $P \in \mathscr{F}_r(X, L^p(U))$ such that

$$\|Tf - Pf\|^p_{p,U} \leqslant \delta \|f\|^p_X \tag{5.9}$$

and, in view of Lemma 5.6, we may suppose that supp $Pf \subset U$ for all $f \in X$. Hence by (5.8) and (5.9),

$$\|Tf - Pf\|^p_{p,\Omega} \leqslant [\Gamma_T(0) + 2\delta]\,\|f\|^p_X,$$

whence $\alpha(T) \leqslant [\Gamma_T(0) + 2\delta]^{1/p}$. Since δ is arbitrary, $[\alpha(T)]^p \leqslant \Gamma_T(0)$ and the proof is complete. □

Remarks 5.8. 1. Suppose that Ω is a bounded domain in $\mathbb{R}^n$ such that for open balls (or cubes) S_r with centre the origin and diameter $2r$, the domains $U_r := \Omega \cap S_r$ $(0 < r < R)$ have boundary of class C. Then each of the embeddings $W^{1,p}(\Omega) \to L^p(U_r)$ is compact and hence, with $A = \partial\Omega\setminus S_R$, the family $\{U_r: 0 < r < R\}$ is A-admissible and I-compact, where I is the natural embedding $W^{1,p}(\Omega) \to L^p(\Omega)$. This is the case in the 'rooms and passages' example discussed in §4.9; in this example we could take the S_r to be cubes and A to be the set consisting of the single point $(l, 0)$.

2. Since $\alpha(I)$, $\tilde{\beta}(I) \leqslant \|I\| \leqslant 1$ for all domains Ω it is natural to ask whether $\tilde{\beta}(I)$, say, can be equal to 1 or whether $\tilde{\beta}(I) < 1$ for all domains Ω no matter what $\partial\Omega$ is like. The ubiquitous 'rooms and passages' example can be used to show that it is possible for $\tilde{\beta}(I)$ to be 1. All we need to do is to take S to be as in Theorem 4.2(ii), with $a > 3$, define the functions u_k as in that part of Theorem 4.2, and observe that given $\varepsilon > 0$, for all large enough k,

$$\begin{aligned}\int_{(\partial S)(\varepsilon)} |u_k(x)|^2\,dx &= \int_S |u_k(x)|^2\,dx \\ &= 1 + \tfrac{1}{3} h_k^{-2}\,(h_{k+1}^{a+1} + h_{k-1}^{a+1}) \to 1\end{aligned}$$

as $k \to \infty$, and

$$\int_S |\nabla u_k(x)|^2\,dx = h_k^{-2}\,(h_{k+1}^{a-1} + h_{k-1}^{a-1}) \leqslant h_k^{a-3}(1 + C^{1-a}).$$

Hence

$$\limsup_{k\to\infty} \|u_k\|_{1,2,S}^2 = 1$$

and thus

$$\Gamma_I(\varepsilon, \partial S) \geqslant \limsup_{k\to\infty}\left(\int_{(\partial S)(\varepsilon)} |u_k(x)|^2\,dx / \|u_k\|_{1,2,S}^2\right) = 1.$$

It follows that $\Gamma_I(0, \partial S) = 1$ and thus, by Theorem 5.7, $\tilde{\beta}(I) = 1$.

5.3. *The Poincaré inequality and* $\alpha(I)$

The Poincaré inequality for $W^{1,p}(\Omega)$, when $|\Omega| < \infty$, is

$$\|f - f_\Omega\|_{p,\Omega} \leqslant K_{p,\Omega,n}\|\nabla f\|_{p,\Omega}, \qquad f_\Omega := |\Omega|^{-1}\int_\Omega f, \tag{5.10}$$

where the constant $K_{p,\Omega,n}$ depends only on p, Ω, and n. We saw in §4.7 that this holds, for instance, if Ω has a boundary of class C and, in particular, when Ω is a cube Q,

$$\|f - f_Q\|_{p,Q} \leqslant K_{p,n}|Q|^{1/n}\|\nabla f\|_{p,Q} \tag{5.11}$$

(cf. Theorem 3.23).

Throughout this section Ω will be a domain.

Lemma 5.9. If (f_m) is a sequence in $W^{1,p}(\Omega)$ which is such that $\|f_m\|_{p,\Omega} = 1$ for all $m \in \mathbb{N}$, and $\|\nabla f_m\|_{p,\Omega} \to 0$ as $m \to \infty$, then there exist a subsequence $(f_{m(j)})$ and a constant c which satisfy $\|f_{m(j)} - c\|_{p,\Omega'} \to 0$ for every $\Omega' \subset\subset \Omega$. If $|\Omega| = \infty$ then $c = 0$. ∎

Proof. By the Whitney decomposition (cf. Stein [1, p. 16]), we may write $\Omega = \cup_{n=1}^\infty Q_n^\circ$ where the Q_n are closed cubes, and for any cube Q in Ω set $Pf = \chi_Q f_Q$, where χ_Q is the characteristic function of Q and $f_Q = |Q|^{-1}\int_Q f$. Then $P \in \mathscr{F}_r(W^{1,p}(\Omega), L^p(\Omega))$ and, by (5.11),

$$\|f - Pf\|_{p,Q} \leqslant K_{p,n}|Q|^{1/n}\|\nabla f\|_{p,Q} \text{ for all } f \in W^{1,p}(\Omega).$$

Thus, by hypothesis, $f_m - Pf_m \to 0$ in $L^p(Q)$. Since $P \in \mathscr{F}_r(W^{1,p}(\Omega), L^p(\Omega))$ and (f_m) is bounded in $W^{1,p}(\Omega)$ there exists a subsequence $(f_{m(k)})$ such that $(Pf_{m(k)})$ converges to a limit ϕ, say, in $L^p(Q)$. Hence $f_{m(k)} \to \phi$ in $L^p(Q)$ and, since P is continuous on $L^p(Q)$, $Pf_{m(k)} \to P\phi$ in $L^p(Q)$. Consequently $\phi = P\phi$, a constant, in $L^p(Q)$. We now proceed by the usual diagonalization procedure. There exists a $\phi_1 : \mathbb{N} \to \mathbb{N}$ which is strictly increasing and such that $\|f_{\phi_1(m)} - c_1\|_{p,Q_1} \to 0$, where c_1 is a constant. Define by induction the sequences (ϕ_m) and (c_m) such that

$$\lim_{k\to\infty} \|f_{\phi_m\circ\phi_{m-1}\circ\ldots\circ\phi_1(k)} - c_m\|_{Q_m} = 0.$$

Let $\psi(m) = \phi_m \circ \phi_{m-1} \circ \ldots \circ \phi_1(m)$. Then $\lim_{m\to\infty} \|f_{\psi(m)} - c_k\|_{Q_k} = 0$ for all k. If $Q_{k_1} \cap Q_{k_2} \neq \varnothing$ then $c_{k_1} = c_{k_2}$ and so by the connectedness of Ω the c_{k_i} are all equal. Since every Ω' is contained in a finite union of the Q_n° the result follows. If $|\Omega| = \infty$ we must have $c = 0$, for otherwise $\|f_m\|_{p,\Omega} = 1$ would be contradicted by choosing $|\Omega'|$ large enough. □

Proposition 5.10. $\alpha(I) = 1$ if $|\Omega| = \infty$. ■

Proof. Let $|\Omega| = \infty$ and suppose that $\alpha(I) < 1$. Then there exist a $P \in \mathscr{F}_r(\mathrm{W}^{1,p}(\Omega), \mathrm{L}^p(\Omega))$ and $k < 1$ such that

$$\|f - Pf\|_{p,\Omega}^p \leqslant k(\|\nabla f\|_{p,\Omega}^p + \|f\|_{p,\Omega}^p).$$

In view of Lemma 5.6 we may suppose that $Pf(x) = 0$ outside a set $\Omega' \subset\subset \Omega$, for all $f \in \mathrm{W}^{1,p}(\Omega)$. Hence

$$\|f\|_{p,\Omega\backslash\Omega'}^p \leqslant k(\|\nabla f\|_{p,\Omega}^p + \|f\|_{p,\Omega\backslash\Omega'}^p + \|f\|_{p,\Omega'}^p)$$

and so

$$\|f\|_{p,\Omega\backslash\Omega'}^p \leqslant [k/(1-k)]\,(\|\nabla f\|_{p,\Omega}^p + \|f\|_{p,\Omega'}^p). \tag{5.12}$$

We now claim that there exists a positive constant K such that

$$\|f\|_{p,\Omega}^p \leqslant K\|\nabla f\|_{p,\Omega}^p \quad \text{for all } f \in \mathrm{W}^{1,p}(\Omega). \tag{5.13}$$

Otherwise there exists a sequence (f_m) satisfying the hypotheses of Lemma 5.9 and so, since $|\Omega| = \infty$, a subsequence $(f_{m(k)})$ exists which converges to zero in $\mathrm{L}^p(\Omega')$. But (5.12) then implies that $f_{m(k)} \to 0$ in $\mathrm{L}^p(\Omega)$, contrary to $\|f_{m(k)}\|_{p,\Omega} = 1$. Thus (5.13) is satisfied. To complete the proof we show that (5.13) is impossible. For $m \in \mathbb{N}$ put $B_m = \{x \in \mathbb{R}^n : |x| \leqslant m\}$ and $\Omega_m = \Omega \cap B_m$, and let $u_m \in \mathrm{C}_0^\infty(\mathbb{R}^n)$ be such that

$$u_m(x) = \begin{cases} 1 & \text{if } x \in B_m, \\ 0 & \text{if } x \notin B_{m+1}, \end{cases}$$

with $0 \leqslant u_m(x) \leqslant 1$ and $|\nabla u_m(x)| \leqslant 2$ for all $x \in \mathbb{R}^n$. Then the restriction of u_m to Ω lies in $\mathrm{W}^{1,p}(\Omega)$, so that if (5.13) holds,

$$|\Omega_m| \leqslant \int_\Omega |u_m(x)|^p\,\mathrm{d}x \leqslant K \int_\Omega |\nabla u_m(x)|^p\,\mathrm{d}x \leqslant 2^p K\,|\Omega_{m+1}\backslash\Omega_m|.$$

Thus, with $\Omega_0 = \varnothing$, we have

$$\sum_{j=0}^{m-1} |\Omega_{j+1}\backslash\Omega_j| \leqslant 2^p K\,|\Omega_{m+1}\backslash\Omega_m|,$$

and hence

$$(1 + 2^{-p}K^{-1}) \sum_{j=0}^{m-1} |\Omega_{j+1}\backslash\Omega_j| \leqslant \sum_{j=0}^{m} |\Omega_{j+1}\backslash\Omega_j|.$$

With $A_m = \sum_{j=0}^{m} |\Omega_{j+1} \backslash \Omega_j|$ this gives

$$A_m \geqslant (1+2^{-p}K^{-1})A_{m-1} \geqslant (1+2^{-p}K^{-1})^{m-1}A_1.$$

But $A_m \leqslant |B_{m+1}| = \omega_n(m+1)^n$, and thus for all $m \in \mathbb{N}$,

$$\omega_n(m+1)^n \geqslant (1+2^{-p}K^{-1})^{m-1}A_1.$$

Since this is clearly impossible (5.13) is contradicted and the proposition is proved. □

Hereafter in this section we assume that $|\Omega| < \infty$.

Theorem 5.11. The Poincaré inequality (5.10) holds if, and only if, $\alpha(I) < 1$. ■

Proof. We saw in the proof of Proposition 5.10 that $\alpha(I) < 1$ implies that (5.12) is satisfied for some $k < 1$ and $\Omega' \subset\subset \Omega$ and for all $f \in W^{1,p}(\Omega)$. Suppose (5.10) is not satisfied. Then there exists a sequence (f_m) in $W^{1,p}(\Omega)$ such that $g_m = f_m - (f_m)_\Omega$ satisfies

$$\|\nabla g_m\|_{p,\Omega} \to 0, \qquad \|g_m\|_{p,\Omega} = 1, \qquad (g_m)_\Omega = 0. \tag{5.14}$$

By Lemma 5.9, (g_m) contains a subsequence $(g_{m(t)})$ which converges to a constant c in $L^p(\Omega')$ and hence, on using (5.12) with $f = g - c$ we see that $g_{m(k)} \to c$ in $L^p(\Omega)$. By (5.14), $c = \lim_{k\to\infty} (g_{m(k)})_\Omega = 0$, which contradicts $\|g_{m(k)}\|_{p,\Omega} = 1$. Therefore (5.10) is valid.

Suppose now that (5.10) is satisfied. The map $f \mapsto f_\Omega$ belongs to $\mathscr{F}_r(W^{1,p}(\Omega), L^p(\Omega))$ and hence by Lemma 5.6, given $\varepsilon > 0$ there exist $R \in \mathscr{F}_r(W^{1,p}(\Omega), L^p(\Omega))$ and $\delta > 0$ such that for all $f \in W^{1,p}(\Omega)$,

$$\|f_\Omega - Rf\|_{p,\Omega}^p \leqslant 2^{1-p}\varepsilon \|f\|_{1,p,\Omega}^p$$

and

$$\operatorname{supp} Rf \subset \Omega \backslash (\partial\Omega)(\delta).$$

Hence, by (5.10),

$$\begin{aligned} \|f - Rf\|_{p,\Omega}^p &\leqslant 2^{p-1}K^p \|\nabla f\|_{p,\Omega}^{p} + \varepsilon \|f\|_{1,p,\Omega}^p \\ &\leqslant (K_1 + \varepsilon)\|\nabla f\|_{p,\Omega}^p + \varepsilon \|f\|_{p,\Omega}^p \end{aligned}$$

where $K_1 = 2^{p-1}K^p$. Therefore

$$\|f\|_{p,(\partial\Omega)(\delta)}^p \leqslant (K_1 + \varepsilon)\|\nabla f\|_{p,\Omega}^p + \varepsilon \|f\|_{p,\Omega}^p$$

and so

$$\|f\|_{p,(\partial\Omega)(\delta)}^p \leqslant [(K_1 + \varepsilon)/(1 + K_1)] \|f\|_{1,p,\Omega}^p.$$

It follows from Theorem 5.7 with $A = \partial\Omega$ that $\alpha(I) \leqslant K_1/(1+K_1) < 1$. □

Finally, the methods of this section lead to the following result.

Theorem 5.12. The map $T: f \mapsto \nabla f$ of $W^{1,p}(\Omega) \to [L^p(\Omega)]^n$ has closed range if, and only if, $\alpha(I) < 1$. ∎

Proof. It is easily seen that T is a closed operator and we know from Theorem I.3.4 that T has closed range if, and only if, its *reduced minimum modulus* $\gamma(T)$ defined by

$$\gamma(T) = \inf\{\|Tf\|_{p,\Omega}/\text{dist}(f, \mathcal{N}(T)) : f \in W^{1,p}(\Omega)\backslash\{0\}\}$$

is positive. Here the null space $\mathcal{N}(T) = \mathbb{C}$ and $\text{dist}(f, \mathcal{N}(T)) = \inf_{\mathbb{C}} \|f - c\|_{1,p,\Omega}$. If $\alpha(I) < 1$, (5.10) is satisfied and for all $f \in W^{1,p}(\Omega)\backslash\{0\}$,

$$\begin{aligned} \|\nabla f\|_{p,\Omega}/\text{dist}(f, \mathcal{N}(T)) &\geqslant \|\nabla f\|_{p,\Omega}/\|f - f_\Omega\|_{1,p,\Omega} \\ &\geqslant (1 + K^p)^{-1/p} \end{aligned}$$

and hence $\gamma(T) > 0$. If $\gamma(T) > 0$

$$\inf \|f - c\|_{1,p,\Omega} \leqslant \gamma(T)^{-1} \|Tf\|_{p,\Omega} \quad \text{for all } f \in W^{1,p}(\Omega).$$

For any $c \in \mathbb{C}$,

$$\begin{aligned} \|f - f_\Omega\|_{p,\Omega} &= \|f - c - (f - c)_\Omega\|_{p,\Omega} \\ &\leqslant 2\|f - c\|_{p,\Omega} \\ &\leqslant 2\|f - c\|_{1,p,\Omega}. \end{aligned}$$

This implies that (5.10) is satisfied and hence, by Theorem 5.11, $\alpha(I) < 1$. The theorem is therefore proved. □

Notes. The case $p = 2$, Ω bounded, $T = I$, and $A = \partial\Omega$ of Theorem 5.7 was proved by Amick in [1]; in that case as the spaces involved are Hilbert spaces it is known beforehand that $\tilde{\beta}(I) = \alpha(I)$. The general result in Theorem 5.7 comes from two sources, namely Edmunds and Evans [3], where it is proved that $[\tilde{\beta}(T)]^p = \Gamma_T(0)$; and Evans and Harris [1], where the result for $\alpha(I)$ is established. The results and methods in §5.3 also come from the latter paper by Evans and Harris. The case $p = 2$ of Proposition 5.10 is proved in Edmunds and Evans [3], while the connection between the Poincaré inequality and the condition $\Gamma_1(0) < 1$ was first discovered by Amick in [1] when $p = 2$.

5.4. *Estimates for $\tilde{\beta}(I_0)$ and $\beta(I_0)$*

We have already seen in §4 that if Ω is a *bounded* open set in $\mathbb{R}^n$ and $p \in [1, \infty)$, the embedding I_0 of $W_0^{1,p}(\Omega)$ in $L^p(\Omega)$ is compact; this embedding is, of course, continuous even if Ω is not bounded, and, as we shall see, is still compact for some unbounded open sets Ω, such as quasi-bounded sets (for which $\lim_{|x| \to \infty, x \in \Omega} \text{dist}(x, \partial\Omega) = 0$). For general unbounded sets the embedding is not compact, and the question arises as to whether it nevertheless may have some property better than mere continuity. We use the set and ball measures of non-compactness to describe these properties, and are able to estimate the measure of non-compactness of the embedding map above; for an interesting class of open sets (including infinite strips) we are able to show that the measure of non-compactness is less than 1.

We begin with some preliminary lemmas. Throughout this sub-section Ω is assumed to be an unbounded open subset of $\mathbb{R}^n$, with $n > 1$.

Lemma 5.13. Let $u \in C_0^1(\Omega)$ and suppose that $d > 0$. Then for all $x \in \Omega$,

$$|u(x)| \leqslant \omega_n^{-1} \int_{\Omega \cap B(x,d)} [d^{-1}|u(y)| + |\nabla u(y)|]\,|x-y|^{1-n}\,dy. \qquad \blacksquare$$

Proof. Let $\theta \in C^1([0, \infty))$ be such that $0 \leqslant \theta \leqslant 1$ and for some $\rho \in (0, \frac{1}{3})$,

$$\theta(r) = \begin{cases} 1 & \text{if } 0 \leqslant r \leqslant \rho d, \\ 0 & \text{if } d(1-\rho) \leqslant r, \end{cases}$$

with $|\theta'(r)| \leqslant d^{-1}(1-3\rho)^{-1}$ for all $r \geqslant 0$. Extend the domain of u to the whole of $\mathbb{R}^n$ by setting u equal to 0 outside Ω, and for all $y \in B(x, d)$ put $y = x + r\xi$, where $0 \leqslant r \leqslant d$ and $|\xi| = 1$, and $u(y) = u(x+r\xi) = \phi(r, \xi)$. Then for $0 < \sigma < \rho d$,

$$\phi(\sigma, \xi) = -\int_\sigma^d \frac{\partial}{\partial r}[\theta(r)\phi(r, \xi)]\,dr.$$

Thus, with standard notation,

$$\left|\int_{|\xi|=1} \phi(\sigma, \xi)\,d\omega(\xi)\right| \leqslant \int_{|\xi|=1}\int_\sigma^d \left|\frac{\partial}{\partial r}[\theta(r)\phi(r, \xi)]\right| dr\,d\omega(\xi)$$

$$\leqslant \int_{|\xi|=1}\int_0^d \left(\left|\frac{\partial \phi}{\partial r}(r, \xi)\right| + d^{-1}(1-3\rho)^{-1}|\phi(r, \xi)|\right) r^{1-n}\,dy$$

$$= \int_{|\xi|=1}\int_0^d [|\nabla u(y)\cdot\xi| + d^{-1}(1-3\rho)^{-1}|u(y)|]\,|x-y|^{1-n}\,dy$$

$$\leqslant \int_{\Omega \cap B(x,d)} [d^{-1}(1-3\rho)^{-1}|u(y)| + |\nabla u(y)|]\,|x-y|^{1-n}\,dy.$$

Now multiply by $\omega^{-1}\sigma^{n-1}$ and integrate with respect to σ over $[0, h]$, where $h < \rho d$: we obtain

$$\omega_n^{-1}\left|\int_0^h\int_{|\xi|=1} \phi(\sigma, \xi)\sigma^{n-1}\,d\sigma\,d\omega(\xi)\right|$$

$$\leqslant h^n/(n\omega_n)\int_{\Omega \cap B(x,d)} [d^{-1}(1-3\rho)^{-1}|u(y)| + |\nabla u(y)|]\,|x-y|^{1-n}\,dy.$$

Since $|B(x, h)| = \omega_n h^n/n$, we obtain from Lebesgue's Differentiation Theorem,

$$|u(x)| = \left| \lim_{h \to 0} |B(x, h)|^{-1} \int_{B(x,h)} u(y)\,dy \right|$$
$$\leqslant \omega_n^{-1} \int_{\Omega \cap B(x,d)} [d^{-1}(1-3\rho)^{-1}|u(y)| + |\nabla u(y)|]\,|x-y|^{1-n}\,dy.$$

Since ρ may be made arbitrarily small the lemma follows. □

Now let $d > 0$ and for all $x \in \mathbb{R}^n$ define

$$\tilde{\eta}_d(x) = |B(x, d) \cap \Omega| / |B(x, d)|. \tag{5.15}$$

Lemma 5.14. Let $b \in (0, n)$ and suppose that $d > 0$. Then for all $x \in \mathbb{R}^n$,

$$\int_{\Omega \cap B(x,d)} |x-y|^{b-n}\,dy \leqslant \omega_n d^b b^{-1} [\tilde{\eta}_d(x)]^{b/n}. \quad ■$$

Proof. Since $|x-y|^{b-n}$ increases as the centre of the ball $B(x, d)$ is approached,

$$\int_{\Omega \cap B(x,d)} |x-y|^{b-n}\,dy \leqslant \int_S |x-y|^{b-n}\,dy,$$

where $S = \{y \in \mathbb{R}^n : |x-y| \leqslant \eta\}$ and $|S| = |\Omega \cap B(x, d)| = \tilde{\eta}_d(x)|B(x, d)|$. It follows that $\eta^n\omega_n/n = \tilde{\eta}_d(x)d^n\omega_n/n$, and so $\eta/d = [\tilde{\eta}_d(x)]^{1/n}$. Thus

$$\int_S |x-y|^{b-n}\,dy = \omega_n\eta^b/b = \omega_n d^b [\tilde{\eta}_d(x)]^{b/n} b^{-1},$$

and the proof is complete. □

In our next lemma we need a new quantity:

$$\tilde{\eta}_{d,\Omega} := \sup_{x \in \Omega} \tilde{\eta}_d(x). \tag{5.16}$$

Lemma 5.15. Let $p \in (1, \infty)$, let $d > 0$ and suppose that $b \in (0, n) \cap (p - n(p-1), p)$. Then for all $u \in W_0^{1,p}(\Omega)$,

$$\|u\|_{p,\Omega} \leqslant [(p-1)/(p-b)]^{(p-1)/p} b^{-1/p} \tilde{\eta}_{d,\Omega}^{1/n} (\|u\|_{p,\Omega} + d\,\|\nabla u\|_{p,\Omega}). \quad ■$$

Proof. Let $u \in C_0^1(\Omega)$ and put

$$v(y) = |u(y)| + d\,|\nabla u(y)| \qquad (y \in \Omega).$$

By Lemma 5.13, followed by Hölder's inequality, we have for any $x \in \Omega$,

$$|u(x)| \leqslant (\omega_n d)^{-1} \left(\int_{\Omega \cap B(x,d)} v^p(y)\,|x-y|^{b-n}\,dy \right)^{1/p}$$
$$\times \left(\int_{\Omega \cap B(x,d)} |x-y|^{(p-b)(p-1)^{-1}-n}\,dy \right)^{1/p'}.$$

Note that $(p-b)(p-1)^{-1}-n<0$ in view of the hypothesis on b, and so, with $c := (p-b)(p-1)^{-1}$, Lemma 5.14 gives

$$|u(x)|^p \leqslant (\omega_n d)^{-p}\{\omega_n d^c c^{-1}[\tilde{\eta}_d(x)]^{c/n}\}^{p/p'} \int_{\Omega \cap B(x,d)} v^p(y)\,|x-y|^{b-n}\,dy.$$

Hence

$$\int_\Omega |u(x)|^p\,dx \leqslant (\omega_n d)^{-p}(\omega_n d^c c^{-1}\,\tilde{\eta}_{d,\Omega}{}^{c/n})^{p/p'} \int_\Omega \int_{\Omega \cap B(x,d)} v^p(y)\,|x-y|^{b-n}\,dy\,dx$$

$$\leqslant (\omega_n d)^{-p}(\omega_n d^c c^{-1}\tilde{\eta}_{d,\Omega}{}^{c/n})^{p/p'}\,\omega_n d^b b^{-1}\,\tilde{\eta}_{d,\Omega}{}^{b/n}\,\|v\|_{p,\Omega}^p,$$

and from this the required inequality follows immediately, for $u \in C_0^1(\Omega)$. Since $C_0^1(\Omega)$ is dense in $W_0^{1,p}(\Omega)$ the proof is complete. □

Theorem 5.16. Let $p \in (1, \infty)$, let $d > 0$, and let I_0 be the natural embedding of $W_0^{1,p}(\Omega)$ in $L^p(\Omega)$. Then

$$\|I_0 u\|_{p,\Omega} \leqslant \tilde{\eta}_{d,\Omega}{}^{1/n}(\|u\|_{p,\Omega}+d\|\nabla u\|_{p,\Omega})$$

for all $u \in W_0^{1,p}(\Omega)$. ■

Proof. From Lemma 5.15 we see that for all $u \in W_0^{1,p}(\Omega)$,

$$\|I_0 u\|_{p,\Omega} \leqslant (p-1)^{(p-1)/p} f(b)\,\tilde{\eta}_{d,\Omega}{}^{1/n}(\|u\|_{p,\Omega}+d\,\|\nabla u\|_{p,\Omega}),$$

where

$$f(b) = (p-b)^{-(p-1)/p}\,b^{-1/p}$$

and $b \in (0, n) \cap (n-p(n-1), p)$. It is easy to see that the minimum value of f occurs when $b = 1$ and $f(1) = (p-1)^{-(p-1)/p}$. The result follows. □

We are now in a position to establish our main result.

Theorem 5.17. Let $p \in (1, \infty)$ and $d \in (0, 1]$ and put

$$\eta_d = \limsup_{\substack{|x| \to \infty \\ x \in \Omega}} \frac{|B(x,d) \cap \Omega|}{|B(x,d)|}.$$

Then the natural embedding I_0 of $W_0^{1,p}(\Omega)$ in $L^p(\Omega)$ is a $\eta_d^{1/n}$-set contraction if $W_0^{1,p}(\Omega)$ is endowed with the norm (equivalent to the usual norm) $\|u\|_{p,\Omega} + \|\nabla u\|_{p,\Omega}$. ■

Proof. Given any $R > 0$, let $\theta_R \in C_0^\infty(B(0, 2R))$ be such that $0 \leqslant \theta_R \leqslant 1$ and $\theta_R(x) = 1$ for all $x \in B(0, R)$; write $I_0 = I_0\theta_R + I_0(1-\theta_R)$. An inspection of the proof of Theorem 5.16 shows that given any $\varepsilon > 0$, there exists $R > 0$ such that

$$\|I_0(1-\theta_R)\| \leqslant \eta_d^{1/n} + \varepsilon.$$

In view of Lemma I.2.8 our result will follow if we can show that $I_0\theta_R$ is compact. To do this, note that if $u \in W_0^{1,p}(\Omega)$, then $\theta_R u \in W_0^{1,p}(\Omega \cap B(0, 2R))$ and

$$\|\theta_R u\|_{1,p,\Omega} \leqslant K\,\|u\|_{1,p,\Omega},$$

where K depends only upon θ_R. However, by Theorem 4.18, $W_0^{1,p}(\Omega \cap B(0, 2R))$ is compactly embedded in $L^p(\Omega \cap B(0, 2R))$: it follows that $I_0\theta_R\colon W_0^{1,p}(\Omega) \to L^p(\Omega)$ is compact, and the proof is complete. □

Remarks 5.18. (1) If instead of the condition $d \in (0, 1]$ we require that $d > 1$, then the proof makes it clear that I_0 is a $\eta_d^{1/n} d$-set contraction.

(2) As the proof uses a decomposition of I_0 into a bounded linear map, the norm of which is estimated, and a compact linear map, I_0 is also an $\eta_d^{1/n}$-ball contraction if $d \in (0, 1]$; similarly for $d > 1$.

(3) If the more usual norm is used on $W_0^{1,p}(\Omega)$, the result is modified by a constant factor. Thus if $p \geqslant 2$,

$$\begin{aligned}\|u\|_{p,\Omega} + \|\nabla u\|_{p,\Omega} &\leqslant 2^{1/p'}\left(\int_\Omega (|u|^p + |\nabla u|^p)\,dx\right)^{1/p}\\ &\leqslant 2^{1/p'}\left(\int_\Omega |u|^p\,dx + n^{(p-2)/2}\int_\Omega \sum_{i=1}^n |D_i u|^p\,dx\right)^{1/p}\\ &\leqslant 2^{1/p'} n^{(p-2)/2p}\,\|u\|_{1,p,\Omega},\end{aligned}$$

so that I_0 is a $2^{1/p'} n^{(p-2)/2p} \eta_d^{1/n}$-set contraction if $d \in (0, 1]$; if $p < 2$,

$$\begin{aligned}\|u\|_{p,\Omega} + \|\nabla u\|_{p,\Omega} &\leqslant 2^{1/p'}\left(\int_\Omega |u|^p\,dx + n^{p/2}\int_\Omega \sum_{i=1}^n |D_i u|^p\,dx\right)^{1/p}\\ &\leqslant 2^{1/p'} n^{1/2}\,\|u\|_{1,p,\Omega},\end{aligned}$$

and thus I_0 is a $2^{1/p'} n^{1/2} \eta_d^{1/n}$-set contraction if $d \in (0, 1]$. The modifications needed if $d > 1$ are obvious.

(4) As a very special case of Theorem 5.17 we see that the embedding I_0 is compact if

$$\limsup_{\substack{|x| \to \infty\\ x \in \Omega}} |B(x, 1) \cap \Omega| = 0,$$

a result obtained by Berger and Schechter [1]. Further remarks on this will be made in §VIII.3 and in §X.6.

(5) A particular case of an unbounded open set for which η_d can be determined explicitly is an infinite strip, of width 2ρ say, in $\mathbb{R}^2$. A routine computation shows that if $\rho < d$ then

$$\eta_d = 1 - (2/\pi)\cos^{-1}(\rho/d) + [2\rho/(\pi d)]\sqrt{(1 - \rho^2/d^2)}.$$

(6) If Ω opens out at infinity, as in the case of a conical region, then $\eta_d = 1$ and our theorem merely tells us that I_0 is a 1-set contraction, an obvious result in view of the fact that $\|I_0\| \leqslant 1$.

(7) Extensions of the theorem to weighted spaces may be made without difficulty. See Edmunds and Evans [1], upon which the treatment of this section is based.

(8) Estimates for I: $W^{1,p}(\Omega) \to L^p(\Omega)$ are possible—see Edmunds and Evans [2].

6. Embeddings and approximation numbers

In the last section we saw how measures of non-compactness can be used to find out whether embedding maps have any properties intermediate between continuity and compactness. Here our point of view is different: we take an embedding map known to be compact, and ask whether it can be classified in some way which will, so to speak, measure the strength of its compactness. There are various ways in which this may be done, and as we are concerned to be illustrative rather than exhaustive (and exhausting) we choose to deal with only two embedding maps and to use the approximation numbers as our main tool: it will be recalled that these numbers measure the closeness by which the map can be approximated by finite-dimensional linear maps. The results obtained are of notable use in the theory of the asymptotic distribution of eigenvalues of elliptic operators.

The first embedding map which we shall analyse in detail is that of $W^{r,p}(\Omega)$ in $L^q(\Omega)$, where Ω is a bounded open subset of $\mathbb{R}^n$, the numbers p and q belong to $[1, \infty)$, and r is a positive integer such that $r/n > 1/p - 1/q$. We know that if $\partial\Omega$ is minimally smooth in the sense of Stein then the embedding is compact. We propose to obtain estimates for the approximation numbers of this embedding map. A key step in our treatment is provided by the following lemma; the proof we give is due to D. J. Harris. In this, and elsewhere in this section, we write $h = r/n - \max\{0, 1/p - 1/q\}$.

Lemma 6.1. Let $a_i, b_i \in \mathbb{R}$, with $a_i < b_i$, for $i = 1, 2, \ldots, n$, let $Q = \{x \in \mathbb{R}^n : a_i < x_i < b_i$ for $i = 1, 2, \ldots, n\}$, and suppose that $1 \leqslant p \leqslant q < \infty$ and $r/n > 1/p - 1/q$, where $r \in \mathbb{N}$. For all $u \in W^{r,p}(Q)$ and all $x \in \mathbb{R}^n$ put

$$(P_{r,Q}u)(x) = \frac{\chi_Q(x)}{|Q|} \sum_{|\alpha| \leqslant r-1} \int_{\mathbb{R}^n} \chi_Q(y) \frac{(x-y)^\alpha}{\alpha!} D^\alpha u(y)\, dy,$$

where χ_Q is the characteristic function of Q. Let Q be subdivided into 2^{nN} congruent boxes Q_j, and set

$$(P_N u)(x) = \sum_{j=1}^{2^{nN}} \chi_{Q_j}(x)(P_{r,Q_j}u)(x) \qquad (x \in \mathbb{R}^n).$$

Then for all $u \in W^{r,p}(Q)$ with $\|u\|_{r,p,Q} = 1$,

$$\|u - P_N u\|_{q,Q} \leqslant C(2^{-nN}|Q|)^h,$$

where C is a constant which depends only upon n, r, p, and q. ∎

Proof. Since $C^r(\bar{Q})$ is dense in $W^{r,p}(Q)$ (by Theorem 4.7) it is enough to prove the lemma when $u \in C^r(\bar{Q})$. For such a function u we have for any $x \in \mathbb{R}^n$, by Taylor's formula, and setting $u = 0$ outside $\bar{\Omega}$,

$$u(x) - (P_{r,Q}u)(x)$$

$$= \chi_Q(x)|Q|^{-1}\sum_{|\alpha|=r}\left(\frac{r}{\alpha!}\int_{\mathbb{R}^n}\chi_Q(y)\int_0^1 (1-\tau)^{r-1}(x-y)^\alpha D^\alpha u(\tau x + y - \tau y)\,d\tau\,dy\right)$$

$$= \sum_{|\alpha|=r}\frac{r}{\alpha!}F_\alpha(x), \text{ say.}$$

Then

$$|F_\alpha(x)| \leqslant \frac{\chi_Q(x)}{|Q|}\int_{\mathbb{R}^n}\int_0^1 \chi_Q\left(x - \frac{z}{\tau}\right)\frac{|z^\alpha|}{\tau^{n+1}}|D^\alpha u(x-z)|\,d\tau\,dz.$$

Let Q_0 be the box centred at 0 and obtained by translation of Q. If $x \in Q$ and $x - z/\tau \in Q$, then $x - z = \tau(x - z/\tau) + (1-\tau)x \in Q$; and if Q has centre $c = (c_i)$ and sides $2l_i$, then if $|x_i - c_i| < l_i$ and $|x_i - z_i/\tau - c_i| < l_i$ it follows that $|z_i/\tau| < 2l_i$. Thus

$$\chi_Q(x)\chi_Q(x - z/\tau) \leqslant \chi_Q(x-z)\chi_{2Q_0}(z/\tau),$$

and hence

$$|F_\alpha(x)| \leqslant |Q|^{-1}\int_{\mathbb{R}^n}\int_0^1 \frac{|z^\alpha|}{\tau^{n+1}}\chi_{2Q_0}\left(\frac{z}{\tau}\right)\chi_Q(x-z)|D^\alpha u(x-z)|\,d\tau\,dz$$

$$= |Q|^{-1}[g_\alpha * (\chi_Q|D^\alpha u|)](x), \text{ say.}$$

Now put $Q_0 = kU$, where $k > 0$ is so chosen that $|U| = 1$, and set $z = k\rho\xi$, with $\xi \in \partial U$. Then

$$g_\alpha(z) = \int_{\rho/2}^1 |z^\alpha|\tau^{-n-1}\chi_{2Q_0}(z/\tau)\,d\tau$$

$$\leqslant (1/n)[(2/\rho)^n - 1](k\rho)^{|\alpha|}|\xi^\alpha|\chi_{2Q_0}(z)$$

$$\leqslant (1/n)(2/\rho)^n(k\rho)^r|\xi^\alpha|\chi_{2Q_0}(z),$$

from which we see that

$$\|g_\alpha\|_{m,\mathbb{R}^n}^m \leqslant (2^n/n)^m k^{rm+n}\int_{\partial U}|\xi^\alpha|^m\int_0^2 \rho^{(r-n)m+n-1}\,d\rho\,d\xi$$

$$= 2^{rm+n}n^{-m}|Q|^{1+mr/n}[(r-n)m+n]^{-1}\int_{\partial U}|\xi^\alpha|^m\,d\xi,$$

provided that $(r-n)m+n>0$. We take m so that $1/m = 1+1/q-1/p$, which ensures that this condition is satisfied since $1/q-1/p > -r/n$. Now use Young's theorem on convolutions (Theorem 1.2) and Hölder's inequality for sums: we have

$$\begin{aligned}\|u-P_{r,Q}u\|_{q,Q} &\leqslant \sum_{|\alpha|=r} (r/\alpha!)\,\|F_\alpha\|_{q,Q}\\ &\leqslant \sum_{|\alpha|=r} (r/\alpha!)|Q|^{-1}\|g_\alpha\|_{m,\mathbb{R}^n}\|\chi_Q D^\alpha u\|_{p,\mathbb{R}^n}\\ &\leqslant C|Q|^h\|u\|_{r,p,Q},\end{aligned}$$

where C is a constant which depends only upon n, r, p, and q. This inequality, with Q_i in place of Q, gives

$$\begin{aligned}\|u-P_N u\|_{q,Q} &= \left(\sum_j \|\chi_{Q_j}(u-P_{r,Q_j}u)\|^q_{q,Q_j}\right)^{1/q}\\ &\leqslant C(|Q|2^{-nN})^h\left(\sum_j \|u\|^q_{r,p,Q_j}\right)^{1/q}\\ &= C(|Q|2^{-nN})^h,\end{aligned}$$

and the proof is complete. □

Corollary 6.2. Under the same conditions as the lemma save that the condition $1 \leqslant p \leqslant q < \infty$ is replaced by $1 \leqslant q \leqslant p < \infty$, we have for all $u \in W^{r,p}(Q)$ with $\|u\|_{r,p,Q} = 1$,

$$\|u-P_N u\|_{q,Q} \leqslant C|Q|^{r/n+1/q-1/p}2^{-hnN}. \qquad \blacksquare$$

Proof. By Hölder's inequality and Lemma 6.1,

$$\begin{aligned}\|u-P_N u\|_{q,Q} &\leqslant \|u-P_N u\|_{p,Q}|Q|^{1/q-1/p}\\ &\leqslant C(2^{-nN}|Q|)^{r/n}|Q|^{1/q-1/p},\end{aligned}$$

as required. □

Remark. Examination of the arguments used in Lemma 6.1 and Corollary 6.2 shows that the constant C may be taken to be of the form

$$C_{r,n,p,q} = K_{r,n}2^{nl}l^l h^{-l},$$

where $l = 1-\max\{0, 1/p-1/q\}$ and $h = r/n - \max\{0, 1/p-1/q\}$.

Lemma 6.1 and Corollary 6.2 will be used below to give upper bounds for the approximation numbers; the following two lemmas help in the estimation from below of these numbers.

Lemma 6.3. Let X be a Banach space with $\dim X = r \in [1, \infty)$, and let $I_X : X \to X$ be the identity map. Then $a_{r-1}(I_X) = 1$. ■

Proof. Let $F \in \mathscr{B}(X)$ be such that $\dim F(X) \leqslant r - 1$. By Riesz's lemma, given any $\theta \in (0, 1)$, there exists $x_\theta \in X$ such that $\|x_\theta\| = 1$ and $\|x_\theta - F(x)\| > \theta$ for all $x \in X$. Hence

$$\|I_X - F\| = \sup\{\|x - F(x)\| : x \in X, \|x\| = 1\} > \theta.$$

Thus $a_{r-1}(I_X) \geqslant 1$. But $a_{r-1}(I_X) \leqslant a_0(I_X) = 1$. □

Lemma 6.4. Let X and Y be Banach spaces such that $X \subset Y$ algebraically and topologically, and let $\imath : X \to Y$ be the natural embedding map. Let X_r be an r-dimensional subspace of X with $r \in [1, \infty)$, and suppose that there is a positive number c such that $\|x\|_X \leqslant c\|x\|_Y$ for all $x \in X_r$. Let P_r be a projection of Y onto $\imath(X_r)$. Then

$$\mathrm{a}_{r-1}(\imath) \geqslant c^{-1}\|P_r\|^{-1}.$$ ■

Proof. Let $I_{X_r} : X_r \to X_r$ be the identity map, let $\imath_r : X_r \to X$ be the natural map and let $j_r : \imath(X_r) \to X_r$ be the identity map. Then $I_{X_r} = j_r \circ P_r \circ \imath \circ \imath_r$, so that by Lemma 6.3,

$$1 = \mathrm{a}_{r-1}(I_{X_r}) \leqslant \|j_r\| \, \|P_r\| \, \|\imath_r\| \mathrm{a}_{r-1}(\imath) \leqslant c\|P_r\| \mathrm{a}_{r-1}(\imath).$$ □

Theorem 6.5. Let Ω be a bounded open set in $\mathbb{R}^n$, let $p, q \in [1, \infty)$ and let $r \in \mathbb{N}$ be such that $r/n > 1/p - 1/q$. Then the sth approximation number $\mathrm{a}_s(I_0)$ of the embedding map $I_0 : \mathrm{W}_0^{r,p}(\Omega) \to \mathrm{L}^q(\Omega)$ is $O(s^{-h})$ as $s \to \infty$, where $h = r/n - \max\{0, 1/p - 1/q\}$. If in addition $p = q$ then $\mathrm{a}_s(I_0) \geqslant \text{const.}\, s^{-r/n}$ for all large enough s. All these results hold also for the embedding $I : \mathrm{W}^{r,p}(\Omega) \to \mathrm{L}^q(\Omega)$ provided that $\partial\Omega$ is minimally smooth. ■

Proof. Lemma 6.1 and Corollary 6.2 give the O-estimates of the theorem almost immediately in the special case when $\Omega = Q$. To see this, observe that the map

$$P_N : u \mapsto \sum_{j=1}^{2^{nN}} \chi_{Q_j} P_{r,Q_j} u$$

is linear and of rank at most

$$2^{nN} \sum_{|\alpha| < r} 1 = 2^{nN} M,$$

say. Thus, since

$$\|I - P_N\|_{q,Q} \leqslant 2^{-nNh} |Q|^{r/n - 1/p + 1/q} C,$$

we see that

$$\mathrm{a}_s(I) \leqslant C|Q|^{r/n + 1/q - 1/p} M^h s^{-h},$$

when s is of the form $2^{nN}M$. However, given any positive integer s, there exists $N \in \mathbb{N}$ such that $2^{nN}M \leqslant s \leqslant 2^{n(N+1)}M$, and

$$a_{2^{n(N+1)}M}(I) \leqslant a_s(I) \leqslant a_{2^{nN}M}(I),$$

from which it follows easily that

$$a_s(I) = O(s^{-h}) \text{ as } s \to \infty.$$

In the more general case in which we merely assume that Ω is bounded and has a minimally smooth boundary we know (Theorem 4.12) that there is an extension map $E\colon W^{r,p}(\Omega) \to W^{r,p}(\mathbb{R}^n)$ such that for all $u \in W^{r,p}(\Omega)$,

$$\|u\|_{r,p,\Omega} \leqslant \|Eu\|_{r,p,\mathbb{R}^n} \leqslant c\|u\|_{r,p,\Omega},$$

where c is a constant independent of u and p, but dependent upon r. Let Q be a box such that $\bar{\Omega} \subset Q$, and observe that $W^{r,p}(Q)$ is the set of restrictions to Q of functions in $W^{r,p}(\mathbb{R}^n)$. Thus given any $u \in W^{r,p}(\Omega)$, define $\tilde{u} = Eu \restriction Q \in W^{r,p}(Q)$; clearly $\tilde{u}(x) = u(x)$ for all $x \in \Omega$, and

$$\|\tilde{u}\|_{r,p,Q} \leqslant \|Eu\|_{r,p,\mathbb{R}^n} \leqslant c\|u\|_{r,p,\Omega}.$$

Hence

$$\left\| u - \sum_{j=1}^{2^{nN}} \chi_{Q_j \cap \Omega} P_{r,Q_j} \tilde{u} \right\|_{q,\Omega} \leqslant \left\| \tilde{u} - \sum_{j=1}^{2^{nN}} \chi_{Q_j} P_{r,Q_j} \tilde{u} \right\|_{q,\Omega}$$
$$\leqslant Cc|Q|^{r/n+1/q-1/p} 2^{-nNh} \|u\|_{r,p,\Omega}, \qquad (6.1)$$

from which we conclude that, as before, $a_s(I) = O(s^{-h})$ as $s \to \infty$. For I_0 this result follows as for I, but without any conditions on $\partial\Omega$ since the extensions needed may be made without them.

To complete the proof it is enough to obtain the lower bounds for the approximation numbers of the map $I_0\colon W^{r,p}(\Omega) \to L^p(\Omega)$ since $a_s(I) \geqslant a_s(I_0)$ for all $s \in \mathbb{N}$. To do this take Q to be the unit cube $(0, 1)^n$, let $s \in \mathbb{N}$, let j be the integer such that $(j-1)^n \leqslant s < j^n$, let $K = \{k \in \mathbb{N}_0^n : 0 \leqslant k_i \leqslant j-1 \text{ for } i = 1, 2, \ldots, n\}$, and for each $k \in K$ put $Q_k = \{x \in \mathbb{R}^n : k_i/j < x_i < (k_i+1)/j \text{ for } i = 1, 2, \ldots, n\}$. Thus $Q \setminus \cup_{k \in K} Q_k$ has zero measure. Let $\phi \in C_0^\infty(Q)$ be such that $\|\phi\|_{2,Q} = 1$; then the functions ϕ_k $(k \in K)$ defined by $\phi_k(x) = j^{\frac{1}{2}n}\phi(jx-k)$ are in $C_0^\infty(Q_k)$ and satisfy

$$(\phi_k, \phi_{k'})_{2,Q} = \delta_{kk'}, \qquad \|\phi_k\|_{q,Q} = j^{n(\frac{1}{2}-1/q)}\|\phi\|_{q,Q} \quad (q \geqslant 1).$$

Let V be the linear space spanned by the ϕ_k $(k \in K)$ and define $\tilde{P}_j\colon L^p(Q) \to V$ by

$$\tilde{P}_j u = \sum_{k \in K} (u, \phi_k)_{2,Q} \phi_k.$$

Then $\tilde{P}_j$ is a projection and

$$\|\tilde{P}_j u\|_{p,Q}^p = j^{n(\frac{1}{2}p-1)} \|\phi\|_{p,Q}^p \sum_{k \in K} |(u, \phi_k)_{2,Q}|^p.$$

Since

$$|(u, \phi_k)_{2,Q}| \leqslant j^{n(1/p-\frac{1}{2})} \|u\|_{p,Q_k} \|\phi\|_{p',Q},$$

it follows easily that

$$\|\tilde{P}_j\| \leqslant \|\phi\|_{p,Q} \|\phi\|_{p',Q}.$$

Next we observe that given any $u \in V$ we have $u = \tilde{P}_j u \in W_0^{r,p}(Q)$ and

$$\begin{aligned}\|u\|_{r,p,Q} &= \left(\sum_{|\alpha| \leqslant r} \sum_{k \in K} |(u, \phi_k)_{2,Q}|^p \|D^\alpha \phi_k\|_{p,Q_k}^p\right)^{1/p} \\ &= j^{n(\frac{1}{2}-1/p)} \left(\sum_{|\alpha| \leqslant r} j^{|\alpha|p} \|D^\alpha \phi\|_{p,Q}^p\right)^{1/p} \left(\sum_{k \in K} |(u, \phi_k)_{2,Q}|^p\right)^{1/p} \\ &= \left(\sum_{|\alpha| \leqslant r} j^{|\alpha|p} \|D^\alpha \phi\|_{p,Q}^p\right)^{1/p} \|u\|_{p,Q} \|\phi\|_{p,Q}^{-1} \\ &\leqslant j^r \|\phi\|_{r,p,Q} \|u\|_{p,Q} \|\phi\|_{p,Q}^{-1} \\ &\leqslant (s^{1/n}+1)^r \|\phi\|_{r,p,Q} \|u\|_{p,Q} \|\phi\|_{p,Q}^{-1}. \end{aligned} \tag{6.2}$$

Since $\dim V = j^n \geqslant s+1$ we may, and shall, choose a set of $s+1$ functions ϕ_k and let X_{s+1} be the linear span of these functions. Define a projection P_{s+1} of $L^p(Q)$ onto X_{s+1} (viewed as a subspace of $L^p(Q)$) by

$$P_{s+1}(u) = \sum_{\phi_k \in X_{s+1}} (u, \phi_k)_{2,Q} \phi_k.$$

Then

$$\|P_{s+1}\| \leqslant \|\tilde{P}_j\| \leqslant \|\phi\|_{p,Q} \|\phi\|_{p',Q},$$

so that by (6.2) and Lemma 6.4,

$$a_s(I_0) \geqslant (\|\phi\|_{r,p,Q} \|\phi\|_{p',Q})^{-1}(s^{1/n}+1)^{-r},$$

which is the required lower bound for $a_s(I_0)$, at least in the case with $\Omega = (0, 1)^n$, and hence for any cube in $\mathbb{R}^n$. To complete the proof, let Ω be any bounded open set in $\mathbb{R}^n$ and let Q be a cube with $\bar{Q} \subset \Omega$. Then the natural embedding map $I_Q: W_0^{r,p}(Q) \to L^p(Q)$ is given by $I_Q = R \circ I_0 \circ E$, where $R: L^p(\Omega) \to L^p(Q)$ is the restriction map and $E: W_0^{r,p}(Q) \to W_0^{r,p}(\Omega)$ is the extension map which extends elements of $W_0^{r,p}(Q)$ by zero in $\Omega \setminus Q$. Hence

$$a_s(I_Q) \leqslant \|R\| \|E\| a_s(I_0) = a_s(I_0),$$

and thus

$$a_s(I_0) \geqslant \text{const}.\, s^{-r/n}.$$

The proof is complete. □

We recall that a map is said to be of type l^p if its approximation numbers form a sequence which is in the sequence space l^p; that is, $\sum_{r=1} a_r^p < \infty$. It is of type c_0 if $\lim_{r\to\infty} a_r = 0$. The results of Theorem 6.5 thus show that the embedding maps I and I_0 are of type l^t for any t such that $ht > 1$.

The techniques used to prove Theorem 6.5 can be used to analyse, from the same point of view, the embedding maps which turn up when we consider unbounded domains Ω, spaces of fractional order, traces on lower-dimensional manifolds and non-isotropic spaces. Instead of proceeding with this analysis, however, we prefer to conclude our discussion in this section by consideration of embeddings into Orlicz spaces.

Theorem 6.6. Let Ω be a bounded open set in $\mathbb{R}^n$ with minimally smooth boundary, let $r \in \mathbb{N}$, suppose that $p := n/r > 1$ and $v \in (1, p')$, and define ϕ_v by $\phi_v(t) = \exp t^v - 1$ $(t \geqslant 0)$. Then the approximation numbers of the embedding map $J : W^{r,p}(\Omega) \to L^{\phi_v}(\Omega)$ satisfy

$$a_s(J) = O\left((\log s)^{1-p^{-1}-v^{-1}}\right) \text{ as } s \to \infty \qquad \blacksquare$$

Proof. Consider first the case $\Omega = Q$, where Q is some box in $\mathbb{R}^n$. Let $u \in W^{k,p}(Q)$ and let Q be subdivided into 2^{nN} congruent boxes Q_j. We set

$$(P_N u)(x) = \sum_{j=1}^{2^{nN}} \chi_{Q_j}(x)(P_{k,Q_j}u)(x) \qquad (x \in \mathbb{R}^n)$$

where

$$(P_{k,Q_j}u)(x) = \chi_{Q_j}(x)|Q_j|^{-1} \sum_{|\alpha| \leqslant k-1} \int_{\mathbb{R}^n} \chi_{Q_j}(y) \frac{(x-y)^\alpha}{\alpha!} D^\alpha u(y)\,dy$$

and χ_{Q_j} denotes the characteristic function of Q_j. Put $U = u - P_N u$. Then, by Lemma 6.1, Corollary 6.2 and the Remark following Lemma 6.3,

$$\|U\|_{q,Q} \leqslant K(k,n) 2^{nl} l^l h^{-l} 2^{-nNh} |Q|^{1/q} \|u\|_{k,p,\Omega},$$

where $h = k/n - \max\{0, 1/p - 1/q\}$ and $l = 1 - \max\{0, 1/p - 1/q\}$. Thus for any $\lambda > 0$,

$$\begin{aligned}\int_Q \phi\left(\frac{|U(x)|}{\lambda}\right) dx &= \int_Q \sum_{j=1}^{\infty} \frac{1}{j!}\left(\frac{|U(x)|}{\lambda}\right)^{jv} dx \\ &= \sum_{j=1}^{\infty} \frac{1}{j!}(\|U\|_{jv,Q}\lambda^{-1})^{jv} \\ &\leqslant |Q| \sum_{j=1}^{\infty} \frac{1}{j!}[K 2^{nl(jv)} 2^{-nNh(jv)} l(jv)^{l(jv)} h(jv)^{-l(jv)} \lambda^{-1} \|u\|_{k,p,Q}]^{jv}\end{aligned}$$

where $l(jv) = 1 - \max\{0, 1/p - 1/jv\}$ and $h(jv) = 1/p - \max\{0, 1/p - 1/jv\}$. First suppose that $p/v \geqslant 1$. Then we have, with the integer part of p/v denoted by $\lfloor p/v \rfloor$,

$$l(jv) = \begin{cases} 1 & \text{if } j < \lfloor p/v \rfloor + 1, \\ 1 - 1/p + 1/jv & \text{if } j \geqslant \lfloor p/v \rfloor + 1, \end{cases}$$

and

$$h(jv) = \begin{cases} 1/p & \text{if } j < \lfloor p/v \rfloor + 1, \\ 1/jv & \text{if } j \geqslant \lfloor p/v \rfloor + 1. \end{cases}$$

Hence

$$\int_Q \phi\left(\frac{|U(x)|}{\lambda}\right) \mathrm{d}x \leqslant |Q| \left\{ \sum_{j=1}^{[p/v]} \frac{1}{j!} (K 2^n 2^{-nN/p} p \lambda^{-1} \|u\|_{k,p,Q})^{jv} \right.$$

$$\left. + \sum_{j=\lfloor p/v \rfloor + 1}^{\infty} \frac{1}{j!} [K 2^{n(\frac{1}{p'} + \frac{1}{jv})} 2^{-\frac{nN}{jv}} \left(\frac{1}{p'} + \frac{1}{jv}\right)^{\frac{1}{p'} + \frac{1}{jv}} (jv)^{\frac{1}{p'} + \frac{1}{jv}} \lambda^{-1} \|u\|_{k,p,Q}]^{jv} \right\}$$

$$\leqslant |Q| 2^{-nN/p} \left[\sum_{j=1}^{\lfloor p/v \rfloor} \frac{1}{j!} (K 2^n p \lambda^{-1} \|u\|_{k,p,Q})^{jv} \right.$$

$$\left. + 2^n \sum_{j=\lfloor p/v \rfloor + 1}^{\infty} \frac{1}{j!} (K 2^{n/p'} \lambda^{-1} \|u\|_{k,p,Q})^{jv} \left(1 + \frac{jv}{p'}\right)^{1 + jv/p'} \right].$$

For $j \in \{1, \ldots, \lfloor p/v \rfloor\}$,

$$(K 2^n p \lambda^{-1} \|u\|_{k,p,Q})^{jv} \leqslant 2^n p^p (K 2^{n/p'} \lambda^{-1} \|u\|_{k,p,Q})^{jv}$$

$$\leqslant 2^n p^p (K 2^{n/p'} \lambda^{-1} \|u\|_{k,p,Q})^{jv} \left(1 + \frac{jv}{p'}\right)^{1 + jv/p'}$$

and so

$$\int_Q \phi\left(\frac{|U(x)|}{\lambda}\right) \mathrm{d}x$$

$$\leqslant |Q| 2^{-nN/p} 2^n p^p \sum_{j=1}^{\infty} \frac{1}{j!} (K 2^{n/p'} \lambda^{-1} \|u\|_{k,p,Q})^{jv} \left(1 + \frac{jv}{p'}\right)^{1 + jv/p'}.$$

If $p/v < 1$, we have

$$\int_Q \phi\left(\frac{|U(x)|}{\lambda}\right) \mathrm{d}x \leqslant |Q| 2^{-nN} 2^n \sum_{j=1}^{\infty} \frac{1}{j!} (K 2^{n/p'} \lambda^{-1} \|u\|_{k,p,Q})^{jv} \left(1 + \frac{jv}{p'}\right)^{1 + jv/p'},$$

and so, in either case,

$$\int_Q \phi\left(\frac{|U(x)|}{\lambda}\right) \mathrm{d}x \leqslant K_2 2^{-nN/p} \sum_{j=1}^{\infty} \frac{1}{j!} (K 2^{n/p'} \lambda^{-1} \|u\|_{k,p,Q})^{jv} \left(1 + \frac{jv}{p'}\right)^{1 + jv/p'},$$

and we need to consider the series

$$S(z) := \sum_{j=1}^{\infty} \frac{1}{j!} (1 + \mu j)^{1 + \mu j} z^j,$$

where $\mu = v/p'$ and $z = (K 2^{n/p'} \lambda^{-1} \|u\|_{k,p,Q})^v$.

Since $\mu < 1$, the series converges for all $z > 0$. Let $\lfloor 1/(1-\mu) \rfloor$ be the integer part of $1/(1-\mu)$. For $j \geqslant \lfloor 1/(1-\mu) \rfloor + 1$ we have $1 + \mu j \leqslant j$, and for $j \leqslant \lfloor 1/(1-\mu) \rfloor$ we have $1 + \mu j < 1/(1-\mu)$. Thus

$$S(z) \leqslant \sum_{j=1}^{\lfloor 1/(1-\mu) \rfloor} \left(\frac{1}{j!}\right)\left(\frac{1}{1-\mu}\right)^{1/(1-\mu)} z^j + \sum_{j=\lfloor 1/(1-\mu) \rfloor + 1}^{\infty} \left(\frac{j^{\mu j}}{j!}\right)(\mathrm{e}z)^j.$$

Evidently the series

$$\sum_{j=1}^{\infty} \left(\frac{j^{\mu j}}{j!}\right)(\mathrm{e}z)^j$$

converges for all $z > 0$ and, as we shall see in a moment, there exists an absolute constant c_0 such that

$$\sum_{j=1}^{\infty} (j^{\mu j}/j!)(\mathrm{e}z)^j \leqslant c_0 \exp[(2\mathrm{e})^{\mu/(1-\mu)}(\mathrm{e}z)^{1/(1-\mu)}].$$

Thus

$$S(z) \leqslant \sum_{j=1}^{\lfloor 1/(1-\mu) \rfloor} (1/j!)(1-\mu)^{-1/(1-\mu)} z^j + c_0 \exp[(2\mathrm{e})^{\mu/(1-\mu)}(\mathrm{e}z)^{1/(1-\mu)}].$$

If $z > 1$, then since it is easily seen that $z + (1-\mu)^{-1} \log(1-\mu)^{-1} \leqslant \mathrm{e}z(2\mathrm{e})^{\mu(1-\mu)}$, we have

$$\sum_{j=1}^{\lfloor 1/(1-\mu) \rfloor} (1-\mu)^{-1/(1-\mu)} z^j/j! \leqslant (1-\mu)^{-1/(1-\mu)} \mathrm{e}^z$$

$$= \exp[z + (1-\mu)^{-1} \log(1-\mu)^{-1}] - \exp[(2\mathrm{e})^{\mu/(1-\mu)}(\mathrm{e}z)^{1/(1-\mu)}]$$
$$+ \exp[(2\mathrm{e})^{\mu/(1-\mu)}(\mathrm{e}z)^{1/(1-\mu)}] \leqslant \exp[(2\mathrm{e})^{\mu/(1-\mu)}(\mathrm{e}z)^{1/(1-\mu)}];$$

and if $z \leqslant 1$,

$$\sum_{j=1}^{\lfloor 1/(1-\mu) \rfloor} (1-\mu)^{-1/(1-\mu)}(z^j/j!) \leqslant (1-\mu)^{-1/(1-\mu)} \mathrm{e};$$

therefore

$$S(z) \leqslant K_3 \exp[(2\mathrm{e})^{\mu/(1-\mu)}(\mathrm{e}z)^{1/(1-\mu)}]$$

and so

$$\int_Q \phi\left(\frac{|U(x)|}{\lambda}\right) \mathrm{d}x \leqslant K_4 2^{-nN/p} \exp[(2\mathrm{e})^{\mu/(1-\mu)}(\mathrm{e}z)^{1/(1-\mu)}]$$
$$\leqslant 1$$

if $(2\mathrm{e})^{\mu/(1-\mu)}(\mathrm{e}z)^{1/(1-\mu)} \leqslant (nN/p)\log 2 - \log K_4$, which is certainly true if

$$\lambda \geqslant K \|u\|_{k,p,Q} [(nN/p)\log 2 - \log K_4]^{-(1-\mu)/\nu} (2\mathrm{e})^{\mu/\nu} \mathrm{e}^{1/\nu} 2^{n/p'}.$$

It now follows easily that as $s \to \infty$;

$$a_s(J) = O\big((\log s)^{1-p^{-1}-\nu^{-1}}\big);$$

when $\Omega = Q$.

To deal with the more general case in which Ω is merely required to be bounded and to have minimally smooth boundary, we know by Theorem 4.12 that there is an extension map E: $W^{k,p}(\Omega) \to W^{k,p}(\mathbb{R}^n)$ such that, for all $u \in W^{k,p}(\Omega)$,

$$\|u\|_{k,p,\Omega} \leqslant \|Eu\|_{k,p,\mathbb{R}^n} \leqslant c\,\|u\|_{k,p,\Omega},$$

where c is a constant independent of u and p, but dependent upon k.

Let Q be an open box such that $\bar{\Omega} \subset Q$. Since $W^{k,p}(Q)$ coincides with the set of restrictions to Q of functions in $W^{k,p}(\mathbb{R}^n)$, given any $u \in W^{k,p}(\Omega)$, we define $\tilde{u} = Eu \upharpoonright Q \in W^{k,p}(Q)$. Clearly, $\tilde{u}(x) = u(x)$ for all $x \in \Omega$ and $\|\tilde{u}\|_{k,p,Q} \leqslant \|Eu\|_{k,p,\mathbb{R}^n} \leqslant c\,\|u\|_{k,p,\Omega}$.

Since

$$\left\| u - \sum_{j=1}^{2^{nN}} \chi_{Q_j \cap \Omega} P_{k,Q_j}\,\tilde{u} \right\|_{\phi,\Omega} \leqslant \left\| \tilde{u} - \sum_{j=1}^{2^{nN}} \chi_{Q_j} P_{k,Q_j}\,\tilde{u} \right\|_{\phi,Q},$$

it follows that

$$\begin{aligned}\left\| u - \sum_{j=1}^{2^{nN}} \chi_{Q_j \cap \Omega} P_{k,Q_j}\,\tilde{u} \right\|_{\phi,\Omega} &\leqslant K_5\,\|\tilde{u}\|_{k,p,Q}[-\log K_4 + (nN/p)\log 2]^{-(1-\mu)/\nu} \\ &\leqslant cK_5\,\|u\|_{k,p,\Omega}[-\log K_4 + (nN/p)\log 2]^{-(1-\mu)/\nu};\end{aligned}$$

and so, as before, $a_s(J) = O\big((\log s)^{1/p' - 1/\nu}\big)$ as $s \to \infty$.

All that remains is to show that for all $z > 0$, if $0 < b < 1$,

$$T(z) := \sum_{k=1}^{\infty} \frac{k^{bk}}{k!} z^k \leqslant c_0 \exp\left[(2\mathrm{e})^{b/(1-b)} z^{1/(1-b)}\right].$$

Stirling's formula

$$k! \asymp \surd(2\pi)\, k^{k+\frac{1}{2}} \mathrm{e}^{-k}[1 + O(k^{-1})] \qquad (k \to \infty)$$

implies that there exist positive constants c_1 and c_2 such that

$$c_1 \leqslant k!/[k^k(k+1)^{\frac{1}{2}}\mathrm{e}^{-k}] \leqslant c_2 \qquad (k > 0).$$

Thus in the series for $T(z)$,

$$\begin{aligned} z^k k^{bk} &= (z/b^b)^k (bk)^{bk} \leqslant (z/b^b)^k c_1^{-1} [\Gamma(bk+1)/(bk+1)^{\frac{1}{2}} \mathrm{e}^{-bk}] \\ &= c_1^{-1} \gamma^k \Gamma(bk+1)/(bk+1)^{\frac{1}{2}}, \end{aligned}$$

where $\gamma = ze^b/b^b$. (Note that $b^b = e^{b \log b} \to 1$ as $b \to 0$.) Hence

$$T(z) \leqslant \sum_{k=1}^{\infty} c_1^{-1} \left(\frac{\gamma^k}{(bk+1)^{\frac{1}{2}}} \right) \frac{\Gamma(bk+1)}{k!}$$

$$= c_1^{-1} \sum_{k=1}^{\infty} (bk+1)^{-\frac{1}{2}} \frac{\gamma^k}{k!} \int_0^\infty e^{-t} t^{bk} \, dt$$

$$< c_1^{-1} \int_0^\infty e^{-t} \sum_{k=1}^{\infty} \frac{\gamma^k t^{bk}}{k!} \, dt$$

$$= c_1^{-1} \left(\int_0^\infty e^{-t+\gamma t^b} \, dt - 1 \right).$$

Define $f(t) = t - \gamma t^b$, so that $f'(t) = 1 - \gamma b t^{b-1}$. Define t_0 by $f'(t_0) = \frac{1}{2}$; then

$$t_0 = (2\gamma b)^{1/(1-b)}$$

and also $\frac{1}{2} \leqslant f'(t) < 1$ for $t \geqslant t_0$. Thus

$$\int_{t_0}^\infty e^{-t+\gamma t^b} \, dt \leqslant 2 \int_{t_0}^\infty e^{-t+\gamma t^b} (1 - \gamma b t^{b-1}) \, dt = 2e^{-t_0 + \gamma t_0^b}.$$

Also

$$\int_0^{t_0} e^{-t+\gamma t^b} \, dt \leqslant e^{\gamma t_0^b}.$$

Hence

$$T(z) < c_1^{-1} \exp(\gamma t_0^b)(1 + 2e^{-t_0})$$

and the result follows. □

Note that Theorem 6.6 enables us to say that the embedding map J is of type c_0.

The results of Theorem 6.5 go back to work of Birman and Solomjak in the 1960s (see Birman and Solomjak [1, 2]); the proof given here is that contained in Edmunds [1]. Theorem 6.6 was first proved in Edmunds and Moscatelli [1] (see also Edmunds [1]). Versions of Theorem 6.5 applicable to certain *unbounded* domains will be found in König [1, 2] and Martins [1]; Edmunds and Edmunds [1] make a corresponding extension of Theorem 6.6. Analogues of Theorems 6.5 and 6.6 for entropy numbers are given in Birman and Solomjak [2], Namasivayam [1], and Edmunds and Edmunds [1].

VI

Generalized Dirichlet and Neumann boundary-value problems

We begin by briefly motivating the generalized or weak interpretation of the Dirichlet and Neumann problems for general elliptic expressions and then use the Lax–Milgram Theorem of Chapter IV to set the problems in the framework of eigenvalue problems for operators acting in a Hilbert space. The results on variational inequalities in Chapter IV enable us to give Stampacchia's weak maximum principle, and this leads to the notion of capacity.

1. Boundary-value problems

1.1. *The weak Dirichlet problem*

The non-homogeneous Dirichlet problem associated with the differential expression τ defined by

$$\tau u = -\sum_{i,j=1}^{n} \mathrm{D}_i(a_{ij}\mathrm{D}_j u) + \sum_{j=1}^{n} b_j \mathrm{D}_j u + qu \tag{1.1}$$

in an open set Ω in $\mathbb{R}^n$, with $n \geqslant 1$, is the following: given functions h and g, to find a solution u of the boundary-value problem

$$\tau u = h \quad \text{in } \Omega, \tag{1.2}$$

$$u = g \quad \text{on } \partial\Omega. \tag{1.3}$$

In the classical problem a solution is taken to mean a function in $\mathrm{C}^2(\Omega) \cap \mathrm{C}(\bar{\Omega})$, say, which satisfies (1.2) and (1.3) identically; this implies matching smoothness conditions on the *coefficients* a_{ij}, b_j, and q in (1.1) and the *data* h and g in (1.2) and (1.3). The existence of such a solution depends not only on τ and the data but also on the regularity of $\partial\Omega$. We shall take a more general view of the problem and interpret the solution in the distributional sense. This allows us to work with discontinuous coefficients and data for which the problem has no meaning in the classical sense. We shall use Hilbert-space methods, these being based on redefining the problem in the so-called variational or weak form and then using the results of §IV.1 for sesquilinear forms. This approach forces us to be rather restrictive as regards (1.3), because it is not enough that g be defined on $\partial\Omega$ since we shall want it to belong to the Sobolev space $W^{1,2}(\Omega)$. If $\partial\Omega$ is sufficiently smooth, a function g in $W^{1,2}(\Omega)$ does have a 'value' on Ω, and conversely a smooth enough g defined on $\partial\Omega$ can be extended off $\partial\Omega$ to a function in $W^{1,2}(\Omega)$ (cf. Proposition V.4.9). We do not investigate these matters in this book; they involve the notion of *trace* and a full treatment may be found in the books of Agmon [1] and Trèves [1]. Suffice it to say that our approach implies some smoothness criteria on g and $\partial\Omega$ in (1.3).

We shall assume initially that Ω is a bounded open set and also that the following conditions are satisfied:

$$h \in L^2(\Omega), \qquad g \in W^{1,2}(\Omega), \tag{1.4}$$

and

(i) the matrix $[a_{ij}(x)]$ is Hermitian for a.e. $x \in \Omega$, $a_{ij} \in L^\infty(\Omega)$, and the smallest eigenvalue of $[a_{ij}(x)]$ is bounded below a.e. on Ω by $\nu > 0$;

(ii) $b_j \in L^{2r}(\Omega)$ where

$$r \in \begin{cases} [\tfrac{1}{2}n, \infty] & \text{if } n \geqslant 3, \\ (1, \infty] & \text{if } n = 2, \\ [1, \infty] & \text{if } n = 1; \end{cases} \tag{1.5}$$

(iii) $q \in L^r(\Omega)$.

If in (1.2) and (1.3) we set $v = u - g$, the problem then becomes

$$\tau v = F \quad \text{in } \Omega, \tag{1.6}$$

$$v = 0 \quad \text{on } \partial\Omega, \tag{1.7}$$

where $F = h - \tau g$ is defined in the distributional sense. Under the above assumptions, τg, and hence F, lies in $W^{-1,2}(\Omega)$, the adjoint of $W_0^{1,2}(\Omega)$. To see this, first note that $a_{ij} D_j g \in L^2(\Omega)$ and so the term $-\Sigma_{i,j=1}^n D_i(a_{ij} D_j g)$ in τg is in $W^{-1,2}(\Omega)$ in view of Corollary IV.1.3. Furthermore, by the Sobolev Embedding Theorem, Theorem V.4.13, if $\partial\Omega$ is minimally smooth, $W^{1,2}(\Omega)$ is continuously embedded in $L^{2r/(r-1)}(\Omega)$ and hence on using Hölder's inequality,

$$\begin{aligned} \left| \int_\Omega b_j D_j g \bar{\phi} \right| &\leqslant \|b_j\|_{2r,\Omega} \|\nabla g\|_{2,\Omega} \|\phi\|_{2r/(r-1),\Omega} \\ &\leqslant \gamma_r(\Omega) \|b_j\|_{2r,\Omega} \|g\|_{1,2,\Omega} \|\phi\|_{1,2,\Omega} \end{aligned} \tag{1.8}$$

and

$$\begin{aligned} \left| \int_\Omega q g \bar{\phi} \right| &\leqslant \|q\|_{r,\Omega} \|g\|_{2r/(r-1),\Omega} \|\phi\|_{2r/(r-1),\Omega} \\ &\leqslant (\gamma_r^2(\Omega) \|q\|_{r,\Omega} \|g\|_{1,2,\Omega}) \|\phi\|_{1,2,\Omega}, \end{aligned} \tag{1.9}$$

where $\gamma_r(\Omega)$ is the norm of the embedding $W^{1,2}(\Omega) \to L^{2r/(r-1)}(\Omega)$. It therefore follows that $\tau g \in W^{-1,2}(\Omega)$ and $F \in W^{-1,2}(\Omega)$ in (1.6). If $v \in W^{1,2}(\Omega)$ the distributional equation (1.6) has the meaning

$$t[\![v,\phi]\!] := \int_\Omega (a_{ij}\mathrm{D}_j v \mathrm{D}_i \bar{\phi} + b_j \mathrm{D}_j v \bar{\phi} + qv\bar{\phi})$$
$$= (F,\phi) \quad \text{for all } \phi \in \mathrm{C}_0^\infty(\Omega), \tag{1.10}$$

and (1.7) is interpreted as $v \in \mathrm{W}_0^{1,2}(\Omega)$. In (1.10) we have used the summation convention, repeated suffices being summed from 1 to n. The sesquilinear form $t[\![\bullet,\bullet]\!]$ defined in (1.10) will be seen to be continuous on $\mathrm{W}_0^{1,2}(\Omega) \times \mathrm{W}_0^{1,2}(\Omega)$ and so (1.10) is satisfied for all $\phi \in \mathrm{W}_0^{1,2}(\Omega)$. This brings us to the following *generalized Dirichlet problem* for τ in Ω: given $F \in \mathrm{W}^{-1,2}(\Omega)$, to find $v \in \mathrm{W}_0^{1,2}(\Omega)$ such that

$$t[\![v,\phi]\!] = (F,\phi) \quad \text{for all } \phi \in \mathrm{W}_0^{1,2}(\Omega). \tag{1.11}$$

This is also referred to as the *weak* or *variational* form of the Dirichlet problem for τ in Ω.

Once the existence of a solution $v \in \mathrm{W}_0^{1,2}(\Omega)$ of (1.11) has been established the next step in this approach is to investigate how the smoothness of the coefficients and data are reflected in the smoothness of the solution v. This is the so-called *regularity problem*. It is extensively treated in many books, e.g. those of Agmon [1] and Friedman [1], and is not included here. Suffice it to say that under an appropriate amount of smoothness the solution v of (1.11) is indeed a classical solution of (1.6) and (1.7) and hence gives rise to a solution u of (1.2) and (1.3). On the other hand, if v is a classical solution of (1.6) and (1.7) an integration by parts leads to (1.10) under appropriate conditions and hence the generalized problem (1.11) is doubly motivated; it truly generalizes the classical problem.

In order to treat the problem (1.11) we need some preliminary lemmas.

Lemma 1.1. Let $f \in \mathrm{L}^p(\Omega)$ for some $p \in [1,\infty)$ and let $\varepsilon > 0$. Then we can write $f = f_1 + f_2$ where $f_1 \in \mathrm{L}^p(\Omega)$, $f_2 \in \mathrm{L}^\infty(\Omega)$ and, for some positive constant K_ε depending on ε,

$$\|f_1\|_{p,\Omega} < \varepsilon, \qquad \|f_2\|_{\infty,\Omega} < K_\varepsilon. \qquad \blacksquare$$

Proof. Define

$$f_2(x) = \begin{cases} f(x) & \text{if } |f(x)| \leqslant k, \\ k & \text{if } |f(x)| > k. \end{cases}$$

Then $|f_2(x)| \leqslant k$ and $f_1 := f - f_2$ satisfies

$$\int_\Omega |f_1|^p = \int_{\Omega \cap \{x : |f(x)| > k\}} |f-k|^p$$
$$\leqslant 2^p \int_{\Omega \cap \{x : |f(x)| > k\}} |f|^p$$
$$< \varepsilon^p$$

for k large enough. □

Proposition 1.2. Let (i)–(iii) be satisfied on a bounded open set Ω in $\mathbb{R}^n$, with $n \geqslant 1$. Then the form t in (1.10) is bounded on $W_0^{1,2}(\Omega) \times W_0^{1,2}(\Omega)$:

$$|t[\![u,v]\!]| \leqslant K\|u\|_{1,2,\Omega}\|v\|_{1,2,\Omega} \quad \text{for all } u,v \in W_0^{1,2}(\Omega), \tag{1.12}$$

where the constant K depends only on $\|a_{ij}\|_{\infty,\Omega}$, $\|b_j\|_{2r,\Omega}$, and $\|q\|_{r,\Omega}$ for $i,j = 1, 2, \ldots, n$. Also, there exist positive constants μ and K such that

$$\operatorname{re} t[\![u]\!] + \mu\|u\|_{2,\Omega}^2 \geqslant K\|\nabla u\|_{2,\Omega}^2 \quad \text{for all } u \in W_0^{1,2}(\Omega), \tag{1.13}$$

where μ and K depend on ν, $\|a_{ij}\|_{\infty,\Omega}$, $\|b_j\|_{2r,\Omega}$, $\|q\|_{r,\Omega}$ $(i,j = 1, 2, \ldots, n)$. Hence $t+\mu$ is coercive on $W_0^{1,2}(\Omega)$. ■

Proof. On using (1.8) and (1.9) we have

$$\begin{aligned}|t[\![u,v]\!]| &\leqslant \left(\max_{1\leqslant i,j\leqslant n} \|a_{ij}\|_{\infty,\Omega}\right)\|\nabla u\|_{2,\Omega}\|\nabla v\|_{2,\Omega} \\ &\quad + \gamma_r(\Omega)\left(\max_{1\leqslant j\leqslant n}\|b_j\|_{2r,\Omega}\right)\|\nabla u\|_{2,\Omega}\|v\|_{1,2,\Omega} \\ &\quad + \gamma_r^2(\Omega)\|q\|_{r,\Omega}\|u\|_{1,2,\Omega}\|v\|_{1,2,\Omega} \\ &\leqslant K\|u\|_{1,2,\Omega}\|v\|_{1,2,\Omega}.\end{aligned}$$

In order to prove (1.13) we decompose each of the coefficients b_j $(j = 1, \ldots, n)$ and q in accordance with Lemma 1.1; we write $b_j = b_j^{(1)} + b_j^{(2)}$ and $q = q^{(1)} + q^{(2)}$, where for a given $\varepsilon > 0$,

$$\begin{aligned}&\|b_j^{(1)}\|_{2r,\Omega} < \varepsilon, \qquad \|b_j^{(2)}\|_{\infty,\Omega} < K_\varepsilon, \\ &\|q^{(1)}\|_{r,\Omega} < \varepsilon, \qquad \|q^{(2)}\|_{\infty,\Omega} < K_\varepsilon.\end{aligned}$$

Then, on using (i),

$$\begin{aligned}\operatorname{re} t[\![u]\!] &\geqslant \nu\|\nabla u\|_{2,\Omega}^2 - \left|\int_\Omega (b_j^{(1)} D_j u\bar{u} + q^{(1)}|u|^2)\right| \\ &\quad - \left|\int_\Omega (b_j^{(2)} D_j u\bar{u} + q^{(2)}|u|^2)\right| \\ &\geqslant \nu\|\nabla u\|_{2,\Omega}^2 - \varepsilon\gamma_r(\Omega)\|\nabla u\|_{2,\Omega}\|u\|_{1,2,\Omega} - \varepsilon\gamma_r^2(\Omega)\|u\|_{1,2,\Omega}^2 \\ &\quad - K_\varepsilon(\|\nabla u\|_{2,\Omega}\|u\|_{2,\Omega} + \|u\|_{2,\Omega}^2) \\ &\geqslant \nu\|\nabla u\|_{2,\Omega}^2 - \varepsilon\gamma_r(\Omega)\|\nabla u\|_{2,\Omega}^2 - \varepsilon\gamma_r^2(\Omega)\|\nabla u\|_{2,\Omega}^2 - K_\varepsilon\|\nabla u\|_{2,\Omega}\|u\|_{2,\Omega} \\ &\quad - K_\varepsilon\|u\|_{2,\Omega}^2 \\ &\geqslant [\nu - \varepsilon\gamma_r(\Omega) - \varepsilon\gamma_r^2(\Omega) - \delta]\|\nabla u\|_{2,\Omega}^2 - K_{\varepsilon,\delta}\|u\|_{2,\Omega}^2\end{aligned}$$

for any $\delta > 0$, on using the inequality $2|ab| \leqslant \delta|a|^2 + \delta^{-1}|b|^2$. The inequality (1.13) therefore follows on choosing ε and δ small enough. □

Theorem 1.3. Let (i)–(iii) be satisfied on a bounded open set Ω in $\mathbb{R}^n$, with $n \geqslant 1$. Then either the generalized non-homogeneous Dirichlet problem

$$t[\![u,\phi]\!] + \lambda(u,\phi)_{2,\Omega} = (F,\phi) \qquad (\phi \in \mathrm{W}_0^{1,2}(\Omega)), \tag{1.14}$$

has a unique solution $u \in \mathrm{W}_0^{1,2}(\Omega)$ for every $F \in \mathrm{W}^{-1,2}(\Omega)$, or else the homogeneous problem

$$t[\![\phi,v]\!] + \lambda(\phi,v)_{2,\Omega} = 0 \qquad (\phi \in \mathrm{W}_0^{1,2}(\Omega)), \tag{1.15}$$

has a non-trivial solution $v \in \mathrm{W}_0^{1,2}(\Omega)$. In the latter case the number of linearly independent solutions of (1.15) is finite and equal to the number of linearly independent solutions of

$$t[\![w,\phi]\!] + \lambda(w,\phi)_{2,\Omega} = 0 \qquad (\phi \in \mathrm{W}_0^{1,2}(\Omega)). \tag{1.16}$$

There exists an at most countable number of values μ_n of λ for which (1.15) and (1.16) have non-trivial solutions; these can accumulate only at infinity and satisfy $\mathrm{re}\,\mu_n < \mu$, where μ is the constant in (1.13). The problem (1.14) has a unique solution if, and only if, $(F,v) = 0$ for all solutions v of (1.15).

If t is symmetric the existence of a countable number of μ_n's is guaranteed. ■

Proof. Let $\hat{A}$ denote the linear bijection in Theorem IV.1.1 associated with the bounded coercive form a defined by

$$a[\![x,y]\!] := t[\![x,y]\!] + \mu(I_0 x, I_0 y)_{2,\Omega} \qquad (x, y \in \mathrm{W}_0^{1,2}(\Omega)),$$

where I_0 is the embedding map $\mathrm{W}_0^{1,2}(\Omega) \to \mathrm{L}^2(\Omega)$. Let I_0^* be the adjoint of I_0, with $\mathrm{L}^2(\Omega)$ identified with its adjoint:

$$(I_0^* x, \phi) = (x, I_0\phi)_{2,\Omega} \quad \text{for all } x \in \mathrm{L}^2(\Omega),\ \phi \in \mathrm{W}_0^{1,2}(\Omega).$$

We can rewrite (1.14) as

$$\begin{aligned} a[\![u,\phi]\!] &= (F,\phi) - (\lambda-\mu)(I_0 u, I_0\phi)_{2,\Omega} \\ &= (F,\phi) - (\lambda-\mu)(Ju,\phi) \end{aligned}$$

where $J = I_0^* I_0 : \mathrm{W}_0^{1,2}(\Omega) \to \mathrm{W}^{-1,2}(\Omega)$. Therefore, by Theorem IV.1.1 there exists a solution u of (1.14) if, and only if, $u = \hat{A}^{-1}F - (\lambda-\mu)\hat{A}^{-1}Ju$, i.e.

$$\hat{A}u + (\lambda-\mu)Ju = F. \tag{1.17}$$

Let C denote the linear isomorphism of $\mathrm{W}_0^{1,2}(\Omega)$ onto $\mathrm{W}^{-1,2}(\Omega)$ defined by $(C\phi,\psi) = (\phi,\psi)_{1,2,\Omega}$ and in (1.17) set $u_0 = C^{-1}\hat{A}u$ and $F_0 = C^{-1}F$. Then (1.17) has a solution if, and only if,

$$u_0 + (\lambda-\mu)Ku_0 = F_0 \tag{1.18}$$

does in $\mathrm{W}_0^{1,2}(\Omega)$, where $K = C^{-1}J\hat{A}^{-1}C$. Since $I_0 \in \mathscr{K}(\mathrm{W}_0^{1,2}(\Omega), \mathrm{L}^2(\Omega))$ we have $K \in \mathscr{K}(\mathrm{W}_0^{1,2}(\Omega))$ and so the Fredholm–Riesz—Schauder theory (see

§I.1) applies to (1.18). The theorem will follow once we have the adjoint K^* of K in $W_0^{1,2}(\Omega)$. Since the adjoint a^* of a also satisfies (1.12) and (1.13) there exists a linear bijection $\hat{B}$ of $W_0^{1,2}(\Omega)$ onto $W^{-1,2}(\Omega)$ satisfying

$$a^*[\![f,\phi]\!] = (\hat{B}f,\phi) \qquad (f,\phi \in W_0^{1,2}(\Omega)).$$

For $\phi, \psi \in W_0^{1,2}(\Omega)$, we have

$$\begin{aligned}(K\phi,\psi)_{1,2,\Omega} &= (C^{-1}J\hat{A}^{-1}C\phi,\psi)_{1,2,\Omega}\\ &= (J\hat{A}^{-1}C\phi,\psi)\\ &= \overline{(J\psi,\hat{A}^{-1}C\phi)}\\ &= \overline{(\hat{B}\hat{B}^{-1}J\psi,\hat{A}^{-1}C\phi)}\\ &= \overline{a^*[\![\hat{B}^{-1}J\psi,\hat{A}^{-1}C\phi]\!]}\\ &= a[\![\hat{A}^{-1}C\phi,\hat{B}^{-1}J\psi]\!]\\ &= (C\phi,\hat{B}^{-1}J\psi)\\ &= (\phi,\hat{B}^{-1}J\psi)_{1,2,\Omega}.\end{aligned}$$

Hence $K^* = \hat{B}^{-1}J$. If θ_n $(\neq 0)$ is an eigenvalue of K^* with eigenvector v_n, then, with $\mu_n = \mu - 1/\bar{\theta}_n$,

$$\hat{B}v_n + (\bar{\mu}_n - \mu)Jv_n = 0$$

and hence, since $a^* = t^* + \mu$,

$$t^*[\![v_n,\phi]\!] + \bar{\mu}_n(I_0v_n, I_0\phi)_{2,\Omega} = 0 \qquad (\phi \in W_0^{1,2}(\Omega)),$$

which is (1.15). Since the θ_n can accumulate only at 0 the μ_n can only accumulate at ∞. Also,

$$\begin{aligned}(1/\theta_n)\|I_0v_n\|_{2,\Omega}^2 &= (1/\theta_n)(Jv_n,v_n)\\ &= (1/\theta_n)a^*[\![K^*v_n,v_n]\!]\\ &= a^*[\![v_n]\!],\end{aligned}$$

so that, by (1.13), $\operatorname{re}\theta_n > 0$ and hence $\operatorname{re}\mu_n < \mu$. Finally, F_0 in (1.18), with $\lambda = \mu_n$, is orthogonal to v_n in $W_0^{1,2}(\Omega)$ if and only if

$$\begin{aligned}(F,v_n) &= (CF_0,v_n)\\ &= (F_0,v_n)_{1,2,\Omega}\\ &= 0.\end{aligned}$$

The theorem is therefore proved for general t.

If t is symmetric the operator $A = (I_0^*)^{-1}\hat{A}I_0^{-1}$ is self-adjoint in $L^2(\Omega)$, by Corollary IV.1.2. Hence, from above, $A^{-1} = I_0\hat{A}^{-1}I_0^* = I_0K^*I_0^{-1}$ is a compact self-adjoint operator in $L^2(\Omega)$. Since A^{-1} is clearly not of finite rank it has an infinite sequence of eigenvalues and these are also eigenvalues of K^*. The final assertion of the theorem is therefore established. □

From Corollary IV.1.2 we also obtain the following theorem; in its statement we suppress the embedding map I_0 for simplicity.

Theorem 1.4. Let (i)–(iii) be satisfied on a bounded open set Ω in $\mathbb{R}^n$, with $n \geqslant 1$. Then there exists a closed, densely defined linear operator T in $L^2(\Omega)$ with the following properties:

(i) $\mathscr{D}(T) = \{u : u \in W_0^{1,2}(\Omega) \text{ and } \tau u \in L^2(\Omega)\}$ and $Tu = \tau u$ for $u \in \mathscr{D}(T)$;
(ii) $a[\![u, \phi]\!] = (Tu, \phi)_{2,\Omega}$ for $u \in \mathscr{D}(T)$ and $\phi \in W_0^{1,2}(\Omega)$;
(iii) $\mathscr{D}(T)$ is dense as a subpace of $W_0^{1,2}(\Omega)$;
(iv) $T + \mu I$ is m-accretive, where μ is the real constant in (1.13);
(v) T^* is the operator associated with t^*;
(vi) T has a compact resolvent; if t is symmetric T is self-adjoint and its spectrum consists of an infinite sequence of eigenvalues λ_n $(n \in \mathbb{N})$, each having finite multiplicity and with re $\lambda_n > -\mu$. ■

Proof. With the notation of Theorem 1.3 and its proof we set $T = A - \mu I$. Since $0 \in \rho(A)$ and, by (1.13),

$$\operatorname{re}(Au, u)_{2,\Omega} = \operatorname{re} a[\![u]\!] > 0 \quad \text{for all } u \in \mathscr{D}(T),$$

it follows from Theorem III.6.5 that $T + \mu I$ is m-accretive. With the exception of (vi) the theorem is an immediate consequence of Corollary IV.1.2. Part (vi) was established at the end of the proof of Theorem 1.3; note that the eigenvalues λ_n are related to the μ_n in Theorem 1.3 by $\lambda_n = -\mu_n$. □

Remark 1.5. The embedding $W_0^{1,2}(\Omega) \to L^{2r/(r-1)}(\Omega)$ is continuous even when Ω is unbounded. Proposition 1.2 therefore remains valid for an arbitrary open set Ω. When Ω is arbitrary we see from (1.13) that

$$\operatorname{re}\{t[\![u]\!] + \mu_1 \|u\|_{2,\Omega}^2\} \geqslant K \|u\|_{1,2,\Omega}^2 \qquad (u \in W_0^{1,2}(\Omega)), \tag{1.19}$$

for some $\mu_1 \geqslant \mu$. Hence $t + \mu_1$ is coercive on $W_0^{1,2}(\Omega)$, (1.14) has a unique solution $u \in W_0^{1,2}(\Omega)$ for every $F \in W^{-1,2}(\Omega)$ whenever re $\lambda \geqslant \mu_1$ and Theorem 1.4(i)–(v) remains true. Furthermore, on replacing μ by μ_1, Theorems 1.3 and 1.4 continue to hold in their entirety if $W_0^{1,2}(\Omega)$ is compactly embedded in $L^2(\Omega)$. By Remark V.5.18(4), this is so if $\lim_{|x| \to \infty, x \in \Omega} |B(x, 1) \cap \Omega| = 0$.

Remark 1.6. In Theorem V.3.22 we saw that the Poincaré inequality

$$\|u\|_{2,\Omega} \leqslant K_\Omega \|\nabla u\|_{2,\Omega}^2 \qquad (u \in W_0^{1,2}(\Omega)),$$

holds for cylindrical open sets Ω in $\mathbb{R}^n$ as well as for bounded Ω. In this case $\|\nabla u\|_{2,\Omega}$ is equivalent to the $W_0^{1,2}(\Omega)$ norm and hence, by (1.13) $a + \lambda$ is coercive on $W_0^{1,2}(\Omega)$ as long as $\text{re}\,\lambda \geqslant \mu$; in Remark 1.5 we therefore have no need to introduce μ_1. Also, with the Poincaré inequality available, it is easy to check from the proof of Theorem 1.2 that (1.13) holds with $\mu = 0$ if the norms $\|a_{ij}\|_{\infty,\Omega}$, $\|b_j\|_{2r,\Omega}$ and $\|q\|_{r,\Omega}$ are sufficiently small.

Remark 1.7. A specially important case of Theorem 1.4, for an arbitrary open set Ω, is when $t[\![u,\phi]\!] = (u,\phi)_{1,2,\Omega}$ and hence the self-adjoint operator T in Theorem 1.4 is the Friedrichs extension of $1-\Delta$ on $C_0^\infty(\Omega)$. We deonote $T-I$ by $-\Delta_{D,\Omega}$ and call it the *Dirichlet Laplacian* on Ω. The operator $\hat{A}$ in Theorem IV.1.1 now satisfies

$$(\hat{A}u,\phi) = (u,\phi)_{1,2,\Omega} \qquad (u,\phi\in W_0^{1,2}(\Omega)),$$

and hence $\hat{A}$ is the canonical linear isometry of $W_0^{1,2}(\Omega)$ onto its adjoint $W^{-1,2}(\Omega)$. Furthermore, for $u\in W_0^{-1,2}(\Omega)$ and $\phi\in C_0^\infty(\Omega)$,

$$(\hat{A}u,\phi) = \int_\Omega u(1-\Delta)\bar{\phi}.$$

Thus, if we regard distributions as conjugate linear functionals on $C_0^\infty(\Omega)$ rather than linear ones as is usually the case, we can identify $\hat{A}$ with the distributional operator $1-\Delta$.

If Ω is bounded, or more generally, if $\|\nabla\bullet\|_{2,\Omega}$ is equivalent to $\|\bullet\|_{1,2,\Omega}$ on $W_0^{1,2}(\Omega)$, the preceding remarks remain valid with $t[\![u,\phi]\!] = (\nabla u,\nabla\phi)_{2,\Omega}$ and consequently $-\Delta$ is the canonical linear isometry of $W_0^{1,2}(\Omega)$, with norm $\|\nabla\bullet\|_{2,\Omega}$, onto $W^{-1,2}(\Omega)$.

We have that $u\in\mathscr{D}(\Delta_{D,\Omega})$ if, and only if, $u\in W_0^{1,2}(\Omega)$ and the distributional Laplacian $\Delta u\in L^2(\Omega)$. Let $\phi\in C_0^\infty(\Omega)$ and set $v=\phi u$ to be zero outside Ω. Then $v\in W^{1,2}(\mathbb{R}^n)$, $\Delta v\in L^2(\mathbb{R}^n)$, and, on taking Fourier transforms, $(1+|\xi|^2)\hat{v}(\xi)\in L^2(\mathbb{R}^n)$. Consequently $\xi_i\xi_j\hat{v}(\xi)\in L^2(\mathbb{R}^n)$ for $i,j=1,2,\ldots,n$ and so $D_iD_jv\in L^2(\mathbb{R}^n)$. This implies that $v\in W^{2,2}(\mathbb{R}^n)$ and

$$\mathscr{D}(\Delta_{D,\Omega}) = \{u : u\in W_0^{1,2}(\Omega)\cap W_{\text{loc}}^{2,2}(\Omega),\ \Delta u\in L^2(\Omega)\}.$$

If we assume that $\partial\Omega$ is of class C^2 then

$$\mathscr{D}(\Delta_{D,\Omega}) = W_0^{1,2}(\Omega)\cap W^{2,2}(\Omega);$$

this result may be found in Gilbarg and Trudinger [1, Theorem 8.12].

If Ω is unbounded the conditions (i)–(iii) are rather restrictive on the coefficients of τ and, except for open sets like those referred to in Remark 1.5, $W_0^{1,2}(\Omega)$ is not compactly embedded in $L^2(\Omega)$; the compactness of $W_0^{1,2}(\Omega) \to L^2(\Omega)$ for unbounded Ω will be discussed in detail in §§VIII.3 and X.6. In general the analogue of the operator T in Theorem 1.3 does not have a wholly discrete spectrum. The nature and spectra of these operators T will be the subject of §VII.1 and Chapter X. □

1.2. *Eigenfunction expansions for the Dirichlet problem*

We now specialize to the case when t is symmetric and hence the operator T in Theorem 1.4 is self-adjoint in $L^2(\Omega)$. This is the case when

$$\operatorname{re} b_j(x) = 0, \qquad 2\mathrm{i} \operatorname{im} q(x) = \sum_{j=1}^{n} D_j b_j(x), \tag{1.20}$$

for a.e. $x \in \Omega$.

Under the conditions in Proposition 1.2 the form a defined by

$$a[\![x, y]\!] = t[\![x, y]\!] + \mu(I_0 x, I_0 y)_{2,\Omega} \qquad (x, y \in W_0^{1,2}(\Omega)),$$

is an inner product on $W_0^{1,2}(\Omega)$ and the associated norm $a^{\frac{1}{2}}[\![\bullet]\!]$ is equivalent to the usual $W_0^{1,2}(\Omega)$ norm $\|\bullet\|_{1,2,\Omega}$. Let us denote $W_0^{1,2}(\Omega)$ with the inner product $a[\![\bullet, \bullet]\!]$ by H_0 and let H denote $L^2(\Omega)$. We therefore have the triplet

$$H_0 \overset{E}{\rightarrow} H \overset{E^*}{\rightarrow} H_0^*$$

where the embedding maps E and E^* are continuous and have dense ranges. With the notation of §1.1 we have for $F \in H_0^*$ that $F = \hat{A}f$ for some $f \in H_0$ and

$$(F, \phi) = (\hat{A}f, \phi) = a[\![f, \phi]\!] = (f, \phi)_{H_0}$$

for all $\phi \in H_0$. Hence

$$\|F\| = \|\hat{A}f\| = \sup_{\|\phi\|_{H_0}=1} |(\hat{A}f, \phi)|$$

$$= \sup_{\|\phi\|_{H_0}=1} |(f, \phi)_{H_0}| = \|f\|_{H_0}.$$

This implies that $\hat{A}$ is a unitary map of H_0 onto H_0^* and can therefore be used to define an inner product on H_0^* which gives rise to a norm equivalent to the usual norm on H_0^*. This inner product is defined as follows: if $F_1, F_2 \in H_0^*$,

$$(F_1, F_2)_{H_0^*} := a[\![f_1, f_2]\!]$$

$$= (f_1, f_2)_{H_0}, \tag{1.21}$$

where $f_i = \hat{A}^{-1} F_i$ $(i = 1, 2)$.

In Theorem 1.4 let the eigenvalues λ_n be repeated according to multiplicity and let ϕ_n denote the eigenvector of T corresponding to λ_n, the set of eigenvectors being orthonormal in H. Then, with $A = T + \mu I$,

$$A\phi_n = (\lambda_n + \mu)\phi_n, \qquad \lambda_n > -\mu. \tag{1.22}$$

If $\psi_n := (\lambda_n + \mu)^{-\frac{1}{2}} E^{-1} \phi_n$ and $G := \hat{A}^{-1} E^* E = E^{-1} A^{-1} E$ (note that we now have $A = (E^*)^{-1} \hat{A} E^{-1}$ since I_0 has been replaced by E),

$$G\phi_n = (\lambda_n + \mu)^{-1} \psi_n \tag{1.23}$$

and

$$\begin{aligned}(\psi_n, \psi_m)_{H_0} &= a[\![\psi_n, \psi_m]\!]\\ &= (\lambda_n + \mu)\, a[\![G\psi_n, \psi_m]\!]\\ &= (\lambda_n + \mu)\, a[\![\hat{A}^{-1}E^*E\psi_n, \psi_m]\!]\\ &= (\lambda_n + \mu)\,(E^*E\psi_n, \psi_m)\\ &= (\lambda_n + \mu)\,(E\psi_n, E\psi_m)_H\\ &= (\phi_n, \phi_m)_H. \end{aligned} \tag{1.24}$$

The eigenvectors ϕ_n $(n \in \mathbb{N})$ form an orthonormal basis of H since A^{-1} has a trivial null set; see Corollary II.5.3. Also, $\{\psi_n : n \in \mathbb{N}\}$ is an orthonormal basis for H_0 since, if $f \in H_0$ is orthogonal to the ψ_n in H_0, we obtain, by the same argument as that leading to (1.24),

$$\begin{aligned}0 = (\psi_n, f)_{H_0} &= a[\![\psi_n, f]\!]\\ &= (\lambda_n + \mu)^{\frac{1}{2}}\,(\phi_n, Ef)_H,\end{aligned}$$

whence $Ef = 0$ and consequently $f = 0$.

We can also obtain an orthonormal basis for H_0^* from the eigenvector ϕ_n. Let $\theta_n = (\lambda_n + \mu)^{\frac{1}{2}} E^*\phi_n$ $(n \in \mathbb{N})$. Then

$$\hat{A}^{-1}\theta_n = (\lambda_n + \mu)\, G\psi_n = \psi_n$$

and

$$\begin{aligned}(\theta_n, \theta_m)_{H_0^*} &= a[\![\psi_n, \psi_m]\!]\\ &= (\psi_n, \psi_m)_{H_0},\end{aligned}$$

so that $\{\theta_n : n \in \mathbb{N}\}$ is orthonormal in H_0^*. It is also complete in H_0^*, because if $F \in H_0^*$ is orthogonal to $\{\theta_n : n \in \mathbb{N}\}$ we have, since $F = \hat{A}f$ for some $f \in H_0$,

$$\begin{aligned}0 = (F, \theta_n)_{H_0^*} &= a[\![f, \psi_n]\!]\\ &= (f, \psi_n)_{H_0},\end{aligned}$$

whence $f = 0$ by the already established completeness of $\{\psi_n : n \in \mathbb{N}\}$ in H_0. This in turn gives $F = 0$ as asserted. We have therefore proved

Theorem 1.8. Let (i)–(iii) and (1.20) be satisfied on a bounded open set Ω in R^n, with $n \geq 1$, and let $\{\phi_n : n \in \mathbb{N}\}$ be the eigenvectors of the operator T in Theorem 1.4. Define

$$\psi_n := (\lambda_n + \mu)^{-\frac{1}{2}} E^{-1}\phi_n \in H_0, \qquad \theta_n := (\lambda_n + \mu)^{\frac{1}{2}} E^*\phi_n \in H_0^*.$$

Then the sequences $\{\phi_n : n \in \mathbb{N}\}$, $\{\psi_n : n \in \mathbb{N}\}$, $\{\theta_n : n \in \mathbb{N}\}$ are orthonormal bases for H, H_0, H_0^* respectively.

The theorem remains true for an unbounded open set $\Omega \subset \mathbb{R}^n$ which is such that $W_0^{1,2}(\Omega)$ is compactly embedded in $L^2(\Omega)$, if μ is replaced by the μ_1 of Remark 1.5 (see also Remark 1.6). ∎

1.3. *The weak Neumann problem*

The problem (1.11) is still meaningful if we replace $W_0^{1,2}(\Omega)$ by $W^{1,2}(\Omega)$ or indeed by any closed subspace of $W^{1,2}(\Omega)$ which contains $W_0^{1,2}(\Omega)$. The problem defined with respect to $W^{1,2}(\Omega)$ will be called the *generalized Neumann problem* for τ in Ω: given $F \in [W^{1,2}(\Omega)]^*$, to find $v \in W^{1,2}(\Omega)$ such that

$$t[\![v, \phi]\!] = (F, \phi) \quad \text{for all } \phi \in W^{1,2}(\Omega). \tag{1.25}$$

This is also referred to as the *weak* or *variational* form of the Neumann problem for τ in Ω.

In order to motivate our description of this as the (generalized) Neumann problem we assume that Ω is smooth enough for $C_0^\infty(\mathbb{R}^n)$, or more precisely, the set of functions which are restrictions to Ω of functions in $C_0^\infty(\mathbb{R}^n)$, to be dense in $W^{1,2}(\Omega)$; by Theorem V.4.7 this is true if $\partial\Omega$ is of class C. Also we assume that our use of Green's Theorem is valid and that $a_{ij} \in C^1(\bar{\Omega})$.

If in (1.25) $v \in C^2(\bar{\Omega})$ and F is given by $(F, \phi) = \int_\Omega f\bar{\phi}$ for all $\phi \in W^{1,2}(\Omega)$ and for some $f \in L^2(\Omega)$, then for all $\phi \in C_0^\infty(\mathbb{R}^n)$,

$$\begin{aligned}\int_\Omega f\bar{\phi} &= (F, \phi) \\ &= t[\![v, \phi]\!] \\ &= \int_\Omega (\tau v)\bar{\phi} - \int_{\partial\Omega} \bar{\phi} a_{ij} m_i D_j v \, d\sigma \end{aligned}\tag{1.26}$$

by Green's Theorem, where $m = (m_1, m_2, \ldots, m_n)$ denotes the exterior normal to $\partial\Omega$ and $d\sigma$ is the Lebesgue measure on $\partial\Omega$. On taking $\phi \in C_0^\infty(\Omega)$ in (1.26) we see that $\tau v = f$ and then, in view of the fact that members of $C_0^\infty(\mathbb{R}^n)$ assume arbitrary values on $\partial\Omega$, we deduce from (1.26) that

$$\tau v = f \text{ in } \Omega, \qquad a_{ij} m_i D_j v = 0 \text{ on } \partial\Omega. \tag{1.27}$$

The term $a_{ij} m_i D_j v$ is called the *co-normal derivative* of v with respect to τ. When $a_{ij} = \delta_{ij}$ the co-normal derivative becomes the normal derivative $\partial v/\partial n = (m \cdot \nabla)v$ and hence (1.27) is the familiar Neumann problem

$$\tau v = f \quad \text{in } \Omega, \qquad \partial v/\partial n = 0 \quad \text{on } \partial\Omega. \tag{1.28}$$

Conversely, under appropriate smootheness conditions, a solution v of (1.27) satisfies (1.25) with $(F, \phi) = \int_\Omega f\bar{\phi}$ and $\phi \in C_0^\infty(\mathbb{R}^n)$, as we see on using Green's formula. Since $C_0^\infty(\mathbb{R}^n)$ is assumed to be dense in $W^{1,2}(\Omega)$, (1.25) then follows by continuity, given that t is continuous on $W^{1,2}(\Omega) \times W^{1,2}(\Omega)$.

In the Dirichlet problem, (1.11) with $\phi \in C_0^\infty(\Omega)$ can be interpreted as the distributional equation $\tau v = F$. Analogously we can give a distributional interpretation to the Neumann problem (1.25) if we regard $C_0^\infty(\mathbb{R}^n)$ as the test-function space instead of $C_0^\infty(\Omega)$ but bear in mind that $\tau v = F$ also incorporates information about the behaviour of v on $\partial\Omega$. The regularity properties of v in and on the boundary of Ω are discussed in detail in the books of Agmon [1] and Friedman [1].

If Ω is such that $W^{1,2}(\Omega)$ is continuously embedded in $L^{2r/(r-1)}(\Omega)$ for r satisfying (1.5), and the embedding $W^{1,2}(\Omega) \to L^2(\Omega)$ is compact, the arguments in §§1.1 and 1.2 go through with $W_0^{1,2}(\Omega)$ replaced by $W^{1,2}(\Omega)$. This is so, for instance, if $\partial\Omega$ is minimally smooth, by Theorem V.4.13. As in Proposition 1.2 we have that t is continuous on $W^{1,2}(\Omega) \times W^{1,2}(\Omega)$ and there exists a positive constant μ such that $t[\![\bullet, \bullet]\!] + \mu(\bullet, \bullet)_{2,\Omega}$ is coercive on $W^{1,2}(\Omega)$; in fact

$$\operatorname{re}(t[\![u]\!] + \mu\|u\|_{2,\Omega}^2) \geqslant K\|u\|_{1,2,\Omega}^2 \qquad (u \in W^{1,2}(\Omega)), \tag{1.29}$$

for some positive constant K depending only on ν, $\|a_{ij}\|_{\infty,\Omega}$, $\|b_j\|_{2r,\Omega}$, and $\|q\|_{r,\Omega}$ for $i, j = 1, 2, \ldots, n$. We therefore have

Theorem 1.9. If $\partial\Omega$ is minimally smooth, Theorems 1.3, 1.4, and 1.8 continue to hold for the Neumann problem, when $W_0^{1,2}(\Omega)$ is replaced by $W^{1,2}(\Omega)$. ■

2. Truncation rules on $W^{1,p}(\Omega)$

In this section Ω is an arbitrary open set in $\mathbb{R}^n$, with $n \geqslant 1$, unless the contrary is explicitly stated, and all the functions and function spaces will be *real*. We shall be investigating the effect of the map $u \mapsto G(u)$ on $W_0^{1,p}(\Omega)$ and $W^{1,p}(\Omega)$ when G is a Lipschitz function defined on $\mathbb{R}$. An important special case of our main theorem can be derived from Proposition V.2.6. To commence, let G be a real-valued, uniformly Lipschitz continuous function on $\mathbb{R}$, i.e. there exists a positive constant γ such that

$$|G(t) - G(s)| \leqslant \gamma|t - s| \quad \text{for all } t, s \in \mathbb{R}. \tag{2.1}$$

Such a function G is locally absolutely continuous on R and has a locally integrable derivative G' such that

$$G(t) - G(s) = \int_s^t G'(z)\,dz.$$

Also, for a.e. $t \in \mathbb{R}$, by Lebesgue's fundamental theorem (see Stein [1, §1.1]),

$$\lim_{s \to t}\left(\frac{1}{s-t}\int_s^t G'(z)\,dz\right) = G'(t)$$

and hence, by (2.1), $G' \in L^\infty(\mathbb{R})$ with $\|G'\|_{\infty,\mathbb{R}} \leqslant \gamma$. Hereafter we shall take G' to be defined and bounded everywhere on $\mathbb{R}$ by putting it zero on the exceptional null set.

Theorem 2.1. Let G be uniformly Lipschitz continuous on $\mathbb{R}$ and let Ω be a bounded open set in $\mathbb{R}^n$, with $n \geqslant 1$. Then $u \mapsto G(u)$ maps the real space $W^{1,p}(\Omega)$ $(1 < p < \infty)$ into itself and the first distributional derivatives of $G(u)$ are given by

$$D_j G(u) = G'(u) D_j u \tag{2.2}$$

with the understanding that the right-hand side vanishes if either factor does. If $G(0) = 0$ then $u \mapsto G(u)$ maps $W_0^{1,p}(\Omega)$ into itself. The theorem remains valid for an unbounded Ω if $G(0) = 0$. ■

Proof. Let Ω be bounded and let $u_m \in C^\infty(\Omega) \cap W^{1,p}(\Omega)$ $(m \in \mathbb{N})$ be such that $u_m \to u$ in $W^{1,p}(\Omega)$; such a sequence exists for each $u \in W^{1,p}(\Omega)$ by virtue of Theorem V.3.2. Each $G(u_m)$ is continuous on Ω and also, for $x, y \in \Omega$,

$$\begin{aligned} |G[u_m(x)]| &\leqslant |G[u_m(y)]| + |G[u_m(x)] - G[u_m(y)]| \\ &\leqslant |G[u_m(y)]| + \gamma |u_m(x) - u_m(y)| \\ &\leqslant K_y + \gamma |u_m(x)|, \end{aligned}$$

where K_y is a positive constant depending on y. Since $u_m \in L^p(\Omega)$ and Ω is bounded we have $G(u_m) \in L^p(\Omega)$. Evidently, by (2.1), $(G(u_m))$ is a Cauchy sequence in $L^p(\Omega)$ and indeed $G(u_m) \to G(u)$ in $L^p(\Omega)$: $G(u)$ is measurable and since

$$|G[u(x)] - G[u_m(x)]| \leqslant \gamma |u(x) - u_m(y)|,$$

we have $G(u) \in L^p(\Omega)$ and $G(u_m) \to G(u)$ in $L^p(\Omega)$.

The next step is to prove (2.2). Let u be an everywhere defined representative of $u \in W^{1,p}(\Omega)$. Let $\phi \in C_0^\infty(\Omega)$ and $\hat{h}_j = (0, 0, \ldots, 0, h_j, 0, \ldots, 0)$, with $h_j \neq 0$ in the jth position. Then for all $j \in \{1, 2, \ldots, n\}$,

$$\begin{aligned} -\int_\Omega G[u(x)] D_j\phi(x)\,dx &= -\lim_{h_j \to 0} \int_\Omega G[u(x)]\, h_j^{-1} [\phi(x+\hat{h}_j) - \phi(x)]\,dx, \\ &= \lim_{h_j \to 0} \int_\Omega h_j^{-1} \big(G[u(x)] - G[u(x-\hat{h}_j)]\big)\phi(x)\,dx. \end{aligned} \tag{2.3}$$

By (2.1), for small enough $h_j \neq 0$,

$$\begin{aligned} h_j^{-1}\big(G[u(x)] - G[u(x-\hat{h}_j)]\big) &= h_j^{-1}\big(G[u(x)] - G[u(x) - h_j D_j u(x)]\big) \\ &\quad + h_j^{-1}\big(G[u(x) - h_j D_j u(x)] - G[u(x-\hat{h}_j)]\big) \\ &= h_j^{-1} \int_{u(x) - h_j D_j u(x)}^{u(x)} G'(z)\,dz + O\big(|h_j^{-1}[u(x) - u(x-\hat{h}_j)] - D_j u(x)|\big). \end{aligned} \tag{2.4}$$

As $h_j \to 0$ the first term on the right-hand side of (2.4) tends to $G'[u(x)]\,D_j u(x)$, with the convention noted in the theorem that the product is taken to be zero if one of the factors vanishes. The second term on the right-hand side of (2.4) satisfies

$$\|h_j^{-1}[u(\bullet)-u(\bullet-h_j)]-D_j u(\bullet)\|_{p,\operatorname{supp}\phi}\to 0$$

as $h_j \to 0$, since this is obviously true for each u_m (in fact the convergence is then uniform on the compact set supp ϕ) and the result follows on allowing $m \to \infty$. We therefore see from (2.4) that

$$-\int_\Omega G(u)\,D_j\phi = \lim_{h_j\to 0}\int_\Omega\left(h_j^{-1}\int_{u(x)-h_jD_ju(x)}^{u(x)} G'(z)\,dz\right)\phi(x)\,dx$$

$$=\int_\Omega [G'(u)\,D_j u]\,\phi,$$

the last step being justified by the Dominated Convergence Theorem, since

$$\left|h_j^{-1}\int_{u(x)-h_jD_ju(x)}^{u(x)} G'(z)\,dz\right| \leqslant \gamma|D_j u(x)|.$$

This proves (2.2).

In proving (2.2) we have shown in particular that $D_j G(u_m) = G'(u_m)D_j u_m$ and consequently $G(u_m) \in W^{1,p}(\Omega)$ and

$$\|D_j G(u_m)\|_{p,\Omega} \leqslant \gamma\|D_j u_m\|_{p,\Omega}$$

$$\leqslant \gamma\|u_m\|_{1,p,\Omega}.$$

Thus $(G(u_m))$ is a bounded sequence in $W^{1,p}(\Omega)$ and we may infer that $(G(u_m))$ contains a subsequence which converges weakly in $W^{1,p}(\Omega)$. Since we have already shown that $G(u_m) \to G(u)$ in $L^p(\Omega)$, the weak limit of this subsequence of $(G(u_m))$ must be $G(u)$ and hence $G(u) \in W^{1,p}(\Omega)$.

If $u \in W_0^{1,p}(\Omega)$, the sequence (u_m) above lies in $C_0^\infty(\Omega)$, and since $G(0) = 0$, the function $G(u_m)$ has compact support in Ω for each m. Thus $G(u_m) \in W_0^{1,p}(\Omega)$, by Theorem V.3.4. Also, from above, $G(u)$ is the weak limit in $W^{1,p}(\Omega)$ of a subsequence, (v_j) say, of $(G(u_m))$. We now invoke Mazur's Theorem (cf. Yosida [1, Theorem V.1.2]) and conclude that since $G(u)$ is the weak limit of (v_j) in $W^{1,p}(\Omega)$ there exists a sequence of averages of the v_j with positive coefficients, (w_j) say, where

$$w_j = \left(\sum_{k=1}^{j} c_{k,j} v_k\right)\Bigg/\sum_{k=1}^{j} c_{k,j} \qquad (c_{k,j} > 0),$$

such that $w_j \to G(u)$ in $W^{1,p}(\Omega)$. Since $w_j \in W_0^{1,p}(\Omega)$, we conclude that $G(u) \in W_0^{1,p}(\Omega)$ and the theorem is proved for a bounded Ω.

If Ω is unbounded and $G(0) = 0$ the above proof requires only minor modifications. Any $u \in W^{1,p}(\Omega)$ is the limit of a sequence (u_m) where

$u_m \in C^\infty(\Omega) \cap W^{1,p}(\Omega)$ and u_m has compact support in $\mathbb{R}^n$, by Corollary V.3.3. Since $G(0) = 0$, it follows that $G(u_m)$ has compact support in $\mathbb{R}^n$ and therefore lies in $L^p(\Omega)$ as before. The previous argument can now be applied without change. □

Remark 2.2. In the proof of Theorem 2.1 we show that if $u_m \in C^\infty(\Omega) \cap W^{1,p}(\Omega)$ and $u_m \to u$ in $W^{1,p}(\Omega)$ (the u_m having compact support in $\mathbb{R}^n$ when Ω is unbounded) then $G(u)$ lies in the closed convex hull $\overline{\text{co}}\{G(u_m): m \in \mathbb{N}\}$ of the sequence $(G(u_m))$ in $W^{1,p}(\Omega)$.

Remark 2.3. Let $u \in W^{1,p}(\Omega)$ and $u_m \to u$ in $W^{1,p}(\Omega)$, where $u_m \in C^\infty(\Omega) \cap W^{1,p}(\Omega)$. Then $G(u_m) \to G(u)$ in $L^p(\Omega)$ and a subsequence of $(G(u_m))$ converges weakly to $G(u)$ in $W^{1,p}(\Omega)$. If $u(\bullet)$ denotes a pointwise representative of u (a function which is unique outside a set of measure zero in Ω) then we can select a subsequence of (u_m), which we continue to denote by (u_m), such that $u_m(x) \to u(x)$ a.e. in Ω and hence $G[u_m(x)] \to G[u(x)]$ a.e. in Ω and $G(u_m) \rightharpoonup G(u)$ in $W^{1,p}(\Omega)$. On applying Mazur's Theorem we have that convex means of the $G(u_m)$ converge pointwise to $G[u(\bullet)]$ a.e. in Ω and strongly to $G(u)$ in $W^{1,p}(\Omega)$. Thus $G[u(\bullet)]$ is a representative of the element $G(u) \in W^{1,p}(\Omega)$, so that $G(u) = G[u(\bullet)]$ a.e. in Ω.

Corollary 2.4. Let $u \in W^{1,p}(\Omega)$ $(W_0^{1,p}(\Omega))$. Then $|u|$, $u^+ := \sup(u, 0)$ and $\bar{u} := -\inf(u, 0)$ also belong to $W^{1,p}(\Omega)$ $(W_0^{1,p}(\Omega))$. If u, $v \in W^{1,p}(\Omega)$ $(W_0^{1,p}(\Omega))$ then $\sup(u, v) := u + (v - u)^+$ and $\inf(u, v) := u - (v - u)^-$ belong to $W^{1,p}(\Omega)$ $(W_0^{1,p}(\Omega))$. ■

Proof. We have $|u| = G(u)$, where $G(t) = |t|$, a uniformly Lipschitz function on $\mathbb{R}$. Thus the assertion about $|u|$ follows from Theorem 2.1. The rest follows since $u^+ = \frac{1}{2}(u + |u|)$ and $u^- = u^+ - u = \frac{1}{2}(|u| - u)$. Note that we can also use Proposition V.2.6 to prove this corollary. □

It is worth observing that Corollary 2.4 has no analogue for spaces $W^{k,p}(\Omega)$ and $W_0^{k,p}(\Omega)$ with $k > 1$.

Remark 2.5. On account of Remark 2.3, $|u| = |u(\bullet)|$ a.e. in Ω and hence $|u| \geqslant 0$ a.e. in Ω. Similarly $u^+ \geqslant 0$ and $u^- \geqslant 0$ a.e. in Ω. Also $\sup(u, v)$ has the representative $\max\{u(\bullet), v(\bullet)\}$ a.e. in Ω. Consequently $\sup(u, v)$ and $\sup(v, u)$ are elements of $W^{1,p}(\Omega)$ which are equal a.e. in Ω. That this means the equality of $\sup(u, v)$ and $\sup(v, u)$ as elements of $W^{1,p}(\Omega)$ will be proved in Proposition 2.7 below. Similarly $\inf(u, v) = \inf(v, u)$ in $W^{1,p}(\Omega)$.

Lemma 2.6. Let $u \in W^{1,p}(\Omega)$. Then $\nabla u = 0$ a.e. on any set where u is constant. ■

Proof. Without loss of generality we may take the constant to be zero. Then from Proposition V.2.6, $\nabla u = \nabla u^+ - \nabla u^- = 0$ a.e. on $\{x : x \in \Omega, u(x) = 0\}$. □

Proposition 2.7. Let $u, v \in W^{1,p}(\Omega)$ have representatives $u(\bullet)$ and $v(\bullet)$ which are equal a.e. on Ω. Then $u = v$ in $W^{1,p}(\Omega)$. ■

Proof. By Lemma 2.6, $\nabla u = \nabla v$ a.e. in Ω and the proposition follows. □

Another useful consequence of Theorem 2.1 (and Proposition V.2.6) is

Corollary 2.8. If $u, v \in W^{1,p}(\Omega)$ and $w = \sup(u, v)$, then for all $j \in \{1, 2, \ldots, n\}$,

$$\|D_j w\|_{p,\Omega} \leqslant \max\{\|D_j u\|_{p,\Omega}, \|D_j v\|_{p,\Omega}\}. \tag{2.5}$$

In particular,

$$\|D_j |u|\,\|_{p,\Omega} \leqslant \|D_j u\|_{p,\Omega}.$$ ■

Proof. Let $u_k, v_k \in C^\infty(\Omega) \cap W^{1,p}(\Omega)$ be such that $u_k \to u$ and $v_k \to v$ in $W^{1,p}(\Omega)$ and define $w_k = \max\{u_k, v_k\}$. Then (w_k) contains a subsequence, which we shall still call (w_k), which converges weakly to w in $W^{1,p}(\Omega)$. Also, w_k is Lipschitz continuous and for a.e. $x \in \Omega$,

$$|D_j w_k(x)| \leqslant \max\{|D_j u_k(x)|, |D_j v_k(x)|\}.$$

Consequently, for all $\phi \in C_0^\infty(\Omega)$,

$$\int_\Omega (D_j w)\phi = \lim_{k \to \infty} \int_\Omega (D_j w_k)\phi$$

and

$$\left|\int_\Omega (D_j w_k)\phi\right| \leqslant \max\{\|D_j u_k\|_{p,\Omega}, \|D_j v_k\|_{p,\Omega}\}\,\|\phi\|_{p',\Omega},$$

where $1/p + 1/p' = 1$. The result therefore follows. □

Corollary 2.9. If $u \in W_0^{1,p}(\Omega)$ and k is a constant then $(u-k)^+ \in W_0^{1,p}(\Omega)$ if $k \geqslant 0$ and $(u-k)^- \in W_0^{1,p}(\Omega)$ if $k \leqslant 0$. ■

Proof. We apply Theorem 2.1 with $G(t) = \max\{t-k, 0\}$ and $-\min\{t-k, 0\}$. □

3. A partial ordering on $W^{1,p}(\Omega)$

As in §2 we shall assume in this section that Ω is an arbitrary open set, unless the contrary is stated, and the function spaces appearing will be real. The following definition presupposes the denseness of $C^\infty(\Omega) \cap W^{1,p}(\Omega)$ in $W^{1,p}(\Omega)$, established in Theorem V.3.2.

Definition 3.1. Let E be a subset of $\bar{\Omega}$ and $u \in W^{1,p}(\Omega)$. We say that u is *non-negative on E in the sense of* $W^{1,p}(\Omega)$, and write $u \succcurlyeq 0$ on E, if there exists a sequence (u_j) in $C^\infty(\Omega) \cap W^{1,p}(\Omega)$ which is such that $u_j \to u$ in $W^{1,p}(\Omega)$ and for each j there exists an open neighbourhood U_j of E in $\mathbb{R}^n$ such that $u_j > 0$ in $U_j \cap \Omega$. If $-u \succcurlyeq 0$ we write $u \preccurlyeq 0$; we write $u \approx 0$ if both $u \preccurlyeq 0$ and $u \succcurlyeq 0$ hold. We define $u \preccurlyeq k$ and $u \succcurlyeq k$ for any real constant k in an analogous way. For u, $v \in W^{1,p}(\Omega)$ we write $u \preccurlyeq v$ to mean that $u - v \preccurlyeq 0$; similarly for $u \succcurlyeq v$. ■

A separate definition of $u \preccurlyeq k$ for a real constant k is necessary in order that unbounded sets Ω may be considered. If Ω is unbounded and E is a bounded subset of $\bar{\Omega}$, the functions u_j in the definition can be taken to have compact supports in $\mathbb{R}^n$. If E is an unbounded set any neighbourhood of E is assumed to be a neighbourhood of the *point at infinity* and consequently is of the form

$$U(\varepsilon, \delta) = \{x \in \mathbb{R}^n : \operatorname{dist}(x, \bar{E}) < \varepsilon \text{ or } |x| > 1/\delta\}$$

for some ε and δ.

The relationship between $u \succcurlyeq 0$ on E and $u(\bullet) \geqslant 0$ a.e. on E is analysed in our first theorem.

Theorem 3.2. Let $u \in W^{1,p}(\Omega)$.

(i) If $u \succcurlyeq 0$ on a subset E of Ω then $u(x) \geqslant 0$ a.e. on E.

(ii) Let E be a compact subset of Ω and $u(x) \geqslant 0$ a.e. on E. Then $u \succcurlyeq 0$ on E.

(iii) If $u(x) \geqslant 0$ a.e. on Ω then $u \succcurlyeq 0$ on Ω.

Thus $u \succcurlyeq 0$ on Ω if, and only if, $u(x) \geqslant 0$ a.e. in Ω. ■

Proof. (i) Let u_j and U_j be as in the Definition 3.1 of $u \succcurlyeq 0$ on E and let $U = \cap_{j=1}^\infty U_j$. Then E is a subset of the measurable set U and $u_j(x) > 0$ on $U \cap \Omega$ for all $j \in \mathbb{N}$. Since a subsequence of $(u_j(x))$ converges to $u(x)$ a.e. in Ω it follows that $u(x) \geqslant 0$ a.e. on E.

(ii) Let $u_j \in C^\infty(\Omega) \cap W^{1,p}(\Omega)$ and $u_j \to u$ in $W^{1,p}(\Omega)$. A subsequence of $(u_j(\bullet))$, which we still denote by $(u_j(\bullet))$, must therefore converge to $u(\bullet)$ a.e. in Ω. Let $\varepsilon > 0$ and let $\phi \in C_0^\infty(\mathbb{R}^n)$ be such that $\phi(x) = 1$ for $x \in E$. Since E is compact and u_j is continuous we can cover E with a finite number of balls B_m such that $|u_j(x) - u_j(y)| < \frac{1}{2}\varepsilon$ for $x, y \in B_m$. Each B_m contains points y for which $u_j(y) \to u(y)$ and $u(y) \geqslant 0$. Hence, for j large enough, $u_j(x) + \varepsilon > 0$ on E. The functions $\phi_{j,\varepsilon} = u_j + \varepsilon\phi$ therefore satisfy $\phi_{j,\varepsilon} > 0$ on an open neighbourhood $U_j(\varepsilon)$ of E and $\|\phi_{j,\varepsilon} - u\|_{1,p,\Omega} = O(\varepsilon)$ as $j \to \infty$. The set $\{\phi_{j,\varepsilon} : j \in \mathbb{N}, \varepsilon \in (0, 1)\}$ therefore contains a sequence which fulfills the requirements of the definition of $u \succcurlyeq 0$ on E.

(iii) From the proof of Theorem V.3.2, establishing the denseness of $C^\infty(\Omega) \cap W^{1,p}(\Omega)$ in $W^{1,p}(\Omega)$, it follows that if $u(\bullet) \geqslant 0$ a.e. in Ω there is a sequence of functions $u_j \in C^\infty(\Omega) \cap W^{1,p}(\Omega)$ which tends to u in $W^{1,p}(\Omega)$ and satisfy $u_j(x) \geqslant 0$ on Ω. Hence, if $\phi \in C^\infty(\Omega) \cap W^{1,p}(\Omega)$ is such that $\phi > 0$ on Ω then $u_j + j^{-1}\phi > 0$ on Ω and $u_j + j^{-1}\phi \to u$ in $W^{1,p}(\Omega)$ as $j \to \infty$. Thus $u \succcurlyeq 0$ on Ω. □

Remark 3.3. By Remark 2.5 and Theorem 3.2, $|u| \succcurlyeq 0$ and $u^+ \succcurlyeq 0$ on Ω for any $u \in W^{1,p}(\Omega)$. Also, if $u, v \in W^{1,p}(\Omega)$, u and v are both $\preccurlyeq \sup(u, v)$ and $\succcurlyeq \inf(v, u)$ in $W^{1,p}(\Omega)$.

Theorem 3.4. Let $u \in W^{1,p}(\Omega)$ and $v \in W_0^{1,p}(\Omega)$ and suppose either $u \preccurlyeq 0$ on $\partial\Omega$ or $u \preccurlyeq 0$ on Ω. Then $\sup(u, v) \in W_0^{1,p}(\Omega)$. ■

Proof. If $u \preccurlyeq 0$ on $\partial\Omega$ there exists a sequence (u_j) in $C^\infty(\Omega) \cap W^{1,p}(\Omega)$ which is such that $u_j \to u$ in $W^{1,p}(\Omega)$ and $u_j < 0$ in $U_j \cap \Omega$, where U_j is an open neighbourhood of $\partial\Omega$ in $\mathbb{R}^n$. Also, as $v \in W_0^{1,p}(\Omega)$, there exists a sequence (v_j) in $C_0^\infty(\Omega)$ which converges to v in $W^{1,p}(\Omega)$. If $\phi_j = u_j - v_j$, the sequence (ϕ_j) satisfies the definition of $u - v \preccurlyeq 0$ on $\partial\Omega$. From Remark 2.2 and Corollary 2.4, $(u-v)^+$ lies in the closed convex hull of $\{\phi_j^+ : j \in \mathbb{N}\}$ in $W^{1,p}(\Omega)$. But ϕ_j^+ has compact support in Ω and therefore belongs to $W_0^{1,p}(\Omega)$. Consequently $(u-v)^+ \in W_0^{1,p}(\Omega)$ and so does $\sup(u, v) = v + (u-v)^+$.

The proof is the same when $u \preccurlyeq 0$ on Ω; indeed $u \preccurlyeq 0$ on Ω clearly implies $u \preccurlyeq 0$ on $\partial\Omega$. □

Theorem 3.5. $u \in W_0^{1,p}(\Omega)$ if, and only if, $u \in W^{1,p}(\Omega)$ and $u \approx 0$ on $\partial\Omega$. ■

Proof. If $u \in W_0^{1,p}(\Omega)$ there exists a sequence (ϕ_j) in $C_0^\infty(\Omega)$ such that $\phi_j \to u$ in $W^{1,p}(\Omega)$. Since ϕ_j vanishes in a neighbourhood of $\partial\Omega$ we have, for a $\phi \in C^\infty(\Omega) \cap W^{1,p}(\Omega)$ which is positive in a neighbourhood of $\partial\Omega$, that the $\phi_j \pm j^{-1}\phi$ $(j \in \mathbb{N})$ satisfy the requirements of the definition for $u \succcurlyeq 0$ and $u \preccurlyeq 0$ on $\partial\Omega$. Thus $u \approx 0$ on $\partial\Omega$.

Let $u \preccurlyeq 0$ on $\partial\Omega$. Then, by Theorem 3.4 with $v = 0$, $u^+ \in W_0^{1,p}(\Omega)$. Similarly $u \succcurlyeq 0$ on $\partial\Omega$ implies that $u^- = \sup(-u, 0) \in W_0^{1,p}(\Omega)$. Hence $u \approx 0$ on $\partial\Omega$ implies $u = u^+ - u^- \in W_0^{1,p}(\Omega)$. □

Theorem 3.6. If $u \in W_0^{1,p}(\Omega)$ and $u(\bullet) \geqslant 0$ a.e. on Ω, there exists a sequence (ϕ_j) in $C_0^\infty(\Omega)$ such that $\phi_j \geqslant 0$ on Ω and $\phi_j \to u$ in $W^{1,p}(\Omega)$. ■

Proof. Since $u(\bullet) \geqslant 0$ a.e. in Ω we see from Remark 2.5 and Proposition 2.7 that $u = |u|$ in $W^{1,p}(\Omega)$. If $u_j \in C_0^\infty(\Omega)$ and $u_j \to u$ in $W^{1,p}(\Omega)$ then $u = |u|$ lies in $\overline{\text{co}}\{|u_j| : j \in \mathbb{N}\}$ in $W^{1,p}(\Omega)$. Each $|u_j|$ has compact support in Ω and so the mollifier $\phi_{j,\varepsilon} = |u_j|_\varepsilon \in C_0^\infty(\Omega)$ if ε is small enough (see §V.1). Also, $\phi_{j,\varepsilon} \geqslant 0$ on Ω and $\phi_{j,\varepsilon} \to |u_j|$ in $W^{1,p}(\Omega)$ as $\varepsilon \to 0$. The set $\overline{\text{co}}\{\phi_{j,\varepsilon} : j \in \mathbb{N}, \varepsilon \in (0, 1)\}$ therefore contains a sequence which converges to u in $W^{1,p}(\Omega)$ and the proof is complete. □

Definition 3.7. Let E be a subset of $\bar{\Omega}$ and $u \in W^{1,p}(\Omega)$. We say that u is *bounded above on E in the sense of* $W^{1,p}(\Omega)$ if $u \preccurlyeq M$ on E for some $M \in \mathbb{R}$. The infimum of all such numbers M is called the *maximum of u on E* and written $\max_E u$. Similarly, we define *boundedness below on E* and $\min_E u$. ■

Theorem 3.8. Let $u \in W^{1,p}(\Omega)$. If $k > \max_{\partial\Omega} u$ then $(u-k)^+ \in W_0^{1,p}(\Omega)$, and if $k < \min_{\partial\Omega} u$ then $(u-k)^- \in W_0^{1,p}(\Omega)$. ■

Proof. If Ω is of finite measure, $k \in W^{1,p}(\Omega)$ and the theorem is a consequence of Theorem 3.4 since $(u-k)^+ = \sup(u-k, 0)$. For a general Ω we argue as follows. As $u \preccurlyeq k$ on $\partial\Omega$, there exists a sequence $(u_j) \subset C^\infty(\Omega) \cap W^{1,p}(\Omega)$ such that $u_j \to u$ in $W^{1,p}(\Omega)$ and $u_j < k$ on $U_j \cap \Omega$, where U_j is an open neighbourhood of $\partial\Omega$ in $\mathbb{R}^n$; recall that if $\partial\Omega$ is unbounded, U_j is a neighbourhood of the point at infinity. Thus $(u_j - k)^+ \in W_0^{1,p}(\Omega)$ and $(u-k)^+ \in \overline{\text{co}}\,\{(u_j-k)^+ : j \in \mathbb{N}\} \subset W_0^{1,p}(\Omega)$. The last part of the theorem follows similarly. □

Remark 3.9. If $u \preccurlyeq 0$ on Ω then $\min_{\partial\Omega} u \leqslant 0$. For if $M = \min_{\partial\Omega} u > 0$ and $0 < k < M$, we conclude from Theorem 3.8 that $(u-k)^- \in W_0^{1,p}(\Omega)$. But $(u-k)^- = k - u > k > 0$ a.e. on Ω since $u \preccurlyeq 0$ on Ω and we therefore have a contradiction. Similarly, $u \succcurlyeq 0$ on Ω implies that $\max_{\partial\Omega} u \geqslant 0$.

4. A weak maximum principle

We continue to assume that Ω is an arbitrary open set in $\mathbb{R}^n$, in general, and that all functions and function spaces appearing are real. Let τ be the differential expression in §1, namely

$$\tau = -\sum_{i,j=1}^{n} D_i(a_{ij} D_j) + \sum_{j=1}^{n} b_j D_j + q, \tag{4.1}$$

where the coefficients a_{ij}, b_j, and q are real-valued and satisfy (i)–(iii) in §1.1. The expression τ gives rise to the bilinear form

$$t[\![u, v]\!] = \int_\Omega (a_{ij} D_j u D_i v + b_j D_j u v + q u v). \tag{4.2}$$

If the embedding $W^{1,2}(\Omega) \to L^{2r/(r-1)}(\Omega)$ is continuous for r satisfying (1.5), then $t[\![\bullet, \bullet]\!]$, is bounded on $W^{1,2}(\Omega) \times W^{1,2}(\Omega)$. Also, there exist positive numbers μ and δ such that

$$t[\![u]\!] + \mu \|u\|_{2,\Omega}^2 \geqslant \delta \|u\|_{1,2,\Omega}^2 \quad \text{for all } u \in W_0^{1,2}(\Omega). \tag{4.3}$$

Furthermore, we observed in Remark 1.6 that we can take $\mu = 0$ in (4.3) if the norms $\|a_{ij}\|_{\infty,\Omega}$, $\|b_j\|_{2r,\Omega}$, and $\|q\|_{r,\Omega}$ are sufficiently small and Ω is such that the Poincaré inequality $\|u\|_{2,\Omega} \leqslant K\|\nabla u\|_{2,\Omega}$ is valid for $u \in W_0^{1,2}(\Omega)$.

Definition 4.1. We say that $u \in \mathscr{D}'(\Omega)$ is a *subsolution* (*supersolution*) of τ if $(\tau u, \phi) \leqslant 0$ ($\geqslant 0$) for all $\phi \in C_0^\infty(\Omega)$ with $\phi \geqslant 0$ on Ω. If $u \in W^{1,2}(\Omega)$ then u is a subsolution (supersolution) of τ if $t[\![u, \phi]\!] \leqslant 0$ ($\geqslant 0$) for all $\phi \in C_0^\infty(\Omega)$ with $\phi \geqslant 0$ on Ω. If $\tau = -\Delta$, subsolutions and supersolutions are called *subharmonic* and *superharmonic* functions respectively. ■

If $u \in W^{1,2}(\Omega)$ is a subsolution (supersolution) of τ we see from Theorem 3.6 that $t[\![u, \phi]\!] \leqslant 0$ ($\geqslant 0$) for all $\phi \in W_0^{1,2}(\Omega)$ with $\phi \succcurlyeq 0$ on Ω.

Theorem 4.2. Suppose there exist positive numbers M and m such that

$$|t[\![u, v]\!]| \leqslant M\|u\|_{1,2,\Omega}\|v\|_{1,2,\Omega} \qquad (u, v \in W^{1,2}(\Omega)) \tag{4.4}$$

and

$$t[\![u]\!] \geqslant m\|u\|_{1,2,\Omega}^2 \qquad (u \in W_0^{1,2}(\Omega)). \tag{4.5}$$

Let $u, v \in W^{1,2}(\Omega)$ be subsolutions of τ. Then $w = \sup(u, v) \in W^{1,2}(\Omega)$ and w is a subsolution of τ. ■

Proof. The assertion that $w \in W^{1,2}(\Omega)$ is a consequence of Corollary 2.4. Let

$$\mathfrak{K}_0 := \{\psi : \psi \in W_0^{1,2}(\Omega), \psi \preccurlyeq 0 \text{ on } \Omega\}, \qquad \mathfrak{K} = \mathfrak{K}_0 + w.$$

Then $\mathfrak{K}_0$ is a closed convex cone in $W_0^{1,2}(\Omega)$; if $\psi_j \in \mathfrak{K}_0$ ($j \in \mathbb{N}$) is such that $\psi_j \to \psi$ in $W_0^{1,2}(\Omega)$, a subsequence of (ψ_j) converges to ψ a.e. in Ω and so by Theorem 3.2, $\psi \preccurlyeq 0$ on Ω. Therefore $\mathfrak{K}$ is a closed, convex, non-empty subset of $W^{1,2}(\Omega)$. The hypotheses of Corollary IV.6.5 are satisfied and so there exists a unique $\eta \in \mathfrak{K}$ such that

$$t[\![\eta, \phi]\!] \geqslant 0 \quad \text{for all } \phi \in \mathfrak{K}_\eta. \tag{4.6}$$

Since $\eta \in \mathfrak{K}$, it follows that $\mathfrak{K} - \eta = \mathfrak{K}_0$ and, for any $\psi \in C_0^\infty(\Omega)$ with $\psi \leqslant 0$ on Ω we have $\psi \in \mathfrak{K}_0 = \mathfrak{K} - \eta$ and so $\psi \in \mathfrak{K}_\eta$. From (4.6) we infer that

$$t[\![\eta, \phi]\!] \leqslant 0 \quad \text{for all } \phi \in C_0^\infty(\Omega) \text{ with } \phi \geqslant 0 \text{ on } \Omega,$$

and η is a subsolution of τ. We shall prove that $\eta = w$. Since $\eta \in \mathfrak{K}$ we already know that $\eta \preccurlyeq w$ on Ω and hence it suffices to prove that $w \preccurlyeq \eta$ on Ω in view of Proposition 2.7 and Theorem 3.2.

Let $\zeta = \sup(u, \eta)$. Then $\zeta - w = \sup(u - w, \eta - w)$; and $u - w \preccurlyeq 0$ on Ω (by definition of w) and $\eta - w \in \mathfrak{K}_0 \subset W_0^{1,2}(\Omega)$. Therefore, Theorem 3.4 yields $\zeta - w \in W_0^{1,2}(\Omega)$. Also, since $u - w \preccurlyeq 0$ and $\eta - w \preccurlyeq 0$ on Ω, in view of Theorem 3.2 we have that $\zeta - w \preccurlyeq 0$ on Ω and hence $\zeta \in \mathfrak{K}$. Thus $\zeta - \eta \in \mathfrak{K}_\eta$ and by (4.6)

$$t[\![\eta, \zeta - \eta]\!] \geqslant 0. \tag{4.7}$$

Since $\zeta = \sup(u, \eta)$, we have $\zeta - u = (\eta - u)^+$ and $\zeta - \eta = -(\eta - u)^-$. Consequently, from Theorem 2.1, for $j, k = 1, 2, \ldots, n$,

$$(\zeta - u)(\zeta - \eta) = 0,$$

$$D_j(\zeta - u)D_k(\zeta - \eta) = 0,$$

$$[D_j(\zeta - u)](\zeta - \eta) = 0,$$

a.e. in Ω. Hence $t[\![\zeta - u, \zeta - \eta]\!] = 0$ and $t[\![\zeta, \zeta - \eta]\!] = t[\![u, \zeta - \eta]\!]$. But $\zeta - \eta = \zeta - w + w - \eta \in W_0^{1,2}(\Omega)$ and $\zeta - \eta \succcurlyeq 0$ on Ω, and since u is a subsolution of τ we conclude that $t[\![u, \zeta - \eta]\!] \leqslant 0$. This in turn gives $t[\![\zeta, \zeta - \eta]\!] \leqslant 0$ and, on using (4.7),

$$t[\![\zeta - \eta, \zeta - \eta]\!] \leqslant 0.$$

We conclude from (4.5) that $\zeta = \eta$ and hence $u \preccurlyeq \eta$ on Ω. Similarly $v \preccurlyeq \eta$ on Ω and so $w \preccurlyeq \eta$ on Ω. The theorem is therefore proved. □

Corollary 4.3. Let (4.4) and (4.5) be satisfied, let Ω have finite measure, and suppose that $q(x) \geqslant 0$ a.e. on Ω. If $u \in W^{1,2}(\Omega)$ is a subsolution of τ so also is $(u-k)^+$ for any constant $k \geqslant 0$. If $q = 0$ on Ω then $(u-k)^+$ is a subsolution for any real constant k. ■

Proof. For all $\phi \in C_0^\infty(\Omega)$ with $\phi \geqslant 0$ on Ω, we have by hypothesis

$$t[\![u-k, \phi]\!] = t[\![u, \phi]\!] - k\int_\Omega q\phi \leqslant 0.$$

Hence $u - k \in W^{1,2}(\Omega)$ is a subsolution of τ and consequently so is $(u-k)^+$ from Theorem 4.2. If $q = 0$ then $t[\![u-k, \phi]\!] = t[\![u, \phi]\!]$ for any constant k and so $(u-k)^+$ is a subsolution if u is one. □

We can now prove the following weak maximum principle.

Theorem 4.4. Let (4.4) and (4.5) be satisfied and $q(x) \geqslant 0$ a.e. on Ω. If $u \in W^{1,2}(\Omega)$ is a subsolution of τ then

$$\max_\Omega u \leqslant \max\{\max_{\partial\Omega} u, 0\}. \tag{4.8}$$

If $q = 0$ then for any subsolution of τ in $W^{1,2}(\Omega)$,

$$\max_\Omega u \leqslant \max_{\partial\Omega} u. \tag{4.9}$$ ■

Proof. We may clearly suppose that $\max_{\partial\Omega} u < \infty$ since otherwise (4.8) is trivially satisfied. The open set Ω is the union of an increasing sequence of bounded open subsets Ω_ν and from Definition 3.7 it is clear that $\max_{\partial\Omega_\nu} u \leqslant \max_{\partial\Omega} u$ for ν large enough. Hence it is sufficient to prove the theorem for a bounded Ω; note that in view of Theorem 3.2 the inequalities (4.8) and (4.9) are satisfied a.e. on Ω.

Let k be any number greater than $\max\{\max_{\partial\Omega} u, 0\}$. Then, by Theorem 3.8, $(u-k)^+ \in W_0^{1,2}(\Omega)$ and is non-negative in the sense of $W^{1,2}(\Omega)$. Furthermore, since $k > 0$, we have that $(u-k)^+$ is a subsolution of τ, by Corollary 4.3, and so

$$t[\![(u-k)^+, \phi]\!] \leqslant 0 \quad \text{for all } \phi \in C_0^\infty(\Omega) \text{ with } \phi \geqslant 0 \text{ on } \Omega.$$

By continuity this holds for all $\phi \in W_0^{1,2}(\Omega)$ which are $\succcurlyeq 0$ on Ω and, in particular,

$$t[\![(u-k)^+, (u-k)^+]\!] \leqslant 0.$$

From the coercivity condition (4.5) we obtain $(u-k)^+ = 0$, whence $u(x) \leqslant k$ a.e. in Ω and this of course implies (4.8). If $q = 0$, k need not be non-negative in Corollary 4.3 and (4.9) is then satisfied. □

Remark 4.5. If $u(x) \geqslant 0$ a.e. in Ω then $\max_{\partial\Omega} u \geqslant 0$, by Remark 3.9. This means that for any subsolution $u \in W^{1,2}(\Omega)$ of τ in Theorem 4.4,

$$\max_{\Omega} u \leqslant \max_{\Omega} u^+ \leqslant \max_{\partial\Omega} u^+. \tag{4.10}$$

Since u is a supersolution of τ if, and only if, $-u$ is a subsolution of τ we also obtain the following result.

Corollary 4.6. Under the hypothesis of Theorem 4.4, if $u \in W^{1,2}(\Omega)$ is a supersolution of τ then

$$\min_{\Omega} u \geqslant \min\{\min_{\partial\Omega} u, 0\}. \tag{4.11}$$

If $q = 0$ in Ω,

$$\min_{\Omega} u \geqslant \min_{\partial\Omega} u. \tag{4.12}$$ ■

Corollary 4.7. Under the hypothesis of Theorem 4.4, if $u \in W^{1,2}(\Omega)$ is a solution of $\tau u = 0$, then

$$\max_{\Omega} |u| \leqslant \max_{\partial\Omega} |u|. \tag{4.13}$$ ■

Proof. Since u and $-u$ are both subsolutions of τ, so are u^+ and $(-u)^+ = u^-$. Consequently (4.10) applies to u^+ and u^- and (4.13) follows since $|u| = u^+ + u^-$. □

Corollary 4.8. Let the hypothesis of Theorem 4.4 hold, let $u \in W^{1,2}(\Omega)$ be a subsolution of τ, and let $h \in W^{1,2}(\Omega)$ be a solution of $\tau h = 0$. If $u - h \in W_0^{1,2}(\Omega)$ then $u(x) \leqslant h(x)$ a.e. in Ω. ■

Proof. By hypothesis, $u-h$ is a subsolution of τ and hence so is $(u-h)^+ = \sup(u-h, 0)$, by Theorem 4.2. Also, from Corollary 2.4, $u-h \in W_0^{1,2}(\Omega)$ implies $(u-h)^+ \in W_0^{1,2}(\Omega)$ and, by virtue of Theorem 3.5, $\max_{\partial\Omega}(u-h)^+ = 0$. Thus, from (4.10), $u-h \preccurlyeq 0$ on Ω and hence $u(x) \leqslant h(x)$ a.e. in Ω. □

Corollary 4.9. If $u \in W_0^{1,2}(\Omega)$ is a subsolution of τ then $u(x) \leqslant 0$ a.e. in Ω. ■

5. Capacity

Let E be a compact subset of Ω, an arbitrary open subset of $\mathbb{R}^n$, and let

$$\mathfrak{K} := \{v : v \in W_0^{1,2}(\Omega),\ v \succcurlyeq 1 \text{ on } E\}; \tag{5.1}$$

$\mathfrak{K}$ is a closed, convex, non-empty subset of $W_0^{1,2}(\Omega)$.

Theorem 5.1. Let t satisfy (4.4) and (4.5) and let $q \geqslant 0$. Then there exists a unique $u \in W_0^{1,2}(\Omega)$ and a positive measure μ having its support in E such that

(i) $t[\![u, \phi]\!] = \displaystyle\int_\Omega \phi \, d\mu$ for all $\phi \in C_0^\infty(\Omega)$,

(ii) $u \approx 1$ on E,

(iii) $u \preccurlyeq 1$ on Ω. ■

Proof. By Theorem IV.6.1 there exists a unique $u \in \mathfrak{K}$ such that

$$t[\![u, w]\!] \geqslant 0 \quad \text{for all } w \in \mathfrak{K}_u, \tag{5.2}$$

where $\mathfrak{K}_u$ is the cone

$$\mathfrak{K}_u = \{w : w \in W_0^{1,2}(\Omega),\ u + \varepsilon w \in \mathfrak{K} \text{ for some } \varepsilon > 0\}.$$

We shall prove that this u satisfies the theorem. Any $\phi \in C_0^\infty(\Omega)$ with $\phi \geqslant 0$ on Ω lies in $\mathfrak{K}_u$ and consequently $t[\![u, \phi]\!] \geqslant 0$; u is therefore a supersolution of τ. The map $\phi \to t[\![u, \phi]\!]$ is therefore a positive linear functional on $C_0^\infty(\Omega)$ and by the Riesz Representation Theorem (cf. Taylor [1, §7.5]), there exists a positive measure μ on Ω such that

$$t[\![u, \phi]\!] = \int_\Omega \phi \, d\mu \quad \text{for all } \phi \in C_0^\infty(\Omega).$$

If $\operatorname{supp}\phi \subset \Omega \setminus E$ then $\pm\phi + u \in \mathfrak{K}$ and $\pm\phi \in \mathfrak{K}_u$. We therefore deduce from (5.2) that $t[\![u, \phi]\!] = 0$ and hence that the support of μ lies in E.

Since $u \succcurlyeq 1$ on E there exists a sequence (u_j) in $C^\infty(\Omega) \cap W^{1,2}(\Omega)$ which is such that $u_j \to u$ in $W^{1,2}(\Omega)$ and $u_j > 1$ in $U_j \cap \Omega$, where U_j is an open neighbourhood of E in $\mathbb{R}^n$, each u_j having compact support in $\mathbb{R}^n$ if Ω is unbounded. The function $\zeta = \inf(u, 1)$ lies in $\overline{\text{co}}\,\{\inf(u_j, 1) : j \in \mathbb{N}\}$ and so ζ is the limit of a sequence (v_j) in $W^{1,2}(\Omega)$ which satisfies $v_j = 1$ on an open neighbourhood of E. On using the denseness of $C^\infty(\Omega) \cap W^{1,2}(\Omega)$ in $W^{1,2}(\Omega)$ and arguing as in the proof of Theorem 3.2 (ii) it readily follows that $\zeta \approx 1$ on E. Parts (ii) and (iii) of the theorem will follow if we prove that $\zeta = u$.

By Corollary 2.9, $\zeta = u - (u-1)^+ \in W_0^{1,2}(\Omega)$. Thus $\zeta \approx 1$ on E implies that $\zeta \in \mathfrak{K}$ with $\zeta - u \in \mathfrak{K}_u$ and, from (5.2),

$$t[\![u, \zeta - u]\!] \geqslant 0.$$

Since $\zeta - u = -(u-1)^+$ and $\zeta - 1 = -(u-1)^-$ in Ω, we have, as in the proof of Theorem 4.2, that $t[\![\zeta - 1, \zeta - u]\!] = 0$ and so

$$\begin{aligned} t[\![\zeta, \zeta - u]\!] &= t[\![1, \zeta - u]\!] \\ &= \int_\Omega q(\zeta - u) \leqslant 0 \end{aligned}$$

as $q(x) \geqslant 0$ by hypothesis and $\zeta(x) \leqslant u(x)$ a.e. in Ω. Thus $t[\![\zeta - u, \zeta - u]\!] \leqslant 0$ and from the coercivity condition (4.5) we deduce that $\zeta = u$. The theorem is therefore proved. □

In Theorem 5.1, since $u(x) \geqslant 1$ a.e. on E and E is a compact subset of Ω, there exist functions $\phi_j \in C_0^\infty(\Omega)$ which are such that $\phi_j \geqslant 1$ on E and $\phi_j \to u$ in $W^{1,2}(\Omega)$. Since the support of μ lies in E it follows that

$$t[\![u]\!] = \lim_{j \to \infty} t[\![u, \phi_j]\!] \geqslant \int_\Omega d\mu.$$

Similarly, as $u(x) \leqslant 1$ a.e. on E, we deduce the reverse inequality and hence

$$t[\![u]\!] = \int_\Omega d\mu.$$

Definition 5.2. The real number $\operatorname{cap}_\tau(E, \Omega) := \int_\Omega d\mu = t[\![u]\!]$ is called the *capacity with respect to τ of E relative to Ω*. The measure $\mu = \mu_\tau$ is called the capacity distribution of E and u the capacity potential of E. We shall usually write $\operatorname{cap}(E, \Omega)$ for $\operatorname{cap}_\tau(E, \Omega)$, the dependence on τ being suppressed unless there is a possibility of confusion. ■

In Chapter VIII we shall define capacity in L^p spaces.

Theorem 5.3. Let t satisfy the hypothesis of Theorem 5.1 and in addition let t be symmetric, i.e. $t[\![u, v]\!] = t[\![v, u]\!]$. Then u in Theorem 5.1 is the unique element of $W_0^{1,2}(\Omega)$ which satisfies

$$t[\![u]\!] = \inf_{v \in \mathfrak{K}} t[\![v]\!]. \tag{5.3}$$

Also $0 \leqslant u(x) \leqslant 1$ a.e. in Ω. Furthermore

$$\operatorname{cap}(E, \Omega) = \inf_{\phi \in \mathfrak{K}_0} t[\![\phi]\!] \tag{5.4}$$

where

$$\mathfrak{K}_0 = \{\phi : \phi \in C_0^\infty(\Omega), 0 \leqslant \phi \leqslant 1 \text{ in } \Omega \text{ and } \phi = 1 \text{ in an open neighbourhood of } E\}.$$

■

Proof. The fact that u is the unique solution of (5.3) follows from (5.2) and Proposition IV.6.3. To prove that $0 \leqslant u(x) \leqslant 1$ a.e. in Ω it is enough to prove that $u = u^+$, in view of Theorem 5.1 (iii). As in the proof of Theorem 4.2, we see that $t[\![u^+, u^-]\!] = 0$ and hence

$$t[\![u]\!] = t[\![u^+]\!] + t[\![u^-]\!].$$

But $u^- \in W_0^{1,2}(\Omega)$ and since $u \approx 1$ on E, we have $u^- \approx 0$ on E and hence $u^- \in \mathfrak{K}_u$. Thus by (5.2),

$$t[\![u^-]\!] = t[\![u, u^-]\!] \geqslant 0,$$

whence

$$t[\![u]\!] \geqslant t[\![u^+]\!].$$

However, $u^+ \in \mathfrak{K}$ and (5.3) imply that $u = u^+$.

$\mathfrak{K}_0$ is clearly a subset of $\mathfrak{K}$ and (5.4) will follow if we can show that u can be approximated in $W^{1,2}(\Omega)$ by functions ϕ from $\mathfrak{K}_0$. Since $u \in W_0^{1,2}(\Omega)$ and $u(x) \geqslant 0$ a.e. on Ω, there exist functions $u_j \in C_0^\infty(\Omega)$ such that $u_j(x) \geqslant 0$ on Ω and $u_j \to u$ in $W^{1,2}(\Omega)$, by Theorem 3.6; also we may suppose that $u_j(x) \to u(x)$ a.e. in Ω. Since $u(x) = 1$ a.e. in E then, given $\varepsilon > 0$ and a $\chi \in C_0^\infty(\Omega)$ which is equal to 1 in a neighbourhood of E and satisfies $0 \leqslant \chi \leqslant 1$ on Ω, it follows that $u_{j,\varepsilon} := u_j + \varepsilon\chi > 1$ on a neighbourhood U_ε of E for j large enough. Also $u_{j,\varepsilon} \geqslant 0$ on Ω, $u_{j,\varepsilon} \in C_0^\infty(\Omega)$, and $\|u_{j,\varepsilon} - u\|_{1,2,\Omega} = O(\varepsilon)$ as $j \to \infty$. Therefore from the set $\{u_{j,\varepsilon} : j \in \mathbb{N}, \varepsilon \in (0, 1)\}$ we can extract a sequence (v_k), say, with the following properties: $v_k \geqslant 0$ on Ω, $v_k > 1$ on a neighbourhood U_k of E, $v_k \in C_0^\infty(\Omega)$ and $v_k \to u$ in $W^{1,2}(\Omega)$.

Let $w_k = \inf(v_k, 1)$. Then $w_k \in W_0^{1,2}(\Omega)$, having compact support in Ω, $0 \leqslant w_k \leqslant 1$ on Ω, and $w_k = 1$ on U_k. Since $u = \inf(u, 1)$ lies in $\overline{\mathrm{co}}\{w_k : k \in \mathbb{N}\}$ there exists a sequence (z_k), consisting of positive averages of the w_k, such that $z_k \to u$ in $W^{1,2}(\Omega)$, $0 \leqslant z_k \leqslant 1$ on Ω and $z_k = 1$ on U_k. Now put $\phi_{k,\varepsilon} = (z_k)_\varepsilon$, a regularization of z_k. For a given k and ε small enough, there exists a neighbourhood $V_{k,\varepsilon}$ of E, with $E \subset V_{k,\varepsilon} \subset\subset U_k \subset \Omega$, such that $\phi_{k,\varepsilon} = 1$ on $V_{k,\varepsilon}$. Also $0 \leqslant \phi_{k,\varepsilon} \leqslant 1$ and $\phi_{k,\varepsilon} \in C_0^\infty(\Omega)$. Since

$$\|u - \phi_{k,\varepsilon}\|_{1,2,\Omega} \leqslant \|u - z_k\|_{1,2,\Omega} + \|z_k - \phi_{k,\varepsilon}\|_{1,2,\Omega}$$

and $z_k \to u$ as $k \to \infty$ and $\phi_{k,\varepsilon} \to z_k$ as $\varepsilon \to 0$, it follows that the set

$\{\phi_{k,\varepsilon}: k \in \mathbb{N}, \varepsilon \in (0, 1)\}$ contains a sequence which approximates u in $W^{1,2}(\Omega)$ and lies in $\mathfrak{K}_0$. The proof is therefore complete. □

Theorem 5.3 also implies the following useful observation:

$$\operatorname{cap}(E, \Omega) = \inf_{v \in \mathfrak{K}_1} t[\![v]\!] \tag{5.5}$$

where

$\mathfrak{K}_1 = \{v : v$ is Lipschitz continuous and has compact support in Ω, $v \geqslant 0$ on Ω and $v \geqslant 1$ on an open neighbourhood of $E\}$.

VII

Second-order differential operators on arbitrary open sets

In this chapter we describe in detail three different methods for obtaining nice operators generated in some L^2 space by second-order differential expressions and either Dirichlet or Neumann boundary conditions. The first method, in §1, is based on sesquilinear forms and uses Theorem IV.2.4 to determine m-sectorial operators. In §2 we produce an m-accretive realization of $-\Delta+q$ by a technique due to Kato [4] in which his distributional inequality plays a leading role. The method in §3 has its roots in the work of Levinson and Titchmarsh and gives operators T which are such that $\mathrm{i}T$ is m-accretive. The class of such operators includes the self-adjoint maps, even those which are unbounded below, and so the results obtained specialize to ones on the essential self-adjointness of the Schrödinger operator. Schrödinger operators whose potentials have strong local singularities require special treatment and a brief outline of known results is given in §4. The quantum-mechanical interpretation of essential self-adjointness of Schrödinger operators points to the type of result possible and furthermore suggests when essential self-adjointness will break down; this is illustrated in §5.

1. Quasi-*m*-sectorial Dirichlet and Neumann operators

We recall that a densely defined linear operator T in a Hilbert space H is quasi-m-sectorial if for some $\beta \in (-\pi,\pi)$ and $\delta \in \mathbb{C}$, the operator $e^{-i\beta}(T+\delta I)$ is m-accretive and its numerical range lies in a sector $\{z \in \mathbb{C}: |\arg z| \leqslant \theta < \frac{1}{2}\pi\}$. In this case the operator $e^{-i\beta}(T+\delta I)$ has no proper accretive extensions and its spectrum lies inside its numerical range and hence inside that sector. In this section we shall study quasi-m-sectorial operators which are generated by second-order differential expressions in $L^2(\Omega)$ with either Dirichlet or Neumann boundary conditions on the boundary $\partial\Omega$ of the open set Ω, Ω being arbitrary in $\mathbb{R}^n$ and $n \geqslant 1$. We begin with the expression $\tau = -\Delta + b\cdot\nabla + q$ where $b = (b_1, \ldots, b_n)$ and q are respectively vector-valued and scalar-valued functions on Ω. Subsequently we shall apply similar techniques to obtain analogous results for general second-order elliptic expressions.

1.1. *The Dirichlet problem on Ω.*

Let Ω be an arbitrary open set in $\mathbb{R}^n$, with $n \geqslant 1$, and set

$$\tau = -\Delta + b\cdot\nabla + q, \qquad b = (b_1, \ldots, b_n), \tag{1.1}$$

where the b_j $(j = 1, 2, \ldots, n)$ and q are complex-valued measurable functions on Ω. In order to make sense of the conditions we shall impose on the b_j and q we first need some notation and some preliminary results.

We saw in Theorem V.4.13 that for any closed cube Q, the Sobolev space $W^{1,1}(Q^\circ)$ is continuously embedded in $L^{s'}(Q)$ for any s' in $[1, n/(n-1)]$ $(n > 1)$. Furthermore, there exists a constant $K > 0$, depending only on s', n, and the volume $|Q|$ of Q such that

$$\|u - u_Q\|_{s',Q} \leqslant K\|\nabla u\|_{1,Q} \qquad (u \in W^{1,1}(Q^\circ)),$$

where $u_Q := |Q|^{-1}\int_\Omega u(x)\,dx$, the integral mean of u on Q; see Theorem V.3.24. If I is the unit open cube $(0,1)^n$ in $\mathbb{R}^n$ we may therefore define

$$\mu(s', n) := \sup\{\|u - u_I\|_{s',I} : u \in W^{1,1}(I),\ \|\nabla u\|_{1,I} = 1\}. \tag{1.2}$$

For any closed cube Q in $\mathbb{R}^n$ it therefore follows by an appropriate change of variable that

$$\sup\{\|u - u_Q\|_{s',Q}/\|\nabla u\|_{1,Q} : u \in W^{1,1}(Q^\circ),\ \nabla u \neq 0\} = \mu(s', n)|Q|^{1/n - 1 + 1/p^*}. \tag{1.3}$$

Lemma 1.1. Let h be a real-valued function in $L^s(Q)$, with $n \leqslant s \leqslant \infty$, and suppose that $h_Q = |Q|^{-1}\int_Q h(x)\,dx = 0$. Then for all $\phi \in W^{1,2}(Q^\circ)$,

$$\left|\int_Q h|\phi|^2\right| \leqslant 2\mu_s|Q|^{1/n - 1/s}\|h\|_{s,Q}\|\phi\|_{2,Q}\|\nabla\phi\|_{2,Q}, \tag{1.4}$$

where

$$\mu_s = \begin{cases} \mu(s', n) \text{ (with } 1/s' = 1 - 1/s) & \text{when } s \geqslant n \geqslant 2, \\ \frac{1}{2} \text{ when } s \geqslant n = 1. \end{cases} \tag{1.5}$$ ■

Proof. Since $h_Q = 0$ we have, for $\phi \in W^{1,1}(Q^\circ)$,

$$\int_Q h(x)|\phi(x)|^2\,dx = \int_Q h(x)\{|\phi(x)|^2 - [|\phi|^2]_Q\}\,dx$$

where $[|\phi|^2]_Q = |Q|^{-1}\int_Q |\phi|^2$. Hence, when $n > 1$, we find, on applying Hölder's inequality, that

$$\left|\int_Q h|\phi|^2\right| \leqslant \|h\|_{s,Q}\|\{|\phi|^2 - [|\phi|^2]_Q\}\|_{s',Q}.$$

Since $\phi \in W^{1,2}(Q^\circ)$, it follows that $|\phi|^2 \in W^{1,1}(Q^\circ)$ and $\nabla(|\phi|^2) = \mathrm{re}(\bar{\phi}\nabla\phi)$. Hence, on using (1.3),

$$\left|\int_Q h|\phi|^2\right| \leqslant \mu(s',n)|Q|^{1/n-1+1/s'} \|h\|_{s,Q} \|\nabla(|\phi|^2)\|_{1,Q}$$

$$\leqslant 2\mu(s',n)|Q|^{1/n-1+1/s'}\|h\|_{s,Q} \|\phi\|_{2,Q}\|\nabla\phi\|_{2,Q}.$$

Thus (1.4) is proved for $n > 1$.

When $n = 1$, the space $W^{1,2}(Q^\circ)$ is continuously embedded in $C^{0,\lambda}(Q)$ $(0 < \lambda < \frac{1}{2})$, the set of Hölder continuous functions with exponent λ on Q. Hence each $\phi \in W^{1,2}(Q^\circ)$ can be identified with a continuous function on Q. Let $x_0 \in Q$ be such that $|\phi(x)|^2$ attains its minimum on Q at x_0. We then have

$$\int_Q h(x)|\phi(x)|^2\,dx = \int_Q h(x)[|\phi(x)|^2 - |\phi(x_0)|^2]\,dx$$

$$\leqslant \int_Q h^+(x)[|\phi(x)|^2 - |\phi(x_0)|^2]\,dx$$

where $h^+ = \max(h,0)$. Also

$$|[|\phi(x)|^2 - |\phi(x_0)|^2]| = \left|2\mathrm{re}\int_{x_0}^{x} \bar{\phi}(t)\phi'(t)\,dt\right|$$

$$\leqslant 2\|\phi\|_{2,Q}\|\phi'\|_{2,Q}$$

and so

$$\int_Q h(x)|\phi(x)|^2\,dx \leqslant 2\|h^+\|_{1,Q}\|\phi\|_{2,Q}\|\phi'\|_{2,Q}.$$

The hypothesis $h_Q = 0$ implies that $\int_Q h^+(x)\,dx = \int_Q h^-(x)\,dx$, where $h^- = -\min(h,0)$, and consequently

$$\int_Q h^+(x)\,dx = \tfrac{1}{2}\|h\|_{1,Q}.$$

On applying Hölder's inequality we therefore obtain

$$\int_Q h(x)|\phi(x)|^2\,dx \leqslant \|h\|_{1,Q}\|\phi\|_{2,Q}\|\phi'\|_{2,Q}$$

$$\leqslant |Q|^{1-1/s}\|h\|_{s,Q}\|\phi\|_{2,Q}\|\phi'\|_{2,Q}.$$

Similarly

$$-\int_Q h(x)|\phi(x)|^2\,dx \leqslant 2\|h^-\|_{1,Q}\|\phi\|_{2,Q}\|\phi'\|_{2,Q}$$

$$\leqslant |Q|^{1-1/s}\|h\|_{s,Q}\|\phi\|_{2,Q}\|\phi'\|_{2,Q}.$$

The proof is therefore complete. □

Let $\mathfrak{F}$ denote a countable set of closed cubes Q having disjoint interiors and which cover Ω, $\Omega \subset \bigcup_{Q \in \mathfrak{F}} Q$. We extend the domain of definition of any function defined on Ω to the whole of $\mathbb{R}^n$ by putting it equal to zero outside Ω. Also if $h \in \mathrm{L}^s(Q)$ we set

$$\rho_s(h, Q) := |Q|^{-1/s} \|h - h_Q\|_{s, Q}. \tag{1.6}$$

The assumptions on the b_j and q in (1.1) can now be given. They are:

I(i) $b_j = b_j^{(1)} + ib_j^{(2)}$ $(j = 1, 2, \ldots, n)$ and there exist positive constants B_1 and B_2 such that, for $k = 1, 2$,

$$\sum_{j=1}^{n} |b_j^{(k)}(x)|^2 \leqslant 4B_k^2 \quad \text{a.e. on } \Omega; \tag{1.7}$$

I(ii) $q = q_0 + q_1$, where $q_0, q_1 \in \mathrm{L}^1_{\mathrm{loc}}(\Omega)$;

I(iii) there exist constants $\beta \in (-\frac{1}{2}\pi, \frac{1}{2}\pi)$, $\delta \in \mathbb{C}$, and $M \geqslant 0$ such that $\mathrm{e}^{\mathrm{i}\beta}(q_0 - \delta) =: f_0 + \mathrm{i}g_0$ satisfies

$$f_0(x) \geqslant 0 \text{ and } |g_0(x)| \leqslant M f_0(x) \quad \text{a.e. on } \Omega, \tag{1.8}$$

i.e. $\mathrm{e}^{\mathrm{i}\beta} q_0(x)$ has values in the sector $\mathscr{S}(\mathrm{e}^{\mathrm{i}\beta}\delta; \tan^{-1} M) = \{z \in \mathbb{C} : |\arg(z - \mathrm{e}^{\mathrm{i}\beta}\delta)| \leqslant \tan^{-1} M\}$;

I(iv) there exists $s \in [n, \infty]$ such that $f + \mathrm{i}g := \mathrm{e}^{\mathrm{i}\beta} q_1 \in \mathrm{L}^s(Q)$ for each $Q \in \mathfrak{F}$ and

$$\left.\begin{aligned} m(f) &:= \inf_{Q \in \mathfrak{F}} f_Q > -\infty, \\ l_s(f^-) &:= \sup_{Q \in \mathfrak{F}} [|Q|^{-1/s} \|f^-\|_{s,Q}(1 + |Q|^{1/n})] < \infty, \\ \alpha_s(f) &:= \mu_s \sup_{Q \in \mathfrak{F}} [|Q|^{1/n} \rho_s(f, Q)] < \infty, \\ p(g) &:= \sup_{Q \in \mathfrak{F}} |g_Q| < \infty, \\ \alpha_s(g) &:= \mu_s \sup_{Q \in \mathfrak{F}} [|Q|^{1/n} \rho_s(g, Q)] < \infty. \end{aligned}\right\} \tag{1.9}$$

The assumption I(iii) is satisfied if the values of q_0 lie in a sector

$$\{z \in \mathbb{C} : \theta \leqslant \arg(z - \delta) \leqslant \pi + \theta - \varepsilon, \ -\pi < \theta < \varepsilon < \pi, \ \varepsilon > 0\}, \tag{1.10}$$

for some θ, the significant point being that the sector is not intersected by the line $\{z : \arg(z - \delta) = \pi\}$. The conditions in I(iv) give the integral inequalities in the next lemma.

Lemma 1.2. Let f and g satisfy (1.9). Then for all $\phi \in \mathrm{W}_0^{1,2}(\Omega)$,

$$\int_\Omega f|\phi|^2 \geqslant m(f) \|\phi\|_{2,\Omega}^2 - 2\alpha_s(f) \|\nabla\phi\|_{2,\Omega} \|\phi\|_{2,\Omega}, \tag{1.11}$$

$$\left|\int_{\Omega} g|\phi|^2\right| \leqslant p(g)\,\|\phi\|_{2,\Omega}^2 + 2\alpha_s(g)\,\|\nabla\phi\|_{2,\Omega}\|\phi\|_{2,\Omega}, \tag{1.12}$$

and for any $\varepsilon \in (0,1)$ there exists a positive constant K, depending only on $l_s(f^-)$, such that

$$\int_{\Omega} f^-\,|\phi|^2 \leqslant \varepsilon\|\nabla\phi\|_{2,\Omega}^2 + \varepsilon^{-1}K\,\|\phi\|_{2,\Omega}^2. \tag{1.13}$$

Also, for any compact subset Ω_0 of Ω there exists a constant K_{Ω_0}, depending on Ω_0, such that

$$\int_{\Omega_0} |g|\,|\phi|^2 \leqslant K_{\Omega_0}\,(\|\nabla\phi\|_{2,\Omega}^2 + \|\phi\|_{2,\Omega}^2) \tag{1.14}$$

and similarly for f. ■

Proof. To prove (1.11) we set $h(x) = f(x) - f_Q$ for $x \in Q$, so that $h_Q = 0$. Lemma 1.1 can therefore be applied to give

$$\begin{aligned}\int_Q f|\phi|^2 &= f_Q\|\phi\|_{2,Q}^2 + \int_Q h|\phi|^2 \\ &\geqslant f_Q\|\phi\|_{2,Q}^2 - 2\mu_s|Q|^{1/n-1/s}\|h\|_{s,Q}\|\nabla\phi\|_{2,Q}\|\phi\|_{2,Q} \\ &\geqslant m(f)\,\|\phi\|_{2,Q}^2 - 2\alpha_s(f)\,\|\nabla\phi\|_{2,Q}\,\|\phi\|_{2,Q}.\end{aligned}$$

On summing over $Q \in \mathfrak{F}$ and noting that $\Sigma_{Q\in\mathfrak{F}}\,\|\nabla\phi\|_{2,Q}\,\|\phi\|_{2,Q} \leqslant (\Sigma_{Q\in\mathfrak{F}}\|\nabla\phi\|_{2,Q}^2)^{\frac{1}{2}}\,(\Sigma_{Q\in\mathfrak{F}}\|\phi\|_{2,Q}^2)^{\frac{1}{2}} = \|\nabla\phi\|_{2,\Omega}\|\phi\|_{2,\Omega}$, we obtain (1.11); recall that $\phi = 0$ outside Ω. The proof of (1.12) is similar. To derive (1.13) when $n > 1$ we apply Holder's inequality, so that with $s' = s/(s-1)$ we have

$$\begin{aligned}\int_Q f^-\,|\phi|^2 &\leqslant \|f^-\|_{s,Q}\,\||\phi|^2\|_{s',Q} \\ &\leqslant \|f^-\|_{s,Q}[\|\,|\phi|^2 - (|\phi|^2)_Q\|_{s',Q} + \|(|\phi|^2)_Q\|_{s',Q}] \\ &\leqslant \|f^-\|_{s,Q}[2\mu(s',n)|Q|^{1/n-1/s}\|\nabla\phi\|_{2,Q}\|\phi\|_{2,Q} + |Q|^{-1/s}\|\phi\|_{2,Q}^2],\end{aligned} \tag{1.15}$$

as in the proof of Lemma 1.1, using the fact that $(|\phi|^2)_Q = |Q|^{-1}\,\|\phi\|_{2,Q}^2$. Hence given any $\varepsilon \in (0,1)$,

$$\begin{aligned}\int_Q f^-\,|\phi|^2\,\mathrm{d}x &\leqslant 2\mu(s',n)\,l_s(f^-)\,\|\nabla\phi\|_{2,Q}\,\|\phi\|_{2,Q} + l_s(f^-)\|\phi\|_{2,Q}^2 \\ &\leqslant \varepsilon\|\nabla\phi\|_{2,Q}^2 + [\varepsilon^{-1}\mu^2(s',n)l_s^2(f^-) + l_s(f^-)]\,\|\phi\|_{2,Q}^2\end{aligned}$$

and (1.13) follows on summing over $Q \in \mathfrak{F}$. The inequality (1.15) continues to hold when f^- is replaced by g or f and (1.14) follows on summing the analogue of (1.15) for g over the finite number of cubes Q which cover Ω_0. The case $n = 1$ is treated similarly. □

We denote the Dirichlet operator generated by τ in (1.1) by T_D. It will be determined by the Dirichlet integral associated with τ and Dirichlet conditions on $\partial\Omega$, this being the sesquilinear form

$$t_D[\![\phi,\psi]\!]:=\int_\Omega [\nabla\phi\cdot\nabla\bar{\psi}+(b\cdot\nabla\phi)\bar{\psi}+q\phi\bar{\psi}] \qquad (\phi,\psi\in\mathfrak{Q}_D). \tag{1.16}$$

The domain $\mathfrak{Q}_D$ of t_D is defined to be the completion of $C_0^\infty(\Omega)$ with respect to the norm $\|\bullet\|_D$ given by

$$\|\phi\|_D^2:=\int_\Omega [|\nabla\phi|^2+(f_0+f^+)|\phi|^2+|\phi|^2], \tag{1.17}$$

where f_0 and f are defined in I(iii) and I(iv). The map which identifies any $\phi\in C_0^\infty(\Omega)$ as a member of $\mathfrak{Q}_D$ with the same function in $W_0^{1,2}(\Omega)$ is continuous since $\|\phi\|_{1,2,\Omega}\leqslant\|\phi\|_D$. It extends uniquely to $\mathfrak{Q}_D$ to give a continuous embedding of $\mathfrak{Q}_D$ into $W_0^{1,2}(\Omega)$ with dense range, since $C_0^\infty(\Omega)$ is dense in $W_0^{1,2}(\Omega)$.

Definition 1.3. We define $v=\Delta u$ to be the *Dirichlet Laplacian of* $u\in W_0^{1,2}(\Omega)$ if $v\in L^1_{\text{loc}}(\Omega)$ and for all $\phi\in C_0^\infty(\Omega)$,

$$\int_\Omega \nabla u\cdot\nabla\phi=-\int_\Omega v\phi. \qquad \blacksquare$$

Theorem 1.4. Let b_j $(j=1,2,\ldots,n)$ and q in (1.1) be complex-valued measurable functions on Ω which satisfy I(i)–I(iv) (p. 329). Then $e^{i\beta}t_D$ is a closed, densely defined and sectorial form in $L^2(\Omega)$. If $e^{i\beta}T_D$ denotes the associated m-sectorial operator, T_D is the Dirichlet operator generated by τ in $L^2(\Omega)$ and has the following properties:

$$(T_D u,\phi)_{2,\Omega}=t_D[\![u,\phi]\!] \qquad (u\in\mathscr{D}(T_D),\ \phi\in\mathfrak{Q}_D),$$

$$\mathscr{D}(T_D)=\{u: u\in\mathfrak{Q}_D,\ \tau u\in L^2(\Omega)\},$$

$$T_D u=\tau u \qquad (u\in\mathscr{D}(T_D)),$$

and Δ in τ is the Dirichlet Laplacian. ■

Proof. Let h_D and i_D denote the real and imaginary parts of $e^{i\beta}t_D$, i.e.

$$h_D:=\tfrac{1}{2}(e^{i\beta}t_D+e^{-i\beta}t_D^*),\qquad i_D:=-\tfrac{1}{2}i(e^{i\beta}t_D-e^{-i\beta}t_D^*).$$

For $\phi\in\mathfrak{Q}_D$,

$$\begin{aligned} h_D[\![\phi]\!]&=\operatorname{re}(e^{i\beta}t_D[\![\phi]\!])\\ &=(\cos\beta)\,\|\nabla\phi\|_{2,\Omega}^2+\operatorname{re}\left(e^{i\beta}\int_\Omega (b\cdot\nabla\phi)\bar{\phi}\right)\\ &\quad+\mu\|\phi\|_{2,\Omega}^2+\int_\Omega (f_0+f)|\phi|^2, \end{aligned} \tag{1.18}$$

where $\mu = \mathrm{re}\,(e^{i\beta}\delta)$. Also

$$\begin{aligned} i_D[\![\phi]\!] &= \mathrm{im}\,(e^{i\beta} t_D[\![\phi]\!]) \\ &= (\sin\beta)\,\|\nabla\phi\|^2_{2,\Omega} + \mathrm{im}\left(e^{i\beta}\int_\Omega (b\cdot\nabla\phi)\bar{\phi}\right) \\ &\quad + \nu\|\phi\|^2_{2,\Omega} + \int_\Omega (g_0+g)\,|\phi|^2, \end{aligned} \tag{1.19}$$

where $\nu = \mathrm{im}\,(e^{i\beta}\delta)$. On using (1.7) and (1.13) in (1.18), we see that, for any $\varepsilon \in (0,1)$,

$$\begin{aligned} h_D[\![\phi]\!] &\geqslant (\cos\beta)\,\|\nabla\phi\|^2_{2,\Omega} - 2(B_1+B_2)\,\|\nabla\phi\|_{2,\Omega}\,\|\phi\|_{2,\Omega} + \mu\|\phi\|^2_{2,\Omega} \\ &\quad + \int_\Omega (f_0+f^+)\,|\phi|^2 - \varepsilon\|\nabla\phi\|^2_{2,\Omega} - \varepsilon^{-1}K\,\|\phi\|^2_{2,\Omega} \\ &\geqslant (\cos\beta - 2\varepsilon)\,\|\nabla\phi\|^2_{2,\Omega} + \int_\Omega (f_0+f^+)|\phi|^2 - K_\varepsilon\,\|\phi\|^2_{2,\Omega} \end{aligned}$$

for some $K_\varepsilon > 0$. Hence there exist positive constants K_0 and K_1 such that

$$(h_D + K_0 + 1)[\![\phi]\!] \geqslant K_1\|\phi\|^2_D \qquad (\phi \in \mathfrak{Q}_D). \tag{1.20}$$

Similarly, there exists a positive constant K_2 such that

$$(h_D + K_0 + 1)\,[\![\phi]\!] \leqslant K_2\,\|\phi\|^2_D \qquad (\phi \in \mathfrak{Q}_D) \tag{1.21}$$

and from (1.7), (1.8), (1.12), and (1.19) we obtain

$$\begin{aligned} |i_D[\![\phi]\!]| &\leqslant |\sin\beta|\;\|\nabla\phi\|^2_{2,\Omega} + 2(B_1+B_2)\,\|\nabla\phi\|_{2,\Omega}\,\|\phi\|_{2,\Omega} + \nu\|\phi\|^2_{2,\Omega} \\ &\quad + M\int_\Omega f_0|\phi|^2 + p(g)\,\|\phi\|^2_{2,\Omega} + 2\alpha_s(g)\,\|\nabla\phi\|_{2,\Omega}\,\|\phi\|_{2,\Omega} \\ &\leqslant K_3\,\|\phi\|^2_D \end{aligned} \tag{1.22}$$

for some positive constant K_3. The two inequalities (1.21) and (1.22) give

$$|i_D[\![\phi]\!]| \leqslant K_1^{-1}K_3\,(h_D + K_0 + 1)\,[\![\phi]\!] \qquad (\phi \in \mathfrak{Q}_D). \tag{1.23}$$

Hence t_D is sectorial with vertex at $-K_0 - 1$. Furthermore (1.20) and (1.21) imply that $(h_D + K_0 + 1)^{\frac{1}{2}}[\![\bullet]\!]$ is a norm on $\mathfrak{Q}_D$ which is equivalent to $\|\bullet\|_D$. Since $\mathfrak{Q}_D$ is complete it follows from §IV.2 that $e^{i\beta}t_D$ is a closed sectorial form in $L^2(\Omega)$ which is densely defined in $L^2(\Omega)$ since $\mathfrak{Q}_D$ contains $C_0^\infty(\Omega)$. We can therefore invoke the First Representation Theorem, Theorem IV.2.4, for such sesquilinear forms to establish the existence of an m-sectorial operator, denoted by $e^{i\beta}T_D$, which satisfies $(T_D u, \phi)_{2,\Omega} = t_D[\![u, \phi]\!]$ for all $u \in \mathscr{D}(T_D)$ and $\phi \in \mathfrak{Q}_D$.

Let $U := \{u : u \in \mathfrak{Q}_D,\ \tau u \in L^2(\Omega)\}$ and let S denote the restriction of τ to U. To complete the proof we need to show that $S = T_D$. For $u \in \mathscr{D}(T_D)$ and

$\phi \in \mathfrak{Q}_D$, we have

$$\int_\Omega \nabla u \cdot \nabla \bar{\phi} = t_D[\![u, \phi]\!] - (b \cdot \nabla u, \phi)_{2,\Omega} - (qu, \phi)_{2,\Omega}$$
$$= (T_D u - b \cdot \nabla u - qu, \phi)_{2,\Omega}$$
$$= (v, \phi)_{2,\Omega}, \tag{1.24}$$

say. Since $u \in \mathscr{D}(T_D) \subset \mathfrak{Q}_D$, we have that $T_D u \in L^2(\Omega) \subset L^1_{loc}(\Omega)$, and $b \cdot \nabla u \in L^1_{loc}(\Omega)$ since $|b| \in L^\infty(\Omega)$ and $|\nabla u| \in L^2(\Omega)$. Moreover, by Lemma 1.2 and (1.17), $f_0^{\frac{1}{2}} u$, $(f^+)^{\frac{1}{2}} u$, $(f^-)^{\frac{1}{2}} u \in L^2(\Omega)$, $|g|^{\frac{1}{2}} u \in L^2_{loc}(\Omega)$ and $|g_0|^{\frac{1}{2}} u \in L^2(\Omega)$ in view of (1.8). All in all, $|qu^2| \in L^1_{loc}(\Omega)$ and since $q \in L^1_{loc}(\Omega)$ and $2|qu| \leqslant |q| + |qu^2|$, it follows that $qu \in L^1_{loc}(\Omega)$. Consequently in (1.24), $v \in L^1_{loc}(\Omega)$ and $-\Delta u = v$ by Definition 1.3 for the Dirichlet Laplacian. We have therefore proved that $T_D u = \tau u$ and so $T_D \subset S$. Conversely, if $u \in U$ then, for all $\phi \in C_0^\infty(\Omega)$,

$$(Su, \phi)_{2,\Omega} = (\tau u, \phi)_{2,\Omega} = t_D[\![u, \phi]\!].$$

From

$$|t_D[\![u, \phi]\!]| \leqslant K \|u\|_D \|\phi\|_D$$

and since $C_0^\infty(\Omega)$ is dense in $\mathfrak{Q}_D$, it follows by continuity that

$$(Su, u) = t_D[\![u, u]\!] \qquad (u \in U),$$

and hence that $e^{i\beta} S$ is sectorial. This implies that $e^{i\beta} S = e^{i\beta} T_D$, and hence $S = T_D$, since the m-sectorial operator $e^{i\beta} T_D$ cannot have a proper sectorial extension. This completes the proof. □

Theorem 1.5. Let the hypotheses of Theorem 1.4 be satisfied and set

$$F := B_1 + B_2 + \alpha_s(f), \tag{1.25}$$
$$G := B_1 + B_2 + \alpha_s(g), \tag{1.26}$$
$$\theta := \mu + m(f) - F^2 \sec \beta, \tag{1.27}$$

where $e^{i\beta}\delta =: \mu + i\nu$. Then $\sigma(T_D)$ lies in the set $e^{-i\beta}\Theta_\beta$, where Θ_β is the set of all $x + iy$ such that $x \geqslant \theta$ and $|y - \nu| - p(g) \leqslant$

$$\begin{cases} [|\tan \beta|(z+F)^2 + 2G(z+F)] \sec \beta & \text{if} \quad M \leqslant |\tan \beta| \\ \quad \text{or if} \quad M > |\tan \beta| \quad \text{and} \quad z \leqslant \dfrac{(G\cos\beta + F|\sin\beta|)}{(M\cos\beta - |\sin\beta|)}, & (1.28) \\ M[x - \mu - m(f)] + \dfrac{(MF+G)^2}{(M\cos\beta - |\sin\beta|)} \quad \text{otherwise} & (1.29) \end{cases}$$

and $z = [(x-\theta)\cos\beta]^{\frac{1}{2}}$. ■

Proof. The fact that $e^{i\beta}T_D$ is m-sectorial implies that its spectrum lies within its numerical range and hence within the numerical range of $e^{i\beta}t_D$. The set $\sigma(T_D)$ must therefore lie within $e^{-i\beta}\Theta(e^{i\beta}t_D)$. With the notation of the proof of Theorem 1.4, we have from (1.18), on using (1.7), (1.8) and (1.11), that for all $\phi\in\mathfrak{Q}_D$,

$$\begin{aligned}
h_D[\![\phi]\!] &\geqslant (\cos\beta)\|\nabla\phi\|_{2,\Omega}^2-2(B_1+B_2)\|\nabla\phi\|_{2,\Omega}\|\phi\|_{2,\Omega}+\mu\|\phi\|_{2,\Omega}^2\\
&\quad+\|f_0^{\frac{1}{2}}\phi\|_{2,\Omega}^2+m(f)\|\phi\|_{2,\Omega}^2-2\alpha_s(f)\|\nabla\phi\|_{2,\Omega}\|\phi\|_{2,\Omega}\\
&=(\cos\beta)\|\nabla\phi\|_{2,\Omega}^2-2F\|\nabla\phi\|_{2,\Omega}\|\phi\|_{2,\Omega}+\|f_0^{\frac{1}{2}}\phi\|_{2,\Omega}^2\\
&\quad+[\mu+m(f)]\|\phi\|_{2,\Omega}^2\\
&\geqslant(\cos\beta-\varepsilon_1)\|\nabla\phi\|_{2,\Omega}^2+\|f_0^{\frac{1}{2}}\phi\|_{2,\Omega}^2+[\mu+m(f)-\varepsilon_1^{-1}F^2]\|\phi\|_{2,\Omega}^2, \qquad (1.30)
\end{aligned}$$

for any $\varepsilon_1>0$. Similarly, from (1.19), on using (1.7), (1.8) and (1.12), we have that

$$\begin{aligned}
|i_D[\![\phi]\!]-\nu\|\phi\|_{2,\Omega}^2| &\leqslant |\sin\beta|\,\|\nabla\phi\|_{2,\Omega}^2+2(B_1+B_2)\|\nabla\phi\|_{2,\Omega}\|\phi\|_{2,\Omega}\\
&\quad+M\|f_0^{\frac{1}{2}}\phi\|_{2,\Omega}^2+p(g)\|\phi\|_{2,\Omega}^2+2\alpha_s(g)\|\nabla\phi\|_{2,\Omega}\|\phi\|_{2,\Omega}\\
&=|\sin\beta|\,\|\nabla\phi\|_{2,\Omega}^2+2G\|\nabla\phi\|_{2,\Omega}\|\phi\|_{2,\Omega}+M\|f_0^{\frac{1}{2}}\phi\|_{2,\Omega}^2+p(g)\|\phi\|_{2,\Omega}^2\\
&\leqslant(|\sin\beta|+\varepsilon_2)\|\nabla\phi\|_{2,\Omega}^2+M\|f_0^{\frac{1}{2}}\phi\|_{2,\Omega}^2+[p(g)+\varepsilon_2^{-1}G^2]\|\phi\|_{2,\Omega}^2, \qquad (1.31)
\end{aligned}$$

for any $\varepsilon_2>0$. From (1.30) and (1.31) it follows that for $0\leqslant t\leqslant M^{-1}$ and $\|\phi\|_{2,\Omega}=1$,

$$\begin{aligned}
h_D[\![\phi]\!]-t|i_D[\phi]-\nu| &\geqslant [(\cos\beta-\varepsilon_1)-t(|\sin\beta|+\varepsilon_2)]\,\|\nabla\phi\|_{2,\Omega}^2\\
&\quad+\{\mu+m(f)-\varepsilon_1^{-1}F^2-t[p(g)+\varepsilon_2^{-1}G^2]\}. \qquad (1.32)
\end{aligned}$$

Hence, on putting $t=0$ and $\varepsilon_1=\cos\beta$ in (1.32), we obtain $h_D[\phi]\geqslant\theta$.

We now choose t, ε_1, ε_2 in (1.32) to satisfy

$$0\leqslant t\leqslant M^{-1},\qquad t<|\cot\beta|, \qquad (1.33)$$

$$\varepsilon_1=:\varepsilon,\qquad 0<\varepsilon<\cos\beta-t|\sin\beta|, \qquad (1.34)$$

$$t\varepsilon_2=\cos\beta-\varepsilon-t|\sin\beta|. \qquad (1.35)$$

On substituting in (1.32), we obtain

$$\begin{aligned}
h_D[\![\phi]\!]-t|i_D[\![\phi]\!]-\nu| &\geqslant \mu+m(f)-tp(g)-\varepsilon^{-1}F^2-t^2G^2(\cos\beta-\varepsilon-t|\sin\beta|)^{-1}\\
&=:\Phi(t,\varepsilon), \qquad (1.36)
\end{aligned}$$

say. For t fixed and satisfying (1.33), $\Phi(t,\bullet)$ attains its maximum value when $\varepsilon=(\cos\beta-t|\sin\beta|)F(F+tG)^{-1}$. This satisfies (1.34) and on substituting in $\Phi(t,\varepsilon)$ we obtain the lower bound

$$h_D[\![\phi]\!]-t|i_D[\![\phi]\!]-\nu|\geqslant\mu+m(f)-tp(g)-(F+tG)^2(\cos\beta-t|\sin\beta|)^{-1}.$$

Since we have already seen that $h_{\mathrm{D}}[\![\phi]\!] \geqslant \theta$, we conclude that $\Theta(\mathrm{e}^{\mathrm{i}\beta} t_{\mathrm{D}})$ lies within the set W_t defined by

$$\left.\begin{aligned} W_t &:= \{x+\mathrm{i}y : x \geqslant \theta, |y-v| \leqslant \Psi(t)\}, \\ \Psi(t) &:= t^{-1}[x-\mu-m(f)+tp(g)+(F+tG)^2(\cos\beta - t|\sin\beta|)^{-1}], \end{aligned}\right\} \tag{1.37}$$

with t satisfying (1.33). We now minimize $\Psi(\bullet)$. It is straight-forward to verify that the minimum occurs when

$$t = t_0 := z(\cos\beta)(G\cos\beta + |\sin\beta| F + z|\sin\beta|)^{-1}$$

where $z^2 = (\cos\beta)(x-\theta)$. Observe that $t_0 < \cot\beta$ and $t_0 \leqslant M^{-1}$ if $M \leqslant |\tan\beta|$. If $M > |\tan\beta|$, then $t_0 < M^{-1}$ as long as

$$z \leqslant (G\cos\beta + F|\sin\beta|)(M\cos\beta - |\sin\beta|)^{-1}.$$

If either of these alternatives prevails, we have $\Psi(t) \geqslant \Psi(t_0)$ and (1.37) with $t = t_0$ gives (1.28). Otherwise, it is seen that $\Psi(t) \geqslant \Psi(M^{-1})$ and (1.37) with $t = M^{-1}$ yields (1.29). The theorem is therefore proved. □

Theorem 1.6. Let the hypotheses of Theorem 1.4 be satisfied and, in addition, suppose that $b_j \in \mathrm{AC}_{\mathrm{loc}}(\Omega)$ $(j = 1, 2, \ldots, n)$ and $\operatorname{div} b \equiv \mathrm{D}_j b_j = 0$. Then $\sigma(T_{\mathrm{D}})$ lies in the set $\mathrm{e}^{-\mathrm{i}\beta}\Theta_\beta$, where Θ_β is defined by (1.28) and (1.29) but with F and G in (1.25) and (1.26) replaced by

$$\left.\begin{aligned} F &= B_1|\sin\beta| + B_2\cos\beta + \alpha_s(f), \\ G &= B_1\cos\beta + B_2|\sin\beta| + \alpha_s(g). \end{aligned}\right\} \tag{1.38}$$ ■

Proof. Since $\mathrm{C}_0^\infty(\Omega)$ is a core of t_{D}, it is sufficient to work with $\phi \in \mathrm{C}_0^\infty(\Omega)$ in the proof of Theorem 1.5. Under the present hypotheses we have, for $\phi \in \mathrm{C}_0^\infty(\Omega)$, that

$$\int_\Omega (b\cdot\nabla\phi)\bar{\phi} = -\int_\Omega \phi(b\cdot\nabla\bar{\phi}),$$

giving

$$\operatorname{re}\int_\Omega (b\cdot\nabla\phi)\bar{\phi} = \mathrm{i}\int_\Omega \left(b^{(2)}\cdot\nabla\phi\right)\bar{\phi},$$

$$\operatorname{im}\int_\Omega (b\cdot\nabla\phi)\bar{\phi} = \int_\Omega (b^{(1)}\cdot\nabla\phi)\bar{\phi},$$

in the notation of I(i). These identities yield the following improved estimates in (1.18) and (1.19):

$$\operatorname{re}\left(\mathrm{e}^{\mathrm{i}\beta}\int_\Omega (b\cdot\nabla\phi)\bar{\phi}\right) \geqslant -2(B_1|\sin\beta| + B_2\cos\beta)\|\nabla\phi\|_{2,\Omega}\|\phi\|_{2,\Omega},$$

$$\left|\operatorname{im}\left(\mathrm{e}^{\mathrm{i}\beta}\int_\Omega (b\cdot\nabla\phi)\bar{\phi}\right)\right| \leqslant 2(B_1\cos\beta + B_2|\sin\beta|)\|\nabla\phi\|_{2,\Omega}\|\phi\|_{2,\Omega}.$$

The rest of the proof of Theorem 1.5 now goes through unchanged on defining the quantities F and G by (1.38). □

Corollary 1.7. Let $\tau = -\Delta + q$, where $q \in L^1_{loc}(\Omega)$ and for some $\beta \in (-\frac{1}{2}\pi, \frac{1}{2}\pi)$, $\delta_0 \in \mathbb{C}$ and $\gamma \in [0, \frac{1}{2}\pi)$, the values of $e^{i\beta} q$ lie in the sector $\mathscr{S}(\delta_0; \gamma) = \{z \in \mathbb{C}: |\arg(z - \delta_0)| \leqslant \gamma\}$ having vertex δ_0 and semi-angle γ. Then

$$\sigma(T_D) \subset e^{-i\beta} \mathscr{S}(\delta_0; \max\{\gamma, |\beta|\}).$$

■

Proof. This follows from Theorem 1.5 on putting $q = q_0$, $f_0 + ig_0 = e^{i\beta} q - \delta_0$, $M = \tan\gamma$, $e^{i\beta}\delta = \delta_0$, and $f = g = 0$. □

Corollary 1.8. Let $\tau = -\Delta + q$ and $q \in L^1_{loc}(\Omega)$ and suppose that, for some $\beta \in (-\frac{1}{2}\pi, \frac{1}{2}\pi)$, the function $e^{i\beta} q$ is real-valued and $e^{i\beta} q = f_0 + f + \mu$, where $\mu \in \mathbb{R}$, $f_0 \geqslant 0$ with $f_0 \in L^1_{loc}(\Omega)$, and $f \in L^s(\Omega)$ for some $s \in [n, \infty]$. Then with μ_s defined by (1.5), the inclusion

$$\sigma(T_D) \subset e^{-i\beta} \Theta_\beta$$

holds, where

$$\Theta_\beta = \{x + iy : x \geqslant \theta, |y| \leqslant |\tan\beta| [(x - \theta)^{\frac{1}{2}} + F(\sec\beta)^{\frac{1}{2}}]\}$$

with

$$\theta - \mu \geqslant \begin{cases} -(8\mu_s^2 n^{-1} \sec\beta)^{n/(2s-n)} (s - \frac{1}{2}n)(s-n)^{-2(s-n)/(2s-n)} \| f \|_{s,\Omega} & \text{if } s > n, \\ -4\mu_n^2 (\sec\beta) \| f \|_{n,\Omega} & \text{if } s = n, \end{cases}$$

and

$$F \leqslant \begin{cases} [2^{s-n} \mu_s^{-n} n^s (s-n)^{-s} (\sec\beta)^{-s} \| f \|_{s,\Omega}^{s-n}]^{1/(2s-n)} & \text{if } s > n, \\ 2\mu_n \| f \|_{n,\Omega} & \text{if } s = n. \end{cases}$$

If $\beta = 0$ then T_D is self-adjoint and $T_D \geqslant \theta I$. ■

Proof. Let $\mathfrak{F}$ be any covering of Ω by closed, congruent cubes having disjoint interiors. On applying Hölder's inequality we obtain from (1.9),

$$\begin{aligned} m(f) &\geqslant -\sup_{Q \in \mathfrak{F}} \left(|Q|^{-1} \int_Q |f| \right) \\ &\geqslant -|Q|^{-1/s} \| f \|_{s,\Omega}, \end{aligned}$$

and

$$\begin{aligned} \alpha_s(f) &= \mu_s |Q|^{1/n - 1/s} \sup_{Q \in \mathfrak{F}} \| f - f_Q \|_{s,Q} \\ &\leqslant 2\mu_s |Q|^{1/n - 1/s} \| f \|_{s,Q}, \end{aligned}$$

since $\|f_Q\|_{s,Q} \leqslant \|f\|_{s,Q}$. Setting $|Q| = d^n$, we get from (1.25)–(1.27),

$$F \leqslant 2\mu_s d^{1-n/s}\|f\|_{s,\Omega}, \qquad G = 0,$$

$$\theta \geqslant \mu - d^{-n/s}\|f\|_{s,\Omega} - 4\mu_s^2 d^{2-2n/s}\|f\|_{s,\Omega}^2 \sec\beta.$$

As a function of d, this lower bound for θ has a maximum when

$$d = [8\mu_s^2 n^{-1}(s-n)\|f\|_{s,\Omega} \sec\beta]^{-s/(2s-n)}$$

if $s > n$, whereas if $s = n$, the supremum is obtained on allowing $d \to \infty$. On substituting these values of d in the above estimates for F and θ and then in Theorem 1.5, we obtain the corollary. □

When $n = 1$, $\beta = 0$ and Ω is $\mathbb{R}$ or $(0, \infty)$, the lower bound θ of T_{D} in Corollary 1.8 agrees with the result obtained by Eastham in [1]. It is similar to, but worse than, the value in Everitt [1]. The optimal value of this lower bound is derived by Veling in [1]: he proves that, when $\Omega = (0, \infty)$,

$$\inf_{q \in L^s(0,\infty)} \inf_{\substack{\phi \in \mathscr{D}(T_{\mathrm{D}}) \\ \|\phi\|_{2,(0,\infty)}=1}} [(T_{\mathrm{D}}\phi, \phi)_{2,(0,\infty)} \|q\|_{s,(0,\infty)}^{-2s/(2s-1)}] = -\ell(s),$$

where $\ell(1) = \frac{1}{4}$ and, for $s > 1$,

$$\ell(s)^{2s-1} = 2^{-2}s^{-2s}(2s-1)^2(s-1)^{2s-4}[\Gamma(\tfrac{1}{2})\Gamma(s-1)/\Gamma(s-\tfrac{1}{2})]^{-2}.$$

In [2], Veling proves an optimal result for the case $\Omega = \mathbb{R}^n$ $(n \geqslant 1)$, $\beta = 0$, $q^+ \in \mathrm{L}^2_{\mathrm{loc}}(\mathbb{R}^n)$ and $q^- \in \mathrm{L}^p(\mathbb{R}^n)$, where $p \in [1, \infty)$ if $n = 1$ and $p \in (\frac{1}{2}n, \infty)$ if $n \geqslant 2$. The result is

$$\inf_{q^- \in \mathrm{L}^p(\mathbb{R}^n)} \inf_{\substack{\phi \in \mathscr{D}(T_{\mathrm{D}}) \\ \|\phi\|_{2,\mathbb{R}^n}=1}} [(T_{\mathrm{D}}\phi, \phi)_{2,\mathbb{R}^n} \|q^-\|_{p,\mathbb{R}^n}^{-1/(1-t)}] = -(1-t)t^{t/(1-t)}\lambda_{n,t}^{-2/(1-t)},$$

where

$$t \in \begin{cases} (0, \frac{1}{2}) & \text{if } n = 1, \\ (0, 1) & \text{if } n \geqslant 2, \end{cases}$$

and

$$\lambda_{n,t} = \inf_{\substack{\psi \in \mathrm{W}^{1,2}(\mathbb{R}^n) \\ \psi \neq 0}} \frac{\|\nabla\psi\|_{2,\mathbb{R}^n}^t \|\psi\|_{2,\mathbb{R}^n}^{1-t}}{\|\psi\|_{2n/(n-2t),\mathbb{R}^n}}.$$

Veling gives estimates for the numbers $\lambda_{n,t}$ in terms of known optimal constants in Sobolev inequalities.

Corollary 1.9. Let $\Omega = \mathbb{R}^n$, $n \geqslant 1$ and $\tau = -\Delta + b \cdot \nabla + q$, where b_j $(j = 1, 2, \ldots, n)$ and q are real constants. Then $\sigma(T_{\mathrm{D}})$ lies in the set

$$\Theta_0 = \left\{x + \mathrm{i}y : x \geqslant q, |y|^2 \leqslant \left(\sum_{j=1}^n b_j^2\right)(x - q)\right\}.$$

The result is precise when $n \geqslant 2$, i.e. $\sigma(T_D) = \Theta_0$, whereas if $n = 1$ then $\sigma(T_D)$ coincides with the boundary of Θ_0, namely

$$\partial\Theta_0 = \{x + iy : x \geqslant q, |y|^2 = b_1^2(x-q)\}. \qquad \blacksquare$$

Proof. That $\sigma(T_D) \subset \Theta_0$ follows from Theorem 1.6 with $b^{(2)} = f = g = 0$, $\beta = 0, f_0 = g_0 = 0$ and $\mu = \delta = q$. The precise result is obtained with the aid of Fourier Theory; a further discussion of constant-coefficient operators will be given later in §§IX.6, 7. We now have $\mathfrak{Q}_D = W_0^{1,2}(\mathbb{R}^n)$ and $\mathscr{D}(T_D) = W^{2,2}(\mathbb{R}^n) \cap W_0^{1,2}(\mathbb{R}^n) = W^{2,2}(\mathbb{R}^n)$ since $W^{2,2}(\mathbb{R}^n) = W_0^{2,2}(\mathbb{R}^n)$ (see §V.3.1). The Fourier Transform $\mathbf{F}: \phi \mapsto \hat{\phi}$ is a unitary map on $L^2(\mathbb{R}^n)$ and T_D is therefore unitarily equivalent to $S = \mathbf{F}T_D\mathbf{F}^{-1}$. We have by Lemma V.1.15,

$$\mathscr{D}(S) = \{\hat{\phi} : \hat{\phi}, |\xi|^2\hat{\phi}(\xi) \in L^2(\mathbb{R}^n)\} \equiv \mathbf{F}W^{2,2}(\mathbb{R}^n),$$
$$(S\hat{\phi})(\xi) = (4\pi^2|\xi|^2 + 2\pi i b\cdot\xi + q)\hat{\phi}(\xi),$$

where $\xi = (\xi_1, \ldots, \xi_n)$. The set $\mathfrak{S} = \{4\pi^2|\xi|^2 + 2\pi i b\cdot\xi + q : \xi \in \mathbb{R}^n\}$ is closed in $\mathbb{C}$ and if $\lambda \notin \mathfrak{S}$, we have for $\hat{\phi} \in \mathscr{D}(S)$,

$$\|(S - \lambda I)\hat{\phi}\|_{2,\mathbb{R}^n}^2 = \int_{\mathbb{R}^n} |\lambda - (4\pi^2|\xi|^2 + 2\pi i b\cdot\xi + q)|^2 |\hat{\phi}(\xi)|^2 \, d\xi$$
$$\geqslant c_\lambda \|\hat{\phi}\|_{2,\mathbb{R}^n}^2$$

for some $c_\lambda > 0$. Consequently $(S - \lambda I)^{-1}$ exists and is bounded on $\mathscr{R}(S - \lambda I)$; the subspace $\mathscr{R}(S - \lambda I)$ is therefore closed since S is a closed operator. The adjoint of S is the operator of multiplication by $4\pi^2|\xi|^2 - 2\pi i b\cdot\xi + q$ on $\mathscr{D}(S)$, and since b and q are real we also have $\|(S^* - \bar{\lambda}I)\hat{\phi}\|_{2,\mathbb{R}^n}^2 \geqslant c_\lambda \|\hat{\phi}\|_{2,\mathbb{R}^n}^2$. It follows that $\sigma(S) \subset \mathfrak{S}$ and $\sigma(T_D) \subset \mathfrak{S}$. It will be shown in §IX.6 that in fact $\sigma(S) = \sigma(T_D) = \mathfrak{S}$. The final assertion of the corollary follows since $\mathfrak{S} = \Theta_0$ when $n \geqslant 2$ and $\mathfrak{S} = \partial\Theta_0$ when $n = 1$. □

Corollary 1.10. Let $\tau = -\Delta + q$, where $\operatorname{re} q$ and $\operatorname{im} q$ are periodic functions with period cube Q^*, and let $\Omega = (\cup_{Q\in\mathfrak{F}} Q)^\circ$, where the cubes Q in $\mathfrak{F}$ are congruent to Q^*. Suppose that $q \in L^s(Q^*)$ for some $s \in [n, \infty]$ and that $(\operatorname{re} q)_{Q^*} > 0$, $(\operatorname{im} q)_{Q^*} = 0$. Then

$$\sigma(T_D) \subset \{x + iy : x \geqslant \theta_s, [|y| - 2\alpha_s(f)\alpha_s(g)]^2 \leqslant 4\alpha_s^2(g)(x - \theta_s)\},$$

where $f = \operatorname{re} q - (\operatorname{re} q)_{Q^*}$, $g = \operatorname{im} q$, $\alpha_s(f) = \mu_s|Q^*|^{1/n - 1/s}\|f\|_{s,Q^*}$ and $\theta_s = (\operatorname{re} q)_{Q^*} - \alpha_s^2(f)$. ■

Proof. In Theorem 1.5 we put $\beta = B_1 = B_2 = 0$, $q_0 = (\operatorname{re} q)_{Q^*}$, $f_0 = g_0 = 0$, $\mu = q_0$, $v = 0$. □

With a little extra care one can improve on Theorems 1.5 and 1.6 in some cases. For instance

Theorem 1.11. Let $\tau = -\Delta + q$ where q is real-valued and $q \in L^s_{\text{loc}}(\Omega)$ for some $s \in [n, \infty]$. Then T_D is a self-adjoint operator which is bounded below by

$$\theta = \inf_{Q \in \mathfrak{F}} (q_Q - \mu_s^2 |Q|^{2/n - 2/s} \|q - q_Q\|^2_{s,Q}),$$

where μ_s is defined in (1.5). ∎

Proof. In the proof of Theorem 1.5 we set $\beta = b = f_0 = \operatorname{im} q = 0$, and $f = q_1 = q$. The value of θ in (1.27) is subsequently improved as follows:

$$h_D[\![\phi]\!] \geqslant \|\nabla\phi\|^2_{2,\Omega} + \sum_{Q \in \mathfrak{F}} [q_Q \|\phi\|^2_{2,Q} - |((q - q_Q)\phi, \phi)_{2,Q}|]$$

$$\geqslant \|\nabla\phi\|^2_{2,\Omega} + \sum_{Q \in \mathfrak{F}} (q_Q \|\phi\|^2_{2,Q} - 2\mu_s |Q|^{1/n - 1/s} \|q - q_Q\|_{s,Q} \|\nabla\phi\|_{2,Q} \|\phi\|_{2,Q})$$

(by Lemma 1.1)

$$\geqslant \|\nabla\phi\|^2_{2,\Omega} + \sum_{Q \in \mathfrak{F}} [(q_Q - \mu_s^2 |Q|^{2/n - 2/s} \|q - q_Q\|^2_{s,Q}) \|\phi\|^2_{2,\Omega} - \|\nabla\phi\|^2_{2,Q}]$$

$$\geqslant \inf_{Q \in \mathfrak{F}} (q_Q - \mu_s^2 |Q|^{2/n - 2/s} \|q - q_Q\|^2_{s,Q}) \|\phi\|^2_{2,\Omega}.$$ □

When $s = n = 1$ and Ω is $\mathbb{R}$ or $(0, \infty)$ the value of Θ in Theorem 1.11 is that obtained by Eastham in [1], namely,

$$\theta = \inf_{Q \in \mathfrak{F}} (q_Q - \tfrac{1}{4} \|q - q_Q\|^2_{1,Q})$$

where the Q's are now closed intervals.

1.2. *The Neumann problem on Ω*

The Neumann problem for (1.1) on Ω can be treated in a similar way to the Dirichlet problem in §1.1 by making some straightforward changes to the assumptions and definitions. We shall now assume the following in place of I(i)–(iv):

II(i) b satisfies I(i);
II(ii) $q = q_0 + q_1$; $q_0, q_1 \in L^1_{\text{loc}}(\bar{\Omega})$;
II(iii) I(iii) holds;
II(iv) (1.9) holds with $\mathfrak{F}$ a covering of Ω such that $\Omega = \cup_{Q \in \mathfrak{F}} Q$, the cubes Q in $\mathfrak{F}$ being closed and having disjoint interiors.

A critical role continues to be played by Lemmas 1.1 and 1.2 but since we cannot extend members ϕ of $W^{1,2}(\Omega)$ to the space $W^{1,2}(\mathbb{R}^n)$ by setting $\phi = 0$ outside Ω it is necessary that the cubes Q in $\mathfrak{F}$ lie within Ω. If Ω is an arbitrary domain, the Whitney covering discussed in Stein [1] fulfills the requirements on $\mathfrak{F}$ in II(iv). In a Whitney covering $\mathfrak{F}$ the diameters of the cubes Q are comparable to their distances from the boundary of Ω. If Ω is a half space or the intersection of half spaces the covering $\mathfrak{F}$ in II(iv) can be a set of congruent cubes.

The Neumann operator T_N is defined by the sesquilinear form

$$t_N[\![\phi,\psi]\!] := \int_\Omega [\nabla\phi \cdot \nabla\bar{\psi} + (b\cdot\nabla\phi)\bar{\psi} + q\phi\bar{\psi}] \qquad (\phi,\psi\in\mathfrak{Q}_N), \quad (1.39)$$

where its domain $\mathfrak{Q}_N$ is the completion of $C_0^\infty(\mathbb{R}^n)$ with respect to the norm defined by (1.17); recall that $C_0^\infty(\mathbb{R}^n)$ is here the space of functions which are restrictions to $\bar{\Omega}$ of functions in $C_0^\infty(\mathbb{R}^n)$. The identification map is a continuous embedding of $\mathfrak{Q}_N$ into $W^{1,2}(\Omega)$ and the embedding has dense range if $C_0^\infty(\mathbb{R}^n)$ is dense in $W^{1,2}(\Omega)$, this being the case if $\partial\Omega$ is smooth enough, for instance, if it is of class C; see Theorem V.4.7. However, we only need to know that the embedding of $\mathfrak{Q}_N$ in $L^2(\Omega)$ has dense range and, of course, this is true. Observe that we require $q\in L^1_{\mathrm{loc}}(\bar{\Omega})$, as assumed in II(ii), in order to ensure that (1.39) is defined on $C_0^\infty(\mathbb{R}^n)$ and hence that t_N is densely defined in $L^2(\Omega)$.

The Dirichlet Laplacian of Definition 1.3 is now replaced by the Neumann Laplacian:

Definition 1.12. We define $v = \Delta u$ to be the *Neumann Laplacian* of $u\in W^{1,2}(\Omega)$ if $v\in L^1_{\mathrm{loc}}(\bar{\Omega})$ and for all $\phi\in C_0^\infty(\mathbb{R}^n)$,

$$\int_\Omega \nabla u \cdot \nabla\phi = -\int_\Omega v\phi. \qquad \blacksquare$$

The results of §1.1 now carry over to the Neumann problem after the above changes have been made. Explicitly, we have the following result.

Theorem 1.13. Let b_j $(j = 1, 2, \ldots, n)$ and q in (1.1) be complex-valued measurable functions on Ω which satisfy II(i)–II(iv). Then $e^{i\beta}t_N$ is a closed, densely defined and sectorial form in $L^2(\Omega)$. If $e^{i\beta}T_N$ denotes the associated m-sectorial operator, T_N is the Neumann operator generated by τ in $L^2(\Omega)$ and has the following properties:

$$(T_N u, \phi)_{2,\Omega} = t_N[\![u,\phi]\!] \qquad (u\in\mathscr{D}(T_N),\ \phi\in\mathfrak{Q}_N),$$

$$\mathscr{D}(T_N) = \{u : u\in\mathfrak{Q}_N,\ \tau u\in L^2(\Omega)\},$$

$$T_N u = \tau u \qquad (u\in\mathscr{D}(T_N)),$$

and Δ in τ is the Neumann Laplacian. The spectrum $\sigma(T_N)$ lies in the set $e^{-i\beta}\Theta_\beta$ where Θ_β is defined in (1.28) and (1.29). $\blacksquare$

1.3. $\tau = -\Delta + q$ *with* $q\in L^{n/2}_{\mathrm{loc}}(\Omega)$ *when* $n\geqslant 3$

In Chapters XI and XII we shall need a description of the Dirichlet and Neumann operators T_D and T_N when $\tau = -\Delta + q$ and $q\in L^{n/2}_{\mathrm{loc}}(\Omega)$ when $n\geqslant 3$.

The above results in §§1.1, 2 are not applicable in this case but the methods used do apply with very little change. The role of Lemma 1.1 is now played by the following lemma.

Lemma 1.14. Let $h \in L^r(Q)$ where

$$r \in \begin{cases} [1, \infty] & \text{if } n = 1, \\ (1, \infty] & \text{if } n = 2, \\ [\tfrac{1}{2}n, \infty] & \text{if } n \geqslant 3. \end{cases} \tag{1.40}$$

Then for all $\phi \in W^{1,2}(Q^\circ)$,

$$\int_Q |h|\,|\phi|^2 \leqslant \gamma_r |Q|^{-1/r} \|h\|_{r,Q} (|Q|^{2/n} \|\nabla\phi\|_{2,Q}^2 + \|\phi\|_{2,Q}^2), \tag{1.41}$$

where

$$\gamma_r = \sup\{\|\phi\|_{2r/(r-1),I}^2 : \phi \in W^{1,2}(I),\ \|\phi\|_{1,2,I} = 1\} \tag{1.42}$$

and $I = (0, 1)^n$. ■

Proof. By Sobolev's Embedding Theorem, Theorem V.4.13, $W^{1,2}(I)$ is continuously embedded in $L^{2r/(r-1)}(I)$ and so γ_r is finite. By a change of variable we have that

$$\sup\{\|\phi\|_{2r/(r-1),Q}^2 / (|Q|^{2/n} \|\nabla\phi\|_{2,Q}^2 + \|\phi\|_{2,Q}^2) : \phi \in W^{1,2}(Q^\circ),\ \phi \neq 0\}$$
$$= |Q|^{-1/r} \gamma_r.$$

Hence, on using Hölder's inequality, we obtain

$$\begin{aligned} \int_Q |h|\,|\phi|^2 &\leqslant \|h\|_{r,Q} \|\phi\|_{2r/(r-1),Q}^2 \\ &\leqslant \gamma_r |Q|^{-1/r} \|h\|_{r,Q} (|Q|^{2/n} \|\nabla\phi\|_{2,Q}^2 + \|\phi\|_{2,Q}^2) \end{aligned}$$

as asserted in (1.41). □

In the next theorem we retain the notation (1.6) and $\mathfrak{F}$ is a covering of Ω by closed cubes with disjoint interiors, $\mathfrak{F}$ being arbitrary for the Dirichlet problem and such that $\Omega = \cup_{Q \in \mathfrak{F}} Q$ in the Neumann problem, as explained in §1.2. Also, t_D and t_N will stand for the Dirichlet and Neumann forms determined by

$$\int_\Omega (\nabla\phi \cdot \nabla\bar{\psi} + q\phi\bar{\psi})$$

and the domains $\mathfrak{Q}_D$ and $\mathfrak{Q}_N$ respectively, these being the completion of $C_0^\infty(\Omega)$ and $C_0^\infty(\mathbb{R}^n)$ respectively with respect to the norm

$$\left(\int_\Omega (|\nabla\phi|^2 + f^+ |\phi|^2 + |\phi|^2)\right)^{\frac{1}{2}}. \tag{1.43}$$

Theorem 1.15. (a) Let q be a complex-valued measurable function on Ω and $q \in L^r_{\text{loc}}(\bar{\Omega})$ for some r satisfying (1.40). Let $q = f + ig$ and suppose that

$$\text{III(i)}\ \theta_0 := \inf_{Q \in \mathfrak{F}} [f_Q - \gamma_r \rho_r(f, Q)] < \infty,$$

$$\text{III(ii)}\ \ \theta_1 := \sup_{Q \in \mathfrak{F}} \{[|g_Q| + \gamma_r \rho_r(g, Q)][f_Q - \gamma_r \rho_r(f, Q) - \theta_0 + 1]^{-1}\} < \infty,$$

$$\text{III(iii)}\ \beta_0 := \gamma_r \sup_{Q \in \mathfrak{F}} [|Q|^{2/n} \rho_r(f, Q)] < 1,$$

$$\text{III(iv)}\ \beta_1 := \gamma_r \sup_{Q \in \mathfrak{F}} [|Q|^{2/n} \rho_r(g, Q)] < \infty,$$

$$\text{III (v)}\ \sup_{Q \in \mathfrak{F}} [|Q|^{-1/r} \|f^-\|_{r,Q} (1 + |Q|^{2/n})] < \infty.$$

Then t_D is a closed, densely defined sectorial form in $L^2(\Omega)$ and T_D, the associated m-sectorial operator, has the following properties:

$$(T_D u, \phi)_{2,\Omega} = t_D[\![u, \phi]\!] \qquad (u \in \mathscr{D}(T_D),\ \phi \in \mathfrak{Q}_D),$$

$$T_D = \tau \restriction \mathscr{D}(T_D), \qquad \mathscr{D}(T_D) = \{u : u \in \mathfrak{Q}_D,\ \tau u \in L^2(\Omega)\}, \tag{1.44}$$

$$\sigma(T_D) \subset \Theta :=$$

$$\left\{x + iy : x \geqslant \theta_0, |y| \leqslant \begin{cases} \theta_1 (x - \theta_0 + 1) & \text{if } \beta_1 \leqslant \theta_1 (1 - \beta_0) \\ [\beta_1/(1 - \beta_0)](x - \theta_0) + \theta_1 & \text{if } \beta_1 > \theta_1\ (1 - \beta_0) \end{cases} \right\}. \tag{1.45}$$

(b) Let $q \in L^1_{\text{loc}}\ (\bar{\Omega}) \cap L^r_{\text{loc}}(\Omega)$ for some r satisfying (1.40) and suppose that III(i)–(v) are satisfied with $\mathfrak{F}$ a covering of Ω appropriate to the Neumann problem. Then (a) holds for the Neumann operator T_N. ■

Proof. (a) We first observe that, on account of III(v), and since in all cases, $r \geqslant n/2$, we may suppose, by subdividing the cubes Q in $\mathfrak{F}$ if necessary, that $\gamma_r |Q|^{2/n - 1/r} \|f_-\|_{r,Q} < \frac{1}{2}$ for all $Q \in \mathfrak{F}$. It then readily follows as in the proof of Theorem 1.4, but using Lemma 1.14 instead of Lemma 1.1, for some $k > 0$ and with $h_D := \text{re}\ t_D$, that $(h_D + k)^{\frac{1}{2}}[\![\bullet]\!]$ is a norm on $\mathfrak{Q}_D$ which is equivalent to (1.43). The remainder of the proof is similar to that of Theorem 1.4 with the following estimates to determine the numerical range of t_D and hence to obtain (1.45). On reverting to the original covering $\mathfrak{F}$, i.e. before the aforementioned subdivision, we find that for all $\phi \in \mathfrak{Q}_D$, on using Lemma 1.14 and setting f and ϕ zero outside Ω,

$$\begin{aligned} h_{\mathrm{D}}[\![\phi]\!] &= \sum_{Q\in\mathfrak{F}} [\|\nabla\phi\|_{2,Q}^2 + f_Q\|\phi\|_{2,Q}^2 + ((f-f_Q)\phi,\phi)_{2,Q}] \\ &\geqslant \sum_{Q\in\mathfrak{F}} [\|\nabla\phi\|_{2,Q}^2 + f_Q\|\phi\|_{2,Q}^2 - \gamma_r\rho_r(f,Q)(|Q|^{2/n}\|\nabla\phi\|_{2,Q}^2 + \|\phi\|_{2,Q}^2)] \\ &\geqslant (1-\beta_0)\|\nabla\phi\|_{2,\Omega}^2 + \sum_{Q\in\mathfrak{F}} [f_Q - \gamma_r\rho_r(f,Q)]\|\phi\|_{2,Q}^2. \end{aligned} \tag{1.46}$$

Similarly, with $i_{\mathrm{D}} := \operatorname{im} t_{\mathrm{D}}$,

$$\begin{aligned} |i_{\mathrm{D}}[\![\phi]\!]| &= \left|\sum_{Q\in\mathfrak{F}} (g\phi,\phi)_{2,Q}\right| \\ &\leqslant \sum_{Q\in\mathfrak{F}} \{|g_Q|\,\|\phi\|_{2,Q}^2 + |((g-g_Q)\phi,\phi)_{2,Q}|\} \\ &\leqslant \sum_{Q\in\mathfrak{F}} [|g_Q|\,\|\phi\|_{2,Q}^2 + \gamma_r\rho_r(g,Q)(|Q|^{2/n}\|\nabla\phi\|_{2,Q}^2 + \|\phi\|_{2,Q}^2)] \\ &\leqslant \beta_1\|\nabla\phi\|_{2,\Omega}^2 + \theta_1\sum_{Q\in\mathfrak{F}} [f_Q - \gamma_r\rho_r(f,Q)]\|\phi\|_{2,Q}^2 + \theta_1(1-\theta_0)\|\phi\|_{2,\Omega}^2 \end{aligned} \tag{1.47}$$

by III(ii). From (1.46), $h_{\mathrm{D}}[\![\phi]\!] \geqslant \theta_0$ when $\|\phi\|_{2,\Omega} = 1$ and, for any positive number l, we obtain from (1.46) and (1.47),

$$\begin{aligned} &h_{\mathrm{D}}[\![\phi]\!] - l|i_{\mathrm{D}}[\![\phi]\!]| \\ &\quad\geqslant (1-\beta_0-l\beta_1)\|\nabla\phi\|_{2,\Omega}^2 + \sum_{Q\in\mathfrak{F}} (1-l\theta_1)[f_Q - \gamma_r\rho_r(f,Q)]\|\phi\|_{2,Q}^2 - l\theta_1(1-\theta_0). \end{aligned}$$

If $\beta_1 \leqslant \theta_1(1-\beta_0)$ and $\theta_1 > 0$ we choose $l = 1/\theta_1$ to obtain

$$h_{\mathrm{D}}[\![\phi]\!] - \theta_1^{-1}|i_{\mathrm{D}}[\![\phi]\!]| \geqslant -(1-\theta_0),$$

i.e.

$$|i_{\mathrm{D}}[\![\phi]\!]| \leqslant \theta_1(h_{\mathrm{D}}[\![\phi]\!] - \theta_0 + 1).$$

This also holds when $\theta_1 = 0$ since then $g = 0$ and hence $i_{\mathrm{D}}[\![\phi]\!] = 0$. If $\beta_1 > \theta_1(1-\beta_0)$ we choose $l = \beta_1^{-1}(1-\beta_0)$. This gives

$$\begin{aligned} &h_{\mathrm{D}}[\![\phi]\!] - \beta_1^{-1}(1-\beta_0)|i_{\mathrm{D}}[\![\phi]\!]| \\ &\quad\geqslant \sum_{Q\in\mathfrak{F}} [1-\theta_1(1-\beta_0)/\beta_1][f_Q - \gamma_r\rho_r(f,Q)]\|\phi\|_{2,Q}^2 - \theta_1(1-\beta_0)(1-\theta_0)/\beta_1 \\ &\quad\geqslant [1-\theta_1(1-\beta_0)/\beta_1]\theta_0 - \theta_1(1-\beta_0)(1-\theta_0)/\beta_1 \\ &\quad= \theta_0 - \theta_1(1-\beta_0)/\beta_1. \end{aligned}$$

The proof of (1.45) is therefore complete. The proof of (b) follows identical lines. □

1.4. *General second-order elliptic operators in* $L^2(\Omega; w)$

We now take τ to be the general second-order elliptic expression

$$\tau = \frac{1}{w}\left(-\sum_{i,j=1}^{n} D_i(a_{ij}D_j) + \sum_{j=1}^{n} b_j D_j + q\right) \quad \left(D_j := \frac{\partial}{\partial x_j}\right),$$
$$\equiv (1/w)\,(-D_i(a_{ij}D_j) + b_j D_j + q), \tag{1.48}$$

on using the summation convention, and we shall obtain quasi-m-sectorial realizations of τ in a weighted space $L^2(\Omega; w)$. We shall be mainly concerned in this section with a method for determining how the relative behaviour of the a_{ij} and q affects the results of the previous sections. The method we adopt imposes radial symmetry on Ω; we take Ω to be a shell $S\{l, m\}$ (with $0 \leqslant l < m \leqslant \infty$) where

$$S\{l, m\} = \{x : x \in \mathbb{R}^n,\ |x| \in \{l, m\}\}$$

and $\{l, m\}$ stands for one of the intervals $[l, m)$, $[l, m]$, $(l, m]$, (l, m). An end point of $\{l, m\}$ will be included to indicate that Neumann boundary conditions are assumed to hold at the corresponding part of $\partial\Omega$ and the coefficients of τ will retain their local properties in bounded sets which include this part of the boundary. The growth of the a_{ij} will be measured in terms of a positive function a and we introduce the function

$$h_\gamma(r) := \left|\int_\gamma^r \frac{dt}{t^{n-1}a(t)}\right| \quad \text{for} \quad r = |x| \in \{l, m\},\ \gamma \in \{l, m\}. \tag{1.49}$$

This function was used by Kalf and Walter [1] in their work on the essential self-adjointness of singular elliptic operators and in [1] by Kalf for characterizing the Friedrichs extension. In (1.49), $\gamma = l$ or $\gamma = m$ is only allowed if the integral is convergent for this value of γ. Let

$$\rho_\gamma(r) := \{a(r)[r^{n-1}h_\gamma(r)]^2\}^{-1}. \tag{1.50}$$

The role of ρ_γ is made apparent in the following generalization of a well-known inequality of Hardy, the result being a modification of one proved by Kalf and Walter in [1].

Lemma 1.16. Let a, $1/a \in L^\infty_{\text{loc}}(S(l, m))$ and define the spherical mean

$$\phi_u(r) := \left(\int_{|\xi|=1} |u(r\xi)|^2\, d\xi\right)^{\frac{1}{2}},$$

where $d\xi$ denotes the surface element of the unit sphere in $\mathbb{R}^n$. Let

$$D := \{u : u \in C^1(S(l, m)),\ a^{\frac{1}{2}}\nabla u \in L^2(S(l, m))\}$$

and

$$D_\gamma = \begin{cases} D \text{ if } \gamma \in (l, m), \\ \{u : u \in D \text{ and } \lim_{r \to m-} \inf \phi_u(r) = 0\} \quad \text{if} \quad h_m(\bullet) < \infty \text{ and } \gamma = m, \\ \{u : u \in D \text{ and } \lim_{r \to l+} \inf \phi_u(r) = 0\} \quad \text{if} \quad h_l(\bullet) < \infty \text{ and } \gamma = l. \end{cases}$$

Then, for $u \in D_\gamma$ with $\gamma \in \{l, m\}$, we have $\rho_\gamma^{\frac{1}{2}} u \in L^2(S(l, m))$ and

$$\int_{S(l,m)} \rho_\gamma |u|^2 \leqslant 4 \int_{S(l,m)} a |\nabla u|^2. \tag{1.51}$$ ■

Proof. We may assume, without loss of generality, that u is real. Let

$$h_\gamma^\circ(r) := \int_\gamma^r \frac{dt}{t^{n-1} a(t)}$$

so that $h_\gamma(r) = |h_\gamma^\circ(r)|$. For $l < R_1 < R_2 < m$, it follows from (1.49) and (1.50) that

$$\begin{aligned} \int_{R_1}^{R_2} \rho_\gamma(r)\, \phi_u^2(r) r^{n-1}\, dr &= \int_{R_1}^{R_2} \phi_u^2(r) [a(r) h_\gamma^2(r) r^{n-1}]^{-1}\, dr \\ &= -\int_{R_1}^{R_2} \phi_u^2(r) \frac{d}{dr}\left(\frac{1}{h_\gamma^\circ(r)}\right) dr \\ &= \frac{\phi_u^2(R_1)}{h_\gamma^\circ(R_1)} - \frac{\phi_u^2(R_2)}{h_\gamma^\circ(R_2)} + 2\int_{R_1}^{R_2} \phi_u(r)\, \phi_u'(r) [h_\gamma^\circ(r)]^{-1}\, dr. \end{aligned} \tag{1.52}$$

In the case $\gamma \in (l, m)$ we choose R_1 and R_2 such that $\gamma \in (R_1, R_2)$. Then $h_\gamma^\circ(R_1) < 0$ and $h_\gamma^\circ(R_2) > 0$, and so

$$\begin{aligned} \int_{R_1}^{R_2} \rho_\gamma(r)\, \phi_u^2(r) r^{n-1}\, dr &\leqslant 2 \int_{R_1}^{R_2} \rho_\gamma^{\frac{1}{2}}(r)\, a^{\frac{1}{2}}(r) |\phi_u(r)\, \phi_u'(r)| r^{n-1}\, dr \\ &\leqslant 2 \left(\int_{R_1}^{R_2} \rho_\gamma(r)\, \phi_u^2(r) r^{n-1}\, dr \right)^{\frac{1}{2}} \times \\ &\quad \left(\int_{R_1}^{R_2} a(r) |\phi_u'(r)|^2 r^{n-1}\, dr \right)^{\frac{1}{2}}. \end{aligned}$$

Hence

$$\begin{aligned} \int_{S(R_1, R_2)} \rho_\gamma |u|^2 &= \int_{R_1}^{R_2} \rho_\gamma\, \phi_u^2 r^{n-1}\, dr \\ &\leqslant 4 \int_{R_1}^{R_2} a(r) |\phi_u'(r)|^2 r^{n-1}\, dr. \\ &\leqslant 4 \int_{S(R_1, R_2)} a |\nabla u|^2 \end{aligned}$$

since

$$|\phi_u(r)\,\phi_u'(r)| = \left|\int_{|\xi|=1} u(r\xi)\frac{\partial}{\partial r}u(r\xi)\,\mathrm{d}\xi\right|$$
$$\leqslant \left(\int_{|\xi|=1} |u(r\xi)|^2\,\mathrm{d}\xi\right)^{\frac{1}{2}}\left(\int_{|\xi|=1} |\frac{\partial}{\partial r}u(r\xi)|^2\,\mathrm{d}\xi\right)^{\frac{1}{2}}$$
$$\leqslant \phi_u(r)\left(\int_{|\xi|=1} |\nabla u(r\xi)|^2\,\mathrm{d}\xi\right)^{\frac{1}{2}}.$$

The result follows.

Suppose now that $h_m(\bullet) < \infty$ and $\gamma = m$. Then $h_m^\circ = -h_m$ and by (1.52) and (1.50),

$$\int_{R_1}^{R_2} \rho_m(r)\,\phi_u^2(r)\,r^{n-1}\,\mathrm{d}r \leqslant \frac{\phi_u^2(R_2)}{h_m(R_2)} + 2\int_{R_1}^{R_2} \rho_m^{\frac{1}{2}}(r)\,a^{\frac{1}{2}}(r)|\phi_u(r)\,\phi_u'(r)|r^{n-1}\,\mathrm{d}r$$
$$\leqslant \frac{\phi_u^2(R_2)}{h_m(R_2)} + \frac{1}{2}\int_{R_1}^{R_2} \rho_m(r)\,\phi_u^2(r)\,r^{n-1}\,\mathrm{d}r + 2\int_{R_1}^{R_2} a(r)\,|\phi_u'(r)|^2\,r^{n-1}\,\mathrm{d}r.$$

Hence

$$\frac{1}{2}\int_{S(R_1,R_2)} \rho_m|u|^2 = \frac{1}{2}\int_{R_1}^{R_2} \rho_m(r)\,\phi_u^2(r)\,r^{n-1}\,\mathrm{d}r$$
$$\leqslant \frac{\phi_u^2(R_2)}{h_m(R_2)} + 2\int_{R_1}^{R_2} a(r)\,|\phi_u'(r)\,|^2\,r^{n-1}\,\mathrm{d}r. \quad (1.53)$$

Also,

$$|\phi_u(r) - \phi_u(r')|^2 \leqslant \left(\int_r^{r'} |\phi_u'(t)|\,\mathrm{d}t\right)^2$$
$$\leqslant \left|\int_r^{r'} \frac{\mathrm{d}t}{t^{n-1}a(t)}\right|\left(\int_r^{r'} a(t)\,|\phi_u'(t)|^2\,t^{n-1}\,\mathrm{d}t\right)$$
$$\leqslant \left|\int_r^{r'} \frac{\mathrm{d}t}{t^{n-1}a(t)}\right|\int_{S(r,r')} a|\nabla u|^2.$$

On allowing $r' \to m-$ through the sequence for which $\phi_u(r') \to 0$, we obtain

$$|\phi_u(r)|^2 \leqslant h_m(r)\int_{S(r,m)} a|\nabla u|^2$$

and consequently

$$\lim_{r\to m-}\,[\phi_u^2(r)/h_m(r)] = 0.$$

It follows from (1.53) that $\rho_m^{\frac{1}{2}}u \in \mathrm{L}^2(S(l,m))$ and (1.51) is satisfied. The case $\gamma = l$ is similar. □

When $a(r) = 1$, $n > 2$, $l = 0$, $m = \infty$ and $\gamma = \infty$ we have

$$h_\infty(r) = (n-2)^{-1} r^{2-n}, \qquad \rho_\infty(r) = (n-2)^2/r^2,$$

and (1.51) becomes, for all $u \in C_0^1(\mathbb{R}^n)$,

$$\int_{\mathbb{R}^n} |x|^{-2} |u(x)|^2 \, dx \leqslant \frac{4}{(n-2)^2} \int_{\mathbb{R}^n} |\nabla u|^2 \tag{1.54}$$

since $\liminf_{r \to \infty} \phi_u(r) = 0$. The case $n = 3$ of (1.54) is the Hardy inequality referred to above. The constant $4/(n-2)^2$ in (1.54) is known to be best possible, with equality if, and only if, $u = 0$ (see Shortley [1] and Kalf, Schmincke, Walter and Wüst [1, Lemma 1]). Note that since the set $C_0^1(\mathbb{R}^n)$ is dense in the Sobolev space $W^{1,2}(\mathbb{R}^n)$, (1.54) holds for all $u \in W^{1,2}(\mathbb{R}^n)$.

In the assumptions made below on the coefficients of τ, the choice of Ω will be determined by the permitted values of γ in (1.49) and also by our wish to have Lemma 1.16 at our disposal for all $u \in C_0^\infty(\Omega)$. If Ω is a half-closed shell, $\Omega = S[l, m)$ say, then any $u \in C_0^\infty(\Omega)$ is understood to be the restriction to Ω of a function in $C_0^\infty(\mathbb{R}^n)$ with support in $B(0, m)$ and hence it need not vanish on $|x| = l$. Also, any $u \in L^1_{\text{loc}}(\Omega)$ is integrable on compact subsets of Ω and such a set may include $|x| = l$ when $\Omega = S[l, m)$. The same remark applies to other function spaces defined by local properties.

The choice of Ω is as follows:

$$\left.\begin{array}{l} \text{(i) if } \gamma \in (l, m) \text{ then } \Omega = S\{l, m\} \text{ is arbitrary;} \\ \text{(ii) if } \gamma = l \text{ then } \Omega = S(l, m] \text{ or } S(l, m); \\ \text{(iii) if } \gamma = m \text{ then } \Omega = S[l, m) \text{ or } S(l, m). \end{array}\right\} \tag{1.55}$$

If, for instance, $\gamma = l$ and hence $\Omega = S(l, m\}$ according to (1.55) (ii), then any $u \in C_0^\infty(\Omega)$ vanishes in a neighbourhood of $|x| = l$ and since $a^{\frac{1}{2}} |\nabla u| \in L^2(\Omega)$ whenever $a \in L^1_{\text{loc}}(\Omega)$ and $u \in C_0^\infty(\Omega)$, Lemma 1.16 is valid for all $u \in C_0^\infty(\Omega)$.

The assumptions to be made on the coefficients of τ in (1.48) can now be stated. They are:

IV (i) $w(x) > 0$ a.e. in Ω, $w \in L^1_{\text{loc}}(\Omega)$ and $1/w \in L^\infty_{\text{loc}}(\Omega)$;

IV (ii) the matrix $[a_{ij}(x)]$ is Hermitian and positive definite for a.e. $x \in \Omega$ and $a_{ij} \in L^\infty_{\text{loc}}(\Omega)$ $(i, j = 1, 2, \ldots, n)$;

IV (iii) if $a_+(x)$ and $a_-(x)$ denote the largest and smallest eigenvalues of $[a_{ij}(x)]$ and

$$a_1(r) := \operatorname*{ess\,inf}_{|x| = r} a_-(x), \qquad a_2(r) := \operatorname*{ess\,sup}_{|x| = r} a_+(x)$$

then there exist a positive function a with a, $1/a \in L^\infty_{\text{loc}}(\Omega)$ and positive constants A_1 and A_2 such that

$$a_1(r) \geqslant A_1 a(r), \quad a_2(r) \leqslant A_2 a_1(r), \quad \text{for a.e. } r \in \{l, m\};$$

IV (iv) $b_j = b_j^{(1)} + i b_j^{(2)}$ $(j = 1, 2, \ldots, n)$ is measurable on Ω and there exist positive constants B_1 and B_2 such that for $k = 1,2$,

$$\sum_{j=1}^{n} |b_j^{(k)}(x)|^2 \leqslant 4B_k^2\, a(|x|)\, w(x) \quad \text{for a.e. } x \in \Omega;$$

IV(v) $q = q_0 + q_1$; $\quad q_0, q_1 \in \mathrm{L}^1_{\mathrm{loc}}(\Omega)$;
IV(vi) there exist constants $\beta \in (-\frac{1}{2}\pi, \frac{1}{2}\pi)$, $\delta \in \mathbb{C}$, $M \geqslant 0$, $\gamma \in \{l, m\}$ such that $\mathrm{e}^{\mathrm{i}\beta}(q_0 - \delta\rho_\gamma) =: f_0 + \mathrm{i}g_0$ satisfies $f_0(x) \geqslant 0$, $|g_0(x)| \leqslant M f_0(x)$, a.e. on Ω, i.e.

$$\mathrm{e}^{\mathrm{i}\beta} q_0(x) - \mathrm{e}^{\mathrm{i}\beta}\delta\rho_\gamma(|x|) \in \mathscr{S}(0; \tan^{-1} M) \quad \text{a.e. on } \Omega;$$

IV(vii) $\mathrm{e}^{\mathrm{i}\beta}\delta =: \mu + \mathrm{i}\nu$ satisfies $A_1 \cos\beta - 4\mu^- > 0$ (with $\mu^- = -\min\{\mu, 0\}$);
IV(viii) there exist constants $m(f)$, $p(g)$, $\alpha(f)$, $\alpha(g)$, depending on Ω and where $f + \mathrm{i}g := \mathrm{e}^{\mathrm{i}\beta} q_1$, such that for all $\phi \in \mathrm{C}_0^\infty(\Omega)$,

$$\int_\Omega f|\phi|^2 \geqslant m(f)\|w^{\frac{1}{2}}\phi\|_{2,\Omega}^2 - 2\alpha(f)\|a^{\frac{1}{2}}\nabla\phi\|_{2,\Omega}\|w^{\frac{1}{2}}\phi\|_{2,\Omega},$$

$$\left|\int_\Omega g|\phi|^2\right| \leqslant p(g)\|w^{\frac{1}{2}}\phi\|_{2,\Omega}^2 + 2\alpha(g)\|a^{\frac{1}{2}}\nabla\phi\|_{2,\Omega}\|w^{\frac{1}{2}}\phi\|_{2,\Omega},$$

and for any $\varepsilon > 0$ and any compact subset Ω_0 of Ω there exist positive constants K_ε and K_{Ω_0} such that

$$\int_\Omega f^-|\phi|^2 \leqslant \varepsilon\|a^{\frac{1}{2}}\nabla\phi\|_{2,\Omega}^2 + K_\varepsilon\|w^{\frac{1}{2}}\phi\|_{2,\Omega}^2,$$

$$\int_{\Omega_0} |g||\phi|^2 \leqslant K_{\Omega_0}(\|a^{\frac{1}{2}}\nabla\phi\|_{2,\Omega}^2 + \|w^{\frac{1}{2}}\phi\|_{2,\Omega}^2).$$

The inequalities in IV(viii) are closely modelled on those in Lemma 1.2. Examples will be given in §1.5 below. We are now in a position to proceed in a similar way to §§1.1 and 1.2. Let

$$t[\![\phi, \psi]\!] = \int_\Omega [a_{ij}\mathrm{D}_j\phi\mathrm{D}_i\bar{\psi} + (b\cdot\nabla\phi)\bar{\psi} + q\phi\bar{\psi}] \qquad (\phi, \psi \in \mathfrak{Q}), \quad (1.56)$$

where the summation convention is understood and where $\mathfrak{Q}$ is the completion of $\mathrm{C}_0^\infty(\Omega)$ with respect to the norm $\|\bullet\|_{\mathfrak{Q}}$ defined by

$$\|\phi\|_{\mathfrak{Q}}^2 = \int_\Omega [a_1|\nabla\phi|^2 + (f_0 + f^+)|\phi|^2 + w|\phi|^2]. \qquad (1.57)$$

Definition 1.17. We define $\mathrm{D}_i(a_{ij}\mathrm{D}_j u)$ for $u \in \mathfrak{Q}$ in the following weak sense: $v = \mathrm{D}_i(a_{ij}\mathrm{D}_j u)$ if $v \in \mathrm{L}^1_{\mathrm{loc}}(\Omega)$ and for all $\phi \in \mathrm{C}_0^\infty(\Omega)$,

$$\int_\Omega a_{ij}\mathrm{D}_j u\mathrm{D}_i\phi = -\int_\Omega v\phi. \qquad \blacksquare$$

This definition implies the Neumann condition $v_i a_{ij} D_j u = 0$ $(v_i = x_i/|x_i|)$ in a generalized sense at $|x| = l$ when $\Omega = S[l, m\}$ and at $|x| = m$ when $\Omega = S\{l, m]$.

Theorem 1.18. Let Ω be determined by (1.55) and let the conditions IV(i)–(viii) be satisfied by the coefficients of τ in (1.48). Then $e^{i\beta}t$ is a closed, densely defined sectorial form in $L^2(\Omega; w)$. If $e^{i\beta}T$ denotes the associated m-sectorial operator, T is quasi-m-sectorial in $L^2(\Omega; w)$ and has the following properties:

$$(wTu, \phi)_{2,\Omega} = t[\![u, \phi]\!] \qquad (u \in \mathscr{D}(T), \phi \in \mathfrak{Q}),$$
$$\mathscr{D}(T) = \{u : u \in \mathfrak{Q},\ \tau u \in L^2(\Omega; w)\},$$
$$Tu = \tau u \qquad (u \in \mathscr{D}(T)).$$

■

Proof. We shall merely indicate the changes which need to be made to the proof of Theorem 1.4 as the underlying method is identical. Let $(\bullet, \bullet)$ and $\|\bullet\|$ denote the $L^2(\Omega)$ inner-product and norm. Denoting the real and imaginary parts of $e^{i\beta}t$ by h_1 and h_2 respectively, we have for all $\phi \in C_0^\infty(\Omega)$,

$$h_1[\![\phi]\!] = (\cos\beta)\int_\Omega a_{ij} D_j\phi D_i\bar{\phi} + \operatorname{re}\left(e^{i\beta}\int_\Omega (b\cdot\nabla\phi)\bar{\phi}\right)$$
$$+ \mu\int_\Omega \rho_\gamma|\phi|^2 + \int_\Omega (f_0 + f)|\phi|^2, \tag{1.58}$$

$$h_2[\![\phi]\!] = (\sin\beta)\int_\Omega a_{ij} D_j\phi D_i\bar{\phi} + \operatorname{im}\left(e^{i\beta}\int_\Omega (b\cdot\nabla\phi)\bar{\phi}\right)$$
$$+ v\int_\Omega \rho_\gamma|\phi|^2 + \int_\Omega (g_0 + g)|\phi|^2. \tag{1.59}$$

Hence on using IV and Lemma 1.16, we have, for any $\varepsilon_1, \varepsilon_2 > 0$,

$$h_1[\![\phi]\!] \geqslant (\cos\beta)\|a_1^{\frac{1}{2}}\nabla\phi\|^2 - 2(B_1 + B_2)\|a^{\frac{1}{2}}\nabla\phi\|\,\|w^{\frac{1}{2}}\phi\| + \mu^+\|\rho_\gamma^{\frac{1}{2}}\phi\|^2$$
$$- 4\mu^-\|a^{\frac{1}{2}}\nabla\phi\|^2 + \|f_0^{\frac{1}{2}}\phi\|^2 + \|(f^+)^{\frac{1}{2}}\phi\|^2 - \|(f^-)^{\frac{1}{2}}\phi\|^2 \tag{1.60}$$
$$\geqslant (\cos\beta - 4\mu^-/A_1)\|a_1^{\frac{1}{2}}\nabla\phi\|^2 - (\varepsilon_1\|a^{\frac{1}{2}}\nabla\phi\|^2 + K_{\varepsilon_1}\|w^{\frac{1}{2}}\phi\|^2) + \|f_0^{\frac{1}{2}}\phi\|^2$$
$$+ \|(f^+)^{\frac{1}{2}}\phi\|^2 - (\varepsilon_2\|a^{\frac{1}{2}}\nabla\phi\|^2 + K_{\varepsilon_2}\|w^{\frac{1}{2}}\phi\|^2)$$
$$\geqslant (\cos\beta - 4\mu^-/A_1 - \varepsilon)\|a_1^{\frac{1}{2}}\nabla\phi\|^2 + \|f_0^{\frac{1}{2}}\phi\|^2 + \|(f^+)^{\frac{1}{2}}\phi\|^2 - K_\varepsilon\|w^{\frac{1}{2}}\phi\|^2$$

for any $\varepsilon > 0$. On choosing $\varepsilon < \cos\beta - 4\mu^-/A_1$ (see IV(vii)) we see that there exist constants K_0 and K_1 such that for all $\phi \in C_0^\infty(\Omega)$,

$$(h_1 + K_0 + 1)[\![\phi]\!] \geqslant K_1\|\phi\|_{\mathfrak{Q}}^2, \tag{1.61}$$

where $1[\![\phi]\!] = \|w^{\frac{1}{2}}\phi\|^2$. Similarly, there exists a positive constant K_2 such that for all $\phi \in C_0^\infty(\Omega)$,

$$(h_1 + K_0 + 1)\,[\![\phi]\!] \leqslant K_2\,\|\phi\|_{\mathfrak{Q}}^2. \tag{1.62}$$

Consequently $(h + K_0 + 1)^{\frac{1}{2}}[\![\bullet]\!]$ is a norm on $C_0^\infty(\Omega)$ which is equivalent to $\|\bullet\|_{\mathfrak{Q}}$. We also see from IV and Lemma 1.16 that

$$\begin{aligned} |h_2[\![\phi]\!]| &\leqslant A_2|\sin\beta|\,\|a_1^{\frac{1}{2}}\nabla\phi\|^2 + 2(B_1 + B_2)\,\|a^{\frac{1}{2}}\nabla\phi\|\,\|w^{\frac{1}{2}}\phi\| \\ &\quad + 4|v|\,\|a^{\frac{1}{2}}\nabla\phi\|^2 + M\,\|f_0^{\frac{1}{2}}\phi\|^2 + \left|\int_\Omega g|\phi|^2\right| \\ &\leqslant K\|\phi\|_{\mathfrak{Q}}^2 \\ &\leqslant K(h_1 + K_0 + 1)\,[\![\phi]\!] \end{aligned} \tag{1.63}$$

in view of (1.61). It follows that $e^{i\beta}t$ is sectorial and closed. The rest follows as in the proof of Theorem 1.4. □

Theorem 1.19. Let IV be satisfied with Ω determined by (1.55) and define

$$\left.\begin{aligned} \rho &= \inf_\Omega[\rho_\gamma(|x|)/w(x)], \\ j &= A_1\cos\beta - 4\mu^- > 0, \\ k &= A_1A_2\,|\sin\beta| + 4|v|, \\ F &= B_1 + B_2 + \alpha(f), \\ G &= B_1 + B_2 + \alpha(|g|), \end{aligned}\right\} \tag{1.64}$$

$$\theta = \begin{cases} \tfrac{1}{4}\rho A_1\cos\beta + \mu\rho + m(f) - \rho^{\frac{1}{2}}F & \text{if } \rho > 0 \text{ and } F \leqslant \tfrac{1}{2}\rho^{\frac{1}{2}}j, \\ \mu^+\rho + m(f) - F^2/j \text{ otherwise.} \end{cases} \tag{1.66}$$

Then the spectrum of the operator T in Theorem 1.18 lies in the set $e^{-i\beta}\Theta_\beta$ where Θ_β is the set of all $x + iy$ such that $x \geqslant \theta$ and

$|y| - p(g) \leqslant$

$$\begin{cases} kj^{-2}(z+F)^2 + 2Gj^{-1}(z+F) \text{ if } M \leqslant k/j \text{ or if } M > k/j \text{ and } z \leqslant \dfrac{jG + kF}{jM - k}, \\ M[x - \mu^+\rho - m(f)] + (MF + G)^2/(jM - k) \quad \text{otherwise}, \end{cases}$$

and

$$z = j^{\frac{1}{2}}[x - \mu^+\rho - m(f) + F^2/j]^{\frac{1}{2}}. \qquad ■$$

Proof. By (1.60), IV(viii) and (1.64), we have for all $\phi \in C_0^\infty(\Omega)$,

$$\begin{aligned} h_1[\![\phi]\!] &\geqslant (\cos\beta)\,\|a_1^{\frac{1}{2}}\nabla\phi\|^2 - 2(B_1+B_2)\,\|a^{\frac{1}{2}}\nabla\phi\|\,\|w^{\frac{1}{2}}\phi\| + \mu^+\rho\|w^{\frac{1}{2}}\phi\|^2 \\ &\quad - 4\mu^-\,\|a^{\frac{1}{2}}\nabla\phi\|^2 + \|f_0^{\frac{1}{2}}\phi\|^2 + m(f)\|w^{\frac{1}{2}}\phi\|^2 - 2\alpha(f)\|\,a^{\frac{1}{2}}\nabla\phi\|\,\|w^{\frac{1}{2}}\phi\| \\ &\geqslant (\cos\beta - 4\mu^-/A_1)\,\|a_1^{\frac{1}{2}}\nabla\phi\|^2 + \|f_0^{\frac{1}{2}}\phi\|^2 + [\mu^+\rho + m(f)]\,\|w^{\frac{1}{2}}\phi\|^2 \\ &\quad - 2F\,\|a^{\frac{1}{2}}\nabla\phi\|\,\|w^{\frac{1}{2}}\phi\| \end{aligned}$$

(on using IV(iii))

$$\geqslant A_1^{-1}(j-\varepsilon)\,\|a_1^{\frac{1}{2}}\nabla\phi\|^2 + \|f_0^{\frac{1}{2}}\phi\|^2 + [\mu^+\rho + m(f) - \varepsilon_1^{-1}F^2]\,\|w^{\frac{1}{2}}\phi\|^2$$

for any $\varepsilon_1 > 0$. Similarly, by (1.63),

$$|h_2[\![\phi]\!]| \leqslant A_1^{-1}(k+\varepsilon_2)\,\|a_1^{\frac{1}{2}}\nabla\phi\|^2 + M\|f_0^{\frac{1}{2}}\phi\|^2 + [p(g) + G^2\varepsilon_2^{-1}]\|w^{\frac{1}{2}}\phi\|^2$$

for any $\varepsilon_2 > 0$. Hereafter the proof follows that of Theorem 1.5. Note that since $C_0^\infty(\Omega)$ is a core of t it suffices to work with functions $\phi \in C_0^\infty(\Omega)$ in the above proof although continuity arguments show that the estimates do in fact hold for all $\phi \in \mathfrak{Q}$. □

As in Theorem 1.6 we obtain the following improvement for the Dirichlet problem.

Theorem 1.20. Let the hypothesis and terminology of Theorem 1.19 hold with $\Omega = S(l,m)$ and suppose also that each $b_j \in AC_{\text{loc}}(\Omega)$ and div $b = D_j b_j = 0$. Then Theorem 1.19 holds with F, G in (1.64) replaced by

$$\left.\begin{aligned} F &= B_1|\sin\beta| + B_2\cos\beta + \alpha(f), \\ G &= B_1\cos\beta + B_2|\sin\beta| + \alpha(g). \end{aligned}\right\} \qquad (1.67)$$ ■

1.5. *Examples*

We begin with an analogue of Lemma 1.2 in which inequalities of the type in IV(viii) are obtained, featuring a possibly singular function a and a weight function w.

Let $R \in \{l, m\}$ with the understanding that $R = l$ if $\Omega = S[l,m)$ and $R = m$ if $\Omega = S(l,m]$. Let $(l_j^\pm)$ be partitions of $[R,m\}$ and $\{l,R]$ respectively with

$$R \leqslant l_1^+ < l_2^+ < \ldots \leqslant m, \qquad R \geqslant l_1^- > l_2^- > \ldots \geqslant l.$$

Set $I_j^+ = (l_j^+, l_{j+1}^+)$ and $I_j^- = (l_{j+1}^-, l_j^-)$, and define the following, with $x = r\xi$ and $|\xi| = 1$, i.e. with (r,ξ) denoting polar coordinates:

$$(M_j^\pm f)(\xi) := \int_{I_j^\pm} f(r\xi)r^{n-1}\,\mathrm{d}r \bigg/ \int_{I_j^\pm} w(r\xi)r^{n-1}\,\mathrm{d}r, \qquad (1.68)$$

$$(N_j^\pm f)(x) := f(x) - (M_j^\pm f)(\xi)\,w(x), \qquad x = r\xi, \qquad r \in I_j^\pm, \qquad (1.69)$$

$$\alpha_j^{\pm}(f) := \sup_{\substack{|\xi|=1 \\ r \in I_j^{\pm}}} \left\{ \left| \int_{I_j^{\pm}}^{r} (N_j^{\pm} f)(t\xi) t^{n-1}\,\mathrm{d}t \right| h_\gamma(r) \rho_\gamma^{\frac{1}{2}}(r) \omega^{-\frac{1}{2}}(r\xi) \right\}, \tag{1.70}$$

$$\alpha^{\pm}(f) := \sup_{j \geqslant 1} \alpha_j^{\pm}(f). \tag{1.71}$$

Lemma 1.21. Let $\Omega = S\{l, m\}$ and let f and g be real-valued $\mathrm{L}^1_{\mathrm{loc}}(\Omega)$ functions satisfying

$$m^{\pm}(f) := \inf_{\substack{|\xi|=1 \\ j \geqslant 1}} (M_j^{\pm} f)(\xi) > -\infty, \qquad p^{\pm}(g) := \sup_{\substack{|\xi|=1 \\ j \geqslant 1}} |(M_j^{\pm} g)(\xi)| < \infty,$$

$$\alpha^{\pm}(f),\ \alpha^{\pm}(g) < \infty.$$

Set $m(f) = \min\{m^+(f), m^-(f)\}$, $p(g) = \max\{p^+(g), p^-(g)\}$, $\alpha(f) = \max\{\alpha^+(f), \alpha^-(f)\}$ and $\alpha(g) = \max\{\alpha^+(g), \alpha^-(g)\}$. Then for all $\phi \in \mathrm{C}_0^\infty(\Omega)$,

$$\int_\Omega f|\phi|^2 \geqslant m(f)\,\|w^{\frac{1}{2}}\phi\|^2_{2,\Omega} - 2\alpha(f)\,\|a^{\frac{1}{2}}\nabla\phi\|_{2,\Omega}\|w^{\frac{1}{2}}\phi\|_{2,\Omega},$$

$$\left| \int_\Omega g|\phi|^2 \right| \leqslant p(g)\,\|w^{\frac{1}{2}}\phi\|^2_{2,\Omega} + 2\alpha(g)\,\|a^{\frac{1}{2}}\nabla\phi\|_{2,\Omega}\,\|w^{\frac{1}{2}}\phi\|_{2,\Omega}. \qquad \blacksquare$$

Proof. From (1.68) and (1.69) we see that

$$\int_{I_j^{\pm}} (N_j^{\pm} f)(r\xi) r^{n-1}\,\mathrm{d}r = 0 \qquad (j \geqslant 1).$$

Hence, on integrating by parts,

$$\left| \int_{I_j^{\pm}} (N_j^{\pm} f)(r\xi) |\phi(r\xi)|^2\, r^{n-1}\,\mathrm{d}r \right|$$

$$= \left| -2\mathrm{re} \int_{I_j^{\pm}} \left(\int_{I_j^{\pm}}^{r} (N_j^{\pm} f)(t\xi) t^{n-1}\,\mathrm{d}t \right) \bar{\phi}(r\xi) \left(\frac{\partial}{\partial r} \right) \phi(r\xi)\,\mathrm{d}r \right|$$

$$\leqslant 2 \int_{I_j^{\pm}} \alpha_j^{\pm}(f)\, a^{\frac{1}{2}}(r\xi)\, |\nabla\phi(r\xi)|\, w^{\frac{1}{2}}(r\xi)\, |\phi(r\xi)|\, r^{n-1}\,\mathrm{d}r$$

from (1.70) and (1.50). On applying the Cauchy-Schwarz inequality we obtain

$$\int_{S(I_j^{\pm})} (N_j^{\pm} f)(x) |\phi(x)|^2\,\mathrm{d}x$$

$$\leqslant 2 \int_{|\xi|=1} \int_{I_j^{\pm}} \alpha_j^{\pm}(f)\, a^{\frac{1}{2}}(r\xi) |\nabla\phi(r\xi)|\, w^{\frac{1}{2}}(r\xi)\, |\phi(r\xi)|\, r^{n-1}\,\mathrm{d}r\,\mathrm{d}\xi$$

$$\leqslant 2\alpha_j^{\pm}(f)\,\|a^{\frac{1}{2}}\nabla\phi\|_{2,S(I_j^{\pm})}\|w^{\frac{1}{2}}\phi\|_{2,S(I_j^{\pm})}.$$

From (1.69) we have

$$\int_{\Omega} f|\phi|^2$$

$$= \sum_{j\geq 1}\left(\int_{|\xi|=1}(M_j^+ f)(\xi)\int_{I_j^+}|\phi(r\xi)|^2\, w(r\xi)r^{n-1}\,dr\,d\xi + \int_{|\xi|=1}\int_{I_j^+} N_j^+ f|\phi|^2\right)$$

$$+ \sum_{j\geq 1}\left(\int_{|\xi|=1}(M_j^- f)(\xi)\int_{I_j^-}|\phi(r\xi)|^2\, w(r\xi)r^{n-1}\,dr\,d\xi\right.$$

$$\left. + \int_{|\xi|=1}\int_{I_j^-}(N_j^- f)|\phi|^2\right)$$

$$\geq m^+(f)\,\|w^{\frac{1}{2}}\phi\|^2_{2,S[R,m\}} - 2\alpha^+(f)\,\|a^{\frac{1}{2}}\nabla\phi\|_{2,S[R,m\}}\|w^{\frac{1}{2}}\phi\|_{2,S[R,m\}}$$

$$+ m^-(f)\|w^{\frac{1}{2}}\phi\|^2_{2,S\{l,R]} - 2\alpha^-(f)\,\|a^{\frac{1}{2}}\nabla\phi\|_{2,S\{l,R]}\,\|w^{\frac{1}{2}}\phi\|_{2,S\{l,R]}$$

$$\geq m(f)\,\|w^{\frac{1}{2}}\phi\|^2_{2,\Omega} - 2\alpha(f)\,\|a^{\frac{1}{2}}\nabla\phi\|_{2,\Omega}\,\|w^{\frac{1}{2}}\phi\|_{2,\Omega}.$$

The inequality for $|\int_{\Omega} g|\phi|^2|$ is proved in the same way. □

The remaining inequalities in IV (viii), that is, those concerning f^- and $|g|$, can also be obtained by the method of Lemma 1.21. For instance, if $p^{\pm}(f^-) < \infty$ and $\alpha^{\pm}(f^-) < \infty$ then we have from Lemma 1.21 that given any $\varepsilon > 0$,

$$\int_{\Omega} f^-|\phi|^2 \leq p(f^-)\|w^{\frac{1}{2}}\phi\|^2_{2,\Omega} + 2\alpha(f^-)\|a^{\frac{1}{2}}\nabla\phi\|_{2,\Omega}\|w^{\frac{1}{2}}\phi\|_{2,\Omega}$$

$$\leq \varepsilon\|a^{\frac{1}{2}}\nabla\phi\|_{2,\Omega} + [p(f^-) + \varepsilon^{-1}\alpha(f^-)]\,\|w^{\frac{1}{2}}\phi\|^2_{2,\Omega}.$$

The inequality for $\int_{\Omega_0}|g|\,|\phi|^2$ in IV(viii) is satisfied for any compact subset Ω_0 of Ω if $\sup_{|\xi|=1}(M_j^{\pm}|g|)(\xi)$ and $\alpha_j^{\pm}(|g|)$ are finite for each value of j. To illustrate Lemma 1.21 we have

Example 1.22. Let $l = 0, m = \infty$, $a(r) = r^{2\alpha}$, $w(x) = r^{2\omega}$ and $|x| = r$, and suppose that for some $\delta \leq 1$ and $\gamma \geq 1$, for any $\varepsilon \in (0,1)$,

$$I_\varepsilon^+(f) := \inf_{\substack{r\in[1,\infty)\\|\xi|=1}}\left(\varepsilon^{-1}r^{-\delta-2\omega-n+1}\int_r^{r+\varepsilon r^\delta} f(t\xi)t^{n-1}\,dt\right) > -\infty,$$

$$I_\varepsilon^-(f) := \inf_{\substack{r\in(0,1]\\|\xi|=1}}\left(\varepsilon^{-1}r^{-\gamma-2\omega-n+1}\int_{r-\varepsilon r^\gamma}^{r} f(t\xi)t^{n-1}\,dt\right) > -\infty,$$

$$J_\varepsilon^+(f) := \sup_{\substack{r\in[1,\infty)\\|\xi|=1}}\left(\varepsilon^{-1}r^{-\alpha-\omega}\int_r^{r+\varepsilon r^\delta}|f(t\xi)|\,dt\right) < \infty,$$

$$J_\varepsilon^-(f) := \sup_{\substack{r\in(0,1]\\|\xi|=1}} \left(\varepsilon^{-1} r^{-\alpha-\omega} \int_{r-\varepsilon r^\gamma}^{r} |f(t\xi)|\,\mathrm{d}t \right) < \infty,$$

$$K_\varepsilon^+(g) := \sup_{\substack{r\in[1,\infty)\\|\xi|=1}} \left(\varepsilon^{-1} r^{-\delta-2\omega-n+1} \int_{r}^{r+\varepsilon r^\delta} |g(t\xi)|\, t^{n-1}\,\mathrm{d}t \right) < \infty,$$

$$K_\varepsilon^-(g) := \sup_{\substack{r\in(0,1]\\|\xi|=1}} \left(\varepsilon^{-1} r^{-\gamma-2\omega-n+1} \int_{r-\varepsilon r^\gamma}^{r} |g(t\xi)|\, t^{n-1}\,\mathrm{d}t \right) < \infty.$$

Then by Lemma 1.21, as $\varepsilon \to 0+$,

$$m^\pm(f) \geqslant I_\varepsilon^\pm(f)\,[1+O(\varepsilon)],$$
$$p^\pm(g) \leqslant K_\varepsilon^\pm(g)\,[1+O(\varepsilon)],$$
$$\alpha^\pm(f) \leqslant \varepsilon J_\varepsilon^\pm(f)\,[1+O(\varepsilon)]. \qquad \blacksquare$$

Proof. In Lemma 1.21 we put $R = 1$ and choose partitions $(l_j^\pm)$ of $[1,\infty)$ and $(0,1]$ to satisfy

$$l_{j+1}^+ = l_j^+ + \varepsilon(l_j^+)^\delta, \quad \delta \leqslant 1, \qquad l_{j+1}^- = l_j^- - \varepsilon(l_j^-)^\gamma, \quad \gamma \geqslant 1.$$

The sequences (l_j^+) and (l_j^-) are increasing and decreasing respectively and $l_j^+ \to \infty$ and $l_j^- \to 0$. Also, any $r \in [l_j^+, l_{j+1}^+]$ satisfies $r = l_j^+[1+O(\varepsilon)]$ uniformly in j and ε, while any $r \in [l_{j+1}^-, l_j^-]$ satisfies $r = l_j^-[1+O(\varepsilon)]$.

For $1 \leqslant r < \infty$ and $2\omega + n \neq 0$,

$$\int_r^{r+\varepsilon r^\delta} w(t)t^{n-1}\,\mathrm{d}t = (2\omega+n)^{-1}\,[(r+\varepsilon r^\delta)^{2\omega+n} - r^{2\omega+n}]$$
$$= \varepsilon r^{2\omega+n+\delta-1}[1+O(\varepsilon)],$$

while, if $2\omega + n = 0$,

$$\int_r^{r+\varepsilon r^\delta} w(t)t^{n-1}\,\mathrm{d}t = \log(1+\varepsilon r^{\delta-1})$$
$$= \varepsilon r^{\delta-1}[1+O(\varepsilon)].$$

Hence, in (1.68),

$$(M_j^+ f)(\xi) = \varepsilon^{-1}(l_j^+)^{-2\omega-n-\delta+1}\left(\int_{l_j^+}^{l_j^+ + \varepsilon(l_j^+)^\delta} f(t\xi)t^{n-1}\,\mathrm{d}t\right)[1+O(\varepsilon)]$$

and this gives

$$m^+(f) \geqslant I_\varepsilon^+(f)\,[1+O(\varepsilon)]$$

and

$$p^+(g) \leqslant K_\varepsilon^+(g)\,[1+O(\varepsilon)].$$

Since $\int_{l_j^+}^{l_{j+1}^+} (N_j^+ f)(t\xi) t^{n-1}\,\mathrm{d}t = 0$, we have

$$\int_{l_j^+}^{l_{j+1}^+} (N_j^+ f)^+ (t\xi) t^{n-1}\,\mathrm{d}t = \int_{l_j^+}^{l_{j+1}^+} (N_j^+ f)^- (t\xi) t^{n-1}\,\mathrm{d}t$$

and consequently, for any $r \in [l_j^+, l_{j+1}^+]$,

$$\begin{aligned}\left|\int_{l_j^+}^{r} (N_j^+ f)(t\xi) t^{n-1}\,\mathrm{d}t\right| &\leqslant \int_{l_j^+}^{l_{j+1}^+} (N_j^+ f)^+ (t\xi) t^{n-1}\,\mathrm{d}t \\ &= \tfrac{1}{2}\int_{l_j^+}^{l_{j+1}^+} |(N_j^+ f)(t\xi)|\, t^{n-1}\,\mathrm{d}t \\ &\leqslant \tfrac{1}{2}\left(\int_{l_j^+}^{l_{j+1}^+} |f(t\xi)| t^{n-1}\,\mathrm{d}t + \int_{l_j^+}^{l_{j+1}^+} |(M_j^+ f)(\xi)|\, w(t\xi) t^{n-1}\,\mathrm{d}t\right) \\ &\leqslant \int_{l_j^+}^{l_{j+1}^+} |f(t\xi)| t^{n-1}\,\mathrm{d}t\end{aligned}$$

by (1.68). Since any $r \in [l_j^+, l_{j+1}^+]$ satisfies $r = l_j^+[1 + O(\varepsilon)]$ we obtain from (1.70) and (1.50) that

$$\begin{aligned}\alpha^+(f) &\leqslant \sup_{\substack{|\xi|=1\\ r\in[l_j, l_{j+1}^+], j\geqslant 1}} \left(r^{-\alpha-\omega-n+1}\int_{l_j^+}^{l_{j+1}^+} |f(t\xi)|\, t^{n-1}\,\mathrm{d}t\right) \\ &\leqslant \sup_{\substack{|\xi|=1\\ r\in[1,\infty)}} \left(r^{-\alpha-\omega}\int_r^{r+\varepsilon r^{\delta}} |f(t\xi)|\,\mathrm{d}t\right)[1 + O(\varepsilon)] \\ &= \varepsilon J_\varepsilon^+(f)[1 + O(\varepsilon)].\end{aligned}$$

The results for $m^-(f)$, $p^-(g)$ and $\alpha^-(f)$ follow similarly. □

We now compute the other quantities in Theorem 1.19 when $l = 0$, $m = \infty$, $a(r) = r^{2\alpha}$ and $w(r) = r^{2\omega}$. In (1.49) we now have

$$h_\gamma(r) = \left|\int_\gamma^r t^{-2\alpha-n+1}\,\mathrm{d}t\right|$$

and the permitted values of γ, and consequently the choice of Ω in (1.55), depend on α. There are three cases.

Case 1: $2\alpha + n < 2$. Here $h_0(r) < \infty$ and $h_\infty(r) = \infty$. Thus in accordance with the convention in (1.55), we choose $\gamma = 0$ when $\Omega = \mathbb{R}_+^n = \mathbb{R}^n \setminus \{0\}$, and $\gamma \in (0, \infty)$ when $\Omega = \mathbb{R}^n$.

If $\Omega = \mathbb{R}_+^n$, the choice $\gamma = 0$ gives

$$h_0(r) = [-(2\alpha + n - 2) r^{2\alpha+n-2}]^{-1}, \qquad \rho_0(r) = (2\alpha + n - 2)^2 r^{2\alpha-2},$$

and in Theorem 1.19,

$$\rho = \begin{cases} 0 & \text{if } \alpha - \omega - 1 \neq 0, \\ (2\alpha + n - 2)^2 & \text{if } \alpha - \omega - 1 = 0. \end{cases}$$

If $\Omega = \mathbb{R}^n$ and $\gamma \in (0, \infty)$, then

$$h_\gamma(r) = |(2\alpha + n - 2)^{-1}(\gamma^{2-n-2\alpha} - r^{2-n-2\alpha})|,$$

$$\rho_\gamma(r) = (2\alpha + n - 2)^2 r^{2-2n-2\alpha} |\gamma^{2-n-2\alpha} - r^{2-n-2\alpha}|^{-2},$$

$$\rho = 0.$$

Case 2: $2\alpha + n > 2$. We now have $h_0(r) = \infty$ and $h_\infty(r) < \infty$ and we can therefore choose $\gamma = \infty$ for $\Omega = \mathbb{R}^n_+$ and $\Omega = \mathbb{R}^n$. This gives

$$h_\infty(r) = [(2\alpha + n - 2)r^{2\alpha+n-2}]^{-1}, \qquad \rho_\infty(r) = (2\alpha + n - 2)^2 r^{2\alpha - 2},$$

$$\rho = \begin{cases} 0 & \text{if } \alpha - \omega - 1 \neq 0, \\ (2\alpha + n - 2)^2 & \text{if } \alpha - \omega - 1 = 0. \end{cases}$$

Case 3: $2\alpha + n = 2$. Since $h_0(r) = h_\infty(r) = \infty$ we are forced to choose $\gamma \in (0, \infty)$ and Ω can be either $\mathbb{R}^n_+$ or $\mathbb{R}^n$. We have

$$h_\gamma(r) = |\log(r/\gamma)|, \qquad \rho_\gamma(r) = \{r^n[\log(r/\gamma)]^2\}^{-1}, \qquad \rho = 0.$$

We now gather together the results in Example 1.22 and the subsequent computations to illustrate Theorem 1.19. For simplicity we take q to be real-valued, $b = 0$ and $\beta = 0$ in Theorem 1.19.

Corollary 1.23. In Theorem 1.19 let $l = 0$, $m = \infty$, $\beta = 0$, $b = 0$, and let $q = q_0 + f$ be real-valued. Suppose that $q_0(x) \geqslant \mu\rho_\gamma(|x|)$, where $\mu > -\frac{1}{4}A_1$ and ρ_γ is given above, and, in the notation of Example 1.22, suppose that

$$\lim_{\varepsilon \to 0+} \varepsilon J^{\pm}_{\varepsilon}(f) = 0, \qquad I(f) := \limsup_{\varepsilon \to 0+} \min\{I^+_\varepsilon(f), I^-_\varepsilon(f)\} > -\infty.$$

Then the operator T in Theorem 1.18 is self-adjoint and bounded below by θ, where θ takes the following values in the three cases examined above:
Case 1: $2\alpha + n < 2$.

$$\theta = \begin{cases} (\frac{1}{4}A_1 + \mu)(2\alpha + n - 2)^2 + I(f) & \text{when } \Omega = \mathbb{R}^n_+ \text{ and } \alpha - \omega - 1 = 0, \\ I(f) & \text{otherwise.} \end{cases}$$

Case 2: $2\alpha + n > 2$.

$$\theta = \begin{cases} (\frac{1}{4}A_1 + \mu)(2\alpha + n - 2)^2 + I(f) & \text{when } \Omega = \mathbb{R}^n_+ \text{ or } \mathbb{R}^n \text{ and } \alpha - \omega - 1 = 0, \\ I(f) & \text{when } \alpha - \omega - 1 \neq 0. \end{cases}$$

Case 3: $2\alpha + n = 2$.

$$\theta = I(f).$$

■

Proof. The only thing to add to the discussion in Example 1.22 and the following remarks is that we allow $\varepsilon \to 0+$ through the sequence which defines $I(f)$. □

2. *M*-accretive realizations of $\tau = -\Delta + q$

The results in §1 depended on showing that the sesquilinear form associated with τ and the given boundary condition was closed and sectorial in $L^2(\Omega)$ or $L^2(\Omega; w)$, and the m-sectorial realization of τ was then determined by the First Representation Theorem, Theorem IV.2.4, for such forms. We now consider the more general problem of determining m-accretive realizations of τ and concern ourselves exclusively with the Dirichlet problem. For simplicity we take $\tau = -\Delta + q$ and assume that $\operatorname{re} q \geqslant 0$ on Ω. The remaining terms in (1.1) are best regarded as perturbations and amenable to the stability result in Theorem III.8.4. The method of §1 is no longer applicable because the Dirichlet form t_D is not necessarily sectorial. Instead we use a technique developed by Kato in [4] which leans heavily on the distributional inequality given in §2.1 below.

2.1. *Kato's inequality*

Definition 2.1. A distribution $T \in \mathscr{D}'(\Omega)$ is said to be non-negative and written $T \geqslant 0$, if $T(\phi) \geqslant 0$ for all $\phi \in C_0^\infty(\Omega)$ such that $\phi \geqslant 0$ on Ω. If $T, S \in \mathscr{D}'(\Omega)$ we write $T \geqslant S$ if $T - S \geqslant 0$. ■

Theorem 2.2. Let $u \in L^1_{\text{loc}}(\Omega)$ and suppose that its distributional Laplacian $\Delta u \in L^1_{\text{loc}}(\Omega)$. Define

$$(\operatorname{sgn} u)(x) = \begin{cases} u(x)/|u(x)| & \text{if } u(x) \neq 0, \\ 0 & \text{if } u(x) = 0, \end{cases}$$

so that $\operatorname{sgn} u \in L^\infty(\Omega)$ and $(\operatorname{sgn} u)\Delta u \in L^1_{\text{loc}}(\Omega)$. Then

$$\Delta|u| \geqslant \operatorname{re}[(\operatorname{sgn} \bar{u})\Delta u], \tag{2.1}$$

that is,

$$\int_\Omega |u|\Delta\phi \geqslant \operatorname{re}\int_\Omega [(\operatorname{sgn} \bar{u}\,\Delta u]\phi \qquad (\phi \in C_0^\infty(\Omega),\ \phi \geqslant 0).$$ ■

In order to prove the inequality (2.1) we need the following approximate result.

Lemma 2.3. Let $u, \Delta u \in L^1_{\text{loc}}(\Omega)$ and for $\varepsilon > 0$ define

$$u_\varepsilon = (|u|^2 + \varepsilon^2)^{\frac{1}{2}}.$$

Then

$$\Delta u_\varepsilon \geqslant \operatorname{re}[(\bar{u}/u_\varepsilon)\Delta u]. \tag{2.2}$$ ■

Proof. We first prove the result for the case when u is smooth, $u \in C^2(\Omega)$ say. From the definition of u_ε, we obtain

$$2u_\varepsilon D_j u_\varepsilon = D_j(|u|^2) = 2\mathrm{re}(\bar{u}D_j u)$$

and

$$2u_\varepsilon D_j^2 u_\varepsilon + 2(D_j u_\varepsilon)^2 = 2\mathrm{re}(\bar{u}D_j^2 u) + 2|D_j u|^2.$$

Hence $|\nabla u_\varepsilon| \leqslant |\bar{u}/u_\varepsilon|\,|\nabla u| \leqslant |\nabla u|$ and

$$u_\varepsilon \Delta u_\varepsilon + |\nabla u_\varepsilon|^2 = \mathrm{re}\,(\bar{u}\Delta u) + |\nabla u|^2.$$

Consequently $u_\varepsilon \Delta u_\varepsilon \geqslant \mathrm{re}\,(\bar{u}\Delta u)$ and the lemma follows for $u \in C^2(\Omega)$ on dividing by u_ε.

For $u \in L^1_{\mathrm{loc}}(\Omega)$ the regularization $u^{(\rho)} = u * j_\rho \in C^\infty(\Omega)$. Therefore from the version of the theorem already established, for all $\phi \in C_0^\infty(\Omega)$ with $\phi \geqslant 0$,

$$\int_\Omega u_\varepsilon^{(\rho)} \Delta\phi \geqslant \mathrm{re} \int_\Omega [(\bar{u}^{(\rho)}/u_\varepsilon^{(\rho)})\Delta u^{(\rho)}]\phi \, dx, \qquad u_\varepsilon^{(\rho)} = (u^{(\rho)})_\varepsilon. \tag{2.3}$$

The lemma will follow if we can justify taking the limit as $\rho \to 0$ in each term in (2.3). We know from Theorem V.1.5 and Lemma V.2.2 that $u^{(\rho)} \to u$ and $\Delta u^{(\rho)} \to \Delta u$ in $L^1_{\mathrm{loc}}(\Omega)$, i.e. in $L^1(\Omega')$ for every $\Omega' \subset\subset \Omega$, and we may therefore conclude that $u^{(\rho)}(x) \to u(x)$ a.e. in Ω as $\rho \to 0$ through a suitable sequence. From the inequality

$$\begin{aligned} |u_\varepsilon^{(\rho)} - u_\varepsilon| &= |(|u^{(\rho)}|^2 + \varepsilon^2)^{\frac{1}{2}} - (|u|^2 + \varepsilon^2)^{\frac{1}{2}}| \\ &= |(|u^{(\rho)}|^2 - |u|^2)|/[(|u^{(\rho)}|^2 + \varepsilon^2)^{\frac{1}{2}} + (|u|^2 + \varepsilon^2)^{\frac{1}{2}}] \\ &\leqslant ||u^{(\rho)}| - |u|| \\ &\leqslant |u^{(\rho)} - u|, \end{aligned}$$

it follows that $u_\varepsilon^{(\rho)} \to u_\varepsilon$ in $L^1_{\mathrm{loc}}(\Omega)$ and also $u_\varepsilon^{(\rho)}(x) \to u_\varepsilon(x)$ a.e. in Ω, as $\rho \to 0$ through this sequence. The integral on the left-hand side of (2.3) therefore converges to $\int_\Omega u_\varepsilon \Delta\phi$ as $\rho \to 0$. Also $(\bar{u}^{(\rho)}/u_\varepsilon^{(\rho)})(x) \to (\bar{u}/u_\varepsilon)(x)$ a.e. in Ω and $|\bar{u}^{(\rho)}/u_\varepsilon^{(\rho)}| \leqslant 1$. Since $\Delta u^{(\rho)} \to \Delta u$ in $L^1_{\mathrm{loc}}(\Omega)$,

$$\int_\Omega (\bar{u}^{(\rho)}/u_\varepsilon^{(\rho)})(\Delta u^{(\rho)} - \Delta u)\phi \to 0,$$

while, by the Dominated Convergence Theorem, since $\Delta u \in L^1_{\mathrm{loc}}(\Omega)$,

$$\int_\Omega (\bar{u}^{(\rho)}/u_\varepsilon^{(\rho)})\phi\Delta u \to \int_\Omega (\bar{u}/u_\varepsilon)\phi\Delta u.$$

Together these results give

$$\mathrm{re}\int_\Omega [(\bar{u}^{(\rho)}/u_\varepsilon^{(\rho)})\Delta u^{(\rho)}]\phi = \mathrm{re}\int_\Omega (\bar{u}^{(\rho)}/u_\varepsilon^{(\rho)})(\Delta u^{(\rho)} - \Delta u)\phi + \mathrm{re}\int_\Omega (\bar{u}^{(\rho)}/u_\varepsilon^{(\rho)})\,\phi\Delta u \to \int_\Omega (\bar{u}/u_\varepsilon)\phi\Delta u$$

as $\rho \to 0$. The lemma therefore follows on allowing $\rho \to 0$ through the chosen sequence in (2.3). □

Proof of Theorem 2.2. As $\varepsilon \to 0$ so $u_\varepsilon \to |u|$ uniformly on Ω and hence $\int_\Omega u_\varepsilon\, \Delta\phi \to \int_\Omega |u|\Delta\phi$. Also $\bar{u}/u_\varepsilon \to \mathrm{sgn}\ \bar{u}$ and since $|(\bar{u}/u_\varepsilon)\Delta u| \leqslant |\Delta u| \in \mathrm{L}^1_{\mathrm{loc}}(\Omega)$ we have, by the Dominated Convergence Theorem, that

$$\int_\Omega \mathrm{re}[(\bar{u}/u_\varepsilon)\Delta u]\phi \to \int_\Omega \mathrm{re}[(\mathrm{sgn}\ \bar{u})\Delta u]\phi \qquad (\phi \in \mathrm{C}_0^\infty(\Omega)).$$

The theorem therefore follows on allowing $\varepsilon \to 0$ in (2.2). □

2.2. *Kato's Theorem*

Let I_0 be the continuous embedding of $\mathrm{W}_0^{1,2}(\Omega)$ into $\mathrm{L}^2(\Omega)$ and $I_0^*: \mathrm{L}^2(\Omega) \to \mathrm{W}^{-1,2}(\Omega)$ its adjoint, $\mathrm{L}^2(\Omega)$ being identified with its adjoint. In Remark VI.1.7 we saw that the conjugate-linear distribution $1-\Delta$ defined by $(1-\Delta)u: \phi \mapsto \int_\Omega u(1-\Delta)\bar{\phi}$ is a linear isometry of $\mathrm{W}_0^{1,2}(\Omega)$ onto $\mathrm{W}^{-1,2}(\Omega)$. Hence, if $f \in \mathrm{W}^{-1,2}(\Omega)$ then $f = (1-\Delta)u$ for some $u \in \mathrm{W}_0^{1,2}(\Omega)$ and for any $\phi \in \mathrm{W}_0^{1,2}(\Omega)$ we can write the duality bracket $(\bullet, \bullet)$ between $\mathrm{W}_0^{1,2}(\Omega)$ and $\mathrm{W}^{-1,2}(\Omega)$ symbolically as

$$\begin{aligned}(f, \phi) = \int_\Omega f\bar{\phi} := \int_\Omega u(1-\Delta)\bar{\phi} \\ = (u, \phi)_{1,2,\Omega}.\end{aligned} \tag{2.4}$$

It follows that

$$\begin{aligned}\|f\|_{-1,2,\Omega} &= \sup_{\|\phi\|_{1,2,\Omega}=1} |(f, \phi)| \\ &= \|u\|_{1,2,\Omega},\ f = (1-\Delta)u,\end{aligned} \tag{2.5}$$

and as in §VI.1.2, $\mathrm{W}^{-1,2}(\Omega)$ is a Hilbert space with inner product

$$(f, g)_{-1,2,\Omega} = (u, v)_{1,2,\Omega}, \qquad f = (1-\Delta)u, \qquad g = (1-\Delta)v. \tag{2.6}$$

Let $\tau = -\Delta + q$, where q is a complex-valued and satisfies

$$\mathrm{re}\, q \geqslant 0, \qquad q \in \mathrm{L}^p_{\mathrm{loc}}(\Omega), \tag{2.7}$$

where

$$p\begin{cases} = 2n/(n+2) & \text{if } n \geqslant 3, \\ > 1 & \text{if } n = 2, \\ = 1 & \text{if } n = 1. \end{cases} \tag{2.8}$$

Since $W_0^{1,2}(\Omega)$ is continuously embedded in $L^{p'}(\Omega)$, where $1/p + 1/p' = 1$, the assumption (2.7) implies that $qu \in L^1_{\text{loc}}(\Omega)$ for all $u \in W_0^{1,2}(\Omega)$ and $\tau u = (-\Delta + q)u$ is defined as a distribution. This prompts the definition

$$A := \tau \upharpoonright \mathscr{D}(A), \qquad \mathscr{D}(A) := \{u : u \in W_0^{1,2}(\Omega),\ \tau u \in W^{-1,2}(\Omega)\}. \tag{2.9}$$

A is the *maximal* realization of τ as a map from $W_0^{1,2}(\Omega)$ into $W^{-1,2}(\Omega)$. If $u \in C_0^\infty(\Omega)$ then $qu \in L^p(\Omega)$ and for all $\phi \in W_0^{1,2}(\Omega)$,

$$\begin{aligned} \left| \int_\Omega qu\bar{\phi} \right| &\leqslant \|qu\|_{p,\Omega}\, \|\phi\|_{p',\Omega} \\ &\leqslant (\gamma \|qu\|_{p,\Omega})\, \|\phi\|_{1,2,\Omega}, \end{aligned}$$

where γ is the norm of the embedding $W_0^{1,2}(\Omega) \to L^{p'}(\Omega)$. It follows that $qu \in W^{-1,2}(\Omega)$ and hence $C_0^\infty(\Omega) \subset \mathscr{D}(A)$. We may therefore define

$$A_0 := A \upharpoonright C_0^\infty(\Omega), \qquad A_0 : C_0^\infty(\Omega) \subset W_0^{1,2}(\Omega) \to W^{-1,2}(\Omega). \tag{2.10}$$

On identifying the Hilbert space $W_0^{1,2}(\Omega)$ with $[W^{-1,2}(\Omega)]^*$, the adjoint A_0^* of A_0 also has domain in $W_0^{1,2}(\Omega)$ and range in $W^{-1,2}(\Omega)$. It is a closed linear operator and we recall from Theorem III.1.5 that A_0^* is densely defined if, and only if, A_0 is closable, in which case the closure of A_0 is A_0^{**}. Note that

$$(A_0\phi, \psi) = (\phi, A_0^*\psi) \equiv \overline{(A_0^*\psi, \phi)} \qquad (\phi \in \mathscr{D}(A_0),\ \psi \in \mathscr{D}(A_0^*)). \tag{2.11}$$

Lemma 2.4. Let A_0^c and A^c be the operators defined by replacing τ in (2.9) and (2.10) by $\bar{\tau} = -\Delta + \bar{q}$. Then $A_0^* = A^c$ and $(A_0^c)^* = A$. ■

Proof. Let $u \in \mathscr{D}(A_0^*)$ and $A_0^* u = f$. Then for all $\phi \in C_0^\infty(\Omega)$,

$$\begin{aligned} (f, \phi) &= (A_0^* u, \phi) = (u, A_0\phi) \\ &= \int_\Omega u\bar{\tau}\bar{\phi}. \end{aligned}$$

Hence the conjugate-linear distribution $\bar{\tau}u = f \in W^{-1,2}(\Omega)$ and consequently $A_0^* \subset A^c$. Conversely, if $u \in \mathscr{D}(A^c)$ and $f = A^c u = \bar{\tau}u$, we have, for all $\phi \in C_0^\infty(\Omega)$,

$$\begin{aligned} (A_0\phi, u) &= \int_\Omega (\tau\phi)\bar{u} \\ &= \int_\Omega \phi\bar{f} = (\phi, f) = \overline{(f, \phi)}. \end{aligned}$$

Since $f \in W^{-1,2}(\Omega)$, we conclude that $u \in \mathscr{D}(A_0^*)$ and $f = A_0^* u$. In other words $A^c \subset A_0^*$ and hence $A_0^* = A^c$. □

Hereafter we suppress the embeddings I_0 and I_0^* since it will be clear from the context when $W_0^{1,2}(\Omega)$ is to be regarded as a subspace of $L^2(\Omega)$ and $L^2(\Omega)$ a subspace of $W^{-1,2}(\Omega)$. The main theorem in this section is the following result proved by Kato in [4].

Theorem 2.5. Let q satisfy (2.7) where p satisfies (2.8). Then the operator T defined by

$$T = \tau \restriction \mathscr{D}(T), \qquad \mathscr{D}(T) := \{u : u \in W_0^{1,2}(\Omega),\ \tau u \in L^2(\Omega)\}$$

is m-accretive in $L^2(\Omega)$. Also $T^* = T^c$ and T is J-self-adjoint with respect to the conjugation $J : u \mapsto \bar{u}$. ■

Proof. For $\phi \in C_0^\infty(\Omega) = \mathscr{D}(A_0)$,

$$\begin{aligned} \operatorname{re}((A_0 + I)\phi, \phi) &= \operatorname{re} \int_\Omega (|\nabla \phi|^2 + q|\phi|^2 + |\phi|^2) \\ &\geqslant \|\phi\|_{1,2,\Omega}^2 \end{aligned}$$

since $\operatorname{re} q \geqslant 0$, and consequently

$$\|\phi\|_{1,2,\Omega} \leqslant \|(A_0 + I)\phi\|_{-1,2,\Omega}.$$

Since $C_0^\infty(\Omega) \subset \mathscr{D}(A^c)$, it follows from Lemma 2.4 that A_0^* is densely defined in $W_0^{1,2}(\Omega)$ and so A_0 has a closure A_0^{**} which therefore satisfies

$$\|u\|_{1,2,\Omega} \leqslant \|(A_0^{**} + I)u\|_{-1,2,\Omega} \qquad (u \in \mathscr{D}(A_0^{**})). \tag{2.12}$$

This inequality implies that $(A_0^{**} + I)^{-1}$ exists and is a bounded, closed operator on the range $\mathscr{R}(A_0^{**} + I)$ of $A_0^{**} + I$ in $W^{-1,2}(\Omega)$. Therefore $\mathscr{R}(A_0^{**} + I)$ is a closed subspace of $W^{-1,2}(\Omega)$. We claim that $\mathscr{R}(A_0^{**} + I) = W^{-1,2}(\Omega)$. Otherwise there exists a $u \in W_0^{1,2}(\Omega)$, with $u \neq 0$, such that $u \in \mathscr{R}(A_0^{**} + I)^\perp$ and hence, for all $\phi \in C_0^\infty(\Omega)$,

$$0 = ((A_0 + I)\phi, u) = \int_\Omega (\tau + 1)\phi \bar{u}.$$

This yields the distributional identity

$$(\bar{\tau} + 1)u = -\Delta u + \bar{q}u + u = 0,$$

and since $\bar{q}u \in L^1_{loc}(\Omega)$ for $u \in W_0^{1,2}(\Omega)$ under the hypothesis (2.7), we conclude that $\Delta u = u + \bar{q}u \in L^1_{loc}(\Omega)$. Kato's inequality, Theorem 2.2, is therefore applicable and we obtain the distributional inequality

$$\begin{aligned} \Delta|u| &\geqslant \operatorname{re}[(\operatorname{sgn} \bar{u})(u + \bar{q}u)] \\ &\geqslant |u|. \end{aligned}$$

In other words,

$$((1-\Delta)|u|,\phi) \leqslant 0 \qquad (\phi \in C_0^\infty(\Omega),\ \phi \geqslant 0). \tag{2.13}$$

From Corollary VI.2.4 and Theorem VI.3.6, $|u| \in W_0^{1,2}(\Omega)$ and there exists a sequence of functions $\phi_k \in C_0^\infty(\Omega)$ such that $\phi_k \geqslant 0$ and $\phi_k \to |u|$ in $W_0^{1,2}(\Omega)$ as $k \to \infty$. Thus $\phi_k \to |u|$ in $L^2(\Omega)$ and

$$\begin{aligned}(|u|,\phi_m)_{1,2,\Omega} &= \lim_{k\to\infty} (\phi_k,\phi_m)_{1,2,\Omega} \\ &= \lim_{k\to\infty} (\phi_k,(1-\Delta)\phi_m)_{2,\Omega}, \\ &= (|u|,(1-\Delta)\phi_m)_{2,\Omega} \\ &= ((1-\Delta)|u|,\phi_m) \leqslant 0\end{aligned}$$

from (2.13). On allowing $m \to \infty$, this gives the contradiction $\| |u| \|_{1,2,\Omega} \leqslant 0$ and so

$$\mathscr{R}(A_0^{**}+I) = W^{-1,2}(\Omega) \tag{2.14}$$

is established. Since $A_0 \subset A$ and A is closed by virtue of Lemma 2.4 we have $A_0^{**} \subset A$ and hence $\mathscr{R}(A+I) = W^{-1,2}(\Omega)$, by (2.14). We next prove that $A \subset A_0^{**}$ and hence $A = A_0^{**}$. Let $u \in \mathscr{D}(A)$. From (2.14) and $A_0^{**} \subset A$, there exists a $v \in \mathscr{D}(A_0^{**})$ such that $(A+I)u = (A_0^{**}+I)v = (A+I)v$, whence $u - v \in \mathscr{N}(A+I)$. But $\mathscr{R}(A+I) = W^{-1,2}(\Omega)$ implies that $\operatorname{nul}(A+I) = \operatorname{def}(A^*+I) = \operatorname{def}(A_0^{c**}+I) = 0$ on using Lemma 2.4 and the analogue of (2.14) for A_0^c. Consequently $u = v \in \mathscr{D}(A_0^{**})$ and $A = A_0^{**}$.

For all $\phi \in C_0^\infty(\Omega)$,

$$\operatorname{re}(A_0\phi,\phi) = \operatorname{re}\int_\Omega (|\nabla\phi|^2 + q|\phi|^2) \geqslant 0$$

and since A was proved in the last paragraph to be A_0^{**}, the closure of A_0, we have

$$\operatorname{re}(A\phi,\phi) \geqslant 0 \qquad (\phi \in \mathscr{D}(A)).$$

This in turn implies that

$$\operatorname{re}(T\phi,\phi)_{2,\Omega} = \operatorname{re}(A\phi,\phi) \geqslant 0 \qquad (\phi \in \mathscr{D}(T))$$

and hence that T is accretive. But $\mathscr{R}(A+I) = W^{-1,2}(\Omega)$ also yields $\mathscr{R}(T+I) = L^2(\Omega)$. We therefore conclude from Theorem III.2.3 that the half-plane $\{\lambda\colon \operatorname{re}\lambda < 0\}$ lies in the resolvent set of T and hence that T is m-accretive.

Finally we show that $T^* = T^c$. For $u \in \mathscr{D}(T^c)$ and $\phi \in \mathscr{D}(T)$,

$$\begin{aligned}(T\phi,u)_{2,\Omega} &= (A\phi,u) \\ &= (\phi,A^c u)\end{aligned}$$

since $A^* = (A_0^c)^{**} = A^c$ from Lemma 2.4 and the last but one paragraph. Thus since $T^c \subset A^c$, we have $(T\phi, u)_{2,\Omega} = (\phi, T^c u)_{2,\Omega}$ and so $T^c \subset T^*$ or $JTJ \subset T^*$. Therefore T is J-symmetric and since it is m-accretive, T is J-self-adjoint from Theorem III.6.7(ii), and $T^* = T^c$. The theorem is proved. □

Theorem 2.6. Let q satisfy (2.7) with $\Omega = \mathbb{R}^n$, $n \geqslant 1$, and define $S = \tau \upharpoonright \mathscr{D}(S)$, $\mathscr{D}(S) := \{u : u \in L^2(\Omega),\ qu \in L^1_{\text{loc}}(\Omega) \text{ and } \tau u \in L^2(\Omega)\}$. Then $S = T$, the m-accretive operator in Theorem 2.5. ■

Proof. Since $qu \in L^1_{\text{loc}}(\mathbb{R}^n)$ for all $u \in W_0^{1,2}(\mathbb{R}^n)$ we have $\mathscr{D}(T) \subset \mathscr{D}(S)$ and hence $T \subset S$. To prove that $S \subset T$ we first recall from Theorem 2.5 that $\mathscr{R}(T+I) = L^2(\mathbb{R}^n)$. Consequently if $u \in \mathscr{D}(S)$ there exists a $v \in \mathscr{D}(T)$ such that $(S+I)u = (T+I)v = (S+I)v$, and so $(S+I)w = 0$ on setting $w = u - v$. It suffices to prove that $w = 0$.

From $(S+I)w = 0$ we have the distributional identity $\Delta w = w + qw$. Also $qw = qu - qv \in L^1_{\text{loc}}(\mathbb{R}^n)$ as $u \in W_0^{1,2}(\mathbb{R}^n)$ and $v \in \mathscr{D}(S)$. Thus $\Delta w \in L^1_{\text{loc}}(\mathbb{R}^n)$ and Kato's inequality gives

$$\Delta|w| \geqslant \operatorname{re}[(\operatorname{sgn}\bar{w})(w + qw)] = |w| + (\operatorname{re} q)|w| \geqslant |w|,$$

i.e. $(1-\Delta)|w| \leqslant 0$. Now put $f = |w|$ and let $f_\varepsilon = f * \rho_\varepsilon$ be the regularization of f. Then $(1-\Delta)f_\varepsilon = [(1-\Delta)f] * \rho_\varepsilon \leqslant 0$ since $\rho_\varepsilon \in C_0^\infty(\mathbb{R}^n)$ and $\rho_\varepsilon \geqslant 0$. Furthermore $f_\varepsilon \geqslant 0$ and hence

$$(f_\varepsilon, (1-\Delta)f_\varepsilon)_{2,\mathbb{R}^n} \leqslant 0.$$

But since $f_\varepsilon \in C^\infty(\mathbb{R}^n) \cap L^2(\mathbb{R}^n)$ and $\Delta f_\varepsilon = f * \Delta\rho_\varepsilon \in L^2(\mathbb{R}^n)$, it follows that $f_\varepsilon \in W^{2,2}(\mathbb{R}^n) = W_0^{2,2}(\mathbb{R}^n)$ and so

$$(f_\varepsilon, (1-\Delta)f_\varepsilon)_{2,\mathbb{R}^n} = \|f_\varepsilon\|^2_{2,\mathbb{R}^n} + \|\nabla f_\varepsilon\|^2_{2,\mathbb{R}^n} \geqslant 0.$$

We conclude that equality must hold in the last equation and hence that $f_\varepsilon = 0$. Since $f_\varepsilon \to f = |w|$ in $L^2(\mathbb{R}^n)$ as $\varepsilon \to 0$, the required conclusion $w = 0$ is obtained. □

Corollary 2.7. Let $\operatorname{re} q$ be bounded below on $\mathbb{R}^n$ and $q \in L^2_{\text{loc}}(\mathbb{R}^n)$, and let $T_0' = \tau \upharpoonright C_0^\infty(\mathbb{R}^n)$. Then T_0' is closable and its closure T_0 is quasi-m-accretive and J-self-adjoint. Moreover $T_0 = T$, where T is the operator in Theorems 2.5 and 2.6. In particular, if q is real, T_0' is essentially self-adjoint and T_0 is self-adjoint. ■

Proof. Since a bounded perturbation has no effect on the properties to be proved on account of Corollary III.8.5, we can add a real constant to q and hence assume that $\operatorname{re} q \geqslant 0$ on $\mathbb{R}^n$. The assumption $q \in L^2_{\text{loc}}(\mathbb{R}^n)$ implies that T_0' is properly defined on $C_0^\infty(\mathbb{R}^n)$ and therefore has a domain dense in $L^2(\mathbb{R}^n)$. Furthermore $T_0' \subset T$ and, as T is closed, T_0' is closable and $T_0 \subset T$.

We first prove that $T_0^* = T^c$. By definition, $u \in \mathscr{D}(T_0^*)$ and $f = T_0^* u$ if, and only if, $f \in L^2(\mathbb{R}^n)$ and for all $\phi \in C_0^\infty(\mathbb{R}^n)$,

$$(\phi, f)_{2,\mathbb{R}^n} = (T_0\phi, u)_{2,\mathbb{R}^n}$$
$$= \int_{\mathbb{R}^n} \tau\phi\bar{u}.$$

It follows that

$$T_0^* = \bar{\tau} \restriction \mathscr{D}(T_0^*), \qquad \mathscr{D}(T_0^*) = \{u : u \in L^2(\mathbb{R}^n), \bar{\tau}u \in L^2(\mathbb{R}^n)\}.$$

Since $qu \in L^1_{loc}(\mathbb{R}^n)$ for any $u \in L^2(\mathbb{R}^n)$ in view of the hypothesis $q \in L^2_{loc}(\mathbb{R}^n)$, we conclude from Theorem 2.6 that $T_0^* = T^c$. The corollary will follow if we prove that $T_0 = T$, and since $T_0 \subset T$ it suffices to prove that $T \subset T_0$.

For all $\phi \in C_0^\infty(\mathbb{R}^n)$,

$$\operatorname{re}((T_0+I)\phi, \phi)_{2,\mathbb{R}^n} = \operatorname{re}\int_{\mathbb{R}^n} (|\nabla\phi|^2 + q|\phi|^2 + |\phi|^2)$$
$$\geqslant \|\phi\|^2_{2,\mathbb{R}^n}$$

and consequently

$$\|\phi\|_{2,\mathbb{R}^n} \leqslant \|(T_0+I)\phi\|_{2,\mathbb{R}^n}.$$

Since T_0 is the closure of T_0', the latter inequality continues to hold for $\phi \in \mathscr{D}(T_0)$ and thus $(T_0+I)^{-1}$ exists and is bounded on its domain $\mathscr{R}(T_0+I)$. Since $(T_0+I)^{-1}$ is also closed it follows that $\mathscr{R}(T_0+I)$ is a closed subspace of $L^2(\mathbb{R}^n)$. Moreover, since $T_0^* = T^c$ we have

$$\mathscr{R}(T_0+I)^\perp = \mathscr{N}(T_0^*+I) = \mathscr{N}(T^c+I) = \{0\},$$

the final step following from Theorem 2.5. We therefore conclude that $\mathscr{R}(T_0+I) = L^2(\mathbb{R}^n)$. Consequently, for any $u \in \mathscr{D}(T)$, there exists a $\phi \in \mathscr{D}(T_0)$ such that

$$(T+I)u = (T_0+I)\phi,$$

and since $T_0 \subset T$ we have $(T+I)(u-\phi) = 0$. But, from Theorem 2.5, $T+I$ is injective with the result that $u = \phi \in \mathscr{D}(T_0)$ and $T \subset T_0$. This completes the proof. □

2.3. *Supplementary results*

We assume throughout this section that

$$\operatorname{re} q \geqslant 0, \qquad q \in L^2_{loc}(\mathbb{R}^n), \tag{2.15}$$

and $\tau = -\Delta + q$. The three lemmas below will be needed in §3. We shall denote the $L^2(\mathbb{R}^n)$, $W^{1,2}(\mathbb{R}^n)$ and $W^{-1,2}(\mathbb{R}^n)$ norms by $\|\bullet\|_2$, $\|\bullet\|_{1,2}$ and $\|\bullet\|_{-1,2}$ respectively.

Lemma 2.8. Let $u \in L^2(\mathbb{R}^n)$ and $\tau u \in W^{-1,2}(\mathbb{R}^n)$. Then $u \in W^{1,2}(\mathbb{R}^n)$ and

$$\|u\|_{1,2} \leqslant \|\tau u\|_{-1,2} + \|u\|_2. \qquad \blacksquare$$

Proof. Let B be the operator from $L^2(\mathbb{R}^n)$ into $W^{-1,2}(\mathbb{R}^n)$ defined by

$$B := \tau \upharpoonright \mathscr{D}(B), \qquad \mathscr{D}(B) = \{u : u \in L^2(\mathbb{R}^n),\ \tau u \in W^{-1,2}(\mathbb{R}^n)\}.$$

Clearly $A \subset B$, where A is the operator defined in (2.9) with $\Omega = \mathbb{R}^n$, and the lemma asserts that $B \subset A$ and hence $B = A$. In the course of proving Theorem 2.5 we showed that $\mathscr{R}(A + I) = W^{-1,2}(\mathbb{R}^n)$ and so, given $u \in \mathscr{D}(B)$, there exists some $v \in \mathscr{D}(A)$ such that $(B + I)u = (A + I)v$, i.e. (since $A \subset B$) such that $(B + I)w = 0$, where $w = u - v$. The last equation implies that $\Delta w = w + qw \in L^1_{\text{loc}}(\mathbb{R}^n)$ and the same argument as that in the proof of Theorem 2.6 gives $w = 0$. Hence $B \subset A$ and $u \in W^{1,2}(\mathbb{R}^n)$ in the lemma.

Since, in the notation of Theorem 2.5, $A_0^{**} = A = B$, we see from (2.12) that

$$\begin{aligned} \|u\|_{1,2} &= \|(B + I)u\|_{-1,2} \\ &\leqslant \|\tau u\|_{-1,2} + \|u\|_{-1,2} \\ &\leqslant \|\tau u\|_{-1,2} + \|u\|_2 \end{aligned}$$

since the embedding $I_0^* : L^2(\mathbb{R}^n) \to W^{-1,2}(\mathbb{R}^n)$ has norm $\leqslant 1$. $\square$

Lemma 2.9. Let $u \in L^2(\mathbb{R}^n)$ and $\tau u \in L^2_{\text{loc}}(\mathbb{R}^n)$. Then $u \in W^{1,2}_{\text{loc}}(\mathbb{R}^n)$ and given any r and R with $0 < r < R$ there exists a constant K depending only on $R - r$ such that

$$\|u\|_{1,2,B_r} \leqslant \|\tau u\|_{2,B_R} + K\|u\|_{2,B_R},$$

where B_r is the ball $\{x \in \mathbb{R}^n : |x| < r\}$. $\blacksquare$

Proof. Let $\phi \in C_0^\infty(B_R)$ be such that $0 \leqslant \phi \leqslant 1$, with $\phi = 1$ on $\bar{B}_r$, and such that the first and second derivatives of ϕ are bounded by a constant depending only on $R - r$. For $\psi \in C_0^\infty(\mathbb{R}^n)$ and u as in the lemma,

$$\begin{aligned} \left| \int (D_j \phi\, D_j u) \bar{\psi} \right| &= \left| \int u D_j (\bar{\psi} D_j \phi) \right| \\ &\leqslant K \|u\|_{2,B_R} \|\psi\|_{1,2}. \end{aligned}$$

This implies that $\nabla\phi \cdot \nabla u \in W^{-1,2}(\mathbb{R}^n)$ and

$$\|\nabla\phi \cdot \nabla u\|_{-1,2} \leqslant K\|u\|_{2,B_R}.$$

Also $\tau(\phi u) = \phi \tau u - 2\nabla\phi \cdot \nabla u - u\Delta\phi \in W^{-1,2}(\mathbb{R}^n)$ and consequently, by Lemma 2.8, $\phi u \in W^{1,2}(\mathbb{R}^n)$ or, equivalently, $u \in W^{1,2}_{\text{loc}}(\mathbb{R}^n)$. Furthermore, since

$\phi = 1$ on B_r,

$$\begin{aligned}\|u\|_{1,2,B_r} &\leqslant \|\phi u\|_{1,2}\\ &\leqslant \|\tau(\phi u)\|_{-1,2} + \|\phi u\|_2\\ &\leqslant \|\phi \tau u\|_{-1,2} + 2\|\nabla\phi\cdot\nabla u\|_{-1,2} + \|u\,\Delta\,\phi\|_{-1,2} + \|u\|_{2,B_R}\\ &\leqslant \|\phi\tau u\|_2 + K\,\|u\|_{2,B_R}\\ &\leqslant \|\tau u\|_{2,B_R} + K\,\|u\|_{2,B_R}.\end{aligned}$$ □

Lemma 2.10. Let $u, \tau u \in \mathrm{L}^2(\mathbb{R}^n)$ and $\operatorname{supp} u \subset B_R$. Then there exists a sequence of functions $u_m \in \mathrm{C}_0^\infty(\mathbb{R}^n)$ such that supp $u_m \subset B_R$ and

$$u_m \to u, \qquad \tau u_m \to \tau u$$

in $\mathrm{L}^2(\mathbb{R}^n)$. ■

Proof. By Lemma 2.8, $u \in \mathscr{D}(T)$ where T is the operator in Theorem 2.5. Thus, in view of Corollary 2.7, $u \in \mathscr{D}(T_0)$, where T_0 is the closure of $T_0' = \tau \upharpoonright \mathrm{C}_0^\infty(\mathbb{R}^n)$. Therefore there exists a sequence of functions $v_m \in \mathrm{C}_0^\infty(\mathbb{R}^n)$ satisfying

$$v_m \to u, \qquad \tau v_m = T_0 v_m \to T_0 u = \tau u$$

in $\mathrm{L}^2(\mathbb{R}^n)$. Now put $u_m = \phi v_m$ where $\phi \in \mathrm{C}_0^\infty(B_R)$ and $\phi(x) = 1$ for $x \in \operatorname{supp} u$. Then supp $u_m \subset B_R$ and $u_m \to \phi u = u$ in $\mathrm{L}^2(\mathbb{R}^n)$. Since $\mathrm{L}^2(\mathbb{R}^n)$ is continuously embedded in $\mathrm{W}^{-1,2}(\mathbb{R}^n)$, we have that $\tau v_m \to \tau u$ in $\mathrm{W}^{-1,2}(\mathbb{R}^n)$ and so $v_m \to u$ in $\mathrm{W}^{1,2}(\mathbb{R}^n)$, by Lemma 2.8. It follows that

$$\begin{aligned}\tau u_m &= \phi\tau v_m - 2\nabla\phi\cdot\nabla v_m - v_m\,\Delta\phi\\ &\to \phi\tau u - 2\nabla\phi\cdot\nabla u - u\,\Delta\phi\\ &= \tau(\phi u) = \tau u\end{aligned}$$

in $\mathrm{L}^2(\mathbb{R}^n)$ and the lemma is proved. □

3. $\tau = -\Delta + q$ with im q semi-bounded

We impose the following conditions throughout this section:

$$\left.\begin{array}{l}\text{(i) re } q \text{ is locally bounded below in } \mathbb{R}^n,\\ \text{(ii) } q \in \mathrm{L}^2_{\mathrm{loc}}(\mathbb{R}^n),\\ \text{(ii) im } q \text{ is semi-bounded on } \mathbb{R}^n.\end{array}\right\} \qquad (3.1)$$

We continue to denote the $\mathrm{W}^{1,2}(\mathbb{R}^n)$ and $\mathrm{L}^2(\mathbb{R}^n)$ norms by $\|\bullet\|_{1,2}$ and $\|\bullet\|_2$ respectively. As in the proof of Corollary 2.7, the adjoint of the operator $T_0' = \tau \upharpoonright \mathrm{C}_0^\infty(\mathbb{R}^n)$ is the operator T^c defined by

$$T^c = \bar{\tau} \upharpoonright \mathscr{D}(T^c), \qquad \mathscr{D}(T^c) = \{u : u \in \mathrm{L}^2(\mathbb{R}^n), \bar{\tau} u \in \mathrm{L}^2(\mathbb{R}^n)\}. \qquad (3.2)$$

In particular T^c, and similarly the operator T defined as in (3.2) with $\bar{\tau}$ replaced by τ, is densely defined and hence T'_0 is closable. Its closure T_0 therefore satisfies

$$T_0^* = T^c. \tag{3.3}$$

Our aim is to obtain conditions on q which are sufficient for T_0 to be J-self-adjoint, i.e. $T_0^* = T_0^c$ or equivalently, in view of (3.3), $T_0 = T$. For real q the problem is that of establishing the essential self-adjointness of T'_0 and this has been the subject of intensive study over the years on account of its importance in quantum mechanics.

If in (3.1), im q is bounded above, im $q \leqslant M$ say, then for all $\phi \in C_0^\infty(\mathbb{R}^n)$,

$$\begin{aligned}\operatorname{im}(T'_0\phi, \phi)_2 &= \operatorname{im}\int_{\mathbb{R}^n} (|\nabla\phi|^2 + q|\phi|^2)\\ &= \int_{\mathbb{R}^n} \operatorname{im} q|\phi|^2\\ &\leqslant M\|\phi\|_2^2.\end{aligned}$$

Hence the numerical range $\Theta(\mathrm{i}T'_0)$ lies in $\{\lambda : \operatorname{re}\lambda \geqslant -M\}$ and $\mathrm{i}T'_0$ is quasi-accretive. Since T_0 is the closure of T'_0, the same is true for iT_0. By Theorem III.2.3, if $\lambda \in \mathbb{C}\setminus\bar{\Theta}(\mathrm{i}T_0)$ then $\operatorname{nul}(\mathrm{i}T_0 - \lambda I) = 0$, $\mathscr{R}(iT_0 - \lambda I)$ is closed and $\operatorname{def}(\mathrm{i}T_0 - \lambda I) = \operatorname{nul}(-\mathrm{i}T_0^* - \bar{\lambda}I)$ is constant for $\operatorname{re}\lambda < \mathrm{M}$. Thus $T_0^* = T_0^c$ implies that $\operatorname{def}(\mathrm{i}T_0 - \lambda I) = \operatorname{nul}(-\mathrm{i}T_0^c - \bar{\lambda}I) = \operatorname{nul}(\mathrm{i}T_0 - \lambda I) = 0$ for $\operatorname{re}\lambda < M$, and consequently that $\mathrm{i}T_0$ is quasi-m-accretive. If im q is bounded below in (3.1) we find similarly that $-\mathrm{i}T_0$ is quasi-m-accretive if $T_0^* = T_0^c$.

The method used in this section relies heavily on some technical lemmas which concern the local behaviour of members of $\mathscr{D}(T)$. The assumptions made ultimately on q allow for widely oscillatory behaviour.

3.1. *Local properties of $\mathscr{D}(T)$*

Lemma 3.1. Let $u \in \mathscr{D}(T)$ and $\phi \in C_0^\infty(B_R)$. Then $\phi u \in \mathscr{D}(T)$ and there exists a sequence $(u_m) \subset C_0^\infty(B_R)$ such that $u_m \to \phi u$ in $W^{1,2}(\mathbb{R}^n)$ and $Tu_m \to T(\phi u)$ in $L^2(\mathbb{R}^n)$. ∎

Proof. Set $q = q_1 + \mathrm{i}q_2$. Since q_1 is locally bounded below by (3.1) (i), there exists a locally bounded function k such that $q_1(x) \geqslant -k(x)$, for a.e. $x \in \mathbb{R}^n$. Let $\tau_R := -\Delta + q_R$, where $q_R = q_{1,R} + \mathrm{i}q_2$ with

$$q_{1,R}(x) = \begin{cases} q_1(x) & \text{for } x \in B_R\\ q_1(x) + k(x) - k_R & \text{otherwise,}\end{cases}$$

and $k_R = \operatorname{ess\,sup}_{x\in B_R} k(x)$. Let $T_{0,R}$ and T_R denote the operators defined by replacing τ in T_0 and T by τ_R. Since $q_{1,R} \geqslant -k_R$ it follows from Corollary 2.7

that $T^*_{0,R} = T^c_R = T^c_{0,R}$. For $u \in \mathscr{D}(T)$ we have $u \in L^2(\mathbb{R}^n)$ and $\tau_R u = \tau u + (q_{1,R} - q_1)u \in L^2_{loc}(\mathbb{R}^n)$ since $q_{1,R} - q_1$ is locally bounded. We therefore infer from Lemma 2.9 that $u \in W^{1,2}_{loc}(\mathbb{R}^n)$ and for $0 < r < R$,

$$\begin{aligned}\|u\|_{1,2,B_r} &\leqslant \|\tau_R u\|_{2,B_R} + K\|u\|_{2,B_R} \\ &= \|\tau u\|_{2,B_R} + K\|u\|_{2,B_R} \end{aligned} \tag{3.4}$$

where K depends on $R - r$. It follows that for $\phi \in C_0^\infty(B_R)$,

$$\tau(\phi u) = \tau_R(\phi u) = \phi\tau u - 2\nabla\phi\cdot\nabla u - u\Delta\phi \in L^2(\mathbb{R}^n).$$

Hence $\phi u \in \mathscr{D}(T)$ and, by Lemma 2.10, there exists a sequence (u_m) in $C_0^\infty(B_R)$ such that $u_m \to \phi u$ and $Tu_m \to T(\phi u)$ in $L^2(\mathbb{R}^n)$. The result $u_m \to \phi u$ in $W^{1,2}(\mathbb{R}^n)$ follows from Lemma 2.9 applied to $u_m - \phi u$. □

In the next two lemmas $K_1, K_2, \ldots$ denote various positive constants and we write $x = r\xi, r = |x|$ and $|\xi| = 1$; (r, ξ) are polar coordinates.

Lemma 3.2. Let ω be a bounded non-negative function in $AC_{loc}(0, \infty)$ and let $q = q_1 + iq_2$, with $q_1 = h_1 + h_2 + h_3$, satisfy (3.1). Given $\theta > 0$ suppose that there exists a non-negative function v (depending on θ) in $AC_{loc}(0, \infty)$ whose support is a compact subset of $(0, \infty)$ and which is such that the following conditions are satisfied for $r = |x| \in \operatorname{supp} v$:

(i) $h_1(x) \geqslant Q_1(r) \geqslant 0, |q_2(x)| \geqslant Q_2(r) \geqslant 0$ and, for some positive constants δ_1 and δ_2,

$$(1 + \delta_1)|\omega'(r)|^2 + \delta_2(n-1)\omega^2(r)/r^2 - Q_1(r)\omega^2(r) \leqslant K_1[1 + Q_2(r)];$$

(ii) $-h_2(x)\omega^2(r) \leqslant K_2 + \gamma^2 Q_2(r)$ for some positive constant γ;

(iii) $\omega^d(r)|Q_3(x)| \leqslant K_3$ for some $d \in [0, 1]$ and $(\partial/\partial r)Q_3(x) \leqslant h_3(x)\omega(r)^{1-d}$;

(iv) $0 \leqslant v(r) \leqslant K_4$;

(v) $|v'(r)| \leqslant \theta v(r)\omega(r)[1 + \omega(r)Q_1^{\frac{1}{2}}(r) + Q_2^{\frac{1}{2}}(r)] + K_5\omega(r)$;

(vi) $\omega(r)|v'(r)| \leqslant K_6$.

Then there exists $\theta > 0$ and a positive constant K depending on θ such that, for all $u \in \mathscr{D}(T)$,

$$\begin{aligned}\|v\omega|\nabla u|\|_2^2 + \|v\omega h_1^{\frac{1}{2}}u\|_2^2 + \|v|q_2|^{\frac{1}{2}}u\|_2^2 \\ \leqslant K(\|vTu\|_2\|u\|_2 + \|u\|_2^2).\end{aligned} \tag{3.5}$$

If $q_2 = 0$ then θ can be any fixed constant and (v) is redundant. ■

Proof. It is sufficient to prove (3.5) for $u \in C_0^\infty(\mathbb{R}^n)$ since the result for any $u \in \mathscr{D}(T)$ will then follow on using Lemma 3.1 with $\phi = 1$ on $\operatorname{supp} v$. To see this we need comment only on the second and third terms in (3.5). If k denotes the locally integrable function $v\omega h_1^{\frac{1}{2}}$ or $v|q_2|^{\frac{1}{2}}$ and $(u_m) \subset C_0^\infty(\mathbb{R}^n)$ is the sequence in Lemma 3.1 which is such that $u_m \to \phi u$ and $Tu_m \to T(\phi u)$, it follows from (3.5) for the u_m that (u_m) is a Cauchy sequence in the weighted spaces $L^2(\mathbb{R}^n; k)$. Since $u_m \to \phi u$ in $L^2(\mathbb{R}^n)$ the limit of (u_m) in $L^2(\mathbb{R}^n; k)$ must be ϕu. Hence (3.5) for $u \in \mathscr{D}(T)$ will follow on allowing $m \to \infty$ in the inequality for u_m.

For $u \in C_0^\infty(\mathbb{R}^n)$ we find, on integration by parts, that

$$\int v^2\omega^2\bar{u}Tu = \int [v^2\omega^2|\nabla u|^2 + \bar{u}\nabla u \cdot \nabla(v^2\omega^2) + v^2\omega^2 q|u|^2].$$

Hence

$$\text{re} \int v^2\omega^2\bar{u}Tu$$
$$\geqslant \|v\omega|\nabla u|\|^2 - 2\int v\omega|\bar{u}\nabla u \cdot \nabla(v\omega)| + \int v^2\omega^2(h_1+h_2+h_3)|u|^2. \quad (3.6)$$

Here and throughout the proof $\|\bullet\|$ denotes the $L^2(\mathbb{R}^n)$ norm. Also K will denote various positive constants which need not be the same on each appearance and $\varepsilon, \varepsilon_1, \varepsilon_2, \ldots$ will denote various small positive constants. Our task is to estimate the integrals on the right-hand side of (3.6). We shall make repeated use of the inequality $2|ab| \leqslant \varepsilon|a|^2 + \varepsilon^{-1}|b|^2$ which is valid for any $\varepsilon > 0$.

From (vi) and the boundedness of ω we have

$$2\int v\omega|\bar{u}\nabla u \cdot \nabla(v\omega)| \leqslant 2K_6\int v\omega|\bar{u}\nabla u| + 2\int v^2\omega|\omega'||\bar{u}\nabla u|$$
$$\leqslant (\varepsilon_1+\varepsilon_2)\|v\omega|\nabla u|\|^2 + \varepsilon_2^{-1}\|v\omega' u\| + K_{\varepsilon_1}\|u\|^2. \quad (3.7)$$

From (ii) and (iv),

$$-\int v^2\omega^2 h_2|u|^2 \leqslant K\|u\|^2 + \gamma^2\|v|q_2|^{\frac{1}{2}}u\|^2. \quad (3.8)$$

On integration by parts and using (iii), (iv) and (vi),

$$-\int v^2\omega^2 h_3|u|^2 \leqslant -\int v^2\omega^{1+d}\frac{\partial}{\partial r}Q_3|u|^2$$
$$= \int Q_3\left(v^2\omega^{1+d}\frac{\partial}{\partial r}|u|^2 + |u|^2\frac{\partial}{\partial r}(v^2\omega^{1+d}) + (n-1)r^{-1}v^2\omega^{1+d}|u|^2\right)$$
$$\leqslant 2K_3\int v^2\omega|u\nabla u| + K\int(v^2|\omega'| + v|v'|\omega)|u|^2 + (n-1)K_3\int r^{-1}v^2\omega|u|^2$$
$$\leqslant \varepsilon_3\|v\omega|\nabla u|\|^2 + K_{\varepsilon_3}\|u\|^2 + \varepsilon_4\|v\omega' u\|^2 + K_{\varepsilon_4}\|u\|^2 + K\|u\|^2$$
$$+ \varepsilon_5(n-1)\|r^{-1}v\omega u\|^2 + K_{\varepsilon_5}\|u\|^2$$
$$\leqslant \varepsilon_3\|v\omega|\nabla u|\|^2 + \varepsilon_4\|v\omega' u\|^2 + \varepsilon_5(n-1)\|r^{-1}v\omega u\|^2 + K_{\varepsilon_3,\varepsilon_4,\varepsilon_5}\|u\|^2. \quad (3.9)$$

On substituting (3.7)–(3.9) in (3.6) we have

$$\operatorname{re}\int v^2\omega^2\bar{u}Tu \geqslant (1-\varepsilon_1-\varepsilon_2-\varepsilon_3)\,\|v\omega|\nabla u|\,\|^2$$

$$+\int v^2[\omega^2 h_1-(\varepsilon_2^{-1}+\varepsilon_4)\,\omega'^2-\varepsilon_5\,(n-1)r^{-2}\omega^2]\,|u|^2$$

$$-\gamma^2\,\|v|q_2|^{\frac{1}{2}}u\|^2-K_{\varepsilon_1,\varepsilon_3,\varepsilon_4,\varepsilon_5}\,\|u\|^2.$$

We now make the choice

$$(1+\delta_1)^{-1}<\varepsilon_2<1,\ \varepsilon_2^{-1}+\varepsilon_4=(1+\delta_1)\eta,\quad \text{where } 0<\eta<1,\ \varepsilon_5=\delta_2\eta,$$

where δ_1 and δ_2 are the positive constants in (i), and choose $\varepsilon_1+\varepsilon_3<1-\varepsilon_2$. On applying (i) we obtain

$$\operatorname{re}\int v^2\omega^2\bar{u}Tu \geqslant K\,\|v\omega|\nabla u|\,\|^2+\int v^2\,[(1-\eta)\,\omega^2 h_1-\eta K_1-\eta K_1\,|q_2|]\,|u|^2$$

$$-\gamma^2\,\|v|q_2|^{\frac{1}{2}}u\|^2-K\,\|u\|^2$$

and hence

$$\|v\omega|\nabla u|\,\|^2+\|v\omega h_1^{\frac{1}{2}}u\|^2 \leqslant K(\|vTu\|\,\|u\|+\|v|q_2|^{\frac{1}{2}}u\|^2+\|u\|^2). \tag{3.10}$$

Next, we consider the identity

$$\operatorname{im}\int v^2\bar{u}Tu=\operatorname{im}\int \bar{u}(\nabla u\cdot\nabla v^2)+\int v^2 q_2\,|u|^2.$$

Suppose that in (3.1) (iii), q_2 is bounded above, $q_2 \leqslant M$ say, for some $M>0$; the case when q_2 is bounded below is treated in the same way. Since $|q_2| = |(M-q_2)-M| \leqslant 2M-q_2$, we have

$$\|v|q_2|^{\frac{1}{2}}u\|^2 \leqslant |\operatorname{im}\int v^2\bar{u}Tu|+2|\int v\bar{u}(\nabla u\cdot\nabla v)|+2M\,\|vu\|^2. \tag{3.11}$$

From (v) we obtain

$$2\left|\int v\bar{u}(\nabla u\cdot\nabla v)\right| \leqslant 2\int[\theta v^2\omega(1+\omega h_1^{\frac{1}{2}}+|q_2|^{\frac{1}{2}})\,|u\nabla u|+K_5 v\omega|u\nabla u|]$$

$$\leqslant 2\theta\|v\omega|\nabla u|\,\|\,\|v(1+\omega h_1^{\frac{1}{2}}+|q_2|^{\frac{1}{2}})u\|+2K_5\,\|v\omega|\nabla u|\,\|\,\|u\|$$

$$\leqslant 2\theta\,\|v\omega|\nabla u|\,\|\,(\|vu\|+\|v\omega h_1^{\frac{1}{2}}u\|+\|v|q_2|^{\frac{1}{2}}u\|)+2K_5\|v\omega|\nabla u|\,\|\,\|u\|$$

$$\leqslant \theta(2\|v\omega|\nabla u|\,\|^2+\|v\omega h_1^{\frac{1}{2}}u\|^2+\|v|q_2|^{\frac{1}{2}}u\|^2)+\varepsilon\|v\omega|\nabla u|\,\|^2+K_\varepsilon\|u\|^2,$$

on using (iv). On substituting in (3.11) and choosing $\theta<\frac{1}{2}$, we obtain

$$\|v|q_2|^{\frac{1}{2}}u\|^2 \leqslant 2\,\|vTu\|\,\|vu\|+2(2\theta+\varepsilon)\,\|v\omega|\nabla u|\,\|^2+2\theta\|v\omega h_1^{\frac{1}{2}}u\|^2+K_\varepsilon\|u\|^2. \tag{3.12}$$

If we now choose θ and ε to be small enough, (3.5) will follow from (3.10) and (3.12).

If $q_2 = 0$, (3.10) gives (3.5) and hence, in this case, θ can have any constant value; in fact (v) is redundant. □

In the next lemma the constant θ in Lemma 3.2 is not arbitrary but depends on the constant γ in (ii). To simplify matters we also assume that the terms in (i) are separately bounded and take $h_1 = 0$.

Lemma 3.3. Let ω and v be bounded non-negative functions in $\mathrm{AC}_{\mathrm{loc}}(0, \infty)$ and let v have compact support in $(0, \infty)$. Let $q = q_1 + iq_2$ and $q_1 = h_2 + h_3$ and suppose that the following conditions are satisfied for $r = |x| \in \operatorname{supp} v$:

(i) $-h_2(x)\omega^2(r) \leqslant K_1 + \gamma^2 Q_2(r)$, where $0 \leqslant Q_2(r) \leqslant |q_2(x)|$ and γ is a positive constant;

(ii) $\omega^d(r)|Q_3(x)| \leqslant K_2$ for some $d \in [0, 1]$ and $(\partial/\partial r) Q_3(x) \leqslant h_3(x)\omega(r)^{1-d}$;

(iii) $|v'(r)| \leqslant \theta v(r)\omega(r)[1 + Q_2^{\frac{1}{2}}(r)] + K_3\omega(r)$, where $\theta < 1/2\gamma$;

(iv) $\omega(r)|v'(r)| \leqslant K_4$;

(v) $|\omega'(r)| \leqslant K_5$.

Then for all $u \in \mathscr{D}(T)$,

$$\|v\omega|\nabla u|\|_2^2 + \|v|q_2|^{\frac{1}{2}}u\|_2^2 \leqslant K(\|vTu\|_2\,\|u\|_2 + \|u\|_2^2). \qquad (3.13)\ ■$$

Proof. As in Lemma 3.2 it is sufficient to prove the result for $u \in \mathrm{C}_0^\infty(\mathbb{R}^n)$. The proof is similar to that of Lemma 3.2, the only difference being that we have to keep track of the constant θ which is no longer arbitrary.

We still have (3.6) with $h_1 = 0$ and (3.8) but the estimates (3.7) and (3.9) are now replaced by the following:

$$2\int v\omega|\bar{u}\nabla u \cdot \nabla(v\omega)| \leqslant K\int v\omega|\bar{u}\nabla u|$$
$$\leqslant \varepsilon_1\|v\omega|\nabla u|\|^2 + K_{\varepsilon_1}\|u\|^2, \qquad (3.14)$$

(on using (iv), (v) and the boundedness of v)

$$-\int v^2\omega^2 h_3|u|^2$$
$$\leqslant \int Q_3\left(v^2\omega^{1+d}\frac{\partial}{\partial r}|u|^2 + |u|^2\frac{\partial}{\partial r}(v^2\omega^{1+d}) + (n-1)r^{-1}v^2\omega^{1+d}|u|^2\right)$$
$$\leqslant K\int(v\omega|u\nabla u| + |u|^2)$$
$$\leqslant \varepsilon_2\|v\omega|\nabla u|\|^2 + K_{\varepsilon_2}\|u\|^2, \qquad (3.15)$$

from (ii), (iv) and (v), since v, ω and $r^{-1}v$ are bounded, the last being so since v is assumed to be supported away from 0. On substituting (3.14), (3.8) and

(3.15) in (3.6) we obtain

$$\mathrm{re}\int v^2\omega^2\bar{u}Tu \geqslant (1-\varepsilon_1-\varepsilon_2)\,\|v\omega|\nabla u|\,\|^2-\gamma^2\,\|v|q_2|^{\frac{1}{2}}u\|^2-K_{\varepsilon_1,\varepsilon_2}\,\|u\|^2. \quad (3.16)$$

From (3.11) and (iii),

$$\begin{aligned}
&\|v|q_2|^{\frac{1}{2}}u\|^2\\
&\leqslant \left|\mathrm{im}\int v^2\bar{u}Tu\right|+2\theta\int v^2\omega(1+|q_2|^{\frac{1}{2}})\,|\bar{u}\nabla u|+2K_3\int v\omega\,|\bar{u}\nabla u|+K\,\|u\|^2\\
&\leqslant \|vTu\|\,\|vu\|+2\theta\|v\omega|\nabla u|\,\|\,\|v|q_2|^{\frac{1}{2}}u\|+K\int v\omega|\bar{u}\nabla u|+K\,\|u\|^2\\
&\leqslant \|vTu\|\,\|vu\|+2\theta^2\,\|v\omega|\nabla u|\,\|^2+\tfrac{1}{2}\|v|q_2|^{\frac{1}{2}}u\|^2+\varepsilon_3\|v\omega|\nabla u|\,\|^2+K_{\varepsilon_3}\,\|u\|^2,
\end{aligned}$$

and this yields

$$\|v|q_2|^{\frac{1}{2}}u\|^2\leqslant 2\|vTu\|\,\|vu\|+(4\theta^2+2\varepsilon_3)\,\|v\omega|\nabla u|\,\|^2+K_{\varepsilon_3}\,\|u\|^2. \quad (3.17)$$

Substitution of (3.17) into (3.16) gives

$$\begin{aligned}
&(1+2\gamma^2)\|vTu\|\,\|vu\|\\
&\qquad\geqslant (1-4\gamma^2\theta^2-2\gamma^2\varepsilon_3-\varepsilon_1-\varepsilon_2)\,\|v\omega|\nabla u|\,\|^2-K_{\varepsilon_1,\varepsilon_2,\varepsilon_3}\,\|u\|^2. \quad (3.18)
\end{aligned}$$

Since $2\gamma\theta<1$, we may choose ε_1, ε_2, ε_3 so small that (3.13) follows from (3.17) and (3.18). □

3.2. *Sufficiency conditions for $T_0 = T$*

Theorem 3.4. Let ω be a bounded non-negative function in $\mathrm{AC}_{\mathrm{loc}}[0,\infty)$ and let $q=q_1+\mathrm{i}q_2$ satisfy (3.1). Suppose that for some $a>0$ and all $r=|x|\geqslant a$ the conditions (i), (ii), (iii) of Lemma 3.2 are satisfied with $q_1=h_1+h_2+h_3$, and also that for any $\theta>0$,

$$\int_a^\infty \omega(r)\exp\left(\theta\int_a^r \omega(t)\,[1+\omega(t)\,Q_1^{\frac{1}{2}}(t)+Q_2^{\frac{1}{2}}(t)]\,\mathrm{d}t\right)=\infty \quad (3.19)$$

and

$$\omega^2(r)\,[1+\omega(r)\,Q_1^{\frac{1}{2}}(r)+Q_2^{\frac{1}{2}}(r)]\leqslant K. \quad (3.20)$$

Then $T_0^*=T_0^{\mathrm{c}}=T^{\mathrm{c}}$ and for any $u\in\mathscr{D}(T)$,

$$\|\omega|\nabla u|\,\|_2^2+\|\,|q_2|^{\frac{1}{2}}u\|_2^2\leqslant K(\|Tu\|_2\,\|u\|_2+\|u\|_2^2). \quad (3.21)$$

If h_1 is locally bounded we can add $\|\omega h_1^{\frac{1}{2}}u\|_2^2$ to the left-hand side of (3.21).

If $q_2 = 0$ we can replace (3.19) and (3.20) by the one condition

$$\int_a^\infty \omega(r)[1+\omega(r)Q_1^{\frac{1}{2}}(r)]\,\mathrm{d}r = \infty. \tag{3.22}$$ ■

Proof. In view of (3.19), $\omega(r)$ does not vanish identically for all large values of r. Suppose that a is such that $\omega(a) \neq 0$ and that $\omega(r) \neq 0$ on $[a, a+1]$; this involves no loss of generality since the length of the interval is of no significance. For $X > a$ we define

$$v(r) = \begin{cases} 0 \ \text{ if } r \in [0, a], \\ r-a \ \text{ if } r \in [a, a+1], \\ 1 \ \text{ if } r \in [a+1, X], \\ \exp[-A(r,\theta)] \displaystyle\int_r^Y \omega(t) \exp A(t,\theta)\,\mathrm{d}t \ \text{ if } r \in [X, Y], \\ 0 \ \text{ if } r \in [Y, \infty), \end{cases} \tag{3.23}$$

where

$$A(r,\theta) = \theta \int_X^r \omega(t)[1+\omega(t)Q_1^{\frac{1}{2}}(t)+Q_2^{\frac{1}{2}}(t)]\,\mathrm{d}t$$

and $Y = Y(X,\theta)$ is chosen such that

$$\int_X^Y \omega(t) \exp A(t,\theta)\,\mathrm{d}t = 1.$$

Such a choice of Y is possible on account of the hypothesis (3.19). It is an easy matter to check that the conditions of Lemma 3.2 are satisfied with this function v and we therefore conclude from (3.5) that

$$\int_{a+1 \leqslant |x| \leqslant X} (\omega^2 |\nabla u|^2 + \omega^2 h_1 |u|^2 + |q_2|\,|u|^2) \leqslant K(\|Tu\|\,\|u\| + \|u\|^2),$$

and since K is independent of X,

$$\int_{|x| \geqslant a+1} (\omega^2 |\nabla u|^2 + \omega^2 h_1 |u|^2 + |q_2|\,|u|^2) \leqslant K(\|Tu\|\,\|u\| + \|u\|^2). \tag{3.24}$$

We next prove that (3.21) is satisfied for all $u \in \mathscr{D}(T)$. Let $\phi \in C_0^\infty(B_r)$ where $r > a+1$ and $\phi = 1$ on B_{a+1}, and let $(u_m) \subset C_0^\infty(B_r)$ be the sequence determined in Lemma 3.1. Since q_1 is locally bounded below we have

$$\begin{aligned} \|\nabla u_m\|^2 &= \mathrm{re}\int \bar{u}_m T u_m - \int q_1 |u_m|^2 \\ &\leqslant \|u_m\|\,\|Tu_m\| + K_r \|u_m\|^2. \end{aligned}$$

On allowing $m \to \infty$ we have, since $\phi = 1$ on B_{a+1},

$$\begin{aligned}\|\nabla u\|_{2,B_{a+1}}^2 &\leqslant \|\nabla(\phi u)\|^2 \\ &\leqslant \|\phi u\| \, \|T(\phi u)\| + K_r \|\phi u\|^2 \\ &\leqslant K_r [\|u\| (\|Tu\| + \|\nabla u\|_{2,B_r} + \|u\|) + \|u\|^2] \\ &\leqslant K_r (\|u\| \, \|Tu\| + \|u\| \, \|\nabla u\|_{2,B_r} + \|u\|^2) \\ &\leqslant K_r (\|u\| \, \|Tu\| + \|u\|^2)\end{aligned}$$

by (3.4). Since ω is bounded, we therefore obtain

$$\|\omega |\nabla u| \|_{2,B_{a+1}}^2 \leqslant K(\|Tu\| \, \|u\| + \|u\|^2). \tag{3.25}$$

If q_2 is bounded above, say $q_2 \leqslant M$ with $M > 0$, we also have

$$\begin{aligned}\| |q_2|^{\frac{1}{2}} u_m \|^2 &\leqslant \int (2M - q_2) |u_m|^2 \\ &= 2M \|u_m\|^2 - \operatorname{im} \int \bar{u}_m T u_m \\ &\leqslant \|Tu_m\| \, \|u_m\| + 2M \|u_m\|^2 .\end{aligned}$$

It therefore follows that $|q_2|^{\frac{1}{2}} u \in \mathrm{L}^2(B_{a+1})$ and

$$\begin{aligned}\| |q_2|^{\frac{1}{2}} u \|_{2,B_{a+1}}^2 &\leqslant \| |q_2|^{\frac{1}{2}} \phi u \|^2 \\ &\leqslant \|T(\phi u)\| \, \|\phi u\| + 2M \|\phi u\|^2 \\ &\leqslant K[\|u\| (\|Tu\| + \|\nabla u\|_{2,B_r} + \|u\|) + \|u\|^2] \\ &\leqslant K(\|Tu\| \, \|u\| + \|u\|^2),\end{aligned} \tag{3.26}$$

on using (3.4). The three estimates (3.24), (3.25) and (3.26) combine to give (3.21). If h_1 is locally bounded, the term $\|\omega h_1^{\frac{1}{2}} u\|^2$ can also be added to the left-hand side of (3.21), in view of (3.24) and since $\|\omega h_1^{\frac{1}{2}} u\|_{2,B_{a+1}}^2 \leqslant K_a \|u\|^2$.

In order to prove that $T_0^* = T_0^c$, and hence $T_0 = T$, by (3.3), we shall prove that $T_0^* - \bar{\lambda} I$ is injective whenever $\operatorname{im}\lambda > M \geqslant q_2$. As noted earlier in §3, such values of λ lie outside the numerical range of T_0 and so, by Theorem III.2.3, if $T_0^* - \bar{\lambda} I$ is shown to be injective then it will follow that $\operatorname{def}(T_0 - \lambda I) = \operatorname{nul}(T_0^* - \bar{\lambda} I) = 0$; the identity $T_0^* = T_0^c$ will therefore be a consequence of Theorem III.5.5. With this strategy in mind, let us suppose that there exists some $u \in \mathscr{D}(T)$, with $u \neq 0$, such that $(T_0^* - \bar{\lambda} I)\bar{u} = 0$ and hence, by (3.3), $(T - \lambda I)u = 0$. Let (u_m) be a sequence in $\mathrm{C}_0^\infty(\mathbb{R}^n)$ which satisfies Lemma 3.1 for some $\phi \in \mathrm{C}_0^\infty(\mathbb{R}^n)$ which will be chosen later. From Green's formula, we obtain for any $t > 0$,

$$-2\mathrm{i}\int_{\bar{B}_t}[\mathrm{im}(\bar{u}_m Tu_m) - q_2|u_m|^2] = \int_{\bar{B}_t}(\bar{u}_m\,\Delta\, u_m - u_m\,\Delta\,\bar{u}_m)$$

$$= 2\mathrm{i}\;\mathrm{im}\int_{|x|=t} u_m(x/|x|\cdot\nabla\bar{u}_m)\,t^{n-1}\,\mathrm{d}\xi \qquad (x = |x|\,\xi).$$

Hence, for any Z in $(a+1, Y)$,

$$\int_Z^Y \omega(t)[1+\omega(t)Q_1^{\frac{1}{2}}(t)+Q_2^{\frac{1}{2}}(t)]\left(\int_{\bar{B}_t}[\mathrm{im}(\bar{u}_m Tu_m)-q_2|u_m|^2]\right)\mathrm{d}t$$

$$\leqslant \int_{\bar{B}_Y\backslash B_Z}\omega(1+\omega Q_1^{\frac{1}{2}}+Q_2^{\frac{1}{2}})\,|u_m\nabla u_m|$$

$$\leqslant K\int_{\bar{B}_Y\backslash B_Z}[|u_m|^2+\omega^2|\nabla u_m|^2+(\omega^2 Q_1+Q_2)\,|u_m|^2], \tag{3.27}$$

on using the Cauchy-Schwarz inequality and the boundedness of ω. We now choose ϕ in Lemma 3.1 to be 1 on $\bar{B}_Y$ so that $u_m \to u$ and $Tu_m \to Tu$ in $\mathrm{L}^2(B_t)$ for any $t \leqslant Y$. From (3.21), $\omega\,|\nabla u_m| \to \omega\,|\nabla u|$ and $|q_2|^{\frac{1}{2}}u_m \to |q_2|^{\frac{1}{2}}u$ in $\mathrm{L}^2(B_t)$, and from (3.24), $\omega h_1^{\frac{1}{2}}u_m \to \omega h_1^{\frac{1}{2}}u$ in $\mathrm{L}^2(B_Y\backslash B_{a+1})$. On allowing $m\to\infty$ in (3.27) it therefore follows that

$$\int_Z^Y \omega(t)[1+\omega(t)Q_1^{\frac{1}{2}}(t)+Q_2^{\frac{1}{2}}(t)]\left(\int_{\bar{B}_t}[\mathrm{im}(\bar{u}Tu)-q_2|u|^2]\right)\mathrm{d}t$$

$$\leqslant K\int_{\bar{B}_Y\backslash B_Z}[|u|^2+\omega^2|\nabla u|^2+(\omega^2 Q_1+Q_2)\,|u|^2],$$

the limit on the left-hand side being justified by the Dominated Convergence Theorem. Since $Tu = \lambda u$, and if $\mathrm{im}\,\lambda - q_2 \geqslant \delta > 0$, we obtain, by (3.24),

$$\delta\int_Z^Y \omega(t)[1+\omega(t)Q_1^{\frac{1}{2}}(t)+Q_2^{\frac{1}{2}}(t)]\,\|u\|_{2,B_t}^2\,\mathrm{d}t \leqslant K\,\|u\|^2,$$

and as K and δ are independent of Y, this gives

$$\int_Z^\infty \omega(t)[1+\omega(t)Q_1^{\frac{1}{2}}(t)+Q_2^{\frac{1}{2}}(t)]\,\|u\|_{2,B_t}^2\,\mathrm{d}t \leqslant K\,\|u\|^2.$$

On choosing Z such that $\|u\|_{2,B_t} > \frac{1}{2}\|u\|$ for $t > Z$ it follows that

$$\int_Z^\infty \omega(1+\omega Q_1^{\frac{1}{2}}+Q_2^{\frac{1}{2}}) < \infty.$$

But this is not possible since it implies that $\omega\in\mathrm{L}^1(0,\infty)$ and hence contradicts (3.19). We must therefore have $u = 0$ and $T_0 = T$.

If $q_2 = 0$, condition (v) in Lemma 3.2 is redundant and in this case we redefine the function v in (3.23) by making it linear between X and Y. The remainder of the proof follows without change. □

Theorem 3.5. Let ω be a bounded non-negative function in $\mathrm{AC}_{\mathrm{loc}}[0, \infty)$ and let $q = q_1 + iq_2$ satisfy (3.1). Suppose that for some $a > 0$ and any $r = |x| \geq a$ the conditions (i), (ii) and (v) in Lemma 3.3 are satisfied with $q_1 = h_2 + h_3$, and also assume that

$$\int_a^\infty \omega(r) \exp\left(\theta \int_a^r \omega(t)[1 + Q_2^{\frac{1}{2}}(t)]\, \mathrm{d}t\right) \mathrm{d}r = \infty, \tag{3.28}$$

and

$$\omega^2(r)[1 + Q_2^{\frac{1}{2}}(r)] \leq K \tag{3.29}$$

for some $\theta < 1/2\gamma$, where γ is the constant in Lemma 3.3(i). Then $T_0^* = T_0^{\mathrm{c}} = T^{\mathrm{c}}$ and for all $u \in \mathscr{D}(T)$,

$$\|\omega|\nabla u|\|_2^2 + \||q_2|^{\frac{1}{2}}u\|_2^2 \leq K(\|Tu\|_2\|u\|_2 + \|u\|_2^2). \tag{3.30}$$ ■

Proof. The proof is identical to that of Theorem 3.4 except that we use Lemma 3.3 instead of Lemma 3.2 and set $Q_1 = h_1 = 0$. □

We can extend the results in Theorems 3.4 and 3.5 to cover the differential expression

$$\nu = -\Delta + ib\cdot\nabla + q \tag{3.31}$$

when $b = (b_1, b_2, \ldots, b_n)$ is real-valued and div $b = 0$.

Theorem 3.6. Let q satisfy the hypotheses of either Theorem 3.4 or 3.5 and let $b = (b_1, b_2, \ldots, b_n)$ be such that each b_j is a real-valued, measurable function in $\mathrm{AC}_{\mathrm{loc}}(\mathbb{R}^n)$ and

$$|b|^2 := \sum_{j=1}^n |b_j|^2 \leq K\omega^2, \qquad \operatorname{div} b := \sum_{j=1}^n \mathrm{D}_j b_j = 0.$$

Then $S_0' := \nu \upharpoonright \mathrm{C}_0^\infty(\mathbb{R}^n)$ is closable in $\mathrm{L}^2(\mathbb{R}^n)$ and its closure S_0 satisfies

$$S_0^* = S^+ = S_0^+ \tag{3.32}$$

where

$$S := \nu \upharpoonright \mathscr{D}(S), \qquad \mathscr{D}(S) = \{u : u, \nu u \in \mathrm{L}^2(\mathbb{R}^n)\} = \mathscr{D}(T)$$

and the superscript + indicates that ν has been replaced by $\nu^+ = \bar{\tau} + ib\cdot\nabla$; also T is the operator generated by $\tau = -\Delta + q$ in Theorems 3.4 and 3.5. If q_2 is bounded above in (3.1), iS_0 is quasi-m-accretive. ■

Proof. The operator $P_0' := ib\cdot\nabla \upharpoonright \mathrm{C}_0^\infty(\mathbb{R}^n)$ is symmetric, and hence closable in

$L^2(\mathbb{R}^n)$, and from (3.21) and (3.30), we see that

$$\|P_0'\phi\|^2 \leqslant K\|\omega|\nabla\phi|\|^2 \leqslant K(\|T_0'\phi\|\|\phi\| + \|\phi\|^2)$$
$$\leqslant \varepsilon\|T_0'\phi\|^2 + K_\varepsilon\|\phi\|^2$$

for any $\varepsilon > 0$. The operator P_0' is therefore T_0'-bounded with relative bound 0. It follows by Lemma III.8.1 and Theorem III.8.2 that the closure P_0 of P_0' has T_0-bound 0 and $S_0 = T_0 + P_0$. Furthermore, on replacing τ by $\bar{\tau}$, we have that P_0 has T_0^c-bound zero and $S_0^+ = T_0^c + P_0$ with $\mathscr{D}(S_0^+) = \mathscr{D}(T_0^c) \subset \mathscr{D}(P_0)$.

Let $q_2 = \text{im}\, q$ be bounded above in (3.1) and suppose, without loss of generality, that $q_2 \leqslant 0$. Then, by Theorem 3.4 or 3.5 and the introductory remarks to §3, $\mathrm{i}T_0$ is m-accretive. As P_0 is symmetric $\mathrm{i}P_0$ is accretive and consequently, by Corollary III.8.5, $\mathrm{i}S_0 = \mathrm{i}T_0 + \mathrm{i}P_0$ is m-accretive. Similarly $-\mathrm{i}S_0^+$ is m-accretive. Moreover $(\mathrm{i}S_0)^*$ is m-accretive by virtue of Theorem III.6.6. But

$$-\mathrm{i}S_0^+ = -\mathrm{i}T_0^c - \mathrm{i}P_0$$
$$= (\mathrm{i}T_0)^* + (\mathrm{i}P_0)^*$$

(since $\mathscr{D}(P_0) \supset \mathscr{D}(T_0^c) = \mathscr{D}(T_0^*)$)

$$\subset (\mathrm{i}S_0)^*.$$

We therefore conclude that $-\mathrm{i}S_0^+ = (\mathrm{i}S_0)^*$, so that $S_0^* = S_0^+$. The proof of $T_0^* = T^c$ in Corollary 2.7 also applies to S_0 to give $S_0^* = S^+$. Consequently $S = S_0$ and in particular $\mathscr{D}(S) = \mathscr{D}(S_0) = \mathscr{D}(T_0) = \mathscr{D}(T)$. The theorem is therefore proved. □

In the first corollary we obtain from Theorem 3.4 the conditions are imposed only on a sequence of disjoint annuli in $\mathbb{R}^n\backslash\{0\}$, apart from the basic assumptions (3.1) which are always present. Outside the annuli q is not restricted and hence highly oscillatory behaviour is allowed.

Corollary 3.7. Let $S_k = \{x: a_k \leqslant |x| \leqslant b_k\}$ $(k = 1, 2, \ldots)$ be pairwise disjoint annuli in $\mathbb{R}^n\backslash\{0\}$ and suppose that in each S_k the following conditions are satisfied with $d_k = b_k - a_k$:

(i) $q_1(x) \geqslant -R_k$, with $R_k \geqslant 0$;

(ii) $|q_2(x)| \geqslant I_k \geqslant 0$;

(iii) $(1 + I_k)^{\frac{3}{2}}(R_k + d_k^{-2})^{-1} \leqslant K$;

(iv) $\sum_{k=1}^{\infty} d_k(1 + I_k)^{\frac{1}{2}}(R_k + d_k^{-2})^{-\frac{1}{2}} = \infty$.

Then $T_0^* = T_0^c = T^c$. ■

Proof. In Theorem 3.4 we choose $\omega = \Sigma_{k=1}^{\infty}\, \omega_k$ where each $\omega_k \in \mathrm{AC_{loc}}(0, \infty)$ has support in $[a_k, b_k]$ and is such that

$$\omega_k(r) = \delta_k := (1 + I_k)^{\frac{1}{2}}(R_k + d_k^{-2})^{-\frac{1}{2}} \quad \text{for } \tfrac{1}{3}d_k \leqslant r - a_k \leqslant \tfrac{2}{3}d_k$$

and

$$\omega_k'(r) = \begin{cases} 3\delta_k d_k^{-1} & \text{for } 0 \leqslant r - a_k < \frac{1}{3}d_k, \\ -3\delta_k d_k^{-1} & \text{for } \frac{2}{3}d_k < r - a_k \leqslant d_k. \end{cases}$$

It is an easy matter to verify that the conditions of Theorem 3.4 are satisfied with $q_1 = h_2$ and $h_1 = h_3 = 0$. □

An interesting special case of Corollary 3.7 is the following result which is a well-known criterion for T_0 to be self-adjoint when $q_2 = 0$; see Titchmarsh [1, Part II, Theorem 12.11].

Corollary 3.8. Let R and I be non-negative functions on $(0, \infty)$ which are respectively non-decreasing and non-increasing and suppose that for some $a > 0$ the following conditions are satisfied whenever $|x| \geqslant a$:

(i) $q_1(x) \geqslant -R(|x|)$;
(ii) $|q_2(x)| \geqslant I(|x|)$;
(iii) $[1 + I(|x|)]^{\frac{3}{2}} \leqslant KR(|x|)$;
(iv) $\displaystyle\int_a^\infty [1 + I(r)]^{\frac{1}{2}}[1 + R(r)]^{-\frac{1}{2}}\,\mathrm{d}r = \infty$.

Then $T_0^* = T_0^c = T^c$. ■

Proof. In Corollary 3.7 we choose annuli S_k to cover $\{x : |x| \geqslant a\}$ and which are such that $d_k \geqslant 2$ and $d_k \leqslant 2d_{k-1}$. Also we set $R_k = R(b_k)$ and $I_k = I(b_k)$. Then from (i) and (ii) we have $q_1(x) \geqslant -R_k$ and $|q_2(x)| \geqslant I_k$ in S_k. Also

$$\begin{aligned}(1 + I_k)^{\frac{3}{2}}(R_k + d_k^{-2})^{-1} &\leqslant [1 + I(b_k)]^{\frac{3}{2}}[R(b_k)]^{-1} \\ &\leqslant K\end{aligned}$$

by (iii), and from (iv) we see that

$$\begin{aligned}\infty &= \int_a^\infty [1 + I(r)]^{\frac{1}{2}}[1 + R(r)]^{-\frac{1}{2}}\,\mathrm{d}r \\ &\leqslant \sum_{k=1}^\infty d_k[1 + I(a_k)]^{\frac{1}{2}}[4d_k^{-2} + R(a_k)]^{-\frac{1}{2}} \\ &\leqslant 2\sum_{k=1}^\infty d_{k-1}[1 + I(b_{k-1})]^{\frac{1}{2}}[d_{k-1}^{-2} + R(b_{k-1})]^{-\frac{1}{2}} \\ &= 2\sum_{k=1}^\infty d_{k-1}(1 + I_{k-1})^{\frac{1}{2}}(d_{k-1}^{-2} + R_{k-1})^{-\frac{1}{2}}.\end{aligned}$$

The corollary therefore follows from Corollary 3.7. □

Another familiar result in the case $q_2 = 0$, due to Levinson [1] when $n = 1$, is the following variant of Corollary 3.8.

Corollary 3.9. Let R and I be non-negative functions on $(0, \infty)$ belonging to $AC_{loc}(0, \infty)$ and suppose that the following conditions are satisfied for $|x| \geqslant a > 0$:

(i) $q_1(x) \geqslant -R(|x|)$;
(ii) $|q_2(x)| \geqslant I(|x|)$;
(iii) $|R'(r)| \leqslant K[1 + R^{\frac{3}{2}}(r)]$ and $|I'(r)| \leqslant K[1 + I(r)]$;
(iv) $R(r) \geqslant K[1 + I(r)]^{\frac{3}{2}}$;
(v) $\int_a^\infty [1 + I(r)]^{\frac{1}{2}} [1 + R(r)]^{-\frac{1}{2}} \mathrm{d}r = \infty$.

Then $T_0^* = T_0^c = T^c$. ■

Proof. This follows from Theorem 3.4 on taking $q_1 = h_2$ and $\omega = (1 + I)^{\frac{1}{2}} (1 + R)^{-\frac{1}{2}}$. □

To demonstrate the relative growths of q_1 and q_2 allowed in Theorems 3.4 and 3.5 we have the following corollary.

Corollary 3.10. Suppose that (3.1) is satisfied and that, for $|x| \geqslant a_0 > 0$, there exist positive constants a and b such that

$$q_1(x) \geqslant -a^2 |x|^\alpha, \qquad |q_2(x)| \geqslant b^2 |x|^\beta.$$

Then $T_0^* = T_0^c = T^c$ in each of the following cases:

(i) $\beta \leqslant 0$ and $\alpha \leqslant 2$,
(ii) $\beta > 0$ and $\alpha < 2\beta + 2$,
(iii) $\beta > 0$, $\alpha = 2\beta + 2$, and $b^2 > \beta a$. ■

Proof. (i) In this case the hypothesis of Corollary 3.8 is satisfied with $R(r) = a^2 r^2$ and $I(r) = b^2 r^\beta$.

(ii) For this we use Theorem 3.4 with $q_1 = h_2$, $Q_2(r) = b^2 r^\beta$ and $\omega(r) = r^{-\frac{1}{2}\beta - \eta}$ in $[1, \infty)$, where $\alpha - 2\beta \leqslant 2\eta < 2$. The integral in (3.19) is divergent if

$$\int_1^\infty r^{-\frac{1}{2}\beta - \eta} \exp[\theta b r^{1-\eta}/(1-\eta)] \mathrm{d}r = \infty$$

for any $\theta > 0$, and this is so since $\eta < 1$. The remaining conditions imposed in Theorem 3.4 are easily checked and (ii) follows.

(iii) Theorem 3.5 is used here for the first time. We set $q_1 = h_2$, $Q_2(r) = b^2 r^\beta$ and $\omega(r) = r^{-\frac{1}{2}\beta - 1}$ in $[1, \infty)$. The integral in (3.28) is divergent if

$$\int_1^\infty r^{-\frac{1}{2}\beta - 1} \exp(\theta b \log r) \mathrm{d}r = \infty,$$

which is the case if $\theta \geqslant \beta/2b$. By Lemma 3.3 we must also have $2\theta\gamma < 1$ where $\gamma = a/b$, in our present terminology. Thus for Theorem 3.5 to be applicable we must be able to choose θ to satisfy $\beta/2b \leqslant \theta < b/2a$. This is possible in view of our assumption that $b^2 > \beta a$. □

When $n = 1$ we have from Corollary III.10.21 that T_0 is J-self-adjoint if, and only if, τ is in Sims' Case I at $\pm\infty$. If $q_1(x) = -a^2(|x|+1)^\alpha$ and $q_2(x) = -b^2(|x|+1)^\beta$ with $\beta > 0$, Theorem III.10.28 and the analogous result for the interval $(-\infty, -1]$ imply that T_0 is J-self-adjoint if, and only if, either $\alpha < 2\beta+2$ or $\alpha = 2\beta+2$ and $b^2 \geqslant \beta a$. When $\beta \leqslant 0$, the operator T_0 is J-self-adjoint if, and only if, $\alpha \leqslant 2$. On comparing these results with Corollary 3.10 we see that the exponents α and β have the best possible range of values in Corollary 3.10 and the only case of the J-self-adjointness of T_0 omitted is when $\alpha = 2\beta+2$ and $b^2 = \beta a$.

Finally we determine the type of oscillatory behaviour that q is allowed to have in Theorems 3.4 and 3.5. For simplicity we take $q_2 = 0$ so that the results give sufficiency criteria for T_0 to be self-adjoint.

Corollary 3.11. If $q(x) = \phi(\xi)r^\alpha \sin r^\beta$ with $x = r\xi$, then T_0 is self-adjoint in each of the following cases:

(i) $\alpha \leqslant 2$, $\beta \in \mathbb{R}$ and ϕ bounded on $|\xi| = 1$;

(ii) $\beta \leqslant 2$, $\alpha \geqslant 0$ and either $\phi \geqslant 0$ or $\phi \leqslant 0$ on $|\xi| = 1$;

(iii) $\alpha \leqslant \beta$ and ϕ bounded on $|\xi| = 1$. ■

Proof. (i) This is a special case of Corollary 3.10.

(ii) We apply Corollary 3.7 with $S_k = \{x: 2k\pi \leqslant |x|^\beta \leqslant (2k+1)\pi\}$, if $\phi(\xi) \geqslant 0$ on $|\xi| = 1$, and $R_k = I_k = 0$. Then $d_k \sim Kk^{1/\beta-1}$ for some positive constant K and Corollary 3.7 is readily verified when $1 \leqslant \beta \leqslant 2$. If $\beta < 1$ we choose any interval of unit length in each S_k and apply Corollary 3.7 to this sequence of intervals. If $\phi(\xi) \leqslant 0$ on $|\xi| = 1$ we choose $S_k = \{x: (2k+1)\pi \leqslant |x|^\beta \leqslant (2k+2)\pi\}$ and repeat the same argument.

(iii) This is obtained from Theorem 3.4 with $q = q_1 = h_3$, with $\omega(r) = r^{-1}$ in $[1, \infty)$, and with $d = 1$ and $(\partial/\partial r)Q_3(x) = \phi(\xi)r^\alpha \sin r^\beta$ in Lemma 3.2 (iii). On integration by parts we get

$$Q_3(x) = \phi(\xi)\textstyle\int^r t^\alpha \sin t^\beta \, dt$$

$$= \beta^{-1}\phi(\xi)\left(-r^{\alpha-\beta+1}\cos t^\beta + (\alpha-\beta+1)\int^r t^{\alpha-\beta}\cos t^\beta \, dt\right).$$

Consequently $\omega(|x|)Q_3(x) \leqslant K(|x|^{\alpha-\beta}) \leqslant K$ for $|x| \geqslant 1$, since $\alpha \leqslant \beta$. The other conditions in Theorem 3.4 are readily verified. □

When q is real, the results of this section relate to the essential-self-adjointness of the operator T_0' in $L^2(\mathbb{R}^n)$. This is a problem which has been worked on extensively over the years, much of the interest being motivated by its importance in quantum mechanics. Many of the important landmarks in this problem, and other related ones like that of the domain of the self-adjoint operator T_0, are given in Kalf [3] and in Chapter X of Reed and Simon [1].

The first comprehensive treatment when τ is a general second-order elliptic expression,

$$\tau = -\sum_{i,j=1}^{n} D_i(a_{ij}D_j) + \sum_{j=1}^{n} b_j D_j + q,$$

was given by Ikebe and Kato [1]. In this important paper they adapted the criterion of Titchmarsh in Corollary 3.8 ($q_2 = 0$) and used some ideas of Stummel [1]. It was shown by Eastham, Evans and McLeod [1] and by Evans [1] that the methods of the present section are effective for general expressions τ, a feature of the technique being that highly oscillatory coefficients are allowed; the significance of this fact is discussed in §5 below. For an up-to-date survey and the most general known result we refer to Kato [6]. An important class of operators are the Schrödinger operators with magnetic potentials. These are generated by expressions of the form

$$\tau = -\sum_{j=1}^{n} (D_j - ic_j)^2 + V \tag{3.33}$$

where the c_j and V are real, and it is of interest to know how badly the c_j can behave if T'_0 is to be essentially self-adjoint. A definitive answer is given by Leinfelder and Simader [1]. They prove that T'_0 is essentially self-adjoint if

$$c_j \in L^4_{\text{loc}}(\mathbb{R}^n), \qquad \operatorname{div} c \in L^2_{\text{loc}}(\mathbb{R}^n), \qquad 0 \leqslant V \in L^2_{\text{loc}}(\mathbb{R}^n), \tag{3.34}$$

where $c = (c_1, c_2, \ldots, c_n)$, these being minimal assumptions for T'_0 to be defined. Another significant question considered by Leinfelder and Simader and which is relevant to our §1 above is the following. If $t_{\max}$ is the sesquilinear form defined by

$$t_{\max}[\![u]\!] = \sum_{j=1}^{n} \|(D_j - ic_j)u\|^2 + \|V^{\frac{1}{2}}u\|^2 \tag{3.35}$$

on the maximal possible domain, consisting of functions $u \in L^2(\mathbb{R}^n)$ for which $t_{\max}[\![u]\!]$ is defined and finite, and $t_{\min}$ is the restriction of $t_{\max}$ to $C_0^\infty(\mathbb{R}^n)$, when is $t_{\max}$ the (form) closure of $t_{\min}$? They prove that this is so if only

$$c_j \in L^2_{\text{loc}}(\mathbb{R}^n), \qquad 0 \leqslant V \in L^1_{\text{loc}}(\mathbb{R}^n). \tag{3.36}$$

Earlier work on this problem was done by Kato [5], Schechter [3] and Simon [2, 4].

When T'_0 is essentially self-adjoint the self-adjoint operator $T_0 = T_0'^*$ has domain

$$\mathscr{D}(T_0) = \{u : u, \tau u \in L^2(\mathbb{R}^n)\}.$$

If, for instance, $\tau = -\Delta + q$, it is of interest to know when T_0 is the operator sum, i.e.

$$\mathscr{D}(T_0) = \mathscr{D}(-\Delta) \cap \mathscr{D}(q);$$

in this case T_0 is said to have the *separation property* or to be *separated.* A special case of a result proved by Evans and Zettl [1] is that separation occurs if on $\mathbb{R}^n$,

$$q(x) \geqslant -K, \qquad |\nabla q(x)| \leqslant \alpha |q(x)|^{\frac{3}{2}} \quad \text{for some } \alpha \in [0,2). \tag{3.37}$$

This is mild as a growth condition on q at infinity and is satisfied by functions like $\exp(|x|^k)$, $\exp[\exp(|x|^k)]$, etc. However, it places a restriction on the oscillatory behaviour of q. Results for functions q which are not bounded below on $\mathbb{R}^n$ will be given at the end of the next section.

4. Schrödinger operators with strongly singular potentials

In non-relativistic quantum mechanics the expression

$$\tau = -\Delta + b \cdot \nabla + q$$

in (1.1) formally represents the Hamiltonian of a finite system of charged particles in the presence of an electromagnetic field. For simplicity we restrict attention to the situation when there is no magnetic field, in which case $\tau = -\Delta + q$, where q is a real-valued function determined by the potential of the electric field and the interaction between the particles. A fundamental problem is to determine the self-adjoint Schrödinger operators generated by τ in $L^2(\mathbb{R}^n)$. It is usual to start with an operator like $T'_0 := \tau \restriction C_0^\infty(\mathbb{R}^n)$ which is obviously symmetric in $L^2(\mathbb{R}^n)$ and then to prove that T'_0 has either a unique self-adjoint extension, namely its closure T_0, or else that it has some self-adjoint extension which is distinguished in some physical sense. The case of q real in §3 is thus of special importance as it concerns the essential self-adjointness of T'_0. The results obtained in §3 are exhaustive for potentials q which are locally bounded below in $\mathbb{R}^n$ and are in $L^2_{\text{loc}}(\mathbb{R}^n)$. However, it is of physical, and mathematical, interest to work with the so-called *strongly singular potentials* q which behave like $|x|^{-2}$ at the origin; for the motivation and further background information on such problems we refer to Kalf, Schmincke, Walter and Wüst [1] and the references therein. If $q(x) \sim c|x|^{-2}$ near zero then $q \notin L^2_{\text{loc}}(\mathbb{R}^n)$ for $n \leqslant 4$ while q is not bounded below near zero if $c < 0$.

To start with, let $q(x) = c|x|^{-2}$ in $\mathbb{R}^n_+ = \mathbb{R}^n \setminus \{0\}$ for some real constant c and define

$$S'_0 := \tau \restriction C_0^\infty(\mathbb{R}^n_+), \qquad \tau = -\Delta + q. \tag{4.1}$$

Note that if $n \leqslant 4$ then T'_0 is not defined since $q \notin L^2_{\text{loc}}(\mathbb{R}^n)$ and so it is imperative to consider S'_0 in place of T'_0. In the Hardy inequality

$$\int_{\mathbb{R}^n} |\nabla\phi|^2 \geqslant (\tfrac{1}{2}n-1)^2 \int_{\mathbb{R}^n} |\phi(x)|^2/|x|^2 \, dx$$

(see (1.54)) the constant $(\frac{1}{2}n-1)^2$ is best possible for all $\phi \in C_0^\infty(\mathbb{R}^n)$ and also for all $\phi \in C_0^\infty(\mathbb{R}_+^n)$; the elegant proof of this by Shortley [1] is reproduced in Kalf, Schmincke, Walter and Wüst [1]. Consequently S_0' is bounded below (in fact non-negative) if, and only if, $c \geqslant -(\frac{1}{2}n-1)^2$ and the same applies to T_0' when it is defined for $n \geqslant 5$. However, $c \geqslant -(\frac{1}{2}n-1)^2$ does not ensure that S_0' is essentially self-adjoint, for we shall see in Proposition 4.1 and Corollary VIII.6.5 that, with $q=0$, $S_0' = -\Delta \restriction C_0^\infty(\mathbb{R}_+^n)$ is not essentially self-adjoint unless $n \geqslant 4$. When $n=1$, the operator S_0' is the orthogonal sum of the operators defined by τ on $C_0^\infty(-\infty, 0)$ and $C_0^\infty(0,\infty)$ and so, in view of Corollary III.10.21, the essential-self-adjointness of S_0' depends on the limit-point, limit-circle classification of τ at 0 and $\pm\infty$. On $(0,\infty)$ the differential equation

$$\tau\phi(x) = -\phi''(x) + cx^{-2}\phi(x) = 0$$

has the linearly independent solutions $\phi_i(x) = x^{\alpha_i}$ $(i=1,2)$ where

$$\alpha_1 = \tfrac{1}{2}[1+\sqrt{(1+4c)}], \qquad \alpha_2 = \tfrac{1}{2}[1-\sqrt{(1+4c)}].$$

Thus near zero $\phi_1 \in L^2$ but $\phi_2 \in L^2$ if, and only if, $\alpha_2 > -\frac{1}{2}$, i.e. $c < \frac{3}{4}$. We conclude that τ is in the limit-point case at zero if and only if $c \geqslant \frac{3}{4}$. Also, τ is clearly limit-point at ∞ (see also Theorem III.10.28). The same applies to the interval $(-\infty, 0)$ and so S_0' is essentially self-adjoint if, and only if, $c \geqslant \frac{3}{4}$.

For $n \geqslant 1$, it follows by separation of variables that S_0' is essentially self-adjoint if, and only if, the one-dimensional operators generated by the expressions

$$-d^2/dx^2 + c/r^2 + [\tfrac{1}{4}(n-1)(n-3) - \kappa_l]r^{-2} \qquad (l = 0, 1, \ldots)$$

on $C_0^\infty(0,\infty)$ are all essentially self-adjoint in $L^2(0,\infty)$, where $r = |x|$ and κ_l are the eigenvalues of the Laplace–Beltrami operator on $L^2(S^{n-1})$ (see Reed and Simon [1, Appendix to X.I, Example 4]). Since it is known that $\kappa_l \leqslant 0$ it follows from the previous paragraph that S_0' is essentially self-adjoint if, and only if, $c + \frac{1}{4}(n-1)(n-3) \geqslant \frac{3}{4}$, i.e. $c \geqslant 1-(\frac{1}{2}n-1)^2$. To summarize, we have the following result.

Proposition 4.1. Let $q(x) = c|x|^{-2}$ in $\mathbb{R}_+^n$. Then S_0' is bounded below (non-negative) if, and only if,

$$c \geqslant -(\tfrac{1}{2}n-1)^2 \tag{4.2}$$

and essentially self-adjoint if, and only if,

$$c \geqslant 1-(\tfrac{1}{2}n-1)^2. \tag{4.3}$$

When $n \geqslant 5$ the same applies to T_0'. ■

The essential-self-adjointness of S_0' for general, strongly singular potentials q have been studied by many authors. The definitive result given in Theorem 4.2 below is due to Simon [1]; in [2] Kalf and Walter give an alternative proof based on Simon's method. This result generalizes earlier work of Kalf and

Walter [1] and Schmincke [1] and is often referred to as the Kalf–Walter–Schmincke–Simon Theorem. Another proof was also given by Simader [1]. In his work on second-order elliptic operators on a general domain, Jörgens also made a significant contribution to this problem.

Theorem 4.2 (Kalf, Walter, Schmincke, Simon). Let $n \geqslant 2$, let $q \in L^2_{loc}(\mathbb{R}^n_+)$ and let $q(x) \geqslant [1-(\frac{1}{2}n-1)^2]|x|^{-2}$ in $\mathbb{R}^n_+$. Then S_0' in (4.1) is essentially self-adjoint. ■

Proof. Since S_0', and hence its closure S_0, is non-negative it is sufficient to show that $S_0'+2I$ has dense range in $L^2(\mathbb{R}^n)$; because then $\bar{S}_0+2I$ has range $L^2(\mathbb{R}^n)$, by Theorem III.2.3, and the essential-self-adjointness of S_0' follows from Theorem III.4.2 (iii). The main tool in Simon's proof is Kato's inequality in Theorem 2.2 with $\Omega = \mathbb{R}^n_+$:

$$u,\ \Delta u \in L^1_{loc}(\mathbb{R}^n_+) \Rightarrow \Delta|u| \geqslant \operatorname{re}[(\operatorname{sgn}\bar{u})\Delta u],$$

that is,

$$\int_{\mathbb{R}^n} |u|\Delta\phi \geqslant \operatorname{re}\int_{\mathbb{R}^n} [(\operatorname{sgn}\bar{u})\Delta u]\phi \qquad (0 \leqslant \phi \in C_0^\infty(\mathbb{R}^n_+)).$$

Suppose that $h \in L^2(\mathbb{R}^n)$ is orthogonal to the range of $S_0'+2I$. Then

$$\int_{\mathbb{R}^n} h(-\Delta+q+2)\phi = 0 \qquad (\phi \in C_0^\infty(\mathbb{R}^n_+))$$

and hence $(-\Delta+q+2)h = 0$ in $\mathscr{D}'(\mathbb{R}^n_+)$. Since $q \in L^2_{loc}(\mathbb{R}^n_+)$, we have $\Delta h = (q+2)h \in L^1_{loc}(\mathbb{R}^n_+)$ and so Kato's inequality applies to give $\Delta|h| \geqslant (q+2)|h|$ or

$$(-\Delta+q+2)|h| \leqslant 0.$$

Setting $Q(|x|) = [1-(\frac{1}{2}n-1)^2]|x|^{-2}$, we therefore have

$$\int_{\mathbb{R}^n} |h|(-\Delta+Q+2)\phi \leqslant \int_{\mathbb{R}^n} |h|(-\Delta+q+2)\phi$$
$$\leqslant 0 \qquad (0 \leqslant \phi \in C_0^\infty(\mathbb{R}^n_+)).$$

The next step is to show that there exist functions Φ_m (m = 1, 2, . . .) in $C_0^\infty(\mathbb{R}^n_+)$ which are non-negative and such that $\{(-\Delta+Q+2)\Phi_m\}$ converges weakly to some positive function Ψ in $L^2(\mathbb{R}^n)$. It will then follow that

$$\int_{\mathbb{R}^n} |h|\Psi \leqslant 0,$$

whence $h = 0$ as required. It is in the construction of Ψ that the proofs of Simon in [1] and Kalf and Walter in [2] differ. Simon obtains Ψ as the solution of a differential equation by cleverly altering the comparison potential Q, whereas the Ψ of Kalf and Walter satisfies a differential inequality. We shall adopt the construction of Kalf and Walter. First let

$$\Phi(x) := r^{\frac{1}{2}(1-n)} \frac{r^{\frac{3}{2}}}{1+r^{\frac{3}{2}}} e^{-r}, \qquad r = |x| \in (0, \infty). \tag{4.4}$$

Clearly $0 < \Phi \in L^2(\mathbb{R}^n)$ and it is readily shown that

$$(-\Delta + Q + 2)\Phi = \Psi$$

where

$$\Psi(x) = \Phi(x)\left\{1 + \frac{3}{r(1+r^{\frac{3}{2}})}\left[1 + \tfrac{1}{4}\sqrt{r}\left(1 + \frac{6}{1+r^{\frac{3}{2}}}\right)\right]\right\}, \qquad r \in (0, \infty).$$

Thus $0 < \Psi \in L^2(\mathbb{R}^n)$. The function Φ is chosen to behave at $r = 0$ and $r = \infty$ like the principal solution of the radial differential equation

$$-y'' - (n-1)r^{-1}y' + Q(r)y = \lambda y, \qquad r \in (0, \infty),$$

for some suitable number λ. We have

$$\Phi(r) = \begin{cases} O(r^{(4-n)/2}) \ \textit{as}\ r \to 0, \\ O(r^{(1-n)/2} e^{-r}) \ \text{as}\ r \to \infty, \end{cases} \tag{4.5}$$

$$\Phi'(r) = \begin{cases} O(r^{(2-n)/2}) \ \text{as}\ r \to 0, \\ O(r^{(1-n)/2} e^{-r}) \ \text{as}\ r \to \infty. \end{cases} \tag{4.6}$$

Let $\eta, \zeta \in C_0^\infty(\mathbb{R}^n)$ satisfy $0 \leqslant \eta, \zeta \leqslant 1$ and

$$\eta(x) = \begin{cases} 1 & \text{if } |x| \geqslant \frac{1}{2}, \\ 0 & \text{if } |x| \leqslant \frac{1}{4}, \end{cases}$$

$$\zeta(x) = \begin{cases} 1 & \text{if } |x| \leqslant 1, \\ 0 & \text{if } |x| \geqslant 2, \end{cases}$$

and define $\Phi_m(x) = \eta(mx)\zeta(x/m)\Phi(x)$ $(m \in \mathbb{N})$. Then $0 \leqslant \Phi_m \in C_0^\infty(\mathbb{R}^n_+)$ and

$$\begin{aligned}(-\Delta + Q + 2)\Phi_m &= \eta(m\bullet)\zeta(\bullet/m)(-\Delta + Q + 2)\Phi + 2\nabla[\eta(m\bullet)\zeta(\bullet/m)]\cdot\nabla\Phi \\ &\qquad + \Phi\Delta[\eta(m\bullet)\zeta(\bullet/m)] \\ &= \eta(m\bullet)\zeta(\bullet/m)\Psi + O\left(m\chi\left[\frac{1}{4m}, \frac{1}{2m}\right]|\nabla\Phi|\right) \\ &\quad + O\left(\frac{1}{m}\chi[m, 2m]\,|\nabla\Phi|\right) \\ &\quad + O\left(m^2\chi\left[\frac{1}{4m}, \frac{1}{2m}\right]|\Phi|\right) + O(m^{-2}\chi[m, 2m]\,|\Phi|) \\ &= \eta(m\bullet)\zeta(\bullet/m)\Psi + A_1(m) + A_2(m) + A_3(m) + A_4(m) \end{aligned} \tag{4.7}$$

say, where $\chi[a, b]$ denotes the characteristic function of $\{x : a \leqslant |x| \leqslant b\}$. From (4.5) and (4.6) it follows easily that as $m \to \infty$ so $A_i(m) \to 0$ in $L^2(\mathbb{R}^n)$. Since $\eta(mx)\zeta(x/m)\Psi(x) \to \Psi(x)$ for each $x \in \mathbb{R}^n_+$ as $m \to \infty$, and also $\Psi \in L^2(\mathbb{R}^n)$, we conclude from (4.7) that

$$(-\Delta + Q + 2)\Phi_m \rightharpoonup \Psi$$

in $L^2(\mathbb{R}^n)$. The Φ_m and Ψ therefore have all the desired properties and the proof is complete. □

Another important result, which we state without proof, is the following.

Theorem 4.3 (Kalf, Simader). Let $q \in L^2_{\text{loc}}(\mathbb{R}^n_+)$ and

$$q(x) \geqslant [1 - (\tfrac{1}{2}n - 1)^2]|x|^{-2} - c|x|^2 \qquad (x \in \mathbb{R}^n_+)$$

for some $c \in \mathbb{R}$. Then S_0' is essentially self-adjoint. ■

In the above form the result is due to Simader [1], Kalf having proved earlier in [2] the result with q assumed to belong to the Stummel space $Q^{\alpha}_{\text{loc}}(\mathbb{R}^n_+)$, i.e.

$$M_\alpha(x) := \int_{|x-y| \leqslant 1} |q(y)|^2 |x-y|^{-n+4-\alpha}\,dy, \qquad \alpha \in (0, 4)$$

is locally bounded in $\mathbb{R}^n_+$. Before Kato's inequality was available most results on the essential-self-adjointness of S_0' (T_0') assumed that $q \in Q^{\alpha}_{\text{loc}}(\mathbb{R}^n_+)$ $(Q^{\alpha}_{\text{loc}}(\mathbb{R}^n))$. This was mainly because of the result proved in the celebrated paper of Ikebe and Kato [1] that this assumption implies

$$\mathscr{D}(S_0'^*) = \{u : u \in W^{2,2}_{\text{loc}}(\mathbb{R}^n_+) \cap L^2(\mathbb{R}^n), \quad \tau u \in L^2(\mathbb{R}^n)\}$$

and similarly for $T_0'^*$ with $\mathbb{R}^n_+$ replaced by $\mathbb{R}^n$. This is valuable information since one way of proving essential-self-adjointness is by showing that the adjoint is symmetric.

If S_0' is bounded below but not essentially self-adjoint the distinguished self-adjoint extension of S_0' mentioned in the introductory paragraph to this section is usually the Friedrichs extension. A complete characterization of the domain of the Friedrichs extension is given by Kalf in [1] and [3].

If

$$q(x) = c|x|^{-\alpha}, \qquad c > 0, \quad \alpha > 0 \tag{4.8}$$

in (4.1) and S_0 is the Friedrichs extension of S_0', it is natural to ask when is S_0 the operator sum of $-\Delta$ and the multiplication operator q, i,e.,

$$\mathscr{D}(S_0) = \mathscr{D}(-\Delta) \cap \mathscr{D}(q). \tag{4.9}$$

This was partially answered by Davies in [1]; he proved that (4.9) holds if $\alpha < \frac{3}{2}$ or $\alpha > 2$ but is false if $\frac{3}{2} \leqslant \alpha < 2$. In the border-line case $\alpha = 2$ he found that (4.9) is false if $0 < c \leqslant \frac{3}{4}$ and true if $c > \frac{3}{2}$; the validity of (4.9) when $\alpha = 2$ and $c > \frac{3}{2}$ had already been established by Robinson [1]. The remaining case $\alpha = 2$

with $\frac{3}{4} < c \leqslant \frac{3}{2}$ was settled by Simon, who proved in [5] that (4.9) is true in this range.

5. Further remarks on self-adjointness and quantum mechanics

In the case when q is real, the theorems in §3 concerning the essential-self-adjointness of $T'_0 = (-\Delta + q)\upharpoonright C_0^\infty(\mathbb{R}^n)$ are linked to the quantum-mechanical interpretation of essential-self-adjointness as the state in which the particle (whose motion is described by the Schrodinger operator $-\Delta + q$) cannot escape to infinity in a finite time. If $q(x)$ is unbounded below as $|x| \to \infty$ then as the particle goes to infinity the potential energy may decrease and the kinetic energy increase so rapidly that the particle reaches infinity in a finite time. In such a case it must be given directions as to its next move and these amount to imposing boundary conditions at infinity so as to select a specific self-adjoint extension. Similar remarks apply if $q(x)$ has singularities at finite points that are too large and negative as in §4, the 'disaster' in this case being that the particle will collide with the centre of attraction. This correspondence between the lack of essential-self-adjointness and a finite escape time is not exact, since one is a phenomenon of quantum mechanics and the other of classical mechanics, and counter-examples to the correspondence can be constructed by investigating potentials which exploit quantum-mechanical effects not found in classical mechanics (see Rauch and Reed [1]). Nonetheless, the correspondence does suggest that the restriction on the negative growth of q in Corollary 3.8, say, is sufficient of a barrier to the outward progress of the particle and Corollary 3.7 implies that essential-self-adjointness is ensured if such a barrier occurs intermittently but sufficiently often.

The above quantum-mechanical interpretation also suggests that if q is sufficiently large and negative in a tube which extends to infinity, then the particle might escape to infinity in finite time along the tube, regardless of the behaviour of q elsewhere in $\mathbb{R}^n$, and so essential-self-adjointness will fail in such circumstances. That this is indeed the case is borne out by Theorem 5.1 below. A fuller and more precise discussion of essential-self-adjointness and its quantum mechanical interpretation can be found in Reed and Simon [1].

Theorem 5.1. Let $\tau = -\Delta + q$, where q is a real function in $L^2_{\mathrm{loc}}(\mathbb{R}^n)$. Assume further that $q(x) = q_1(x_1)$ in the tube $\mathbb{R}^+ \times \Omega_{n-1} = \{x = (x_1, \ldots, x_n) : x_1 \geqslant 0, (x_2, \ldots, x_n) \in \Omega_{n-1}\}$, where Ω_{n-1} is a bounded open set in $\mathbb{R}^{n-1}$. Then if $q_1(x_1)$ is such that the equation

$$-\frac{d^2}{dx_1^2}u + q_1(x_1)u = \lambda u$$

is in the limit-circle case at ∞, the operator $T'_0 = \tau \upharpoonright C_0^\infty(\mathbb{R}^n)$ is not essentially self-adjoint. ■

Proof. By Theorem III.4.1, T'_0 is essentially self-adjoint if, and only if, its adjoint is symmetric in $L^2(\mathbb{R}^n)$. From (3.3), $(T'_0)^*$ is the operator $T = \tau \restriction \mathscr{D}(T)$, where

$$\mathscr{D}(T) = \{u : u \in L^2(\mathbb{R}^n) \text{ and } \tau u \in L^2(\mathbb{R}^n)\}.$$

Thus to prove the theorem it is sufficient to construct a function $u \in \mathscr{D}(T)$ which is such that

$$\int_{Q_N} (u\tau\bar{u} - \bar{u}\tau u) \to L \neq 0$$

as $N \to \infty$, where Q_N is the cube in $\mathbb{R}^n$ with centre at the origin and side N.

Since $q_1(x_1)$ gives the limit-circle case, we know that every solution of the equation

$$-\frac{d^2y}{dx_1^2} + q_1(x_1)y = iy$$

lies in $L^2(0, \infty)$. Let h be a (non-trivial) solution with $h(1) = 0$ and set

$$u(x) = h^*(x_1)g(x_2, \ldots, x_n),$$

where

$$h^*(x_1) = \begin{cases} h(x_1) & \text{for } x_1 \geq 1, \\ 0 & \text{for } x_1 < 0, \end{cases}$$

with otherwise $h^* \in C^2(0, 1)$, and where g is a non-trivial real function in $C_0^\infty(\Omega_{n-1})$. Clearly $u \in \mathscr{D}(T)$ and

$$\int_{Q_N} (u\tau\bar{u} - \bar{u}\tau u) = -2i \int_{1 \leq x_1 \leq N} |u|^2$$

$$\not\to 0,$$

as required. □

From Theorem III.10.28, $q_1(x_1) = -Kx_1^{2+\alpha}$, with $\alpha > 0$, will do in Theorem 5.1 and this strong negative growth is typical of functions $q_1(x_1)$ which are such that $-d^2/dx_1^2 + q(x_1)$ is limit-circle at infinity. Theorem 5.1 is merely a sample of many results of this kind. For instance, if in the same tube $q(x) = q_1(x_1) + O(1)$, where $q_1(x_1)$ is the same as before, then T'_0 will not be essentially self-adjoint. Other results are possible; if, for example, $q(x) = q(|x|)$ in some cone with vertex at the origin and is such that, as a function of $r = |x|$, it gives rise to the limit-circle case in the radial equation obtained by separation of variables. In this case we choose for the test function u a suitable function of r multiplied by a function of the angular co-ordinates.

VIII

Capacity and compactness criteria

The main results of this chapter are necessary and sufficient conditions for the Poincaré inequality to hold, for the embedding $\mathrm{W}_0^{1,p}(\Omega) \to \mathrm{L}^p(\Omega)$ to be compact and for a self-adjoint realization of $-a_{ij}\mathrm{D}_i\mathrm{D}_j + q$ to have a wholly discrete spectrum when q is real and bounded below. The latter result is a notable one obtained by Molcanov [1]. We shall prove these results by the method of Maz'ja in [1]; see also Adams [1, Chapter VI].

1. Capacity and its basic properties

Let E be a compact subset of a non-empty open set Ω in $\mathbb{R}^n$, with $n \geqslant 1$, and let $p \in [1, \infty)$. By analogy with Definition VI.5.2 and Theorem VI.5.3 we define the *p-capacity* (with respect to $-\Delta$) of E relative to Ω by

$$p\text{-cap}\,(E, \Omega) := \inf\{\|\nabla v\|_{p,\Omega}^p : v \in \mathfrak{K}(E, \Omega)\} \tag{1.1}$$

where

$$\mathfrak{K}(E, \Omega) = \{v \in \mathrm{W}_0^{1,p}(\Omega) : v \succcurlyeq 1 \text{ on } E\}. \tag{1.2}$$

We shall continue to denote 2-cap (E, Ω) by cap (E, Ω). The proof of (5.4) in Theorem VI.5.3 continues to hold in $\mathrm{W}_0^{1,p}(\Omega)$ and we get

$$p\text{-cap}\,(E, \Omega) = \inf\{\|\nabla \phi\|_{p,\Omega}^p : \phi \in \mathfrak{K}_0(E, \Omega)\} \tag{1.3}$$

where

$\mathfrak{K}_0(E, \Omega) = \{\phi : \phi \in \mathrm{C}_0^\infty(\Omega),\ 0 \leqslant \phi \leqslant 1$ on Ω, and $\phi = 1$ in an open neighbourhood of E in $\Omega\}$.

By implication, functions in the convex sets $\mathfrak{K}(E, \Omega)$ and $\mathfrak{K}_0(E, \Omega)$ are real.

Lemma 1.1. Let E, E_1, E_2 be compact subsets of the non-empty open sets Ω, Ω_1, Ω_2 (in $\mathbb{R}^n$) respectively. Then the following results hold:

(i) p-cap $(\varnothing, \Omega) = 0$.

(ii) (Monotonicity). If $E_1 \subset E_2$ and $\Omega_1 \supset \Omega_2$ then

$$p\text{-cap}\,(E_1, \Omega_1) \leqslant p\text{-cap}\,(E_2, \Omega_2).$$

(iii) (Right continuity). For each $\varepsilon > 0$ there exists an open neighbourhood U of E, with $U \subset \Omega$, such that for any compact subset E' satisfying $E \subset E' \subset U$,

$$p\text{-cap}(E', \Omega) \leqslant p\text{-cap}(E, \Omega) + \varepsilon.$$

(iv) If $\Omega_2 \subset \Omega_1$ and Ω_1 is bounded then

$$p\text{-cap}(E_2, \Omega_1) \leqslant p\text{-cap}(E_2, \Omega_2) \leqslant K(1 + |\Omega_1|^{p/n}) p\text{-cap}(E_2, \Omega_1),$$

where K is a constant depending on p, n, and the distance between E_2 and Ω_2.

(v) If E_1 and E_2 are compact subsets of Ω,

$$p\text{-cap}(E_1 \cup E_2, \Omega) + p\text{-cap}(E_1 \cap E_2, \Omega) \leqslant p\text{-cap}(E_1, \Omega) + p\text{-cap}(E_2, \Omega).$$

(vi) If $\{E_j : j = 1, 2, \ldots\}$ is a decreasing sequence of compact subsets of Ω and $E = \cap_{j=1}^{\infty} E_j$ then

$$p\text{-cap}(E, \Omega) = \lim_{j \to \infty} [p\text{-cap}(E_j, \Omega)].$$ ■

Proof. (i) and (ii). These follow immediately from the definition.

(iii) Let $\varepsilon > 0$ be given and let $\phi \in \mathfrak{R}_0(E, \Omega)$ be such that $\phi(x) = 1$ on an open neighbourhood U of E in Ω and

$$\|\nabla \phi\|_{p,\Omega}^p \leqslant p\text{-cap}(E, \Omega) + \varepsilon.$$

If $E \subset E' \subset U$ then $\phi \in \mathfrak{R}_0(E', \Omega)$ and

$$\begin{aligned} p\text{-cap}(E', \Omega) &\leqslant \|\nabla \phi\|_{p,\Omega}^p \\ &\leqslant p\text{-cap}(E, \Omega) + \varepsilon. \end{aligned}$$

(iv) The first inequality is a special case of (ii). Let $\phi \in \mathfrak{R}_0(E_2, \Omega_1)$ and $\alpha \in \mathfrak{R}_0(E_2, \Omega_2)$. Then $\alpha\phi \in \mathfrak{R}_0(E_2, \Omega_2)$ and

$$\begin{aligned} p\text{-cap}(E_2, \Omega_2) &\leqslant \|\nabla(\alpha\phi)\|_{p,\Omega_2}^p \\ &\leqslant K_\alpha(\|\nabla\phi\|_{p,\Omega_1}^p + \|\phi\|_{p,\Omega_1}^p) \\ &\leqslant K_\alpha(1 + |\Omega_1|^{p/n} \|\nabla\phi\|_{p,\Omega_1}^p), \end{aligned}$$

the final step following from the Poincaré inequality (V.3.20), the constant K_α depending on bounds for $|\alpha|$ and $|\nabla\alpha|$ and hence on the distance between E_2 and Ω_2. This yields the result since $\phi \in \mathfrak{R}_0(E_2, \Omega_1)$ is arbitrary.

(v) Let $u_i \in \mathfrak{R}_0(E_i, \Omega)$ $(i = 1, 2)$ and set $\phi = \max(u_1, u_2)$ and $\psi = \min(u_1, u_2)$. Then ϕ and ψ are Lipschitz continuous and have compact support in Ω. Also $\phi(x) = 1$ in an open neighbourhood of $E_1 \cup E_2$ and $\psi(x) = 1$ in an open neighbourhood of $E_1 \cap E_2$. The set $\{x : u_1(x) \neq u_2(x)\}$ is the union of the open sets $\{x : u_1(x) < u_2(x)\}$ and $\{x : u_1(x) > u_2(x)\}$; also $\nabla u_1(x) = \nabla u_2(x)$ a.e. on the set $\{x : u_1(x) = u_2(x)\}$ (see Lemma VI.2.6). Consequently, on writing Ω as the union of the sets in which $u_1(x) < u_2(x)$, $u_1(x) > u_2(x)$, and

$u_1(x) = u_2(x)$ respectively, we obtain

$$\int_\Omega |\nabla\phi|^p + \int_\Omega |\nabla\psi|^p = \int_\Omega |\nabla u_1|^p + \int_\Omega |\nabla u_2|^p.$$

Given $\varepsilon > 0$, choose u_1 and u_2 such that $\|\nabla u_i\|_{p,\Omega}^p \leqslant p\text{-cap}\,(E_i, \Omega) + \varepsilon$ for $i = 1, 2$. Then

$$\|\nabla\phi\|_{p,\Omega}^p + \|\nabla\psi\|_{p,\Omega}^p \leqslant p\text{-cap}\,(E_1, \Omega) + p\text{-cap}\,(E_2, \Omega) + 2\varepsilon$$

and the result follows since $\phi \in \mathfrak{K}(E_1 \cup E_2, \Omega)$ and $\psi \in \mathfrak{K}(E_1 \cap E_2, \Omega)$.

(vi) This follows easily from (ii) and (iii). □

A function defined on the set of compact subsets of Ω and satisfying (ii)–(v) is called a *Choquet capacity*. We refer the reader to Maz'ja [1, 2, 3] for a fuller treatment of capacity and in particular of the capacity of non-compact subsets of Ω (see also §5 below). Maz'ja in [1, 2, 3] also works with a notion of capacity which allows him to deal with function spaces other than L^p and with derivatives of arbitrary order.

Lemma 1.1(iv) has the following extension when $n > p$.

Lemma 1.2. Let E be a compact subset of a bounded, non-empty, open set B in $\mathbb{R}^n$ with $n > p$, and let Ω be an open subset of $\mathbb{R}^n$ containing B. Then

$$p\text{-cap}\,(E, \Omega) \leqslant p\text{-cap}\,(E, B) \leqslant K(1 + |B|^{p/n})\,[p\text{-cap}\,(E, \Omega)]$$

where K is a positive constant which depends only on n, p and the distance between E and B. ■

Proof. The first inequality follows from Lemma 1.1(ii). Let $\phi \in \mathfrak{K}_0(E, \Omega)$ and $\alpha \in \mathfrak{K}_0(E, B)$. Then $\alpha\phi \in \mathfrak{K}_0(E, B)$ and

$$\begin{aligned} p\text{-cap}\,(E, B) &\leqslant \|\nabla(\alpha\phi)\|_{p,B}^p \\ &\leqslant K_\alpha(\|\nabla\phi\|_{p,B}^p + \|\phi\|_{p,B}^p) \\ &\leqslant K_\alpha(\|\nabla\phi\|_{p,B}^p + |B|^{p/n}\,\|\phi\|_{p^*,B}^p) \end{aligned}$$

(on using Hölder's inequality with $1/p^* = 1/p - 1/n$)

$$\begin{aligned} &\leqslant K_\alpha(\|\nabla\phi\|_{p,\Omega}^p + |B|^{p/n}\,\|\phi\|_{p^*,\Omega}^p) \\ &\leqslant K_\alpha(1 + |B|^{p/n})\,\|\nabla\phi\|_{p,\Omega}^p \end{aligned}$$

by Theorem V.3.6. □

Lemma 1.3. Let E be a non-empty compact subset of a bounded open set $\Omega \subset \mathbb{R}^n$ with $n < p$. Then

$$p\text{-cap}\,(E, \Omega) \geqslant K|\Omega|^{1-p/n}$$

where K is a positive constant depending only on p and n. ■

Proof. From (V.3.15) it follows that for any $\phi \in \mathfrak{K}_0(E, \Omega)$,

$$1 = \max_{x \in \Omega} |\phi(x)| \leqslant K |\Omega|^{1/n - 1/p} \|\nabla \phi\|_{p, \Omega},$$

whence the result. □

A significant consequence of Lemma 1.3 is that if $n < p$, even sets E consisting of single points have positive p-capacity. In this case only the empty set has zero, or even very small, p-capacity. When $n \geqslant p$, one can have sets E of arbitrarily small p-capacity and the p-capacity of E is related to its Lebesgue measure $|E|$ as follows.

Lemma 1.4. Let E be a compact subset of a bounded, non-empty, open set $\Omega \subset \mathbb{R}^n$ and let $n \geqslant p$. Then

$$p\text{-cap}\,(E, \Omega) \geqslant K |\Omega|^{1 - p/n - p/r} |E|^{p/r},$$

for any $r \in [1, pn/(n-p)]$ if $n > p$, and any $r \in [1, \infty)$ if $n = p$; the real number K is a constant depending only on p and n. ■

Proof. For any $\phi \in \mathfrak{K}_0(E, \Omega)$, we see from §§V.3.2 and V.3.3 that

$$\begin{aligned} |E|^{p/r} &\leqslant \|\phi\|_{r,\Omega}^p \\ &\leqslant K |\Omega|^{p/r - p/p^*} \|\nabla \phi\|_{p,\Omega}^p \end{aligned}$$

where $1/p^* = 1/p - 1/n$ with $n \geqslant p$, which yields the result. □

Finally we give two simple inequalities which will be used in subsequent sections. Firstly, let Q_d denote an open cube of side d and define

$$\mu_{n,p} := \sup \{ \|u\|_{p,Q_1} / \|\nabla u\|_{p,Q_1} : u \in C_0^\infty(Q_1),\ \|\nabla u\|_{p,Q_1} \neq 0 \}. \tag{1.4}$$

Then, by a similarity transformation,

$$\|\phi\|_{p,Q_d}^p \leqslant d^p \mu_{n,p}^p \|\nabla \phi\|_{p,Q_d}^p \qquad (\phi \in C_0^\infty(Q_d)) \tag{1.5}$$

and consequently, for any compact subset E of Q_d,

$$p\text{-cap}\,(E, Q_d) \geqslant d^{-p} \mu_{n,p}^{-p} |E|. \tag{1.6}$$

The constant $\mu_{n,2}^{-2}$ is the first eigenvalue in the Dirichlet problem for $-\Delta$ on Q_1 and we shall see later in §XI.2.3 that $\mu_{n,2}^{-2} = \pi^2 n$.

We also readily obtain, for any compact subset E of $\bar{Q}_d$,

$$\begin{aligned} p\text{-cap}\,(E, Q_{2d}) &\leqslant p\text{-cap}\,(\bar{Q}_d, Q_{2d}) \\ &= d^{n-p}\, p\text{-cap}\,(\bar{Q}_1, Q_2) \\ &\leqslant 2^{n+p} d^{n-p}, \end{aligned} \tag{1.7}$$

where Q_{2d} is the cube of side $2d$ which is concentric to Q_d.

2. Some integral inequalities

Most of the results in this section are primarily tools which will subsequently be used to obtain a necessary and sufficient condition for the compactness of the embedding $W_0^{1,p}(\Omega) \to L^p(\Omega)$ and also to prove Molcanov's criterion for a discrete spectrum. However the main theorems are of intrinsic interest. Theorems 2.8 and 2.9 give a necessary and sufficient condition for the validity of an inequality of the form

$$\|u\|_{p,\Omega}^p \leqslant K\left(\|\nabla u\|_{p,\Omega}^p + \int_\Omega q\,|u|^p\right) \qquad (u \in C_0^\infty(\Omega)),$$

when Ω is an arbitrary open subset of $\mathbb{R}^n$, with $n \geqslant 1$, and q is a real $L^1_{\mathrm{loc}}(\Omega)$ function which is positive a.e. on Ω. In Theorem 2.10 we obtain a necessary and sufficient condition for the Poincaré inequality

$$\|u\|_{p,\Omega}^p \leqslant K\,\|\nabla u\|_{p,\Omega}^p \qquad (u \in C_0^\infty(\Omega))$$

to hold.

First we need some preliminary lemmas. We denote by Q_d an open cube of side d in $\mathbb{R}^n$, and let $\bar{Q}_d$ stand for the closure of this cube and Q_{cd} the concentric cube of side cd. Without loss of generality, we assume that the functions in this section are real.

Lemma 2.1. Let E be a closed subset of $\bar{Q}_d$ and let $u \in C^\infty(\bar{Q}_d)$ be such that $u \geqslant 1$ on E. Then

$$p\text{-cap}\,(E, Q_{2d}) \leqslant K(\|\nabla u\|_{p,Q_d}^p + d^{-p}\,\|u\|_{p,Q_d}^p) \tag{2.1}$$

where K is a positive constant depending only on p and n. ■

Proof. It suffices to prove the result for $d = 1$ since the stated result will then follow by a similarity transformation. From the proof of Theorem V.4.11 there is an extension of u to a function $v \in C^\infty(\mathbb{R}^n) \cap W^{1,p}(\mathbb{R}^n)$ such that $v = u$ on $\bar{Q}_1$ and

$$\|v\|_{1,p,\mathbb{R}^n} \leqslant K\,\|u\|_{1,p,Q_1}.$$

Let $\theta \in C_0^\infty(Q_2)$ be such that $\theta = 1$ on $\bar{Q}_1$ and $0 \leqslant \theta \leqslant 1$. Then $\theta v \in W_0^{1,p}(Q_2)$ and $\theta v \geqslant 1$ on E. By Theorem VI.3.2, it follows that $\theta v \in \mathfrak{K}(E, Q_2)$ in the notation of (1.2). Consequently

$$\begin{aligned} p\text{-cap}\,(E, Q_2) &\leqslant \|\nabla(\theta v)\|_{p,Q_2}^p \\ &\leqslant K(\|\nabla v\|_{p,Q_2}^p + \|v\|_{p,Q_2}^p) \\ &\leqslant K(\|\nabla u\|_{p,Q_1}^p + \|u\|_{p,Q_1}^p) \end{aligned}$$

as asserted. □

The next lemma is interesting in itself in that it implies (see Corollary 2.3) a necessary and sufficient condition for Poincaré's inequality to hold for functions in $C^\infty(\bar{Q}_d)$ which vanish on some closed subset E of $\bar{Q}_d$. If E contains an open neighbourhood of the boundary of Q_d the functions under consideration are in $C_0^\infty(Q_d)$ and hence we know from Theorem V.3.22 that the inequality is valid. It is the criterion for the size of E which is the significant feature in the following lemma.

Lemma 2.2. Let E be a closed subset of $\bar{Q}_d$.

(i) If $u \in C^\infty(\bar{Q}_d)$ is such that $u(x) \leqslant 0$ on E, the integral mean $u_{Q_d} = |Q_d|^{-1} \int_{Q_d} u \geqslant 0$ and p-cap $(E, Q_{2d}) > 0$, then

$$\|u\|^p_{r,Q_d} \leqslant A\|\nabla u\|^p_{p,Q_d} \tag{2.2}$$

for any r satisfying

$$r \in \begin{cases} [1, pn/(n-p)] & (n > p), \\ [1, \infty) & (n = p), \\ [1, \infty] & (n < p), \end{cases}$$

and any A satisfying

$$A^{-1} \geqslant Kd^{-pn/r}[p\text{-cap}\,(E, Q_{2d})], \tag{2.3}$$

where $K = K(n, p)$.

(ii) Suppose that (2.2) holds for all $u \in C^\infty(\bar{Q}_d)$ which vanish in an open neighbourhood of E in $\bar{Q}_d$ and let $p\text{-cap}\,(E, Q_{2d}) \leqslant \gamma d^{n-p}$, where $\gamma \leqslant 2^{-2p-1}\mu_{n,p}^{-p}$ and $\mu_{n,p}$ is the constant in (1.4). Then

$$A^{-1} \leqslant 2^p d^{-pn/r}[p\text{-cap}\,(E, Q_{2d})]. \tag{2.4}$$ ∎

Proof. It suffices to prove the lemma for $d = 1$; the factors $d^{-pn/r}$ in (2.3) and (2.4), and d^{n-p} in $p\text{-cap}\,(E, Q_{2d})$, are the scaling factors obtained from a similarity transformation.

(i) Let $N = \|u\|_{p,Q_1}$ and $\phi = 1 - N^{-1}u$. Then $\phi \in C^\infty(\bar{Q}_1)$ and $\phi(x) \geqslant 1$ on E. Therefore by Lemma 2.1,

$$\begin{aligned} N^p[p\text{-cap}\,(E, Q_2)] &\leqslant KN^p(\|\nabla\phi\|^p_{p,Q_1} + \|\phi\|^p_{p,Q_1}) \\ &= K(\|\nabla u\|^p_{p,Q_1} + \|N - u\|^p_{p,Q_1}). \end{aligned} \tag{2.5}$$

Since $0 \leqslant u_{Q_1} \leqslant \|u\|_{p,Q_1} = N$, we have

$$\begin{aligned} |N - u_{Q_1}| &= N - u_{Q_1} \\ &\leqslant \|u - u_{Q_1}\|_{p,Q_1} \end{aligned}$$

and

$$\begin{aligned}\|N-u\|_{p,Q_1} &\leqslant \|N-u_{Q_1}\|_{p,Q_1}+\|u-u_{Q_1}\|_{p,Q_1}\\ &\leqslant 2\,\|u-u_{Q_1}\|_{p,Q_1}\\ &\leqslant K\,\|\nabla u\|_{p,Q_1},\end{aligned}$$

on using the Poincaré inequality of Theorem V.3.23. Thus, by (2.5),

$$\|u\|_{p,Q_1}^p \, p\text{-cap}\,(E, Q_2) \leqslant K\,\|\nabla u\|_{p,Q_1}^p. \tag{2.6}$$

Next we apply the Sobolev Embedding Theorem V.4.13, from which we have $\|u\|_{r,Q_1} \leqslant K\,\|u\|_{1,p,Q_1}$, and (2.6) to obtain

$$\begin{aligned}\|u\|_{r,Q_1}^p &\leqslant K\,(\|\nabla u\|_{p,Q_1}^p+\|u\|_{p,Q_1}^p)\\ &\leqslant K\,\{1+[p\text{-cap}\,(E,Q_2)]^{-1}\}\,\|\nabla u\|_{p,Q_1}^p\\ &\leqslant K\,[p\text{-cap}\,(E,Q_2)]^{-1}\,\|\nabla u\|_{p,Q_1}^p,\end{aligned}$$

by (1.7).

(ii) Let $\psi \in \mathfrak{K}_0(E, Q_2)$ be such that for $0 < \varepsilon < \gamma$,

$$\|\nabla\psi\|_{p,Q_2}^p \leqslant p\text{-cap}\,(E,Q_2)+\varepsilon < 2\gamma$$

and set $u = 1-\psi$. On using the Poincaré inequality and (1.4),

$$\begin{aligned}0 \leqslant \psi_{Q_1} &= \int_{Q_1}\psi\\ &\leqslant \|\psi\|_{p,Q_1} \leqslant \|\psi\|_{p,Q_2}\\ &\leqslant 2\mu_{n,p}\,\|\nabla\psi\|_{p,Q_2}\\ &\leqslant 2^{1+1/p}\mu_{n,p}\gamma^{1/p} \leqslant \tfrac{1}{2}.\end{aligned}$$

Hence $u_{Q_1} = 1-\psi_{Q_1} \geqslant \frac{1}{2}$ and

$$1 \leqslant 2u_{Q_1} \leqslant 2\,\|u\|_{r,Q_1}.$$

Also,

$$\begin{aligned}\|\nabla u\|_{p,Q_1}^p &\leqslant \|\nabla\psi\|_{p,Q_2}^p\\ &\leqslant p\text{-cap}\,(E,Q_2)+\varepsilon.\end{aligned}$$

On substituting these last two inequalities in (2.2) we find that

$$A^{-1} \leqslant 2^p\,[p\text{-cap}\,(E,Q_2)+\varepsilon]$$

and (2.4) follows when $d = 1$ since ε is arbitrary. The proof of the lemma is therefore complete when $d = 1$ and hence for all values of d. □

Corollary 2.3. Let E be a closed subset of $\bar{Q}_d$. Then (2.2) is satisfied by all $u \in C^\infty(\bar{Q}_d)$ which vanish on E if, and only if, $p\text{-cap}\,(E, Q_{2d}) > 0$. ■

Proof. Since u vanishes on E the requirement $u_{Q_d} \geqslant 0$ in Lemma 2.2(i) is redundant, since we could replace u by $-u$ if necessary. The sufficiency of p-cap $(E, Q_{2d}) > 0$ is therefore immediate.

On the other hand, if (2.2) is satisfied by the functions in the corollary it is satisfied for all functions in Lemma 2.2(ii). Moreover, if p-cap $(E, Q_{2d}) = 0$, then p-cap $(E, Q_{2d}) < \gamma d^{n-p}$ for any $\gamma > 0$ and (2.4) provides the contradiction sought. □

Corollary 2.4. Let $\Omega^c = \mathbb{R}^n \setminus \Omega$ and define any $v \in C_0^\infty(\Omega)$ to be zero outside Ω.

(i) If p-cap $(\bar{Q}_d \cap \Omega^c, Q_{2d}) > 0$ then for any $v \in C_0^\infty(\Omega)$,

$$\|v\|_{r, Q_d}^p \leqslant A \|\nabla v\|_{p, Q_d}^p \tag{2.7}$$

where for some $K = K(n, p)$,

$$A^{-1} \geqslant K d^{-pn/r} \, p\text{-cap}\, (\bar{Q}_d \cap \Omega^c, Q_{2d})$$

and r is as in Lemma 2.2.

(ii) If $n < p$ and $\bar{Q}_d \cap \Omega^c \neq \varnothing$, then there exists $K = K(n, p)$ such that for all $v \in C_0^\infty(\Omega)$,

$$\max_{\bar{Q}_d} |v(x)|^p \leqslant K d^{n-p} \|\nabla v\|_{p, Q_d}^p. \tag{2.8}$$ ■

Proof. Any $v \in C_0^\infty(\Omega)$ vanishes on $E = \bar{Q}_d \cap \Omega^c$ and (i) is an immediate consequence of Lemma 2.2(i) for the reason noted in the first part of the proof of Corollary 2.3.

When $n < p$, Lemma 1.3 yields p-cap $(\bar{Q}_d \cap \Omega^c, Q_{2d}) \geqslant K d^{n-p}$ and (2.8) follows from (2.7) with $r = \infty$. □

In subsequent lemmas we shall be concerned with closed subsets E of $\bar{Q}_d$ which are such that p-cap $(E, Q_{2d}) < \varepsilon d^{n-p}$ for small values of ε. When $n \geqslant p$ we see from (1.6) that

$$|E| \leqslant \varepsilon 2^p d^n \mu_{n,p}^p \tag{2.9}$$

and hence

$$\begin{aligned} |\bar{Q}_d \setminus E| &\geqslant (1 - \varepsilon 2^p \mu_{n,p}^p) d^n \\ &> 0 \end{aligned}$$

if $\varepsilon < 1/(2\mu_{n,p})^p$.

Lemma 2.5. Let q be a real integrable function on Q_d which is positive a.e. on Q_d. Given $\varepsilon \in (0, 2^{-p}\mu_{n,p}^{-p})$ there exists a positive constant K, depending only on p and n, such that for all $u \in C^\infty(\bar{Q}_d)$,

$$\|u\|_{p, Q_d}^p \leqslant K \left[\varepsilon^{-1} d^p \|\nabla u\|_{p, Q_d}^p + \left(d^n / \inf \int_{\bar{Q}_d \setminus E} q \right) \int_{Q_d} q |u|^p \right], \tag{2.10}$$

where the infimum is taken over all the compact subsets $E \subset \bar{Q}_d$ which are such that

$$p\text{-cap}\,(E, Q_{2d}) \leqslant \varepsilon d^{n-p}. \tag{2.11}$$ ■

Proof. We may assume, without loss of generality, that $u_{Q_d} \geqslant 0$ since otherwise we work with $-u$. Set

$$2\tau = d^{-n/p}\|u\|_{p,Q_d}, \qquad E_\tau = \{x \in \bar{Q}_d : u(x) \leqslant \tau\}.$$

Then

$$\begin{aligned}\|u\|_{p,Q_d} &\leqslant \|u-\tau\|_{p,Q_d} + \tau d^{n/p} \\ &\leqslant \|u-\tau\|_{p,Q_d} + \tfrac{1}{2}\|u\|_{p,Q_d},\end{aligned}$$

and

$$\|u\|_{p,Q_d} \leqslant 2\|u-\tau\|_{p,Q_d}.$$

We first consider the case when $p\text{-cap}\,(E_\tau, Q_{2d}) > \varepsilon d^{n-p}$. If $u_{Q_d} \geqslant \tau$ we find, on applying Lemma 2.2(i) to $u-\tau$ and E_τ, that

$$\begin{aligned}\|u\|^p_{p,Q_d} &\leqslant 2^p\|u-\tau\|^p_{p,Q_d} \\ &\leqslant A\|\nabla u\|^p_{p,Q_d}\end{aligned}$$

where

$$\begin{aligned}A &\leqslant K d^n[p\text{-cap}\,(E_\tau, Q_{2d})]^{-1} \\ &\leqslant K\varepsilon^{-1}d^p.\end{aligned}$$

If $u_{Q_d} < \tau$ we argue as follows. The Poincaré inequality yields

$$\|u-u_{Q_d}\|_{p,Q_d} \leqslant Kd\|\nabla u\|_{p,Q_d}.$$

Since

$$\|u\|_{p,Q_d} - 2d^{n/p}u_{Q_d} > \|u\|_{p,Q_d} - 2d^{n/p}\tau = 0,$$

we have

$$\begin{aligned}\|u\|_{p,Q_d} &\leqslant 2(\|u\|_{p,Q_d} - d^{n/p}u_{Q_d}) \\ &\leqslant 2(\|u-u_{Q_d}\|_{p,Q_d} + \|u_{Q_d}\|_{p,Q_d} - d^{n/p}u_{Q_d}) \\ &= 2\|u-u_{Q_d}\|_{p,Q_d} \\ &\leqslant Kd\|\nabla u\|_{p,Q_d}.\end{aligned}$$

Hence if $p\text{-cap}(E_\tau, Q_{2d}) > \varepsilon d^{n-p}$ we have shown that

$$\|u\|^p_{p,Q_d} \leqslant K\varepsilon^{-1}d^p\|\nabla u\|^p_{p,Q_d}. \tag{2.12}$$

If, on the other hand, $p\text{-cap}(E_\tau, Q_{2d}) \leqslant \varepsilon d^{n-p}$ then

$$\begin{aligned}\|u\|^p_{p,Q_d} &= 2^p d^n \tau^p \\ &\leqslant \left(2^p d^n \Big/ \int_{\bar{Q}_d\setminus E_\tau} q\right)\int_{\bar{Q}_d\setminus E_\tau} q|u|^p \\ &\leqslant \left(2^p d^n \Big/ \inf \int_{\bar{Q}_d\setminus E} q\right)\int_{Q_d} q|u|^p\end{aligned}$$

where the infimum is over all the compact subsets E of $\bar{Q}_d$ which satisfy (2.11). The lemma is therefore proved. □

The next result is a converse of Lemma 2.5.

Lemma 2.6. Let E be a closed subset of $\bar{Q}_d$ satisfying

$$p\text{-cap}(E, Q_{2d}) < \varepsilon d^{n-p}, \quad 0 < \varepsilon < c_0 := 1/2^{n+p+1}\mu_{n,p}^p, \tag{2.13}$$

and let q be non-negative and integrable on Q_d. Suppose that there exist positive constants A and B such that for all $u \in C^\infty(\bar{Q}_d)$ which vanish in an open neighbourhood of E in $\bar{Q}_d$,

$$\|u\|_{p,Q_{d/2}}^p \leqslant A\|\nabla u\|_{p,Q_d}^p + B\int_{Q_d} q|u|^p. \tag{2.14}$$

Then there exists $K = K(n, p)$ such that

$$d^n \leqslant K\left(A\varepsilon d^{n-p} + B\inf\int_{\bar{Q}_d\setminus F} q\right)$$

where the infimum is over all the closed subsets of $\bar{Q}_d$ which satisfy (2.13). ■

Proof. It suffices to consider the case $d = 1$. Let $G = E \cup F$, where F, as well as E, satisfies (2.13). From the Choquet inequality in Lemma 1.1(v),

$$\begin{aligned} p\text{-cap}(G, Q_2) &\leqslant p\text{-cap}(E, Q_2) + p\text{-cap}(F, Q_2) \\ &< 2\varepsilon. \end{aligned}$$

Let $\psi \in \mathfrak{K}_0(G, Q_2)$ be such that

$$\|\nabla\psi\|_{p,Q_2}^p < 2\varepsilon \tag{2.15}$$

and set $u = 1 - \psi$. Then u vanishes in an open neighbourhood of G, and hence of E, has values in $[0, 1]$ and lies in $C^\infty(\bar{Q}_1)$. Hence, from the hypothesis and (2.15),

$$\begin{aligned} \|1-\psi\|_{p,Q_{1/2}}^p &\leqslant A\|\nabla\psi\|_{p,Q_1}^p + B\int_{Q_1} q|1-\psi|^p \\ &\leqslant 2A\varepsilon + B\int_{\bar{Q}_1\setminus F} q. \end{aligned} \tag{2.16}$$

But $\psi \in C_0^\infty(Q_2)$ implies that

$$\begin{aligned} \|1-\psi\|_{p,Q_{1/2}} &\geqslant 2^{-n/p} - \|\psi\|_{p,Q_{1/2}} \\ &\geqslant 2^{-n/p} - \|\psi\|_{p,Q_2} \\ &\geqslant 2^{-n/p} - 2\mu_{n,p}\|\nabla\psi\|_{p,Q_2} \end{aligned}$$

(from (1.5))

$$\geqslant 2^{-n/p} - 2\mu_{n,p}(2\varepsilon)^{1/p}$$
$$> 0.$$

The lemma therefore follows from (2.16). □

We infer from Lemma 1.3 that when $n < p$ there exists an $\varepsilon_0 > 0$, depending only on n and p, such that for all non-empty compact subsets E of $\bar{Q}_d$,

$$p\text{-cap}\,(E, Q_{2d}) > \varepsilon_0 d^{n-p}. \tag{2.17}$$

Let $e_0 < \min\{\varepsilon_0, c_0\}$, where c_0 is the constant $1/2^{n+p+1}\mu_{n,p}^p$ in (2.13).

Definition 2.7. A closed subset E of $\bar{Q}_d$ is said to be *inessential* if $p\text{-cap}\,(E, Q_{2d}) < e_0 d^{n-p}$. ■

In view of (2.17) and our choice of e_0 there are no non-empty inessential sets when $n <$ p.

Theorem 2.8. Let q be locally integrable and positive a.e. on an open subset Ω of $\mathbb{R}^n$ and let $n \geqslant p$. Then

$$\|v\|_{p,\Omega}^p \leqslant A\left(\|\nabla v\|_{p,\Omega}^p + \int_\Omega q|v|^p\right) \qquad (v \in C_0^\infty(\Omega)) \tag{2.18}$$

if, and only if, there exist positive constants d and k such that for all cubes $\bar{Q}_d$ having an inessential intersection with Ω^c, i.e. $p\text{-cap}\,(\bar{Q}_d \cap \Omega^c, Q_{2d}) < e_0 d^{n-p}$, we have

$$\int_{Q_d \setminus E} q \geqslant k \qquad (q = 0 \quad \text{in} \quad \Omega^c) \tag{2.19}$$

for all inessential closed subsets E of $\bar{Q}_d$. ■

Proof. To prove sufficiency we consider a tesselation of $\mathbb{R}^n$ by cubes Q_d where d is such that (2.19) is satisfied. Whenever $\bar{Q}_d \cap \Omega^c$ is inessential we therefore have from Lemma 2.5, setting $v \in C_0^\infty(\Omega)$ to be zero outside Ω,

$$\|v\|_{p,Q_d}^p \leqslant K\left(e_0^{-1} d^p \|\nabla v\|_{p,Q_d}^p + k^{-1} d^n \int_{Q_d} q|u|^p\right). \tag{2.20}$$

If $\bar{Q}_d \cap \Omega^c$ is not inessential, i.e. $p\text{-cap}\,(\bar{Q}_d \cap \Omega^c, Q_{2d}) \geqslant e_0 d^{n-p}$, then by Corollary 2.4(i) with $r = p$,

$$\|v\|_{p,Q_d}^p \leqslant K e_0^{-1} d^p \|\nabla v\|_{p,Q_d}^p.$$

On summing over all the cubes Q_d in the tesselation of $\mathbb{R}^n$ we therefore obtain

$$\|v\|_{p,\Omega}^p \leqslant K\left(e_0^{-1} d^p \|\nabla v\|_{p,\Omega}^p + k^{-1} d^n \int_\Omega q|u|^p\right)$$

and hence (2.18).

In order to prove necessity we suppose that (2.18) is satisfied and let $\bar{Q}_d$ have an inessential intersection with Ω^c. Let $\theta \in \mathfrak{K}_0(\bar{Q}_{d/2}, Q_d)$ be such that $|\nabla\theta| \leqslant 5/d$. If $u \in C^\infty(\bar{Q}_d)$ vanishes in an open neighbourhood of $\bar{Q}_d \cap \Omega^c$ then $\theta u \in C_0^\infty(\Omega)$ and by (2.18),

$$\|\theta u\|_{p,\Omega}^p \leqslant A\left(\|\nabla(\theta u)\|_{p,\Omega}^p + \int_\Omega q|\theta u|^p\right)$$

$$\leqslant 2^{p-1}A\left(\|\nabla u\|_{p,Q_d}^p + (5d^{-1})^p\|u\|_{p,Q_d}^p + \int_{Q_d} q|u|^p\right). \tag{2.21}$$

By the Poincaré inequality,

$$\|u - u_{Q_d}\|_{p,Q_d} \leqslant Kd\|\nabla u\|_{p,Q_d}$$

and so

$$\|u\|_{p,Q_d} \leqslant \|u - u_{Q_d}\|_{p,Q_d} + |Q_d|^{1/p}|u_{Q_d}|$$

$$\leqslant Kd\|\nabla u\|_{p,Q_d} + |Q_d|^{1/p-1}\left(\int_{Q_{d/2}} |u| + \int_{Q_d\setminus Q_{d/2}} |u|\right)$$

$$\leqslant Kd\|\nabla u\|_{p,Q_d} + 2^{-n(1-1/p)}\|u\|_{p,Q_{d/2}} + (1-2^{-n})^{1-1/p}\|u\|_{p,Q_d},$$

whence

$$\|u\|_{p,Q_d} \leqslant K(d\|\nabla u\|_{p,Q_d} + \|u\|_{p,Q_{d/2}}). \tag{2.22}$$

On substituting (2.22) in (2.21) we obtain

$$\|u\|_{p,Q_{d/2}}^p \leqslant K\left(\|\nabla u\|_{p,Q_d}^p + d^{-p}\|u\|_{p,Q_{d/2}}^p + \int_{Q_d} q|u|^p\right)$$

and, for d large enough, this yields

$$\|u\|_{p,Q_{d/2}}^p \leqslant K\left(\|\nabla u\|_{p,Q_d}^p + \int_{Q_d} q|u|^p\right).$$

We now invoke Lemma 2.6 with $E = \bar{Q}_d \cap \Omega^c$ to conclude that

$$d^n \leqslant K\left(d^{n-p} + \int_{\bar{Q}_d\setminus F} q\right)$$

for all inessential subsets F of $\bar{Q}_d$. On choosing d large enough, (2.19) follows and the theorem is proved. □

The above proof continues to hold even when $n < p$. However, in this case there are no non-empty inessential subsets of Q_d. Thus the only cubes $\bar{Q}_d$ which have an inessential intersection with Ω^c are those which lie in Ω. Theorem 2.8 therefore becomes

Theorem 2.9. Let q be locally integrable and positive a.e. on the open subset Ω of $\mathbb{R}^n$ and let $n < p$. Then (2.18) is valid if, and only if, there exist positive constants d and k such that for all $\bar{Q}_d \subset \Omega$,

$$\int_{Q_d} q \geqslant k. \tag{2.23}$$ ■

The methods of this section also give the following necessary and sufficient condition for the validity of the Poincaré inequality.

Theorem 2.10. There exists a constant $A > 0$ such that

$$\|v\|_{p,\Omega}^p \leqslant A\,\|\nabla v\|_{p,\Omega}^p \qquad (v \in C_0^\infty(\Omega)), \tag{2.24}$$

if, and only if, there exist positive numbers d and k such that

$$p\text{-cap}(\bar{Q}_d \cap \Omega^c, Q_{2d}) \geqslant k d^{n-p} \tag{2.25}$$

for all cubes Q_d. The constant A in (2.24) satisfies

$$A \leqslant K d^p k^{-1} \tag{2.26}$$

where K is a constant depending only on n and p. ■

Proof. If (2.25) is satisfied for some d and k, we find from Corollary 2.4(i) that

$$\|v\|_{p,Q_d}^p \leqslant A\,\|\nabla v\|_{p,Q_d}^p \qquad (v \in C_0^\infty(\Omega)),$$

where

$$\begin{aligned} A &\leqslant K d^n [p\text{-cap}(\bar{Q}_d \cap \Omega^c, Q_{2d})]^{-1} \\ &\leqslant K d^p k^{-1}. \end{aligned}$$

On summing over a tesselation of $\mathbb{R}_n$ by cubes Q_d we obtain (2.24) and (2.26).

Suppose that (2.24) is satisfied and let $\bar{Q}_d$ be assumed to have an inessential intersection with Ω^c, i.e.

$$p\text{-cap}(\bar{Q}_d \cap \Omega^c, Q_{2d}) < e_0 d^{n-p}.$$

Let $\theta \in C_0^\infty(Q_d)$ be such that $0 \leqslant \theta \leqslant 1$, $\theta = 1$ on $Q_{d/2}$ and $|\nabla\theta| \leqslant 5/d$. Then if $u \in C^\infty(\bar{Q}_d)$ vanishes in an open neighbourhood of $\bar{Q}_d \cap \Omega^c$, we have $\theta u \in C_0^\infty(\Omega)$ and by (2.24),

$$\begin{aligned} \|u\|_{p,Q_{d/2}}^p &\leqslant \|\theta u\|_{p,Q}^p \\ &\leqslant A\,\|\nabla(\theta u)\|_{p,Q_d}^p \\ &\leqslant AK(\|\nabla u\|_{p,Q_d}^p + d^{-p}\,\|u\|_{p,Q_d}^p). \end{aligned}$$

Lemma 2.6 with $q = 1$ and $F = \varnothing$ now implies

$$d^n \leqslant KA(e_0 d^{n-p} + d^{n-p})$$

which is impossible for large enough d. Our assumption that $\bar{Q}_d$ has an inessential intersection with Ω^c is therefore contradicted and the necessity of (2.25) is established. □

By (1.6), (2.25) is satisfied and hence the Poincaré inequality is valid if, for some γ and d, and all Q_d,

$$|\bar{Q}_d \cap \Omega^c| \geqslant \gamma d^n \tag{2.27}$$

This is so if Ω is a quasi-cylindrical or quasi-bounded domain in the sense of Definition X.6.1 below. If $n < p$, (2.25) is satisfied if, and only if, $\bar{Q}_d \cap \Omega^c \neq \emptyset$, on account of Lemma 1.3. Hence, if $n = 1$ and $p > 1$ then Ω must either be a bounded interval or the union of bounded intervals whose closures are disjoint.

3. Compactness criterion for the embedding $W_0^{1,p}(\Omega) \to L^p(\Omega)$

This section concerns unbounded open sets Ω since we already know that the embedding $W_0^{1,p}(\Omega) \to L^p(\Omega)$ is compact for any bounded Ω. In Theorem V.5.17 we showed that the embedding is compact if $\lim_{|x|\to\infty} |\Omega \cap B(x,1)| = 0$. Furthermore, it is clear that the embedding is not compact if Ω contains a sequence of disjoint balls (or cubes) of equal diameter; in this case there exists a sequence of functions (ϕ_m) with disjoint supports which are bounded in $W_0^{1,p}(\Omega)$ and satisfy $\|\phi_m - \phi_k\|_{p,\Omega}^p = 2$. Therefore a necessary condition for compactness is that

$$\lim_{|x|\to\infty} \operatorname{dist}(x, \Omega^c) = 0.$$

It is known that this condition is not sufficient for compactness; see Adams [1, §6.14].

Theorem 3.1. The embedding $W_0^{1,p}(\Omega) \to L^p(\Omega)$ is compact if, and only if,

$$\inf_{d>0} \left(d^{p-n} \liminf_{\rho\to\infty} \{p\text{-cap}(\bar{Q}_d \cap \Omega^c, Q_{2d}) : Q_d \subset B_\rho^c\} \right) > 0, \tag{3.1}$$

where $B_\rho = B(0,\rho)$. ■

Proof. In order to establish the sufficiency of (3.1) we shall prove that the set $F = \{u : u \in C_0^\infty(\Omega) \text{ and } \|u\|_{1,p,\Omega} \leqslant 1\}$ is relatively compact in $L^p(\Omega)$. Since F is relatively compact in $L^p(\Omega \cap B_\rho)$ for any $\rho > 0$, it suffices to show that, given $\varepsilon > 0$, there exists a ρ such that

$$\|u\|_{p,\Omega\cap B_\rho^c} < \varepsilon \qquad (u \in F); \tag{3.2}$$

see Corollary V.5.5. By (3.1), for any $d > 0$, there exists a ρ such that for all Q_d

which intersect B_ρ^c we have

$$\frac{p\text{-cap}(\bar{Q}_d \cap \Omega^c, Q_{2d})}{d^{n-p}} \geqslant \delta > 0$$

where δ is independent of d. Hence, by Corollary 2.4(i) with $r = p$,

$$\begin{aligned}\|u\|_{p,Q_d}^p &\leqslant K\, d^n [p\text{-cap}(\bar{Q}_d \cap \Omega^c, Q_{2d})]^{-1} \|\nabla u\|_{p,Q_d}^p \\ &\leqslant K\, d^p \delta^{-1} \|\nabla u\|_{p,Q_d}^p.\end{aligned}$$

On summing over a tesselation of $\mathbb{R}^n$ by cubes Q_d, we therefore obtain

$$\|u\|_{p,\Omega\cap B_\rho^c}^p \leqslant K\, d^p \delta^{-1}.$$

Since d is arbitrary (3.2) follows and the sufficiency of (3.1) is proved.

For necessity we suppose that F is relatively compact in $L^p(\Omega)$. Then, for any $\varepsilon > 0$, there exists a ρ such that

$$\|u\|_{p,\Omega\cap B_\rho^c} < \varepsilon \qquad (u \in F). \tag{3.3}$$

With $d > 0$ fixed, let $Q_d \subset B_\rho^c$ and suppose that

$$p\text{-cap}(\bar{Q}_d \cap \Omega^c, Q_{2d}) < e_0\, d^{n-p}$$

where e_0 is the constant in Definition 2.7; in other words, $\bar{Q}_d$ has an inessential intersection with Ω^c. The argument now follows that in the necessity part of the proof of Theorem 2.10. Let $\theta \in C_0^\infty(Q_d)$ be such that $0 \leqslant \theta \leqslant 1$, $|\nabla\theta| \leqslant 5d^{-1}$, and $\theta = 1$ on $Q_{d/2}$. Then, if $v \in C^\infty(\bar{Q}_d)$ vanishes in a neighbourhood of $\Omega^c \cap \bar{Q}_d$, it follows that $\theta v \in C_0^\infty(\Omega)$ and, by (3.3),

$$\begin{aligned}\|v\|_{p,Q_{d/2}}^p &\leqslant \|\theta v\|_{p,Q_d}^p \\ &\leqslant \varepsilon^p \|\theta v\|_{1,p,Q_d}^p \\ &\leqslant K\varepsilon^p [\|\nabla v\|_{p,Q_d}^p + (1 + d^{-p}) \|v\|_{p,Q_d}^p].\end{aligned}$$

Lemma 2.6 with $q = 1$ and $F = \varnothing$ now yields

$$d^n \leqslant K\, \varepsilon^p [e_0 d^{n-p} + d^n(1 + d^{-p})].$$

This is a contradiction since ε may be chosen arbitrarily. Therefore for any $d > 0$,

$$p\text{-cap}(\bar{Q}_d \cap \Omega^c, Q_{2d}) \geqslant e_0\, d^{n-p}$$

for all $Q_d \subset B_\rho^c$ when ρ is sufficiently large. The proof is complete. □

By (1.6) and (1.7)

$$\mu_{n,p}^{-p} d^{-n} |\bar{Q}_d \cap \Omega^c| \leqslant d^{p-n} [p\text{-cap}(\bar{Q}_d \cap \Omega^c, Q_{2d})] \leqslant 2^{n+p}. \tag{3.4}$$

The first inequality in (3.4) and Theorem 3.1 imply that the embedding

$W_0^{1,p}(\Omega) \to L^p(\Omega)$ is compact if

$$\inf_{d>0} \lim_{\rho\to\infty} \{d^{-n}|\bar{Q}_d \cap \Omega^c| : Q_d \subset B_\rho^c\} > 0 \tag{3.5}$$

which is equivalent to the criterion in Theorem V.5.17. Note also that (3.1) implies that for a compact embedding, Ω does not contain a sequence of disjoint cubes Q_d for any value of d.

4. Molcanov's criterion for a discrete spectrum

Let

$$t_0[\![u,v]\!] = \sum_{i,j=1}^{n} \int_\Omega a_{ij} D_j u D_i \bar{v}, \tag{4.1}$$

$$t'[\![u,v]\!] = t_0[\![u,v]\!] + \int_\Omega q u \bar{v} \tag{4.2}$$

on $C_0^\infty(\Omega)$, where Ω is an arbitrary open set in $\mathbb{R}^n$, $n \geqslant 1$, and suppose that

(i) the matrix $[a_{ij}(x)]$ is Hermitian for a.e. $x \in \Omega$ and $a_{ij} \in L^\infty_{\mathrm{loc}}(\Omega)$, and there are positive constants C_1 and C_2 such that the largest and smallest eigenvalues, $\lambda^+(x)$ and $\lambda^-(x)$ respectively, of $[a_{ij}(x)]$ satisfy, for a.e. x,

$$0 < C_1 \leqslant \lambda^-(x) \leqslant \lambda^+(x) \leqslant C_2;$$

(ii) q is a real $L^1_{\mathrm{loc}}(\Omega)$ function which is bounded below on Ω; for the discussion in this section we may assume, without loss of generality, that $q \geqslant 1$ since a positive displacement does not alter the nature of the spectrum of the linear operators considered.

It follows from (i) and (ii) that t_0 and t' are positive symmetric forms and also that for all $u \in C_0^\infty(\Omega)$,

$$C_1 \|\nabla u\|^2_{2,\Omega} \leqslant t_0[\![u]\!] \leqslant C_2 \|\nabla u\|^2_{2,\Omega}. \tag{4.3}$$

Furthermore, t' is closable in $L^2(\Omega)$. To see this, suppose that $(\phi_k) \subset C_0^\infty(\Omega)$ is a Cauchy sequence with respect to the norm $(t'+1)^{\frac{1}{2}}[\![\bullet]\!]$, and converges to zero in $L^2(\Omega)$. From (4.3), $(D_j\phi_k)$ converges to some limit ψ_j in $L^2(\Omega)$ for $j = 1, 2, \ldots, n$ and, for $\phi \in C_0^\infty(\Omega)$,

$$\begin{aligned}\int_\Omega \psi_j \phi &= \lim_{k\to\infty} \int_\Omega (D_j\phi_k)\phi \\ &= -\lim_{k\to\infty} \int_\Omega \phi_k D_j \phi \\ &= 0.\end{aligned}$$

Hence $\psi_j = 0$ and $\phi_k \to 0$ in $W_0^{1,2}(\Omega)$. Since (ϕ_k) is also a Cauchy sequence in the weighted space $L^2(\Omega; q)$ it converges to a limit in $L^2(\Omega; q)$ and this limit must be zero a.e in Ω since the null sets of $L^2(\Omega; q)$ and $L^2(\Omega)$ coincide. We have therefore proved that $t'[\![\phi_k]\!] \to 0$ as $k \to \infty$ and hence that t' is closable. We denote the closure of t' by t. Its domain is the completion of $C_0^\infty(\Omega)$ with respect to the norm given by

$$|\!|\!| u |\!|\!|_t^2 = \|\nabla u\|_{2,\Omega}^2 + \|u\|_{2,\Omega}^2 + \int_\Omega q|u|^2. \tag{4.4}$$

Let T be the positive self-adjoint operator associated with t (see Theorem IV.2.4). From Theorem IV.2.8, the domain of t is $\mathscr{D}(T^{\frac{1}{2}})$ and

$$\begin{aligned} t[\![u, v]\!] &= (T^{\frac{1}{2}}u, T^{\frac{1}{2}}v)_{2,\Omega} \qquad (u, v \in \mathscr{D}(T^{\frac{1}{2}})), \\ |\!|\!| u |\!|\!|_t^2 &= (t+1)[\![u]\!] \\ &= \|(T^{\frac{1}{2}} \pm iI)u\|_{2,\Omega}^2. \end{aligned}$$

If the $a_{ij} \in C^1(\Omega)$ and $q \in L^2_{\text{loc}}(\Omega)$, T is the Friedrichs extension of the operator T' defined by

$$\tau = -D_i a_{ij} D_j + q$$

on $C_0^\infty(\Omega)$. When $\Omega = \mathbb{R}^n$ we proved in Corollary VII.2.7 that T' is essentially self-adjoint and hence T is the closure of T'.

By Theorem IV.2.9, T has a compact resolvent if, and only if, the Hilbert space $H(t)$ determined by the domain of t and the inner-product norm $|\!|\!| \bullet |\!|\!|_t$ is compactly embedded in $L^2(\Omega)$. Since $C_0^\infty(\Omega)$ is dense in $H(t)$, T has a compact resolvent, and hence a discrete spectrum consisting of positive isolated eigenvalues of finite multiplicity which tend to infinity, if, and only if, the set

$$\Gamma = \{u : u \in C_0^\infty(\Omega) \text{ and } |\!|\!| u |\!|\!|_t \leqslant 1\} \tag{4.5}$$

is relatively compact in $L^2(\Omega)$.

We are now in a position to give Molcanov's Theorem.

Theorem 4.1. Let T be the operator defined above under the conditions (i) and (ii) and let $n \geqslant 2$. Then T has a discrete spectrum if, and only if, for any positive number d,

$$\lim_{\substack{|c_d| \to \infty \\ \bar{Q}_d \in \mathfrak{F}_d}} \inf_{E \in \mathfrak{E}_d} \int_{\bar{Q}_d \setminus E} q = \infty \tag{4.6}$$

where $\mathfrak{F}_d$ is the set of all cubes $\bar{Q}_d$, centre c_d, which are such that $\bar{Q}_d \cap \Omega^c$ is inessential, and $\mathfrak{E}_d$ is the set of all inessential closed subsets of $\bar{Q}_d$. ■

Proof. Suppose that (4.6) is satisfied for all d. Since $q \geqslant 1$ on Ω the set Γ in (4.5) is bounded in $W_0^{1,2}(\mathbb{R}^n)$ and hence relatively compact in $L^2(B_\rho)$ for any ρ. From Corollary V.5.5, Γ is compactly embedded in $L^2(\Omega)$ if, and only if, for any given $\delta > 0$, there exists ρ such that

$$\|u\|_{2,B_\rho^c} < \delta \qquad (u \in \Gamma). \tag{4.7}$$

We set $d = \delta$ and choose ρ to have the property that, for every cube Q_δ which meets B_ρ^c and is such that $\bar{Q}_\delta \cap \Omega^c$ is inessential,

$$\int_{Q_\delta \setminus E} q\,dx \geqslant \delta^{n-2}$$

for every closed inessential subset E of $\bar{Q}_\delta$; such a choice of ρ is possible by our hypothesis. By Lemma 2.5 with $p = 2$,

$$\|u\|_{2,Q_\delta}^2 \leqslant K\delta^2 \left(\|\nabla u\|_{2,Q_\delta}^2 + \int_{Q_\delta} q|u|^2 \right) \tag{4.8}$$

for $u \in \Gamma$. If Q_δ meets B_ρ^c and $\bar{Q}_\delta \cap \Omega^c$ is not inessential, i.e. cap $(\bar{Q}_\delta \cap \Omega^c, Q_{2\delta})$ $> e_0 \delta^{n-2}$, then, from Corollary 2.4(i),

$$\|u\|_{2,Q_\delta}^2 \leqslant K\delta^2 \|\nabla u\|_{2,Q_\delta}^2 \qquad (u \in \Gamma). \tag{4.9}$$

On summing over all the cubes Q_δ of a tesselation of $\mathbb{R}^n$ which intersect B_ρ^c we therefore have from (4.8) and (4.9) that

$$\|u\|_{2,B_\rho^c} < K\delta.$$

The sufficiency of (4.6) is therefore established.

If Γ is relatively compact in $L^2(\Omega)$, then for each $\delta > 0$ there exists ρ such that (4.7) is satisfied for all $u \in \Gamma$. Hence, for all $u \in C_0^\infty(\Omega)$ with support in B_ρ^c,

$$\|u\|_{2,\Omega}^2 \leqslant \delta^2 \|\!|u|\!\|_t^2$$

and consequently, if $\delta^2 < \frac{1}{2}$,

$$\|u\|_{2,\Omega}^2 \leqslant 2\delta^2 \left(\|\nabla u\|_{2,\Omega}^2 + \int_\Omega q|u|^2 \right). \tag{4.10}$$

From here the argument proceeds as in the proof of necessity in Theorem 2.8. Let d be any positive number and let $Q_d \subset B_\rho^c$ be such that $\bar{Q}_d \cap \Omega^c$ is inessential. Then, if $\theta \in \mathfrak{R}_0(\bar{Q}_{d/2}, Q_d)$, the set in (1.3), and $u \in C^\infty(\bar{Q}_d)$ vanishes in a neighbourhood of $\bar{Q}_d \cap \Omega^c$, we have that $\theta u \in C_0^\infty(\Omega)$ and satisfies (4.10). On using (2.21) and (2.22) with $A = 2\delta^2$ and $p = 2$,

$$\|u\|_{2,Q_{d/2}}^2 \leqslant K\delta^2 \left(\|\nabla u\|_{2,Q_d}^2 + d^{-2}\|u\|_{2,Q_{d/2}}^2 + \int_{Q_d} q|u|^2 \right).$$

Consequently, if δ is small enough, we obtain

$$\|u\|_{2,Q_{d/2}}^2 \leqslant K\delta^2 \left(\|\nabla u\|_{2,Q_d}^2 + \int_{Q_d} q|u|^2 \right).$$

From Lemma 2.6 we deduce that

$$d^n \leqslant K\delta^2 \left(d^{n-2} + \int_{\bar{Q}_d \setminus E} q \right)$$

for any closed inessential subset E of $\bar{Q}_d$. The necessity of (4.6) is therefore proved. □

In the case $n = 1$, due to the absence of non-empty inessential sets, the theorem becomes

Theorem 4.2. When $n = 1$, the operator T has a discrete spectrum if, and only if, for any $d > 0$,

$$\lim_{\substack{|c_d| \to \infty \\ Q_d \in \mathfrak{F}_d}} \int_{Q_d} q = \infty$$

where $\mathfrak{F}_d$ is the set of all intervals $\bar{Q}_d \subset \Omega$, and c_d is the centre of Q_d. ■

5. Continuous representatives of $W^{1,p}(\Omega)$

It is a fundamental property of the Lebesgue integral that any $u \in L^p(\Omega)$ $(1 \leqslant p < \infty)$ has a pointwise representative which is continuous on the complement in Ω of a set of zero measure and is unique if we identify functions which are equal almost everywhere in Ω. If (ϕ_j) is a sequence of $C_0^\infty(\Omega)$ functions which converges to u in $L^p(\Omega)$ then a subsequence converges uniformly on $\Omega \setminus E$, where E is a subset of arbitrarily small measure and the limit is the aforementioned representative of u, being continuous on the complement in Ω of a set of zero measure. It is the analogue of this result which we now discuss for $W^{1,p}(\Omega)$; the result is due to Lewy and Stampacchia [1].

To proceed we need the notion of capacity of arbitrary subsets of $\bar{\Omega}$. In this section Ω can be assumed to be a bounded, non-empty, open subset of $\mathbb{R}^n$ and we shall denote by Ω' any fixed open set satisfying $\bar{\Omega} \subset \Omega' \subset \mathbb{R}^n$. For an arbitrary subset A of $\bar{\Omega}$ we define the p-capacity (with respect to $-\Delta$ and relative to Ω') by

$$p\text{-cap}(A) := \sup_{E \subset A} [p\text{-cap}(E, \Omega')] \tag{5.1}$$

where the supremum is over all compact subsets E of A. It is more usual perhaps to refer to (5.1) as the *inner capacity*, written $p\text{-}\underline{\text{cap}}\,(A)$ and then define the *outer capacity* $p\text{-}\overline{\text{cap}}\,(A)$ by

$$p\text{-}\overline{\text{cap}}\,(A) := \inf_{G \supset A} [p\text{-}\underline{\text{cap}}\,(G)] \tag{5.2}$$

where the infimum is over all open subsets G of Ω' which contain A. The set A is then said to be *capacitable* if its inner and outer capacities coincide and the common value is referred to as the capacity of A. However, the definition (5.1) will serve our purpose. It clearly agrees with that in §1 when A is compact. From Lemma 1.1 it follows that p-cap$(\bullet)$ is a non-negative, non-decreasing set function on subsets of $\bar{\Omega}$. The set A has zero p-capacity if, and only if, the p-capacity of every compact subset of A is zero; also $p\text{-cap}(A) = 0$ if A is contained in subsets of $\bar{\Omega}$ having arbitrarily small p-capacity.

Lemma 5.1. Let $A \subset \bar{\Omega}$ and $A = \cup_{j=1}^{\infty} E_j$ where each E_j is compact. Then

$$p\text{-cap}(A) \leqslant \sum_{j=1}^{\infty} p\text{-cap}(E_j). \tag{5.3} \blacksquare$$

Proof. Let $u_j \in \mathfrak{K}(E_j, \Omega')$, the convex set defined in (1.2), and assume that for a given $\varepsilon > 0$,

$$\|\nabla u_j\|_{p,\Omega'}^p \leqslant p\text{-cap}(E_j) + \varepsilon 2^{-j} \qquad (j = 1,2, \ldots).$$

Let E be any compact subset of A; then $E \subset \cup_{j=1}^{N} E_j$ for some integer N. By Corollary VI.2.4, $u := \sup_{1 \leqslant j \leqslant N} (u_j) \in W_0^{1,p}(\Omega')$ and in fact it is readily seen that $u \in \mathfrak{K}(E, \Omega')$. Furthermore, by Corollary VI.2.8,

$$\|\nabla u\|_{p,\Omega'}^p \leqslant \max_{1 \leqslant j \leqslant N} \|\nabla u_j\|_{p,\Omega'}^p.$$

Consequently

$$\begin{aligned} p\text{-cap}(E) &\leqslant \|\nabla u\|_{p,\Omega'}^p \\ &\leqslant \sum_{j=1}^{N} \|\nabla u_j\|_{p,\Omega'}^p \\ &\leqslant \sum_{j=1}^{N} [p\text{-cap}(E_j) + \varepsilon 2^{-j}] \\ &\leqslant \left(\sum_{j=1}^{\infty} p\text{-cap}(E_j) \right) + \varepsilon. \end{aligned}$$

The lemma is therefore proved. □

To set the scene for the next theorem we recall that any $u \in W^{1,p}(\Omega) \subset L^p(\Omega)$ has a representative which is continuous on the complement of a set of measure zero; indeed this representative is absolutely continuous in the sense explained in Theorem V.3.11. When $n < p$ and Ω has a minimally smooth boundary, $u \in W^{1,p}(\Omega)$ can be identified with an everywhere continuous function and it is significant that in this case there are no non-empty subsets of Ω with zero p-capacity. If $n \geqslant p$, sets of zero p-capacity have zero measure.

Theorem 5.2. Any $u \in W^{1,p}(\Omega)$ can be represented by a pointwise function which is continuous on the complement in Ω of a set of zero p-capacity and is unique if we identify functions which are equal outside subsets of Ω of zero p-capacity. ■

Proof. Let E be any compact subset of Ω, let Ω_0 be a bounded open subset of Ω such that $E \subset \Omega_0 \subset \Omega$, and let $\theta \in C_0^\infty(\Omega_0)$ with $\theta = 1$ on E. Then if $u \in W^{1,p}(\Omega)$, we have $\theta u \in W_0^{1,p}(\Omega_0)$ and $\theta u = u$ on E. It therefore suffices to prove the theorem for $u \in W_0^{1,p}(\Omega)$ and Ω bounded.

Let $\phi_k \in C_0^\infty(\Omega)$ $(k \in \mathbb{N})$ and suppose that $\phi_k \to u$ in $W_0^{1,p}(\Omega)$. We extract a subsequence from (ϕ_k), which we continue to denote by (ϕ_k), which satisfies

$$\sum_{k=1}^{\infty} 2^{pk} \|\nabla(\phi_{k+1} - \phi_k)\|_{2,\Omega}^p < \infty. \tag{5.4}$$

The sets

$$E_k = \{x : |\phi_{k+1}(x) - \phi_k(x)| \geqslant 2^{-k}\}$$

are compact subsets of Ω and clearly $2^k|\phi_{k+1} - \phi_k| \in \mathfrak{R}(E_k, \Omega')$. Hence

$$\begin{aligned} p\text{-cap}(E_k) &\leqslant 2^{pk} \|\nabla|\phi_{k+1} - \phi_k|\,\|_{p,\Omega}^p \\ &\leqslant 2^{pk} \|\nabla(\phi_{k+1} - \phi_k)\|_{p,\Omega}^p. \end{aligned}$$

If $W_j = \cup_{k=j}^{\infty} E_k$, we have $W_j \supset W_{j+1}$ and by Lemma 5.1,

$$\begin{aligned} p\text{-cap}(W_j) &\leqslant \sum_{k=j}^{\infty} p\text{-cap}(E_k) \\ &\leqslant \sum_{k=j}^{\infty} 2^{pk} \|\nabla(\phi_{k+1} - \phi_k)\|_{p,\Omega}^p \\ &\to 0 \end{aligned}$$

as $j \to \infty$, on using (5.4). In $\Omega \setminus W_j$, each ϕ_k is uniformly continuous since it has compact support in Ω, and for $k \geqslant j$,

$$|\phi_{k+1}(x) - \phi_k(x)| < 2^{-k}.$$

Consequently, (ϕ_k) converges uniformly to a function ϕ which is continuous on $\Omega \setminus W_j$. Since $W = \cap_{j=1}^{\infty} W_j$ has zero p-capacity, the sequence (ϕ_k) converges to ϕ except possibly on the set W of zero p-capacity. This function ϕ is unique if we identify functions which are equal outside sets of zero p-capacity. To see this, suppose that (ϕ'_k) is another sub-sequence of the original sequence (ϕ_k) which also satisfies (5.4). Since (ϕ_k) and (ϕ'_k) converge to the same limit in $W_0^{1,p}(\Omega)$ we can choose subsequences which also satisfy

$$\sum_{k=1}^{\infty} 2^{pk} \|\nabla(\phi_k - \phi'_k)\|_{p,\Omega}^p < \infty.$$

We now define $V_j = \cup_{k=j}^{\infty} F_k$, where

$$F_k = E_k \cup E'_k \cup \{x : |\phi'_{k+1} - \phi_{k+1}| \geqslant 2^{-k}\} \cup \{x : |\phi_k - \phi'_k| \geqslant 2^{-k}\}$$

and E'_k denotes E_k with ϕ_k replaced by ϕ'_k. We readily see that p-cap $(V_j) \to 0$ as $j \to \infty$, and in $\Omega \setminus V_j$ the sequence (ϕ'_k) converges uniformly to a limit ϕ', say, as well as $\phi_k \to \phi$ and $\phi_k - \phi'_k \to 0$. Consequently $\phi' = \phi$ outside the set $\bigcap_{j=1}^{\infty} V_j$ which has zero p-capacity. The proof is therefore complete. □

From Theorem 5.2 it follows that two elements of $W^{1,p}(\Omega)$ are identical in $W^{1,p}(\Omega)$ only if their representatives are equal outside a set of zero p-capacity, such sets being of zero measure when $n \geqslant p$. This complements the result in Theorem VI.2.7 that two elements of $W^{1,p}(\Omega)$ are identical if their representatives are equal outside a set of zero measure.

6. Density results for $C_0^\infty(\Omega)$

For what domains Ω in $\mathbb{R}^n$ is $C_0^\infty(\Omega)$ dense in $W^{m,p}(\mathbb{R}^n)$, or, in other words, when is $W_0^{m,p}(\Omega) = W^{m,p}(\mathbb{R}^n)$? This is the question discussed in this section. The answer is given in terms of the notion of (m, p')-polar sets introduced by Hörmander and Lions [1] and also in terms of a generalized capacity defined by Maz'ja [3].

We assume throughout that $1 < p < \infty$ and p' is the conjugate of p, that is, $1/p' = 1 - 1/p$. The adjoint $W^{-m,p'}(\mathbb{R}^n)$ of $W^{m,p}(\mathbb{R}^n)$ consists of conjugate linear distributions; recall that $F \in W^{-m,p'}(\mathbb{R}^n)$ is said to have its support supp F in a closed subset E of $\mathbb{R}^n$ if $(F, \phi) = 0$ for all $\phi \in C_0^\infty(\mathbb{R}^n)$ whose support lies outside E.

Definition 6.1. A closed set E is said to be *(m, p')-polar* if $F \in W^{-m,p'}(\mathbb{R}^n)$ and supp $F \subset E$ imply $F = 0$. ■

If E is (m, p')-polar it must be of zero Lebesgue measure, since otherwise the characteristic function of any closed bounded subset of E belongs to $L^{p'}(\mathbb{R}^n)$ and hence to $W^{-m,p'}(\mathbb{R}^n)$. Furthermore, if $n < mp$, an (m, p')-polar set must be empty. To see this, recall that $W^{m,p}(\mathbb{R}^n)$ is continuously embedded in $C_B(\mathbb{R}^n)$ when $n < mp$ (see page 265) and hence $|\phi(x)| \leqslant K\|\phi\|_{m,p,\mathbb{R}^n}$ for some positive constant K and all $x \in \mathbb{R}^n$. If δ_x is the (conjugate) Dirac delta function defined by $\delta_x(\phi) = \bar{\phi}(x)$ we have for all $\phi \in C_0^\infty(\mathbb{R}^n)$ that

$$|\delta_x(\phi)| \leqslant K\|\phi\|_{m,p,\mathbb{R}^n},$$

and so δ_x can be extended by continuity to a member of $W^{-m,p'}(\mathbb{R}^n)$. Since supp $\delta_x = \{x\}$, an (m, p')-polar set must be empty when $n < mp$.

Theorem 6.2. $C_0^\infty(\Omega)$ is dense in $W^{m,p}(\mathbb{R}^n)$ if, and only if, Ω^c is (m, p')-polar. ■

Proof. Suppose $C_0^\infty(\Omega)$ is dense in $W^{m,p}(\mathbb{R}^n)$ and let $F \in W^{-m,p'}(\mathbb{R}^n)$ have its support in Ω^c. If $\phi \in W^{m,p}(\mathbb{R}^n)$ there is a sequence $(\phi_k) \subset C_0^\infty(\Omega)$ which converges to ϕ in $W^{m,p}(\mathbb{R}^n)$ and hence since supp $F \subset \Omega^c$,

$$(F, \phi) = \lim_{k\to\infty} (F, \phi_k) = 0.$$

Consequently $F = 0$ and Ω^c is (m, p')-polar.

If $C_0^\infty(\Omega)$ is not dense in $W^{m,p}(\mathbb{R}^n)$ there exists a non-zero $u \in W^{m,p}(\mathbb{R}^n)$ and a positive constant δ such that $\|u-\phi\|_{m,p,\mathbb{R}^n} \geqslant \delta$ for all $\phi \in C_0^\infty(\Omega)$. By the Hahn–Banach Theorem it follows that there exists $F \in W^{-m,p'}(\mathbb{R}^n)$ such that $(F, u) \neq 0$ and $(F, \phi) = 0$ for all $\phi \in C_0^\infty(\Omega)$. Thus $F \neq 0$ and supp $F \subset \Omega^c$, whence Ω^c is not (m, p')-polar. □

Another criterion for $C_0^\infty(\Omega)$ to be dense in $W^{m,p}(\mathbb{R}^n)$ is given in terms of a generalized capacity. For a compact set E in $\mathbb{R}^n$ we define

$$\operatorname{Cap}(E, W^{m,p}(\mathbb{R}^n)) = \inf\{\|u\|_{m,p,\mathbb{R}^n} : u \in \mathfrak{K}_1(E)\} \tag{6.1}$$

where

$$\mathfrak{K}_1(E) = \{u \in C_0^\infty(\mathbb{R}^n) : u = 1 \text{ in an open neighbourhood of } E\}.$$

When $m = 1$, (6.1) is related to the p-capacity of E relative to any bounded set Ω' containing E; there exist positive constants $K_{1,\Omega'}$ and $K_{2,\Omega'}$, depending on Ω', such that

$$K_{1,\Omega'} \operatorname{Cap}(E, W^{1,p}(\mathbb{R}^n)) \leqslant p\text{-cap}(E, \Omega') \leqslant K_{2,\Omega'} \operatorname{Cap}(E, W^{1,p}(\mathbb{R}^n)). \tag{6.2}$$

The first inequality in (6.2) is a consequence of the Poincaré inequality

$$\|\phi\|_{p,\Omega'} \leqslant K_{\Omega'} \|\nabla\phi\|_{p,\Omega} \qquad (\phi \in C_0^\infty(\Omega')).$$

To prove the second inequality take any $u \in \mathfrak{K}_1(E)$ and let $\phi \in \mathfrak{K}_1(E)$ have its support inside Ω'. Then, in the notation of (1.2), $\phi u \in \mathfrak{K}(E, \Omega')$ and

$$\|\nabla(\phi u)\|_{p,\Omega'} \leqslant K_{\Omega'} \|u\|_{1,p,\mathbb{R}^n},$$

whence the required inequality.

For an arbitrary subset A of $\mathbb{R}^n$ we define

$$\operatorname{Cap}(A, W^{m,p}(\mathbb{R}^n)) := \sup_{\substack{E\subset A\\ E \text{ compact}}} \operatorname{Cap}(E, W^{m,p}(\mathbb{R}^n)).$$

Theorem 6.3. The set $C_0^\infty(\Omega)$ is dense in $W^{m,p}(\mathbb{R}^n)$ if, and only if, $\operatorname{Cap}(\Omega^c, W^{m,p}(\mathbb{R}^n)) = 0$. ■

Proof. By Theorem 6.2 it suffices to prove that $\operatorname{Cap}(\Omega^c, W^{m,p}(\mathbb{R}^n)) = 0$ if, and only, if Ω^c is (m, p')-polar.

First let $\operatorname{Cap}(\Omega^c, W^{m,p}(\mathbb{R}^n)) = 0$ and suppose there exists $F \in W^{-m,p'}(\mathbb{R}^n)$ with supp $F \subset \Omega^c$. We may assume, without loss of generality, that supp F is compact; otherwise consider αF for some $\alpha \in C_0^\infty(\mathbb{R}^n)$. By hypothesis, there exists a sequence $(u_k) \subset \mathfrak{K}_1$ (supp F) which is such that $\|u_k\|_{m,p,\mathbb{R}^n} \to 0$ as

$k \to \infty$. For any $\phi \in C_0^\infty(\mathbb{R}^n)$, we have $\phi(1-u_k) = 0$ in a neighbourhood of supp F and so $(F, \phi(1-u_k)) = 0$. This gives

$$(F, \phi) = \lim_{k\to\infty} (F, \phi u_k) = 0.$$

Since $C_0^\infty(\mathbb{R}^n)$ is dense in $W^{m,p}(\mathbb{R}^n)$, we conclude that $F = 0$ and Ω^c is (m, p')-polar.

Conversely, let Ω^c be (m, p')-polar. Then each compact subset E of Ω^c is also (m, p')-polar and $C_0^\infty(\mathbb{R}^n \setminus E)$ is dense in $W^{m,p}(\mathbb{R}^n)$ by Theorem 6.2. Thus, for each $u \in \mathfrak{K}_1(E)$, there exists a sequence $(u_k) \subset C_0^\infty(\mathbb{R}^n \setminus E)$ which converges to u in $W^{m,p}(\mathbb{R}^n)$. Since $u - u_k \in \mathfrak{K}_1(E)$ and $\|u-u_k\|_{m,p,\mathbb{R}^n} \to 0$ it follows that $\operatorname{Cap}(E, W^{m,p}(\mathbb{R}^n)) = 0$ and hence $\operatorname{Cap}(\Omega^c, W^{m,p}(\mathbb{R}^n)) = 0$. □

Corollary 6.4. $C_0^\infty(\mathbb{R}^n \setminus \{0\})$ is dense in $W^{m,p}(\mathbb{R}^n)$ if, and only if, $n \geq mp$. ■

Proof. For $n > mp$ define $\theta \in C_0^\infty[0, \infty)$ such that

$$\theta(r) = \begin{cases} 1 & (0 \leq r \leq \frac{1}{2}) \\ 0 & (r \geq 1), \end{cases}$$

and set $\theta_k(x) = \theta(k|x|)$. Then $|D^\alpha \theta_k| \leq K k^{|\alpha|}$ and

$$\begin{aligned} \|\theta_k\|_{m,p,\mathbb{R}^n}^p &\leq K \int_{|x|\leq 1/k} \sum_{0\leq|\alpha|\leq m} k^{|\alpha|p}\, dx \\ &= O(k^{mp-n}) = o(1) \end{aligned}$$

as $k \to \infty$. Hence $\operatorname{Cap}(\{0\}, W^{m,p}(\mathbb{R}^n)) = 0$ snd $C_0^\infty(\mathbb{R}^n \setminus \{0\})$ is dense in $W^{m,p}(\mathbb{R}^n)$ by Theorem 6.3.

For $n = mp$ we proceed as follows. Let $\phi \in C^\infty[0, 1]$ be such that $\phi = 0$ in a right neighbourhood of 0 and $\phi = 1$ in a left neighbourhood of 1; let $\delta \in (0, \frac{1}{2})$ and set

$$\psi_\delta(x) = \begin{cases} \phi(\log_\delta |x|) & \text{if } \delta < |x| < 1, \\ 1 & \text{if } |x| \leq \delta, \\ 0 & \text{if } x \geq 1. \end{cases}$$

Then $\psi_\delta \in C_0^\infty(B(0, 1))$ and $\psi_\delta = 1$ in a neighbourhood of $B(0, \delta)$. For $\delta \leq |x| \leq 1$,

$$|\nabla^m \psi_\delta(x)| \leq K[|x|^m \log(1/\delta)]^{-1}$$

and so, since $n = mp$,

$$\begin{aligned} \|\nabla^m \psi_\delta\|_{p,B(0,1)}^p &\leq K[\log(1/\delta)]^{-p} \int_\delta^1 r^{-mp+n-1}\, dr \\ &\leq K[\log(1/\delta)]^{-p+1}. \end{aligned}$$

Also by the Poincaré inequality,

$$\|\nabla^j \psi_\delta\|_{p,B(0,1)}^p \leqslant K\|\nabla^m \psi_\delta\|_{p,B(0,1)}^p \qquad (j = 0, 1, \ldots, m-1),$$

and consequently

$$\|\psi_\delta\|_{m,p,\mathbb{R}^n} = O([\log(1/\delta)]^{-1/p'}) \to 0$$

as $\delta \to 0$. It follows that Cap $(\{0\}; W^{m,p}(\mathbb{R}^n)) = 0$ and $C_0^\infty(\mathbb{R}^n \setminus \{0\})$ is dense in $W^{m,p}(\mathbb{R}^n)$ by Theorem 6.3.

If $n < mp$ an (m, p')-polar set must be empty as we saw in the paragraph preceding Theorem 6.2. Hence $\{0\}$ is not (m, p')-polar in this case and $C_0^\infty(\mathbb{R}^n \setminus \{0\})$ is not dense in $W^{m,p}(\mathbb{R}^n)$ by Theorem 6.2. □

Corollary 6.5. The restriction of $-\Delta$ to $C_0^\infty(\mathbb{R}^n \setminus \{0\})$ is essentially self-adjoint in $L^2(\mathbb{R}^n)$ if, and only if, $n \geqslant 4$. ■

Proof. The operator $S := -\Delta \restriction C_0^\infty(\mathbb{R}^n \setminus \{0\})$ is symmetric with a self-adjoint extension $T := -\Delta \restriction W^{2,2}(\mathbb{R}^n)$; therefore S is essentially self-adjoint if, and only if, T is the closure of S and this is equivalent to $C_0^\infty(\mathbb{R}^n \setminus \{0\})$ being dense in $W^{2,2}(\mathbb{R}^n)$ since the norm of $W^{2,2}(\mathbb{R}^n)$ is equivalent to the graph norm of T. The corollary therefore follows from Corollary 6.4. □

Another interesting question is: when is $C_0^\infty(\Omega)$ dense in $W^{m,p}(\Omega)$ or, in other words, when is $W_0^{m,p}(\Omega) = W^{m,p}(\Omega)$? This is discussed by Adams in [1, Chapter III], following the approach of Lions in [1], and we have nothing to add. Adams proves in particular in [1, Theorem 3.31] that if $p \geqslant 2$ then $W_0^{m,p}(\Omega) = W^{m,p}(\Omega)$ if, and only if, Ω^c is (m, p')-polar.

Theorem 6.3 may be found in Maz'ja [3, Kapitel 4, Sätzen 2.1, 2.2] and we also adopt Maz'ja's ideas in our proof of Corollary 6.4. The case $p = 2$ of Corollary 6.4 is proved by Faris in [1, Example 2, p 33] using Fourier Theory and the result that a distribution with support at a single point is a linear combination of derivatives of a delta function.

IX
Essential spectra

The theme of this chapter is a study of the various essential spectra for a closed operator and the changes that occur when the operator is perturbed. Constant-coefficient differential operators on $\mathbb{R}^n$ and $[0, \infty)$ are investigated in detail.

1. General remarks

Let X be a complex Banach space and let $T \in \mathscr{C}(X)$ be densely defined. The various essential spectra of T are defined as in §I.4 to be the sets $\sigma_{ek}(T) = \mathbb{C} \setminus \Delta_k(T)$, $k = 1, 2, 3, 4, 5$ where with

$$\Phi_{\pm}(T) = \{\lambda \in \mathbb{C} : T - \lambda I \in \mathscr{F}_{\pm}(X)\}$$

we have

$$\begin{aligned}
\Delta_1(T) &= \Phi_+(T) \cup \Phi_-(T),\\
\Delta_2(T) &= \Phi_+(T),\\
\Delta_3(T) &= \{\lambda \in \mathbb{C} : T - \lambda I \in \mathscr{F}(X)\} = \Phi_+(T) \cap \Phi_-(T),\\
\Delta_4(T) &= \{\lambda \in \mathbb{C} : \lambda \in \Delta_3(T) \quad \text{and} \quad \operatorname{ind}(T - \lambda I) = 0\},\\
\Delta_5(T) &= \text{union of all the components of } \Delta_1(T) \text{ which}\\
&\qquad \text{intersect the resolvent set } \rho(T) \text{ of } T.
\end{aligned}$$

Each of the sets $\sigma_{ek}(T)$ is closed since the $\Delta_k(T)$ are open, as proved in Theorems I.3.18 and I.3.25. Also $\sigma_{ek}(T) \subset \sigma_{el}(T)$ for $k < l$, the inclusion between any pair being strict in general (see Gustafson and Weidmann [1]).

By Theorem I.3.7, $\mathscr{R}(T - \lambda I)$ is closed if, and only if, $\mathscr{R}(T^* - \bar{\lambda} I)$ is closed, and in this case $\operatorname{nul}(T - \lambda I) = \operatorname{def}(T^* - \bar{\lambda} I)$ and $\operatorname{def}(T - \lambda I) = \operatorname{nul}(T^* - \bar{\lambda} I)$. A consequence of this and the fact that

$$\rho(T) = \{\lambda \in \mathbb{C} : (T - \lambda I) \in \mathscr{F}(X), \operatorname{nul}(T - \lambda I) = \operatorname{def}(T - \lambda I) = 0\} \tag{1.1}$$

is that $\rho(T) = \overline{\rho(T^*)}$. Also, if $\sigma_{e2}^*(T) = \mathbb{C} \setminus \Phi_-(T)$, then $\lambda \in \sigma_{e2}^*(T)$ if, and only if, $\bar{\lambda} \in \sigma_{e2}(T^*)$.

Theorem 1.1. Let $T \in \mathscr{C}(X)$ be densely defined. For $k = 1, 3, 4, 5, \lambda \in \sigma_{ek}(T)$ if, and only if, $\bar{\lambda} \in \sigma_{ek}(T^*)$. ■

Proof. The result for $k = 1, 3, 4$ follows from the remarks in the preceding paragraph. If $\lambda \in \Delta_5(T)$ it lies in a component $\Delta_1^{(n)}$ of $\Delta_1(T)$ which interesects $\rho(T)$. Therefore $\rho(T^*)$ must interesect the conjugate set $\bar{\Delta}_1^{(n)} = \{\bar{\mu} : \mu \in \Delta_1^{(n)}\}$ which is a component of $\Delta_1(T^*)$ and hence $\bar{\lambda} \in \Delta_5(T^*)$. Since $(T^*)^* = T$ the converse also holds and the theorem is proved. □

In applications it is often convenient to work with the equivalent definitions of the essential spectra given in Theorems 1.3–1.5 below. First we need the notion of a singular sequence.

Definition 1.2. A sequence $(u_n)_{n \in \mathbb{N}}$ in $\mathscr{D}(T)$ is called a *singular sequence* of T corresponding to $\lambda \in \mathbb{C}$ if it contains no convergent subsequence in X and satisfies $\|u_n\|_X = 1 \quad (n \in \mathbb{N})$ and $(T - \lambda I)u_n \to 0$ in X as $n \to \infty$. ■

If X is reflexive there exists a singular sequence of T corresponding to λ if and only if there exists a sequence $(v_n)_{n \in \mathbb{N}}$ in $\mathscr{D}(T)$ with the following properties:

$$\|v_n\|_X = 1 \quad (n \in \mathbb{N}), \qquad v_n \rightharpoonup 0 \text{ and } (T - \lambda I)v_n \to 0 \quad \text{in } X. \tag{1.2}$$

To see this, suppose that (u_n) is a singular sequence of T corresponding to λ. Then (u_n) is weakly compact if X is reflexive and so it contains a subsequence, (w_n) say, which converges weakly to a limit w in X and is such that $\|w_n - w_m\|_X \geq \delta > 0$ for all m, $n \in \mathbb{N}$. Hence $v_n = (w_n - w_{n-1})/\|w_n - w_{n-1}\|_X$ satisfies (1.2). Conversely, if (v_n) satisfies (1.2) it cannot contain a convergent subsequence without contradicting the two requirements $\|v_n\|_X = 1$ and $v_n \rightharpoonup 0$, and hence (v_n) is a singular sequence of T corresponding to λ.

Theorem 1.3. Let $T \in \mathscr{C}(X)$ be densely defined. Then

(i) $\lambda \in \sigma_{e2}(T)$ if, and only if, there exists a singular sequence of T corresponding to λ,

(ii) $\lambda \in \sigma_{e2}^*(T)$ if, and only if, there exists a singular sequence of T^* corresponding to $\bar{\lambda}$. ■

Proof. The first part follows from Corollary I.4.7 (which continues to hold for closed maps; see also Kato [1, Theorems IV-5.10, 5.11]) and the second part is a consequence of the remarks preceding Theorem 1.1. □

The Hilbert-space version of Theorem 1.3 is due to Wolf [1]. The general Banach-space case was proved by Balslev and Schubert in an unpublished report in 1964 (see Schechter [2, Chapter 11]).

Theorem 1.4. Let $T \in \mathscr{C}(X)$ be densely defined. Then

$$\sigma_{e4}(T) = \bigcap_{P \in \mathscr{K}(X)} \sigma(T + P). \qquad ■$$

Proof. Let $\lambda \notin \cap_{P \in \mathscr{K}(X)} \sigma(T+P)$. Then there exists an operator $P \in \mathscr{K}(X)$ such that $\lambda \in \rho(T+P)$ and hence $\lambda \in \Delta_4(T+P)$. It follows from Theorem I.3.17 that $\lambda \in \Delta_4(T)$ and hence $\lambda \notin \sigma_{e4}(T)$. Conversely, suppose that $\lambda \in \Delta_4(T)$. We have to show, for some $P \in \mathscr{K}(X)$, that $\lambda \in \rho(T+P)$. Since $\lambda \in \Delta_4(T)$, the subspace $\mathscr{R}(T-\lambda I)$ is closed in X and $\text{nul}(T-\lambda I) = \text{def}(T-\lambda I) = n$ say. Let $\{x_1, x_2, \ldots, x_n\}, \{y'_1, y'_2, \ldots, y'_n\}$ be bases for $\mathscr{N}(T-\lambda I)$ and $\mathscr{R}(T-\lambda I)^{\circ} = \mathscr{N}[(T-\lambda I)^*]$ respectively. Without loss of generality we may put $\lambda = 0$. Define $x'_j \in X^*$ and $y_j \in X$ $(j = 1, 2, \ldots, n)$ by

$$(x'_j, x_k) = \delta_{jk}, \qquad (y'_j, y_k) = \delta_{jk}$$

and for each $x \in X$ put

$$Px = \sum_{k=1}^{n} (x'_k, x) y_k.$$

The operator P is of finite rank and is bounded since

$$\|Px\| \leqslant \left(\sum_{k=1}^{n} \|x'_k\| \, \|y_k\| \right) \|x\|.$$

Hence P is compact and $T+P \in \mathscr{F}(X)$ with $\text{ind}(T+P) = \text{ind}\, T = 0$ by Theorem I.3.17. Let $x \in \mathscr{N}(T+P)$ so that $Tx = -Px$. Then since $Tx \in {}^{\circ}\mathscr{N}(T^*)$ we have $(y'_k, Tx) = 0$ $(k = 1, 2, \ldots, n)$ and hence $(y'_k, Px) = 0$. Consequently from the definition of P,

$$(x'_k, x) = 0 \qquad (k = 1, 2, \ldots, n)$$

and $Px = 0$. Therefore $Tx = 0$ and $x = \Sigma_{k=1}^{n} \alpha_k x_k$ for some $\alpha_k \in \mathbb{C}$. But $\alpha_k = (x'_k, x) = 0$ and hence $x = 0$; in other words we have proved that $\text{nul}(T+P) = 0$. This in turn gives $\text{def}(T+P) = 0$ since $T+P$ has zero index. We therefore conclude that $(T+P)^{-1}$ exists and its domain is X. Since $(T+P)^{-1}$ is also closed $0 \in \rho(T+P)$ and the theorem is proved. □

Theorem 1.5. We have that $\lambda \notin \sigma_{e5}(T)$ if, and only if, $\lambda \in \Delta_4(T)$ and a deleted neighbourhood of λ lies in the resolvent set of T. ■

Proof. Let $\lambda \notin \sigma_{e5}(T)$, i.e. $\lambda \in \Delta_5(T)$. Then from the definition of $\Delta_5(T)$, λ lies in a component $\Delta_1^{(n)}$ of $\Delta_1(T)$ which intersects $\rho(T)$. Since this intersection is an open set in $\mathbb{C}$ it follows from the stability results of §I.3, as noted in §I.4, that for all $\mu \in \Delta_1^{(n)}$ we have $\text{ind}(T-\mu I) = 0$ and furthermore $\text{nul}(T-\mu I) = \text{def}(T-\mu I) = 0$, and hence $\mu \in \rho(T)$, except possibly at some isolated values of μ. Thus if $\lambda \in \Delta_5(T)$ it is either in the resolvent set or else is one of these isolated points in $\Delta_4(T)$ with a deleted neighbourhood in $\rho(T)$. The converse is immediate. □

If $\lambda \in \sigma_{e4}(T) \backslash \sigma_{e2}(T) = \Delta_2(T) \backslash \Delta_4(T)$ then $\mathscr{R}(T-\lambda I)$ is closed and either λ is an eigenvalue of T of finite geometric multiplicity or else $\text{nul}(T-\lambda I) = 0$

$\neq \text{def}(T-\lambda I)$, i.e. λ lies in the residual spectrum $\sigma_r(T)$ of T. Furthermore

$$\{\lambda : \lambda \in \sigma_r(T) \text{ and } \mathscr{R}(T-\lambda I) \text{ closed}\} \subset \sigma_{e4}(T)\backslash\sigma_{e2}(T).$$

For $\lambda \notin \sigma_{e4}(T)$ we have $\text{nul}(T-\lambda I) = \text{def}(T-\lambda I) < \infty$ and hence the residual spectrum lies inside $\sigma_{e4}(T)$.

Theorem 1.6. Let X be a Hilbert space. Then

(i) if T is self-adjoint the sets $\sigma_{ek}(T)$ $(k = 1, 2, 3, 4, 5)$ are identical and $\lambda \in \sigma(T)\backslash\sigma_{ek}(T)$ if, and only if, it is an isolated eigenvalue of finite multiplicity,
(ii) If T is J-self-adjoint, the sets $\sigma_{ek}(T)$ $(k = 1, 2, 3, 4)$ are identical. ∎

Proof. (i) It suffices to show that $\Delta_1(T) \subset \Delta_5(T)$ or that every connected component of $\Delta_1(T)$ intersects $\rho(T)$. But since $\sigma(T) \subset \mathbb{R}$ every neighbourhood of every $\lambda \in \mathbb{C}$ intersects $\rho(T)$ and hence the result follows.

(ii) Since T is J-self-adjoint, $\text{nul}(T-\lambda I) = \text{def}(T-\lambda I)$ for any $\lambda \in \mathbb{C}$ if $\mathscr{R}(T-\lambda I)$ is closed, by Lemma III.5.4. Hence $\Delta_1(T) \subset \Delta_4(T)$ and $\sigma_{e1}(T) = \sigma_{e4}(T)$. □

Theorem 1.7. Let T be a closed, densely defined operator in a reflexive Banach space X and define

$$d(\lambda) = \text{dist}\,(\lambda, \sigma_{e2}(T)),$$

$$\rho(\lambda) = \inf\{\liminf_{n\to\infty} \|(T-\lambda I)f_n\| : (f_n) \in \mathscr{E}\},$$

where $\mathscr{E}$ is the set of all sequences (f_n) in $\mathscr{D}(T)$ such that $\|f_n\| = 1$ and $f_n \rightharpoonup 0$. Then $\rho(\lambda) \leqslant d(\lambda)$. If X is a Hilbert space and T is self-adjoint, $\rho(\lambda) = d(\lambda)$. ∎

Proof. Since $\sigma_{e,2}(T)$ is closed there exists $\mu \in \sigma_{e,2}(T)$ such that $d(\lambda) = |\lambda - \mu|$, and by Theorem 1.3 there is a singular sequence of T corresponding to μ:

$$\|(T-\mu I)f_n\| \to 0, \qquad \|f_n\| = 1, \quad f_n \rightharpoonup 0.$$

Hence

$$\|(T-\lambda I)f_n\| \leqslant \|(T-\mu I)f_n\| + |\lambda - \mu|$$

and consequently $\rho(\lambda) \leqslant d(\lambda)$; also $\rho(\lambda) = d(\lambda) = 0$ if $\lambda \in \sigma_{e2}(\lambda)$, i.e. $d(\lambda) = 0$.

Suppose that $d(\lambda) > 0$ and that T is self-adjoint in the Hilbert space X. Then $\sigma_{e2}(T) = \sigma_e(T)$ and given any ε with $0 < \varepsilon < \frac{1}{2}d(\lambda)$, the interval $I_\lambda = (\lambda - d(\lambda) + \varepsilon,\ \lambda + d(\lambda) - \varepsilon)$ lies outside $\sigma_e(T)$ and consequently application of Theorem 1.6 shows that this interval intersects $\sigma(T)$ only in a finite set of isolated eigenvalues, each eigenvalue having finite multiplicity. Let $\lambda_1, \lambda_2, \ldots, \lambda_k$ denote these eigenvalues (counted according to multiplicity) and let E denote the linear span of the associated eigenvectors $e_1, e_2, \ldots, e_k$. Then

X has the orthogonal sum decomposition, $X = E \oplus E^{\perp}$. If T_1 and T_2 denote the restrictions of T to E and $E^{\perp}$ respectively, then $T_1 E \subset E$ and $T_2 E^{\perp} \subset E^{\perp}$ (in the standard terminology—see §I.1—the decomposition reduces T). Also, it is an easy matter to verify that $\sigma(T_1) = \{\lambda_1, \ldots, \lambda_k\}$ and $\sigma(T_2) \cap I_\lambda = \emptyset$. Let $(f_n) \in \mathscr{E}$ and set

$$\phi_n = f_n - \sum_{j=1}^{k} (f_n, e_j) e_j.$$

Then $\phi_n \in E^{\perp}$, $\|\phi_n\| \to 1$ and,

$$\|(T-\lambda I) f_n - (T_2 - \lambda I)\phi_n\| \to 0.$$

Furthermore, since $I_\lambda \subset \rho(T_2)$, we have $\|(T_2 - \lambda I)\phi_n\| \geqslant [d(\lambda) - \varepsilon]\|\phi_n\|$. It follows that $d(\lambda) \leqslant \rho(\lambda)$ and hence $d(\lambda) = \rho(\lambda)$. □

2. Invariance of the essential spectra under perturbations

The first theorem is a generalization, to arbitrary closed operators densely defined in a Banach space X, of a celebrated result of H. Weyl [1] for self-adjoint operators in Hilbert space.

Theorem 2.1. Let $T \in \mathscr{C}(X)$ be densely defined in X and let P be T-compact. Then $T + P \in \mathscr{C}(X)$ and

$$\sigma_{ek}(T+P) = \sigma_{ek}(T) \qquad (k = 1, 2, 3, 4). \tag{2.1}$$ ■

Proof. Since P is T-compact it has T-bound zero by Corollary III.7.7, and hence $T + P \in \mathscr{C}(X)$ by Theorem III.8.2. Let $\lambda \in \Delta_k(T)$ for some $k \in \{1, 2, 3, 4\}$. By Theorems I.3.17 and I.3.21, a relatively compact perturbation P takes the sets $\mathscr{F}(X)$ and $\mathscr{F}_{\pm}(X)$ into themselves and preserves the index; note that $P \in \mathscr{K}(X(T), X)$ and so we are using Remark I.3.27 here. Hence $\lambda \in \Delta_k(T+P)$ and so $\sigma_{ek}(T+P) \subset \sigma_{ek}(T)$. The converse follows by the same argument once we have shown that P is $(T+P)$-compact. Since P has T-bound zero, for any $\varepsilon \in (0, 1)$ there exists a positive constant $K = K(\varepsilon)$ such that

$$\|Pu\| \leqslant \varepsilon \|Tu\| + K\|u\| \qquad (u \in \mathscr{D}(T)).$$

Hence

$$\|(T+P)u\| \geqslant (1-\varepsilon)\|Tu\| - K\|u\|.$$

Thus, given a sequence (u_n) in $\mathscr{D}(T)$ such that $\|(T+P)u_n\| + \|u_n\| \leqslant c$, it follows that $\|Tu_n\| + \|u_n\| \leqslant c'$ and since P is T-compact, (Pu_n) is precompact in X. This proves that P is $(T+P)$-compact as required to complete the proof of the theorem. □

The following example shows that Theorem 2.1 does not hold for $k = 5$. Additional hypotheses are required on the spectra of T and P for the result to be true for σ_{e5}; this problem is treated fully in Reed and Simon [2, §XIII.4].

Example 2.2. Let $X = l^2(\mathbb{Z})$, the Hilbert space of complex bilateral sequences $u = (\xi_j)_{j\in\mathbb{Z}}$ satisfying $\|u\|^2 = \sum_{-\infty}^{\infty} |\xi_j|^2 < \infty$, and let (e_n) be the canonical basis for X, that is, $e_n = (\delta_{jn})_{j\in\mathbb{Z}}$. Let T be the bounded linear operator defined by

$$Te_0 = 0, \qquad Te_j = e_{j-1} \quad (j\in\mathbb{Z}\backslash\{0\}).$$

Since $\|T\| = 1$, we have that $\sigma(T) \subset \{\lambda : |\lambda| \leqslant 1\}$. We claim that $\text{nul}\,(T-\lambda I) = 1$ if $0 \leqslant |\lambda| < 1$ and hence that the unit open disk $\{\lambda : |\lambda| < 1\}$ lies in $\sigma_{e5}(T)$. To see this, suppose that $u = (\xi_j) = \sum_{-\infty}^{\infty} \xi_j e_j$ satisfies $(T-\lambda I)u = 0$. Then

$$\sum_{j\neq 0} \xi_j e_{j-1} - \lambda \Sigma \xi_j e_j = 0$$

and this is satisfied if, and only if, $\xi_{j+1} = \lambda\xi_j$ for $j \neq -1$ and $\lambda\xi_{-1} = 0$. Hence if $\lambda = 0$ then $\xi_j = 0 \ (j\neq 0)$, and if $0 < |\lambda| < 1$ then $\xi_j = \lambda^j \xi_0 \ (j = 1, 2, \ldots)$ with $\xi_j = 0 \ (j = -1, -2, \ldots)$; the subspace $\mathcal{N}(T-\lambda I)$ is therefore spanned by e_0 when $\lambda = 0$ and by $\Sigma_{j=0}^{\infty} \lambda^j e_j$ when $0 < |\lambda| < 1$.

Let P be the bounded linear operator of rank 1 defined by $Pe_j = 0 \ (j \neq 0)$ with $Pe_0 = e_{-1}$. Therefore P is compact and hence T-compact since T is bounded on X. Moreover $S = T+P$ is the left shift operator: $Se_j = e_{j-1} (j\in\mathbb{Z})$. Therefore S is a unitary operator mapping X onto itself and so its spectrum lies on the unit circle $\{\lambda : |\lambda| = 1\}$; in fact $\sigma(S)$ coincides with the unit circle. Consequently $\sigma_{e5}(S) \neq \sigma_{e5}(T)$.

In this example, it is easily shown that $\text{def}\,(T-\lambda I) = 1$ if $0 \leqslant |\lambda| < 1$, and hence $\{\lambda : |\lambda| < 1\} \subset \Delta_4(T)$ since $\mathcal{R}(T-\lambda I)$ is closed, by Theorem I.3.2. Thus the essential spectra $\sigma_{ek}(T)$ $(k = 1, 2, 3, 4)$ lie on the unit circle. This example also works if $X = l^p(\mathbb{Z})$, with $p \geqslant 1$.

If $\lambda \in \rho(T)$, the operator $T-\lambda I$ is continuous and continuously invertible as a map from $\mathcal{D}(T)$ endowed with the graph norm into X. It therefore follows that P is T-compact if, and only if, $P(T-\lambda I)^{-1}$ is compact in X for some (and consequently for all) $\lambda\in\rho(T)$. If $S = T+P$ and $\lambda \in \rho(\mathrm{T}) \cap \rho(S)$,

$$S-\lambda I = [I+P(T-\lambda I)^{-1}](T-\lambda I)$$

and hence

$$(T-\lambda I)^{-1} - (S-\lambda I)^{-1} = (S-\lambda I)^{-1} P(T-\lambda I)^{-1}.$$

It follows that if P is T-compact, $(T-\lambda I)^{-1} - (S-\lambda I)^{-1}$ is compact. This property is also sufficient in itself to guarantee that $\sigma_{ek}(S) = \sigma_{ek}(T)$ for $k = 1$, 2, 3, 4. Before proving this in Theorem 2.4 below we need the following Spectral Mapping Theorem.

Theorem 2.3. Let $T\in\mathscr{C}(X)$ and $\xi\in\rho(T)$, and set $T_\xi = (T-\xi I)^{-1}$. Then for $\lambda \neq \xi$,

(i) $\lambda\in\sigma(T)$ if, and only if, $1/(\lambda-\xi)\in\sigma(T_\xi)$,

(ii) λ is an eigenvalue of T if, and only if, $1/(\lambda-\xi)$ is an eigenvalue of T_ξ with the same algebraic and geometric multiplicities,
(iii) $\lambda\in\sigma_{ek}(T)$ if, and only if, $1/(\lambda-\xi)\in\sigma_{ek}(T_\xi)$ $(k = 1,2,3,4,5)$. ■

Proof. The first two parts are consequences of the following identities on $\mathscr{D}(T)$:

$$T-\lambda I = (\lambda-\xi)[(\lambda-\xi)^{-1}I-T_\xi](T-\xi I) \tag{2.2}$$

$$= (\lambda-\xi)(T-\xi I)[(\lambda-\xi)^{-1}I-T_\xi]. \tag{2.3}$$

(iii) Since $\xi\in\rho(T)$ it follows that $T-\xi I$ is in $\mathscr{F}(X)$ and has zero index. Hence if $(\lambda-\xi)^{-1}\in\Delta_1(T_\xi)$, both operators on the right-hand side of (2.2) are semi-Fredholm and consequently so is $T-\lambda I$, by Theorem I.3.19. Therefore $\lambda\in\Delta_1(T)$. Furthermore if $(\lambda-\xi)^{-1}\in\Delta_2(T_\xi)$ then $T-\lambda I\in\mathscr{F}_+(X)$ and $\lambda\in\Delta_2(T)$. If $(\lambda-\xi)^{-1}\in\Delta_3(T_\xi)$, we see from Theorem I.3.16 that $T-\lambda I\in\mathscr{F}(X)$ and

$$\operatorname{ind}(T-\lambda I) = \operatorname{ind}[T_\xi-(\lambda-\xi)^{-1}I]+\operatorname{ind}(T-\xi I)$$
$$= \operatorname{ind}[T_\xi-(\lambda-\xi)^{-1}I].$$

Thus $(\lambda-\xi)^{-1}\in\Delta_k(T_\xi)$ implies $\lambda\in\Delta_k(T)$ for $k = 3,4$ and, in view of (i) and Theorem 1.5, we also conclude that the implication is valid for $k = 5$. We have therefore proved that $\lambda\in\Delta_k(T)$ if $(\lambda-\xi)^{-1}\in\Delta_k(T_\xi)$ for $k = 1,2,3,4,5$.

For the converse we use Theorem I.3.20 in (2.2) and (2.3). We find that $T_\xi-(\lambda-\xi)^{-1}I\in\mathscr{F}_\pm(X)$ $(\mathscr{F}(X))$ whenever $T-\lambda I\in\mathscr{F}_\pm(X)$ $(\mathscr{F}(X))$ and it follows that $(\lambda-\xi)^{-1}\in\Delta_k(T_\xi)$ if $\lambda\in\Delta_k(T)$. The proof is therefore complete. □

An immediate consequence of Theorems 2.1 and 2.3 is the following.

Theorem 2.4. Let $T, S\in\mathscr{C}(X)$ and suppose there exists $\xi\in\rho(T)\cap\rho(S)$ such that $(T-\xi I)^{-I}-(S-\xi I)^{-1}\in\mathscr{K}(X)$. Then $\sigma_{ek}(S) = \sigma_{ek}(T)$ for $k = 1, 2, 3, 4$. ■

In relation to the hypothesis in Theorem 2.4 it is useful to note the following result.

Proposition 2.5. Let $T, S\in\mathscr{C}(X)$. If $(T-\lambda I)^{-1}-(S-\lambda I)^{-1}$ is compact for some $\lambda\in\rho(T)\cap\rho(S)$ it is compact for all $\lambda\in\rho(T)\cap\rho(S)$. ■

Proof. Let T_λ and S_λ denote $(T-\lambda I)^{-1}$ and $(S-\lambda I)^{-1}$ respectively and suppose that $T_\mu-S_\mu\in\mathscr{K}(X)$ for some $\mu\in\rho(T)\cap\rho(S)$. For $\lambda\in\rho(T)\cap\rho(S)$ we have the resolvent identities

$$T_\lambda = T_\mu+(\lambda-\mu)T_\lambda T_\mu = T_\mu+(\lambda-\mu)T_\mu T_\lambda,$$
$$S_\lambda = S_\mu+(\lambda-\mu)S_\lambda S_\mu = S_\mu+(\lambda-\mu)S_\mu S_\lambda.$$

On using these, we obtain

$$[I+(\lambda-\mu)T_\lambda]\,(T_\mu-S_\mu)[I+(\lambda-\mu)S_\lambda] = T_\lambda - S_\lambda.$$

Since $I+(\lambda-\mu)T_\lambda$ and $I+(\lambda-\mu)S_\lambda$ are in $\mathscr{B}(X)$ it follows that $T_\lambda - S_\lambda \in \mathscr{K}(X)$ as asserted. □

In applications it is often important to weaken the hypothesis in Theorems 2.1 and 2.4. We first specialize to self-adjoint operators T acting in a Hilbert space H. By analogy with Definition III.7.3 for a T-compact operator, an operator P is said to be *T^2-compact* if $\mathscr{D}(P) \supset \mathscr{D}(T^2)$ and, for any sequence $(u_n)_{n\in\mathbb{N}}$ in $\mathscr{D}(T^2)$ such that $\|T^2 u_n\| + \|u_n\|$ $(n \in \mathbb{N})$ is bounded, (Pu_n) contains a subsequence which converges in H.

If T is self-adjoint and $\lambda \in \rho(T)$ then $(T-\lambda I)^2$ is an isomorphism of $\mathrm{H}(T^2)$, the Hilbert space determined by $\mathscr{D}(T^2)$ and the graph inner product of T^2, onto H. Hence, in this case P is T^2-compact if, and only if, $P(T-\lambda I)^{-2}$ is compact in H for some (and hence all) $\lambda \in \rho(T)$.

The following theorem is a special case of a result of Schechter [1]. We now write $\sigma_e(T)$ for the coincident essential spectra of a self-adjoint operator T.

Theorem 2.6. Let T be a self-adjoint operator in a Hilbert space H, let P be a symmetric operator defined on $\mathscr{D}(T)$ which is T^2-compact and suppose that $S = T + P$ is self-adjoint. Then $(T+\mathrm{i}I)^{-1} - (S+\mathrm{i}I)^{-1} \in \mathscr{K}(H)$ and hence $\sigma_e(S) = \sigma_e(T)$. ■

Proof. From Proposition III.7.2, since $\mathscr{D}(S) = \mathscr{D}(T)$, the graph norms $\|\bullet\|_S$ and $\|\bullet\|_T$ are equivalent on $\mathscr{D}(T)$ and hence $P = S - T$ is T-bounded. Also $T \pm \mathrm{i}I$ are unitary maps of $\mathrm{H}(T)$ onto H and so $(T+\mathrm{i}I)^{-1}P \in \mathscr{B}(\mathrm{H}(T))$. Since P is T^2-compact, $P(T-\mathrm{i}I)^{-1} = P(T-\mathrm{i}I)^{-2}(T-\mathrm{i}I) \in \mathscr{K}(\mathrm{H}(T), H)$ and therefore $P(T-\mathrm{i}I)^{-1}(T+\mathrm{i}I)^{-1}P \in \mathscr{K}(\mathrm{H}(T), H)$. From this and the fact that

$$\|(T+\mathrm{i}I)^{-1}Pu\|^2 = (P(T-\mathrm{i}I)^{-1}(T+\mathrm{i}I)^{-1}Pu, u) \leqslant \|P(T-\mathrm{i}I)^{-1}(T+\mathrm{i}I)^{-1}Pu\|\,\|u\|_T$$

it follows that $(T+\mathrm{i}I)^{-1}P \in \mathscr{K}(\mathrm{H}(T), H)$ and $(T+\mathrm{i}I)^{-1}P(T+\mathrm{i}I)^{-1} \in \mathscr{K}(H)$. If we now write

$$(T+\mathrm{i}I)^{-1} - (S+\mathrm{i}I)^{-1} = (S+\mathrm{i}I)^{-1}P(T+\mathrm{i}I)^{-1} = (S+\mathrm{i}I)^{-1}(T+\mathrm{i}I)[(T+\mathrm{i}I)^{-1}P(T+\mathrm{i}I)^{-1}]$$

and use the fact that $(S+\mathrm{i}I)^{-1}(T+\mathrm{i}I) \in \mathscr{B}(H)$, we obtain the result. □

For a self-adjoint operator T one can use the Spectral Theorem to define arbitrary powers of T, with $T \geqslant 0$ when necessary, and indeed $f(T)$ for suitable functions f. Thus, in analogy with Definition III.7.3, one can define T^p-compact operators and even $f(T)$-compact operators P. However, surprisingly, this leads to no weakening of the hypothesis of Theorem 2.6, a fact proved by Gustafson and Weidmann [1]; see also Weidmann [1, Theorem 9.11].

On using Theorem 2.4 in Theorems IV.4.2 and IV.5.1 we obtain results for the invariance of the essential spectra of arbitrary self-adjoint and m-sectorial operators under perturbations of the associated sesquilinear forms. In particular, from Corollaries IV.4.3 and IV.5.2 respectively we have the following two results whose usefulness in practice will be demonstrated in Chapter X.

Theorem 2.7. Let A and S be self-adjoint operators with forms a and s and coincident form domains $\mathfrak{Q}$ and let $p = s - a$ be such that $\hat{P}(A - \mathrm{i}zI)^{-1} \in \mathscr{K}(H, Q^*)$ for some $z \in \mathbb{R}$. Then $\sigma_e(S) = \sigma_e(A)$. ∎

Theorem 2.8. Let A and S be m-sectorial operators with forms a and s and coincident form domains $\mathfrak{Q}$ and let $p = s - a$ be such that $\hat{P}(A + z^2 I)^{-1} \in \mathscr{K}(H, Q^*)$ for some $z \in \mathbb{R}$. Then $\sigma_{ek}(S) = \sigma_{ek}(A)$ $(k = 1, 2, 3, 4)$. ∎

Another class of perturbations P worth investigating in the context of this section consists of operators with a specified measure of non-compactness. More explicitly, in the following theorem P is a T-bounded perturbation of a self-adjoint operator T and as a map from $\mathrm{H}(T)$ into H, P has ball measure of non-compactness $\tilde{\beta}(P) = a < 1$. Note that since $T + \mathrm{i}I$ is an isometry of $\mathrm{H}(T)$ onto H, $\tilde{\beta}(P)$ is equal to the ball measure of non-compactness of the operator $P(T + \mathrm{i}I)^{-1}$ in H.

Theorem 2.9. Let T be a self-adjoint operator in a Hilbert space H and let P be a symmetric T-bounded operator with T-bound less than 1. Then $S = T + P$ is self-adjoint and, if $\tilde{\beta}(P) = a < 1$,

$$(1-a)^2 \leqslant \frac{l_S^2 + 1}{l_T^2 + 1} \leqslant (1+a)^2$$

where

$$l_S := \inf\{|\lambda| : \lambda \in \sigma_e(S)\}$$

and similarly for l_T. ∎

Proof. The operator S is self-adjoint by Theorem III.8.5(ii). Since $(T + \mathrm{i}I)^{-1}$ and $(S + \mathrm{i}I)^{-1}$ are bounded normal operators on H it follows from Theorems I.2.19, I.4.8 and I.4.10 that the radii of their essential spectra coincide with their ball measures of non-compactness which we denote by β_0 and β_1 respectively. Also, $\|(T + \mathrm{i}I)^{-1}\| \leqslant 1$ and $\|(S + \mathrm{i}I)^{-1}\| \leqslant 1$ imply that $\beta_0, \beta_1 \leqslant 1$. By

Theorem 2.3(iii), $\lambda \in \sigma_e(T)$ if, and only if, $(\lambda + i)^{-1} \in \sigma_e[(T+iI)^{-1}]$ and so

$$\begin{aligned}\sigma_e(T) &= \{\lambda \in \mathbb{R} : (\lambda+i)^{-1} \in \sigma_e[(T+iI)^{-1}]\} \\ &\subset \{\lambda \in \mathbb{R} : |(\lambda+i)^{-1}| \leqslant \beta_0\} \\ &= \{\lambda \in \mathbb{R} : |\lambda| \geqslant (\beta_0^{-2}-1)^{\frac{1}{2}}\}.\end{aligned}$$

Similarly

$$\sigma_e(S) \subset \{\lambda \in \mathbb{R} : |\lambda| \geqslant (\beta_1^{-2}-1)^{\frac{1}{2}}\}.$$

Also, $l_T = (\beta_0^{-2}-1)^{\frac{1}{2}}$ and $l_S = (\beta_1^{-2}-1)^{\frac{1}{2}}$. Since

$$S + iI = [I + P(T+iI)^{-1}](T+iI)$$

we have

$$(T+iI)^{-1} = (S+iI)^{-1} + (S+iI)^{-1}P(T+iI)^{-1}.$$

Hence by Lemmas I.2.5 and I.2.8, $\beta_0 \leqslant \beta_1 + a\beta_1$ and also $\beta_1 \leqslant \beta_0 + a\beta_1$ giving

$$(1+a)^{-1}\beta_0 \leqslant \beta_1 \leqslant (1-a)^{-1}\beta_0.$$

Since $\beta_1^{-2} = l_S^2 + 1$ and $\beta_0^{-2} = l_T^2 + 1$ the theorem is proved. □

3. Operators with a compact resolvent

If $P \in \mathscr{K}(X)$ then P is P-compact and hence by Theorem 2.1 $\sigma_{ek}(P) = \sigma_{ek}(P-P) = \sigma_{ek}(O_X)$ $(k = 1,2,3,4)$ where O_X is the zero operator on X. Since O_X has null space and deficiency space X we have, for $k = 1, 2, 3, 4$, that $\sigma_{ek}(P) = \{0\}$ if X is infinite-dimensional and $\sigma_{ek}(P) = \varnothing$ if X is of finite dimension. This is also true for $k = 5$ since the only non-zero points in $\mathbb{C}$ which can lie in the spectrum of a compact operator are isolated eigenvalues of finite algebraic multiplicity; this result follows from Theorem I.1.9 and Remark I.1.19(i).

Theorem 3.1. Let X be an infinite-dimensional Banach space, $T \in \mathscr{C}(X)$ and $(T-\xi I)^{-1} \in \mathscr{K}(X)$ for some $\xi \in \rho(T)$. Then $(T-\lambda I)^{-1} \in \mathscr{K}(X)$ for all $\lambda \in \rho(T)$, and $\sigma_{ek}(T) = \varnothing$ for $k = 1,2,3,4,5$. The spectrum of T consists entirely of isolated eigenvalues of finite algebraic multiplicity.

If X is a Hilbert space, T is self-adjoint with a real $\lambda \in \rho(T)$ and $(T-\lambda I)^{-1} \in \mathscr{K}(X)$, then the eigenvectors of T form an orthonormal basis for X. Conversely if T is self-adjoint, $\sigma_e(T) = \varnothing$ and the eigenvectors of T form an orthonormal basis for X, then $(T-\lambda I)^{-1} \in \mathscr{K}(X)$ for every $\lambda \in \rho(T)$. ■

Proof. The first part is an immediate consequence of the resolvent equation

$$T_\lambda = T_\xi + (\lambda - \xi)T_\lambda T_\xi \qquad (\lambda, \xi \in \rho(T))$$

where $T_\lambda = (T-\lambda I)^{-1}$ and similarly for T_ξ. The spectrum of T_ξ in $\mathbb{C}\backslash\{0\}$ consists only of eigenvalues μ_n which are isolated points in $\sigma(T_\xi)$ and have finite algebraic multiplicity. Hence, by Theorem 2.3, $\sigma(T)$ consists only of eigenvalues $\xi + 1/\mu_n$ which are isolated points in $\sigma(T)$ and have finite algebraic multiplicities.

If X is a Hilbert space and T is self-adjoint, T_λ is compact and self-adjoint for any real $\lambda \in \rho(T)$. Since nul $T_\lambda = 0$ the eigenvectors of T_λ, and hence of T, form an orthonormal basis of X on account of Corollary II.5.3. For the converse we first observe from the hypothesis that $\sigma(T)$ consists of a sequence of isolated real eigenvalues (λ_n) having finite multiplicities and such that $|\lambda_n| \to \infty$ as $n \to \infty$. Suppose, without loss of generality that $0 \in \rho(T)$. Then for $u \in X$,

$$T^{-1}u = \sum_{n=1}^{\infty} \lambda_n^{-1}(u, \phi_n)\phi_n$$

where the ϕ_n $(n \in \mathbb{N})$ are the eigenvectors of T. Since $1/\lambda_n \to 0$ we have $T^{-1} \in \mathscr{K}(X)$ because the operators S_m defined by $S_m u = \Sigma_{n=1}^{m} \lambda_n^{-1}(u, \phi_n)\phi_n$ are bounded and of finite rank and also $\|(T^{-1} - S_m)u\|^2 = \Sigma_{n=m+1}^{\infty} \lambda_n^{-2}|(u, \phi_n)|^2 \leqslant \lambda_{m+1}^{-2}\|u\|^2$ (see Proposition I.1.2). □

Definition 3.2. Let X be a Banach space. An operator $T \in \mathscr{C}(X)$ is said to have a *compact resolvent* if $\rho(T) \neq \varnothing$ and $(T - \lambda I)^{-1} \in \mathscr{K}(X)$ for some, and hence all, $\lambda \in \rho(T)$. ■

4. Finite-dimensional extensions

Let X be a Banach space and let S and T be operators acting in X with S an extension of T. We say that S is an *m-dimensional extension* of T if the quotient space $\mathscr{D}(S)/\mathscr{D}(T)$ is of dimension m, i.e. there is an m-dimensional subspace F of $\mathscr{D}(S)$ such that $\mathscr{D}(S) = \mathscr{D}(T) \dot{+} F$.

Theorem 4.1. Let S be a closed m-dimensional extension of the closed, densely defined operator T in X. Then

(i) nul $T \leqslant$ nul $S \leqslant$ nul $T + m$,

(ii) def $S \leqslant$ def $T \leqslant$ def $S + m$,

(iii) $T \in \mathscr{F}(X)$ if, and only if, $S \in \mathscr{F}(X)$ and ind S = ind $T + m$. ■

Proof. (i) Since $T \subset S$, we have $\mathscr{N}(T) \subset \mathscr{N}(S)$ and hence nul $T \leqslant$ nul S. For the second inequality we may suppose that nul $T < \infty$, in which case $\mathscr{N}(T)$ has a complementary subspace in X; let $\mathscr{N}(S) = \mathscr{N}(T) \dot{+} K$, say. Then dim K = nul S − nul T and $K \cap \mathscr{D}(T) = K \cap \mathscr{N}(T) = \{0\}$. We conclude that $K \dot{+} \mathscr{D}(T) \subset \mathscr{D}(S)$ and hence that dim$K \leqslant m$ and nul $S \leqslant$ nul $T + m$.

(ii) Since $\mathscr{R}(S) \supset \mathscr{R}(T)$ we have def $S \leqslant$ def T. If def $S < \infty$ there exists a subspace R, say, complementary to $\mathscr{R}(S)$ in X with dim R = def $S < \infty$. From the hypothesis, $\mathscr{D}(S) = \mathscr{D}(T) \dot{+} F$ for an m-dimensional subspace F of X and we obtain

$$\begin{aligned} X &= \mathscr{R}(S) \dot{+} R \\ &= \mathscr{R}(T) + (R + SF), \end{aligned}$$

whence $\operatorname{def} T \leqslant \dim(R+SF)$. But $R \cap SF \subset R \cap \mathscr{R}(S) = \{0\}$ and so $\dim(R+SF) = \dim R + \dim SF \leqslant \operatorname{def} S + m$. This concludes the proof of (ii).

(iii) Let $\mathrm{X}(T)$ and $\mathrm{X}(S)$ denote, as usual, the Banach spaces determined by $\mathscr{D}(T)$ and $\mathscr{D}(S)$ respectively with the appropriate graph norms. Then $S \in \mathscr{B}(\mathrm{X}(S), X)$, $T \in \mathscr{B}(\mathrm{X}(T), X)$ and $\mathrm{X}(S) = \mathrm{X}(T) \dot{+} F$ for an m-dimensional subspace F of $\mathrm{X}(S)$. Let P be the projection of $\mathrm{X}(S)$ onto $\mathrm{X}(T)$. Then $P \in \mathscr{B}(\mathrm{X}(S), \mathrm{X}(T))$ and moreover $P \in \mathscr{F}(\mathrm{X}(S), \mathrm{X}(T))$ with $\operatorname{ind} P = \dim F - \operatorname{codim} \mathrm{X}(T) = m$. Also, in the decomposition

$$S = TP + S(I-P),$$

$S(I-P) \in \mathscr{B}(\mathrm{X}(S), X)$ is of rank $\leqslant m$ and is hence in $\mathscr{K}(\mathrm{X}(S), X)$. If $T \in \mathscr{F}(X)$ it is in $\mathscr{F}(\mathrm{X}(T), X)$ with the same index and, on account of Theorems I.3.16 and I.3.17, we deduce that $S \in \mathscr{F}(\mathrm{X}(S), X)$ and

$$\operatorname{ind} S = \operatorname{ind}(TP) = \operatorname{ind} T + \operatorname{ind} P = \operatorname{ind} T + m.$$

If $S \in \mathscr{F}(X)$ then $\operatorname{nul} T < \infty$ and $\operatorname{def} T < \infty$ by (i) and (ii). Hence by Theorem I.3.2, $T \in \mathscr{F}(X)$ and (iii) is proved. □

Corollary 4.2. Let S be a closed m-dimensional extension of the closed, densely defined operator T in X. Then,

$$\sigma_{ek}(T) = \sigma_{ek}(S) \qquad (k = 1, 2, 3), \tag{4.1}$$

$$\Delta_k(T) \cap \Delta_k(S) = \varnothing \qquad (k = 4, 5). \tag{4.2}$$

Thus $\sigma_{ek}(T) \neq \sigma_{ek}(S)$ for $k = 4, 5$ unless $\sigma_{ek}(T) = \sigma_{ek}(S) = \mathbb{C}$. ■

Proof. If $\lambda \in \Phi_+(T)$ then $\mathscr{R}(T-\lambda I)$ is closed and $\operatorname{nul}(T-\lambda I) < \infty$. Since $\mathscr{R}(S-\lambda I) = \mathscr{R}(T-\lambda I) + (S-\lambda I)F$, where F is an m-dimensional subspace of X, the subspace $\mathscr{R}(S-\lambda I)$ is therefore closed and $\lambda \in \Phi_+(S)$ by Theorem 4.1(i). Thus $\Phi_+(T) \subset \Phi_+(S)$ and $\sigma_{e2}(S) \subset \sigma_{e2}(T)$. Conversely, if $\lambda \in \sigma_{e2}(T)$ there exists a singular sequence of T corresponding to λ. As this is also a singular sequence of S corresponding to λ we have $\sigma_{e2}(T) \subset \sigma_{e2}(S)$ and thus $\sigma_{e2}(T) = \sigma_{e2}(S)$.

If $\lambda \in \Phi_-(T)$ then $\mathscr{R}(S-\lambda I)$ is closed as before and $\operatorname{def}(S-\lambda I) < \infty$ by Theorem 4.1(ii); thus $\Phi_-(T) \subset \Phi_-(S)$. Conversely, if $\lambda \in \Phi_-(S)$ then $\operatorname{def}(T-\lambda I) < \infty$ by Theorem 4.1(ii) and since $T - \lambda I \in \mathscr{C}(X)$ we can invoke Theorem I.3.2 to obtain $\lambda \in \Phi_-(T)$. We have therefore proved $\Phi_-(T) = \Phi_-(S)$ as well as $\Phi_+(T) = \Phi_+(S)$ and hence $\sigma_{ek}(T) = \sigma_{ek}(S)$ for $k = 1, 2, 3$.

The identity (4.2) is an immediate consequence of Theorem 4.1(iii). □

5. Direct and orthogonal sums of operators

Let T be a linear operator acting in a Hilbert space H and suppose that H and T have the orthogonal sum decompositions $H = H_1 \oplus H_2$ and $T = T_1 \oplus T_2$, i.e.

H_1 and H_2 are invariant subspaces of T, and T_i is the restriction of T to H_i $(i = 1, 2)$. If $x_n \in H$ and $x_n = x_n^{(1)} + x_n^{(2)}$, where $x_n^{(i)} \in H_i$, then (x_n) is a Cauchy sequence in H if, and only if, the sequences $(x_n^{(i)})$ $(i = 1, 2)$ are both Cauchy, and furthermore $x_n \to y$ if, and only if, $x_n^{(i)} \to y^{(i)}$ where $y = y^{(1)} + y^{(2)}$. Therefore H_1 and H_2 are closed subspaces of H and are Hilbert spaces with respect to the inner product of H. Also, $\mathscr{D}(T)$ is dense in H if, and only if, $\mathscr{D}(T_i) = \mathscr{D}(T) \cap H_i$ is dense in H_i for $i = 1, 2$; and T is closable if, and only if, T_1 and T_2 are both closable, in which case $\bar{T} = \bar{T}_1 \oplus \bar{T}_2$. If $T \in \mathscr{C}(H)$ the following are immediate:

$$\left.\begin{aligned} \operatorname{nul} T &= \operatorname{nul} T_1 + \operatorname{nul} T_2, \\ \operatorname{def} T &= \operatorname{def} T_1 + \operatorname{def} T_2, \\ \operatorname{ind} T &= \operatorname{ind} T_1 + \operatorname{ind} T_2, \\ \mathscr{R}(T) \text{ is closed if, and only if, } &\mathscr{R}(T_1) \text{ and } \mathscr{R}(T_2) \text{ are closed.} \end{aligned}\right\} \tag{5.1}$$

In the notation of §1 we therefore have

$$\left.\begin{aligned} \Phi_{\pm}(T) &= \Phi_{\pm}(T_1) \cap \Phi_{\pm}(T_2), \\ \sigma_{ek}(T) &= \sigma_{ek}(T_1) \cup \sigma_{ek}(T_2) \quad (k = 2, 3), \\ \sigma_{e2}^*(T) &= \sigma_{e2}^*(T_1) \cup \sigma_{e2}^*(T_2), \\ \sigma_{e1}(T) &\supset \sigma_{e1}(T_1) \cup \sigma_{e1}(T_2), \\ \sigma_{ek}(T) &\subset \sigma_{ek}(T_1) \cup \sigma_{ek}(T_2) \quad (k = 4, 5). \end{aligned}\right\} \tag{5.2}$$

These identities for the essential spectra σ_{ek} are the only ones available in general. When $k = 1$ for instance, $\Delta_1(T) \subset \Delta_1(T_1) \cap \Delta_1(T_2)$ but the reverse inclusion is not true in general; if $\lambda \in \Delta_1(T_1) \cap \Delta_1(T_2)$ with $\operatorname{nul}(T_1 - \lambda I) = \infty$ and $\operatorname{def}(T_2 - \lambda I) = \infty$ then $\lambda \notin \Delta_1(T)$, by (5.1). Also, when $k = 4$ we have $\Delta_4(T) \supset \Delta_4(T_1) \cap \Delta_4(T_2)$ but $\operatorname{ind}(T - \lambda I) = 0$ does not imply that $\operatorname{ind}(T_1 - \lambda I) = \operatorname{ind}(T_2 - \lambda I) = 0$. Since $\lambda \in \rho(T)$ if, and only if, $\operatorname{nul}(T - \lambda I) = \operatorname{def}(T - \lambda I) = 0$, by Theorem I.3.2, we have $\rho(T) = \rho(T_1) \cap \rho(T_2)$. We therefore have

$$\left.\begin{aligned} \sigma(T) &= \sigma(T_1) \cup \sigma(T_2), \\ \sigma_p(T) &= \sigma_p(T_1) \cup \sigma_p(T_2), \\ \sigma_r(T) &= \sigma_r(T_1) \cup \sigma_r(T_2). \end{aligned}\right\} \tag{5.3}$$

When used in conjunction with (5.2) and (5.3) the next lemma is useful, especially in the determination of the essential spectra of operators generated by ordinary differential expressions.

Lemma 5.1. Let X be a Banach space and let $T \in \mathscr{C}(X)$ be such that the identity operator I on X is T-compact. Then for $k = 1, 2, 3, 4$, either $\sigma_{ek}(T) = \varnothing$ or $\sigma_{ek}(T) = \mathbb{C}$. ■

Proof. For any constant c, the operator cI is T-compact and so by Theorem 2.1,

$$\sigma_{ek}(T+cI) = \sigma_{ek}(T) \quad (k = 1, 2, 3, 4).$$

Hence, since c is arbitrary, $\sigma_{ek}(T)$ is either empty or the whole of $\mathbb{C}$. □

In Lemma 5.1, I is T-compact if, and only if, the Banach space $X(T)$ determined by $\mathscr{D}(T)$ and the graph norm is compactly embedded in X. This is a situation which often prevails; for instance if T is an mth-order differential operator in $L^2(\Omega)$ with smooth enough coefficients, then $X(T)$ is topologically isomorphic to a closed subspace of the Sobolev space $W^{m,2}(\Omega)$ which is compactly embedded in $L^2(\Omega)$ when Ω is bounded and has smooth boundary. Of particular relevance to ordinary differential operators is the next theorem, the so-called *Decomposition Principle*. H will denote a Hilbert space with $H = H_1 \oplus H_2$, and I_i will be the identity operator on H_i; also $\Theta(T_1)$ will stand for the numerical range of T_1.

Theorem 5.2. Let $T = T_1 \oplus T_2 \in \mathscr{C}(H)$, let $\sigma_{ek}(T_1) = \varnothing$ for $k = 2$ (3), and let I_1 be T_1-compact. Then if S and S_2 are closed finite-dimensional extensions of T and T_2 respectively, $\sigma_{ek}(S) = \sigma_{ek}(S_2)$ for $k = 2$ (3). If $T_1 - \lambda I_1 \in \mathscr{F}(H_1)$ for all $\lambda \in \mathbb{C}$, and so $\sigma_{e1}(T_1) = \varnothing$, we also have $\sigma_{e1}(S) = \sigma_{e1}(S_2)$. If $\Theta(T_1)$ is not dense in $\mathbb{C}$, then $\sigma_{ek}(T_1) = \varnothing$ for $k = 1, 2$. ■

Proof. The first part is an immediate consequence of Corollary 4.2 and (5.2). If $T_1 - \lambda I \in \mathscr{F}(H_1)$ for all $\lambda \in \mathbb{C}$ we have from (5.1),

$$\begin{aligned}\sigma_{e1}(T) &= \sigma_{e1}(T_1) \cup \sigma_{e1}(T_2)\\ &= \sigma_{e1}(T_2).\end{aligned}$$

Thus $\sigma_{e1}(S) = \sigma_{e1}(S_2)$ by Corollary 4.2.

For the final part we invoke Theorem III.2.3. We have $T_1 - \lambda I_1 \in \mathscr{F}_+(X_1)$ for all $\lambda \in \mathbb{C} \backslash \Theta(T_1)$, so $\Delta_2(T_1)$ contains this set. Hence $\Delta_k(T_1) \neq \varnothing$ $(k = 1, 2)$ and Lemma 5.1 implies that $\sigma_{ek}(T_1) = \varnothing$ $(k = 1, 2)$. □

The significance of the Decomposition Principle is in the fact that the essential spectra σ_{ek} $(k = 1, 2, 3)$ of T and its finite-dimensional extensions do not depend on the behaviour of T in H_1. The utility of this result will be demonstrated in §9 below in locating the essential spectra of operators generated by ordinary differential expressions. We cannot expect the result to be generally true for σ_{e4} because, in view of Theorem 4.1, the index of an m-dimensional extension of T depends on m as well as on the index of T. However, we do have the following version of the Decomposition Principle for σ_{e4} by matching finite-dimensional extensions of T and T_1.

Theorem 5.3. Let $T = T_1 \oplus T_2 \in \mathscr{C}(H)$ and suppose there exists a closed m-dimensional extension S_1 of T_1 with a compact resolvent. Then if S is a closed m-dimensional extension of T we have $\sigma_{e4}(S) = \sigma_{e4}(T_2)$. ■

Proof. Since S_1 has a compact resolvent, I_1 is S_1-compact and $\Delta_4(S_1) \supset \rho(S_1) \neq \emptyset$. We therefore conclude from Lemma 5.1 that $\sigma_{e4}(S_1) = \emptyset$; in fact this result is established in Theorem 3.1. Also on assuming, without loss of generality, that $0 \in \rho(S_1)$, we have that S_1 is a Fredholm operator with zero index and moreover nul S_1 = def $S_1 = 0$. Furthermore, for $\lambda \neq 0$, the operator $S_1^{-1} - \lambda^{-1} I_1 \in \mathscr{F}(H_1)$ has zero index, since $S_1^{-1} \in \mathscr{K}(H_1)$ and $I_1 \in \mathscr{F}(H_1)$ has zero index. From

$$S_1 - \lambda I = -\lambda (S_1^{-1} - \lambda^{-1} I_1) S_1$$

and Theorem I.3.16 it follows that $S_1 - \lambda I_1 \in \mathscr{F}(H_1)$ with zero index, and this holds for all $\lambda \in \mathbb{C}$ since it is assumed to be so when $\lambda = 0$. Consequently, by (5.1), if $R = S_1 \oplus T_2$ then $\sigma_{e4}(R) = \sigma_{e4}(T_2)$. This operator R is an m-dimensional extension of T since $\mathscr{D}(S_1)/\mathscr{D}(T_1)$ is m-dimensional. Theorem 4.1 therefore implies that $R - \lambda I$ and $T - \lambda I$ are Fredholm operators for the same values of λ with

$$\text{ind}\,(R - \lambda I) = \text{ind}\,(T - \lambda I) + m.$$

But, since S is an m-dimensional extension of T, we have $\Delta_3(S) = \Delta_3(T) = \Delta_3(R)$ and

$$\begin{aligned} \text{ind}\,(S - \lambda I) &= \text{ind}\,(T - \lambda I) + m \\ &= \text{ind}\,(R - \lambda I) \end{aligned}$$

whenever $\lambda \in \Delta_3(S) = \Delta_3(R)$. Consequently $\Delta_4(S) = \Delta_4(R)$ and $\sigma_{e4}(S) = \sigma_{e4}(R) = \sigma_{e4}(T_2)$. The theorem is therefore proved. □

If X is a Banach space and $T \in \mathscr{C}(X)$ has the direct sum decomposition $T = T_1 \dotplus T_2$ we still have

$$\left.\begin{aligned} \text{nul}\, T &= \text{nul}\, T_1 + \text{nul}\, T_2, \\ \text{def}\, T &= \text{def}\, T_1 + \text{def}\, T_2, \\ \text{ind}\, T &= \text{ind}\, T_1 + \text{ind}\, T_2. \end{aligned}\right\} \tag{5.4}$$

However, the last property in (5.1) is no longer valid. If we assume that T, T_1, T_2 are closed and $T - \lambda I$, $T_1 - \lambda I_1$, $T_2 - \lambda I_2$ have closed ranges for all $\lambda \in \mathbb{C}$, then the results in (5.2) and Theorems 5.2 and 5.3 continue to hold. However, this assumption is unnecessary for results to do with σ_{ek} when $k = 3, 4, 5$. This is because, by (5.4), def $T < \infty$ if, and only if, def $T_1 < \infty$ and def $T_2 < \infty$, and in view of Theorem I.3.2 an operator has finite deficiency only if its range is closed. To summarize, we have the following theorem.

Theorem 5.4. Let X be a Banach space, let $T = T_1 \dotplus T_2 \in \mathscr{C}(X)$ with $T_i \in \mathscr{C}(X_i)$, let $\sigma_{e3}(T_1) = \emptyset$, and let I_1 be T_1-compact. Then if S and S_2 are closed finite-dimensional extensions of T and T_2 respectively, we have $\sigma_{e3}(S) = \sigma_{e3}(T_2)$.

If S is a closed m-dimensional extension of T and T_1 has a closed m-dimensional extension with a compact resolvent, then $\sigma_{e4}(S) = \sigma_{e4}(T_2)$.

The remaining conclusions of Theorem 5.2 hold if $T - \lambda I, T_1 - \lambda I_1, T_2 - \lambda I_2$ have closed ranges for all $\lambda \in \mathbb{C}$. ■

6. Constant-coefficient operators in $L^2(\mathbb{R}^n)$

A polynomial in n variables and of degree $m \geqslant 1$ has the form

$$P(\xi) = \sum_{|\alpha| \leqslant m} c_\alpha \xi^\alpha \qquad (\xi \in \mathbb{R}^n),$$

where α is a multi-index, $c_\alpha \in \mathbb{C}$, and $c_\alpha \neq 0$ for at least one α satisfying $|\alpha| = m$. Such a polynomial determines the differential polynomial

$$P\left(\frac{1}{2\pi i} D\right) = \sum_{|\alpha| \leqslant m} \left(\frac{1}{2\pi i}\right)^{|\alpha|} c_\alpha D^\alpha$$

and the differential operator T_0' defined in $L^2(\mathbb{R}^n)$ by

$$\mathscr{D}(T_0') = C_0^\infty(\mathbb{R}^n), \qquad T_0'\phi = P\left(\frac{1}{2\pi i} D\right)\phi.$$

If $\mathbf{F}$ denotes the Fourier transform, we have from Lemma V.1.15 that $\mathbf{F}\left(\frac{1}{(2\pi i)^{|\alpha|}} D^\alpha \phi\right)(\xi) = \xi^\alpha (\mathbf{F}\phi)(\xi)$ and so

$$\mathbf{F}(T_0'\phi)(\xi) = P(\xi)(\mathbf{F}\phi)(\xi),$$

i.e.

$$(T_0'\phi) = [\mathbf{F}^{-1} M(P) \mathbf{F}]\phi \qquad (\phi \in C_0^\infty(\Omega)),$$

where $M(P)$ denotes the maximal operator of multiplication by $P(\xi)$, having domain

$$\mathscr{D}(M(P)) = \left\{u : u \in L^2(\mathbb{R}^n) \text{ and } \int_{\mathbb{R}^n} |P(\xi)u(\xi)|^2 \, d\xi < \infty\right\}$$

(see Definition III.9.1). By Theorem III.9.2, $M(P)$ is a closed, densely defined operator whose adjoint is $M(\bar{P})$, the maximal operator of multiplication by the conjugate polynomial $\bar{P}(\xi) = \Sigma_{0 \leqslant |\alpha| \leqslant m} \bar{c}_\alpha \xi^\alpha$. Before proceeding we need the result

Lemma 6.1. The set $\{\xi \in \mathbb{R}^n : P(\xi) = 0\}$ has Lebesgue measure zero. ■

Proof. Let $\Lambda = \{\xi \in \mathbb{R}^n : P(\xi) = 0\}$ and denote by Λ_α the set of points ξ in Λ which are such that $D^\alpha P(\xi) = 0$ but $\nabla D^\alpha P(\xi) \neq 0$. Then Λ is the union of a finite number of such sets Λ_α. If $\xi_0 \in \Lambda_\alpha$ and $D_k[D^\alpha P(\xi_0)] \neq 0$, the Jacobian of the map $\xi \mapsto \phi(\xi) = x$ defined by

$$x_k = \mathrm{D}^\alpha P(\xi), \qquad x_j = \xi_k \quad \text{for } j \neq k$$

is non-zero at ξ_0 and it follows from the Inverse Function Theorem that there exists a neighbourhood $U(\xi_0)$ of ξ_0 in $\mathbb{R}^n$ such that $U(\xi_0) \cap \Lambda_\alpha$ is of measure zero. From the fact that Λ is covered by such sets $U(\xi_0) \cap \Lambda_\alpha$ one readily proves that Λ is of measure zero. □

Theorem 6.2.

(i) T_0' is closable and its closure is the operator $T_0 = \mathbf{F}^{-1}\mathrm{M}(\bar{P})\mathbf{F}$:

$$\mathscr{D}(T_0) = \left\{u : u \in \mathrm{L}^2(\mathbb{R}^n), \int_{\mathbb{R}^n} |P(\xi)\hat{u}(\xi)|^2 \,\mathrm{d}\xi < \infty\right\},$$

$$\mathbf{F}(T_0 u)(\xi) = P(\xi)\hat{u}(\xi),$$

where $\hat{u} = \mathbf{F}u$.

(ii) $T_0^* = \mathbf{F}^{-1}\mathrm{M}(\bar{P})\mathbf{F}$.

(iii) $\sigma_{ek}(T_0) = \sigma(T_0) = \overline{\{P(\xi) : \xi \in \mathbb{R}^n\}}$ $(k = 1, 2, 3, 4, 5)$ and T_0 has no eigenvalues.

(iv) If $\lambda \in \rho(T_0)$ then $(T_0 - \lambda I)^{-1} = \mathbf{F}^{-1}\mathrm{M}((P-\lambda)^{-1})\mathbf{F}$, where $\mathrm{M}((P-\lambda)^{-1})$ is the maximal operator of multiplication by $(P(\xi)-\lambda)^{-1}$. ■

Proof. (i) Let P_0 be the restriction of $\mathrm{M}(P)$ to $\mathscr{D}(P_0) = \{\hat{\phi} : \phi \in \mathrm{C}_0^\infty(\mathbb{R}^n)\}$ so that $T_0' = \mathbf{F}^{-1} P_0 \mathbf{F}$. Since $\mathrm{M}(P)$ is closed, P_0 is closable and $\bar{P}_0 \subset \mathrm{M}(P)$. Furthermore, as T_0' and P_0 are unitarily equivalent (i.e. $T_0' = \mathbf{F}^{-1} P_0 \mathbf{F}$ and $\mathbf{F}$ is unitary), T_0' is closable and $\bar{T}_0' = \mathbf{F}^{-1}\bar{P}_0\mathbf{F}$. We therefore need to prove that $\bar{P}_0 = \mathrm{M}(P)$ or equivalently that $\mathscr{D}(P_0)$ is dense in $\mathrm{H}(P)$, the Hilbert space defined by $\mathscr{D}[\mathrm{M}(P)]$ and the graph norm

$$\|u\|_p = [\|\mathrm{M}(P)u\|^2 + \|u\|^2]^{\frac{1}{2}}$$

where $\|\bullet\|$ denotes the $\mathrm{L}^2(\mathbb{R}^n)$ norm.

Let χ_N denote the characteristic function of the ball B_N with centre 0 and radius N and set $v_N = \chi_N v$ for $v \in \mathrm{H}(P)$. Then

$$\|v - v_N\|_p^2 = \int_{B_N^{\mathrm{C}}} [|P(\xi)v(\xi)|^2 + |v(\xi)|^2] \,\mathrm{d}\xi \to 0$$

as $N \to \infty$. Also for each N the regularization $(v_N)_\varepsilon \in \mathrm{C}_0^\infty(\mathbb{R}^n)$ and $(v_N)_\varepsilon \to v_N$ in $\mathrm{H}(P)$ as $\varepsilon \to 0+$. Thus $\mathrm{C}_0^\infty(\mathbb{R}^n)$ is dense in $\mathrm{H}(P)$ and consequently, so is the Schwartz space $\mathfrak{S}$ $(= \mathscr{S}(\mathbb{R}^n))$ of rapidly decreasing functions; see §V.1 for all we need to know about $\mathfrak{S}$. Also $\mathbf{F}^{-1}\mathfrak{S} = \mathfrak{S} \subset \mathrm{W}_0^{m,2}(\mathbb{R}^n)$ and hence for each $\hat{w} \in \mathfrak{S}$ there exists a sequence (ϕ_k) of $\mathrm{C}_0^\infty(\mathbb{R}^n)$ functions which converge to w in $\mathrm{W}_0^{m,2}(\mathbb{R}^n)$. Consequently $\phi_k \to w$ and $P[(2\pi\mathrm{i})^{-1}\mathrm{D}]\phi_k \to P[(2\pi\mathrm{i})^{-1}\mathrm{D}]w$ in

$L^2(\mathbb{R}^n)$ or, equivalently $\hat{\phi}_k \to \hat{w}$ in $H(P)$. All in all we have proved that $\mathsf{F}C_0^\infty(\mathbb{R}^n)$ is dense in $H(P)$ and hence that $M(P) = \bar{P}_0$.

(ii) Since $T_0 = \mathsf{F}^{-1} M(P)\mathsf{F}$ we have $T_0^* = \mathsf{F}^{-1}[M(P)]^* \mathsf{F} = \mathsf{F}^{-1} M(\bar{P})\mathsf{F}$ by Theorem III.9.2.

(iii) Let $\mathfrak{P}$ denote the closure of the set $\{P(\xi): \xi \in \mathbb{R}^n\}$. If $\lambda \notin \mathfrak{P}$ there exists a $\delta > 0$ such that $|\lambda - P(\xi)| > \delta$ for all $\xi \in \mathbb{R}^n$ and hence, for all $u \in \mathscr{D}[M(P)]$,

$$\|[M(P) - \lambda I]u\|^2 = \int_{\mathbb{R}^n} |P(\xi) - \lambda|^2 \, |u(\xi)|^2 \, d\xi$$
$$\geqslant \delta^2 \|u\|^2.$$

For any $\lambda \in \mathbb{C}$, the set $\{\xi: P(\xi) = \lambda\}$ has zero measure, by Lemma 6.1, and so $\mathscr{R}[M(P) - \lambda I]$ is dense in $L^2(\mathbb{R}^n)$ on account of Theorem III.9.2. Since $[M(P) - \lambda I]^{-1}$ is closed and bounded on $\mathscr{R}[M(P) - \lambda I]$ for $\lambda \notin \mathfrak{P}$, it must have a closed domain, and consequently $\lambda \in \rho[M(P)]$ or $\sigma[M(P)] \subset \mathfrak{P}$. If $\lambda = P(\xi_0)$ for some $\xi_0 \in \mathbb{R}^n$, then given any $\varepsilon > 0$ there exists a $\delta > 0$ such that $|P(\xi) - \lambda| < \varepsilon$ for $|\xi - \xi_0| < \delta$. Any $u \in \mathscr{D}[M(P)]$ with support in the ball $B(\xi_0, \delta)$ therefore satisfies

$$\|[M(P) - \lambda I]u\| \leqslant \varepsilon \|u\|,$$

proving that every such λ lies in $\sigma[M(P)]$. Since the spectrum is a closed set we conclude that $\sigma[M(P)] = \mathfrak{P}$ and consequently $\sigma(T_0) = \mathfrak{P}$, because T_0 and $M(P)$, being unitarily equivalent, have the same spectrum. We have also shown that $\mathscr{R}[M(P) - \lambda I]$ is dense in $L^2(\mathbb{R}^n)$ for any $\lambda \in \mathbb{C}$ and so, since $M(P) - \lambda I$ is injective by Theorem III.9.2, it follows that $\mathscr{R}[M(P) - \lambda I]$ is not closed for any $\lambda \in \mathfrak{P}$. This in turn implies that $\mathfrak{P} \subset \sigma_{ek}[M(P)] = \sigma_{ek}(T_0)$ for $k = 1, 2, 3, 4, 5$. Furthermore, since $M(P) - \lambda I$ is injective for all $\lambda \in \mathbb{C}$, neither $M(P)$ nor T_0 has eigenvalues.

(iv) If $\lambda \in \rho(T_0) = \rho[M(P)]$ then $P(\xi) - \lambda$ is nowhere zero and hence $[M(P) - \lambda I]^{-1}$ is the maximal operator of multiplication by $[P(\xi) - \lambda]^{-1}$, by Theorem III.9.2. The theorem is therefore proved. □

In general the set $\{P(\xi): \xi \in \mathbb{R}^n\}$ is not closed. For instance, let

$$P(\xi) = (\xi_1 \xi_2 - 1)^2 + \xi_2^2 + \cdots + \xi_n^2.$$

Then $P(\xi) \neq 0$ for all $\xi \in \mathbb{R}^n$. However, if we take $\xi_1 \neq 0$, $\xi_2 = 1/\xi_1$, $\xi_3 = \cdots = \xi_n = 0$, and let $|\xi_1| \to \infty$, then $P(\xi) \to 0$ and hence 0 lies in the closure of $\{P(\xi): \xi \in \mathbb{R}^n\}$.

If we restrict ourselves to polynomials P such that

$$|P(\xi)| \to \infty \quad \text{as} \quad |\xi| \to \infty \qquad (\xi \in \mathbb{R}^n), \tag{6.1}$$

the set $\{P(\xi): \xi \in \mathbb{R}^n\}$ is indeed closed. This is because, if for all $\xi \in \mathbb{R}^n$ $\lambda \neq P(\xi)$, then $|\lambda - P(\xi)| \geqslant C_1 > 0$ for all ξ in any compact subset of $\mathbb{R}^n$, while from (6.1),

$|\lambda - P(\xi)| \geqslant C_2 > 0$ for $|\xi|$ large enough. Hence

$$|\lambda - P(\xi)| \geqslant C > 0 \qquad (\xi \in \mathbb{R}^n)$$

and our assertion is proved.

We now consider polynomials P which have the property (6.1).

Definition 6.3. If $P(\xi) = \Sigma_{|\alpha| \leqslant m} c_\alpha \xi^\alpha$, the homogeneous polynomial $P_m(\xi) = \Sigma_{|\alpha| = m} c_\alpha \xi^\alpha$ of degree m is called the *principal symbol* of the differential polynomial $P((1/2\pi \mathrm{i})\mathrm{D})$; $P((1/2\pi \mathrm{i})\mathrm{D})$ (or $P(\xi)$) is said to be *elliptic* if for all $\xi \in \mathbb{R}^n \setminus \{0\}$, $P_m(\xi) \neq 0$. ■

Proposition 6.4. Let $T_0 = \mathbf{F}^{-1} \mathrm{M}(P) \mathbf{F}$, where $P(\xi) = \Sigma_{|\alpha| \leqslant m} c_\alpha \xi^\alpha$ is of degree m. Then the following are equivalent:

(i) $P((1/2\pi \mathrm{i})\mathrm{D})$ is elliptic.

(ii) There exists a positive constant c such that $1 + |P(\xi)| \geqslant c(1 + |\xi|^2)^{m/2}$ for all $\xi \in \mathbb{R}^n$.

(iii) $\mathscr{D}(T_0) = \mathrm{W}^{m,2}(\mathbb{R}^n)$.

If any (and hence all) of the conditions are satisfied, the graph norm $\|\bullet\|_{T_0}$ of T_0 is equivalent to the $\mathrm{W}^{m,2}(\mathbb{R}^n)$ norm $\|\bullet\|_{m,2,\mathbb{R}^n}$. ■

Proof. (i) *implies* (ii). *Let* $\delta = \min\{|P_m(\xi)| : |\xi| = 1\}$. Then $\delta > 0$ by (i), and for all $\xi \in \mathbb{R}^n$,

$$\begin{aligned} |P(\xi)| &= \left| P_m(\xi) + \sum_{|\alpha| \leqslant m-1} c_\alpha \xi^\alpha \right| \\ &\geqslant \delta |\xi|^m - K(1 + |\xi|^{m-1}) \\ &\geqslant \tfrac{1}{2} \delta |\xi|^m - K_\delta \end{aligned}$$

since $|\xi|^{m-1} \leqslant \frac{1}{2}\delta|\xi|^m + K_\delta$ for some positive constant K_δ depending on δ. Hence there exists a positive constant $K_{1,\delta}$ such that

$$\begin{aligned} K_{1,\delta} + |P(\xi)| &\geqslant \tfrac{1}{2}\delta(1 + |\xi|^m) \\ &\geqslant c(\delta)\,(1 + |\xi|^2)^{m/2}; \end{aligned}$$

(ii) follows from this.

(ii) *implies* (iii). Since $|P(\xi)| \leqslant K(1 + |\xi|^2)^{m/2}$ for some positive constant K, (ii) implies that $\mathscr{D}(\mathrm{M}(P)) = \mathscr{D}(\mathrm{M}(k_m))$ where $k_m(\xi) = (1 + |\xi|^2)^{m/2}$. But $\mathscr{D}(\mathrm{M}(k_m)) = \mathrm{W}^{m,2}(\mathbb{R}^n)$ by Lemma V.1.15.

(iii) *implies* (i). We shall prove first that (iii) implies (ii) and then that (ii) implies (i). From (iii) we infer that $\mathscr{D}(\mathrm{M}(k_m)) = \mathscr{D}(\mathrm{M}(P)) = \mathscr{D}(\mathrm{M}(|P| + 1))$ and hence, in view of Proposition III.7.2, these domains, with the appropriate graph norms, are topologically isomorphic. Consequently $\mathrm{M}(k_m/(|P| + 1))$ is bounded above and away from zero, implying (ii). If $P_m(\xi_0) = 0$ for some

$\xi_0 \neq 0$, then for any t we may conclude from (ii) that

$$c(1+t^2|\xi_0|^2)^{m/2} \leqslant 1 + P(t\xi_0)$$
$$= 1 + \left| \sum_{|\alpha| \leqslant m-1} c_\alpha (t\xi_0)^{m-1} \right|,$$

which is evidently contradicted on allowing $t \to \infty$. Therefore (ii) implies (i) and (iii) implies (i). The equivalence of the norms $\|\bullet\|_T$ and $\|\bullet\|_{m,2,\mathbb{R}^n}$ is clearly seen from (ii) and the inequality $|P(\xi)| \leqslant K(1+|\xi|^2)^{m/2}$. □

If the coefficients of P_m are real and $n \geqslant 2$ then P is elliptic only if m is even. The latter condition is, however, not necessary since the Cauchy–Riemann operator $D_1 + iD_2$ in $\mathbb{R}^2$ is elliptic. It is true that when $n \geqslant 3$, the operator P is elliptic only if m is even (see Agmon [1, Theorem 4.1]). When $n = 1$, any P is elliptic.

By Proposition 6.4, $|P(\xi)| \to \infty$ as $|\xi| \to \infty$ if P is elliptic and hence, in view of the remarks made about (6.1) we have the following result.

Corollary 6.5. If P is elliptic, the set $\{P(\xi)\colon \xi \in \mathbb{R}^n\}$ is closed in $\mathbb{C}$. ■

Corollary 6.6. If P has real coefficients then T_0 is self-adjoint and $\sigma(T_0) = \mathbb{R}$, $[\gamma, \infty)$ or $(-\infty, \gamma]$ according to whether P is unbounded above and below, bounded below and bounded above, respectively. ■

Corollary 6.7. Let $n \geqslant 2$ and let $\mathrm{re}\, P(\xi) = \Sigma_{|\alpha| \leqslant m} \mathrm{re}\, c_\alpha \xi^\alpha$ be elliptic (and hence of even degree m). Then $\mathrm{re}\, P(\xi)$ is semi-bounded. If $\gamma = \inf\{\mathrm{re}\, P(\xi)\colon \xi \in \mathbb{R}^n\} > -\infty$, then $T_0 - \gamma I$ is m-accretive, T_0 being self-adjoint with $T_0 \geqslant \gamma$ and $\sigma(T_0) = [\gamma, \infty)$ if $P(\xi)$ has real coefficients. A similar result holds when $\mathrm{re}\, P(\xi)$ is bounded above. ■

Proof. By Proposition 6.4(ii), $|\mathrm{re}\, P(\xi)| \to \infty$ as $|\xi| \to \infty$. Since $\mathrm{re}\, P(\xi)$ is elliptic and of course continuous, we infer that for some $c_0 > 0$, either $\mathrm{re}\, P(\xi) > 0$ for $|\xi| > c_0$ or $\mathrm{re}\, P(\xi) < 0$ for $|\xi| > c_0$ and hence $\mathrm{re}\, P(\xi)$ tends either to $+\infty$ or $-\infty$ as $|\xi| \to \infty$. The polynomial $\mathrm{re}\, P$ is therefore either bounded above or bounded below. The rest follows from Theorem 6.2. □

Example 6.8. If $P(\xi) = \xi_1 + i\xi_2$ for $\xi = (\xi_1, \xi_2) \in \mathbb{R}^2$, then T_0 is the Cauchy-Riemann operator $(1/2\pi i)(D_1 + iD_2)$ with domain $W^{1,2}(\mathbb{R}^2)$. By Theorem 6.2, $\sigma_{ek}(T_0) = \mathbb{C}$ for $k = 1, 2, 3, 4, 5$.

Example 6.9. Let $P(\xi) = c_0 + 2\pi i c \cdot \xi + 4\pi^2 |\xi|^2$, where $\xi \in \mathbb{R}^n$, $c_0 \in \mathbb{R}$, and $c \in \mathbb{R}^n$. Then T_0 is defined by $c_0 + c \cdot \nabla - \Delta$ with domain $W^{2,2}(\mathbb{R}^n)$ and for

$k = 1, 2, 3, 4, 5,$

$$\sigma_{ek}(T) = \sigma(T) = \{z \in \mathbb{C}: z = c_0 + 2\pi i c\cdot\xi + 4\pi^2|\xi|^2, \xi \in \mathbb{R}^n\}$$
$$= \begin{cases} \{x+iy: x \geqslant c_0 \text{ and } |y|^2 \leqslant |c|^2(x-c_0)\} & \text{if } n \geqslant 2, \\ \{x+iy: x \geqslant c_0 \text{ and } |y|^2 = |c|^2(x-c_0)\} & \text{if } n = 1. \end{cases}$$

$T_0 - c_0 I$ is m-accretive; in fact T_0 is m-sectorial.

7. Constant-coefficient operators in $L^2(0, \infty)$

We now have

$$P(\xi) = \sum_{j=0}^{m} c_j \xi^j \quad (\xi \in \mathbb{C}), \qquad c_m \neq 0,$$

$$P\left(\frac{1}{2\pi i} D\right) = \sum_{j=0}^{m} \left(\frac{1}{2\pi i}\right)^j c_j D^j, \qquad D = \frac{d}{dx},$$

and T_0' is the restriction of $P\left(\frac{1}{2\pi i} D\right)$ to $C_0^\infty(0, \infty)$:

$$\mathscr{D}(T_0') = C_0^\infty(0, \infty), \qquad T_0'\phi = P\left(\frac{1}{2\pi i} D\right)\phi \qquad (\phi \in C_0^\infty(\Omega)).$$

We denote the $L^2(0, \infty)$ inner product and norm by $(\bullet, \bullet)$ and $\|\bullet\|$ respectively. We shall identify functions which are equal almost everywhere without comment.

Lemma 7.1.

$$W^{m,2}(0, \infty) = \left\{u : u^{(m-1)} \in AC_{loc}[0, \infty);\ u, P\left(\frac{1}{2\pi i} D\right) u \in L^2(0, \infty)\right\} \tag{7.1}$$

$$W_0^{m,2}(0, \infty) = \left\{u : u^{(m-1)} \in AC_{loc}[0, \infty), u^{(j)}(0) = 0 \ (j = 0, 1, \ldots, m-1), \right.$$
$$\left. u, P\left(\frac{1}{2\pi i} D\right) u \in L^2(0, \infty)\right\} \tag{7.2} \blacksquare$$

Proof. By Corollary V.3.12 any $u \in W^{m,2}(0, \infty)$ can be identified with a function u which is such that $u^{(m-1)} \in AC_{loc}[0, \infty)$, and evidently $u, P((1/2\pi i)D)u \in L^2(0, \infty)$. Hence $W^{m,2}(0, \infty)$ is contained in the set on the right-hand side of (7.1). Conversely let $u^{(m-1)} \in AC_{loc}[0, \infty)$ and $u, P((1/2\pi i)D)u \in L^2(0, \infty)$. If $d \geqslant 1$ we see from Theorem V.4.14 and a similarity transformation that there exists a positive constant γ such that

$$\|u^{(j)}\|_{2,[0,d]} \leqslant \gamma \|u\|_{m,2,[0,d]}^{j/m} \|u\|_{2,[0,d]}^{(m-j)/m} \quad (j = 0, 1, \ldots, m-1). \tag{7.3}$$

On using Young's inequality, $|ab| \leqslant |a|^p/p + |b|^{p'}/p'$ (with $1/p + 1/p' = 1$), we find that for any $\delta > 0$,

$$\|u^{(j)}\|^2_{2,[0,d]} \leqslant \gamma^2 (\delta^{2m-2j}\|u\|^2_{m,2,[0,d]})^{j/m} (\delta^{-2j}\|u\|^2_{2,[0,d]})^{(m-j)/m}$$
$$\leqslant \gamma^2\left(\frac{j}{m}\delta^{2m-2j}\|u\|^2_{m,2,[0,d]} + \frac{m-j}{m}\delta^{-2j}\|u\|^2_{2,[0,d]}\right). \tag{7.4}$$

Therefore, given any $\varepsilon > 0$ there exists a positive constant K_ε independent of d such that

$$\|u^{(j)}\|^2_{2,[0,d]} \leqslant \varepsilon\|u\|^2_{m,2,[0,d]} + K_\varepsilon\|u\|^2_{2,[0,d]} \qquad (j = 0, 1, \ldots, m-1). \tag{7.5}$$

These inequalities yield

$$\|u\|^2_{m,2,[0,d]} \leqslant K\left[\left\|P\left(\frac{1}{2\pi \mathrm{i}}\mathrm{D}\right)u\right\|^2_{2,[0,d]} + \|u\|^2_{2,[0,d]}\right] \tag{7.6}$$

and hence $u \in W^{m,2}(0, \infty)$. This completes the proof of (7.1).

It follows as in Corollary V.3.21 that $W_0^{m,2}(0,\infty)$ can be identified with a subset of the set in (7.2). Conversely, if u lies in the latter set then $u \in W^{m,2}(0,\infty)$ from (7.1). In view of Theorem V.3.4, that $u \in W_0^{m,2}(0,\infty)$ will follow if we prove that $x^{-m}u \in L^2(0,\infty)$. This is a consequence of the generalized Hardy inequality

$$\int_0^\infty x^{-2m}|u(x)|^2\,\mathrm{d}x \leqslant \left(2^{2m}\frac{m!}{(2m)!}\right)^2 \int_0^\infty |u^{(m)}|^2\,\mathrm{d}x. \tag{7.7}$$

Firstly we observe that we may suppose, without loss of generality, that u has compact support in $[0,\infty)$. This is because, if $\theta \in C^\infty[0,\infty)$ satisfies $\theta(x) = 1$ for $x \leqslant 1$ and $\theta(x) = 0$ for $x \geqslant 2$, then the functions $u_k(\bullet) = u(\bullet)\theta(\bullet/k)$ converge to u in $W^{m,2}(0,\infty)$ as $k \to \infty$ and hence if we prove (7.7) for the u_k it will also follow for u. Secondly, successive integration by parts yields

$$u^{(j)}(t) = \int_0^t \frac{(t-s)^{m-j-1}}{(m-j-1)!}u^{(m)}(s)\,\mathrm{d}s, \qquad (0 \leqslant j \leqslant m-1).$$

Hence as $t \to 0$, since $u^{(m)} \in L^2(0,\infty)$,

$$u^{(j)}(t) = O\left[t^{m-j-\frac{1}{2}}\left(\int_0^t |u^{(m)}(s)|^2\,\mathrm{d}s\right)^{\frac{1}{2}}\right]$$
$$= o(t^{m-j-\frac{1}{2}}).$$

On integration by parts, for $\varepsilon \to 0+$,

$$\int_\varepsilon^\infty x^{-2m}|u(x)|^2\,\mathrm{d}x = o(1) + \frac{2}{(2m-1)}\int_\varepsilon^\infty x^{-2m+1}\,\mathrm{re}(\bar{u}u')(x)\,\mathrm{d}x,$$

whence $x^{-m}u \in L^2(0,\infty)$ and

$$\int_0^\infty x^{-2m}|u(x)|^2\,dx \leqslant \frac{4}{(2m-1)^2}\int_0^\infty x^{-2m+2}|u'(x)|^2\,dx.$$

Successive application of this yields (7.7) and hence (7.2). □

Theorem 7.2. T_0' is closable with closure T_0 and adjoint T_0^* defined by

$$\mathscr{D}(T_0) = W_0^{m,2}(0,\infty), \qquad T_0 u = P\left(\frac{1}{2\pi i}D\right)u \quad (u \in \mathscr{D}(T_0)), \tag{7.8}$$

$$\mathscr{D}(T_0^*) = W^{m,2}(0,\infty),$$

$$T_0^* v = \bar{P}\left(\frac{1}{2\pi i}D\right)v = \sum_{j=1}^{m}\left(\frac{1}{2\pi i}\right)^j \bar{c}_j D^j v \quad (v \in \mathscr{D}(T_0^*)). \tag{7.9}$$ ■

Proof. By Theorem V.3.8 there exists a positive constant γ such that for all $\phi \in C_0^\infty(0,\infty)$,

$$\|\phi^{(j)}\| \leqslant \gamma \|\phi^{(m)}\|^{j/m}\|\phi\|^{(m-j)/m} \qquad (j = 0,1,\ldots,m-1). \tag{7.10}$$

On repeating the argument leading to (7.5) we find that for any $\varepsilon > 0$ there exists a positive constant K_ε such that for all $\phi \in C_0^\infty(0,\infty)$,

$$\|\phi^{(j)}\|^2 \leqslant \varepsilon\|\phi^{(m)}\|^2 + K_\varepsilon\|\phi\|^2 \qquad (j = 0,1,\ldots,m-1). \tag{7.11}$$

These inequalities yield

$$\|\phi^{(m)}\|^2 \leqslant K_1\left[\left\|P\left(\frac{1}{2\pi i}D\right)\phi\right\|^2 + \|\phi\|^2\right], \quad \|P(D)\phi\|^2 \leqslant K_2(\|\phi^{(m)}\|^2 + \|\phi\|^2),$$

for some positive constants K_1 and K_2. Consequently the graph norm $\|\bullet\|_{T_0'}$ is equivalent to the $W_0^{m,2}(0,\infty)$ norm on $C_0^\infty(0,\infty)$, whence T_0' is closable and $\mathscr{D}(T_0) = W_0^{m,2}(0,\infty)$ with $\|\bullet\|_{T_0}$ and $\|\bullet\|_{m,2,(0,\infty)}$ equivalent. Thus (7.8) is proved.

Let $v \in \mathscr{D}(T_0^*)$ and set $v^* = T_0^* v$. We shall prove, for any $d > 0$, that $v^{(m-1)} \in AC[0,d]$ and $v^* = \bar{P}((1/2\pi i)D)v$; Lemma 7.1 will then imply that $\mathscr{D}(T_0^*) \subset W^{m,2}(0,\infty)$ and $T_0^* v = \bar{P}((1/2\pi i)D)v$. For all $u \in \mathscr{D}(T_0)$,

$$(T_0 u, v) = (u, v^*). \tag{7.12}$$

In particular this holds for all $u \in \mathfrak{D}_0 = \{f : f \in \mathscr{D}(T_0) \text{ and } \operatorname{supp} f \subset [0,d]\}$. For $u \in \mathfrak{D}_0$, successive integration by parts yields

$$u^{(j)}(t) = \int_0^t \frac{(t-s)^{m-j-1}}{(m-j-1)!}u^{(m)}(s)\,ds \qquad (0 \leqslant j \leqslant m-1).$$

On substituting in (7.12) we see that

$$\int_0^d \left((2\pi i)^{-m} c_m D^m u(t) + \sum_{j=0}^{m-1} (2\pi i)^{-j} c_j \int_0^t \frac{(t-s)^{m-j-1}}{(m-j-1)!} u^{(m)}(s)\, ds \right) \bar{v}(t)\, dt$$
$$= \int_0^d \bar{v}^*(t) \int_0^t \frac{(t-s)^{m-1}}{(m-1)!} u^{(m)}(s)\, ds\, dt.$$

By Fubini's Theorem we may interchange the order of integration in the repeated integrals to obtain

$$0 = \int_0^d (2\pi i)^{-m} c_m u^{(m)} \bar{v} dt +$$
$$\sum_{j=0}^{m-1} (2\pi i)^{-j} c_j \int_0^d \left(\int_s^d \frac{(t-s)^{m-j-1}}{(m-j-1)!} \bar{v}(t)\, dt \right) u^{(m)}(s)\, ds$$
$$- \int_0^d \left(\int_s^d \frac{(t-s)^{m-1}}{(m-1)!} \bar{v}^*(t)\, dt \right) u^{(m)}(s)\, ds$$
$$= \int_0^d u^{(m)} \bar{f}\, dt, \tag{7.13}$$

where

$$\bar{f}(t) = (2\pi i)^{-m} c_m \bar{v}(t) + \sum_{j=0}^{m-1} (2\pi i)^{-j} c_j \int_t^d \frac{(z-t)^{m-j-1}}{(m-j-1)!} \bar{v}(z)\, dz$$
$$- \int_t^d \frac{(z-t)^{m-1}}{(m-1)!} \bar{v}^*(z)\, dz.$$

Let $\mathscr{P}_{m-1}$ denote the set of polynomials of degree $\leqslant m-1$ and let $\mathscr{P}_{m-1}^{\perp}$ be its orthogonal complement in $L^2(0,d)$. If $w \in \mathscr{P}_{m-1}^{\perp}$, the function u defined by

$$u(t) = \begin{cases} \displaystyle\int_0^t \frac{(t-s)^{m-1}}{(m-1)!} w(s)\, ds & \text{if } 0 \leqslant t \leqslant d, \\ 0 & \text{if } t > d, \end{cases}$$

lies in $\mathscr{D}_0$ and $u^{(m)} = w$ on $[0,d]$. Thus from (7.13), and since $\mathscr{P}_{m-1}$ is finite-dimensional, $f \in \mathscr{P}_{m-1}^{\perp\perp} = \mathscr{P}_{m-1}$. We conclude that there exists a polynomial p of degree $\leqslant m-1$ such that

$$(2\pi i)^{-m} c_m \bar{v}(t) = p(t) - \sum_{j=1}^{m-1} (2\pi i)^{-j} c_j \int_t^d \frac{(z-t)^{m-j-1}}{(m-j-1)!} \bar{v}(z)\, dz$$
$$+ \int_t^d \frac{(z-t)^{m-1}}{(m-1)!} \bar{v}^*(z)\, dz.$$

On repeated differentiation it readily follows that $v^{(m-1)} \in AC[0,d]$ and

$\bar{P}((1/2\pi i)D)v = v^*$ on $[0, d]$. Since d is arbitrary, $\mathscr{D}(T_0^*) \subset W^{m,2}(0, \infty)$, as explained earlier, and $T_0^* v = \bar{P}((1/2\pi i)D)v$. The reverse inclusion $W^{m,2}(0,\infty) \subset \mathscr{D}(T_0^*)$ is immediate and so the proof of (7.9) is complete. □

Theorem 7.3. Let $\mathfrak{P} = \{P(\xi): \xi \in \mathbb{R}\}$ and $\mathfrak{R} = \{P(\xi): \xi \in \mathbb{C}, \operatorname{im} \xi \leqslant 0\}$. Then
(i) $\sigma(T_0) = \mathfrak{R}$,
(ii) $\sigma_{ek}(T_0) = \mathfrak{P}$ $(k = 1, 2, 3)$,
(iii) $\sigma_{ek}(T_0) = \mathfrak{R}$ $(k = 4, 5)$,
(iv) $\mathfrak{R} \setminus \mathfrak{P} = \sigma_r(T_0) \cap \Delta_3(T_0)$; if $\lambda \in \mathfrak{R} \setminus \mathfrak{P}$ then $\operatorname{def}(T_0 - \lambda I)$ is the number of zeros of $P(\xi) - \lambda$ for im $\xi < 0$. ■

Proof. It is well known (see Coddington and Levinson [1, Chapter 3, Theorem 6.5]) that a fundamental set of solutions of the equation $\{\bar{P}((1/2\pi i)D) - \bar{\lambda} I\}u = 0$ is given by the m functions

$$u_{k_j}(x) = x^{k_j} \exp(2\pi i x \xi_j) \qquad (k_j = 0, 1, \ldots, m_j - 1, \quad j = 0, 1, \ldots, s), \tag{7.14}$$

where m_j is the multiplicityof the root ξ_j of $\bar{P}(\xi) = \bar{\lambda}$, i.e. $P(\bar{\xi}) = \lambda$. If $\lambda \notin \mathfrak{R}$ each ξ_j satisfies im $\xi_j < 0$ and so the functions in (7.14) and their linear span lie outside $L^2(0, \infty)$. This implies that $\operatorname{nul}(T_0^* - \bar{\lambda} I) = 0$ for all $\lambda \notin \mathfrak{R}$.

If $\lambda \notin \mathfrak{P}$ there exists a $\delta > 0$ such that $|\lambda - P(\xi)| \geqslant \delta$ for all $\xi \in \mathbb{R}$; recall that $\mathfrak{P}$ is a closed set as P is elliptic. On setting any $u \in L^2(0, \infty)$ to be zero outside $[0, \infty)$ we obtain

$$\begin{aligned} \|(T_0 - \lambda I)u\|^2 &= \int_{-\infty}^{\infty} |P(\xi) - \lambda|^2 \, |\hat{u}(\xi)|^2 \, d\xi \\ &\geqslant \delta^2 \, \|u\|^2 \qquad (u \in \mathscr{D}(T_0)). \end{aligned} \tag{7.15}$$

Thus $(T_0 - \lambda I)^{-1}$ exists and is a closed and bounded operator on $\mathscr{R}(T_0 - \lambda I)$ which must therefore be a closed subspace of $L^2(0, \infty)$. From the preceding paragraph we therefore conclude that for $\lambda \notin \mathfrak{R}$, the operator $T_0 - \lambda I$ is Fredholm with zero nullity and deficiency. Consequently $\lambda \in \rho(T)$ and $\sigma(T_0) \subset \mathfrak{R}$. The proof of (i) will be complete if we prove (ii) and (iii).

Since $\operatorname{nul}(T_0^* - \lambda I) \leqslant m$ for all $\lambda \in \mathbb{C}$ and, for $\lambda \notin \mathfrak{P}$, the space $\mathscr{R}(T_0 - \lambda I)$ is closed and $\operatorname{nul}(T_0 - \lambda I) = 0$ we see that $\sigma_{e3}(T_0) \subset \mathfrak{P}$. To prove $\sigma_{e2}(T_0) = \sigma_{e3}(T_0) = \mathfrak{P}$ it suffices to prove that $\mathfrak{P} \subset \sigma_{e2}(T_0)$, because $\sigma_{e2}(T_0) \subset \sigma_{e3}(T_0)$, and this will follow if we exhibit a singular sequence of T_0 corresponding to any $P(\xi)$ in $\mathfrak{P}$; see Theorem 1.3. Let $\eta \in C_0^\infty(0, \infty)$ satisfy the conditions $\eta(x) = 1$ for $|x| < \frac{1}{2}$, $\eta(x) = 0$ for $|x| > 1$, and $0 \leqslant \eta \leqslant 1$ on $(0, \infty)$. Let $I_k = \{x : |x - x_k| \leqslant l_k\}$ $(k = 1, 2, \ldots)$ be mutually disjoint intervals on $[0, \infty)$ with $l_k \to \infty$ as $k \to \infty$, and define

$$u_k(x) = l_k^{-\frac{1}{2}} \eta[(x - x_k) l_k^{-1}] \exp[2\pi i \xi (x - x_k)] \qquad (k = 1, 2, \ldots),$$

where $\xi \in \mathbb{R}$. Then the sequence (u_k) is orthogonal and

$$1 = \int_{|x-x_k|<\frac{1}{2}l_k} |u_k(x)|^2 \, dx \leqslant \|u_k\|^2$$

$$\leqslant l_k^{-1} \int_{|x-x_k|<l_k} dx \leqslant 2.$$

Furthermore, $u_k \in C_0^\infty(I_k) \subset \mathscr{D}(T_0)$ and

$$\left[P\left(\frac{1}{2\pi i} D\right) - P(\xi) \right] u_k(x) = O\left(l_k^{-\frac{1}{2}} \sum_{t=1}^{m} |D^t \eta[(x-x_k)l_k^{-1}]| \right).$$

Thus

$$\|[T_0 - P(\xi)I]u_k\|^2 = O(l_k^{-1}) \to 0,$$

proving that (u_k) is a singular sequence of T_0 corresponding to $P(\xi)$ and hence that $\mathfrak{P} \subset \sigma_{e2}(T_0)$ and $\sigma_{e2}(T_0) = \sigma_{e3}(T_0) = \mathfrak{P}$. The sequence (u_k) is similarly a singular sequence of T_0^* corresponding to $\bar{P}(\xi)$ $(\xi \in \mathbb{R})$, so that on using Theorem 1.3(ii) we see that $\mathfrak{P} \subset \sigma_{e1}(T_0) = \sigma_{e2}(T_0) \cap \sigma_{e2}^*(T_0)$ and consequently $\sigma_{e1}(T_0) = \mathfrak{P}$. The proof of (ii) is therefore concluded.

By (7.15), as remarked earlier, for any $\lambda \notin \mathfrak{P}$, the subspace $\mathscr{R}(T_0 - \lambda I)$ is closed, $\operatorname{nul}(T_0 - \lambda I) = 0$ and also $\operatorname{def}(T_0 - \lambda I) = \operatorname{nul}(T_0^* - \bar{\lambda} I) \leqslant m$. If $\lambda \in \mathfrak{R} \setminus \mathfrak{P}$, $\lambda = P(\xi)$ for some ξ with $\operatorname{im} \xi < 0$ and thus $x \mapsto \exp(2\pi i x \bar{\xi})$ lies in the null space of $T_0^* - \bar{\lambda} I$. Consequently $\operatorname{def}(T_0 - \lambda I) \neq 0$, $\mathfrak{R} \setminus \mathfrak{P} \subset \sigma_{e4}(T_0)$ and (iii) follows from (ii) and $\sigma(T_0) \subset \mathscr{R}$; (i) is therefore established.

If $\lambda \in \mathfrak{R} \setminus \mathfrak{P}$ then $\operatorname{nul}(T_0 - \lambda I) = 0$ and, by (7.14), $\operatorname{def}(T_0 - \lambda I)$ $(\neq 0)$ is the number of zeros of $P(\xi) - \lambda$ in $\operatorname{im} \xi < 0$. This concludes the proof of the theorem. □

Let $T(P)$ denote the restriction of $P(D)$ to $W^{m,2}(0, \infty)$ so that by Theorem 7.2, $T(\bar{P}) = T_0^*$. Theorem 1.1 therefore yields the following result.

Corollary 7.4. If $T = T(P)$,
(i) $\sigma(T) = \{P(\xi) \colon \operatorname{im} \xi \geqslant 0\}$,
(ii) $\sigma_{ek}(T) = \{P(\xi) \colon \xi \in \mathbb{R}\}$ $(k = 1, 2, 3)$,
(iii) $\sigma_{ek}(T) = \{P(\xi) \colon \operatorname{im} \xi \geqslant 0\}$ $(k = 4, 5)$. ■

Example 7.5. If $P(\xi) = \xi$ then T_0 is defined by

$$\mathscr{D}(T_0) = \{u \colon u \in AC_{loc}[0, \infty);\ u, u' \in L^2(0, \infty);\ u(0) = 0\}, \qquad T_0 u = (1/2\pi i)u'.$$

Also $\mathfrak{R}$ is the lower half-plane $\{\xi \colon \operatorname{im} \xi \leqslant 0\}$ and $\mathfrak{P} = \mathbb{R}$; T_0 is maximal symmetric with deficiency indices $(0, 1)$, $\sigma_{ek}(T_0) = \mathbb{R}$ for $k = 1, 2, 3$; $\sigma_{ek}(T_0) = \mathfrak{R}$ for $k = 4, 5$, and $\rho(T_0) = \mathbb{C}_+ = \{\lambda \colon \operatorname{im} \lambda > 0\}$.

Example 7.6. Let $P(\xi) = \xi^2$ so that $\mathscr{D}(T_0) = W_0^{2,2}(0, \infty)$ and $T_0\phi = -(1/4\pi^2)\phi^{(2)}$. Then $\mathfrak{R} = \mathbb{C}$ and $\mathfrak{P} = [0, \infty)$; T_0 is non-negative and symmetric but not self-adjoint.

Example 7.7. Let $P(\xi) = c_0 + ic_1\xi + c_2\xi^2$, where c_0, c_1, c_2 are real constants and $c_2 \neq 0$. Then $\mathfrak{R} = \mathbb{C}$ and $\mathfrak{P}$ is the parabola

$$\{x + iy : y^2 = c_1^2 c_2^{-1}(x - c_0),\ x \geqslant c_0\}.$$

8. Relatively bounded and relatively compact perturbations of constant-coefficient operators

Let $\Omega = \mathbb{R}^n$ when $n \geqslant 2$ and either $\Omega = \mathbb{R}$ or $(0, \infty)$ when $n = 1$. In each case we shall write $\Omega = \cup_{Q \in \mathfrak{F}} Q$, where $\mathfrak{F}$ is a set of closed unit cubes with disjoint interiors. Let T_0 be the closed operator defined in either Theorem 6.2 or Theorem 7.2 in terms of the *elliptic* differential polynomial

$$P\left(\frac{1}{2\pi i}\mathrm{D}\right) = \sum_{|\alpha| \leqslant m} \left(\frac{1}{2\pi i}\right)^{|\alpha|} c_\alpha \mathrm{D}^\alpha.$$

By Proposition 6.4 and Theorem 7.2, the assumed ellipticity of P implies that $\mathscr{D}(T_0) = W_0^{m,2}(\Omega)$ and the Hilbert space $H(T_0)$ determined by $\mathscr{D}(T_0)$ and the graph norm of T_0,

$$\|\phi\|_{T_0} = (\|T_0\phi\|_{2,\Omega}^2 + \|\phi\|_{2,\Omega}^2)^{\frac{1}{2}},$$

is topologically isomorphic to $W_0^{m,2}(\Omega)$. In particular there exists a positive constant γ such that

$$\|\phi\|_{m,2,\Omega} \leqslant \gamma \|\phi\|_{T_0} \qquad (\phi \in \mathscr{D}(T_0) = W_0^{m,2}(\Omega)). \tag{8.1}$$

We also know, from Theorem V.4.14 and the argument leading to (7.5), that given any $\varepsilon > 0$ there exists a constant $K_\varepsilon > 0$ such that for all $\phi \in W_0^{m,2}(\Omega)$,

$$\|\mathrm{D}^\alpha \phi\|_{2,\Omega}^2 \leqslant \varepsilon \|\phi\|_{m,2,\Omega}^2 + K_\varepsilon \|\phi\|_{2,\Omega}^2 \qquad (|\alpha| \leqslant m-1). \tag{8.2}$$

Also, by Theorem V.4.14, there is a constant $K_\varepsilon > 0$ such that for all $\phi \in W_0^{m,2}(Q)$,

$$\|\mathrm{D}^\alpha \phi\|_{2,Q}^2 \leqslant \varepsilon \|\phi\|_{m,2,Q}^2 + K_\varepsilon \|\phi\|_{2,Q}^2 \qquad (|\alpha| \leqslant m-1). \tag{8.3}$$

We shall now consider perturbations of T_0 by operators B defined by differential expressions

$$v = \sum_{|\alpha| \leqslant m-1} b_\alpha \mathrm{D}^\alpha \tag{8.4}$$

as follows:

$$\mathscr{D}(B) = \{\phi : \phi, v\phi \in L^2(\Omega)\}, \qquad B\phi = v\phi, \tag{8.5}$$

where, under suitable conditions on the functions b_α, we understand $v\phi$ in the distributional sense.

Theorem 8.1. Let Ω, T_0 and B be defined as above and suppose that in (8.4), $b_\alpha \in L^{2n}(\Omega)$ for each unit cube Q in the covering $\mathfrak{F}$ of Ω and for $|\alpha| \leqslant m-1$. If

$$\sup_{Q\in\mathfrak{F}} \|b_\alpha\|_{2n,Q} < \infty \qquad (|\alpha| \leqslant m-1), \tag{8.6}$$

B is T_0-bounded with T_0-bound zero. Conversely, if B is T_0-bounded then

$$\sup_{Q\in\mathfrak{F}} \|b_\alpha\|_{2,Q} < \infty \qquad (|\alpha| \leqslant m-1). \tag{8.7}$$

In particular when $n = 1$, (8.7) is necessary and sufficient for B to be T_0-bounded and in this case B has T_0-bound zero. ■

Proof. From Lemma VII.1.1 (see the proof of Lemma VII.1.2) it follows that if $f \in L^n(Q)$ then given $\varepsilon > 0$,

$$\int_Q |f||\phi|^2 \leqslant \varepsilon\|\nabla\phi\|^2_{2,Q} + [\varepsilon^{-1}(\mu_s\|f\|_{n,Q})^2 + |Q|^{-1/n}\|f\|_{n,Q}]\|\phi\|^2_{2,Q} \tag{8.8}$$

for all $\phi \in W^{1,2}(Q^\circ)$, the constant μ_s being given in (VII.1.5) with $s = n$. On applying this inequality to $f = b_\alpha^2$ and making use of (8.6), we see that for any $\varepsilon_0 > 0$ there exists a constant K_{ε_0} depending only on ε_0 such that

$$\|b_\alpha D^\alpha\phi\|^2_{2,Q} \leqslant \varepsilon_0\|\phi\|^2_{|\alpha|+1,2,Q} + K_{\varepsilon_0}\|D^\alpha\phi\|^2_{2,Q} \qquad (|\alpha| \leqslant m-1) \tag{8.9}$$

for all $\phi \in W_0^{m,2}(\Omega)$. If we now sum (8.9) over all the cubes Q in $\mathfrak{F}$ and use (8.2) we obtain, for any $\varepsilon_1 > 0$ and all $\phi \in W_0^{m,2}(\Omega)$,

$$\|b_\alpha D^\alpha\phi\|^2_{2,\Omega} \leqslant \varepsilon_1\|\phi\|^2_{m,2,\Omega} + K_{\varepsilon_1}\|\phi\|^2_{2,\Omega} \qquad (|\alpha| \leqslant m-1). \tag{8.10}$$

This inequality implies that $\mathscr{D}(T_0) \subset \mathscr{D}(B)$ and, for any $\varepsilon > 0$ and $\phi \in \mathscr{D}(T_0)$,

$$\begin{aligned}\|B\phi\|_{2,\Omega} &\leqslant \sum_{|\alpha|\leqslant m-1} \|b_\alpha D^\alpha\phi\|_{2,\Omega}\\ &\leqslant \gamma^{-1}\varepsilon\|\phi\|_{m,2,\Omega} + K_\varepsilon\|\phi\|_{2,\Omega}\\ &\leqslant \varepsilon\|T_0\phi\|_{2,\Omega} + K_\varepsilon\|\phi\|_{2,\Omega},\end{aligned}$$

by (8.1). Thus (8.6) implies that B is T_0-bounded with T_0-bound zero.

Suppose now that B is T_0-bounded. Let $\theta \in C_0^\infty(\mathbb{R}^n)$ be 1 in the unit cube $[-\frac{1}{2}, \frac{1}{2}]^n$ and zero outside $(-1, 1)^n$, and define $h_\beta(x) = (x^\beta/\beta!)\theta(x)$ and $h_{\beta,Q}(x) = h_\beta(x - x_Q)$, where β is a multi-index and x_Q is the centre of Q. Except when

Q is the first interval $[0, 1]$ in the case $\Omega = [0, \infty)$, we have $h_{\beta,Q} \in C_0^\infty(\Omega) \subset \mathscr{D}(T_0)$. Since $b_\alpha \in L^2(0, 1)$ by hypothesis, it is sufficient to consider cubes Q which are such that $h_{\beta,Q} \in C_0^\infty(\Omega) \subset \mathscr{D}(T)$ in order to establish (8.7). Henceforth this will be understood.

Since $\theta(x - x_Q) = 1$ when $x \in Q$ we have for all $x \in Q$,

$$D^\alpha h_{\beta,Q}(x) = \begin{cases} (x - x_Q)^{\beta-\alpha}/(\beta-\alpha)! & \text{if } \beta \geqslant \alpha, \text{ i.e. } \beta_j \geqslant \alpha_j \text{ for all } j, \\ 0 & \text{otherwise,} \end{cases}$$

and hence

$$Bh_{\beta,Q}(x) = b_0(x)(x - x_Q)^\beta/\beta! + \sum_{\substack{\alpha<\beta \\ 1 \leqslant |\alpha| \leqslant |\beta|-1}} b_\alpha(x)\,(x - x_Q)^{\beta-\alpha}/(\beta-\alpha)! + b_\beta(x). \tag{8.11}$$

If $|\beta| = 0$ then $Bh_{0,Q}(x) = b_0(x)$ for $x \in Q$, and since B is assumed to be T_0-bounded,

$$\begin{aligned} \|b_0\|_{2,Q} &\leqslant \|Bh_{0,Q}\|_{2,Q} \leqslant \|Bh_{0,Q}\|_{2,\Omega} \\ &\leqslant K\|h_{0,Q}\|_{T_0} \\ &\leqslant K\|h_{0,Q}\|_{m,2,\Omega} \\ &= K\|\theta\|_{m,2,\Omega}. \end{aligned}$$

Thus (8.7) is proved for the case $|\alpha| = 0$. Suppose now that

$$\sup \|b_\alpha\|_{2,Q} < \infty \qquad (|\alpha| \leqslant k-1 < m-1), \tag{8.12}$$

and let β be a multi-index with $|\beta| = k$. By (8.11)

$$\begin{aligned} \|b_\beta\|_{2,Q} &\leqslant \|Bh_{\beta,Q}\|_{2,\Omega} + \sum_{|\alpha| \leqslant k-1} \|b_\alpha D^\alpha h_{\beta,Q}\|_{2,Q} \\ &\leqslant K\|h_{\beta,Q}\|_{T_0} + M \sum_{|\alpha| \leqslant k-1} \|b_\alpha\|_{2,Q} \end{aligned}$$

where

$$M = \sup_{\substack{x \in Q \in \mathfrak{F} \\ |\alpha| \leqslant k-1}} |D^\alpha h_{\beta,Q}(x)| \leqslant \sup_{\substack{|\alpha| \leqslant k-1 \\ \alpha<\beta}} \frac{2^{\alpha-\beta}}{(\beta-\alpha)!}.$$

On using the induction hypothesis (8.12) we obtain

$$\|b_\beta\|_{2,Q} \leqslant K(\|h_{\beta,Q}\|_{m,2,\Omega} + 1) = K(\|h_\beta\|_{m,2,\Omega} + 1).$$

Consequently (8.12) holds for $|\alpha| \leqslant k \leqslant m-1$ and the necessity of (8.7) follows by induction. □

Theorem 8.2. Let Ω, T_0 and B be as in Theorem 8.1 with $b_\alpha \in L^{2n}(Q)$ for $Q \in \mathfrak{F}$

and $|\alpha| \leqslant m-1$. If

$$\limsup_{|x_Q| \to \infty} \|b_\alpha\|_{2n,Q} = 0 \qquad (|\alpha| \leqslant m-1), \tag{8.13}$$

then B is T_0-compact. Conversely, if B_0 is T_0-compact then

$$\limsup_{|x_Q| \to \infty} \|b_\alpha\|_{2,Q} = 0 \qquad (|\alpha| \leqslant m-1). \tag{8.14}$$

Therefore, when $n = 1$, (8.14) is necessary and sufficient for B to be T_0-compact. ■

Proof. For $N \in \mathbb{N}$, let χ_N be the characteristic function of $\Omega_N = \{x \in \Omega \colon |x_i| \leqslant N \ (i = 1, 2, \ldots, n)\}$ and define

$$B_N \phi = \chi_N B\phi = \sum_{|\alpha| \leqslant m-1} \chi_N b_\alpha \mathrm{D}^\alpha \phi \qquad (\phi \in \mathscr{D}(B)). \tag{8.15}$$

For all $\phi \in \mathscr{D}(T_0)$ we therefore have

$$\begin{aligned}
\|(B-B_N)\phi\|_{2,\Omega} &\leqslant \sum_{|\alpha| \leqslant m-1} \|(1-\chi_N) b_\alpha \mathrm{D}^\alpha \phi\|_{2,\Omega} \\
&\leqslant \sum_{|\alpha| \leqslant m-1} \Big(\sum_{Q \not\subset \Omega_N} \|b_\alpha \mathrm{D}^\alpha \phi\|_{2,Q}^2 \Big)^{\frac{1}{2}} \\
&\leqslant \sum_{|\alpha| \leqslant m-1} \sum_{Q \not\subset \Omega_N} \|b_\alpha \mathrm{D}^\alpha \phi\|_{2,Q}.
\end{aligned}$$

From (8.8) it follows that given any $\varepsilon_1 > 0$ there exists a constant $\mathrm{K}_{\varepsilon_1} > 0$ such that

$$\begin{aligned}
\sum_{Q \not\subset \Omega_N} \|b_\alpha \mathrm{D}^\alpha \phi\|_{2,Q}^2 &\leqslant \sum_{Q \not\subset \Omega_N} [\varepsilon_1 \|\phi\|_{|\alpha|+1,2,Q}^2 + K_{\varepsilon_1} \|b_\alpha\|_{2n,Q} \|\mathrm{D}^\alpha \phi\|_{2,Q}^2] \\
&\leqslant \varepsilon_1 \|\phi\|_{|\alpha|+1,2,\Omega}^2 + K_{\varepsilon_1} \Big(\sup_{Q \not\subset \Omega_N} \|b_\alpha\|_{2n,Q} \Big) \|\mathrm{D}^\alpha \phi\|_{2,\Omega}^2 \\
&\leqslant \Big(\varepsilon_1 + K_{\varepsilon_1} \sup_{Q \not\subset \Omega_N} \|b_\alpha\|_{2n,Q} \Big) \|\phi\|_{m,2,\Omega}^2
\end{aligned}$$

If we now assume (8.13) we see that for any $\varepsilon > 0$ there exists an $N_0 \in \mathbb{N}$ such that for all $\phi \in \mathscr{D}(T_0)$ and all $N > N_0$,

$$\begin{aligned}
\|(B-B_N)\phi\|_{2,\Omega} &\leqslant \varepsilon\gamma^{-1} \|\phi\|_{m,2,\Omega} \\
&\leqslant \varepsilon \|\phi\|_{T_0},
\end{aligned}$$

by (8.1). Hence $\|B - B_N \colon \mathrm{H}(T_0) \to \mathrm{L}^2(\Omega)\| \to 0$ as $N \to \infty$. Thus if we can prove that each B_N is T_0-compact, it will follow that B is T_0-compact. From (8.8) and (8.3) it follows that for any positive constants ε_1, ε_2, ε and all $\phi \in \mathscr{D}(T_0)$,

$$\begin{aligned}\|B_N\phi\|_{2,\Omega}^2 &\leqslant K \sum_{|\alpha| \leqslant m-1} \sum_{Q \subset \Omega_{N+1}} \|b_\alpha \mathrm{D}^\alpha \phi\|_{2,Q}^2 \\ &\leqslant \sum_{|\alpha| \leqslant m-1} [\varepsilon_1 \|\phi\|_{|\alpha|+1,2,\Omega_{N+1}}^2 + K_{\varepsilon_1} \|\mathrm{D}^\alpha \phi\|_{2,\Omega_{N+1}}^2] \\ &\leqslant \varepsilon_2 \|\phi\|_{m,2,\Omega}^2 + K_{\varepsilon_2} \|\phi\|_{2,\Omega_{N+1}}^2 \\ &\leqslant \varepsilon \|\phi\|_{T_0}^2 + K_\varepsilon \|\phi\|_{2,\Omega_{N+1}}^2. \end{aligned} \tag{8.16}$$

Let (ϕ_j) be a bounded sequence in $\mathrm{H}(T_0)$, say $\|\phi_j\|_{T_0} \leqslant 1$. By (8.1), (ϕ_j) is bounded in $\mathrm{W}_0^{m,2}(\Omega)$ and, since the embedding $\mathrm{W}_0^{m,2}(\Omega) \to \mathrm{L}^2(\Omega_{N+1})$ is compact, (ϕ_j) contains a subsequence, which we continue to denote by (ϕ_j), which converges in $\mathrm{L}^2(\Omega_{N+1})$. On using this information in (8.16) we conclude that

$$\|B_N\phi_j - B_N\phi_k\|_{2,\Omega}^2 \leqslant 4\varepsilon + K_\varepsilon \|\phi_j - \phi_k\|_{2,\Omega_{N+1}}^2$$

and

$$\limsup_{j,k \to \infty} \|B_N\phi_j - B_N\phi_k\|_{2,\Omega}^2 \leqslant 4\varepsilon.$$

Since ε is arbitrary, $(B_N\phi_j)$ is a Cauchy sequence in $\mathrm{L}^2(\Omega)$ and hence B_N is T_0-compact. Therefore B is T_0-compact if (8.13) is satisfied.

To prove the necessity of (8.14) if B is T_0-compact we use the functions $h_{\beta,Q}$ defined in the proof of Theorem 8.1. Contrary to (8.14), suppose that there exists $\varepsilon > 0$ and a sequence of cubes $Q_k \in \mathfrak{F}$ such that as $k \to \infty$,

$$|x_{Q_k}| \to \infty, \qquad \|b_0\|_{2,Q_k} \geqslant \varepsilon.$$

Since $Bh_{0,Q} = b_0$ on Q we have for all $k \in \mathbb{N}$,

$$\|Bh_{0,Q_k}\|_{2,\Omega} \geqslant \|b_0\|_{2,Q_k} \geqslant \varepsilon. \tag{8.17}$$

But $(h_{0,Q_k})_{k\in\mathbb{N}}$ is a bounded sequence in $\mathrm{H}(T_0)$ and since B is assumed to be T_0-compact $(Bh_{0,Q_k})_{k\in\mathbb{N}}$ contains a subsequence which converges to some limit, say y, in $\mathrm{L}^2(\Omega)$. For any bounded set Ω_0 in Ω, the support of Bh_{0,Q_k} lies outside Ω_0 for k large enough and consequently for such k,

$$\begin{aligned}\|y\|_{2,\Omega_0} &= \|y - Bh_{0,Q_k}\|_{2,\Omega_0} \\ &\leqslant \|y - Bh_{0,Q_k}\|_{2,\Omega} \to 0\end{aligned}$$

as $k \to \infty$ through suitable values. We conclude that $y = 0$ a.e. on Ω_0, and as Ω_0 is arbitrary, $y = 0$ a.e. in Ω. But this contradicts (8.17). Thus (8.14) must hold for $|\alpha| = 0$. Suppose that

$$\limsup_{|x_Q| \to \infty} \|b_\alpha\|_{2,Q} = 0 \qquad (|\alpha| \leqslant k-1 < m-1). \tag{8.18}$$

Let $|\beta| = k$ and suppose that there exists an $\varepsilon > 0$ and a sequence of cubes $Q_k \in \mathfrak{F}$ such that

$$|x_{Q_k}| \to \infty, \qquad \|b_\beta\|_{2,Q_k} \geqslant \varepsilon.$$

It then follows from (8.11) and our inductive hypothesis (8.18) that as $k \to \infty$,

$$\|Bh_{\beta,Q_k}\|_{2,\Omega} \geqslant \tfrac{1}{2}\|b_\beta\|_{2,Q_k} \geqslant \tfrac{1}{2}\varepsilon.$$

The inductive step is established by the same argument as when $|\alpha| = 0$. The necessity of (8.14) therefore follows and the theorem is proved. □

9. The Decomposition Principle for operators in L^2 (a, ∞)

Let I be an arbitrary interval in $\mathbb{R}$ and consider the differential expression

$$\tau\phi(x) = \sum_{j=0}^{m} a_j(x)\phi^{(j)}(x) \qquad (x \in I),$$

where the coefficients a_j satisfy the following conditions:

(i) $a_j \in C^j(I)$ $(j = 0, 1, \ldots, m)$,
(ii) $a_m \neq 0$ on I and $1/a_m \in L^\infty(I)$,
(iii) $a_j^{(k)} \in L^\infty(I)$ $(k = 0, 1, \ldots, j; j = 0, 1, \ldots, m)$.

The formal adjoint of τ is the expression

$$\tau^+\psi = \sum_{j=0}^{m} (-1)^j(\bar{a}_j\psi)^{(j)},$$

the expressions τ and τ^+ being related, on integration by parts, by Green's formula

$$\int_a^b \bar{\psi}\tau\phi - \int_a^b \phi\,\overline{\tau^+\psi} = \left[\sum_{k=1}^{m}\sum_{j=0}^{k-1} (-1)^j(a_k\bar{\psi})^{(j)}\phi^{(k-1-j)}\right]_a^b \tag{9.1}$$

for ϕ and ψ having absolutely continuous $(m-1)$th derivatives on $[a, b]$. It can be shown that $\tau^{++} = \tau$.

It follows as in the proof of Theorem 7.2 that for $\phi \in C_0^\infty(I)$ there exist positive constants K_1 and K_2 such that

$$\|\phi^{(m)}\|^2 \leqslant K_1(\|\tau\phi\|^2 + \|\phi\|^2), \tag{9.2}$$

$$\|\tau\phi\|^2 \leqslant K_2(\|\phi^{(m)}\|^2 + \|\phi\|^2), \tag{9.3}$$

and hence that the closure T_0 of the restriction of τ to $C_0^\infty(I)$ is the restriction of τ to $W_0^{m,2}(I)$. Furthermore, if $I = [a, \infty)$ then

$$\begin{aligned}\mathscr{D}(T_0) &= W_0^{m,2}(a, \infty)\\ &= \{u : u^{(m-1)} \in AC_{\text{loc}}[a, \infty), u^{(j)}(a) = 0 \ (j = 0, 1, \ldots, m-1);\\ &\qquad u, \tau u \in L^2(a, \infty)\},\end{aligned} \tag{9.4}$$

while if $I = [a,b]$,

$$\begin{aligned}\mathscr{D}(T_0) &= \mathrm{W}_0^{m,2}(a,b)\\ &= \{u : u^{(m-1)} \in \mathrm{AC}[a,b], u^{(j)}(a) = u^{(j)}(b) = 0 \ \ (j = 0,1,\ldots,m-1);\\ &\qquad\qquad u, \tau u \in \mathrm{L}^2(a,b)\}. \end{aligned} \tag{9.5}$$

The argument in Theorem 7.2 also carries through to prove that T_0^* is the restriction of the formal adjoint τ^+ to

$$\mathscr{D}(T_0^*) = \mathrm{W}^{m,2}(I) = \{u : u^{(m-1)} \in \mathrm{AC}_{\mathrm{loc}}(I);\ u, \tau^+ u \in \mathrm{L}^2(I)\}. \tag{9.6}$$

Results of the type (9.4)–(9.6) can also be obtained without the stringent assumptions made in (i), (ii) and (iii). We saw this when $m = 2$ in §III.10 in our discussion of quasi-differential operators, and similar results can be achieved for higher-order operators with minimal smoothness assumptions on the coefficients. However, our concern here is to illustrate the results in §§4,5 and the extra complications that would have to be overcome to meet the demands of weak smoothness conditions on the coefficients would only serve to obscure the theme of this section. It will be clear that the method used may be adapted to deal with assumptions weaker than (i), (ii) and (iii).

Lemma 9.1. If $I = [a,b]$, with $-\infty < a < b < \infty$, then for any $\lambda \in \mathbb{C}$, the operator $T_0 - \lambda I$ has closed range, zero nullity and deficiency m. Hence

$$\sigma_{ek}(T_0) = \begin{cases} \varnothing & (k = 1,2,3),\\ \mathbb{C} & (k = 4,5). \end{cases}$$ ■

Proof. For any $f \in \mathrm{L}^2(a,b)$ and $\lambda \in \mathbb{C}$ there exists a solution $u \in \mathscr{D}(T_0^*)$ of the differential equation $(\bar{\tau} - \bar{\lambda})u = f$. Hence the range of $T_0^* - \bar{\lambda} I$ is $\mathrm{L}^2(a,b)$. Consequently $\mathscr{R}(T_0 - \lambda I)$ is closed, by Theorem I.3.7. Furthermore, the equations

$$(\tau - \lambda)u = 0 \quad (u \in \mathscr{D}(T_0)), \qquad (\tau^+ - \bar{\lambda})v = 0 \quad (v \in \mathscr{D}(T_0^*)),$$

have respectively no non-trivial solution and precisely m linearly independent solutions, for any $\lambda \in \mathbb{C}$. In other words $\mathrm{nul}\,(T_0 - \lambda I) = 0$ and $\mathrm{def}\,(T_0 - \lambda I) = \mathrm{nul}\,(T_0^* - \bar{\lambda} I) = m$. The rest is immediate. □

Lemma 9.2. Let $I = [a,b]$, with $-\infty < a < b < \infty$, and let T_1 be the restriction of τ to

$$\mathscr{D}(T_1) = \{u : u^{(m-1)} \in \mathrm{AC}[a,b],\ u^{(i)}(a) = 0 \ \ (i = 0,1,\ldots,m-1)\}.$$

Then T_1 is a closed injective m-dimensional extension of T_0; further, $0 \in \rho(T_1)$ and T_1 has a compact resolvent. ■

Proof. Since T_1 is clearly the closure of the restriction of τ to $C_0^\infty(a,b]$ it is therefore closed. It is injective since the equation $\tau u = 0$ has no non-trivial solution in $\mathscr{D}(T_1)$.

Let $\phi \in C_0^\infty(\mathbb{R})$ be 1 in a neighbourhood of b and 0 in a neighbourhood of a, and define the m linearly independent functions

$$\theta_j(x) = \phi(x)(x-b)^j/j! \qquad (j = 0, 1, \ldots, m-1).$$

Then $\theta_j \in C_0^\infty(\mathbb{R})$, $\theta_j^{(i)}(a) = 0$ $(i = 0, 1, \ldots, m-1)$, and $\theta_j^{(i)}(b) = \delta_{ij}$, the Kronecker delta. Thus, if $u \in \mathscr{D}(T_1)$,

$$u - \sum_{j=0}^{m-1} u^{(j)}(b)\theta_j \in \mathscr{D}(T_0),$$

i.e.

$$\mathscr{D}(T_1) \subset \Theta \dotplus \mathscr{D}(T_0)$$

where Θ is the linear span of $\{\theta_0, \theta_1, \ldots, \theta_{m-1}\}$; note that $\Theta \cap \mathscr{D}(T_0) = \{0\}$. Since Θ and $\mathscr{D}(T_0)$ are subspaces of $\mathscr{D}(T_1)$ we therefore have

$$\mathscr{D}(T_1) = \Theta \dotplus \mathscr{D}(T_0) \tag{9.7}$$

and hence

$$\dim[\mathscr{D}(T_1)/\mathscr{D}(T_0)] = \dim \Theta = m.$$

By Theorem 4.1 and Lemma 9.1, T_1 is a Fredholm operator and $\operatorname{ind} T_1 = \operatorname{ind} T_0 + m = -\operatorname{def} T_0 + m = 0$. Since $\operatorname{nul} T_1 = 0$ it follows that $\operatorname{def} T_1 = 0$ and hence that $0 \in \rho(T_1)$. Finally, T_1 has compact resolvent if, and only if, I is T_1-compact. But I is T_1-compact since $H(T_1) \subset W^{m,2}(a,b)$ is continuous [cf. (7.6)] and the embedding $W^{m,2}(a,b) \to L^2(a,b)$ is compact. Consequently T_1 has a compact resolvent and the lemma is proved. □

Armed with the above preliminary results we are now in a position to apply Theorems 5.2 and 5.3 to the differential operators generated by τ in $L^2(a,\infty)$. First we need some notation. When the interval is I we shall write $T_0(\tau, I)$ and $T_1(\tau, I)$ for the operators T_0 and T_1 defined above. Also we define $T(\tau, I)$ to be the restriction of τ to $W^{m,2}(I)$. Hence we have

$$T_0^*(\tau, I) = T(\tau^+, I). \tag{9.8}$$

The operators $T_0(\tau, I)$ and $T(\tau, I)$ are the so-called minimal and maximal operators generated by τ in $L^2(I)$. Finally, for $-\infty < a < b < \infty$, define $T_{0b}(\tau, [a,\infty))$ to be the restriction of τ to

$$\begin{aligned}&\mathscr{D}[T_{0b}(\tau, [a,\infty))] \\ &= \{u : u^{(m-1)} \in AC_{\text{loc}}[a,\infty), u^{(j)}(a) = u^{(j)}(b) = 0 \ (j = 0, 1, \ldots, m-1); \\ &\qquad\qquad u, \tau u \in L^2(a,\infty)\}.\end{aligned} \tag{9.9}$$

We can write $T_{0b}(\tau, [a,\infty))$ as the orthogonal sum

$$T_{0b}(\tau, [a,\infty)) = T_0(\tau, [a,b]) \oplus T_0(\tau, [b,\infty)) \tag{9.10}$$

in $L^2(a,\infty) = L^2(a,b) \oplus L(b,\infty)$. We can now prove the following Decomposition Principles concerning all operators $S(\tau, I)$ generated by τ in $L^2(I)$ whose domain contains $C_0^\infty(I)$.

Theorem 9.3. Let $S(\tau,[a,\infty))$ be a closed operator satisfying $T_0(\tau,[a,\infty)) \subset S(\tau,[a,\infty)) \subset T(\tau,[a,\infty))$ for any $a > -\infty$. Then the sets $\sigma_{ek}[S(\tau,[a,\infty))]$ ($k = 1, 2, 3$) are coincident and, for any $b \in (a,\infty)$,

$$\sigma_{ek}[T_0(\tau,[a,\infty))] = \sigma_{ek}[S(\tau,[a,\infty))] = \sigma_{ek}[T(\tau,[a,\infty))] \quad (k = 1,2,3), \tag{9.11}$$

$$\sigma_{ek}[S(\tau,[a,\infty))] = \sigma_{ek}[S(\tau,[b,\infty))] \quad (k = 1,2,3). \tag{9.12}$$

Furthermore

$$\sigma_{e4}[T_0(\tau,[a,\infty))] = \sigma_{e4}[T_0(\tau,[b,\infty))], \tag{9.13}$$

$$\sigma_{e4}[T(\tau,[a,\infty))] = \sigma_{e4}[T(\tau,[b,\infty))], \tag{9.14}$$

and

$$\sigma_{e4}[T_0(\tau,[a,\infty))] \neq \sigma_{e4}[T(\tau,[a,\infty))], \tag{9.15}$$

unless both sets in (9.15) coincide with $\mathbb{C}$. ■

Proof. By an argument similar to that given in the proof of Lemma 9.2 it follows that

$$\dim\left(\mathscr{D}[T(\tau,[a,\infty))]/\mathscr{D}[T_0(\tau,[a,\infty))]\right) = m, \tag{9.16}$$

$$\dim\left(\mathscr{D}[T_0(\tau,[a,\infty))]/\mathscr{D}[T_{0b}(\tau,[a,\infty))]\right) = m. \tag{9.17}$$

Hence any closed operator $S(\tau,[a,\infty))$ satisfying $T_0(\tau,[a,\infty)) \subset S(\tau,[a,\infty)) \subset T(\tau,[a,\infty))$ is a finite-dimensional extension of $T_0(\tau,[a,\infty))$ and of $T_{0b}(\tau,[a,\infty))$. Also, whenever $\mathscr{R}[S(\tau,[a,\infty)) - \lambda I]$ is closed, $S(\tau,[a,\infty)) - \lambda I$ has finite nullity and deficiency so that the sets σ_{ek} ($k = 1,2,3$) are coincident. We therefore obtain (9.11) from Corollary 4.2 and furthermore, (9.12) is obtained on applying (5.2) and Lemma 9.1 to (9.10).

To prove (9.13) we appeal to Theorem 5.3. By (9.17), $T_0(\tau,[a,\infty))$ is an m-dimensional extension of $T_{0b}(\tau,[a,\infty))$ while by Lemma 9.2, $T_0(\tau,[a,b])$ has the closed m-dimensional extension $T_1(\tau,[a,b])$ with a compact resolvent. The identity (9.13) therefore follows from Theorem 5.3 applied to (9.10). Since $T(\tau,I) = T_0^*(\tau^+,I)$ when $I = [a,\infty)$ or $[b,\infty)$, (9.14) is a consequence of (9.13) and Theorem 1.1; recall our earlier observation that $\tau^{++} = \tau$.

Finally (9.15) is obtained from (9.16) and (4.2) in Corollary 4.2. □

As an application of the Decomposition Principles in Theorem 9.3 we have

Corollary 9.4. Let $\tau = \sum_{j=0}^{m} (2\pi i)^{-j} a_j D^j$ in $[a,\infty)$ where the a_j satisfy (i), (ii) and (iii), and a_m is a non-zero constant. Suppose that

$$\lim_{x\to\infty} a_j(x) = c_j \quad (j = 0,1,\ldots,m-1). \tag{9.18}$$

Then with $a_m = c_m$ and $P(\xi) = \sum_{j=0}^{m} c_j \xi^j$ we have, for any closed operator $S(\tau, [a, \infty))$ satisfying $T_0(\tau, [a, \infty)) \subset S(\tau, [a, \infty)) \subset T(\tau, [a, \infty))$, that

$$\sigma_{ek}[S(\tau, [a, \infty))] = \{P(\xi) : \xi \in \mathbb{R}\} \qquad (k = 1, 2, 3). \tag{9.19}$$

Also

$$\sigma_{e4}[T_0(\tau, [a, \infty))] = \{P(\xi) : \text{im}\, \xi \leqslant 0\}, \tag{9.20}$$

$$\sigma_{e4}[T(\tau, [a, \infty))] = \{P(\xi) : \text{im}\, \xi \geqslant 0\}. \tag{9.21}$$ ■

Proof. In view of Theorem 9.3 it is enough to prove (9.19)–(9.21) for any interval $[b, \infty)$, with $a < b < \infty$. Given $\varepsilon > 0$, we choose b such that

$$|a_j(x) - c_j| < \varepsilon \qquad (x \in [b, \infty);\ j = 0, 1, \ldots, m-1).$$

Let us write A and B for the operators $T_0(\tau, [b, \infty))$ and $T_0[P((1/2\pi \mathrm{i})\mathrm{D}), [b, \infty)]$ respectively. Then $A = B + C$ in $L^2(b, \infty)$, where $\mathscr{D}(B) = \mathscr{D}(C) = W_0^{m,2}(b, \infty)$, $C\phi = \Sigma_{j=0}^{m-1} b_j \mathrm{D}^j \phi$ and $|b_j(x)| < \varepsilon$ for $x \in [b, \infty)$ $(j = 0, 1, \ldots, m-1)$. On using (7.5) and (7.6) we see that for any $\lambda \in \mathbb{C}$ there exists a constant K, depending on λ, such that

$$\|(C - \lambda I)\phi\|^2 \leqslant \varepsilon K[\|(A - \lambda I)\phi\|^2 + \|\phi\|^2] \qquad (\phi \in \mathscr{D}(C) = \mathscr{D}(A)),$$

$\|\bullet\|$ denoting the $L^2(b, \infty)$ norm. We therefore conclude from Theorem I.3.18 (see Remark I.3.27) that there exists a value of b such that $A - \lambda I$ is Fredholm if, and only if, $B - \lambda I$ is Fredholm and the two operators have a common index. Consequently $\sigma_{ek}(A) = \sigma_{ek}(B)$ $(k = 1, 2, 3, 4)$ and (9.19) and (9.20) follow from (9.12), (9.13) and Theorem 7.3. For (9.21) we can either repeat the argument and appeal to Corollary 7.4 or else use (9.20) and take adjoints. □

If $S(\tau, [a, \infty))$ is a closed operator satisfying

$$T_0(\tau, [a, \infty)) \subset S(\tau, [a, \infty)) \subset T(\tau, [a, \infty)), \tag{9.22}$$

then on taking adjoints and using (9.8) we have

$$T_0(\tau^+, [a, \infty)) \subset S^*(\tau, [a, \infty)) \subset T(\tau^+, [a, \infty)). \tag{9.23}$$

It is therefore of interest to seek operators $S(\tau, [a, \infty))$ which have the property that

$$S^*(\tau, [a, \infty)) = S(\tau^+, [a, \infty)). \tag{9.24}$$

One important example is the following so-called Dirichlet operator in $L^2(a, \infty)$ when m is even. Let $m = 2l$ and define $S(\tau, [a, \infty))$ to be the restriction of τ to

$$\mathscr{D}[S(\tau, [a, \infty))] = W_0^{l,2}(a, \infty) \cap W^{2l,2}(a, \infty) \tag{9.25}$$
$$= \{u : u^{(2l-1)} \in AC_{\text{loc}}[a, \infty);\ u^{(j)}(a) = 0\ \ (j = 0, 1, \ldots l-1);\ u,\ \tau u \in L^2(a, \infty)\}$$

(see Lemma 7.1).

Theorem 9.5. The operator $S(\tau, [a, \infty))$ with $m = 2l$ and domain (9.25) has the following properties:

(i) $S(\tau, [a, \infty))$ is a closed, l-dimensional extension of $T_0(\tau, [a, \infty))$;

(ii) $S^*(\tau, [a, \infty)) = S(\tau^+, [a, \infty))$;

(iii) $S(\tau, [a, \infty))$ is self-adjoint if $\tau^+ = \tau$; and J-self-adjoint, with J the conjugation map $\phi \mapsto \bar{\phi}$, if $\tau^+ = \bar{\tau}$. ■

Proof. (i) Let us denote the operators in (9.22) by T_0, S and T, with S defined by (9.25). The Hilbert space $H(T)$ determined by $\mathscr{D}(T)$ and the graph norm of T is topologically isomorphic to the Sobolev space $W^{2l,2}(a, \infty)$. Thus S, being the restriction of T to the closed subspace $W_0^{l,2}(a, \infty) \cap W^{2l,2}(a, \infty)$ of $W^{2l,2}(a, \infty)$, and hence of $H(T)$, is a closed operator in $L^2(a, \infty)$. The proof of the fact that S is an l-dimensional extension of T_0 follows by an argument similar to that in Lemma 9.2.

(ii) Since (9.23) is satisfied, any $u \in \mathscr{D}(S^*)$ lies in $\mathscr{D}[T(\tau^+, [a, \infty))] = W^{2l,2}(a, \infty)$. Also, if $\phi \in \mathscr{D}(S)$, there exists a sequence $(\phi_j) \subset \mathscr{D}(S)$ such that $\phi_j = \phi$ in a right neighbourhood of a, the function ϕ_j has compact support in $[a, \infty)$, and $\phi_j \to \phi$ and $S\phi_j \to S\phi$ in $L^2(a, \infty)$. To see this, take any $\theta \in C^\infty(\mathbb{R})$ that satisfies $\theta(x) = 1$ for $x \leqslant 1$ and $\theta(x) = 0$ for $x \geqslant 2$ and define $\phi_j(x) = \theta((x-a)/j)\,\phi(x)$. Then $\phi_j \in \mathscr{D}(S)$ and $\phi_j \to \phi$ in $W^{2l,2}(a, \infty)$, and hence $\phi_j \to \phi$ in $H(T)$. Since S is closed, $\phi_j \to \phi$ in $H(S)$ and our claim is established. Thus if $u \in \mathscr{D}(S^*)$, for each $\phi \in \mathscr{D}(S)$ there exists a sequence (ϕ_j) with the above properties, and on integration by parts,

$$\begin{aligned}(\phi, S^*u) &= (S\phi, u)\\ &= \lim_{j\to\infty} (S\phi_j, u)\\ &= \lim_{j\to\infty} (\tau\phi_j, u)\\ &= -[\phi, u](a) + \lim_{j\to\infty} (\phi_j, \tau^+ u)\\ &= -[\phi, u](a) + (\phi, S^*u),\end{aligned} \tag{9.26}$$

on using (9.23). In (9.26) we have used (9.1) and the notation

$$\begin{aligned}[\phi, u](a) &= \sum_{k=1}^{2l} \sum_{n=0}^{k-1} (-1)^n (a_k\bar{u})^{(n)}(a)\phi^{(k-1-n)}(a)\\ &= \sum_{k=l+1}^{2l} \sum_{n=0}^{k-1-l} (-1)^n (a_k\bar{u})^{(n)}(a)\phi^{(k-1-n)}(a)\end{aligned}$$

(since $\phi(a) = \phi^{(1)}(a) = \ldots = \phi^{(l-1)}(a) = 0$ for $\phi \in \mathscr{D}(S)$)

$$= \sum_{k=l+1}^{2l} \sum_{n=0}^{k-1-l} (-1)^n \sum_{j=0}^{n} \binom{n}{j} a_k^{(n-j)}(a)\bar{u}^{(j)}(a)\phi^{(k-1-n)}(a)$$

$$= \sum_{j=0}^{l-1} \alpha_j \bar{u}^{(j)}(a), \tag{9.27}$$

where

$$\alpha_j = \sum_{n=j}^{l-1} \sum_{k=n+l+1}^{2l} (-1)^n \binom{n}{j} a_k^{(n-j)}(a)\phi^{(k-1-n)}(a).$$

The function $\phi \in \mathscr{D}(S)$ can be chosen so that the derivatives $\phi^{(i)}(a)$ $(i = l, \ldots, 2l-1)$ which appear in the above α_j, have any prescribed values; to obtain such a function take a solution of $\tau v = 0$ which is such that $v^{(i)}(a) = 0$ $(i = 0, 1, \ldots, l-1)$ and the $v^{(i)}(a)$ $(i = l, l+1, \ldots, 2l-1)$ coincide with the prescribed values, and multiply by a function $\theta \in C_0^\infty[a, \infty)$ which is 1 in a right neighbourhood of a. The constants α_j $(j = 0, \ldots, l-1)$ can therefore assume arbitrary values by choice of $\phi \in \mathscr{D}(S)$, and so from (9.26) and (9.27) we must have $u^{(j)}(a) = 0$ $(j = 0, 1, \ldots, l-1)$, i.e. $u \in \mathscr{D}(S)$. We have therefore proved that $S^* \subset S(\tau^+, [a, \infty))$. Integration by parts shows that the reverse inclusion also holds and hence (ii) is proved.

(iii) This is an immediate consequence of (ii). □

Theorem 9.6. Let $S(\tau, [a, \infty))$ be the operator in Theorem 9.5 and assume that $\mathrm{re}[(-1)^l a_{2l}(x)] \geqslant \delta > 0$ for $x \in [a, \infty)$. Then $S(\tau, [a, \infty))$ is m-sectorial and is the Friedrichs extension of $T_0(\tau, [a, \infty))$, and $\sigma_{ek}[S(\tau, [a, \infty))] = \sigma_{e3}[T_0(\tau, [a, \infty))]$ $(k = 1, 2, 3)$.

If τ is the differential polynomial $P((1/2\pi i)D) = \Sigma_{j=0}^{2l} (1/2\pi i)^j c_j D^j$ with $c_{2l} > 0$, then $\sigma_{ek}[S(\tau, [a, \infty))] = \{P(\xi): \xi \in \mathbb{R}\}$ $(k = 1, 2, 3)$; and if $\Lambda = \inf\{P(\xi): \xi \in \mathbb{R}\}$, the numerical range $\Theta(S)$ and hence the spectrum $\sigma(S)$ lies in the half plane $\{\lambda: \mathrm{re}\,\lambda \geqslant \Lambda\}$. In particular, when the c_j $(j = 0, 1, \ldots, 2l)$ are real, $S = S(\tau, [a, \infty))$ is self-adjoint and $\sigma(S) = \sigma_e(S) = \{P(\xi): \xi \in \mathbb{R}\} = [\Lambda, \infty)$. ■

Proof. For any $\phi \in \mathscr{D}(S)$, with $S \equiv S(\tau, [a, \infty))$, we have

$$(S\phi, \phi) = \sum_{j=0}^{2l} (a_j \phi^{(j)}, \phi)$$

$$= \sum_{j=l}^{2l} (-1)^l (\phi^{(j-l)}, (\bar{a}_j \phi)^{(l)}) + \sum_{j=0}^{l-1} (a_j \phi^{(j)}, \phi). \tag{9.28}$$

The inequalities (7.5) imply that given any $\varepsilon_0 > 0$ there exists a constant K_{ε_0}

such that

$$\|\phi^{(j)}\| \leqslant \varepsilon_0 \|\phi^{(l)}\| + K_{\varepsilon_0} \|\phi\| \qquad (j = 0, 1, \ldots, l-1).$$

Use of these estimates in (9.28) shows that for any $\varepsilon > 0$,

$$|(S\phi, \phi) - (-1)^l (\phi^{(l)}, \bar{a}_{2l}\phi^{(l)})| \leqslant \varepsilon \|\phi^{(l)}\|^2 + K_\varepsilon \|\phi\|^2$$

for some constant K_ε. Hence, since we are assuming that $\mathrm{re}[(-1)^l a_{2l}] \geqslant \delta > 0$,

$$\mathrm{re}(S\phi, \phi) \geqslant (\delta - \varepsilon) \|\phi^{(l)}\|^2 - K_\varepsilon \|\phi\|^2$$

and hence for some positive constant γ and choosing $\varepsilon = \frac{1}{2}\delta$,

$$\mathrm{re}(S\phi, \phi) + \gamma \|\phi\|^2 \geqslant \tfrac{1}{2}\delta(\|\phi^{(l)}\|^2 + \|\phi\|^2).$$

We also have for some constant K_1,

$$\begin{aligned} |\mathrm{im}(S\phi, \phi)| &\leqslant K_1 (\|\phi^{(l)}\|^2 + \|\phi\|^2) \\ &\leqslant 2\delta^{-1} K_1 [\mathrm{re}(S\phi, \phi) + \gamma \|\phi\|^2]. \end{aligned}$$

This proves that S is sectorial. Since $T_0 = T_0(\tau, [a, \infty)) \subset S$, it follows that T_0 is also sectorial. The form domain $\mathscr{Q}(T_0)$ of T_0 is $\mathrm{W}_0^{l,2}(a, \infty)$; it is the completion of $\mathscr{D}(T_0)$ with respect to the norm

$$\mathrm{re}(T_0\phi, \phi) + \gamma \|\phi\|^2$$

and this is equivalent to the $\mathrm{W}^{l,2}(a, \infty)$ norm as is established above. Since $\mathscr{D}(S) \subset \mathrm{W}_0^{l,2}(a, \infty)$, it will follow from Theorem IV.2.6 that S is the Friedrichs extension of T_0 if we show that S is m-sectorial. To see this we first observe that since $\mathscr{D}(S) = \mathscr{D}(S^*) = \mathrm{W}_0^{l,2}(a, \infty) \cap \mathrm{W}^{2l,2}(a, \infty)$ by Theorem 9.5(ii), then $\lambda \in \Theta(S)$ if, and only if, $\bar{\lambda} \in \Theta(S^*)$. Also, by Theorem III.2.3, if $\lambda \notin \bar{\Theta}(S)$ then $\mathrm{nul}(S - \lambda I) = 0$ and $\mathscr{R}(S - \lambda I)$ is closed. Thus for $\lambda \notin \bar{\Theta}(S)$, we have $\mathrm{def}(S - \lambda I) = \mathrm{nul}(S^* - \bar{\lambda} I) = 0$; note that $\bar{\lambda} \notin \bar{\Theta}(S^*)$, by our earlier observation. Consequently any $\lambda \notin \bar{\Theta}(S)$ lies in $\rho(S)$ and S is therefore m-sectorial. The fact that $\sigma_{ek}(S) = \sigma_{e3}(T_0)$ follows from Theorem 9.5(i) and Corollary 4.2.

If $\tau = P((1/2\pi\mathrm{i})\mathrm{D})$, we have for all $\phi \in \mathscr{D}(T_0)$,

$$(T_0\phi, \phi) = \int_{\mathbb{R}} P(\xi) |\hat{\phi}(\xi)|^2 \, \mathrm{d}\xi$$

where $\hat{\phi}$ is the Fourier transform of the function which is equal to ϕ on $[a, \infty)$ and zero elsewhere on $\mathbb{R}$. Hence $\mathrm{re}(T_0\phi, \phi) \geqslant \Lambda \|\hat{\phi}\|^2 = \Lambda \|\phi\|^2$. The numerical range of T_0 and consequently that of S (by Theorem IV.2.4(v)) lie in the half-plane $\{\lambda : \mathrm{re}\,\lambda \geqslant \Lambda\}$. Since we have shown S to be m-sectorial, $\sigma(S)$ also lies in this half-plane. Furthermore, by Theorem 7.3, $\sigma_{ek}(S) = \{P(\xi) : \xi \in \mathbb{R}\}$ $(k = 1, 2, 3)$.

Finally, when the c_j are real, S is self-adjoint, in which case all the essential spectra coincide with the interval $[\Lambda, \infty)$. Any other λ lies outside $\bar{\Theta}(S)$ and is hence in $\rho(S)$. The theorem is therefore proved. □

Example 9.7. Let $\tau = P((1/2\pi i)D)$ and $P(\xi) = \xi^2 + \xi + 1$. Then the operator $S = S(\tau, [a, \infty))$ with domain $W_0^{1,2}(a, \infty) \cap W^{2,2}(a, \infty)$ $= \{u : u^{(1)} \in AC_{loc}[a, \infty);\ u(a) = 0;\ u,\ \tau u \in L^2(a, \infty)\}$ is self-adjoint and

$$\sigma(S) = \sigma_e(S) = [\tfrac{3}{4}, \infty).$$

Remark 9.8. In the proof of Theorem 9.6 we made use of the property $\mathscr{D}(S) = \mathscr{D}(S^*)$, which is satisfied under our assumptions (i)–(iii), to establish that S is m-sectorial. In the remaining results in this section we only needed the strength of our assumptions (i)–(iii) to simplify our task of characterizing the operators T_0, S, T_0^*. These characterizations are possible, and indeed well-known, under much weaker hypotheses; examples of this can be found in Goldberg's book [1, Chapter VI] under the hypothesis $a_j \in C^j(a, \infty)$ $(j = 0, 1, \ldots, m)$ and $a_m \neq 0$, while in Akhiezer and Glazman [1, Volume II, Appendix 2] and Naimark [1, Chapter V] the problem is treated under minimal conditions on the coefficients. In these instances the domains are no longer the spaces $W_0^{m,2}(a, \infty)$, $W_0^{l,2}(a, \infty) \cap W^{2l,2}(a, \infty)$, and $W^{m,2}(a, \infty)$ but the results do not depend on this in any substantial way and alternative paths can readily be found.

X

Essential spectra of general second-order differential operators

The operators under consideration in this chapter will mainly be those quasi-m-sectorial differential operators in § VII.1 and the essential spectrum will be any one of the sets σ_{ek} ($k = 1, 2, 3, 4$) which remain invariant under compact perturbations of the operator. Our main tools will be a type of Decomposition Principle (Theorem 1.1) and a perturbation result (Theorem 4.1), both of these results being consequences of the abstract results for forms given in §IV.4, 5.

1. A Decomposition Principle

In the Decomposition Principle enunciated in §IX.5, an operator S was decomposed as the sum of an operator with empty essential spectrum and an operator T whose essential spectrum coincided with that of S. The abstract result was found to be ideal for application to ordinary differential operators in §IX.9 but is clearly of no use when the operators are generated by partial differential expressions since the deficiencies of such operators are in general infinite. To overcome this obstacle, and indeed to cover more general circumstances, we shall obtain in this section a decomposition principle in which a partial differential operator S is decomposed as a sum in the sense of forms. More precisely, S will be a quasi-m-sectorial operator whose form s is written as $s = t + p$, where t is the form of another quasi-m-sectorial operator T and p satisfies the compactness criterion in Theorem IV.5.1(v), thus ensuring that $\sigma_{ek}(S) = \sigma_{ek}(T)$ ($k = 1, 2, 3, 4$) in view of Theorem IX.2.4 (see Theorem IX.2.8). The implication of the results obtained will be the same as before, namely, that these essential spectra of the operators S considered do not depend on the behaviour of the coefficients of the differential expression in any compact subset of the open set Ω but only on their behaviour at the boundary $\partial\Omega$ and at infinity.

Let τ be the differential expression

$$\tau = (1/w)\,(-\mathrm{D}_i(a_{ij}\mathrm{D}_j) + b_j\mathrm{D}_j + q) \tag{1.1}$$

where the summation convention is understood. We shall be dealing with those differential operators generated by τ in $L^2(\Omega; w)$ which are studied in §VII.1; if $a_{ij} = \delta_{ij}$, the Kronecker delta, Ω is an arbitrary open set in $\mathbb{R}^n$ ($n \geq 1$), but otherwise Ω is one of the shells $S\{l, m\}$ defined in §VII.1.4. We shall assume throughout this section that the coefficients w, a_{ij}, b_j and q in (1.1) satisfy the hypothesis of any one of Theorems VII.1.4, 1.13, 1.15, 1.18; the form and associated operator under consideration will be denoted by t_Ω and T_Ω respectively. The properties we need are listed in (i)–(iv) below and these are satisfied in each of the theorems cited in the previous sentence.

(i) $$t_\Omega[\![\phi, \psi]\!] := \int_\Omega (a_{ij} D_j \phi D_i \bar{\psi} + b_j D_j \phi \bar{\psi} + q\phi\bar{\psi}) \qquad (\phi, \psi \in \mathfrak{Q}(\Omega)), \tag{1.2}$$

where $\mathfrak{Q}(\Omega)$ is the completion of $C_0^\infty(\Omega)$ in the Dirichlet problem and $C_0^\infty(\mathbb{R}^n)$ in the Neumann problem with respect to the norm $\|\bullet\|_{\mathfrak{Q}(\Omega)}$ given by

$$\|\phi\|^2_{\mathfrak{Q}(\Omega)} = \int_\Omega [a_1|\nabla\phi|^2 + (f_0 + f^+)|\phi|^2 + w|\phi|^2]. \tag{1.3}$$

(ii) $e^{i\beta} t_\Omega$ is a closed, densely defined sectorial form in $L^2(\Omega; w)$.

(iii) If $h_\Omega := \mathrm{re}\,(e^{i\beta} t_\Omega)$ and $z \in \mathbb{R}$ is large enough, there exist positive constants K_1 and K_2 such that for all $\phi \in \mathfrak{Q}(\Omega)$,

$$K_1 \|\phi\|^2_{\mathfrak{Q}(\Omega)} \leq (h_\Omega + z^2)[\![\phi]\!] \leq K_2 \|\phi\|^2_{\mathfrak{Q}(\Omega)}. \tag{1.4}$$

(iv) $$\mathscr{D}(T_\Omega) = \{u : u \in \mathfrak{Q}(\Omega),\ \tau u \in L^2(\Omega; w)\}, \qquad T_\Omega u = \tau u; \tag{1.5}$$

$$(w T_\Omega u, \phi)_{2,\Omega} = t_\Omega[\![u, \phi]\!] \qquad (u \in \mathscr{D}(T_\Omega),\ \phi \in \mathfrak{Q}(\Omega)). \tag{1.6}$$

In (1.5) τ is defined in the weak sense with respect to $\mathfrak{Q}(\Omega)$: for $u \in \mathfrak{Q}(\Omega)$, $v = D_i(a_{ij} D_j u)$ if $v \in L^1_{\mathrm{loc}}(\Omega)$ and

$$\int_\Omega a_{ij} D_j u D_i \phi = -\int_\Omega v\phi \qquad (\phi \in \mathfrak{Q}(\Omega)). \tag{1.7}$$

Theorem 1.1. Let $\Omega_1 \subset \subset \Omega$ and let $\tilde{a}_{ij}$, $\tilde{b}_j$, $\tilde{q}$, $\tilde{w}$, $\tilde{a}$ $(i, j = 1, 2, \ldots, n)$, be functions which are equal to a_{ij}, b_j, q, w, a respectively outside Ω_1 and have the same properties as their counterparts throughout Ω. In addition let

$$w, \tilde{w} \in L^\infty_{\mathrm{loc}}(\Omega), \qquad a_{ij} \in AC_{\mathrm{loc}}(\Omega), \qquad w^{-1} D_j a_{ij} \in L^\infty_{\mathrm{loc}}(\Omega), \tag{1.8}$$

and, for some constant K_{Ω_1} depending on Ω_1,

$$\int_{\Omega_1} |\tilde{q} - q|\,|u|^2 \leq K_{\Omega_1} \|u\|^2_{\mathfrak{Q}(\Omega)} \qquad (u \in \mathfrak{Q}(\Omega)). \tag{1.9}$$

Define

$$s_\Omega[\![u, v]\!] := \int_\Omega [\tilde{a}_{ij} D_j u D_i \bar{v} + \tilde{b}_j (D_j u)\bar{v} + \tilde{q} u \bar{v}] \qquad (u, v \in \mathfrak{Q}(\Omega)). \tag{1.10}$$

Then $e^{i\beta}s_\Omega$ is a closed, densely defined, sectorial form in $L^2(\Omega; w)$ and if $e^{i\beta}S_\Omega$ is the associated m-sectorial operator, S_Ω and T_Ω have the same essential spectra σ_{ek} $(k = 1, 2, 3, 4)$. ■

Proof. The functions a, $\tilde{a}$, w, $\tilde{w}$ are all bounded above and away from zero on $\bar{\Omega}_1$, and so for $x \in \bar{\Omega}_1$,

$$\tilde{a}(x) \asymp a(x) \asymp a_1(x), \qquad w(x) \asymp \tilde{w}(x), \tag{1.11}$$

and

$$\sum_{j=1}^{n} |b_j(x) - \tilde{b}_j(x)|^2 = O(a_1(x)w(x)), \tag{1.12}$$

with constants depending, of course, on Ω_1. If $\tilde{\mathfrak{Q}}(\Omega)$ denotes the Hilbert space determined by (1.3) but with the new coefficients $\tilde{a}$, $\tilde{q}$, $\tilde{w}$, it is readily seen from the assumptions that $\tilde{\mathfrak{Q}}(\Omega)$ and $\mathfrak{Q}(\Omega)$ are topologically isomorphic. Furthermore $e^{i\beta}s_\Omega$ is closed, densely defined, and sectorial for precisely the same reasons that $e^{i\beta}t_\Omega$ has these properties.

We shall prove the theorem by applying Corollary IV.5.2 to $e^{i\beta}s_\Omega$, $e^{i\beta}t_\Omega$, and $p = e^{i\beta}(s_\Omega - t_\Omega)$, that is,

$$p[\![u, v]\!] = e^{i\beta} \int_{\Omega_1} [(\tilde{a}_{ij} - a_{ij})D_j u D_j \bar{v} + (\tilde{b}_j - b_j)(D_j u)\bar{v} + (\tilde{q} - q)u\bar{v}] \qquad (u, v \in \mathfrak{Q}(\Omega)). \tag{1.13}$$

From (1.9), (1.11) and (1.12) it follows that

$$\begin{aligned} |p[u, v]| &\leqslant K_{\Omega_1} \|u\|_{\mathfrak{Q}(\Omega_1)} \|v\|_{\mathfrak{Q}(\Omega_1)} \\ &\leqslant [K_{\Omega_1} \|u\|_{\mathfrak{Q}(\Omega_1)}] \|v\|_{\mathfrak{Q}(\Omega)}. \end{aligned}$$

Consequently, with $\mathfrak{Q} \equiv \mathfrak{Q}(\Omega)$ and $\hat{P}: \mathfrak{Q} \to \mathfrak{Q}^*$ defined by $(\hat{P}u, v) = p[\![u, v]\!]$,

$$\|\hat{P}u\|_{\mathfrak{Q}^*} \leqslant K_{\Omega_1} \|u\|_{\mathfrak{Q}(\Omega_1)}. \tag{1.14}$$

Let Ω_2 be a bounded open subset of Ω with boundary of class C and also $\Omega_1 \subset\subset \Omega_2 \subset\subset \Omega$. Let $\zeta \in C_0^\infty(\Omega)$ be such that $\zeta(x) = 1$ if $x \in \Omega_1$, $\zeta(x) = 0$ if $x \notin \Omega_2$, and $0 \leqslant \zeta(x) \leqslant 1$. For $u \in \mathfrak{Q}$ it is readily shown that $\zeta u \in \mathfrak{Q}$ and

$$\|u\|_{\mathfrak{Q}(\Omega_1)} \leqslant \|\zeta u\|_{\mathfrak{Q}}. \tag{1.15}$$

Moreover, for $v \in \mathscr{D}(T_\Omega)$, we see from (1.7) that $w\tau(\zeta v) \in L^1_{\text{loc}}(\Omega)$ and

$$\begin{aligned} w\tau(\zeta v) = w\zeta\tau v - a_{ij}D_i\zeta D_j v - a_{ij}D_i v D_j\zeta \\ - (D_i a_{ij} D_j \zeta)v - a_{ij} v D^2_{ij}\zeta + b_j v D_j \zeta. \end{aligned} \tag{1.16}$$

Indeed $\tau(\zeta v) \in L^2(\Omega; w)$, and on using the Cauchy–Schwarz inequality for the inner product $a_{ij}\xi_i\bar{\xi}_j$ in (1.16) we have

$$\|w^{\frac{1}{2}}\tau(\zeta v)\|_{2,\Omega} \leqslant \|w^{\frac{1}{2}}\tau v\|_{2,\Omega} + K_{\Omega_2}\|v\|_{\mathfrak{Q}(\Omega_2)}. \tag{1.17}$$

Consequently

$$v \in \mathscr{D}(T_\Omega) \Rightarrow \zeta v \in \mathscr{D}(T_\Omega). \tag{1.18}$$

Next we prove that $\hat{P}(e^{i\beta}T_\Omega + z^2 I)^{-1} \in \mathscr{K}(L^2(\Omega; w); \mathfrak{Q}^*)$ for z real and $|z|$ large enough. Let z^2 be so large that $-z^2$ is in the resolvent sets of $e^{i\beta}T_\Omega$ and $e^{i\beta}S_\Omega$ and such that

$$\|u\|_{\mathfrak{Q}}^2 \leqslant K(h_\Omega + z^2)[\![u]\!] \qquad (u \in \mathfrak{Q}), \tag{1.19}$$

(see (1.4)). Let (u_n) be a sequence in $L^2(\Omega; w)$ with $\|u_n\|_{L^2(\Omega;w)} = 1$ and set $v_n = (e^{i\beta}T_\Omega + z^2 I)^{-1}u_n$. Then $v_n \in \mathscr{D}(T_\Omega)$, the sequence (v_n) is bounded in $L^2(\Omega; w)$, and by (1.18), $\zeta v_n \in \mathscr{D}(T_\Omega)$. On using (1.14), (1.15) and (1.19) we have

$$\begin{aligned}
\|\hat{P}v_n\|_{\mathfrak{Q}^*}^2 &\leqslant K_{\Omega_1}(h_\Omega + z^2)[\![\zeta v_n]\!] \\
&\leqslant K_{\Omega_1}|(e^{i\beta}t_\Omega + z^2)[\![\zeta v_n]\!]| \\
&= K_{\Omega_1}|(w(e^{i\beta}T_\Omega + z^2 I)\zeta v_n, \zeta v_n)_{2,\Omega}| \\
&\leqslant K_{\Omega_1}\|w^{\frac{1}{2}}(e^{i\beta}T_\Omega + z^2 I)\zeta v_n\|_{2,\Omega}\|w^{\frac{1}{2}}v_n\|_{2,\Omega_2} \\
&\leqslant K_{\Omega_1,\Omega_2}[\|w^{\frac{1}{2}}(e^{i\beta}T_\Omega + z^2 I)v_n\|_{2,\Omega} + \|v_n\|_{\mathfrak{Q}}]\,\|v_n\|_{2,\Omega_2},
\end{aligned} \tag{1.20}$$

using (1.17) and the boundedness of w on Ω_2. Also,

$$\begin{aligned}
\|v_n\|_{\mathfrak{Q}}^2 &\leqslant K(h_\Omega + z^2)[\![v_n]\!] \\
&\leqslant K|(e^{i\beta}t_\Omega + z^2)[\![v_n]\!]| \\
&= K|(w(e^{i\beta}T_\Omega + z^2)v_n, v_n)_{2,\Omega}| \\
&= K|(wu_n, v_n)_{2,\Omega}| \\
&\leqslant K.
\end{aligned} \tag{1.21}$$

Therefore, by (1.20) and (1.21),

$$\|\hat{P}v_n\|_{\mathfrak{Q}^*}^2 \leqslant K_{\Omega_1,\Omega_2}\|v_n\|_{2,\Omega_2}. \tag{1.22}$$

Moreover, since $1/a$ and $1/w$ are bounded on Ω_2, we see from (1.21) that

$$\begin{aligned}
\|\nabla v_n\|_{2,\Omega_2}^2 + \|v_n\|_{2,\Omega_2}^2 &\leqslant K_{\Omega_2}\|v_n\|_{\mathfrak{Q}}^2 \\
&\leqslant K_{\Omega_2}.
\end{aligned} \tag{1.23}$$

The sequence (v_n) is therefore bounded in $W^{1,2}(\Omega_2)$ and hence precompact in $L^2(\Omega_2)$, since $\partial\Omega_2$ is assumed to be of class C. It follows from (1.22) that $(\hat{P}v_n)$ is precompact in $\mathfrak{Q}^*$ and hence that $\hat{P}(e^{i\beta}T_\Omega + z^2 I)^{-1} \in \mathscr{K}(L^2(\Omega; w); \mathfrak{Q}^*)$. The theorem is therefore a consequence of Corollary IV.5.2. □

Remark 1.2. If $\Omega = S(l,m)$ and $w, 1/a, 1/w \in L^{\infty}_{\text{loc}}(S[l,m))$, then Ω_1 in Theorem 1.1 can be $S[l,N)$, with $l < N < m$; similarly we can choose $S(N,m]$ for Ω_1 when $w, 1/a, 1/w \in L^{\infty}_{\text{loc}}(S(l,m])$. For an arbitrary open set Ω in the Dirichlet problem, if $w, 1/a, 1/w \in L^{\infty}_{\text{loc}}(\bar{\Omega})$ then Ω_1 can be any bounded open subset of Ω, i.e. $\bar{\Omega}_1$ may intersect $\partial\Omega$. However, in the Neumann problem, (1.18) fails to hold if $\bar{\Omega}_1$ intersects $\partial\Omega$, excepting of course the cases discussed above when Ω is a shell.

In the applications of Theorem 1.1 to the problems associated with the differential operators in §VII.1 we choose $\Omega_1 = \Omega_N^c$ where $\Omega_N^c \to \Omega$ as $N \to \infty$. The essential spectra are then seen to depend only on the behaviour of the coefficients on $\partial\Omega$. In the circumstances described in Remark 1.2 for an unbounded Ω, the boundary of Ω can be included in Ω_N^c and hence it is only the behaviour of the coefficients at infinity which have any effect in these cases.

2. Essential spectra of realizations of $\tau = -\Delta + q$ in $L^2(\Omega)$

We first apply Theorem 1.1 to the operators in Theorems VII.1.4–1.13 in the special case when $b_j = 0$ $(j = 1, 2, \ldots, n)$, so that the operators are generated by $\tau = -\Delta + q$ in $L^2(\Omega)$. We shall continue to use the notation of Chapter VII and, to summarize, the assumptions made on q are as follows:

(i) $q = q_0 + q_1$, where $q_0, q_1 \in L^1_{\text{loc}}(\Omega)$ in the Dirichlet problem and $L^1_{\text{loc}}(\bar{\Omega})$ in the Neumann problem;

(ii) there exist constants $\beta \in (-\frac{1}{2}\pi, \frac{1}{2}\pi)$, $\mu + i\nu \in \mathbb{C}$, and $M \geq 0$, such that for a.e. $x \in \Omega$,

$$e^{i\beta} q_0(x) \in \mathscr{S}(\mu + i\nu;\ \tan^{-1} M),$$

the sector in $\mathbb{C}$ with vertex $\mu + i\nu$, semi-angle $\tan^{-1} M$ and axis $\{\xi + i\nu : \xi \geq \mu\}$;

(iii) $e^{i\beta} q_1 =: f + ig \in L^s(Q)$ for some $s \in [n, \infty]$ and all cubes Q in the covering $\mathfrak{F}$ of Ω;

(iv) $m(f) := \inf_{Q \in \mathfrak{F}} f_Q > -\infty, \qquad f_Q := |Q|^{-1} \int_Q f,$

$$l_s(f^-) := \sup_{Q \in \mathfrak{F}} [|Q|^{-1/s} \| f^- \|_{s,Q} (1 + |Q|^{1/n})] < \infty,$$

$$\alpha_s(f) := \mu_s \sup_{Q \in \mathfrak{F}} [|Q|^{1/n} \rho_s(f, Q)] < \infty, \quad \rho_s(f, Q) := |Q|^{-1/s} \| f - f_Q \|_{s,Q},$$

where μ_s is defined in (VII.1.5),

$$p(g) := \sup_{Q \in \mathfrak{F}} |g_Q| < \infty, \qquad \alpha_s(g) < \infty.$$

Theorems VII.1.4, 1.5 for the Dirichlet problem and Theorem VII.1.13 for the Neumann problem are then satisfied, the spectrum of each of the operators being within the set $e^{-i\beta}\Theta_\beta$, where Θ_β is the set of all $x+iy$ such that $x \geqslant \theta$ and

$$|y-v| - p(g) \leqslant \begin{cases} [|\tan\beta|(z+F)^2 + 2G(z+F)]\sec\beta \\ \qquad \text{if} \quad M \leqslant |\tan\beta| \text{ or } M > |\tan\beta| \text{ and } z \leqslant \dfrac{G\cos\beta + F|\sin\beta|}{M\cos\beta - |\sin\beta|}, \\ M[x-\mu-m(f)] + \dfrac{(MF+G)^2}{M\cos\beta - |\sin\beta|} \quad \text{otherwise,} \end{cases} \tag{2.1}$$

with the notation

$$\theta = \mu + m(f) - \alpha_s^2(f)\sec\beta,$$

$$F = \alpha_s(f), \qquad G = \alpha_s(G), \qquad z^2 = (\cos\beta)(x-\theta).$$

In the next theorem and thereafter the sets Ω_N are the complements in Ω of $\Omega_N^c = \cup_{Q\in\mathfrak{F}_N} Q$ $(N\in\mathbb{N})$, where $\Omega_N^c \subset \Omega_{N+1}^c$ and $\mathfrak{F}_N$ denotes a subset of $k(N)$ cubes Q in the covering $\mathfrak{F}$ of Ω with $k(N)\to\infty$ and $\Omega\cap\Omega_N^c \to \Omega$ as $N\to\infty$, i.e. $\cup_{N\in\mathbb{N}}(\Omega\cap\Omega_N^c) = \Omega$. In the Dirichlet problem $\mathfrak{F}$ can be a tesselation of $\mathbb{R}^n$ and the sets Ω_N^c can be chosen to be cubes with centre at the origin. Also, in the Neumann problem when Ω is an interval, $(0,\infty)$ say, we can take $\mathfrak{F}$ to consist of non-negative intervals Q which are such that $\bar{\Omega} = \cup_{Q\in\mathfrak{F}} Q$ and $\Omega_N = (k(N), \infty)$. However, on account of the point made in Remark 1.2 concerning the Neumann problem for a general Ω, the covering $\mathfrak{F}$ in such a problem will always be understood to be the Whitney covering, the subsets Ω_N^c then being relatively compact subsets of Ω. When we do not have to distinguish between the Dirichlet and Neumann operators $T_{\mathrm{D},\Omega}$ and $T_{\mathrm{N},\Omega}$ we shall write T_Ω for the operator under consideration.

Theorem 2.1. Let q satisfy (i)–(iv) and let Ω_N be the sets defined above. For $x\in\Omega_N$ suppose there are constants $\beta_N \in (-\frac{1}{2}\pi, \frac{1}{2}\pi)$, $\mu_N + iv_N \in \mathbb{C}$, and $M_N > 0$ such that

$$e^{i\beta_N} q_0(x) \in \mathcal{S}(\mu_N + iv_N; \tan^{-1} M_N), \tag{2.2}$$

and define

$$m_{\Omega_N}(f) = \inf_{Q\subset\Omega_N} f_Q, \tag{2.3}$$

$$\alpha_{s,\Omega_N}(f) = \mu_s \sup_{Q\subset\Omega_N} [|Q|^{1/n} \rho_s(f,Q)], \tag{2.4}$$

$$p_{\Omega_N}(g) = \sup_{Q\subset\Omega_N} |g_Q|, \tag{2.5}$$

Then $\sigma_{e4}(T_\Omega)$ lies in the set $e^{-i\beta_N}\Theta_{\beta,N}$ where $\Theta_{\beta,N}$ is defined by (2.1), with the constants in (ii) and (iv) replaced by their counterparts in (2.2)–(2.5). ■

Proof. In Theorem 1.1 we take $\bar{\Omega}_1 = \Omega_N^c$ and

$$\tilde{q}(x) = \begin{cases} q(x) & (x \in \Omega_N = \Omega \setminus \Omega_N^c), \\ h(x) & (x \in \Omega_N^c), \end{cases} \tag{2.6}$$

where $h \in L^s(\Omega_N^c)$ but is otherwise arbitrary. In view of (VII.1.14), (1.9) is satisfied and the theorem is a consequence of Theorem 1.1 on account of the arbitrary choice of h; for instance, if $e^{i\beta}h$ is chosen to be a large enough positive constant, the numbers $m(\tilde{f})$, $\alpha_s(\tilde{f})$, $p(\tilde{g})$, and $\alpha_s(\tilde{g})$ in Theorem 2.1 are equal to those in (2.3)–(2.5). □

Corollary 2.2. (i) Suppose there exist $\beta \in (-\frac{1}{2}\pi, \frac{1}{2}\pi)$, $\mu + i\nu \in \mathbb{C}$ and $\gamma \in [0, \frac{1}{2}\pi)$ such that $e^{i\beta}q(x) \in \mathscr{S}(\mu + i\nu; \gamma)$ as $|x| \to \infty$. Then

$$\sigma_{e4}(T_{D,\Omega}) \subset e^{-i\beta}\mathscr{S}(\mu + i\nu; \max[\gamma, |\beta|]). \tag{2.7}$$

If $\lim_{\substack{|x|\to\infty \\ x\in\Omega}} |q(x)| = \infty$ and, for large values of $|x|$, $\arg q(x)$ lies inside some angle $\delta < \arg q(x) < \delta + \pi - \varepsilon$, with $\varepsilon > 0$, which does not contain the negative real axis, then $\sigma_{e4}(T_{D,\Omega}) = \varnothing$.

The same result holds for $T_{N,\Omega}$ when Ω is an interval in $\mathbb{R}$.

(ii) In the Neumann problem with Whitney covering $\mathfrak{F}$, if $e^{i\beta}q_0(x) \in \mathscr{S}(\mu + i\nu; \gamma)$ as $x \to \partial\Omega$ then

$$\sigma_{e4}(T_{N,\Omega}) \subset e^{-i\beta}\mathscr{S}(\mu + i\nu; \max\{\gamma, |\beta|\}). \tag{2.8}$$

If $\lim_{x\to\partial\Omega} |q(x)| = \infty$ and $\arg q(x)$ lies inside an angle $(\delta, \delta + \pi - \varepsilon)$ which does not contain the negative real axis then $\sigma_{e4}(T_{N,\Omega}) = \varnothing$. ■

Proof. (i) Given $\varepsilon' > 0$ we may choose N so large that for $x \in \Omega_N$ in Theorem 2.1, $e^{i\beta_N}q(x) \in \mathscr{S}(\mu_N + i\nu_N; \gamma_N)$ where $\mu_N > \mu - \varepsilon'$, $|\nu_N - \nu| < \varepsilon'$, $|\beta_N - \beta| < \varepsilon'$ and $0 \leqslant \gamma_N \leqslant \gamma + \varepsilon'$. By Theorem 2.1 with $q = q_0$ in Ω_N it follows that $\sigma_{e4}(T_{D,\Omega})$ lies in the set $e^{-i\beta_N}\Theta_{\beta,N}$, where

$$\begin{aligned} \Theta_{\beta,N} &= \{x + iy : x \geqslant \mu_N, |y - \nu_N| \leqslant (x - \mu_N)\max\{|\tan\beta_N|, \tan\gamma\}\} \\ &= \mathscr{S}(\mu_N + i\nu_N; \max\{|\beta_N|, \gamma\}). \end{aligned}$$

Since ε' is arbitrary we have (2.7).

If $\arg q(x)$ lies inside the stated angle, there exist $\beta \in (-\frac{1}{2}\pi, \frac{1}{2}\pi)$ and $\gamma \in [0, \frac{1}{2}\pi)$ such that $e^{i\beta}q(x) \in \mathscr{S}(0, \gamma)$ for $|x|$ large enough. Since $|q(x)| \to \infty$ it follows, for any $\gamma' \in (\gamma, \frac{1}{2}\pi)$, that $e^{i\beta}q(x) \in \mathscr{S}(\mu_N; \gamma')$ for $|x| \geqslant N$, where $\mu_N \to \infty$ as $N \to \infty$. That $\sigma_{e4}(T_{D,\Omega}) = \varnothing$ is therefore a consequence of the first part already proved.

The same argument applies to the Neumann problem when Ω is an interval.

(ii) In the Neumann problem for a general Ω, the domains Ω_N in Theorem 2.1 are spread around the boundary $\partial\Omega$ of Ω and so we must have $x \to \partial\Omega$ in the hypothesis. □

Corollary 2.3. If $\lim_{N\to\infty} \inf_{Q\subset\Omega_N} f_Q = \infty$ then $\sigma_{e4}(T_\Omega) = \varnothing$ in the Dirichlet and Neumann problems. ■

Proof. The θ_{Ω_N} in Theorem 2.1 can be made arbitrarily large by choice of N, and this implies the corollary. □

In the case when q is real, Corollary 2.3 is reminiscent of Molcanov's criterion in Theorem VIII. 4.1 which is necessary and sufficient for the Dirichlet operator generated by τ to have a wholly discrete spectrum when q is bounded below. We no longer assume that q is bounded below in Corollary 2.3, but instead we assume that $\alpha_s(q) < \infty$ (f now being q in (iv)) and this has the effect of regulating the oscillations of q.

Corollary 2.4. In the notation of Theorem 2.1 let

$$\lim_{N\to\infty} \sup_{Q\subset\Omega_N} g_0(x) = 0, \quad \lim_{N\to\infty} p_{\Omega_N}(g) = 0, \quad \lim_{N\to\infty} \alpha_{s,\Omega_N}(g) = 0,$$

and define

$$\mu_0 = \lim_{N\to\infty} \inf_{Q\subset\Omega_N} f_0(x), \quad m_0 = \lim_{N\to\infty} \mathrm{m}_{\Omega_N}(f), \quad \alpha_0 = \lim_{N\to\infty} \alpha_{s,\Omega_N}(f).$$

Then

$$\sigma_{e4}(T_\Omega) \subset [\mu_0 + m_0 - \alpha_0^2, \infty). \tag{2.9}$$

If q is real, $\sigma_{e4}(T_\Omega) \subset [\theta_e, \infty)$ where

$$\theta_e = \lim_{N\to\infty} \inf_{Q\subset Q_N} [f_0(x) + f_Q - |Q|^{2/n} \rho_s^2(f, Q)]. \tag{2.10}$$ ■

Proof. For large enough N the $\beta_N, \nu_N, M_N, p_{\Omega_N}(g), \alpha_{s,\Omega_N}(g)$ in Theorem 2.1 are within an arbitrary ε (> 0) of zero, and μ_N, $m_{\Omega_N}(f), \alpha_{s,\Omega_N}(f)$ are within ε of μ_0, m_0, α_0 respectively. The corollary therefore follows from Theorem 2.1, (2.10) being a consequence of Corollary VII.1.11. □

Corollary 2.5. Let $q = q_0 + q_1$, where $q_1 \in L^s(\Omega)$ for some $s \in [n, \infty]$, and with $q_0 = f_0 + ig_0$, let

$$\lim_{N\to\infty} \sup_{Q\subset\Omega_N} g_0(x) = 0, \qquad \mu_0 := \lim_{N\to\infty} \inf_{Q\subset\Omega_N} f_0(x).$$

Then, if the cubes in $\mathfrak{F}$ are congruent, $\sigma_{e4}(T_{D,\Omega}) \subset [\mu_0, \infty]$.

The same result holds for $T_{N,\Omega}$ if Ω is an interval in $\mathbb{R}$. ■

Proof. In Theorem 2.1, β_N, ν_N and M_N are close to zero for large enough N, and μ_N is near μ_0. Moreover, with $q_1 = f + ig$, we have

$$|f_Q| \leqslant |Q|^{-1/s} \|f\|_{s,Q}, \qquad |g_Q| \leqslant |Q|^{-1/s} \|g\|_{s,Q},$$

and both these quantities tend to zero as $|x_Q| \to \infty$. Also $m_{\Omega_N}(f)$, $p_{\Omega_N}(g)$, $\alpha_{s,\Omega_N}(f)$ and $\alpha_{s,\Omega_N}(g)$ tend to zero and the result follows from Theorem 2.1. □

In Corollaries 2.4 and 2.5 the implication is that the essential spectrum of T_Ω coincides with that of the lower semi-bounded self-adjoint operator re T_Ω in these circumstances. That this is indeed so will be proved in Theorem 4.2 below.

Finally in this section we apply the Decomposition Principle to the operators in §VII.1.3 generated by $\tau = \Delta + q$ when $q \in L^r_{\text{loc}}(\bar{\Omega})$ for some r satisfying (VII.1.40), i.e.

$$r \in \begin{cases} [1, \infty] & \text{if } n = 1, \\ (1, \infty] & \text{if } n = 2, \\ [\tfrac{1}{2}n, \infty] & \text{if } n \geqslant 3. \end{cases}$$

The proof is similar to that of Theorem 2.1.

Theorem 2.6. Let q satisfy the hypothesis of Theorem VII.1.15 and define

$$\phi_0 := \lim_{N\to\infty} \inf_{Q \subset \Omega_N} [f_Q - \gamma_r \rho_r(f, Q)], \qquad \delta_0 := \gamma_r \lim_{N\to\infty} \sup_{Q \subset \Omega_N} [|Q|^{2/n} \rho_r(f, Q)],$$

$$\phi_1 := \lim_{N\to\infty} \sup_{Q \subset \Omega_N} \{[|g_Q| + \gamma_r \rho_r(f, Q)]\,[f_Q - \gamma_r \rho_r(f, Q) - \theta_0 + 1]^{-1}\},$$

$$\delta_1 := \gamma_r \lim_{N\to\infty} \sup_{Q \subset \Omega_N} [|Q|^{2/n} \rho_r(g, Q)].$$

Then $\sigma_e(T_\Omega)$ is contained in the set of all $x + iy$ such that $x \geqslant \phi_0$ and

$$|y| \leqslant \begin{cases} \phi_1(x + 1 - \phi_0) & \text{if } \delta_1 \leqslant \phi_1(1 - \delta_0), \quad (2.11) \\ \dfrac{\delta_1}{1 - \delta_0}(x - \phi_0) + \phi_1 & \text{if } \delta_1 > \phi_1(1 - \delta_0), \quad (2.12) \end{cases}$$

where T_Ω is the Dirichlet or Neumann operator in Theorem VII.1.15. ■

3. Essential spectra of general second-order operators in shells

Theorem 1.1. will now be applied to the operators defined in §VII.1.4 by the expressions

$$\tau = (1/w)(-D_i(a_{ij}D_j) + b_jD_j + q) \qquad (3.1)$$

in shells $S\{l,m\}$. We refer to §VII.1.4 for the notation.

Theorem 3.1. Let (1.8) and the hypotheses of Theorem VII.1.18 be satisfied in $\Omega = S\{l,m\}$ (see (VII.1.55)) and let $\Omega_N = \Omega \setminus S[l_N, m_N]$, where $l_N \to l+$ and $m_N \to m-$ as $N \to \infty$. Define

$$\left.\begin{aligned} &\rho(N) := \inf_{\Omega_N} [\rho_\gamma(|x|)/w(x)], \\ &A_1(N) := \inf_{\Omega_N} [a_1(r)/a(r)], \qquad A_2(N) := \sup_{\Omega_N} [a_2(r)/a_1(r)], \\ &B_j(N) := \frac{1}{2} \sup_{\Omega_N} \left[\left(\sum_{i=1}^{n} |b_i^{(j)}(x)|^2 \right)^{1/2} a^{-1/2}(|x|) w^{-1/2}(x) \right] \quad (j = 1, 2). \end{aligned}\right\} \tag{3.2}$$

Suppose there exist constants $\beta_N \in (-\frac{1}{2}\pi, \frac{1}{2}\pi)$, $\mu_N + i\nu_N \in \mathbb{C}$ and $M_N > 0$ such that for $x \in \Omega_N$,

$$e^{i\beta_N} q_0(x) - (\mu_N + i\nu_N)\rho_\gamma(|x|) \in \mathscr{S}(0; \tan^{-1} M_N) \tag{3.3}$$

where

$$\mu_N > -\tfrac{1}{4} A_1(N) \cos \beta_N. \tag{3.4}$$

Let $m_{\Omega_N}(f)$, $\alpha_{\Omega_N}(f)$, $p_{\Omega_N}(g)$, $\alpha_{\Omega_N}(g)$ be the constants defined in IV (viii) of §VII.1.4 when $\Omega = \Omega_N$ and set

$$\left.\begin{aligned} &j(N) := A_1(N)\cos\beta_N - 4\mu_N^-, \\ &k(N) := A_1(N)A_2(N)|\sin\beta_N| + 4|\nu_N|, \\ &F(N) := B_1(N) + B_2(N) + \alpha_{\Omega_N}(f), \\ &G(N) := B_1(N) + B_2(N) + \alpha_{\Omega_N}(g), \end{aligned}\right\} \tag{3.5}$$

$$\theta(N) = \begin{cases} \frac{1}{4}\rho(N)A_1(N)\cos\beta_N + \mu_N\rho(N) + m_{\Omega_N}(f) - \rho^{\frac{1}{2}}(N)F(N) \\ \quad \text{if } \rho(N) > 0 \text{ and } F(N) \leqslant \frac{1}{2}\rho^{\frac{1}{2}}(N)j(N), \\ \mu_N^+\rho(N) + m_{\Omega_N}(f) - F^2(N)/j(N) \quad \text{otherwise}. \end{cases} \tag{3.6}$$

Then $\sigma_{e,4}(T_\Omega) \subset e^{-i\beta_N}\Theta_\beta(N)$, where $\Theta_\beta(N)$ is obtained from Θ_β in Theorem VII.1.19 by replacing the numbers (VII.1.64–1.66) by their counterparts defined above in Ω_N.

If either $\Omega = S(l,m)$ with $w, 1/w, a \in L^\infty_{\mathrm{loc}}(S[l,m))$ or $\Omega = \mathbb{R}^n$, and also $b_j(x) = b_j(|x|)$ with $D_j b_j = 0$, then $F(N)$ and $G(N)$ in (3.5) can be replaced by

$$\left.\begin{aligned} &F(N) := B_1(N)|\sin\beta_N| + B_2(N)\cos\beta_N + \alpha_{\Omega_N}(f), \\ &G(N) := B_1(N)\cos\beta_N + B_2(N)|\sin\beta_N| + \alpha_{\Omega_N}(g), \end{aligned}\right\} \tag{3.7}$$

where now $\Omega_N = S(m_N, m)$, $m_N \to m$, $m \leqslant \infty$. ■

Proof. For $R \in (l_N, m_N)$ we define the operation $h \mapsto \bar{h}$ by

$$\bar{h}(x) = \bar{h}(r\xi) = \begin{cases} h(l_N\xi) & \text{if } r \in (l_N, R], \\ h(m_N\xi) & \text{if } r \in (R, m_N), \\ h(r\xi) & \text{if } r \notin (l_N, m_N). \end{cases}$$

In Theorem 1.1 we choose $\Omega_1 = S(l_N, m_N)$ and define $\tilde{a}_{ij}$, $\bar{b}_j$, etc. as indicated above. It is an easy matter to check that the conditions of Theorem 1.1 are satisfied. We also have, for instance,

$$\tilde{a}_1(r) \geqslant A_1(N)\tilde{a}(r) \quad (x = r\xi \in \Omega),$$

and in this way the numbers in Theorem VII.1.19 are replaced by their counterparts in this theorem. The first part of the theorem follows.

In the second part of the theorem we choose $\Omega_1 = S[l, m_N)$ in Theorem 1.1 and define

$$\bar{h}(r\xi) = \begin{cases} h(r\xi) & \text{if } r \in [m_N, m), \\ h(m_N\xi) & \text{if } r \in [l, m_N). \end{cases}$$

The condition $D_j\bar{b}_j = 0$ is satisfied almost everywhere in Ω and the result follows from Theorem VII.1.20. □

To illustrate Theorem 3.1 we consider an example in which $\Omega = S\{1, \infty)$ and $a(r) = r^{2\alpha}$, with $2\alpha > 2 - n$. It is based on Example VII.1.22. There are analogous results for $2\alpha \leqslant 2 - n$ and also when $\Omega = S\{0, 1\}$, $\mathbb{R}^n_+$ or $\mathbb{R}^n$. We shall continue to suppose that (1.8) and the hypotheses of Theorem VII.1.18 are satisfied.

Corollary 3.2. Let $\Omega = S\{1, \infty)$, $a(r) = r^{2\alpha}$, $2\alpha > 2 - n$, and $w(r) = r^{2\omega}$, and for $r \geqslant N > 1$ suppose that

(i) $a_1(r) \geqslant A_1(N)r^{2\alpha}$, $\quad a_2(r) \leqslant A_2 a_1(r)$;

(ii) (3.3) and (3.4) are satisfied with $\rho_\gamma(r) = (n - 2 + 2\alpha)^2 r^{2\alpha - 2}$;

(iii) $|q_1(x)| \leqslant k(|x|)$ and for all $\varepsilon > 0$ and some $\delta \leqslant 1$,

$$\alpha_{\varepsilon,N} := \sup_{r \geqslant N} \left(r^{-\alpha-\omega} \int_r^{r+\varepsilon r^\delta} k(t)\mathrm{d}t \right) < \infty; \tag{3.8}$$

(iv) $e^{i\beta N} q_1 = f + ig$ and

$$m_{\varepsilon,N}(f) := \inf_{\substack{r \geqslant N \\ |\xi| = 1}} \left(\frac{1}{\varepsilon r^{\delta + 2\omega + n - 1}} \int_r^{r+\varepsilon r^\delta} f(t\xi)t^{n-1}\mathrm{d}t \right) > -\infty, \tag{3.9}$$

$$p_{\varepsilon,N}(g) := \sup_{\substack{r \geqslant N \\ |\xi| = 1}} \left(\frac{1}{\varepsilon r^{\delta + 2\omega + n - 1}} \int_r^{r+\varepsilon r^\delta} |g(t\xi)|t^{n-1}\mathrm{d}t \right) < \infty; \tag{3.10}$$

(v) $\sum_{j=1}^{n} |b_j^{(l)}(x)|^2 |x|^{-2\alpha - 2\omega} \leqslant 4B_l^2 \quad (l = 1, 2).$

Then in Theorem 3.1,

$$\rho(N) = \begin{cases} (n-2+2\alpha)^2 N^{2\alpha-2-2\omega} \text{ if } \alpha-\omega \geqslant 1. \\ 0 \quad \text{otherwise,} \end{cases} \tag{3.11}$$

and as $\varepsilon \to 0+$,

$$\left.\begin{array}{ll} m_{\Omega_N}(f) \geqslant m_{\varepsilon, N}(f)[1+O(\varepsilon)], & p_{\Omega_N}(g) \leqslant p_{\varepsilon, N}(g)[1+O(\varepsilon)], \\ \alpha_{\Omega_N}(f), \alpha_{\Omega_N}(g) \leqslant \alpha_{\varepsilon, N}[1+O(\varepsilon)]. & \end{array}\right\} \tag{3.12}$$

■

Proof. In (VII.1.49) we have, since $2\alpha > 2-n$, that $h_\infty(\bullet) < \infty$ and, according to our convention, we naturally choose $\gamma = \infty$ rather than $\gamma \in (1, \infty)$ which is also permissible. Thus

$$h_\infty(r) = [(n-2+2\alpha)r^{n-2+2\alpha}]^{-1}, \qquad \rho_\infty(r) = (n-2+2\alpha)^2 r^{2\alpha-2}.$$

The rest follows from Example VII.1.22. □

Corollary 3.3. In Corollary 3.2 let $b_j = 0$ $(j = 1, 2, \ldots, n)$ and suppose that $\lim_{\varepsilon \to 0+} \lim_{N \to \infty} \alpha_{\varepsilon, N} = 0$. Then $\sigma_{e4}(T_\Omega) = \varnothing$ in each of the following cases:

(a) $\lim\limits_{\varepsilon \to 0+} \lim\limits_{N \to \infty} m_{\varepsilon, N}(f) = \infty$;

(b) $\alpha - \omega \geqslant 1$, $\liminf\limits_{N \to \infty} \dfrac{\mu_N}{[A_1(N) \cos \beta_N]} > -\frac{1}{4}$, and

$$\lim_{N \to \infty} [N^{2\alpha-2-2\omega} A_1(N)] = \infty;$$

(c) $\alpha - \omega \geqslant 1$,

$$\lim_{N \to \infty} \inf_{|x| \geqslant N} |q_0(x)|\, |x|^{2-2\alpha} N^{2\alpha-2-2\omega} = \infty,$$

and when $|x|$ is large, arg $q(x)$ lies inside an angle $(\delta', \delta' + \pi - \varepsilon')$, with $\varepsilon' > 0$, which does not contain the negative real-axis. ■

Proof. From (3.6), (3.11) and (3.12) we have

$$\theta(N) \geqslant \begin{cases} \frac{1}{4}(n-2+2\alpha) N^{2\alpha-2-2\omega}[A_1(N)\cos\beta_N + 4\mu_N] + m_{\varepsilon, N}(f)[1+O(\varepsilon)] \\ \qquad\qquad\qquad\qquad\qquad\qquad \text{if} \quad \alpha-\omega \geqslant 1, \\ m_{\varepsilon,N}[1+O(\varepsilon)] \quad \text{if} \quad \alpha - \omega < 1, \end{cases} \tag{3.13}$$

and this yields (a) and (b) immediately.

In (c) there exist $\beta \in (-\frac{1}{2}\pi, \frac{1}{2}\pi)$ and $\gamma \in [0, \frac{1}{2}\pi)$ such that $e^{i\beta} q_0(x)/\rho_\infty(|x|) = e^{i\beta} q_0(x)|x|^{2-2\alpha}/(n-2+2\alpha)^2 \in \mathscr{S}(0; \gamma)$ for $|x| \geqslant N$, say. Thus there exists $\gamma'' \in (\gamma, \frac{1}{2}\pi)$ such that $e^{i\beta} q_0(x)/\rho_\infty(|x|) \in \mathscr{S}(\mu_N; \gamma'')$, where

$\mu_N = \frac{1}{2}\inf_{|x|\geqslant N}[|q_0(x)|/\rho_\infty(|x|)]$. As $\mu_N\, N^{2\alpha-2-2\omega} \to \infty$ we have $\theta(N) \to \infty$ in (3.13) and (c) is therefore proved. □

The criterion in Corollary 3.3(a), namely

$$\lim_{\varepsilon\to 0+}\ \lim_{N\to\infty}\ \inf_{\substack{r\geqslant N\\ |\xi|=1}} \left(\frac{1}{\varepsilon r^{\delta+2\omega+n-1}} \int_r^{r+\varepsilon r^\delta} f(t\xi)t^{n-1}\mathrm{d}t \right) = \infty, \qquad \delta \leqslant 1 \tag{3.14}$$

is the analogue of the Molcanov criterion for discreteness given in Theorem VIII.4.1 for the case when q is real and bounded below; see also Corollary 2.3. Molcanov's result was extended to complex q by Fortunato [1] (see also Evans [2]). For $n = 1$, extensions were obtained by Hinton and Lewis in [1, 2] to higher-order real differential expressions of the form $\tau y = \sum_{j=1}^m (-1)^j (p_j y^{(j)})^{(j)}$. The criterion of Hinton and Lewis in [1] is the case $\delta = -1$ of (3.14); however, they do not require the assumption (3.8). Similar results are also given in Evans, Kwong and Zettl [1] for the case $n = 1$ and in Lewis [1] when $n \geqslant 1$.

When $\alpha - \omega < 1$ in Corollary 3.3(c), the method used in this section can be adapted to prove that $\sigma_{e4}(T_\Omega) = \varnothing$ if $|x|^{-2\omega}|q_0(x)| \to \infty$ and $\arg q(x)$ eventually lies in a sector which excludes the negative real axis (see Evans [3]). The case $\alpha = \omega = 0$ of this is given in Glazman [1, §35, Theorem 30]; it is interesting to note that the spectrum need not be discrete if q is real and $q(x) \to -\infty$.

4. Perturbation results

We shall now apply the abstract perturbation result in Theorem IV.5.1 and Theorem IX.2.4 to perturbations of the various forms t_Ω considered in §§1,3 by a sesquilinear form p defined on the form domain $\mathfrak{Q}(\Omega)$ appropriate to the problem under consideration by

$$p[\![u, v]\!] = \int_\Omega (c_j \mathrm{D}_j u + du)\bar{v} \qquad (u, v \in \mathfrak{Q}(\Omega)), \tag{4.1}$$

the summation suffix j being summed from 1 to n. The following assumptions are made:

(A) Under Neumann boundary conditions, $c_j = 0$ $(j = 1, 2, \ldots n)$.

(B) Each of the functions

$$\sum_{j=1}^n |c_j|^2/a, \qquad \left(\sum_{j=1}^n |\mathrm{D}_j c_j|^2\right)^{\frac{1}{2}}, \qquad d$$

satisfies the following conditions on h: given any $\varepsilon > 0$ there exist bounded open subsets Ω_1 and Ω_2 of Ω, with $\Omega_1 \subset\subset \Omega_2 \subset\subset \Omega$, and a constant $K_{\varepsilon, \Omega_2}$ such that for all $u \in \mathfrak{Q}(\Omega)$,

$$\int_{\Omega_1} |h|\,|u|^2 \leqslant \varepsilon \|u\|^2_{\mathfrak{Q}(\Omega)} + K_{\varepsilon,\Omega_2} \|w^{\frac{1}{2}} u\|^2_{2,\Omega_2}, \tag{4.2}$$

$$\int_{\Omega \setminus \Omega_1} |h|\,|u|^2 \leqslant \varepsilon \|u\|^2_{\mathfrak{Q}(\Omega)}. \tag{4.3}$$

Examples of functions h satisfying (B) will be given in Examples 4.3–4.7 below. The main result is

Theorem 4.1. Let t_Ω be any one of the forms in §1 satisfying §1 (i)–(iv) and let T_Ω be the associated operator defined by (1.5) and (1.6). Let p be defined by (4.1), where $c_j \in \mathrm{AC}_{\mathrm{loc}}(\Omega)$ and $d \in \mathrm{L}^1_{\mathrm{loc}}(\Omega)$ satisfy (A) and (B). Then $\mathrm{e}^{\mathrm{i}\beta} s_\Omega := \mathrm{e}^{\mathrm{i}\beta} t_\Omega + p$ is a closed, densely defined sectorial form in $\mathrm{L}^2(\Omega; w)$ and if $\mathrm{e}^{\mathrm{i}\beta} S_\Omega$ is the associated m-sectorial operator, S_Ω and T_Ω have the same essential spectrum σ_{e4}. ■

Proof. Let $|c|^2 := \sum_{j=1}^{n} |c_j|^2$. From (B), for any $\varepsilon_1, \varepsilon > 0$ and $u \in \mathfrak{Q}(\Omega)$,

$$\begin{aligned}
|p[\![u]\!]| &\leqslant \int_\Omega (|c|\,|\nabla u| + |d|\,|u|)|u| \\
&\leqslant \|a^{\frac{1}{2}} \nabla u\|_{2,\Omega} \left(\int_\Omega a^{-1} |c|^2 |u|^2 \right)^{\frac{1}{2}} + \int_\Omega |d|\,|u|^2 \\
&\leqslant \|a^{\frac{1}{2}} \nabla u\|_{2,\Omega} (\varepsilon_1 \|u\|^2_{\mathfrak{Q}(\Omega)} + K_{\varepsilon_1,\Omega_2} \|w^{\frac{1}{2}} u\|^2_{2,\Omega_2} + \varepsilon_1 \|u\|^2_{\mathfrak{Q}(\Omega)}) \\
&\quad + \varepsilon_1 \|u\|^2_{\mathfrak{Q}(\Omega)} + K_{\varepsilon_1,\Omega_2} \|w^{\frac{1}{2}} u\|^2_{2,\Omega_2} + \varepsilon_1 \|u\|^2_{\mathfrak{Q}(\Omega)} \\
&\leqslant \varepsilon \|u\|^2_{\mathfrak{Q}(\Omega)} + K_{\varepsilon,\Omega_2} \|w^{\frac{1}{2}} u\|^2_{2,\Omega_2}.
\end{aligned} \tag{4.4}$$

Hence (IV.5.9) is satisfied and all that remains for us to prove is that, in the notation of Theorem IV.5.1, $\hat{P}(\mathrm{e}^{\mathrm{i}\beta} T_\Omega + z^2 I)^{-1} \in \mathscr{K}(\mathrm{L}^2(\Omega; w), \mathfrak{Q}^*)$ for large enough z^2. Let (u_n) be a bounded sequence in $\mathrm{L}^2(\Omega; w)$ and put $v_n = (\mathrm{e}^{\mathrm{i}\beta} T_\Omega + z^2 I)^{-1} u_n$. Then, if z^2 is in the resolvent set of $\mathrm{e}^{\mathrm{i}\beta} T_\Omega$, the sequence (v_n) is bounded in $\mathrm{L}^2(\Omega; w)$ and of course $v_n \in \mathscr{D}(T_\Omega)$. The proof now proceeds along the lines of that of Theorem 1.1. For z^2 large enough, we see from (1.4) that, with $h_\Omega = \mathrm{re}(\mathrm{e}^{\mathrm{i}\beta} t_\Omega)$,

$$\begin{aligned}
\|v_n\|^2_{\mathfrak{Q}(\Omega)} &\leqslant K(h_\Omega + z^2)[\![v_n]\!] \\
&\leqslant K|(\mathrm{e}^{\mathrm{i}\beta} t_\Omega + z^2)[\![v_n]\!]| \\
&= K|(w(\mathrm{e}^{\mathrm{i}\beta} T_\Omega + z^2 I)v_n, v_n)_{2,\Omega}| \\
&= K|(wu_n, v_n)_{2,\Omega}| \\
&\leqslant K.
\end{aligned}$$

Also, since $1/a$ and $1/w$ are in $L^{\infty}_{\text{loc}}(\Omega)$, for any bounded open set $G \subset\subset \Omega$,

$$\|\nabla v_n\|^2_{2,G} + \|v_n\|^2_{2,G} \leqslant K_G \|v_n\|^2_{\mathfrak{Q}(\Omega)} \leqslant K_G.$$

Thus as long as G has a smooth enough boundary (if it is of class C, say) (v_n) is precompact in $L^2(G)$. Given $\varepsilon > 0$ we now choose Ω_1 and Ω_2, as in (B) and let G satisfy $\Omega_2 \subset G \subset \Omega$ (see Theorem V.4.20). From (B), for the Dirichlet problem, we have for all $\phi \in \mathfrak{Q}(\Omega)$,

$$\begin{aligned}(\hat{P}v_n, \phi) &= p[\![v_n, \phi]\!] \\ &= \int_\Omega (c_j D_j v_n + d v_n)\bar{\phi} \\ &= -\int_\Omega v_n(c_j D_j \bar{\phi} + \bar{\phi} D_j c_j) + \int_\Omega d v_n \bar{\phi}.\end{aligned}$$

This last equation holds trivially for the Neumann problem since each $c_j = 0$ in that case. On using (B) we obtain

$$\begin{aligned}&|(\hat{P}v_n, \phi)| \\ &\leqslant \int_\Omega [(a^{\frac{1}{2}}|\nabla\phi|)(a^{-\frac{1}{2}}|c||v_n|) + (|D_j c_j|^{\frac{1}{2}}|v_n|)(|D_j c_j|^{\frac{1}{2}}|\phi|) + (|d|^{\frac{1}{2}}|v_n|)(|d|^{\frac{1}{2}}|\phi|)] \\ &\leqslant (\varepsilon \|v_n\|^2_{\mathfrak{Q}(\Omega)} + K_{\varepsilon,\Omega_2} \|w^{\frac{1}{2}} v_n\|^2_{2,\Omega_2})^{\frac{1}{2}} (\|a^{\frac{1}{2}}\nabla\phi\|^2_{2,\Omega_2} + \|w^{\frac{1}{2}}\phi\|^2_{2,\Omega} + K\|\phi\|^2_{\mathfrak{Q}(\Omega)})^{\frac{1}{2}} \\ &\leqslant K(\varepsilon\|v_n\|^2_{\mathfrak{Q}(\Omega)} + K_{\varepsilon,\Omega_2} \|v_n\|^2_{2,G})^{\frac{1}{2}} \|\phi\|_{\mathfrak{Q}(\Omega)}.\end{aligned}$$

Therefore

$$\|\hat{P}v_n\|^2_{\mathfrak{Q}(\Omega)^*} \leqslant \varepsilon K \|v_n\|^2_{\mathfrak{Q}(\Omega)} + K_{\varepsilon,\Omega_2} \|v_n\|^2_{2,G}.$$

Since (v_n) is precompact in $L^2(G)$ it contains a subsequence, which we still denote by (v_n), such that

$$\begin{aligned}\limsup_{m,n\to\infty} \|\hat{P}(v_n - v_m)\|^2_{\mathfrak{Q}(\Omega)^*} &\leqslant \varepsilon K \limsup_{m,n\to\infty} \|v_n - v_m\|^2_{\mathfrak{Q}(\Omega)} \\ &\leqslant \varepsilon K.\end{aligned}$$

As ε is arbitrary, $(\hat{P}v_n)$ is precompact in $\mathfrak{Q}(\Omega)^*$ and the theorem follows from Theorem IV.5.1(v) and Theorem IX.2.4. □

The special case $p = -\mathrm{i}\,\mathrm{im}\, t_\Omega$ and $\beta = 0$ of Theorem 4.1 is of special interest, for then $s_\Omega = \mathrm{re}\, t_\Omega$ and $S_\Omega = \mathrm{re}\, T_\Omega$.

Corollary 4.2. Let t_Ω be as in Theorem 4.1 with $\beta = 0$, and in addition assume that $h = \mathrm{im}\, q$ satisfies (4.2) and (4.3), that $b_j = 0$ in the Neumann problem and that in the Dirichlet problem $b_j \in AC_{\text{loc}}(\Omega)$ and $h = (\sum_{j=1}^{n} |D_j b_j|^2)^{\frac{1}{2}}$ satisfies (4.2) and (4.3). Then $\sigma_{e4}(T_\Omega) = \sigma_{e4}(\mathrm{re}\, T_\Omega)$. ■

Proof. In Theorem 4.1 we take $p = -\mathrm{i}\,\mathrm{im}\,t_\Omega$ so that

$$p[\![u,v]\!] = -\int_\Omega \{\tfrac{1}{2}[b_j(D_j u)\bar{v} - \bar{b}_j u D_j \bar{v}] + \mathrm{i}(\mathrm{im}\,q)u\bar{v}\}$$

$$= \int_\Omega [-(\mathrm{re}\,b_j)(D_j u)\bar{v} - \tfrac{1}{2}(D_j \bar{b}_j)u\bar{v} - \mathrm{i}(\mathrm{im}\,q)u\bar{v}]$$

under the given hypothesis. The result follows from Theorem 4.1 with $c_j = -\mathrm{re}\,b_j$ and $d = -\frac{1}{2}D_j\bar{b}_j - \mathrm{i}\,\mathrm{im}\,q$. □

On using (VII.1.12) it is readily seen that the hypothesis of Corollary 4.2 is satisfied in Corollaries 2.4 and 2.5. The Decomposition Principle allows us to take $\beta = 0$ in these last corollaries.

Examples of functions h which satisfy (4.2) and (4.3) are now given.

Example 4.3. Let $a \geqslant 1$ and $w \geqslant 1$, and let Ω be an arbitrary open set in $\mathbb{R}^n$. Let $h \in L^t_{\mathrm{loc}}(\Omega)$ where

$$t \in \begin{cases} [1,\infty] \text{ if } n = 1, \\ (1,\infty] \text{ if } n = 2, \\ [\tfrac{1}{2}n, \infty] \text{ if } n \geqslant 3, \end{cases} \tag{4.5}$$

and if $\mathfrak{F}$ is a tesselation of $\mathbb{R}^n$ by congruent cubes Q, let

$$\lim_{|x_Q| \to \infty} \|h\|_{t,Q\cap\Omega} = 0. \tag{4.6}$$

Then (4.2) and (4.3) are satisfied for any $u \in W_0^{1,2}(\Omega)$. ■

Proof. We put $u = 0$ outside Ω. By the Sobolev Embedding Theorem V.4.13, $W^{1,2}(Q)$ is continuously embedded in $L^{2t'}(Q)$, where $1/t' = 1 - 1/t$. Hence, on using Hölder's inequality

$$\int_{Q\cap\Omega} |h||u|^2 \leqslant \|h\|_{t,Q\cap\Omega}\|u\|^2_{2t',Q}$$

$$\leqslant K\|h\|_{t,Q\cap\Omega}\|a^{\frac{1}{2}}\nabla u\|^2_{2,Q} + \|w^{\frac{1}{2}}u\|^2_{2,Q}.$$

In view of (4.6), it follows that for any $\varepsilon > 0$ there exists a bounded open subset Ω_1 of Ω such that

$$\int_{\Omega\setminus\Omega_1} |h||u|^2 \leqslant \varepsilon\|u\|^2_{\mathfrak{D}(\Omega)},$$

which is (4.3). For any $\varepsilon' > 0$ we can decompose h on Ω_1 as $h = h_1 + h_2$ where

$$\|h_1\|_{t,\Omega_1} < \varepsilon', \qquad \|h_2\|_{\infty,\Omega_1} < K_{\varepsilon',\Omega_1}.$$

This is proved in Lemma VI.1.1.

On using this decomposition we have

$$\begin{aligned}\int_{\Omega_1} |h|\,|u|^2 &\leqslant \varepsilon' \|u\|^2_{2t',\Omega_1} + K_{\varepsilon',\Omega_1} \|u\|^2_{2,\Omega_1} \\ &\leqslant \varepsilon \|u\|^2_{1,2,\Omega_1} + K_{\varepsilon,\Omega_1} \|u\|^2_{2,\Omega_1} \\ &\leqslant \varepsilon \|u\|^2_{\mathfrak{D}(\Omega)} + K_{\varepsilon,\Omega_1} \|u\|^2_{2,\Omega_1}\end{aligned}$$

so that (4.2) is satisfied. □

Example 4.4. Let $\Omega = S\{0, \infty\}$, $a(r) = r^{2\alpha}$, $w(x) = w(r) = r^{2\omega}$ $(r = |x|)$, and

$$|h(x)| \leqslant \begin{cases} H_1(r) & (0 < r \leqslant 1), \\ H_2(r) & (1 < r < \infty), \end{cases}$$

where for some $\gamma \geqslant \max\{1, \alpha - \omega\}$ and $\delta \leqslant \min\{1, \alpha - \omega\}$, and for all $\varepsilon > 0$, $R_1 \in (0, 1]$ and $R_2 \geqslant 1$, we suppose

$$J^-_{\varepsilon,R_1} := \sup_{0<r\leqslant R_1} \left(\frac{1}{\varepsilon r^{\gamma+2\omega}} \int_{r-\varepsilon r^\gamma}^{r} H_1(t)\,\mathrm{d}t \right) < \infty,$$

$$J^+_{\varepsilon,R_2} := \sup_{R_2\leqslant r<\infty} \left(\frac{1}{\varepsilon r^{\delta+2\omega}} \int_{r}^{r+\varepsilon r^\delta} H_2(t)\,\mathrm{d}t \right) < \infty,$$

and $J^-_{\varepsilon,R_1} = o(1)$ and $J^+_{\varepsilon,R_2} = o(1)$ as $\varepsilon \to 0$, $R_1 \to 0$ and $R_2 \to \infty$.
Then (4.2) and (4.3) are satisfied. ■

Proof. In the notation of §VII.1.5 we choose $R = 1$, $R_1 = l_k^-$ and $R_2 = l_k^+$ and define $p_k(|h|) = \max\{p_k^+(|h|), p_k^-(|h|)\}$ and $\alpha_k(|h|) = \max\{\alpha_k^+(|h|), \alpha_k^-(|h|)\}$, where the subscript k indicates that the suprema in the definitions of $p^\pm(|h|)$ and $\alpha^\pm(|h|)$ in §VII.1.5 are taken over $j \geqslant k$. With $\Omega_1 = S(R_1, R_2)$, we have as in Lemma VII.1.21 that for all $u \in C_0^\infty(\Omega)$,

$$\int_{\Omega\setminus\Omega_1} |h|\,|u|^2 \leqslant p_k(|h|)\, \|w^{\frac{1}{2}} u\|^2_{2,\Omega\setminus\Omega_1} + 2\alpha_k(|h|)\, \|a^{\frac{1}{2}} \nabla u\|_{2,\Omega\setminus\Omega_1} \|w^{\frac{1}{2}} u\|_{2,\Omega\setminus\Omega_1}.$$

Since $\gamma \geqslant \alpha - \omega$ and $\delta \leqslant \alpha - \omega$ it follows as in Lemma VII.1.22 that

$$p_k^-(|h|) = O(J^-_{\varepsilon,R_1}), \qquad p_k^+(|h|) = O(J^+_{\varepsilon,R_2}),$$
$$\alpha_k^-(|h|) = O(\varepsilon J^-_{\varepsilon,R_1}), \qquad \alpha_k^+(|h|) = O(\varepsilon J^+_{\varepsilon,R_2}).$$

Hence, given any $\varepsilon' > 0$, we can choose k and ε such that

$$\int_{\Omega\setminus\Omega_1} |h|\,|u|^2 \leqslant \varepsilon' \|u\|^2_{\mathfrak{D}(\Omega)},$$

which is (4.3). Also, by Lemma VII.1.21,

$$\int_{\Omega_1} |h|\,|u|^2 \leqslant K_{1,k}\,\|w^{\frac{1}{2}}u\|_{2,\Omega_1}^2 + 2K_{2,k}\,\|a^{\frac{1}{2}}\nabla u\|_{2,\Omega_1}\,\|w^{\frac{1}{2}}u\|_{2,\Omega_1}$$
$$\leqslant \varepsilon'\,\|u\|_{\mathfrak{Q}(\Omega)}^2 + K_{\varepsilon',\Omega_1}\,\|w^{\frac{1}{2}}u\|_{2,\Omega_1}^2$$

and (4.2) is satisfied. □

Example 4.5. Let $\Omega = S\{1,\infty)$, $a(r) \geqslant r^{2\alpha}$, $w(x) \geqslant r^{2\alpha-2}$ $(r = |x|)$ and $h \in \mathrm{L}^t_{\mathrm{loc}}(\Omega)$, where t satisfies (4.5). Furthermore, suppose that

$$I_t(k) := \left(\int_{k<|x|<2k} |h(x)|^t\,|x|^{-(2\alpha-2+nt^{-1})t}\,\mathrm{d}x\right)^{1/t} \to 0 \qquad (4.7)$$

as $k \to \infty$. Then (4.2) and (4.3) are satisfied with $\Omega_1 = S\{1,N)$ and N large enough. ■

Proof. Since $\mathrm{W}^{1,2}(S(1,2))$ is continuously embedded in $\mathrm{L}^{2t'}(S(1,2))$, where $1/t' = 1 - 1/t$, we have, for some constant $K > 0$ and all $u \in \mathfrak{Q}(\Omega)$,

$$\int_{S(k,2k)} (a|\nabla u|^2 + w|u|^2) \geqslant \int_{S(k,2k)} |x|^{2\alpha-2}(|x|^2\,|\nabla u(x)|^2 + |u(x)|^2)\,\mathrm{d}x$$
$$\geqslant \int_{S(k,2k)} r^{2\alpha-2}\left(r^2\left|\frac{\partial}{\partial r}\,u(r\xi)\right|^2 + |u(r\xi)|^2\right) r^{n-1}\,\mathrm{d}r\,\mathrm{d}\xi$$
$$= k^{2\alpha+n-2}\int_{S(1,2)} r^{2\alpha-2}\left(r^2\left|\frac{\partial u}{\partial r}\,(kr\xi)\right|^2 + |u(kr\xi)|^2\right) r^{n-1}\,\mathrm{d}r\,\mathrm{d}\xi$$
$$\geqslant K\,k^{2\alpha+n-2}\int_{S(1,2)} \left(\left|\frac{\partial u}{\partial r}\,(kr\xi)\right|^2 + |u(kr\xi)|^2\right) r^{n-1}\,\mathrm{d}r\,\mathrm{d}\xi$$
$$\geqslant K\,k^{2\alpha+n-2}\,\|u(k\,\bullet)\|_{2t',\,S(1,2)}^2$$
$$\geqslant K\left(\int_{S(k,2k)} |x|^{t'(2\alpha+n-2-nt'^{-1})}\,|u(x)|^{2t'}\,\mathrm{d}x\right)^{1/t'}$$
$$= K\,\|\,|\bullet|^{\alpha-1+n/2t}\,u(\bullet)\|_{2t',S(k,2k)}^2.$$

Hence, by Hölder's inequality,

$$\int_{S(k,2k)} |h|\,|u|^2 \leqslant \|\,|\bullet|^{-2(\alpha-1+n/2t)}h(\bullet)\|_{t,S(k,2k)}\|\,|\bullet|^{2(\alpha-1+n/2t)}u^2(\bullet)\|_{t',S(k,2k)}$$
$$= I_t(k)\,\|\,|\bullet|^{\alpha-1+n/2t}\,u(\bullet)\|_{2t',S(k,2k)}^2$$
$$\leqslant K I_t(k)\,(\|a^{\frac{1}{2}}\nabla u\|_{2,S(k,2k)}^2 + \|w^{\frac{1}{2}}u\|_{2,S(k,2k)}^2)\,.$$

It follows from (4.7) that given any $\varepsilon > 0$ there exists an $N > 1$ such that

$$\int_{S(N,\infty)} |h|\,|u|^2 < \varepsilon \|u\|^2_{\mathfrak{D}(\Omega)} \qquad (u \in \mathfrak{D}(\Omega_1)),$$

which proves (4.3) with $\Omega \setminus \Omega_1 = S(N, \infty)$. The proof of (4.2) follows as in Example 4.3 since $h \in \mathrm{L}^t(\Omega_1)$. □

A similar argument gives the following example.

Example 4.6. Let $\Omega = S(0, 1\}$, $a(r) \geqslant r^{2\alpha}$, $w(x) \geqslant r^{2\alpha-2}$ and $h \in \mathrm{L}^t_{\mathrm{loc}}(\Omega)$, where t satisfies (4.5). Then, if

$$J_t(k) := \left(\int_{S(k,2k)} |h(x)|^t |x|^{(2-2\alpha-nt^{-1})t}\, \mathrm{d}x \right)^{1/t} \to 0 \tag{4.8}$$

as $k \to 0+$, (4.2) and (4.3) are satisfied with $\Omega_1 = S(N, 1\}$ and N small enough.

Example 4.7. The equations (4.7) and (4.8) are satisfied if

$$|\bullet|^{-(2\alpha-2+nt^{-1})} h(\bullet) \in \mathrm{L}^t(\Omega). \tag{4.9}$$

The efficacy of Theorem 4.1 and the above examples (Example 4.3 in particular) is illustrated in the next theorem.

Theorem 4.8. Let $\tau = -\Delta + q$ where q is real, $q \in \mathrm{L}^r_{\mathrm{loc}}(\Omega)$ for some r satisfying (4.5), and $q = 0$ outside Ω. Suppose also that for a tesselation $\mathfrak{F}$ of $\mathbb{R}^n$ by congruent cubes Q having centres x_Q,

$$\lim_{|x_Q| \to \infty} q_Q = \mu \in \mathbb{R}, \tag{4.10}$$

$$\lim_{|x_Q| \to \infty} \|q - q_Q\|_{r,Q} = 0, \tag{4.11}$$

$$\sup_{Q \in \mathscr{F}} \left(|Q|^{-1/r} \|q^-\|_{r,Q} \right) < \infty \tag{4.12}$$

where $q^- = -\min(0, q)$. Then the lower semi-bounded self-adjoint operator $T_{\mathrm{D},\Omega}$ in Theorem VII.1.15(a) is defined and has the same essential spectrum as $-\Delta_{\mathrm{D},\Omega} + \mu$. In particular, if $\Omega = \mathbb{R}^n$ then $\sigma_{\mathrm{e}}(T_{\mathrm{D},\mathbb{R}^n}) = [\mu, \infty)$. ■

Proof. We first observe that by subdividing the cubes Q if necessary we have for each $Q \in \mathfrak{F}$ and with γ_r as defined in (VII.1.42)

$$\gamma_r |Q|^{2/n - 1/r} \|q - q_Q\|_{r,Q} \leqslant 2\gamma_r |Q|^{2/n - 1/r} \|q\|_{r,Q} < 1$$

since $r \geqslant \frac{1}{2}n$. Hence, on account of (4.11),

$$\gamma_r \sup_{Q \in \mathfrak{F}} (|Q|^{2/n-1/r} \|q - q_Q\|_{r,Q}) < 1$$

and the hypotheses of Theorem VII.1.15(a) are satisfied for q real; $T_{D,\Omega}$ is therefore defined. Next, we set, for each $Q \in \mathfrak{F}$,

$$q_0(x) = q_Q \quad (x \in Q), \quad h(x) = q(x) - q_Q \quad (x \in Q).$$

By (4.11), h satisfies (4.6) and so, by Theorem 4.1, $T_{D,\Omega}$ has the same essential spectrum as the operator generated by $-\Delta + q_0$. But in view of the Decomposition Principle the essential spectrum of the latter operator coincides with that of $-\Delta_{D,\Omega} + \mu$. Indeed, since (4.10) implies, for any $\varepsilon > 0$, that $|q_0(x) - \mu| < \varepsilon$ for $|x|$ large enough, we may assume that $q_0(x) \in (\mu - \varepsilon, \mu + \varepsilon)$ throughout Ω, by the Decomposition Principle. Putting $q_0 = \mu + (q_0 - \mu)$ and observing that since $|q_0(x) - \mu| < \varepsilon$, the function $q_0 - \mu$ satisfies (4.2) and (4.3) trivially, we see that the result follows from Theorem 4.1.

If $\Omega = \mathbb{R}^n$ then $\sigma_e(-\Delta_{D,\mathbb{R}^n} + \mu) = [\mu, \infty)$ as we saw in Theorem IX.6.2. □

The essential spectrum of $-\Delta_{D,\Omega} + \mu$ for arbitrary Ω will be the subject of §7 below.

5. A result of Persson

A useful tool for examining the least point of the essential spectrum of a semi-bounded self-adjoint operator is provided by the following theorem. It yields a generalization of a result due to Persson[1] for operators generated by $-\Delta + q$ in $L^2(\Omega)$. In this section Ω will be a domain in $\mathbb{R}^n$ and we shall assume familiarity with the concepts and notation in §V.5.2.

Theorem 5.1. Let $S \geqslant 1$ be a selfadjoint operator in $L^2(\Omega)$ and denote by X its form domain $\mathscr{D}(S^{\frac{1}{2}})$ with norm $\|u\|_X = \|S^{\frac{1}{2}}u\|_{2,\Omega}$. Let E be a dense subspace of X and, with A a closed subset of $\bar{\Omega}$, suppose there exists an A-admissible family $\mathfrak{U}_A$ of domains in Ω which is compact with respect to the embedding $I_0: X \to L^2(\Omega)$. Finally, if $0 < \varepsilon_2 < \varepsilon_1$ assume there is a $\phi \in C_0^\infty(\mathbb{R}^n \setminus A)$ which satisfies $0 \leqslant \phi \leqslant 1$, $\phi(x) = 0$ in $A(\varepsilon_2)$ and $\phi(x) = 1$ in $\Omega \setminus A(\varepsilon_1)$, and is such that $u \mapsto \phi u$ maps E into E and $\|(1-\phi)v_n\|_X \leqslant 1 + o(1)$ as $n \to \infty$ whenever $(v_n) \subset E$ satisfies $\|v_n\|_X = 1$ and $v_n \rightharpoonup 0$ in X.

Then, if

$$\Gamma_{I_0}^*(\varepsilon, A) := \sup\{\|I_0 u\|_{2,A(\varepsilon)}^2 : u \in E, \|u\|_X = 1, \operatorname{supp} u \subset A(\varepsilon)\}$$

and

$$\Gamma_{I_0}^*(0, A) := \lim_{\varepsilon \to 0+} \Gamma_{I_0}^*(\varepsilon, A)$$

we have

$$\mathrm{r_e}(S^{-1}) = \tilde{\beta}(S^{-\frac{1}{2}})^2 = \beta(S^{-\frac{1}{2}})^2 = \Gamma^*_{I_0}(0, A) = \Gamma_{I_0}(0, A)$$

in the notation of §V.5, $\mathrm{r_e}(S^{-1}) = \sup\{|\lambda| : \lambda \in \sigma_e(S^{-1})\}$ being the radius of the essential spectrum of S^{-1}. ■

Proof. By Theorem V.5.7, Theorem I.2.17, Corollary I.4.9 and Theorem I.4.10,

$$\begin{aligned}\Gamma^*_{I_0}(0, A) \leqslant \Gamma_{I_0}(0, A) &= \tilde{\beta}(I_0)^2 = \tilde{\beta}(S^{-\frac{1}{2}})^2 \\ &= \beta(S^{-\frac{1}{2}})^2 = \mathrm{r}_e^2(S^{-\frac{1}{2}}) \qquad (5.1)\end{aligned}$$

since $S^{-\frac{1}{2}}$ is the composition of an isometry of $\mathrm{L}^2(\Omega)$ onto X and I_0.

Let r_e stand for $\mathrm{r_e}(S^{-\frac{1}{2}})$. Given $\delta > 0$ there exist $\lambda \in \sigma_e(S^{-\frac{1}{2}})$ such that

$$r_e^2 - \delta < |\lambda|^2 \leqslant r_e^2$$

and a corresponding singular sequence $(u_n) \subset \mathrm{L}^2(\Omega)$ satisfying $\|u_n\|_{2,\Omega} = 1$, $u_n \rightharpoonup 0$, and $(S^{-\frac{1}{2}} - \lambda I)u_n \to 0$ in $\mathrm{L}^2(\Omega)$. On setting $v_n = S^{-\frac{1}{2}}u_n \in \mathscr{D}(S^{\frac{1}{2}}) = X$, we have $\|v_n\|_X = 1$, $v_n \rightharpoonup 0$ in X, and $I_0 v_n - \lambda u_n \to 0$ in $\mathrm{L}^2(\Omega)$. Thus for n large enough, $n \geqslant n_0$ say,

$$\begin{aligned}\|I_0 v_n\|^2_{2,\Omega} &= |\lambda| \, \|u_n\|^2_{2,\Omega} + o(1) \\ &\geqslant r_e^2 - \tfrac{3}{2}\delta.\end{aligned}$$

Since E is dense in X, there therefore exists a sequence $(w_n) \subset E$ satisfying $\|w_n\|_X = 1$, $w_n \rightharpoonup 0$ in X, and $\|I_0 w_n\|^2_{2,\Omega} \geqslant r_e^2 - 2\delta$ for $n \geqslant n_0$. Choose $\varepsilon_1 > 0$ such that $\Gamma^*_{I_0}(\varepsilon_1, A) \leqslant \Gamma^*_{I_0}(0, A) + \delta$ and let $0 < \varepsilon_2 < \varepsilon$. Since $\mathfrak{U}_A$ is I_0-compact we have

$$\|I_0 w_n\|_{2, \Omega \setminus A(\varepsilon_2)} \to 0$$

as $n \to \infty$, and so for $n \geqslant n_1$ say,

$$\|I_0 w_n\|^2_{2, A(\varepsilon_2)} \geqslant r_e^2 - 3\delta.$$

With ϕ as in the statement of the theorem we obtain

$$\begin{aligned}r_e^2 - 3\delta &\leqslant \|I_0 w_n\|^2_{2, A(\varepsilon_2)} \\ &= \|I_0[(1-\phi)w_n]\|^2_{2, A(\varepsilon_2)} \\ &\leqslant [\Gamma^*_{I_0}(0, A) + \delta] \, \|(1-\phi)w_n\|^2_X \\ &\leqslant [\Gamma^*_{I_0}(0, A) + \delta](1+\delta)^2\end{aligned}$$

for large enough n. Since δ is arbitrary we have $r_e^2 \leqslant \Gamma^*_{I_0}(0, A)$ and when this is coupled with (5.1) the result follows as $r_e^2 = \mathrm{r_e}(S^{-1})$. □

Corollary 5.2. Let $q \in \mathrm{L}^1_{\mathrm{loc}}(\Omega)$ be real and with $A = \partial\Omega$ or $\varnothing$ (when Ω is unbounded) let $\mathfrak{U}_A = \{U_r : 0 < r < \infty\}$ be an A-admissible family of domains in Ω which is such that $U_r \subset \bar{U}_r \subset U_s$ whenever $r < s$. Suppose that

(i) $\lim\limits_{\varepsilon \to 0} \inf\limits_{x \in A(\varepsilon)} q(x) =: q_0 > -\infty$,

(ii) given $\delta > 0$ and $0 < r < s$ there exists a constant $K_{\delta,r,s} > 0$ such that

$$(q^- u, u)_{2,U_r} \leqslant \delta \|\nabla u\|_{2,U_s}^2 + K_{\delta,r,s} \|u\|_{2,U_s}^2 \qquad (u \in C_0^\infty(\Omega)).$$

Then the Dirichlet form t_0 given by

$$t_0[\![u, v]\!] = \int_\Omega (\nabla u \cdot \nabla \bar{v} + q u \bar{v}) \qquad (u, v \in C_0^\infty(\Omega))$$

is bounded below and closable in $L^2(\Omega)$. If T is the semi-bounded self-adjoint operator associated with the closure t of t_0, then

$$\theta_e := \inf\{\lambda : \lambda \in \sigma_e(T)\}$$

$$= \lim_{\varepsilon \to 0+} \inf\left\{\int_{A(\varepsilon)} (|\nabla u|^2 + q|u|^2) : u \in C_0^\infty(\Omega), \|u\|_{2,\Omega} = 1, \text{supp } u \subset A(\varepsilon)\right\}. \quad \blacksquare$$

Proof. If r is so large that $q(x) > q_0 - 1$ in $\Omega \setminus U_r$, we see from (i) and (ii) that for any $u \in C_0^\infty(\Omega)$ and any $\delta > 0$,

$$t_0[\![u]\!] \geqslant \int_\Omega |\nabla u|^2 - |q_0 - 1| \int_{\Omega \setminus U_r} |u|^2 - \int_{U_r} q^- |u|^2 + \int_\Omega q_+ |u|^2$$

$$\geqslant (1-\delta) \int_\Omega |\nabla u|^2 - (|q_0 - 1| + K_{\delta,r,s}) \|u\|_{2,\Omega}^2 + \int_\Omega q^+ |u|^2.$$

Thus t_0 is bounded below and there exist positive constants K_1, K_2, K_3 such that

$$K_1 \int_\Omega (|\nabla u|^2 + q^+ |u|^2 + |u|^2) \leqslant (t_0 + K_2)[\![u]\!]$$

$$\leqslant K_3 \int_\Omega (|\nabla u|^2 + q^+ |u|^2 + |u|^2). \tag{5.2}$$

It follows that t_0 is closable in $L^2(\Omega)$ and the domain $\mathfrak{Q}_D$ of its closure t is the completion of $C_0^\infty(\Omega)$ with respect to the norm

$$\|u\|_D = \left(\int_\Omega (|\nabla u|^2 + q^+ |u|^2 + |u|^2)\right)^{\frac{1}{2}}.$$

Since the addition of a constant has no influence on our result, we may suppose, without loss of generality, that the operator T associated with t satisfies $T \geqslant 1$ and also that $q_0 > 0$.

We now apply Theorem 5.1 with $S = T$ and $E = C_0^\infty(\Omega)$. Then X is continuously embedded in $W_0^{1,2}(\Omega)$ in view of (5.2), and so $\mathfrak{U}_A$ is I_0-compact. Suppose that $0 < \varepsilon_2 < \varepsilon_1$ and $\phi \in C_0^\infty(\mathbb{R}^n \setminus A)$ is such that $0 \leqslant \phi \leqslant 1$, $\phi(x) = 0$ in $A(\varepsilon_2)$, and $\phi(x) = 1$ in $U_r \in \mathfrak{U}_A$ with $\Omega \setminus A(\varepsilon_1) \subset U_r \subset \bar{U}_r \subset \bar{\Omega} \setminus A(\varepsilon_2)$. If ε_1 is sufficiently small, $q(x) > 0$ in $A(\varepsilon_1)$ and for all $u \in C_0^\infty(\Omega)$,

$$\begin{aligned}\|(1-\phi)u\|_X^2 &= \int_{\Omega\setminus U_r} (|\nabla[(1-\phi)u]|^2 + q|(1-\phi)u|^2)\\ &\leqslant \int_{\Omega\setminus U_r} (1-\phi)^2(|\nabla u|^2 + q|u|^2) + 2\int_{U_s} (1-\phi)|\nabla\phi|\,|u\nabla u|\\ &\quad + \int_{U_s} |\nabla\phi|^2|u|^2,\end{aligned}$$

(where $U_s \in \mathfrak{U}_A$ and $U_s \supset \Omega \setminus A(\varepsilon_2)$),

$$\begin{aligned}&\leqslant \int_\Omega (|\nabla u|^2 + q|u|^2) - \int_{U_r} q|u|^2 + \delta\int_{U_s}|\nabla u|^2 + K_\delta\int_{U_s}|u|^2\\ &\leqslant \|u\|_X^2 + 2\delta\|\nabla u\|_{2,U_s}^2 + K_\delta\|u\|_{2,U_s}^2\end{aligned}$$

(on using (ii)),

$$\leqslant (1+\delta K)\|u\|_X^2 + K_\delta\|u\|_{2,U_s}^2,$$

by (5.2) and the assumption $T \geqslant 1$. Consequently, if $(v_n) \subset C_0^\infty(\Omega)$ satisfies $\|v_n\|_X = 1$ and $v_n \rightharpoonup 0$ in X, we conclude that

$$\|(1-\phi)v_n\|_X^2 \leqslant 1 + \delta K + o(1)$$

since $\mathfrak{U}_A$ is I_0-compact. As δ is arbitrary,

$$\|(1-\phi)v_n\|_X \leqslant 1 + o(1)$$

and Theorem 5.1 applies. The corollary is then a consequence of the Spectral Mapping Theorem IX.2.3. □

One immediate consequence of Corollary 5.2 is that $\theta_e \geqslant q_0$. Other applications will be made in the next section.

6. The essential spectrum of $-\Delta_{D,\Omega}$

We shall show how $\sigma_e(-\Delta_{D,\Omega})$ depends on the ultimate nature of the domain Ω. The lower bounds we obtain for the least point in $\sigma_e(-\Delta_{D,\Omega})$ are equivalent to upper bounds for the ball and set measures of non-compactness of the embedding $W_0^{1,2}(\Omega) \to L^2(\Omega)$. We have two ways of describing the relevant properties of Ω.

6.1. *Quasi-conical, quasi-cylindrical, and quasi-bounded domains*

In this subsection the significant features of Ω as regards $\sigma_e(-\Delta_{D,\Omega})$ are described in the terms introduced by Glazman[1, §49]. We differ from Glazman in that we use cubes instead of balls in the definitions.

Definition 6.1. Let Ω be a domain in $\mathbb{R}^n$, with $n \geqslant 1$.
(i) Ω is *quasi-conical* if it contains arbitrarily large cubes.
(ii) Ω is *quasi-cylindrical* if it is not quasi-conical but there exists a $d > 0$ such that Ω contains a sequence of disjoint cubes $\bar{Q}_d$ of side d.
(iii) Ω is *quasi-bounded* if it is neither quasi-conical nor quasi-cylindrical.
We also define

$$k_d := d^{2-n} \lim_{\rho\to\infty} \inf\{\mathrm{cap}(\bar{Q}_d \cap \Omega^c, Q_{2d}) \colon Q_d \subset B_\rho^c\} \qquad (d > 0), \tag{6.1}$$

$$d_0 := \inf\{d : k_d > 0\}, \tag{6.2}$$

$$d_1 := \sup\{d : \Omega \text{ contains a sequence of disjoint cubes } \bar{Q}_d\}. \tag{6.3}$$ ■

Note that in (6.1) we have written cap for 2-cap.

Clearly Ω is quasi-conical, quasi-cylindrical or quasi-bounded according to whether $d_1 = \infty$, $0 < d_1 < \infty$, or $d_1 = 0$ respectively. Also $d_0 \geqslant d_1$ since $\mathrm{cap}(\bar{Q}_d \cap \Omega^c, Q_{2d}) = 0$ whenever $\bar{Q}_d \subset \Omega$. Consequently if Ω is quasi-conical $d_0 = \infty$, while if $d_0 = \infty$ all we can say is that there exist cubes $\bar{Q}_d$ of arbitrarily large side d which have an inessential intersection with Ω^c.

An important ingredient in what follows is the Poincaré inequality proved in Theorem VIII.2.10, namely, that there exists a constant γ_0, depending only on n, such that

$$\|v\|_{2,\Omega}^2 \leqslant \gamma_0 (d^2/k) \|\nabla v\|_{2,\Omega}^2 \qquad (v \in C_0^\infty(\Omega)) \tag{6.4}$$

if, and only if,

$$d^{2-n} \mathrm{cap}(\bar{Q}_d \cap \Omega^c, Q_{2d}) \geqslant k \tag{6.5}$$

for all cubes Q_d. If (6.5) holds for all $Q_d \subset B_\rho^c$ then (6.4) is satisfied for all $v \in C_0^\infty(\Omega \setminus B_\rho)$ with the same constant γ_0.

Theorem 6.2. Let $\theta_e = \inf\{\lambda : \lambda \in \sigma_e(-\Delta_{D,\Omega})\}$. Then
(i) if $d_1 = \infty$ then $\theta_e = 0$;
(ii) if $0 < d_0 < \infty$ then $\theta_e \geqslant k_d/\gamma_0 d^2 > 0$ for all $d > d_0$;
(iii) if $d_1 > 0$ then $\theta_e \leqslant \pi^2 n/d_1^2$;
(iv) $\theta_e = \infty$ if, and only if, $d_0 = 0$ and $\inf_{d>0} k_d > 0$. ■

Proof. From Corollary 5.2 with $A = \varnothing$ (Ω unbounded), $q = 0$ and

$U_r = B_r \cap \Omega$, we have

$$\theta_e = \lim_{\rho \to \infty} \theta_\rho$$

where

$$\theta_\rho = \inf\left\{\int_\Omega |\nabla u|^2 : u \in C_0^\infty(\Omega \setminus B_\rho),\ \|u\|_{2,\Omega} = 1\right\}. \tag{6.6}$$

We also need (see (VIII.1.4))

$$\inf\{\|\nabla\phi\|_{2,Q_d}^2 : \phi \in C_0^\infty(Q_d),\ \|\phi\|_{2,Q_d} = 1\} = \pi^2 n/d^2. \tag{6.7}$$

(i) If $Q_d \subset \Omega \setminus B_\rho$ it follows from (6.6) and (6.7) that $\theta_\rho \leqslant \pi^2 n/d^2$ and this implies $\theta_e = 0$ since d is arbitrary.

(ii) For any $d \in (d_0, \infty)$, $k_d > 0$ and hence, given any $\varepsilon \in (0, k_d)$, there exists a ρ such that

$$d^{2-n} \operatorname{cap}(\bar{Q}_d \cap \Omega^c, Q_{2d}) > k_d - \varepsilon \qquad (Q_d \subset B_\rho^c).$$

Theorem VIII.2.10 now yields

$$\|u\|_{2,\Omega}^2 \leqslant [\gamma_0 d^2/(k_d - \varepsilon)]\ \|\nabla u\|_{2,\Omega}^2 \qquad (u \in C_0^\infty(\Omega \setminus B_\rho)),$$

and so $\theta_\rho \geqslant (k_d - \varepsilon)/\gamma_0 d^2$. Since $\varepsilon > 0$ is arbitrary, $\theta_e \geqslant k_d/\gamma_0 d^2$.

(iii) This is an immediate consequence of (6.6) and (6.7).

(iv) We have that $\theta_e = \infty$ if, and only if, $-\Delta_{D,\Omega}$ has a compact resolvent, and this is equivalent to the compactness of the embedding $W^{1,2}(\Omega) \to L^2(\Omega)$, by Theorem IV.2.9. This part therefore follows from Theorem VIII.3.1. □

The space $W_0^{1,2}(\Omega)$ is the form domain of $-\Delta_{D,\Omega} + I$ and $(-\Delta_{D,\Omega} + I)^{\frac{1}{2}}$ is an isometry of $W_0^{1,2}(\Omega) \to L^2(\Omega)$, by the Second Representation Theorem IV.2.8. Since the embedding $I_0 : W_0^{1,2}(\Omega) \to L^2(\Omega)$ is the composite of this isometry and the bounded self-adjoint operator $(-\Delta_{D,\Omega} + I)^{-\frac{1}{2}}$ in $L^2(\Omega)$ we see from Theorem I.2.17 and Theorem I.4.8 and its corollaries that

$$\begin{aligned}\beta(I_0) &= \beta\left((-\Delta_{D,\Omega} + I)^{-\frac{1}{2}}\right) \\ &= \beta^{\frac{1}{2}}\left((-\Delta_{D,\Omega} + I)^{-1}\right) = (\theta_e + 1)^{-\frac{1}{2}},\end{aligned}$$

where β stands for the set or ball measure of non-compactness of I_0. Thus we have the following result.

Corollary 6.3. Let $\beta(I_0)$ denote the set or ball measure of non-compactness of the embedding $I_0 : W_0^{1,2}(\Omega) \to L^2(\Omega)$. Then:
(i) if $d_1 = \infty$ then $\beta(I_0) = 1$;
(ii) if $0 < d_0 < \infty$ then $\beta^2(I_0) \leqslant [1 + k_d/(\gamma_0 d^2)]^{-1} < 1$ for all $d > d_0$;
(iii) if $d_1 > 0$ then $\beta^2(I_0) \geqslant [1 + \pi^2 n/d_1^2]^{-1}$;
(iv) $\beta(I_0) = 0$ if, and only if, $d_0 = 0$ and $\inf_{d>0} k_d > 0$. ■

Corollary 6.4. Let Ω be unbounded, $A = \varnothing$ and $U_r = B_r \cap \Omega$ in Corollary

5.2. Then $\theta_e \geqslant q_0$; further, $\theta_e = q_0$ if $\lim_{|x|\to\infty,\, x\in\Omega_0} q(x) = q_0$ on a quasi-conical subdomain Ω_0 of Ω. ■

Proof. It is obvious that $\theta_e \geqslant q_0$. Since Ω_0 is quasi-conical there exists a cube Q_d of arbitrarily large side d in $\Omega \setminus B_\rho$ for any ρ. Given $\varepsilon > 0$, if ρ is so large that $q(x) < q_0 + \varepsilon$ in $\Omega_0 \setminus B_\rho$, we see from (6.7) that there is a $\phi \in C_0^\infty(\Omega \setminus B_\rho)$ such that $\|\phi\|_{2,\Omega} = 1$ and

$$\int_\Omega (|\nabla\phi|^2 + q|\phi|^2) < \pi^2 n/d^2 + q_0 + 2\varepsilon.$$

The result follows on allowing $d \to \infty$ and $\varepsilon \to 0+$. □

Theorem 6.2(i) is contained in the following precise result.

Theorem 6.5. If Ω is quasi-conical, $\sigma(-\Delta_{D,\Omega}) = \sigma_e(-\Delta_{D,\Omega}) = [0, \infty)$. ■

Proof. Since $-\Delta_{D,\Omega} \geqslant 0$ we have $\sigma_e(-\Delta_{D,\Omega}) \subset \sigma(-\Delta_{D,\Omega}) \subset [0,\infty)$ and so we need to prove that every positive real number lies in the essential spectrum of $-\Delta_{D,\Omega}$. We shall do this by constructing a singular sequence for $-\Delta_{D,\Omega}$ corresponding to each $\lambda \in (0, \infty)$.

For $\rho > 0$ and $j \in \mathbb{N}$, let θ_j be a $C^\infty(\mathbb{R})$ function which is such that $\theta_j(x) = 1$ for $|x| < \rho$, $\theta_j(x) = 0$ for $|x| > (j+1)\rho$, $|\theta_j'(x)| = O((j\rho)^{-1})$ and $|\theta_j''(x)| = O((j\rho)^{-2})$, the latter bounds being uniform as $j \to \infty$ and for $\rho \in (0,\infty)$. Let

$$\psi(x) = \rho^{-\frac{1}{2}} \sin(\pi x \rho^{-1}) \qquad (x \in \mathbb{R})$$

and

$$\psi_j = C_j \theta_j \psi,$$

where the constants C_j are chosen such that $\int_\mathbb{R} |\psi_j|^2 = 1$; note that this implies $|C_j| \leqslant 1$ since

$$|C_j|^2 = |C_j|^2 \int_{-\rho}^{\rho} |\psi(x)|^2 \,dx \leqslant \int_\mathbb{R} |\psi_j|^2 = 1.$$

Thus $\psi_j \in C_0^\infty(-(j+1)\rho, (j+1)\rho)$ and

$$\begin{aligned}\int_\mathbb{R} |-\psi_j'' - (\pi^2/\rho^2)\psi_j|^2 &= |C_j|^2 \int_{-(j+1)\rho}^{(j+1)\rho} |2\theta_j'\psi' + \theta_j''\psi|^2 \\ &= O\left((j\rho)^{-2}(j+1)\int_{-\rho}^{\rho} |\psi'|^2 + (j\rho)^{-4}(j+1)\int_{-\rho}^{\rho} |\psi|^2\right) \\ &= O((j\rho^4)^{-1})\end{aligned}$$

as $j \to \infty$. Since Ω is quasi-conical, for any $\rho > 0$ there exists a sequence of disjoint cubes Q_j in Ω having sides of length $2(j+1)\rho$ and parallel to the co-ordinate axes; let $a_j = (a_j^{(1)}, \ldots, a_j^{(n)})$ denote the centre of Q_j. Let

$$\Psi_j(x) = \prod_{i=1}^{n} \psi_j(x_i - a_j^{(i)}).$$

Then (Ψ_j) is orthonormal in $L^2(\Omega)$, and as $j \to \infty$,

$$\|(-\Delta - n\pi^2/\rho^2)\Psi_j\|_{2,\Omega}^2 = O\left((j\rho^4)^{-1}\right) \to 0.$$

Since $\Psi_j \in C_0^\infty(Q_j) \subset \mathscr{D}(-\Delta_{D,\Omega})$ it follows that $n\pi^2\rho^{-2} \in \sigma_e(-\Delta_{D,\Omega})$. But $\rho \in (0,\infty)$ is arbitrary and hence $(0,\infty) \subset \sigma_e(-\Delta_{D,\Omega})$. The theorem is therefore proved. □

For quasi-cylindrical domains the precise location of $\sigma_e(-\Delta_{D,\Omega})$ is difficult and depends on the regularity of Ω. A detailed analysis of this problem may be found in Glazman's book [1].

Theorem 6.6. (i) If Ω contains the cylinder $\Omega_z \times (a,\infty)$, with $a \geqslant -\infty$, where Ω_z is a bounded domain in $\mathbb{R}^{n-1}$, and if λ_z is the smallest eigenvalue of $-\Delta_{D,\Omega_z}$, then $[\lambda_z,\infty) \subset \sigma_e(-\Delta_{D,\Omega})$.

(ii) If $\Omega_z = \Omega \cap \{x = (x_1, \ldots, x_n)\colon x_n = z\}$ and $\gamma = \liminf_{z\to\infty} \lambda_z$ then $\sigma_e(-\Delta_{D,\Omega}) \subset [\gamma,\infty)$.

Hence if $\Omega = \{x = (x_1, \ldots, x_n)\colon |x_i| < \rho \text{ for } i = 1,2,\ldots,m \text{ and } x_i \in (a,\infty) \text{ for } i = m+1,\ldots,n\}$, then $\sigma_e(-\Delta_{D,\Omega}) = [\pi^2 m\rho^{-2},\infty)$. ■

Proof. (i) Let ϕ_0 be the normalized eigenvector of $-\Delta_{D,\Omega_z}$ corresponding to the smallest eigenvalue λ_z, and let ψ_j $(j \geqslant j_0)$ be the functions constructed in the proof of Theorem 6.5 having their supports in disjoint subintervals I_j of (a,∞) of lengths $2(j+1)\rho$, with $\rho \in (0,\infty)$ arbitrary. Let

$$\Psi_j(x) = \phi_0(x')\psi_j(x_n), \qquad x' = (x_1, \ldots, x_{n-1}), \qquad x = (x', x_n).$$

Then

$$[-\Delta - (\lambda_z + \pi^2/\rho^2)]\Psi(x) = [-\psi_j''(x_n) - (\pi^2/\rho^2)\psi_j(x_n)]\phi_0(x')$$

and as $j \to \infty$,

$$\|[-\Delta - (\lambda_z + \pi^2/\rho^2)]\Psi_j\|^2 = O\left((j\rho^4)^{-1}\right) \to 0.$$

Since $(\Psi_j) \subset \mathscr{D}(-\Delta_{D,\Omega})$ and is orthonormal, it follows that $\lambda_z + \pi^2/\rho^2 \in \sigma_e(-\Delta_{D,\Omega})$ and hence $[\lambda_z,\infty) \subset \sigma_e(-\Delta_{D,\Omega})$ since $\rho \in (0,\infty)$ is arbitrary.

(ii) We shall prove that every $\lambda < \gamma$ lies outside $\sigma_e(-\Delta_{D,\Omega})$. Let $\lambda' \in (\lambda,\gamma)$ and choose z_0 large enough that $\lambda' < \lambda_z$ for every $z \geqslant z_0$. If $\lambda \in \sigma_e(-\Delta_{D,\Omega})$ we conclude from the Decomposition Principle that $\lambda \in \sigma_e(-\Delta_{D,\Omega} + q)$ for any bounded function q which is zero near infinity. Hence by Theorem III.4.4, here exists $u \in \mathscr{D}(-\Delta_{D,\Omega})$ such that

$$((-\Delta_{D,\Omega}+q-\lambda')u,u)_{2,\Omega} = \int_\Omega [|\nabla u|^2 + (q-\lambda')|u|^2] < 0$$

Since the choice of q in any bounded subdomain of Ω is arbitrary, it follows that

$$\int_z^\infty \int_{\Omega_z} [|\nabla u(x',x_n)|^2 - \lambda' |u(x',x_n)|^2]\, dx'\, dx_n < 0.$$

From this we can infer that for some $z \geqslant z_0$,

$$\int_{\Omega_z} [|\nabla u(x',z)|^2 - \lambda' |u(x',z)|^2]\, dx' < 0$$

and so

$$\int_{\Omega_z} [|\nabla' u(x',z)|^2 - \lambda' |u(x',z)|^2]\, dx' < 0,$$

where $\nabla' \equiv (\partial/\partial x_1, \ldots, \partial/\partial x_{n-1})$. Since $u(\bullet,z) \in \mathscr{D}(-\Delta_{D,\Omega_z})$ this implies that $\lambda' > \lambda_z$, contrary to assumption. This proves (ii).

The final assertion of the theorem follows by repeated application of the first two parts and the observation that the smallest eigenvalue of $-\Delta_{D,\Omega}$ on the m-dimensional cube $(-\rho,\rho)^m$ is $\pi^2 m/\rho^2$. □

Since the Laplacian is invariant under translations and rotations and these transformations in $\mathbb{R}^n$ give rise to unitary maps on L^2 and $W^{1,2}$, Theorem 6.5 remains valid for domains Ω which can be transformed by translation and rotation to satisfy the requirements of the theorem.

6.2. *A mean-distance function*

Let $\xi \in \mathbb{R}^n$ with $n \geqslant 1$ and $|\xi| = 1$, and for $x \in \Omega$ define

$$\rho_\xi(x) = \min\{|t| : x + t\xi \notin \Omega\},$$

the distance from x to $\mathbb{R}^n \setminus \Omega$ in the direction of ξ. If $\{e(i): i = 1,2,\ldots,n\}$ is an orthonormal basis of $\mathbb{R}^n$ we set

$$\rho_i(x) = \rho_{e(i)}(x) \qquad (i = 1,2,\ldots,n).$$

The *mean distance* $m(x)$ from x to $\partial\Omega$ is defined by

$$\frac{1}{m(x)^2} = \frac{1}{n\omega_n} \int_{|\xi|=1} \frac{1}{\rho_\xi^2(x)}\, d\sigma(\xi) \tag{6.8}$$

where $d\sigma(\xi)$ is the usual measure on the unit sphere S^{n-1} in $\mathbb{R}^n$ and $n\omega_n$ is the measure of S^{n-1}. This function was introduced by Davies in [2] where the following important inequality is proved.

Theorem 6.7. For all $f \in C_0^\infty(\Omega)$,

$$\frac{n}{4}\int_\Omega \frac{|f(x)|^2}{m^2(x)}\,dx \leqslant \int_\Omega |\nabla f(x)|^2\,dx. \tag{6.9}$$ ■

Proof. The root of the result is the one-dimensional inequality

$$\int_a^b \frac{|\phi(t)|^2}{\rho^2(t)}\,dt \leqslant 4\int_a^b |\phi'(t)|^2\,dt \qquad (\phi \in C_0^\infty(a,b)), \tag{6.10}$$

where $\rho(t) = \min\{|t-a|, |t-b|\}$. To prove (6.10) we obtain an inequality in each half of (a,b) separately. With $c = \frac{1}{2}(a+b)$,

$$\begin{aligned}
\int_a^c \frac{|\phi(t)|^2}{(t-a)^2}\,dt &= \int_a^c (t-a)^{-2}\left(\int_a^t [|\phi(x)|^2]'\,dx\right)dt \\
&= \int_a^c [|\phi(x)|^2]'\left(\int_x^c (t-a)^{-2}\,dt\right)dx \\
&= \int_a^c \frac{[|\phi(x)|^2]'}{(x-a)}\,dx - \frac{1}{(c-a)}\int_a^c [|\phi(x)|^2]'\,dx \\
&= \int_a^c \frac{[|\phi(x)|^2]'}{(x-a)}\,dx - |\phi(c)|^2/(c-a) \\
&\leqslant 2\int_a^c \frac{|\phi(x)|\,|\phi'(x)|}{(x-a)}\,dx.
\end{aligned}$$

Similarly

$$\int_c^b \frac{|\phi(t)|^2}{(b-t)^2}\,dt \leqslant 2\int_c^b \frac{|\phi(x)|\,|\phi'(x)|}{(b-x)}\,dx$$

and the two inequalities combine to give

$$\begin{aligned}
\int_b^a \frac{|\phi|^2}{\rho^2} &\leqslant 2\int_a^b \frac{|\phi|}{\rho}|\phi'| \\
&\leqslant 2\left(\int_a^b \frac{|\phi|^2}{\rho^2}\right)^{\frac{1}{2}}\left(\int_a^b |\phi'|^2\right)^{\frac{1}{2}}
\end{aligned}$$

and hence (6.10).

If $x \in \Omega$ has co-ordinates (x_i) with respect to an orthonormal basis $\{e(i)\}$ of $\mathbb{R}^n$ we have from (6.10),

$$\int \frac{|f(x)|^2}{\rho_i^2(x)}\,dx_i \leqslant 4\int |D_i f(x)|^2\,dx_i$$

and hence

$$\int_\Omega \frac{|f(x)|^2}{\rho_i^2(x)}\,dx \leqslant 4\int_\Omega |D_i f(x)|^2\,dx.$$

This gives

$$n\int_\Omega \left(\frac{1}{n}\sum_{i=1}^{n} \rho_i^{-2}(x)\right)|f(x)|^2\,dx \leqslant 4\int_\Omega |\nabla f(x)|^2\,dx.$$

The inequality (6.9) follows on averaging over all the orthonormal bases of $\mathbb{R}^n$.

□

Remark 6.8. Theorem 6.7 implies that $f/m \in L^2(\Omega)$ for every $f \in W_0^{1,2}(\Omega)$ and also that (6.9) is satisfied for all $f \in W_0^{1,2}(\Omega)$. This complements the result in Theorem V.3.4 that

$$f \in W^{1,2}(\Omega),\ f/d \in L^2(\Omega) \quad \Rightarrow \quad f \in W_0^{1,2}(\Omega)$$

where $d(x) = \min\{|x-y| : y \notin \Omega\}$, the distance from x to $\partial\Omega$.

It is clear that $d(x) \leqslant m(x)$. If $\partial\Omega$ is sufficiently regular an inequality in the reverse direction is also available, in which case

$$d(x) \leqslant m(x) \leqslant c\,d(x) \tag{6.11}$$

for some $c > 1$. An example of this is given by Davies [2, Theorem 18]. The boundary $\partial\Omega$ is said to satisfy a *θ-cone condition* if every $x \in \partial\Omega$ is the vertex of a circular cone C_x of semi-angle θ which lies entirely in $\mathbb{R}^n \setminus \Omega$. Let $\omega(\alpha)$ denote the solid angle subtended at the origin by a ball of radius $\alpha < 1$ whose centre is at a distance 1 from the origin. Explicitly

$$\omega(\alpha) = \tfrac{1}{2}\int_0^{\arcsin\alpha} \sin^{n-2} t\,dt \Big/ \int_0^{\frac{1}{2}\pi} \sin^{n-2} t\,dt.$$

Proposition 6.9. If $\partial\Omega$ satisfies a θ-cone condition then

$$d(x) \leqslant m(x) \leqslant 2\omega^{-\frac{1}{2}}(\tfrac{1}{2}\sin\theta)\,d(x).$$ ■

Proof. Let $x \in \Omega$, $y \in \partial\Omega$ and $d(x) = |x-y|$. If ξ is the unit vector directed along the axis of the cone C_y in Ω^c, then the ball with centre $y + d(x)\xi$ and radius $d(x)\sin\theta$ lies inside C_y and hence outside Ω. The solid angle Λ subtended by this ball at x is at least $\omega(\frac{1}{2}\sin\theta)$ and every line from x within this solid angle meets $\partial\Omega$ at a distance at most $2d(x)$ from x. Consequently

$$\frac{1}{m^2(x)} \geqslant \frac{1}{n\omega_n}\int_\Lambda \frac{1}{[2d(x)]^2}\,d\sigma(\xi) \geqslant \frac{\omega(\frac{1}{2}\sin\theta)}{[2d(x)]^2}$$

and the lemma is proved. □

Remark 6.10. If we define the *mean inradius* of Ω by

$$\mu = \sup\{m(x) : x \in \Omega\},$$

Theorem 6.7 yields the lower bound $n/4\mu^2$ for the least point λ_0 of $\sigma(-\Delta_{D,\Omega})$. An upper bound can be obtained in terms of the ordinary *inradius* of Ω, namely

$$\delta = \sup\{d(x) : x \in \Omega\}.$$

The reason is that if $\rho < \delta$ then Ω contains a ball B_ρ of radius ρ and

$$\begin{aligned}\lambda_0 &\leqslant \inf\{\|\nabla f\|^2_{2,B_\rho} : f \in C_0^\infty(B_\rho), \|f\|_{2,B_\rho} = 1\}\\ &= (1/\rho^2)\inf\{\|\nabla f\|^2_{2,B_1} : f \in C_0^\infty(B_1), \|f\|_{2,B_1} = 1\}\\ &= (1/\rho^2)\, e_0,\end{aligned}$$

where e_0 is the smallest eigenvalue of $-\Delta_{D,B_1}$. We therefore have

$$n/4\mu^2 \leqslant \lambda_0 \leqslant e_0/\delta^2. \tag{6.12}$$

It follows from Corollary VIII.6.4 that when $n \geqslant 2$, the value of λ_0 is unchanged when Ω is punctured by a finite number of points. However while μ is unaffected the value of δ is reduced. Thus the mean inradius is of greater significance than the inradius δ in the determination of λ_0 when $n \geqslant 2$.

In the rest of this section we estimate the least point θ_e of $\sigma_e(-\Delta_{D,\Omega})$ in terms of the mean distance function m. Since $\theta_e = \infty$ when Ω is bounded we may assume that Ω is unbounded. First we need some preliminary results.

Lemma 6.11. Let G be an open subset of $\mathbb{R}^n$, $n \geqslant 1$. Then $W^{1,2}(\mathbb{R}^n)$ is compactly embedded in $L^2(G)$ if, and only if,

$$\lim_{|x|\to\infty} |G \cap B(x)| = 0, \tag{6.13}$$

where $B(x)$ is the unit ball with centre x. ■

Proof. If (6.13) is not satisfied there exists a sequence of unit balls $(B(x_m) : m \in \mathbb{N})$ with $B_2(x_m) = \{x : |x - x_m| < 2\}$ $(m \in \mathbb{N})$ pairwise disjoint and $|G \cap B(x_m)| \geqslant \delta$ for some $\delta > 0$. Let $\phi_m \in C_0^\infty(B_2(x_m))$ be such that $0 \leqslant \phi_m \leqslant 1$ and $\phi_m(x) = 1$ on $B(x_m)$. Then (ϕ_m) is bounded in $W^{1,2}(\mathbb{R}^n)$ while

$$\begin{aligned}\|\phi_m - \phi_k\|^2_{2,G} &= \|\phi_m\|^2_{2,B_2(x_m)\cap G} + \|\phi_k\|^2_{2,B_2(x_k)\cap G}\\ &\geqslant |B(x_m) \cap G| + |B(x_k) \cap G|\\ &\geqslant 2\delta.\end{aligned}$$

The embedding $W^{1,2}(\mathbb{R}^n) \to L^2(G)$ is therefore not compact and the necessity of (6.13) is established.

To prove (6.13) is sufficient, we consider a tesselation of $\mathbb{R}^n$ by cubes Q of side d and use the Poincaré inequality (V.3.22) on each Q. For any $v \in W^{1,2}(\mathbb{R}^n)$,

$$\|v - v_Q\|_{2,Q} \leqslant Kd\,\|\nabla v\|_{2,Q}, \qquad v_Q = |Q|^{-1}\int_Q v,$$

for some positive constant K. By (6.13), given any $\varepsilon > 0$ there exists $R > 0$ such that $|G \cap Q| < \varepsilon$ for all Q which meet $Q_R^c = \{x : |x_i| \geq R \ (i = 1,2, \ldots, n)\}$. Thus if $v \in W^{1,2}(\mathbb{R}^n)$ and $Q \cap Q_R^c \neq \emptyset$

$$\begin{aligned}\|v\|_{2,G\cap Q} &\leq \|v - v_Q\|_{2,Q} + |v_Q|\,|G \cap Q|^{\frac{1}{2}}\\ &\leq Kd\|\nabla v\|_{2,Q} + |Q|^{-\frac{1}{2}}\,|G \cap Q|^{\frac{1}{2}}\,\|v\|_{2,Q}\\ &\leq K(d + d^{-n/2}\,\varepsilon^{\frac{1}{2}})\,\|v\|_{1,2,Q}.\end{aligned}$$

On summing over the cubes and choosing $d = \varepsilon^{1/(n+2)}$, we obtain

$$\|v\|_{2,G\backslash Q_R} \leq K\,\varepsilon^{1/(n+2)}\|v\|_{1,2,\mathbb{R}^n}. \tag{6.14}$$

Since $W^{1,2}(\mathbb{R}^n) \to L^2(G \cap Q_R)$ is compact for every R it follows from (6.14) that $W^{1,2}(\mathbb{R}^n) \to L^2(G)$ is compact and the proof is complete. □

The next lemma is due to Benci and Fortunato [1].

Lemma 6.12. Let M be a measurable subset of $\mathbb{R}^n$ which is such that $\lim_{|x|\to\infty} |M \cap B(x)| = 0$. Then there exists an open set $G \supset M$ satisfying

$$\lim_{|x|\to\infty} |G \cap B(x)| = 0.$$ ■

Proof. Let B_r denote the open ball with radius r and centre at the origin and choose open sets G_k $(k \in \mathbb{N})$ to satisfy

$$G_1 = B_1, \qquad M \cap (\bar{B}_{k+1} \backslash B_k) \subset G_{k+1} \quad (k \in \mathbb{N}),$$

and

$$|G_{k+1} \backslash [M \cap (\bar{B}_{k+1} \backslash B_k)]| < 1/k \qquad (k \in \mathbb{N}). \tag{6.15}$$

We shall show that $G = \bigcup_{k\in\mathbb{N}} G_k$ has the required properties. Obviously $M \subset G$ and

$$|G \cap B(x)| \leq |M \cap B(x)| + |(G \backslash M) \cap B(x)|.$$

The result will follow if we prove that $|(G \backslash M) \cap B(x)| \to 0$ as $|x| \to \infty$. Let $x \notin \bar{B}_2$ and let m be the greatest integer such that $m \leq |x| - 1$. Then, by (6.15), as $|x| \to \infty$,

$$|(G \backslash M) \cap B(x)| \leq \left|\bigcup_{k=m+1}^{m+3} G_k \backslash M\right| < \frac{3}{m} < \frac{3}{|x| - 2}$$

and the lemma follows. □

Theorem 6.13. Let $M_\varepsilon = \{y \in \Omega : m(y) > \varepsilon\}$ and $\varepsilon_0 = \inf\{\varepsilon : \lim_{|x|\to\infty} |M_\varepsilon \cap B(x)| = 0\}$. Then $\theta_e \geq n/4\varepsilon_0^2$, $\sigma(-\Delta_{D,\Omega})$ being discrete if $\varepsilon_0 = 0$. ■

Proof. From Lemmas 6.11 and 6.12, for any $\varepsilon > \varepsilon_0$ there exists an open set $G_\varepsilon \supset M_\varepsilon$ such that $W^{1,2}(\mathbb{R}^n)$ is compactly embedded in $L^2(G_\varepsilon)$. Thus, given any $\delta \in (0,1)$ there exists a ρ such that

$$\|\phi\|^2_{2, G_\varepsilon \setminus B_\rho} \leqslant \delta \|\phi\|^2_{1,2,\mathbb{R}^n} \qquad (\phi \in W^{1,2}(\mathbb{R}^n)). \tag{6.16}$$

For $\phi \in C_0^\infty(\Omega \setminus B_\rho)$,

$$\begin{aligned} \|\phi\|^2_{2,\Omega} &= \int_{\Omega \cap M_\varepsilon} |\phi|^2 + \int_{\Omega \setminus M_\varepsilon} |\phi|^2 \\ &\leqslant \int_{\Omega \cap G_\varepsilon} |\phi|^2 + \varepsilon^2 \int_\Omega \frac{|\phi|^2}{m^2} \\ &\leqslant \delta \|\phi\|^2_{1,2,\Omega} + (4\varepsilon^2/n) \|\nabla\phi\|^2_{2,\Omega}, \end{aligned}$$

on using (6.16) and (6.9). This yields

$$(\delta + 4n^{-1}\varepsilon^2) \|\nabla\phi\|^2_{2,\Omega} \geqslant (1-\delta) \|\phi\|^2_{2,\Omega}$$

and so

$$\theta_\rho \geqslant (1-\delta)/(\delta + 4n^{-1}\varepsilon^2)$$

in the notation of (6.6). Since $\varepsilon > \varepsilon_0$ is arbitrary the result follows. □

Corollary 6.14. The spectrum of $-\Delta_{D,\Omega}$ is discrete if $\lim_{|x|\to\infty} m(x) = 0$. Thus if (6.11) is satisfied, $\sigma(-\Delta_{D,\Omega})$ is discrete if, and only if, $\lim_{|x|\to\infty} d(x) = 0$. ■

Corollary 6.15. If $\limsup_{|x|\to\infty} m(x) = \mu_0$ and $\limsup_{|x|\to\infty} d(x) = \delta_0$, then

$$\frac{n}{4\mu_0^2} \leqslant \theta_e \leqslant \frac{e_0}{\delta_0^2},$$

where e_0 is the smallest eigenvalue of $-\Delta_{D,B_1}$. ■

Proof. The hypothesis clearly implies $\varepsilon_0 \leqslant \mu_0$ in Theorem 6.13 and hence $\theta_e \geqslant n/4\mu_0^2$. The upper bound is obtained easily from Corollary 5.2 by an argument similar to that used for (6.12). □

Since the ball (and set) measure of non-compactness $\beta(I_0)$ of the embedding I_0: $W_0^{1,2}(\Omega) \to L^2(\Omega)$ satisfies $\beta^2(I_0) = 1/(\theta_e + 1)$, we also obtain from Theorem 6.13

Corollary 6.16. In the notation of Theorem 6.13,

$$\beta^2(I_0) \leqslant e_0^2/(\tfrac{1}{4}n + \varepsilon_0^2).$$

If $\limsup_{|x|\to\infty} m(x) = \mu_0$ and $\limsup_{|x|\to\infty} d(x) = \delta_0$, then

$$\delta_0^2/(e_0 + \delta_0^2) \leqslant \beta^2(I_0) \leqslant \mu_0^2/(\tfrac{1}{4}n + \mu_0^2).$$ ■

The method used in the proof of Theorem 6.13 also gives the following result for the lower semi-bounded self-adjoint operator T associated with the form

$$t[\![u,v]\!] = \int_{\Omega} (\nabla u \cdot \nabla \bar{v} + qu\bar{v}) \qquad (u,v \in C_0^{\infty}(\Omega)),$$

under the assumptions $q \geqslant 1$ and $q \in L^1_{loc}(\Omega)$.

Theorem 6.17. If T is the operator defined above, then $\sigma_e(T) \subset [\theta_e, \infty)$ where

$$\theta_e \geqslant n/4\varepsilon_0^2$$

with

$$\varepsilon_0 = \inf\{\varepsilon\colon \lim_{|x|\to\infty} |M_\varepsilon \cap B(x)| = 0\}$$

and

$$M_\varepsilon = \{y \in \Omega\colon 4n^{-1}q(y) + m^{-2}(y) < \varepsilon^{-2}\}.$$

In particular, $\sigma(T)$ is discrete if

$$\lim_{|x|\to\infty} \int_{\Omega \cap B(x)} [q(y) + \tfrac{1}{4}nm^{-2}(y)]^{-1}\, dy = 0. \qquad \blacksquare$$

XI

Global and asymptotic estimates for the eigenvalues of $-\Delta + q$ when q is real

This chapter is devoted to the study of the Schrödinger operator $-\Delta + q$ where the potential q is real, our concern being largely with the distribution of eigenvalues of this operator. For example, if q is a positive smooth function defined on $\mathbb{R}^n$ and $q(x) \to \infty$ as $|x| \to \infty$, it is known that the spectrum of the self-adjoint realization of $-\Delta + q$ in $L^2(\mathbb{R}^n)$ is discrete, and that the number $N(\lambda)$ of eigenvalues of this realization which are less than λ satisfies

$$\lim_{\lambda \to \infty} \left(N(\lambda) \Big/ \int_{\mathbb{R}^n} (\lambda - q)_+^{n/2} \right) = \omega_n (2\pi)^{-n}.$$

(In this and the final chapter it is convenient to denote $f^{\pm}$ by $f_{\pm}$. Also we use $(\bullet, \bullet)_S$ and $\|\bullet\|_S$ as a shorthand for the $L^2(S)$ inner product and norm.) Again, if instead q is a smooth negative function on $\mathbb{R}^n$ which goes to zero at infinity in an appropriate manner, then the spectrum of the Schrödinger operator is known to be continuous to the right of the origin, discrete to the left; and the behaviour of $N(-\mu)$ as $\mu \to 0+$ may be estimated. We establish these results and the analogues for Dirichlet and Neumann realizations of $-\Delta + q$ on general unbounded open sets $\Omega \subset \mathbb{R}^n$; and also prove the celebrated Cwikel–Lieb–Rosenbljum inequality

$$N(0) \leqslant c \int_{\mathbb{R}^N} q_-^{n/2},$$

which holds for all potentials q in $L^{n/2}(\mathbb{R}^n)$, when $n \geqslant 3$. All this is done in a unified manner by means of a general theorem (Theorem 2.7) which can be specialized to a variety of particular cases. This theorem provides upper and lower bounds for $N(\lambda)$ when λ lies below the essential spectrum, and is obtained by standard techniques: localizations to cubes, the Max–Min Principle and the systematic use of the mean value of q on cubes. Theorem 2.7 (and its specializations to the asymptotic problems $\lambda \to \infty$ and $\mu \to 0+$ mentioned above) is obtained under conditions on the mean values of q on cubes, and it is this which makes it not directly comparable with the strongest theorems for those problems known to us (Rosenbljum [1, 2, 3, 4], Tamura [1, 2, 3]), since Rosenbljum and Tamura resort, partly, to pointwise conditions on q.

We begin with some preparatory material.

1. The Max–Min Principle for semi-bounded, self-adjoint operators

Throughout this section H will stand for a (complex) Hilbert space, and eigenvalues of self-adjoint operators acting in H will be counted according to their multiplicity and arranged in increasing order.

Lemma 1.1. Let $A:\mathscr{D}(A)\to H$, with $\mathscr{D}(A)\subset H$, be self-adjoint and bounded below and let $\lambda_1, \lambda_2, \ldots, \lambda_k$ be eigenvalues of A with corresponding orthonormal eigenvectors $e_1, e_2, \ldots, e_k$ and such that

$$\lambda_1 \leqslant \lambda_2 \leqslant \ldots \leqslant \lambda_k < \lambda_e(A) := \inf\{\lambda : \lambda\in\sigma_e(A)\}.$$

Then

$$\inf\{(A\psi,\psi) : \|\psi\| = 1, \psi\in\mathscr{D}(A)\cap[e_1,\ldots,e_k]^{\perp}\} = \inf\{\lambda : \lambda\in\sigma(A)\backslash\{\lambda_1,\ldots,\lambda_k\}\}. \qquad \blacksquare$$

Proof. Denote $[e_1, \ldots, e_k]$, the linear span of $e_1, \ldots, e_k$, by G, and observe that $H = G\oplus G^{\perp}$; let A_1 and A_2 be the restrictions of A to G and $G^{\perp}$ respectively. Clearly A_1 and A_2 are self-adjoint operators acting in the Hilbert spaces G and $G^{\perp}$ respectively, and $A = A_1\oplus A_2$. Given any $\lambda\notin\{\lambda_1, \lambda_2, \ldots, \lambda_k\}$ and any $\psi\in G$, an easy computation shows that

$$\|(A_1-\lambda I)\psi\| \geqslant \varepsilon\|\psi\|,$$

where $\varepsilon = \min\{|\lambda-\lambda_j| : 1\leqslant j\leqslant k\}$. Thus $\sigma(A_1) = \{\lambda_1, \lambda_2, \ldots, \lambda_k\}$. It is just as easy to verify that $\{\lambda_1, \lambda_2, \ldots, \lambda_k\}\cap\sigma(A_2) = \varnothing$ and that $\sigma(A) = \sigma(A_1)\cup\sigma(A_2)$. Hence $\sigma(A_2) = \sigma(A)\backslash\{\lambda_1, \lambda_2, \ldots, \lambda_k\}$, and since by Theorem III.4.4,

$$\inf\{(A_2\psi,\psi) : \psi\in\mathscr{D}(A_2), \|\psi\| = 1\} = \inf\{\lambda : \lambda\in\sigma(A_2)\},$$

the result follows. $\square$

Theorem 1.2. Let A be a self-adjoint operator acting in H which is bounded below, and for each $n\in\mathbb{N}$ define

$$\mu_n(A) = \sup_{M_{n-1}} \inf_{\psi\in\mathscr{D}(A)\cap M_{n-1}^{\perp}, \|\psi\|=1} (A\psi,\psi), \tag{1.1}$$

where the supremum is taken over all linear subspaces M_{n-1} of dimension at most $n-1$. Then for each $n\in\mathbb{N}$ we have:

(i) $\mu_n(A) < \lambda_e(A) := \inf\{\lambda\in\sigma_e(A)\}$ if, and only if, A has at least n eigenvalues less than $\lambda_e(A)$. In this case, $\mu_n(A)$ is the nth eigenvalue of A, and the infimum in (1.1) is attained when $M_{n-1} = [e_1, \ldots, e_{n-1}]$, where e_j is the jth eigenvector of A $(j = 1, 2, \ldots, n-1)$ corresponding to the jth eigenvalue; (ii) $\mu_n(A) = \lambda_e(A)$ if, and only if, A has at most $n-1$ eigenvalues less than $\lambda_e(A)$, and in this case $\mu_m(A) = \mu_n(A)$ for all $m > n$. ■

Proof. The spectrum of A is discrete in $(-\infty, \lambda_e(A))$ and so, given any $\varepsilon > 0$, the operator A has only a finite number k (depending upon ε) of eigenvalues in $(-\infty, \lambda_e(A) - \varepsilon)$. Denote them by $\lambda_1, \lambda_2, \ldots, \lambda_k$: thus $\lambda_1 \leqslant \lambda_2 \leqslant \ldots \leqslant \lambda_k < \lambda_e(A) - \varepsilon$. We first prove that A has at most $n-1$ eigenvalues less than $\mu_n(A)$. Suppose the contrary; then for some $j \geqslant n$ we have $\lambda_j < \mu_n(A)$; and then given any linear subspace M_{n-1}, there exists $\psi \in [e_1, \ldots, e_j]$ with $\psi \perp M_{n-1}$ and $\psi \neq 0$. Thus $\psi = \sum_{i=1}^{j} (\psi, e_i) e_i$ and

$$(A\psi, \psi) = \sum_{i=1}^{j} \lambda_i |(\psi, e_i)|^2 < \mu_n(A) \|\psi\|^2.$$

Since M_{n-1} is an arbitrary linear subspace of dimension at most $n-1$, it follows that $\mu_n(A) < \mu_n(A)$, which is absurd.

We now deal with the two parts of the theorem separately.

(i) Suppose that A has at least n eigenvalues in $(-\infty, \lambda_e(A))$. Then as shown above, $\mu_n(A) \leqslant \lambda_n < \lambda_e(A)$. Conversely, let us assume that $\mu_n(A) < \lambda_e(A)$. If the number k of eigenvalues less than $\lambda_e(A)$ is such that $k \leqslant n-1$ we have the contradiction

$$\begin{aligned}\mu_n(A) \geqslant \mu_{k+1}(A) &\geqslant \inf\{(A\psi, \psi) : \psi \in \mathscr{D}(A) \cap [e_1, \ldots, e_k]^\perp, \|\psi\| = 1\} \\ &= \lambda_e(A),\end{aligned}$$

the final step following from Lemma 1.1. Thus $k \geqslant n$, as asserted. Since at most $n-1$ eigenvectors are less than $\mu_n(A)$ we must have $\lambda_n \geqslant \mu_n(A)$. But from Lemma 1.1,

$$\mu_n(A) \geqslant \inf\{(A\psi, \psi) : \psi \in \mathscr{D}(A) \cap [e_1, \ldots, e_{n-1}]^\perp, \|\psi\| = 1\} = \lambda_n,$$

whence $\mu_n(A) = \lambda_n$. Moreover, in (1.1), $\mu_n(A)$ is attained when $M_{n-1} = [e_1, \ldots, e_{n-1}]$ and $\psi = e_n$.

(ii) Suppose that $\mu_n(A) = \lambda_e(A)$. Then as shown in the first paragraph of the proof, A has at most $n-1$ eigenvalues in $(-\infty, \lambda_e(A))$. Conversely, suppose that A has at most $n-1$ eigenvalues in $(-\infty, \lambda_e(A))$. If $\lambda_j \geqslant \mu_n(A)$ for some $j \leqslant n-1$, then $\mu_n(A) \geqslant \mu_{j+1}(A) \geqslant \lambda_j \geqslant \mu_n(A)$, whence $\lambda_j = \mu_n(A)$. Thus from Lemma 1.1, $\mu_n(A) \geqslant \lambda_e(A)$. We claim that $\mu_m(A) \leqslant \lambda_e(T)$ if $m \geqslant n$. Granted this, and since $\mu_m(A) \geqslant \mu_n(A)$ if $m \geqslant n$, it will follow that $\mu_m(A) = \mu_n(A) = \lambda_e(T)$.

To justify our claim, let $\lambda \in \sigma_e(A)$ and let M_{m-1} be any linear subspace of H with $\dim M_{m-1} = r \leqslant m-1$. Since A is bounded below, there exists $c < \lambda$ such that $(A-cI)^{-1} \in \mathscr{B}(H)$, and by the Spectral Mapping Theorem IX.2.3, $(\lambda-c)^{-1} \in \sigma_e((A-cI)^{-1})$. Hence, as $(A-cI)^{-1}$ is self-adjoint, it follows from Theorem IX.1.3 that there is a singular sequence (ψ_k) of $(A-cI)^{-1}$ corresponding to $(\lambda-c)^{-1}$:

$$\left.\begin{aligned} \|\psi_k\| = 1 \quad (k \in \mathbb{N}), \\ \psi_k \rightharpoonup 0 \quad \text{and} \quad [(A-cI)^{-1} - (\lambda-c)^{-1} I]\psi_k \to 0 \quad \text{as } k \to \infty. \end{aligned}\right\} \tag{1.2}$$

Let $N_{m-1} = (A-cI)^{-1}M_{m-1}$. Then $\dim N_{m-1} = \dim M_{m-1} = r$ and we may choose an orthonormal basis $\{u_1, \ldots, u_r\}$ for N_{m-1}. Let θ_k be the orthogonal projection of ψ_k on $N_{m-1}^{\perp}$:

$$\theta_k = \psi_k - \sum_{j=1}^{r} (\psi_k, u_j) u_j \qquad (k \in \mathbb{N}).$$

From (1.2) it follows that as $k \to \infty$,

$$\theta_k \rightharpoonup 0, \qquad \|\theta_k\| \to 1, \qquad [(A-cI)^{-1} - (\lambda-c)^{-1} I]\,\theta_k \to 0. \tag{1.3}$$

Now put $\eta_k = (\lambda-c)(A-cI)^{-1}\theta_k$. Then $\eta_k \in \mathscr{D}(A) \cap M_{m-1}^{\perp}$, and from (1.3) we see that as $k \to \infty$,

$$\|\eta_k\| \to 1, \qquad (A-\lambda I)\eta_k = (\lambda-c)\,\theta_k - (\lambda-c)^2 (A-cI)^{-1}\,\theta_k \to 0.$$

We conclude that

$$\inf\{(A\psi, \psi) : \psi \in \mathscr{D}(A) \cap M_{m-1}^{\perp}, \|\psi\| = 1\} \leqslant \lambda$$

and, since λ and M_{m-1} are arbitrary, $\mu_m(A) \leqslant \lambda_e(A)$. The proof is therefore complete. □

It will be convenient to have a version of Theorem 1.2 in which the form and form domain of A appear, and we give this next.

Theorem 1.3. Let A, $\mu_n(A)$ and M_{n-1} be as in Theorem 1.2. Then for each $n \in \mathbb{N}$,

$$\mu_n(A) = \sup_{M_{n-1}} \inf_{\psi \in \mathscr{Q}(A) \cap M_{n-1}^{\perp}, \|\psi\| = 1} a[\![\psi]\!], \tag{1.4}$$

where $a[\![\bullet]\!]$ and $\mathscr{Q}(A)$ are the form and form domain of A respectively. ■

Proof. Denote the right-hand side of (1.4) by $\tilde{\mu}_n(A)$. Since $\mathscr{D}(A) \subset \mathscr{Q}(A)$ it follows that $\tilde{\mu}_n(A) \leqslant \mu_n(A)$. With the notation of Theorem 1.2 put $M_k = [e_1, \ldots, e_k]$; we claim that

$$\inf\{a[\![\psi]\!] : \psi \in M_k^{\perp} \cap \mathscr{Q}(A), \|\psi\| = 1\} = \inf\{(A\psi, \psi) : \psi \in M_k^{\perp} \cap \mathscr{D}(A), \|\psi\| = 1\}. \tag{1.5}$$

To justify this first note that if $\psi \in M_k^{\perp} \cap \mathscr{Q}(A)$, there is a sequence (ψ_j) in $\mathscr{D}(A)$

such that $\psi_j \to \psi$ in $\mathcal{Q}(A)$; and since $\mathcal{Q}(A)$ is continuously embedded in H we also have $\psi_j \to \psi$ in H. For each $j \in \mathbb{N}$ put

$$\phi_j = \psi_j - \sum_{i=1}^{k} (\psi_j, e_i) e_i.$$

Clearly $\phi_j \in M_k^\perp \cap \mathcal{D}(A)$, and with $\mathcal{Q} = \mathcal{Q}(A)$,

$$\|\phi_j - \psi\|_{\mathcal{Q}} \leqslant \|\psi_j - \psi\|_{\mathcal{Q}} + \left(\sum_{i=1}^{k} |(\psi_j, e_i)|^2 \|e_i\|_{\mathcal{Q}}^2 \right)^{\frac{1}{2}} \to 0$$

since $(\psi_j, e_i) \to (\psi, e_i) = 0$. It follows that $\phi_j \to \psi$ in H, and also

$$(A\phi_j, \phi_j)/\|\phi_j\|^2 = a[\![\phi_j]\!]/\|\phi_j\|^2 \to a[\![\psi]\!]/\|\psi\|^2.$$

Hence

$$\inf\{a[\![\psi]\!] : \psi \in M_k^\perp \cap \mathcal{Q}(A), \|\psi\| = 1\} \geqslant \inf\{(A\psi, \psi) : \psi \in M_k^\perp \cap \mathcal{D}(A), \|\psi\| = 1\}.$$

Since $\mathcal{D}(A) \subset \mathcal{Q}(A)$, the reverse inequality is obvious, and our claim (1.5) is established.

If $\mu_n(A) < \lambda_e(A)$, we know from Theorem 1.2 that $\mu_n(A)$ is attained with $M_{n-1} = [e_1, \ldots, e_{n-1}]$. Hence by (1.5),

$$\begin{aligned} \mu_n(A) &= \inf\{(A\psi, \psi) : \psi \in M_{n-1}^\perp \cap \mathcal{D}(A), \|\psi\| = 1\} \\ &= \inf\{a[\![\psi]\!] : \psi \in M_{n-1}^\perp \cap \mathcal{Q}(A), \|\psi\| = 1\} \\ &\leqslant \tilde{\mu}_n(A). \end{aligned}$$

Since the reverse inequality holds, it follows that $\mu_n(A) = \tilde{\mu}_n(A)$. If $\mu_n(A) = \lambda_e(A)$, A has at most $n-1$ eigenvalues to the left of $\lambda_e(A)$, with corresponding eigenvectors $e_1, \ldots, e_j$ $(j \leqslant n-1)$. With $M_j = [e_1, \ldots, e_j]$ we see with the aid of Lemma 1.1 that

$$\begin{aligned} \tilde{\mu}_n(A) &\geqslant \inf\{a[\![\psi]\!] : \psi \in M_j^\perp \cap \mathcal{Q}(A), \|\psi\| = 1\} \\ &= \inf\{(A\psi, \psi) : \psi \in M_j^\perp \cap \mathcal{D}(A), \|\psi\| = 1\} = \lambda_e(A). \end{aligned}$$

This shows that again $\tilde{\mu}_n(A) = \mu_n(A)$, and the theorem is proved. □

To conclude this section we give the well-known Rayleigh–Ritz lemma and use it to discuss the bound states of the Schrödinger operator.

Lemma 1.4 (Rayleigh–Ritz). Let A be a lower semi-bounded self-adjoint operator in H. Let V be an n-dimensional subspace of $\mathcal{D}(A)$ and let P be the orthogonal projection of H onto V. Let $A_V = \mathrm{PAP}$ and let $\hat{\mu}_1, \hat{\mu}_2, \ldots, \hat{\mu}_n$ be the eigenvalues of $A_V \restriction V$, with $\hat{\mu}_1 \leqslant \hat{\mu}_2 \leqslant \ldots \leqslant \hat{\mu}_n$. Then

$$\mu_m(A) \leqslant \hat{\mu}_m \qquad (m = 1, 2, \ldots, n),$$

where $\mu_n(A)$ is defined in (1.1). In particular, if A has eigenvalues (counting multiplicity) $\lambda_1, \lambda_2, \ldots, \lambda_k$ below its essential spectrum, with $\lambda_1 \leqslant \lambda_2 \leqslant \ldots \leqslant \lambda_k$, then

$$\lambda_m \leqslant \hat{\mu}_m \qquad (m = 1, 2, \ldots, \min\{k, n\}). \qquad \blacksquare$$

Proof. By Theorem 1.2, the eigenvalues of $A_V \restriction V$ satisfy

$$\hat{\mu}_m = \sup_{\phi_1, \ldots, \phi_{m-1} \in V} \inf_{\substack{\psi \in V; \|\psi\| = 1 \\ \psi \in [\phi_1, \ldots, \phi_{m-1}]^\perp}} (A\psi, \psi)$$

$$= \sup_{\phi_1, \ldots, \phi_{m-1} \in H} \inf_{\substack{\psi \in V; \|\psi\| = 1 \\ \psi \in [P\phi_1, \ldots, P\phi_{m-1}]^\perp}} (A\psi, \psi)$$

$$= \sup_{\phi_1, \ldots, \phi_{m-1} \in H} \inf_{\substack{\psi \in V; \|\psi\| = 1 \\ \psi \in [\phi_1, \ldots, \phi_{m-1}]^\perp}} (A\psi, \psi)$$

(since $(\psi, P\phi_i) = (P\psi, \phi_i) = (\psi, \phi_i)$)

$$\geqslant \sup_{\phi_1, \ldots, \phi_{m-1} \in H} \inf_{\substack{\psi \in \mathcal{D}(A), \|\psi\| = 1 \\ \psi \in [\phi_1, \ldots, \phi_{m-1}]^\perp}} (A\psi, \psi)$$

$$= \mu_m(A). \qquad \square$$

If q is a real function which satisfies the conditions of Theorem X.4.8 with $\mu = 0$, the Dirichlet operator T defined by $-\Delta + q$ in $L^2(\mathbb{R}^n)$ is bounded below and its essential spectrum coincides with $[0, \infty)$. The question arises whether T has a finite or infinite number of negative eigenvalues? These negative eigenvalues are called the *bound states* of T. The answer depends on the behaviour of the potential q. To investigate the problem we use the Rayleigh–Ritz method of Lemma 1.4.

Theorem 1.5. Let q satisfy the hypothesis of Theorem X.4.8 in $L^2(\mathbb{R}^n)$ with $\mu = 0$ and let T denote the lower-semi-bounded Dirichlet operator determined by $-\Delta + q$.

(a) Suppose that $q(x) \leqslant -a|x|^{-2+\varepsilon}$ if $|x| > R_0$, for some positive numbers a, ε, and R_0. Then T has an infinite number of negative eigenvalues.

(b) Suppose that $n \geqslant 3$ and $q(x) \geqslant -\frac{1}{4}(n-2)^2 b|x|^{-2}$ if $|x| > R_0$, for some positive R_0 and some $b \in [0, 1)$. Then T has only a finite number of negative eigenvalues. $\blacksquare$

Proof. (a) Since $\sigma_e(T) = [0, \infty)$, it is sufficient to show that $\mu_m(T) < 0$ for each $m \in \mathbb{N}$ in Theorem 1.2. Let $\psi \in C_0^\infty(\mathbb{R}^n)$ be a function with support in $\{x : 1 < |x| < 2\}$ and such that $\|\psi\| = 1$. Then $\psi_R(x) = R^{-n/2}\psi(xR^{-1})$ has support in $\{x: R < |x| < 2R\}$ and satisfies $\|\psi_R\| = 1$. Furthermore, if $R > R_0$,

$$\begin{aligned}(T\psi_R, \psi_R) &= (-\Delta\psi_R, \psi_R) + (q\psi_R, \psi_R)\\ &\leqslant (-\Delta\psi_R, \psi_R) - a\int |x|^{-2+\varepsilon}|\psi_R(x)|^2\,dx\\ &= R^{-2}(-\Delta\psi, \psi) - aR^{-2+\varepsilon}\int |y|^{-2+\varepsilon}|\psi(y)|^2\,dy.\end{aligned}$$

Since $\varepsilon > 0$, the last expression is negative for R large enough, $R > R_1$ say. Now let $\phi_m = \psi_{2^m R_1}$ $(m = 1, 2, \ldots)$. The ϕ_m are orthonormal with disjoint supports and so $(T\phi_m, \phi_k) = 0$ if $m \neq k$. Thus, given N and defining V_N to be the linear span of $\{\phi_1, \ldots, \phi_N\}$ we have that if P_N is the orthogonal projection onto V_N then $P_N T P_N \restriction V_N$ has eigenvalues $\{(T\phi_m, \phi_m): m = 1, 2, \ldots, N\}$. By Lemma 1.4 it therefore follows that

$$\mu_N(T) \leqslant \sup\{(T\phi_m, \phi_m): m = 1, 2, \ldots, N\} < 0.$$

Since N is arbitrary, T has an infinite number of negative eigenvalues. (b) For $\phi \in \mathscr{D}(T)$, we have from the Hardy inequality (VII.1.54)

$$\begin{aligned}\int (|\nabla\phi|^2 + q|\phi|^2) &= \int [(1-b)\,|\nabla\phi(x)|^2 + q(x) + \tfrac{1}{4}(n-2)^2\, b|x|^{-2})\,|\phi(x)|^2]\,dx\\ &\quad + b\int [|\nabla\phi(x)|^2 - \tfrac{1}{4}(n-2)^2\,|x|^{-2}\,|\phi(x)|^2]\,dx\\ &\geqslant \int [(1-b)|\nabla\phi(x)|^2 + q(x) + \tfrac{1}{4}(n-2)^2\, b|x|^{-2})\,|\phi(x)|^2]\,dx\\ &\geqslant \int [(1-b)|\nabla\phi|^2 + \tilde{q}|\phi|^2]\end{aligned}$$

where $\tilde{q} = \chi_{R_0} q$, with χ_{R_0} the characteristic function of the ball B_{R_0}. Thus, by the Max–Min Principle of Theorem 1.3,

$$0 \geqslant \mu_m(T) \geqslant \mu_m(-(1-b)\Delta + \tilde{q}) = (1-b)\mu_m(-\Delta + (1-b)^{-1}\tilde{q}).$$

Since $q \in L^{n/2}_{\text{loc}}(\mathbb{R}^n)$ and $\tilde{q}$ has compact support, it follows that $\tilde{q} \in L^{n/2}(\mathbb{R}^n)$ and as we shall prove in Theorem 5.4 below, this implies that $-\Delta + (1-b)^{-1}\tilde{q}$ has only a finite number, N_0 say, of negative eigenvalues. Thus, for $m \geqslant N_0$,

$$\mu_m(-\Delta + (1-b)^{-1}\tilde{q}) = 0 \quad \text{for } m \geqslant N_0$$

and consequently

$$\mu_m(T) = 0 \qquad (m \geqslant N_0).$$

Therefore T has at most N_0 negative eigenvalues. □

Let q be continuous and define

$$\check{q}(r) = \min_{|x|=r} q(x), \qquad \hat{q}(r) = \max_{|x|=r} q(x).$$

In Glazman [1, §48 Theorem 6] it is proved that if

$$\limsup_{r\to\infty} r^2\hat{q}(r) < -\tfrac{1}{4}(n-2)^2$$

then T has an infinite number of negative eigenvalues, while if

$$\liminf_{r\to\infty} r^2 q(r) > -\tfrac{1}{4}(n-2)^2$$

there is only a finite number of negative eigenvalues. The method of proof is similar to that used in Theorem 1.5 above and is linked with a well known criterion due to Kneser for an ordinary differential equation of second-order to be oscillatory. The intimate connection between the negative spectrum of T and the oscillatory behaviour of solutions of associated equations is investigated by Allegretto in [1].

2. Bounds for N(λ, T_D, Ω) and N(λ, T_N, Ω)

2.1. *The operators $T_{D,\Omega}$ and $T_{N,\Omega}$*

Let Ω be an open set in $\mathbb{R}^n$ and let q be a real-valued function on Ω. Throughout, $\mathfrak{F}$ will stand for a family of closed cubes in $\mathbb{R}^n$ with disjoint interiors, sides parallel to the coordinate axes and such that $\Omega \subset \cup_{Q\in\mathfrak{F}} Q$. The choice of $\mathfrak{F}$ will depend upon the problem studied: for the Dirichlet problem the cubes in $\mathfrak{F}$ may be assumed to be congruent in appropriate cases, although in important examples (§§4, 5) this is not so; for the Neumann problem, unless Ω can be written as the interior of the union of congruent cubes (as when Ω is a half-space) it will be assumed that $\mathfrak{F}$ is a *Whitney covering* of Ω in which $|Q| \to 0$ as the centre x_Q of Q approaches $\partial\Omega$ and $\Omega = \cup_{Q\in\mathfrak{F}} Q$. The existence of such Whitney coverings is ensured by Theorem 3 in Stein [1, Chapter I].

We shall frequently use a number s such that

$$s \in [\tfrac{1}{2}n, \infty] \text{ if } n \geqslant 3, \quad s \in (1, \infty] \text{ if } n = 2, \quad s \in [1, \infty] \text{ if } n = 1. \tag{2.1}$$

As in Chapter VII we write

$$t[\![u, v]\!] = \int_\Omega (\nabla u \cdot \nabla \bar{v} + qu\bar{v}) \tag{2.2}$$

for appropriate functions u and v, and if $q \in L^1_{\mathrm{loc}}(\Omega)$ we denote by $\mathfrak{D}_{D,\Omega}$ the completion of $C_0^\infty(\Omega)$ with respect to the metric induced by the norm $|\!|\!|\bullet|\!|\!|_{t,\Omega}$,

where

$$|||u|||_{t,\Omega}^2 = \|\nabla u\|_{2,\Omega}^2 + \|u\|_{2,\Omega}^2 + (q^+u, u)_{2,\Omega}; \tag{2.3}$$

if $q \in L^1_{\text{loc}}(\bar{\Omega})$ then $\mathfrak{Q}_{N,\Omega}$ is defined to be the completion, with respect to the metric induced by $|||\bullet|||_{t,\Omega}$, of the set of all functions on Ω which are restrictions to Ω of functions in $C_0^\infty(\mathbb{R}^n)$. It has already been shown (cf. Chapter VII, §1) that under appropriate conditions on q the sesquilinear forms t_D and t_N defined by t on $\mathfrak{Q}_{D,\Omega}$ and $\mathfrak{Q}_{N,\Omega}$ respectively are closed and semibounded; these are the forms associated with the Dirichlet and Neumann realizations of the Schrödinger operator in $L^2(\Omega)$. More precisely, it has been shown that if $q \in L^s_{\text{loc}}(\bar{\Omega})$ for some s satisfying (2.1), and

$$\sup_{Q \in \mathfrak{F}} |Q|^{-1/s} \|q^-\|_{s,Q} < \infty, \tag{2.4}$$

$$\inf_{Q \in \mathfrak{F}} (q_Q - \gamma_s |Q|^{-1/s} \|f_Q\|_{s,Q}) > -\infty, \tag{2.5}$$

$$\delta := \gamma_s \sup_{Q \in \mathfrak{F}} |Q|^{2/n - 1/s} \|f_Q\|_{s,Q} < 1, \tag{2.6}$$

where

$$q_Q = |Q|^{-1} \int_Q q, \qquad f_Q = q - q_Q,$$

and γ_s is the norm of the embedding of $W^{1,2}((0,1)^n)$ in $L^{2s/(s-1)}((0,1)^n)$, then t_D is a densely defined, closed, lower semi-bounded, sesquilinear form on $L^2(\Omega)$ and the associated self-adjoint operator $T_{D,\Omega}$ satisfies

$$(T_{D,\Omega}u, \phi)_{2,\Omega} = t_D[\![u, \phi]\!] \tag{2.7}$$

for all u in $\mathscr{D}(T_{D,\Omega})$, the domain of $T_{D,\Omega}$, and all ϕ in $\mathfrak{Q}_{D,\Omega}$, where

$$\mathscr{D}(T_{D,\Omega}) = \{u \in \mathfrak{Q}_{D,\Omega}\colon -\Delta u + qu \in L^2(\Omega)\}, \qquad T_{D,\Omega}u = -\Delta u + qu, \tag{2.8}$$

Δ being the Dirichlet Laplacian, and

$$T_{D,\Omega} \geqslant \inf_{Q \in \mathfrak{F}} (q_Q - \gamma_s |Q|^{-1/s} \|f_Q\|_{s,Q}) I, \tag{2.9}$$

I being the identity map of $L^2(\Omega)$ to itself. Moreover, if instead we suppose that $q \in L^s_{\text{loc}}(\Omega) \cap L^1_{\text{loc}}(\bar{\Omega})$ for some s satisfying (2.1), and that (2.4), (2.5) and (2.6) hold, $\mathfrak{F}$ being a covering of the kind used for the Neumann problem, then (2.7), (2.8), and (2.9) hold for the Neumann problem, that is, with the subscript D replaced by N and with the Neumann rather than the Dirichlet Laplacian. If the cubes in $\mathfrak{F}$ are congruent, these latter results hold if $q \in L^s_{\text{loc}}(\bar{\Omega})$ and (2.4), (2.5) and (2.6) are satisfied.

It has also been shown (Theorem X.2.6) that if $\mathfrak{F}$ is a congruent covering of $\mathbb{R}^n$ and the above conditions relating to $T_{D,\Omega}$ hold, then the essential spectrum

$\sigma_e(T_{D,\Omega})$ of $T_{D,\Omega}$ is contained in $[\theta_{D,\Omega}, \infty)$, where

$$\theta_{D,\Omega} = \liminf_{|x_Q| \to \infty} (q_Q - \gamma_s |Q|^{-1/s} \|f_Q\|_{s,Q}). \tag{2.10}$$

Under the conditions above relating to $T_{N,\Omega}$, we have seen (Theorem X.2.6) that $\sigma_e(T_{N,\Omega}) \subset [\theta_{N,\Omega}, \infty)$, where

$$\theta_{N,\Omega} = \liminf_{x_Q \to \partial\Omega} (q_Q - \gamma_s |Q|^{-1/s} \|f_Q\|_{s,Q}), \tag{2.11}$$

with the understanding that if $\Omega = \mathbb{R}^n$ then $\liminf\limits_{|x_Q| \to \infty}$ is intended.

Now let us suppose that $\mathfrak{F}$ is a Whitney covering of Ω unless Ω is the whole of $\mathbb{R}^n$ or is expressible as the interior of the union of a family of closed congruent cubes with disjoint interiors; and even in these latter cases $\mathfrak{F}$ may on occasions consist of cubes which are not congruent. Although we shall retain the same covering $\mathfrak{F}$ for the Dirichlet and Neumann problems in Theorem 2.7 below, we could, alternatively, take one covering $\mathfrak{F}_1$ for the Dirichlet problem on Ω and a different (Whitney) covering $\mathfrak{F}_2$ for the Neumann problem.

Suppose that $q \in L^s_{\mathrm{loc}}(\bar{\Omega})$ for some s satisfying (2.1), suppose that (2.4), (2.5) and (2.6) hold, and let $N(\lambda, T_D, \Omega)$ ($N(\lambda, T_N, \Omega)$) be the number of eigenvalues λ_0 less than λ of the problem $T_{D,\Omega}u = \lambda_0 u$ ($T_{N,\Omega}u = \lambda_0 u$); that is, of the Dirichlet (Neumann) problem on Ω, with zero boundary data, for $-\Delta + q$. For any $Q \in \mathfrak{F}$ we have that $q \in L^s(Q)$: thus by Lemma VII.1.14, the form domains of $T_{D,Q}$ and $T_{N,Q}$ (that is, the domains of the corresponding sesquilinear forms t_D and t_N) are $W^{1,2}(Q^\circ)$ and $W^{1,2}(Q^\circ)$ respectively. We also see that, as in the proof of Theorem VII.1.15, for any ϕ in $W^{1,2}(Q^\circ)$,

$$\int_Q (|\nabla\phi|^2 + q|\phi|^2) \leqslant (1+\delta)\|\nabla\phi\|^2_{2,Q} + (q_Q + \gamma_s|Q|^{-1/s}\|f_Q\|_{s,Q})\|\phi\|^2_{2,Q} \tag{2.12}$$

and

$$\int_Q (|\nabla\phi|^2 + q|\phi|^2) \geqslant (1-\delta)\|\nabla\phi\|^2_{2,Q} + (q_Q - \gamma_s|Q|^{-1/s}\|f_Q\|_{s,Q})\|\phi\|^2_{2,Q}. \tag{2.13}$$

The Max–Min Principle (Theorem 1.3), (2.12) and (2.13) now show that, with $-\Delta_D$ and $-\Delta_N$ written for T_D and T_N respectively when $q = 0$, we have

$$N(\lambda, T_D, Q^\circ) \geqslant N(\mu_Q, -\Delta_D, Q^\circ), \tag{2.14}$$

where

$$\mu_Q = (1+\delta)^{-1}(\lambda - q_Q - \gamma_s|Q|^{-1/s}\|f_Q\|_{s,Q}); \tag{2.15}$$

and

$$N(\lambda, T_N, Q^\circ) \leqslant N(\eta_Q, -\Delta_N, Q^\circ), \tag{2.16}$$

where

$$\eta_Q = (1-\delta)^{-1}(\lambda - q_Q + \gamma_s|Q|^{-1/s}\|f_Q\|_{s,Q}). \tag{2.17}$$

Now let $\lambda < \theta_{N,\Omega}$, our lower bound for points of $\sigma_e(T_{N,\Omega})$, given by (2.11), and put

$$I_1 = \{Q \in \mathfrak{F}: q_Q + \gamma_s |Q|^{-1/s} \|f_Q\|_{s,Q} < \lambda\}, \tag{2.18}$$

$$I_2 = \{Q \in \mathfrak{F}: q_Q - \gamma_s |Q|^{-1/s} \|f_Q\|_{s,Q} \leqslant \lambda\}. \tag{2.19}$$

Our object is to compare $N(\lambda, T_D, \Omega)$ and $N(\lambda, T_N, \Omega)$ with sums of corresponding expressions involving the Laplacian on cubes in I_1 and I_2. To do this some preparatory work on orthogonal sums of operators is useful.

2.2. *Variation of eigenvalues*

Let H be a Hilbert space which is decomposed into the orthogonal sum of two closed linear subspaces H_1 and H_2, i.e. $H = H_1 \oplus H_2$; let T_1 and T_2 be self-adjoint maps acting in H_1 and H_2 respectively and with domains $\mathscr{D}(T_1)$ and $\mathscr{D}(T_2)$, and let $T = T_1 \oplus T_2$ be the orthogonal sum of T_1 and T_2, with domain $\mathscr{D}(T) = \{x_1 + x_2 : x_1 \in \mathscr{D}(T_1),\ x_2 \in \mathscr{D}(T_2)\}$. It is clear that T is a self-adjoint map acting in H. Let $E_\lambda(T)$, $E_\lambda(T_1)$, $E_\lambda(T_2)$ be resolutions of the identity corresponding to T, T_1, T_2 respectively, and given any Borel subset Ω of $\mathbb{R}$ write $P_\Omega(T) = \int_\Omega dE_\lambda(T)$ and $P_\Omega(T_j) = \int_\Omega dE_\lambda(T_j)$ $(j = 1, 2)$. Then $P_\Omega(T) = P_\Omega(T_1) \oplus P_\Omega(T_2)$; and with $N(\lambda, T) = \operatorname{rank} P_{(-\infty,\lambda)}(T)$ and similar expressions for $N(\lambda, T_1)$ and $N(\lambda, T_2)$, it follows that

$$N(\lambda, T) = N(\lambda, T_1) + N(\lambda, T_2). \tag{2.20}$$

For more details about direct sums of operators see Taylor [1, Chapter 5, §4].

Proposition 2.1. Let Ω_1 and Ω_2 be disjoint open subsets of $\mathbb{R}^n$, so that $L^2(\Omega_1 \cup \Omega_2) = L^2(\Omega_1) \oplus L^2(\Omega_2)$. Then for the corresponding orthogonal sum of operators we have

$$-\Delta_{D,\Omega_1 \cup \Omega_2} = -\Delta_{D,\Omega_1} \oplus -\Delta_{D,\Omega_2} \tag{2.21}$$

and

$$-\Delta_{N,\Omega_1 \cup \Omega_2} = -\Delta_{N,\Omega_1} \oplus -\Delta_{N,\Omega_2}. \tag{2.22}$$ ■

Proof. Given any $f \in C_0^\infty(\Omega_1 \cup \Omega_2)$, put $f_j = f \restriction \Omega_j$ $(j = 1, 2)$; since $\Omega_1 \cap \Omega_2 = \varnothing$ it follows that $f_j \in C_0^\infty(\Omega_j)$ $(j = 1, 2)$. Moreover, for all $f, g \in C_0^\infty(\Omega_1 \cup \Omega_2)$,

$$\int_{\Omega_1 \cup \Omega_2} \nabla f \cdot \nabla \bar{g} = \int_{\Omega_1} \nabla f_1 \cdot \nabla \bar{g}_1 + \int_{\Omega_2} \nabla f_2 \cdot \nabla \bar{g}_2,$$

and from this it is clear that the forms corresponding to the two sides of (2.21) are equal. Thus (2.21) holds. The proof of (2.22) is similar. □

Corollary 2.2. Let $\Omega_1, \Omega_2, \ldots, \Omega_k$ be disjoint open subsets of $\mathbb{R}^n$ and let

$$N(\lambda, -\Delta_D, \Omega) = \operatorname{rank} P_{[0,\lambda)}(-\Delta_{D,\Omega}),$$

with $N(\lambda, -\Delta_N, \Omega)$ defined analogously. Then

$$N\left(\lambda, -\Delta_D, \bigcup_1^k \Omega_j\right) = \sum_{j=1}^{k} N(\lambda, -\Delta_D, \Omega_j)$$

and

$$N\left(\lambda, -\Delta_N, \bigcup_1^k \Omega_j\right) = \sum_{j=1}^{k} N(\lambda, -\Delta_N, \Omega_j). \qquad \blacksquare$$

Proof. The result follows immediately from (2.20), (2.21) and (2.22). □

Let H_1 be a closed linear subspace of a Hilbert space H, let $\mathscr{D}(A)$ and $\mathscr{D}(B)$ be dense linear subspaces of H and H_1 respectively, let $A: \mathscr{D}(A) \to H$ and $B: \mathscr{D}(B) \to H_1$ be both self-adjoint and non-negative, and let $\mathscr{Q}(A)$ and $\mathscr{Q}(B)$ be the form domains of A and B respectively. We shall write $0 \leqslant A \leqslant B$ if, and only if, $\mathscr{Q}(B) \subset \mathscr{Q}(A)$ and for all $u \in \mathscr{Q}(B)$ we have $0 \leqslant a[\![u]\!] \leqslant b[\![u]\!]$ where a and b are the forms of A and B.

Lemma 2.3. Let A and B be as above and suppose that $0 \leqslant A \leqslant B$. Then $\lambda_n(A) \leqslant \lambda_n(B)$ for all $n \in \mathbb{N}$, and $N(\lambda, A) \geqslant N(\lambda, B)$. $\blacksquare$

Proof. Since $A \leqslant B$ we have for any $\psi_1, \psi_2, \ldots, \psi_n \in H$,

$$\begin{aligned} &\min\{a[\![\phi]\!] : \phi \in \mathscr{Q}(A),\ \phi \perp \psi_j \quad \text{for } j = 1, 2, \ldots, n\} \\ &\leqslant \min\{b[\![\phi]\!] : \phi \in \mathscr{Q}(B),\ \phi \perp \psi_j \quad \text{for } j = 1, 2, \ldots, n\}. \end{aligned}$$

The lemma follows directly from this and the Max–Min Principle (Theorem 1.3). □

Proposition 2.4. (i) Let Ω and Ω' be open subsets of $\mathbb{R}^n$ with $\Omega \subset \Omega'$. Then $0 \leqslant -\Delta_{D,\Omega'} \leqslant -\Delta_{D,\Omega}$, with the understanding that any element of $L^2(\Omega)$ is to be regarded as an element of $L^2(\Omega')$ by setting it equal to 0 in $\Omega' \backslash \Omega$.

(ii) For any open subset Ω of $\mathbb{R}^n$ we have $0 \leqslant -\Delta_{N,\Omega} \leqslant -\Delta_{D,\Omega}$.

(iii) Let Ω_1 and Ω_2 be disjoint open subsets of an open set Ω in $\mathbb{R}^n$ such that $\Omega_1 \cup \Omega_2$ has interior equal to Ω and $\Omega \backslash (\Omega_1 \cup \Omega_2)$ has Lebesgue n-measure 0. Then $0 \leqslant -\Delta_{D,\Omega} \leqslant -\Delta_{D,\Omega_1 \cup \Omega_2}$ and $0 \leqslant -\Delta_{N,\Omega_1 \cup \Omega_2} \leqslant -\Delta_{N,\Omega}$. $\blacksquare$

Proof. (i) Since $C_0^\infty(\Omega) \subset C_0^\infty(\Omega')$ and the restriction of the form of $-\Delta_{D,\Omega'}$ to functions in $C_0^\infty(\Omega)$ coincides with the form of $-\Delta_{D,\Omega}$ evaluated at such functions, (i) follows.

(ii) This is obvious since $C_0^\infty(\Omega) \subset C_0^\infty(\mathbb{R}^n)$.

(iii) The inequalities relating to the Dirichlet Laplacian are just a special case of (i). As for the Neumann inequalities, denote the form domain of $-\Delta_{N,\Omega}$ by

$\tilde{W}^{1,2}(\Omega)$: the space $\tilde{W}^{1,2}(\Omega)$ is simply the closure in $W^{1,2}(\Omega)$ of the set of all functions on Ω which are restrictions to Ω of functions in $C_0^\infty(\mathbb{R}^n)$; in general, $\tilde{W}^{1,2}(\Omega) \underset{\neq}{\subset} W^{1,2}(\Omega)$, although by Theorem V.4.7 these two spaces coincide if $\partial\Omega$ is of class C, and hence in particular when Ω is a cube. Given any $f \in \tilde{W}^{1,2}(\Omega)$, it is clear that $f \upharpoonright \Omega_1 \cup \Omega_2 \in \tilde{W}^{1,2}(\Omega_1) \oplus \tilde{W}^{1,2}(\Omega_2)$; moreover, since $\Omega \backslash (\Omega_2 \cup \Omega_2)$ has zero measure,

$$0 \leqslant \int_\Omega |\nabla f|^2 = \int_{\Omega_1 \cup \Omega_2} |\nabla f|^2,$$

and so the desired inequalities follow. □

Propositions 2.1 and 2.4, together with (2.14) and (2.16), now give with the aid of Lemma 2.3,

$$\sum_{Q \in I_1} N(\mu_Q, -\Delta_D, Q^\circ) \leqslant \sum_{Q \in I_1} N(\lambda, T_D, Q^\circ) \leqslant N(\lambda, T_D, \Omega) \leqslant N(\lambda, T_N, \Omega). \tag{2.23}$$

Moreover, if we let U be the interior of $\cup_{Q \in I_2} Q$ and Ω' be the interior of $\Omega \backslash U$, then by (2.13), $T_{N,\Omega'} \geqslant \lambda I$, and so $(-\infty, \lambda)$ contains no point of the spectrum of $T_{N,\Omega'}$. Since I_2 is a finite set ($\lambda < \theta_{N,\Omega}$), it follows from Proposition 2.4 and Corollary 2.2 that

$$N(\lambda, T_N, \Omega) \leqslant N(\lambda, T_N, \Omega') + N(\lambda, T_N, U) = N(\lambda, T_N, U)$$

$$= \sum_{Q \in I_2} N(\lambda, T_N, Q^\circ) \leqslant \sum_{Q \in I_2} N(\eta_Q, -\Delta_N, Q^\circ), \tag{2.24}$$

the final step being a consequence of (2.16). From (2.23) and (2.24) we obtain

$$\sum_{Q \in I_1} N(\mu_Q, -\Delta_D, Q^\circ) \leqslant N(\lambda, T_D, \Omega) \leqslant N(\lambda, T_N, \Omega) \leqslant \sum_{Q \in I_2} N(\eta_Q, -\Delta_N, Q^\circ). \tag{2.25}$$

Note that $\mu_Q > 0$ (and hence $N(\mu_Q, -\Delta_D, Q^\circ) \neq 0$) if, and only if, $Q \in I_1$; while $\eta_Q \geqslant 0$ (and hence $N(\eta_Q, -\Delta_N, Q^\circ) \neq 0$) if, and only if, $Q \in I_2$. The next task is to estimate the terms $N(\mu_Q, -\Delta_D, Q^\circ)$ and $N(\eta_Q, -\Delta_N, Q^\circ)$, and fortunately, explicit computations of the eigenvalues of both Dirichlet and Neumann Laplacians on cubes is possible. Before doing this, we give a helpful characterization of operator cores for $-\Delta_D$ and $-\Delta_N$ on a cube which links the weak and classical formulations of the Dirichlet and Neumann problems on a cube Q.

2.3. *The eigenvalues of $-\Delta_{D,Q}$ and $-\Delta_{N,Q}$*

Proposition 2.5. Let Q be an open cube in $\mathbb{R}^n$, let $\partial Q'$ be ∂Q minus the vertices of Q, and let $\partial/\partial n$ be the normal derivative outwards from Q at $\partial Q'$.

(i) Let $\mathfrak{D}_D := \{f \in C^\infty(\bar{Q}) : f \upharpoonright \partial Q = 0\}$ and $A = -\Delta \upharpoonright \mathfrak{D}_D$. Then $\bar{A} = -\Delta_{D,Q}$; that is, $\mathfrak{D}_D$ is an (operator) core of $-\Delta_{D,Q}$.
(ii) Let $\mathfrak{D}_N = \{f \in C^\infty(\bar{Q}) : (\partial f/\partial n) \upharpoonright \partial Q' = 0\}$ and $B = -\Delta \upharpoonright \mathfrak{D}_N$. Then $\bar{B} = -\Delta_{N,Q}$. ■

Proof. (i) Without loss of generality, assume that $Q = (-1, 1)^n$. First note that $-\Delta_{D,Q}$ is the unique self-adjoint extension of $-\Delta \upharpoonright C_0^\infty(Q)$ with domain in the form domain of $-\Delta_{D,Q}$, namely $W_0^{1,2}(Q)$; that is, $-\Delta_{D,Q}$ is the Friedrichs extension of $-\Delta \upharpoonright C_0^\infty(Q)$. Since A is a symmetric extension of $-\Delta \upharpoonright C_0^\infty(Q)$, if we can prove that $\bar{A}$ is self-adjoint and has domain $\mathscr{D}(\bar{A})$ contained in $W_0^{1,2}(Q)$, it will follow from the First Representation Theorem (cf. Theorem IV.2.6) that $\bar{A} = -\Delta_{D,Q}$, as required.

To prove that $\bar{A}$ is self-adjoint, observe that the functions Φ_m, where $\Phi_m(x) = \prod_{i=1}^n \phi_{m_i}(x_i)$ (with $m = (m_j) \in \mathbb{N}^n$) and $\phi_k(t) = \cos(\frac{1}{2}k\pi t)$ (k odd), $\phi_k(t) = \sin(\frac{1}{2}k\pi t)$ (k even, $k \neq 0$), are eigenfunctions of A and form an orthonormal basis of $L^2(Q)$. If $f \in \mathscr{D}(\bar{A})$, there is a sequence (f_k) in $\mathscr{D}(A)$ such that $f_k \to f$ and $Af_k \to \bar{A}f$. Hence

$$(\bar{A}f, \Phi_m)_Q = \lim_{k\to\infty} (Af_k, \Phi_m)_Q = \lim_{k\to\infty} (f_k, A\Phi_m)_Q = \lambda_m (f, \Phi_m)_Q$$

for all $m \in \mathbb{N}^n$, and

$$\|\bar{A}f\|_Q^2 = \sum_m \lambda_m^2 |(f, \Phi_m)_Q|^2 < \infty.$$

Conversely, if $f \in L^2(Q)$ is such that $\Sigma_m \lambda_m^2 |(f, \Phi_m)_Q|^2 < \infty$, then with $f_k = \Sigma_{|m| \leq k} (f, \Phi_m)_Q \Phi_m$, we see that $\|f - f_k\|_Q \to 0$ and $\|Af_k - Af_j\|_Q \to 0$ as $j, k \to \infty$; thus $f \in \mathscr{D}(\bar{A})$. Hence $f \in \mathscr{D}(\bar{A})$ if, and only if, $\sum_m \lambda_m^2 |(f, \Phi_m)_Q|^2 < \infty$; and for $f \in \mathscr{D}(\bar{A})$,

$$\bar{A}f = \sum_m \lambda_m (f, \Phi_m)_Q \Phi_m.$$

It follows that $\bar{A}^* = \bar{A}$, i.e. $\bar{A}$ is self-adjoint.

All that remains is to show that $\mathscr{D}(\bar{A}) \subset W_0^{1,2}(Q)$. But $\mathscr{D}(\bar{A})$ is contained in the form domain $\mathscr{Q}(\bar{A})$ of $\bar{A}$, and $\mathscr{Q}(\bar{A})$ is the completion of $\mathscr{D}(A) = \mathfrak{D}_D$ with respect to the norm

$$(Af, f)_Q = \int_Q |\nabla f|^2,$$

that is, the $W_0^{1,2}(Q)$ norm; moreover, by Theorem V.3.4, $\mathscr{D}(A) \subset W_0^{1,2}(Q)$. It follows that $\mathscr{D}(\bar{A}) \subset \mathscr{Q}(\bar{A}) \subset W_0^{1,2}(Q)$ and the proof is complete.

(ii) As before, assume that $Q = (-1, 1)^n$, without loss of generality. That $\bar{B}$ is self-adjoint is established just as $\bar{A}$ was shown above to have this property. For any $f \in \mathscr{D}(B)$ and any g in the form domain $W^{1,2}(Q)$ of $-\Delta_{N,Q}$,

$$(g, Bf)_Q = \int_Q g(-\Delta \bar{f}) = \int_Q \nabla g \cdot \nabla \bar{f}$$

and so

$$(g, (B+I)f)_Q = (g, f)_{1,2,Q};$$

$-\Delta_{N,Q} + I$ is the unique self-adjoint operator in $L^2(Q)$ associated with the semi-bounded form $(\bullet, \bullet)_{1,2,Q}$ and hence, if we can prove that $\mathscr{D}(B)$ is dense in $W^{1,2}(Q)$ it will follow from the First Representation Theorem that $B + I \subset -\Delta_{N,Q} + I$, so that $\bar{B} = -\Delta_{N,Q}$. Since $\mathscr{D}(B) \subset W^{1,2}(Q)$ and the form domain $\mathscr{Q}(\bar{B})$ of $\bar{B}$ is the completion of $\mathscr{D}(B)$ with respect to the metric induced by

$$(f, (B+I)f)_Q = (f, f)_{1,2,Q},$$

$\mathscr{D}(B)$ is dense in $W^{1,2}(Q)$ if, and only if, $\mathscr{Q}(\bar{B}) = W^{1,2}(Q)$; since $\mathscr{D}(B) \subset W^{1,2}(Q)$ we see that $\mathscr{Q}(\bar{B}) \subset W^{1,2}(Q)$, and hence it is enough to show that $W^{1,2}(Q) \subset \mathscr{Q}(\bar{B})$. Let $f \in W^{1,2}(Q)$ and suppose that $g \in C^\infty(\bar{Q})$ is such that $g(\pm 1, x_2, \ldots, x_n) = 0$ whenever $(\pm 1, x_2, \ldots, x_n) \in \bar{Q}$. Then we claim that

$$(D_1 f, g)_Q = -(f, D_1 g)_Q.$$

To justify this, first suppose that $g \restriction \partial Q = 0$. By Theorem V.3.4, $g \in W_0^{1,2}(Q)$; hence there is a sequence $(g_m) \subset C_0^\infty(Q)$ such that $g_m \to g$ and $\nabla g_m \to \nabla g$ in $L^2(Q)$; and as our claim is obviously true for each g_m it must also hold for g. To remove the assumption that $g \restriction \partial Q = 0$, let (η_m) be a sequence of functions in $C^\infty(\bar{Q})$, each depending upon $x_2, \ldots, x_n$ only and with compact support in $(-1+m^{-1}, 1-m^{-1})^{n-1}$, such that for each $x \in Q$ we have $\eta_m(x) \uparrow 1$ as $m \to \infty$. The claim holds for each $g\eta_m$, and since $D_1(g\eta_m) = \eta_m D_1 g$ it must also be true for g.

Now let (Ψ_m) be the sequence of eigenvectors of B given by

$$\Psi_m(x) = \prod_{i=1}^{n} \psi_{m_i}(x_i),$$

where $m = (m_i) \in \mathbb{N}_0^n$ and

$$\psi_k(t) = \begin{cases} \sin(\frac{1}{2}k\pi t) & (k \text{ odd}), \\ \cos(\frac{1}{2}k\pi t) & (k \text{ even}, k \neq 0), \\ \frac{1}{2}\sqrt{2} & (k = 0), \end{cases}$$

and put $\Phi_m(x) = \prod_{i=1}^{n} \phi_{m_i}(x_i)$, where $\phi_{m_i} = \psi_{m_i}$ if $i \neq 1$ and

$$\phi_{m_1}(t) = \begin{cases} \cos(\frac{1}{2}m_1\pi t) & (m_1 \text{ odd}), \\ \sin(\frac{1}{2}m_1\pi t) & (m_1 \text{ even}, m_1 \neq 0). \end{cases}$$

Then (Φ_m) is orthonormal in $L^2(Q)$ and $D_1\Phi_m = \pm\frac{1}{2}\pi m_1 \Psi_m$; also $\Phi_m \in C^\infty(\bar{Q})$ and $\Phi_m(\pm 1, x_2, \ldots, x_n) = 0$. In view of our claim, established above, and the orthonormality of the Φ_m it follows that

$$\sum_m |(f, \Psi_m)_Q|^2 m_1^2 = \left(\frac{2}{\pi}\right)^2 \sum_m |(D_1 f, \Phi_m)_Q|^2 \leqslant \left(\frac{2}{\pi}\right)^2 \|D_1 f\|_Q^2,$$

and repetition of this process for each variable shows that

$$\sum_m (1+|m|^2)|(f, \Psi_m)_Q|^2 \leqslant \|f\|_Q^2 + \left(\frac{2}{\pi}\right)^2 \|\nabla f\|_Q^2.$$

where $|m|^2 = \sum_{i=1}^n m_i^2$. We see that $f_k := \sum_{|m| \leqslant k} (f, \Psi_m)_Q \Psi_m \to f$ in $L^2(Q)$, and also (f_k) converges in $\mathcal{Q}(\bar{B})$; note that the eigenvalues of B are $\frac{1}{4}\pi^2|m|^2$. Since $f_k \in \mathcal{D}(B)$ and $\mathcal{Q}(\bar{B})$ is continuously embedded in $L^2(Q)$, we have $f \in \mathcal{Q}(\bar{B})$ and the proof is complete. □

Theorem 2.6. Let λ, $a > 0$ and put $Q = (-a, a)^n$. Then there is a constant C, independent of a and λ, such that

$$|N(\lambda, -\Delta_D, Q) - \omega_n(2\pi)^{-n}|Q|\lambda^{n/2}| \leqslant C[1 + (|Q|\lambda^{n/2})^{(n-1)/n}] \tag{2.26}$$

and

$$|N(\lambda, -\Delta_N, Q) - \omega_n(2\pi)^{-n}|Q|\lambda^{n/2}| \leqslant C[1 + (|Q|\lambda^{n/2})^{(n-1)/n}], \tag{2.27}$$

where ω_n is the volume of the unit ball in $\mathbb{R}^n$. ■

Proof. On account of Proposition 2.5 the eigenvalues of $-\Delta_{D,Q}$ and $-\Delta_{N,Q}$ may be computed explicitly; and a routine argument using separation of variables shows that the eigenvectors of $-\Delta_{D,Q}$ are precisely the functions Φ_α $(\alpha \in \mathbb{N}^n)$, where

$$\Phi_\alpha(x) = a^{-n/2} \prod_{j=1}^n \phi_{\alpha_j}(x_j/a), \tag{2.28}$$

and

$$\phi_k(y_k) = \begin{cases} \cos(\frac{1}{2}k\pi y_k) & (k \text{ odd}), \\ \sin(\frac{1}{2}k\pi y_k) & (k \text{ even}). \end{cases} \tag{2.29}$$

The corresponding eigenvalues are

$$\eta_\alpha = \left(\frac{\pi}{2a}\right)^2 \sum_{j=1}^n \alpha_j^2 \qquad (\alpha \in \mathbb{N}^n). \tag{2.30}$$

For $-\Delta_{N,Q}$ the eigenvectors are the functions Ψ_α $(\alpha \in \mathbb{N}_0^n)$, where

$$\Psi_\alpha(x) = a^{-n/2} \prod_{j=1}^n \psi_{\alpha_j}(x_j/a), \tag{2.31}$$

and

$$\psi_k(y_k) = \begin{cases} \sin(\frac{1}{2}k\pi y_k) & (k \text{ odd}), \\ \cos(\frac{1}{2}k\pi y_k) & (k \text{ even}, k \neq 0), \\ \frac{1}{2}\sqrt{2} & (k = 0). \end{cases} \tag{2.32}$$

The corresponding eigenvalues are given by (2.30), but with α running through $\mathbb{N}_0^n$ rather than $\mathbb{N}^n$.

So far as the estimates (2.26) and (2.27) are concerned it is merely necessary to deal with the case in which $a = 1$, because $-\Delta_{D,(-1,1)^n}$ is unitarily equivalent to $a^2(-\Delta_{D,Q})$, as a change of scale shows, and $N(\lambda, -\Delta_D, Q) = N(a^2\lambda, -\Delta_D, Q_1)$, where $Q_1 = (-1, 1)^n$.

Let $\tilde{B}_\lambda = B(0, 2\lambda^{\frac{1}{2}}/\pi)$. In view of (2.30), $N(\lambda, -\Delta_D, Q_1)$ is the number of points of $\mathbb{N}^n$ in $\tilde{B}_\lambda$; $N(\lambda, -\Delta_N, Q_1)$ is the number of points of $\mathbb{N}_0^n$ in $\tilde{B}_\lambda$. For each $\alpha \in \mathbb{N}^n \cap \tilde{B}_\lambda$,

$$\begin{aligned} Q_\alpha := \{x \in \mathbb{R}^n : \alpha_i - 1 \leqslant x_i \leqslant \alpha_i \text{ for } i = 1, 2, \ldots, n\} \\ \subset \tilde{B}_\lambda \cap \{x \in \mathbb{R}^n : x_i \geqslant 0 \text{ for } i = 1, 2, \ldots, n\}, \end{aligned}$$

and so

$$N(\lambda, -\Delta_D, Q_1) \leqslant 2^{-n}\omega_n(2\lambda^{\frac{1}{2}}/\pi)^n. \tag{2.33}$$

Moreover, $\tilde{B}_\lambda \cap \{x \in \mathbb{R}^n : x_i \geqslant 0 \text{ for } i = 1, 2, \ldots, n\}$ is covered by the family of cubes $\{x \in \mathbb{R}^n : \alpha_i \leqslant x_i < \alpha_i + 1 \text{ for } i = 1, 2, \ldots, n\}$ obtained by letting α run through $\tilde{B}_\lambda \cap \mathbb{N}_0^n$, and so

$$2^{-n}\omega_n(2\lambda^{\frac{1}{2}}/\pi)^n \leqslant N(\lambda, -\Delta_N, Q_1). \tag{2.34}$$

Also,

$$N(\lambda, -\Delta_N, Q_1) - N(\lambda, -\Delta_D, Q_1) = \#\left(\tilde{B}_\lambda \cap (\mathbb{N}_0^n \setminus \mathbb{N}^n)\right) = \# \bigcup_{k=0}^{n-1} M_k(\lambda),$$

where

$$M_k(\lambda) = \{\alpha \in \tilde{B}_\lambda \cap \mathbb{N}_0^n : \text{exactly } k \text{ of the } \alpha_i \text{ are non-zero}\}.$$

Just as in (2.33), $2^k \# M_k(\lambda)$ is majorized by the volume in $\mathbb{R}^k$ of $\binom{n}{k}$ balls (in $\mathbb{R}^k$) of radius $2\lambda^{\frac{1}{2}}/\pi$, and thus

$$\sum_{k=1}^{n-1} \# M_k(\lambda) \leqslant \text{const} \cdot \sum_{k=1}^{n-1} \lambda^{k/2} \leqslant C(1 + \lambda^{(n-1)/2}).$$

This shows that

$$N(\lambda, -\Delta_N, Q_1) - N(\lambda, -\Delta_D, Q_1) \leqslant C(1 + \lambda^{(n-1)/2}),$$

which together with (2.33) and (2.34) gives (2.26) and (2.27). □

2.4. *The main estimates*

Note that in view of the translation-invariance of the Laplacian, the conclusions of Theorem 2.6 hold for any cube in our covering $\mathfrak{F}$. The terms in (2.25) may thus be estimated as follows:

$$N(\mu_Q, -\Delta_D, Q^\circ) \geq \omega_n(2\pi)^{-n}|Q|\mu_Q^{n/2} - O\big(1 + (|Q|\mu_Q^{n/2})^{(n-1)/n}\big) \quad (2.35)$$

and

$$N(\eta_Q, -\Delta_N, Q^\circ) \leq \omega_n(2\pi)^{-n}|Q|\eta_Q^{n/2} + O\big(1 + (|Q|\eta_Q^{n/2})^{(n-1)/n}\big), \quad (2.36)$$

where the constants implicit in the O terms are independent of Q, μ_Q, and η_Q

To estimate $\mu_Q^{n/2}$ we use the inequality

$$(A-B)^{n/2} \geq A^{n/2} - KB^{1/2}(A^{(n-1)/2} + B^{(n-1)/2}), \quad (2.37)$$

valid for all A and B with $A \geq B \geq 0$, where K is a suitable positive constant independent of A and B. This is obvious if $n = 1$, and if $n > 1$ it follows because

$$\begin{aligned}(A-B)^{n/2} \geq (A^{1/2} - B^{1/2})^n &= A^{n/2} - O\Big(\sum_{j=1}^{n} A^{(n-j)/2}B^{j/2}\Big)\\ &= A^{n/2} - O\big(B^{1/2}(A^{(n-1)/2} + B^{(n-1)/2})\big),\end{aligned}$$

since by Young's inequality,

$$A^{(n-j)/2}B^{(j-1)/2} \leq \frac{n-j}{n-1}A^{(n-1)/2} + \frac{j-1}{n-1}B^{(n-1)/2}.$$

Together with the definition of μ_Q in (2.15) and (2.6) this gives, for all $Q \in I_1$,

$$\begin{aligned}&|Q|\mu_Q^{n/2} \geq (1+\delta)^{-n/2}|Q|(\lambda - q_Q)^{n/2}\\ &-O\big(|Q|^{1/n}(|Q|^{-1/s}\|f_Q\|_{s,Q})^{\frac{1}{2}}[|Q|(\lambda - q_Q)^{n/2}]^{(n-1)/n} + |Q|^{1-n/2s}\|f_Q\|_{s,Q}^{n/2}\big)\\ &\qquad\geq (1+\delta)^{-n/2}\Lambda_Q - O(\Lambda_Q{}^{(n-1)/n} + 1),\end{aligned} \quad (2.38)$$

where

$$\Lambda_Q := |Q|(\lambda - q_Q)_+^{n/2},$$

since $\lambda > q_Q$ if $Q \in I_1$. Also, from (2.6) and the definition of μ_Q, if $Q \in I_1$ then

$$\begin{aligned}(|Q|\mu_Q^{n/2})^{(n-1)/n} &\leq \{|Q|[(1+\delta)^{-1}(\lambda - q_Q)_+]^{n/2}\}^{(n-1)/n}\\ &= O(\Lambda_Q^{(n-1)/n}).\end{aligned}$$

It now follows from (2.35) that if $Q \in I_1$,

$$N(\mu_Q, -\Delta_D, Q^\circ) \geq \omega_n(2\pi)^{-n}(1+\delta)^{-n/2}\Lambda_Q - O(\Lambda_Q^{(n-1)/n} + 1). \quad (2.39)$$

Sum this over all Q in I_1 and apply Hölder's inequality:

$$\sum_{Q\in I_1} \mathrm{N}(\mu_Q, -\Delta_{\mathrm{D}}, Q^\circ)$$

$$\geqslant \omega_n(2\pi)^{-n}(1+\delta)^{-n/2} \sum_{Q\in I_1} \Lambda_Q - O\left[\left(\sum_{Q\in I_1} \Lambda_Q\right)^{(n-1)/n}\left(\sum_{Q\in I_1} 1\right)^{1/n} + \sum_{Q\in I_1} 1\right]$$

$$\geqslant \omega_n(2\pi)^{-n}(1+\delta)^{-n/2}\left(\sum_{q_Q<\lambda} \Lambda_Q - \sum_{0<\lambda-q_Q\leqslant \gamma_s|Q|^{-1/s}\|f_Q\|_{s,Q}} \Lambda_Q\right)$$

$$- O\left[\left(\sum_{q_Q<\lambda} \Lambda_Q\right)^{(n-1)/n} N_1^{1/n} + N_1\right]$$

$$= \omega_n(2\pi)^{-n}(1+\delta)^{-n/2}\Lambda - O(\Lambda^{(n-1)/n}N_1^{1/n} + N_1), \tag{2.40}$$

where

$$\Lambda = \sum_{q_Q<\lambda} \Lambda_Q = \sum_{q_Q<\lambda} |Q|(\lambda - q_Q)_+^{n/2} \tag{2.41}$$

and N_1 is the number of cubes Q for which $q_Q<\lambda$: that the term involving

$$(1+\delta)^{-n/2} \sum_{0<\lambda-q_Q\leqslant \gamma_s|Q|^{-1/s}\|f_Q\|_{s,Q}} \Lambda_Q$$

is $O(N_1)$ follows from (2.6). For the upper bound in (2.25) we argue similarly, using this time the inequalities

$$(A+B)^{n/2} \leqslant (A^{1/2}+B^{1/2})^n \leqslant A^{n/2} + O\left(B^{1/2}(A^{(n-1)/2} + B^{(n-1)/2})\right)$$

and

$$(A+B)^{(n-1)/2} = O(A^{(n-1)/2} + B^{(n-1)/2})$$

valid for all $A, B \geqslant 0$. Using the definition of η_Q in (2.17) it follows, for $Q\in I_2$, that

$$|Q|\eta_Q^{n/2} \leqslant (1-\delta)^{-n/2}\Lambda_Q + O(\Lambda_Q^{(n-1)/n}+1), \quad (|Q|\eta_Q^{n/2})^{(n-1)/n} = O(\Lambda_Q^{(n-1)/n}+1).$$

With (2.36) these give

$$\mathrm{N}(\eta_Q, -\Delta_{\mathrm{N}}, Q^\circ) \leqslant \omega_n(2\pi)^{-n}(1-\delta)^{-n/2}\Lambda_Q + O(\Lambda_Q^{(n-1)/n}+1), \tag{2.42}$$

and on summing this over the cubes in I_2 and noting that $\sum_{Q\in I_2} \Lambda_Q = \Lambda$ since $\Lambda_Q = 0$ if $Q \in I_2\backslash\{Q : q_Q < \lambda\}$, we find that

$$\sum_{Q\in I_2} \mathrm{N}(\eta_Q, -\Delta_{\mathrm{N}}, Q^\circ) \leqslant \omega_n(2\pi)^{-n}(1-\delta)^{-n/2}\Lambda + O(\Lambda^{(n-1)/n}N_2^{1/n} + N_2), \tag{2.43}$$

where N_2 is the number of cubes $Q \in \mathfrak{F}$ in I_2. Since every cube $Q \in \mathfrak{F}$ for which $q_Q < \lambda$ is in I_2, (2.25) finally yields the following:

Theorem 2.7. Suppose that $q \in L^s_{loc}(\bar{\Omega})$ for some s satisfying (2.1), and assume that (2.4), (2.5) and (2.6) hold. Then for $T = T_{D,\Omega}$ or $T_{N,\Omega}$, and $\lambda < \theta_{N,\Omega}$,

$$|N(\lambda, T, \Omega) - \omega_n(2\pi)^{-n}\Lambda| \leqslant \tfrac{1}{2}n\delta(1-\delta)^{-n/2-1}\omega_n(2\pi)^{-n}\Lambda + O(\Lambda^{(n-1)/n}N_2^{1/n} + N_2), \tag{2.44}$$

where Λ is given by (2.41) and N_2 is the number of cubes Q in I_2 (cf. equation (2.19)). ■

Further approximations may now be made, aimed at the introduction of integrals rather than sums. Let $\Omega_1 = \{x \in \Omega : q(x) < \lambda\}$, let N_3 be the number of cubes $Q \in \mathfrak{F}$ which lie in Ω_1, and let U_3 be the union of all such cubes; put

$$L_Q = \int_Q (\lambda - q)_+^{n/2}, \qquad L_3 = \int_{U_3} (\lambda - q)_+^{n/2}.$$

Lemma 2.8. Suppose that $n \geqslant 2$ and let $q \in L^s_{loc}(\bar{\Omega})$ for some s satisfying (2.1). Then:

(i) for all $Q \in \mathfrak{F}$ we have $\Lambda_Q \leqslant L_Q$;

(ii) for all $Q \in \mathfrak{F}$ with $Q \subset \Omega_1$,

$$0 \leqslant L_Q - \Lambda_Q \leqslant \tfrac{1}{2}n \| f_Q \|_{s,Q} L_Q^{(n-2)/n} |Q|^{2/n - 1/s}.$$

Proof. (i) By Hölder's inequality,

$$\Lambda_Q = |Q|(\lambda - q_Q)_+^{n/2} \leqslant |Q|\left(|Q|^{-1} \int_Q (\lambda - q)_+\right)^{n/2} \leqslant \int_Q (\lambda - q)_+^{n/2} = L_Q.$$

(ii) Let $Q \in \mathfrak{F}$ with $Q \subset \Omega_1$. Then $q(x) < \lambda$ for all x in Q and $\Lambda_Q = |Q|(\lambda - q_Q)^{n/2}$. Hence

$$\Lambda_Q = |Q|\left(|Q|^{-1} \int_Q (\lambda - q)\right)^{n/2}.$$

By the Mean-Value Theorem,

$$[\lambda - q(x)]^{n/2} - (\lambda - q_Q)^{n/2} = -\tfrac{1}{2}n[q(x) - q_Q][z(x)]^{n/2-1},$$

where $|z(x)| \leqslant \max\{\lambda - q(x), \lambda - q_Q\}$. Hence

$$\int_Q |z|^{n/2} \leqslant \max\left\{\int_Q (\lambda - q)^{n/2}, |Q|(\lambda - q_Q)^{n/2}\right\} = \max\{L_Q, \Lambda_Q\} = L_Q.$$

Also, $q - q_Q = f_Q$, and so by Hölder's inequality,

$$\left| \int_Q (\lambda - q)^{n/2} - |Q|(\lambda - q_Q)^{n/2} \right| \leqslant \tfrac{1}{2}n \int_Q |f_Q|\,|z|^{(n/2)-1}$$
$$\leqslant \tfrac{1}{2}n \|f_Q\|_{n/2,Q} \|z\|_{n/2,Q}^{(n-2)/2}$$
$$\leqslant \tfrac{1}{2}n|Q|^{2/n-1/s} L_Q^{(n-2)/n} \|f_Q\|_{s,Q}.$$

The proof is complete. □

To proceed, we assume that the conditions of Theorem 2.7 hold and that $n \geqslant 2$, and use (2.39) together with Lemma 2.8 to show that for all $Q \subset U_3$ which are in I_1,

$$\mathrm{N}(\mu_Q, -\Delta_{\mathrm{D}}, Q^\circ) \geqslant \omega_n(2\pi)^{-n}(1+\delta)^{-n/2}(L_Q - \tfrac{1}{2}n|Q|^{2/n-1/s}\|f_Q\|_{s,Q} L_Q^{(n-2)/n})$$
$$- O(L_Q^{(n-1)/n} + 1) \geqslant \omega_n(2\pi)^{-n}(1+\delta)^{-n/2} L_Q - O(L_Q^{(n-1)/n} + 1),$$

the final step following because, by Young's inequality,

$$L_Q^{(n-2)/n} = O(L_Q^{(n-1)/n} + 1),$$

and of course (2.6) holds. Now sum over all these cubes; (2.25) and (2.6) give (cf. (2.40))

$$\mathrm{N}(\lambda, T_{\mathrm{D}}, \Omega) \geqslant \omega_n(2\pi)^{-n}(1+\delta)^{-n/2} L_3 - O(N_3 + N_3^{1/n} L_3^{(n-1)/n}). \quad (2.45)$$

Finally, let N_4 be the number of cubes $Q \in \mathfrak{F}$ which have non-empty intersection with Ω_1, let U_4 be the union of all such cubes and put

$$L_4 = \int_{U_4} (\lambda - q)_+^{n/2}.$$

From (2.42) and the inequality $\Lambda_Q \leqslant L_Q$ we have

$$\mathrm{N}(\eta_Q, -\Delta_{\mathrm{N}}, Q^\circ) \leqslant \omega_n(2\pi)^{-n}(1-\delta)^{-n/2} L_Q + O(1 + L_Q^{(n-1)/n}),$$

which gives, as in (2.24),

$$\mathrm{N}(\lambda, T_{\mathrm{N}}, \Omega) \leqslant \omega_n(2\pi)^{-n}(1-\delta)^{-n/2} L_4 + O(N_4 + N_4^{1/n} L_4^{(n-1)/n}).$$

These conclusions may be summarized as follows:

Theorem 2.9. Under the conditions of Theorem 2.7 and with, in addition, $n \geqslant 2$, we have for $T = T_{\mathrm{D},\Omega}$ or $T_{\mathrm{N},\Omega}$, and $\lambda < \theta_{\mathrm{N},\Omega}$,

$$\omega_n(2\pi)^{-n}(1+\delta)^{-n/2} L_3 - O(N_3 + N_3^{1/n} L_3^{(n-1)/n}) \leqslant \mathrm{N}(\lambda, T, \Omega)$$
$$\leqslant \omega_n(2\pi)^{-n}(1-\delta)^{-n/2} L_4 + O(N_4 + N_4^{1/n} L_4^{(n-1)/n}).$$
■

3. The case $\lambda \to \infty$: an example

Let $\Omega = \mathbb{R}^n$, let $\lambda_0 > 1$ and $\alpha \geqslant 1$, let β be a positive number such that $1 - \alpha^{-1} \leqslant 2\beta < 1$, and for any $\lambda > \lambda_0$, let $\mathfrak{F}_\beta(\lambda)$ be a tesselation of $\mathbb{R}^n$ by congruent cubes of side $\lambda^{-\beta}$. Let $\mathfrak{F}(\lambda)$ be a family of cubes covering $\mathbb{R}^n$ and obtained from $\mathfrak{F}_\beta(\lambda)$ by subdividing the cubes in $\mathfrak{F}_\beta(\lambda)$ when necessary so that for all $\lambda > \lambda_0$ and all Q in $\mathfrak{F}(\lambda)$ with centre x_Q,

$$|Q|^{2/n}|x_Q|^{\alpha-1} = O(1). \tag{3.1}$$

Henceforth we shall write $\mathfrak{F}$ and $\mathfrak{F}_\beta$ for $\mathfrak{F}(\lambda)$ and $\mathfrak{F}_\beta(\lambda)$. Note that for the cubes Q in $\mathfrak{F}_\beta$ such that $|x_Q| \leqslant K\lambda^{1/\alpha}$ we have

$$\lambda^{-2\beta}|x_Q|^{\alpha-1} \leqslant K^{\alpha-1}\lambda^{1-\alpha^{-1}-2\beta} = O(1),$$

and hence no subdivision of these cubes of $\mathfrak{F}_\beta$ is necessary; that is,

$$|Q|^{1/n} = \lambda^{-\beta} \quad \text{for all } Q \in \mathfrak{F} \text{ with } |x_Q| \leqslant K\lambda^{1/\alpha}. \tag{3.2}$$

We now assume that there are positive constants c_1, c_2, c_3, c_4 and θ such that for any $\lambda > \lambda_0$ and any $Q \in \mathfrak{F}$,

$$c_1|x_Q|^\alpha - c_2 \leqslant q_Q \leqslant c_3(|x_Q|^\alpha + 1) \tag{3.3}$$

and

$$\|f_Q\|_{s,Q} \leqslant c_4|Q|^{\theta/n+1/s}(|x_Q|^{\alpha-1}+1). \tag{3.4}$$

Here s satisfies (2.1) and in addition we assume that (2.4) is satisfied by q^-. Note that from (3.3) and (3.4) it follows that $\theta_{N,\mathbb{R}^n} = \infty$, and so the spectrum is discrete. Now observe the following: for all large enough λ,

$$\begin{aligned}\inf_{Q\in\mathfrak{F}}(q_Q - \gamma_s|Q|^{-1/s}\|f_Q\|_{s,Q}) &\geqslant \inf_{Q\in\mathfrak{F}}[c_1|x_Q|^\alpha - c_2 - \gamma_s c_4|Q|^{\theta/n}(|x_Q|^{\alpha-1}+1)] \\ &> -\infty,\end{aligned}$$

and also

$$\begin{aligned}\sup_{Q\in\mathfrak{F}}(|Q|^{2/n-1/s}\|f_Q\|_{s,Q}) &\leqslant c_4 \sup_{Q\in\mathfrak{F}}[|Q|^{(2+\theta)/n}(|x_Q|^{\alpha-1}+1)] \\ &= O(\lambda^{-\theta\beta}).\end{aligned}$$

Hence for large λ, (2.5) and (2.6) are satisfied and

$$\delta = O(\lambda^{-\theta\beta}). \tag{3.5}$$

Next, we estimate the terms Λ and N_2 which appear in Theorem 2.7. For $Q \in I_2$, we have from (2.19), (3.3) and (3.4),

$$\lambda \geqslant q_Q - \gamma_s|Q|^{-1/s}\|f_Q\|_{s,Q} \geqslant \tfrac{1}{2}c_1|x_Q|^\alpha - 2c_2$$

for large enough λ. Hence any Q in I_2 lies in the ball $\mathrm{B}(0, r)$, where

$$r = [2(\lambda + 2c_2)/c_1]^{1/\alpha} + \lambda^{-\beta} = O(\lambda^{1/\alpha}).$$

Thus $|Q|^{1/n} = \lambda^{-\beta}$ for $Q \in I_2$, and the number N_2 of cubes in I_2 satisfies

$$N_2 = O(\lambda^{n/\alpha + n\beta}). \tag{3.6}$$

From (3.3),

$$\Lambda = \sum_{q_Q < \lambda} |Q|(\lambda - q_Q)^{n/2} \geqslant c_3^{n/2} \sum_{|x_Q| < v} |Q|(v^\alpha - |x_Q|^\alpha)^{n/2},$$

where $v = (\lambda/c_3 - 1)^{1/\alpha}$ with $\lambda > c_3$. For $|x_Q| < v$ and $x \in Q$,

$$\begin{aligned} \big||x|^\alpha - |x_Q|^\alpha\big| &= \alpha\big|(|x| - |x_Q|)w^{\alpha-1}\big| \qquad \text{(with } |w| \leqslant \max\{|x|, |x_Q|\}) \\ &= O\big(|Q|^{1/n}(|x_Q| + |Q|^{1/n})^{\alpha-1}\big) \\ &= O(\lambda^{-\beta+(\alpha-1)/\alpha}). \end{aligned}$$

Hence with the aid of the inequality for $(A+B)^{n/2}$ used in §2.4,

$$\begin{aligned} \int_Q |v^\alpha - |x|^\alpha|^{n/2}\,\mathrm{d}x &= \int_Q [(v^\alpha - |x_Q|^\alpha) + O(\lambda^{-\beta+(\alpha-1)/\alpha})]^{n/2}\,\mathrm{d}x \\ &= |Q|(v^\alpha - |x_Q|^\alpha)^{n/2} \\ &\quad + O\big(|Q|\lambda^{-\beta/2+(\alpha-1)/2\alpha}(\lambda^{(n-1)/2} + \lambda^{-\beta(n-1)/2+(\alpha-1)(n-1)/2\alpha})\big) \\ &= |Q|(v^\alpha - |x_Q|^\alpha)^{n/2} + O(\lambda^{n/2-1/2\alpha-\beta(n+\frac{1}{2})}), \end{aligned}$$

and thus, with $S(v) = \bigcup_{|x_Q| < v} Q$, we have

$$\begin{aligned} \Lambda &\geqslant c_3^{n/2} \int_{S(v)} |v^\alpha - |x|^\alpha|^{n/2}\,\mathrm{d}x - O\Big(\lambda^{n/2-1/2\alpha-\beta(n+\frac{1}{2})} \sum_{|x_Q| < v} 1\Big) \\ &\geqslant c_3^{n/2} \int_{|x| < v} (v^\alpha - |x|^\alpha)^{n/2}\,\mathrm{d}x - O(\lambda^{n/2-1/2\alpha-\beta(n+\frac{1}{2})+n/\alpha+\beta n}), \end{aligned}$$

since the number M of cubes in $\mathfrak{F}$ with $|x_Q| < v$ satisfies

$$M|Q| = O\big((v + |Q|^{1/n})^n\big) = O(\lambda^{n/\alpha}),$$

and so $M = O(\lambda^{n/\alpha+n\beta})$ since $|Q|^{1/n} = \lambda^{-\beta}$ for $|x_Q| < v = O(\lambda^{1/\alpha})$. Note that passage from the integral over $S(v)$ to that over $B(0, v)$ is achieved by means of the estimate

$$\begin{aligned} \int_{v - c\lambda^{-\beta} < |x| < v} (v^\alpha - |x|^\alpha)^{n/2}\,\mathrm{d}x &= O(\lambda^{n/2+n/2\alpha-\beta(\frac{1}{2}n+1)-1/\alpha}) \\ &= O(\lambda^{n/2-1/2\alpha-\beta(n+\frac{1}{2})+n/\alpha+\beta n}). \end{aligned}$$

It follows that

$$\begin{aligned} \Lambda &\geqslant c_3^{n/2} v^{\alpha n/2+n} \omega_n \int_0^1 (1-r^\alpha)^{n/2} r^{n-1}\,\mathrm{d}r - O(\lambda^{n/2+n/\alpha-1/2\alpha-\beta/2}) \\ &= c_3^{-n/\alpha} \lambda^{n/2+n/\alpha} \omega_n \int_0^1 (1-r^\alpha)^{n/2} r^{n-1}\,\mathrm{d}r\,[1 - O(\lambda^{-1/2\alpha-\beta/2})]. \end{aligned} \tag{3.7}$$

We therefore obtain from Theorem 2.7, (3.5), (3.6) and (3.7),

$$\begin{aligned}|\mathrm{N}(\lambda, T, \mathbb{R}^n) - \omega_n (2\pi)^{-n} \Lambda| &= O(\lambda^{-\theta\beta}\Lambda + \lambda^{\beta-\frac{1}{2}}\Lambda + \lambda^{n(\beta-\frac{1}{2})}\Lambda) \\ &= O(\lambda^{-\theta\beta}\Lambda + \lambda^{\beta-\frac{1}{2}}\Lambda),\end{aligned}$$

or

$$\mathrm{N}(\lambda, T, \mathbb{R}^n) = \omega_n (2\pi)^{-n} \Lambda [1 + O(\lambda^{-\theta\beta} + \lambda^{\beta-\frac{1}{2}})]. \tag{3.8}$$

It remains to establish an appropriate upper bound for Λ. From (3.3),

$$\Lambda \leqslant c_1^{n/2} \sum_{|x_Q| < \mu} |Q| (\mu^\alpha - |x_Q|^\alpha)^{n/2},$$

where $\mu = [(\lambda + c_2)/c_1]^{1/\alpha}$.

By proceeding as in our derivation of the lower bound (3.7) we obtain

$$\begin{aligned}\Lambda &\leqslant c_1^{n/2} \int_{|x_Q| < \mu} |v^\alpha - |x|^\alpha|^{n/2} \, \mathrm{d}x + O(\lambda^{n/2 + n/\alpha - 1/2\alpha - \beta/2}) \\ &= c_1^{n/2} \int_{|x| < \mu} (\mu^\alpha - |x|^\alpha)^{n/2} \, \mathrm{d}x + c_1^{n/2} \int_{\mu < |x| < \mu + \lambda^{-\beta}} |\mu^\alpha - |x|^\alpha|^{n/2} \, \mathrm{d}x \\ &\qquad + O(\lambda^{n/2 + n/\alpha - 1/2\alpha - \beta/2}) \\ &= c_1^{n/2} \mu^{n\alpha/2 + n} \omega_n \int_0^1 (1 - r^\alpha)^{n/2} r^{n-1} \, \mathrm{d}r + O([(\mu + \lambda^{-\beta})^n - \mu^n][\lambda^{-\beta}\mu^{\alpha-1}]^{n/2}) \\ &\qquad + O(\lambda^{n/2 + n/\alpha - 1/2\alpha - \beta/2}) \\ &= c_1^{n/2} \lambda^{n/2 + n/\alpha} \omega_n \int_0^1 (1 - r^\alpha)^{n/2} r^{n-1} \, \mathrm{d}r \, [1 + O(\lambda^{-1/2\alpha - \beta/2})].\end{aligned}$$

We have therefore proved the following theorem.

Theorem 3.1. Let $\alpha \geqslant 1$, let β be a positive number such that $1 - \alpha^{-1} \leqslant 2\beta < 1$, let q satisfy (2.4) for some s satisfying (2.1), and suppose that (3.3) and (3.4) hold for all cubes Q in a tesselation of $\mathbb{R}^n$ satisfying (3.1) and (3.2) and arising out of a tesselation $\mathfrak{F}_\beta$ of $\mathbb{R}^n$ by cubes of side $\lambda^{-\beta}$, for all sufficiently large λ. Then for all large enough λ,

$$\mathrm{N}(\lambda, T, \mathbb{R}^n) = \omega_n (2\pi)^{-n} \Lambda \, [1 + O(\lambda^{-\theta\beta} + \lambda^{\beta-\frac{1}{2}})], \tag{3.9}$$

where

$$\Lambda = \sum_{q_Q < \lambda} |Q| \, (\lambda - q_Q)^{n/2}$$

and

$$\omega_n c_3^{-n/3} J[1 + o(1)] \leqslant \Lambda / \lambda^{n/2 + n/\alpha} \leqslant \omega_n c_1^{-n/2} J[1 + o(1)], \tag{3.10}$$

with

$$J = \int_0^1 (1-r^\alpha)^{n/2}\, r^{n-1}\, dr = \alpha^{-1} B(n/\alpha, 1+\tfrac{1}{2}n)$$

($B(p,q)$ here standing for, in traditional notation, the Eulerian integral of the first kind). The $o(1)$ terms are in fact $O(\lambda^{-1/2\alpha-\beta/2})$. ∎

To conclude this section we try to replace the sum Λ in (3.9) by an integral. Recall that $\Omega_1 = \{x \in \mathbb{R}^n : q(x) < \lambda\}$ and $U_3 = \cup_{Q \subset \Omega_1} Q$; write

$$L = \int_{\mathbb{R}^n} (\lambda - q)_+^{n/2}, \qquad L_3 = \int_{U_3} (\lambda - q)_+^{n/2}.$$

By Lemma 2.8, $\Lambda \leqslant L$ and

$$\begin{aligned} \Lambda &\geqslant \sum_{Q \subset U_3} |Q|\, (\lambda - q_Q)^{n/2} \\ &\geqslant \int_{U_3} (\lambda - q)_+^{n/2} - O\left(\sum_{Q \subset U_3} \| f_Q \|_{s,Q} |Q|^{2/n - 1/s} \left(\int_Q (\lambda - q)_+^{n/2} \right)^{(n-2)/n} \right) \\ &= L_3 - O(\lambda^{-\theta\beta} L_3^{(n-2)/n} N_3^{2/n}). \end{aligned}$$

Since $U_3 \subset I_2$, the number N_3 of cubes in U_3 is $O(\lambda^{n/\alpha + n\beta})$, by (3.6). Hence

$$\Lambda \geqslant L_3 - O(\lambda^{-\theta\beta + 2/\alpha + 2\beta} L^{(n-2)/n}).$$

If now we suppose that, as $\lambda \to \infty$,

$$0 \leqslant L - L_3 = \int_{\substack{q(x) < \lambda \\ x \notin U_3}} [\lambda - q(x)]^{n/2}\, dx = o(L), \tag{3.11}$$

then using (3.10) and the fact that $\beta < \frac{1}{2}$, we find that

$$\begin{aligned} \Lambda &\geqslant L[1 + o(1) + O(\lambda^{(2-\theta)\beta + 2/\alpha} L^{-2/n})] \\ &= L[1 + o(1) + O(\lambda^{(2-\theta)\beta + 2/\alpha - 1 - 2/\alpha})] \\ &= L[1 + o(1)]. \end{aligned}$$

We summarize our conclusions as follows:

Theorem 3.2. Let $n \geqslant 2$, suppose that the hypotheses of Theorem 3.1 hold and also assume that (3.11) is satisfied. Then as $\lambda \to \infty$,

$$N(\lambda, T, \mathbb{R}^n) = \omega_n (2\pi)^{-n} L[1 + o(1)], \tag{3.12}$$

where

$$L = \int_{\mathbb{R}^n} (\lambda - q)_+^{n/2}.$$ ∎

A particular case in which Theorem 3.2 holds is that given in Reed and Simon [2, p. 275]. Suppose that there is a constant $\alpha > 1$ and positive constants a_1, a_2, a_3 such that for all x, y in $\mathbb{R}^n$,

$$a_1(|x|^\alpha - 1) \leqslant q(x) \leqslant a_2(|x|^\alpha + 1) \tag{3.13}$$

and

$$|q(x) - q(y)| \leqslant a_3|x-y|(\max\{|x|, |y|\})^{\alpha-1}. \tag{3.14}$$

Then if $\mathfrak{F}$ is a tesselation as in Theorem 3.1,

$$\begin{aligned} |q_Q - q(x_Q)| &= |Q|^{-1}\left|\int_Q [q(x) - q(x_Q)]\,\mathrm{d}x\right| \\ &\leqslant a_3|Q|^{1/n}(|x_Q| + |Q|^{1/n})^{\alpha-1} \\ &= O\left(|Q|^{1/n}(|x_Q|+1)^{\alpha-1}\right). \end{aligned}$$

It follows, with the aid of (3.13), that (3.3) is satisfied. Also, from (3.14),

$$\begin{aligned} |f_Q(x)| &\leqslant |Q|^{-1}\int_Q |q(x) - q(y)|\,\mathrm{d}y \\ &\leqslant a_3|Q|^{1/n}(|x_Q| + |Q|^{1/n})^{\alpha-1} \\ &= O\left(|Q|^{1/n}(|x_Q|+1)^{\alpha-1}\right). \end{aligned}$$

Hence

$$\|f_Q\|_{s,Q} = O\left(|Q|^{1/s+1/n}(|x_Q|^{\alpha-1}+1)\right),$$

so that (3.4) is satisfied with $\theta = 1$.

Finally we verify that (3.11) holds. From (3.13), Ω_1 is contained in the ball with centre the origin and radius $(\lambda/a_1 + 1)^{1/\alpha}$, and so $|Q|^{1/n} = \lambda^{-\beta}$ for all Q in U_3, in view of (3.2). Also, since the distance between U_3 and the surface $q(x) = \lambda$ is less than $\lambda^{-\beta}$, it follows that for any $x \in \Omega_1 \setminus U_3$, there exists y such that $q(y) = \lambda$ and $|x - y| \leqslant \lambda^{-\beta}$. Thus

$$\begin{aligned} |\lambda - q(x)| = |q(y) - q(x)| &\leqslant a_3|x-y|\,[\max\{|x|, |y|\}]^{\alpha-1} \\ &= O(\lambda^{-\beta+(\alpha-1)/\alpha}). \end{aligned}$$

Hence

$$\begin{aligned} \int_{\Omega_1\setminus U_3} (\lambda - q)^{n/2} &= O\left((\lambda^{-\beta+1-1/\alpha})^{n/2}\,\lambda^{n/\alpha}\right) \\ &= O(\lambda^{-n\beta/2 - n/2\alpha}\,L), \end{aligned}$$

the final step following from (3.10) and the fact that $L \geqslant \Lambda$. Thus (3.11) is satisfied and (3.12) holds with the $o(1)$ terms actually $O(\lambda^{-n\beta/2-n/2\alpha} + \lambda^{\beta-1})$. The conclusion of Theorem 3.2 therefore holds for any potential q which satisfies the pointwise estimates (3.13) and (3.14). These pointwise estimates are, of course, decidedly stronger than the conditions involving mean values over cubes which are required for Theorem 3.2 to hold.

Remark 3.3. Results of the form of (3.12) have been established by many authors, under various conditions on the potential q. The first in the field seems to have been de Wet and Mandl [1]; and for similar results we refer to Fleckinger [1], Pham Thé Lai [1], Reed and Simon [2, Chapter XIII, §15], Robert [1], Rosenbljum [1,2,3], Tamura [1,2] and Titchmarsh [2, Chapter XVII, §17]. Tamura [2] even obtains a sharp estimate for the error committed in estimating $\mathrm{N}(\lambda, T, \mathbb{R}^n)$ by

$$\omega_n (2\pi)^{-n} \int_{\mathbb{R}^n} (\lambda - q)_+^{n/2}.$$

Under the assumptions that (i) there exists $m > 0$ such that $q(x)/(1+|x|^2)^{m/2}$ is bounded above and below by positive numbers independent of x, (ii) $|\mathrm{D}^\alpha q(x)| \leqslant C_\alpha (1+|x|^2)^{(m-|\alpha|)/2}$, and $x \cdot \nabla q(x) \geqslant C(1+|x|^2)^{m/2}$ for $|x| > R$, he shows that

$$\mathrm{N}(\lambda, T, \mathbb{R}^n) = \omega_n (2\pi)^{-n} [1 + O(\lambda^{-1/2-1/m})] \int_{\mathbb{R}^n} [\lambda - q(x)]_+^{n/2} \, \mathrm{d}x.$$

The sharpness of this estimate is illustrated by the case in which $q(x) = |x|^2$ and $n = 3$, for then the eigenvalues are $\lambda_j = 2j + 3$ with multiplicities $\frac{1}{2}(j+1)(j+2)$ $(j \in \mathbb{N}_0)$, so that

$$\mathrm{N}(\lambda, T, \mathbb{R}^n) = \sum_{j=0}^{[(\lambda-3)/2]} \tfrac{1}{2}(j+1)(j+2) = \tfrac{1}{48}\lambda^3 + O(\lambda^2).$$

As $\mathrm{N}(\lambda, T, \mathbb{R}^n)$ has a jump of $O(\lambda_j^2)$ at $\lambda = \lambda_j$ it follows that it cannot be approximated by a continuous function with an error estimate $o(\lambda^2)$ as $\lambda \to \infty$; and so the estimate is sharp. An extensive bibliography dealing with problems of this nature is given in Birman and Solomjak [2]. The variations played by these authors on the basic theme contained in (3.12) include the replacement of $\mathbb{R}^n$ and the Laplacian by unbounded subsets of $\mathbb{R}^n$ and more general elliptic operators, respectively, plus of course changes in the hypotheses made upon q. The treatment given here is based upon that of Edmunds and Evans [4].

4. The case $\lambda \to 0-$: an example

What we have in mind in this section is that the potential q should be negative, and that as $|x| \to \infty$, the quantity $q(x)$ should go to zero like a suitable negative power of $|x|$, giving a partly discrete, partly continuous spectrum. As before, our intention is to use Theorem 2.7, but this time the tesselation $\mathfrak{F}$ of $\mathbb{R}^n$ has to be chosen with considerably more care than in §3, where $q(x)$ was large for large $|x|$.

4.1. *The tesselation*

We begin with a lemma which establishes the existence of tesselations of the kind we shall need: in the lemma, and elsewhere, we shall write $[a]$ for the integer part of a; and given two positive functions f and g defined on an interval $(0, \mu_0)$, we shall write $f(\mu) \asymp g(\mu)$ as a convenient shorthand for the statement that there are positive absolute constants c_1 and c_2, such that for all μ in $(0, \mu_0)$,

$$c_1 \leqslant f(\mu)/g(\mu) \leqslant c_2.$$

Lemma 4.1. Let $0 < \alpha < 2$, let

$$(2-\alpha)/3\alpha < \gamma < 1/\alpha - \tfrac{1}{2} \tag{4.1}$$

and let $0 < \varepsilon < 1 - (2-\alpha)/3\alpha\gamma$. Then given any sufficiently small $\mu > 0$, say $\mu < \mu_0 < 1$, there is a tesselation $\mathfrak{F}$ of $\mathbb{R}^n$ by (non-congruent) closed cubes with disjoint interiors, with the following properties:

(i) $\mathfrak{F} = \cup_{k=1}^{\infty} \mathfrak{F}_k$, where each $\mathfrak{F}_k$ is a union of $O(\mu^{-\gamma n})$ congruent cubes; $\mathfrak{F}_1$ is a tesselation of a cube $\hat{Q}_1$ with centre the origin; $\mathfrak{F}_k$ $(k \geqslant 2)$ is a tesselation of $\hat{Q}_k \setminus \hat{Q}_{k-1}$, where $\hat{Q}_k$ is a cube with centre the origin and for all cubes Q_k in $\mathfrak{F}_k$,

$$|\hat{Q}_k|^{1/n} = |\hat{Q}_{k-1}|^{1/n} + 2[\mu^{-\gamma}]\,|Q_k|^{1/n} \qquad (k \geqslant 2) \tag{4.2}$$

$$|\hat{Q}_1|^{1/n} = 2[\mu^{-\gamma}]|Q_1|^{1/n}. \tag{4.3}$$

(ii) There is an integer m, which depends only upon ε and α, such that

$$|\hat{Q}_m|^{1/n}\mu^{1/\alpha} \to \infty \quad \text{as} \quad \mu \to 0+. \tag{4.4}$$

(iii) For each $k \in \mathbb{N}$, each cube Q_k in $\mathfrak{F}_k$, and each $\hat{Q}_k$, we have

$$\begin{aligned} &|Q_1|^{1/n} \asymp \mu^{\varepsilon\gamma/2} \quad \text{and} \quad |\hat{Q}_1|^{1/n} \asymp \mu^{-(1-\frac{1}{2}\varepsilon)\gamma}, \\ &|Q_k|^{1/n} \asymp \mu^{-\gamma a(k-1)} \text{ and } |\hat{Q}_k|^{1/n} \asymp \mu^{-\gamma[a(k-1)+1]} \qquad (2 \leqslant k \leqslant m), \end{aligned} \tag{4.5}$$

where

$$a(k) = \sum_{j=1}^{k} (1-\varepsilon^j)\left(\frac{\alpha+1}{3}\right)^j, \; |Q_k| = |Q_m| \text{ for all } k > m. \tag{4.6}$$

(iv) Any cube in $\mathbb{R}^n$ with centre the origin and sides of length $O(\mu^{-1/\alpha})$ is covered by $O(\mu^{-\gamma n})$ cubes in $\mathfrak{F}$.

(v) For each $k \geqslant 2$ and each cube Q_k in $\mathfrak{F}_k$, with centre x_{Q_k}, there is a positive constant c such that

$$|x_{Q_k}| \geqslant \tfrac{1}{2}|\hat{Q}_{k-1}|^{1/n} \geqslant \begin{cases} c\mu^{-[a(k-2)+1]\gamma} & \text{if } k > 2, \\ c\mu^{-(1-\frac{1}{2}\varepsilon)\gamma} & \text{if } k = 2. \end{cases} \tag{4.7}$$

Moreover,

$$|Q_k|^{3/n}/|x_{Q_k}|^{\alpha+1} = O(\mu^{3\theta_k\gamma}), \tag{4.8}$$

where

$$\theta_k = \begin{cases} \sum_{j=1}^{k-1} \varepsilon^j \left(\frac{\alpha+1}{3}\right)^j - \sum_{j=1}^{k-2} \varepsilon^j \left(\frac{\alpha+1}{3}\right)^{j+1} \quad (>0) \quad \text{if } k>2; \\ \frac{1}{2}\varepsilon \frac{\alpha+1}{3} \quad \text{if } k=2. \end{cases}$$ ■

Proof. We construct $\mathfrak{F}_1$ out of cubes Q_1, each of side $\mu^{\varepsilon\gamma/2}$, and take $|\hat{Q}_1|^{1/n} = 2\lfloor \mu^{-\gamma} \rfloor |Q_1|^{1/n} \asymp \mu^{-(1-\frac{1}{2}\varepsilon)\gamma}$, so that $\hat{Q}_1$ is composed of $O(\mu^{-\gamma n})$ such cubes Q_1.

The choice of $Q_2 \in \mathfrak{F}_2$ is initially determined by the need to satisfy (4.8). The centre x_{Q_2} of any such cube Q_2 must be such that $|x_{Q_2}| > \frac{1}{2}|\hat{Q}_1|^{1/n}$, and thus (4.8) will hold if

$$|Q_2|^{1/n} = O\left(\exp_\mu\left\{-\left[\left(1-\frac{\varepsilon}{2}\right)\frac{(\alpha+1)}{3} - \theta_2\right]\gamma\right\}\right) = O(\mu^{-a(1)\gamma}),$$

where $\exp_\mu(\bullet) := \mu^\bullet$, since $\theta_2 = \varepsilon(\alpha+1)/6$. Now put

$$b_\mu = |\hat{Q}_1|^{1/n} \mu^{\gamma a(1)} \asymp \exp_\mu\left[-\left((1-\varepsilon)\frac{(2-\alpha)}{3} + \tfrac{1}{2}\varepsilon\right)\gamma\right],$$

and make the choice

$$|Q_2|^{1/n} = |\hat{Q}_1|^{1/n}/\lfloor b_\mu \rfloor \asymp \mu^{-a(1)\gamma}.$$

Thus (4.5) is satisfied for $|Q_2|$. As for $|\hat{Q}_2|$, since we require (4.2) to hold, we have

$$\begin{aligned} |\hat{Q}_2|^{1/n} &= |\hat{Q}_1|^{1/n} + 2\lfloor \mu^{-\gamma} \rfloor |Q_2|^{1/n} \\ &= 2\lfloor \mu^{-\gamma} \rfloor (|Q_1|^{1/n} + |Q_2|^{1/n}) \\ &\asymp \mu^{-[a(1)+1]\gamma}, \end{aligned}$$

and so the rest of (4.5) holds for $k=2$. It is also clear that $\hat{Q}_2 \setminus \hat{Q}_1$ is the union of $O(\mu^{-\gamma n})$ cubes Q_2 in $\mathfrak{F}_2$.

Now suppose that $\hat{Q}_k$ has been constructed with the essential property

$$|\hat{Q}_k|^{1/n} \asymp \mu^{-[a(k-1)+1]\gamma},$$

and put

$$c_\mu = |\hat{Q}_k|^{1/n} \mu^{a(k)\gamma} \asymp \exp_\mu\left\{-\left[1-\left(\frac{\alpha+1}{3}\right)^k (1-\varepsilon^k)\right]\gamma\right\};$$

$c_\mu \to \infty$ as $\mu \to 0+$ since $\alpha < 2$. The choice of $|Q_{k+1}|$ such that

$$|Q_{k+1}|^{1/n} = |\hat{Q}_k|^{1/n}/\lfloor c_\mu \rfloor \asymp \mu^{-a(k)\gamma}$$

ensures that (4.5) is satisfied. Also, since $|x_{Q_{k+1}}| > \frac{1}{2}|\hat{Q}_k|^{1/n}$,

$$|Q_{k+1}|^{3/n}/|x_{Q_{k+1}}|^{\alpha+1} = O\left(\exp_\mu\left\{-3\left[a(k)-\left(\frac{\alpha+1}{3}\right)a(k-1)-\frac{\alpha+1}{3}\right]\gamma\right\}\right)$$
$$= O(\mu^{3\theta_{k+1}\gamma})$$

as required by (4.8). The inductive step is therefore established.

Next we turn our attention to the integer m mentioned in (ii). Since

$$\sum_{j=0}^{\infty}\left(\frac{\alpha+1}{3}\right)^j = \frac{3}{2-\alpha},$$

there is an integer m, depending only upon α and ε, such that

$$\sum_{j=0}^{m-1}\left(\frac{\alpha+1}{3}\right)^j > \frac{3}{2-\alpha} - \varepsilon.$$

Of course, there are infinitely many such m, but for the sake of definiteness we take the least one. From (4.5) we see that there is a positive constant K such that

$$|\hat{Q}_m|^{1/n} \geqslant K\mu^{-\gamma[a(m-1)+1]}$$
$$\geqslant K\exp_\mu\left\{-\gamma\left[\frac{3}{2-\alpha}-\varepsilon-\sum_{j=1}^{\infty}\varepsilon^j\left(\frac{\alpha+1}{3}\right)^j\right]\right\},$$

and since

$$\sum_{j=1}^{\infty}\varepsilon^j\left(\frac{\alpha+1}{3}\right)^j = \varepsilon\left(\frac{\alpha+1}{3}\right)\Big/\left[1-\varepsilon\left(\frac{\alpha+1}{3}\right)\right] < \varepsilon\left(\frac{\alpha+1}{3}\right)\Big/\left(1-\frac{\alpha+1}{3}\right)$$
$$= \varepsilon(\alpha+1)/(2-\alpha),$$

it follows that $|\hat{Q}_m|^{1/n} \geqslant K\mu^{-3\gamma(1-\varepsilon)/(2-\alpha)}$. Since $\varepsilon < 1-(2-\alpha)/3\alpha\gamma$, we have $3(1-\varepsilon)/(2-\alpha) > 1/\alpha\gamma$. Hence $|\hat{Q}_m|^{1/n}\mu^{1/\alpha} \to \infty$ as $\mu \to 0+$, and (4.4) holds.

All that remains is to show that (iv) holds. However, this is immediate from (4.4), which makes it plain that for all small enough μ, the given cube may be covered by the cubes in $\cup_{k=1}^{m}\mathfrak{F}_k$, and there are evidently $O(\mu^{-\gamma n})$ such cubes. □

Remarks 4.2. With a view to the use shortly to be made of the tesselation $\mathfrak{F}$, we make a few helpful observations here about what happens to various quantities as $\mu \to 0+$.

(1) For $k = 2, 3, \cdots, m$, and all cubes Q_{k-1} and Q_k in $\mathfrak{F}_{k-1}$ and $\mathfrak{F}_k$

respectively we have

$$|Q_{k-1}|^{1/n}/|Q_k|^{1/n} \asymp \exp_\mu\left[(1-\varepsilon^{k-1})\left(\frac{\alpha+1}{3}\right)^{k-1}\gamma\right] = o(1). \tag{4.9}$$

This is immediate from (4.5).

(2) Given any cube Q in $\mathfrak{F}$, with centre x_Q, there is a positive integer τ such that

$$|Q|^{1/n}/(|x_Q|+1) = O(\mu^\tau); \tag{4.10}$$

in fact $\tau = \frac{1}{2}\varepsilon\gamma$ for all Q in $\mathfrak{F}_1$ and $\tau = (1-\varepsilon^{k-1})\left(\frac{2-\alpha}{3}\right)+\varepsilon^{k-1}$ for all Q in $\mathfrak{F}_k$ $(k \geqslant 2)$.

The proof of this follows directly from (4.5) for cubes in $\mathfrak{F}_1$, while for a cube Q in $\mathfrak{F}_k$ $(k \geqslant 3)$ we again use (4.5) together with (4.7) and the estimate

$$\begin{aligned}
|Q|^{1/n}/(|x_Q|+1) &= O(\mu^{-[a(k-1)-a(k-2)-1]\gamma}) \\
&= O\left(\exp_\mu\left\{-\left[(1-\varepsilon^{k-1})\left(\frac{\alpha+1}{3}\right)^{k-1}-1\right]\gamma\right\}\right) \\
&= O\left(\exp_\mu\left\{-\left[(1-\varepsilon^{k-1})\left(\frac{\alpha+1}{3}\right)-1\right]\gamma\right\}\right) \\
&= O\left(\exp_\mu\left\{\left[(1-\varepsilon^{k-1})\left(\frac{2-\alpha}{3}\right)+\varepsilon^{k-1}\right]\gamma\right\}\right) \\
&= O(\mu^\tau).
\end{aligned}$$

The argument for cubes in $\mathfrak{F}_2$ is similar.

(3) For all x in an arbitrary cube Q of $\mathfrak{F}$, with centre x_Q,

$$|x|+1 \asymp |x_Q|+1 = (|x|+1)[1+O(\mu^\tau)], \tag{4.11}$$

where τ is as in Remark 2.

This follows directly from the observation that

$$|x_Q|-|Q|^{1/n}+1 \leqslant |x|+1 \leqslant |x_Q|+|Q|^{1/n}+1.$$

For then by Remark 2,

$$|(|x|+1)-(|x_Q|+1)| \leqslant |Q|^{1/n} \leqslant (|x_Q|+1)\,O(\mu^\tau) = (|x|+1)\,O(\mu^\tau).$$

(4) For all x in an arbitrary cube Q belonging to $\mathfrak{F}_k$ $(k \geqslant 2)$,

$$|x_Q| = |x|[1+O(\mu^\tau)]. \tag{4.12}$$

This follows since, by Remarks 2 and 3, and the fact that $|\hat{Q}_1|^{1/n} \geqslant 1$,

$$||x_Q|-|x|| \leqslant |Q|^{1/n} = (|x|+1)\,O(\mu^\tau) = |x|\,O(\mu^\tau).$$

(5) For all cubes Q in $\mathfrak{F}_k$ $(k \geqslant 2)$,

$$|Q|^{1/n}/|x_Q|^{\alpha/2} = O\left(\exp_\mu\left[-\tfrac{1}{2}\gamma + \tfrac{1}{2}\gamma\left(\frac{\alpha+1}{3}\right)^{k-1}\right]\right). \tag{4.13}$$

To establish this, first note that if $Q \in \mathfrak{F}_k$ and $k \geqslant 3$, then by (4.5) and (4.7),

$$|Q|^{1/n}/|x_Q|^{\alpha/2} = O\left(\exp_\mu\left[-\left(a(k-1) - \frac{\alpha}{2}a(k-2) - \frac{\alpha}{2}\right)\gamma\right]\right)$$

$$= O\left(\exp_\mu\left\{-\left[(1-\varepsilon^{k-1})\left(\frac{\alpha+1}{3}\right)^{k-1} + \left(\frac{2-\alpha}{\alpha}\right)a(k-2) - \frac{\alpha}{2}\right]\gamma\right\}\right).$$

Summation of the series for $a(k-2)$ now shows that the right-hand side is

$$o\left(\exp_\mu\left\{-\left[\tfrac{1}{2} - \tfrac{1}{2}\left(\frac{\alpha+1}{3}\right)^{k-1} - \varepsilon^{k-1}\left(\frac{\alpha+1}{3}\right)^{k-1}\right]\gamma\right\}\right),$$

which is in turn

$$o\left(\exp_\mu\left\{-\left[\tfrac{1}{2} - \tfrac{1}{2}\left(\frac{\alpha+1}{3}\right)^{k-1}\right]\gamma\right\}\right),$$

as required. The proof for cubes in $\mathfrak{F}_2$ is similar.

(6) Since $\sum_{j=1}^{\infty}\left(\frac{\alpha+1}{3}\right)^j = \frac{\alpha+1}{2-\alpha}$, it follows from (4.5) that for all Q in $\mathfrak{F}$,

$$|Q|^{1/n} = o(\mu^{-\beta}) \quad \text{where} \quad \beta = \left(\frac{\alpha+1}{2-\alpha}\right)\gamma. \tag{4.14}$$

Note that if

$$\gamma < \min\left\{\frac{2-\alpha}{2\alpha}, \frac{2-\alpha}{\alpha(\alpha+1)}\right\}, \tag{4.15}$$

then

$$\tfrac{1}{2} < \beta < 1/\alpha;$$

the left-hand inequality holds because, since $\alpha < 2$,

$$\beta > \left(\frac{\alpha+1}{2-\alpha}\right)\left(\frac{2-\alpha}{3\alpha}\right) = \frac{\alpha+1}{3\alpha} > \tfrac{1}{2};$$

the right-hand inequality is clear.

4.2. *The spectrum*

We may now embark on our study of the spectrum. Suppose there are positive constants a_1, a_2, a_3 and α, with $\alpha < 2$, such that for all x and y in $\mathbb{R}^n$,

$$-a_1(|x|+1)^{-\alpha} \leqslant q(x) \leqslant -a_2(|x|+1)^{-\alpha} \tag{4.16}$$

and

$$|q(x)-q(y)| \leqslant a_3|x-y|(\min\{|x|,|y|\}+1)^{-\alpha-1}. \tag{4.17}$$

Let $\mathfrak{F}$ be the tesselation in Lemma 4.1. From (4.11) and (4.16) it is clear that for all x in an arbitrary cube Q of $\mathfrak{F}$,

$$-q(x) \asymp (|x_Q|+1)^{-\alpha},$$

and so there are positive constants c_1 and c_2 such that

$$-c_1(|x_Q|+1)^{-\alpha} \leqslant q_Q \leqslant -c_2(|x_Q|+1)^{-\alpha}. \tag{4.18}$$

Moreover, (4.17) and (4.11) show that

$$\begin{aligned} |f_Q(x)| &\leqslant a_3|Q|^{-1}\int_Q |x-y|(\min\{|x|,|y|\}+1)^{-\alpha-1}\,dy \\ &= O\big(|Q|^{1/n}(|x_Q|+1)^{-\alpha-1}\big). \end{aligned}$$

Hence there is a positive constant c_3 such that for all cubes Q in $\mathfrak{F}$,

$$\gamma_s|Q|^{-1/s}\|f_Q\|_{s,Q} \leqslant c_3|Q|^{1/n}(|x_Q|+1)^{-\alpha-1}. \tag{4.19}$$

These inequalities (4.18) and (4.19) are, in general, better adapted to our techniques than (4.16) and (4.17), and we shall make heavy use of them in what follows. Note that from (4.18) and (4.19) it follows that $\theta_{N,\mathbf{R}^n}=0$ and hence $\sigma_e \subset [0,\infty)$. In fact, Theorem X.4.8 can be used to show that $\sigma_e=[0,\infty)$ under these conditions.

We now set about the verification of the hypotheses under which our basic results in §2 hold. First we note that, by (4.16) and (4.11),

$$|q^-(x)| \leqslant \text{const}\,.\,(|x_Q|+1)^{-\alpha}$$

for all x in an arbitrary cube Q in $\mathfrak{F}$; thus for all such Q,

$$|Q|^{-1/s}\|q^-\|_{s,Q} = O(|x_Q|+1)^{-\alpha},$$

and (2.4) is satisfied. As for (2.5), we have, by (4.18) and (4.19),

$$q_Q-\gamma_s|Q|^{-1/s}\|f_Q\|_{s,Q} \geqslant -c_1(|x_Q|+1)^{-\alpha}-c_3|Q|^{1/n}(|x_Q|+1)^{-\alpha-1}.$$

If $Q\in\mathfrak{F}_k$ for some $k\geqslant 2$, then by (4.8),

$$-|Q|^{1/n}(|x_Q|+1)^{-\alpha-1} \geqslant \text{const}\,.\,|Q|^{-2/n}\mu^{3\theta_k\gamma},$$

and from this (2.5) follows immediately. Moreover, for all Q in $\mathfrak{F}$, (4.19) and (4.8) show that

$$|Q|^{2/n-1/s}\|f_Q\|_{s,Q} = O(|Q|^{3/n}(|x_Q|+1)^{-\alpha-1}) = o(1),$$

and so (2.6) holds.

Our next objective is the estimation of N_2, the number of cubes in I_2 (cf. (2.19), with $\lambda=-\mu$). For any cube Q in I_2 we have

$$q_Q < -\mu+\gamma_s|Q|^{-1/s}\|f_Q\|_{s,Q},$$

and hence, by (4.18), (4.19) and Remark 2,

$$-c_1(|x_Q|+1)^{-\alpha} < -\mu + O\big(|Q|^{1/n}/(|x_Q|+1)\big)(|x_Q|+1)^{-\alpha}$$
$$= -\mu + o\big((|x_Q|+1)^{-\alpha}\big).$$

Thus for sufficiently small μ we have $2c_1(|x_Q|+1)^{-\alpha} > \mu$. Together with Remark 6 this implies that

$$U_2 = \bigcup_{Q \in I_2} Q$$

is contained in the ball with centre 0 and radius r given by

$$r = \left(\frac{\mu}{2c_1}\right)^{-1/\alpha} - 1 + o(\mu^{-\beta}).$$

If γ is chosen to satisfy (4.15) we therefore have $r = O(\mu^{-1/\alpha} + \mu^{-\beta}) = O(\mu^{-1/\alpha})$, so that, by (iv) of Lemma 4.1,

$$N_2 = O(\mu^{-\gamma n}). \tag{4.20}$$

It remains to estimate

$$\Lambda = \sum_{q_Q < -\mu} |Q|(-\mu - q_Q)^{n/2},$$

both from above and from below. First note that

$$\Lambda \leqslant \sum_{-c_1(|x_Q|+1)^{-\alpha} < -\mu} |Q|[-\mu + c_1(|x_Q|+1)^{-\alpha}]^{n/2},$$

so that if we write $\xi = (\mu/c_1)^{-1/\alpha} - 1$ we find that

$$\Lambda \leqslant c_1^{n/2}(\xi+1)^{-\alpha n/2} \sum_{|x_Q| < \xi} |Q|[(\xi+1)^{\alpha} - (|x_Q|+1)^{\alpha}]^{n/2}(|x_Q|+1)^{-\alpha n/2}$$
$$= c_1^{n/2}(\xi+1)^{-\alpha n/2} J, \text{ say.}$$

Now write the sum J as an integral:

$$J = \int_{|x_Q| < \xi} [(\xi+1)^{\alpha} - (|x_Q|+1)^{\alpha}]^{n/2}(|x_Q|+1)^{-\alpha n/2}\,dx,$$

and divide the region of integration into two parts, the first over all cubes in $\mathfrak{F}_1$, the second over all cubes in $\mathfrak{F}$ not in $\mathfrak{F}_1$ but with $|x_Q| < \xi$. This gives two integrals, J_1 and J_2, say. The first of these is easily estimated, and we find that

$$J_1 = O(\xi^{\alpha n/2}|\hat{Q}_1|) = O(\mu^{-\frac{1}{2}n - (1-\frac{1}{2}\varepsilon)\gamma n})$$
$$= o(\mu^{-n(\frac{1}{2}+\gamma)}).$$

Since $\gamma + \frac{1}{2} < 1/\alpha$ this shows that

$$J_1 = o(\mu^{-n/\alpha}). \tag{4.21}$$

More effort is needed to deal with J_2. Let $Q \in \mathfrak{F} \setminus \mathfrak{F}_1$ be such that $|x_Q| < \xi$. Then with the aid of (4.12) we find that

$$(\xi+1)^\alpha - (|x_Q|+1)^\alpha = \xi^\alpha(1+\xi^{-1})^\alpha - |x|^\alpha(1+|x|^{-1})^\alpha[1+O(\mu^\tau)]$$
$$= \xi^\alpha - |x|^\alpha + O(\mu^{1/\alpha - 1}) + O\big(|x|^\alpha(|x_Q|^{-1}+\mu^\tau)\big).$$

By (4.3) and (4.5),

$$|x_Q| > |\hat{Q}_1|^{1/n} > \text{const}\,.\,\mu^{-(1-\frac{1}{2}\varepsilon)\gamma},$$

and so for $|x_Q| < \xi$, and hence $|x| = O(\xi)$, we have

$$(\xi+1)^\alpha - (|x_Q|+1)^\alpha = \xi^\alpha - |x|^\alpha + O(\mu^{\tau-1}) + O(\mu^{\gamma-1-\frac{1}{2}\varepsilon\gamma}),$$

since $1/\alpha > \gamma + \frac{1}{2} > \gamma(1-\frac{1}{2}\varepsilon)$. It follows that

$$[(\xi+1)^\alpha - (|x_Q|+1)^\alpha]^{n/2} \leqslant |\xi^\alpha - |x|^\alpha|^{n/2}$$
$$+O\big((\mu^{(\tau-1)/2} + \mu^{(\gamma-1-\frac{1}{2}\varepsilon\gamma)/2})(\mu^{-(n-1)/2} + \mu^{(n-1)(\tau-1)/2} + \mu^{(n-1)(\gamma-1-\frac{1}{2}\varepsilon\gamma)/2})\big)$$
$$= |\xi^\alpha - |x|^\alpha|^{n/2} + O(\mu^{(\tau-n)/2} + \mu^{(\gamma-n-\frac{1}{2}\varepsilon\gamma)/2}).$$

Since in addition

$$|x_Q|+1 = |x|(1+|x|^{-1})[1+O(\mu^\tau)] = |x|[1+O(|x_Q|^{-1})][1+O(\mu^\tau)]$$
$$= |x|[1+O(\mu^{(1-\frac{1}{2}\varepsilon)\gamma})][1+O(\mu^\tau)]$$
$$= |x|[1+O(\mu^{(1-\frac{1}{2}\varepsilon)\gamma}) + O(\mu^\tau)],$$

we ultimately find that

$$[(\xi+1)^\alpha - (|x_Q|+1)^\alpha]^{n/2}(|x_Q|+1)^{-\alpha n/2}$$
$$= |\xi^\alpha - |x|^\alpha|^{n/2}|x|^{-\alpha n/2}[1+O(\mu^{(1-\frac{1}{2}\varepsilon)\gamma}) + O(\mu^\tau)]$$
$$+O\big(|x_Q|^{-\alpha n/2}(\mu^{(\tau-n)/2} + \mu^{(\gamma-n-\frac{1}{2}\varepsilon\gamma)/2})\big).$$

Thus for $Q \in \mathfrak{F}_k$ $(k \geqslant 2)$ and $|x_Q| < \xi$, we have, with the aid of (4.13),

$$\int_Q [(\xi+1)^\alpha - (|x_Q|+1)^\alpha]^{n/2}(|x_Q|+1)^{-\alpha n/2}\,dx = \int_Q |\xi^\alpha - |x|^\alpha|^{n/2}|x|^{-\alpha n/2}\,dx$$
$$+O\big(\mu^{-n/2}(\mu^{(1-\frac{1}{2}\varepsilon)\gamma} + \mu^\tau)|Q|\,|x_Q|^{-\alpha n/2}\big)$$
$$+O\big(|Q|\,|x_Q|^{-\alpha n/2}(\mu^{(\tau-n)/2} + \mu^{(\gamma-n-\frac{1}{2}\varepsilon\gamma)/2})\big)$$
$$= \int_Q |\xi^\alpha - |x|^\alpha|^{n/2}|x|^{-\alpha n/2}\,dx$$
$$+o\left((\mu^{\frac{1}{2}(\tau-n)} + \mu^{\frac{1}{2}(\gamma-n-\frac{1}{2}\varepsilon\gamma)})\exp_\mu\left[-\tfrac{1}{2}n\gamma + \tfrac{1}{2}n\gamma\left(\frac{\alpha+1}{3}\right)^{k-1}\right]\right)$$
$$= \int_Q |\xi^\alpha - |x|^\alpha|^{n/2}|x|^{-\alpha n/2}\,dx + o(\mu^{\frac{1}{2}(\tau-n-n\gamma)} + \mu^{\frac{1}{2}(\gamma-n-n\gamma-\frac{1}{2}\varepsilon\gamma)}).$$

Since I_2 contains $O(\mu^{-\gamma n})$ cubes of $\mathfrak{F}$ (cf. (4.20)), and by (4.14), $|x| < \xi + K\mu^{-\beta}$ if $|x_Q| < \xi$, with K a positive constant, it follows that

$$J_2 \leqslant \int_{|x| < \xi + K\mu^{-\beta}} |\xi^\alpha - |x|^\alpha|^{n/2} |x|^{-\alpha n/2}\, dx + o(\mu^{\frac{1}{2}(\tau - n - 3n\gamma)} + \mu^{\frac{1}{2}(\gamma - n - 3n\gamma - \frac{1}{2}\varepsilon\gamma)}).$$

We now split the integral on the right-hand side into two parts, one over the region $|x| < \xi$ and the other over $\xi \leqslant |x| < \xi + K\mu^{-\beta}$, change to polar coordinates in the first integral and observe that the second is $O(\xi^{n-1}\mu^{-\beta}) = o(\xi^n)$; then it turns out that

$$J_2 \leqslant \xi^n \omega_n \int_0^1 (1 - r^\alpha)^{n/2} r^{n-1-\frac{1}{2}\alpha n}\, dr + o(\xi^n) + o(\mu^{\frac{1}{2}(\tau - n - 3n\gamma)} + \mu^{\frac{1}{2}(\gamma - n - 3n\gamma - \frac{1}{2}\varepsilon\gamma)}).$$

At this stage we make a particular choice of γ. Write

$$\gamma = \tfrac{2}{3}\alpha^{-1} - \tfrac{1}{3} + \delta,$$

where δ is some positive number small enough for (4.1) and (4.15) to hold. Then

$$\tfrac{1}{2} - \alpha^{-1} + \tfrac{3}{2}\gamma = \tfrac{3}{2}\delta,$$

and so

$$\tfrac{1}{2}(\tau - n - 3n\gamma) = \tfrac{1}{2}\tau - n\alpha^{-1} - \tfrac{3}{2}n\delta > -n\alpha^{-1} \tag{4.22}$$

if $\delta < \frac{1}{3}\tau/n$; while

$$\tfrac{1}{2}(\gamma - n - 3n\gamma - \tfrac{1}{2}\varepsilon\gamma) = \tfrac{1}{2}\gamma - n\alpha^{-1} - \tfrac{3}{2}n\delta - \tfrac{1}{4}\varepsilon\gamma > -n\alpha^{-1} \tag{4.23}$$

if $\varepsilon + 6n\delta\gamma^{-1} < 2$. Note that by (4.10), $\tau > 0$. Thus (4.22) and (4.23) may be satisfied simply by choosing δ and ε sufficiently small: henceforth we assume that such choices have been made. This implies that

$$J_2 \leqslant \xi^n \omega_n [1 + o(1)] \int_0^1 (1 - r^\alpha)^{n/2} r^{n-1-\frac{1}{2}\alpha n}\, dr,$$

which together with (4.21) yields the estimate

$$\Lambda \leqslant c_1^{n/\alpha} \mu^{-n/\alpha + n/2} \omega_n [1 + o(1)] \int_0^1 (1 - r^\alpha)^{n/2} r^{n-1-\frac{1}{2}\alpha n}\, dr. \tag{4.24}$$

Next we derive a lower bound for Λ. With $v = (\mu/c_2)^{-1/\alpha} - 1$ it follows easily that

$$\Lambda \geqslant c_2^{n/2} (v+1)^{-\alpha n/2} \sum_{|x_Q| < v} [(v+1)^\alpha - (|x_Q| + 1)^\alpha]^{n/2} (|x_Q| + 1)^{-\alpha n/2}$$
$$= c_2^{n/2} (v+1)^{-\alpha n/2} S, \text{ say.}$$

Evidently,

$$S \geqslant \int_{\substack{Q \in \mathfrak{F}\setminus\mathfrak{F}_1 \\ |x_Q| < \nu}} [(\nu+1)^\alpha - (|x_Q|+1)^\alpha]^{n/2} (|x_Q|+1)^{-\alpha n/2}\, dx.$$

By proceeding as in the derivation of the upper bound for Λ we find that if $Q \in \mathfrak{F}\setminus\mathfrak{F}_1$ and $|x_Q| < \nu$, then

$$[(\nu+1)^\alpha - (|x_Q|+1)^\alpha]^{n/2} \geqslant |\nu^\alpha - |x|^\alpha|^{n/2} - O(\mu^{\frac{1}{2}(\tau-n)} + \mu^{\frac{1}{2}(\gamma-n-\frac{1}{2}\varepsilon\gamma)})$$

and

$$\begin{aligned}[(\nu+1)^\alpha &- (|x_Q|+1)^\alpha]^{n/2} (|x_Q|+1)^{-\alpha n/2} \\ &\geqslant |\nu^\alpha - |x|^\alpha|^{n/2} |x|^{-\alpha n/2} [1 + O(\mu^{(1-\frac{1}{2}\varepsilon)\gamma} + \mu^\tau)] \\ &\quad - O\big((\mu^{\frac{1}{2}(\tau-n)} + \mu^{\frac{1}{2}(\gamma-n-\frac{1}{2}\varepsilon\gamma)}) |x_Q|^{-\alpha n/2}\big),\end{aligned}$$

whence

$$\begin{aligned}\int_Q [(\nu+1)^\alpha &- (|x_Q|+1)^\alpha]^{n/2} (|x_Q|+1)^{-\alpha n/2}\, dx \\ &\geqslant \int_Q |\nu^\alpha - |x_Q|^\alpha|^{n/2} |x|^{-\alpha n/2}\, dx - o(\mu^{\frac{1}{2}(\tau-n-n\gamma)} + \mu^{\frac{1}{2}(\gamma-n-n\gamma-\frac{1}{2}\varepsilon\gamma)}).\end{aligned}$$

Since the sum S involves $O(\mu^{-\gamma n})$ cubes and $|x_Q| < \nu$ if $|x| < \nu - |Q|^{1/n}$, we find, just as before, that

$$\Lambda \geqslant c_2^{n/\alpha} \mu^{-n/\alpha + n/2} \omega_n [1 - o(1)] \int_0^1 (1 - r^\alpha)^{n/2} r^{n-1-\frac{1}{2}n\alpha}\, dr.$$

Moreover, $N_2 = O(\mu^{-\gamma n}) = o(\mu^{-n/\alpha + n/2})$ since $\gamma < 1/\alpha - \frac{1}{2}$, and the derivation of (4.18) makes it plain that as $\mu \to 0+$, so $c_i = a_i[1 + o(1)]$ $(i = 0, 1)$. We have therefore proved the following.

Theorem 4.3. Let q satisfy (4.16) and (4.17). Then as $\mu \to 0+$,

$$\mathrm{N}(-\mu, T, \mathbb{R}^n) = \omega_n (2\pi)^{-n} \Lambda [1 + o(1)],$$

where

$$\begin{aligned}a_2^{n/\alpha}[1 + o(1)] &\leqslant \Lambda \mu^{n/\alpha - n/2} \omega_n^{-1} \Big/ \int_0^1 (1 - r^\alpha)^{n/2} r^{n-1-\frac{1}{2}n\alpha}\, dr \\ &\leqslant a_1^{n/\alpha}[1 + o(1)].\end{aligned}$$

∎

Note that

$$\int_0^1 (1 - r^\alpha)^{n/2} r^{n-1-\frac{1}{2}n\alpha}\, dr = \alpha^{-1} \mathrm{B}\left(\frac{n}{\alpha} - \frac{n}{2}, 1 + \frac{n}{2}\right)$$

in standard notation.

To complete our analysis it is desirable to represent $\mathrm{N}(-\mu, T, \mathbb{R}^n)$ in integral form. This is now a comparatively simple matter. Put

$$\Omega_1 = \{x \in \mathbb{R}^n : q(x) < -\mu\}, \qquad U_1 = \bigcup_{Q \in \mathfrak{F}, Q \subset \Omega_1} Q,$$

$$L = \int_{\mathbb{R}^n} (-\mu - q)_+^{n/2}, \quad L_1 = \int_{U_1} (-\mu - q)_+^{n/2}, \quad L_Q = \int_Q (-\mu - q)_+^{n/2}.$$

From Lemma 2.8,

$$\begin{aligned} L \geqslant \Lambda &\geqslant \sum_{Q \subset U_1} |Q| (-\mu - q_Q)_+^{n/2} \\ &\geqslant L_1 - O\Big(\sum_{Q \subset U_1} \|f_Q\|_{s,Q} |Q|^{2/n - 1/s} L_Q^{(n-2)/n} \Big). \end{aligned}$$

From (4.16) and Lemma 4.1 we see that there are $O(\mu^{-\gamma n})$ cubes in U_1. Also, as in Theorem 4.3, it follows that $\mathrm{L}_1 \asymp \mu^{-n/\alpha + n/2}$. Thus, with (4.19) and (4.8), we find that

$$\begin{aligned} L \geqslant \Lambda &\geqslant L_1 - o\big(L_1^{(n-2)/n} (\mu^{-\gamma n})^{2/n}\big) \\ &= L_1 [1 - o\big((\mu^{-\gamma n + n/\alpha - n/2})^{2/n}\big)] \\ &= L_1 [1 - o(1)], \end{aligned}$$

the final step following because $\gamma < \alpha^{-1} - \frac{1}{2}$.

Now let $x \in \Omega_1 \setminus U_1$. Recalling the properties of the integer m introduced in Lemma 4.1 we see that, provided that μ is small enough, there is an element y of $\mathbb{R}^n$ such that $q(y) = -\mu$ and $|x - y| < |Q_m|^{1/n}$. Thus

$$\begin{aligned} |-\mu - g(x)| &= O\big(|Q_m|^{1/n} / (\min\{|x|, |y|\} + 1)^{\alpha+1}\big) \\ &= O\big(|Q_m|^{1/n} / (|x_{Q_m}| + 1)^{\alpha+1}\big) \end{aligned}$$

since by (4.8),

$$\begin{aligned} \min_{Q_m} \{|x|, |y|\} &\geqslant |x_{Q_m}| - |Q_m|^{1/n} \geqslant |x_{Q_m}| - o(|x_{Q_m}|^{(\alpha+1)/3}) \\ &= [1 - o(1)] |x_{Q_m}|. \end{aligned}$$

Thus since $q(x) < -\mu$ if, and only if, $|x| < \text{const}.\, \mu^{-1/\alpha}$, we have

$$\begin{aligned} 0 \leqslant L - L_1 &= O\Big(\int_{|x| \leqslant O(\mu^{-1/\alpha})} [|Q_m|^{1/n} / (|x_{Q_m}| + 1)^{\alpha+1}]^{n/2} \, dx \Big) \\ &= O\big([|Q_m|^{3/n} / (|x_{Q_m}| + 1)^{\alpha+1}]^{n/2} \mu^{-n/\alpha} |Q_m|^{-1}\big). \end{aligned}$$

By (4.4), for any cube Q_m in $\hat{Q}_m$ we must have

$$\Big[\sum_{j=1}^{m-1} (1 - \varepsilon^j) \Big(\frac{\alpha+1}{3} \Big)^j + 1 \Big] \gamma > \frac{1}{\alpha},$$

so that from (4.5),

$$|Q_m|^{1/n} > \text{const}.\, \mu^{\gamma - 1/\alpha}.$$

Thus by (4.8),

$$\begin{aligned} 0 \leqslant L - L_1 &= O(\mu^{3\theta_m \gamma - n\gamma}) = o(\mu^{-n\gamma}) \\ &= o(\mu^{-n/\alpha + n/2}) \\ &= o(L_1) = o(L), \end{aligned}$$

and so $L_1 = L[1 + o(1)]$ and $\Lambda = L[1 + o(1)]$.

We summarize our conclusions:

Theorem 4.4. Under the hypotheses of Theorem 4.3,

$$\mathrm{N}(-\mu, T, \mathbb{R}^n) = \omega_n (2\pi)^{-n} [1 + o(1)] \int_{\mathbb{R}^n} (-\mu - q)_+^{n/2}.$$

■

This is the result stated by Reed and Simon ([2], Theorem XIII.82). They do not, however, give the additional information contained in our Theorem 4.3. Related results were obtained by Brownell and Clark [1], Feigin [1], McLeod [1], Rosenbljum [4], Schmincke [2] and Tamura [1].

5. The case $\Omega = \mathbb{R}^n$, $q \in \mathrm{L}^{n/2}(\mathbb{R}^n)$, $n \geqslant 3$

In this section we derive the Cwikel–Lieb–Rosenbljum bound for $\mathrm{N}(0, T, \mathbb{R}^n)$ from Theorem 2.7. Throughout we shall assume that $n \geqslant 3$ and $q \in \mathrm{L}^{n/2}(\mathbb{R}^n)$. In this case, since $\|f_Q\|_{n/2,Q} \leqslant 2\|q\|_{n/2,Q}$ and $|q_Q| \leqslant |Q|^{-2/n}\|q\|_{n/2,Q}$, we have that $\theta_{\mathrm{D},\mathbb{R}^n} = \theta_{\mathrm{N},\mathbb{R}^n} = 0$, so that $\sigma_e(T) \subset [0, \infty)$. In fact, it follows from Theorem X.4.8 that $\sigma_e(T) = [0, \infty)$ and that $-\Delta + q$ is bounded below.

Proposition 5.1. Suppose that $\|q\|_{n/2,\mathbb{R}^n} < 1/(2\gamma_{n/2})$. Then $\mathrm{N}(0, T, \mathbb{R}^n) = 0$. ■

Proof. We first observe that for any cube Q,

$$\gamma_{n/2}\|f_Q\|_{n/2,Q} \leqslant 2\gamma_{n/2}\|q\|_{n/2,\mathbb{R}^n} := \delta < 1.$$

Hence, given any $\varepsilon > 0$, there is a tesselation $\mathfrak{F}$ of $\mathbb{R}^n$ by congruent cubes Q such that for all $Q \in \mathfrak{F}$,

$$\gamma_{n/2}\|f_Q\|_{n/2,Q} \leqslant \delta < 1, \qquad |Q|^{-2/n}\|q\|_{n/2,Q} < \varepsilon.$$

Now apply Theorem 2.7 with $\mu = -\lambda > (1 + 2\gamma_{n/2})\varepsilon$:

$$\Lambda = \Sigma |Q| (-\mu - q_Q)_+^{n/2} \leqslant \Sigma |Q| [-\mu + (q_Q)_-]_+^{n/2} = 0$$

since

$$(q_Q)_- \leqslant |Q|^{-2/n}\|q\|_{n/2,Q} < \varepsilon < \mu,$$

and

$$N_2 = \#\{Q \in \mathfrak{F}: q_Q - \gamma_{n/2} |Q|^{-2/n} \|f_Q\|_{n/2,Q} \leqslant -\mu\} = 0$$

since

$$\begin{aligned} q_Q - \gamma_{n/2} |Q|^{-2/n} \|f_Q\|_{n/2,Q} &\geqslant -|Q|^{-2/n} \|q\|_{n/2,Q} - 2\gamma_{n/2} |Q|^{-2/n} \|q\|_{n/2,Q} \\ &\geqslant -(1+2\gamma_{n/2})\varepsilon > -\mu. \end{aligned}$$

It follows that $\mathrm{N}(-\mu, T, \mathbb{R}^n) = 0$. Since ε, and hence μ, may be chosen arbitrarily small, the result follows. □

Proposition 5.2. Let $n \geqslant 3$, let $q \in L^{n/2}(\mathbb{R}^n)$, and suppose that q has compact support. Then

$$\mathrm{N}(0, T, \mathbb{R}^n) \leqslant c_n \int_{\mathbb{R}^n} |q|^{n/2},$$

where c_n is a constant depending only on n. ■

Proof. Let the support of q lie inside a cube Q_0, and set

$$J(Q) = (2\gamma_{n/2})^{n/2} \int_Q |q|^{n/2}$$

for any cube Q. Let $\varepsilon \in (0, \frac{1}{2})$ and let M be the integer part of $\varepsilon^{-1} J(Q_0)$; note that in view of Proposition 5.1 we may, and shall, assume that $J(Q_0) \geqslant 1$. We now need some results concerning *Besicovich coverings* of the cube $\bar{Q}_0$, by which is meant a covering of $\bar{Q}_0$ by open cubes $Q \subset \mathbb{R}^n$ with edges parallel to those of $\bar{Q}_0$, and such that given any $x \in \bar{Q}_0$, there is a Q in the covering with x as its centre. The *linkage* $\mathrm{K}(\mathfrak{H})$ of a covering $\mathfrak{H}$ is the least integer k such that $\mathfrak{H}$ can be split into k subsets $\mathfrak{H}_1, \mathfrak{H}_2, \ldots, \mathfrak{H}_k$ where the cubes in each $\mathfrak{H}_j$ are disjoint. We claim that there is a covering $\mathfrak{H}$ of $\bar{Q}_0$ by cubes $Q \subset \bar{Q}_0$ which are open in $\bar{Q}_0$ and satisfy the following conditions:

(i) $J(Q) \leqslant M^{-1} J(Q_0)$;
(ii) $\mathrm{K}(\mathfrak{H}) \leqslant c_n$, a constant which depends only on n;
(iii) $\# \{Q : Q \in \mathfrak{H}\} \leqslant M c_n$.

This claim will be proved in a moment. Accepting it for the time being, we take as our family $\mathfrak{F}$ of cubes the cubes in the covering $\mathfrak{H}$ of $\bar{Q}_0$ together with those in an arbitrary (disjoint) covering of $\mathbb{R}^n \setminus \bar{Q}_0$. Of course, the cubes in $\mathfrak{F}$ are no longer disjoint: this would affect matters only to the extent of introducing a factor c_n in the right-hand side of (2.9), (2.10) and (2.11). Since $\theta_{\mathrm{D},\mathbb{R}^n} = \theta_{\mathrm{N},\mathbb{R}^n} = 0$ in the case under consideration, our results are unaffected and, in particular, the conclusion of Theorem 2.7 is unchanged.

From the above we see that for each $Q \in \mathfrak{H}$,

$$\begin{aligned}\gamma_{n/2} \|f_Q\|_{n/2,Q} &\leqslant J^{2/n}(Q) \leqslant [M^{-1}J(Q_0)]^{2/n} \\ &\leqslant \{J(Q_0)/[\varepsilon^{-1}J(Q_0)-1]\}^{2/n} \\ &\leqslant [\varepsilon/(1-\varepsilon)]^{2/n} < 1\end{aligned}$$

since $J(Q_0) \geqslant 1$ and $\varepsilon < \frac{1}{2}$. Let $\lambda = -\mu < 0$. Then

$$\begin{aligned}\Lambda = \sum_{Q \in \mathfrak{F}} |Q| (-\mu - q_Q)_+^{n/2} &\leqslant \sum_{Q \in \mathfrak{H}} |Q| [(q_Q)_-]^{n/2} \\ &\leqslant c_n \|q\|_{n/2, \mathbb{R}^n}^{n/2},\end{aligned}$$

and

$$\begin{aligned}N_2 &= \# \{Q : q_Q - \gamma_{n/2} |Q|^{-2/n} \|f_Q\|_{n/2,Q} \leqslant -\mu\} \\ &\leqslant \# \{Q : Q \in \mathfrak{H}\} \leqslant c_n M \\ &\leqslant c_n J(Q_0) \leqslant c_n \|q\|_{n/2, \mathbb{R}^n}^{n/2}.\end{aligned}$$

The Proposition now follows from Theorem 2.7. It remains, however, to establish the claim that a covering $\mathfrak{H}$ exists which satisfies (i), (ii) and (iii), and we now set about this, following de Guzman [1]. □

Theorem 5.3. There is a constant k_n, which depends only upon n, such that given any Besicovich covering $\mathfrak{H}$ of $\bar{Q}_0$, there is a finite subcover $\mathfrak{H}_0$ of $\bar{Q}_0$ with linkage $\mathrm{K}(\mathfrak{H}_0) \leqslant k_n$. ■

Proof. Given any cube Q, let $\delta(Q)$ be the diameter of Q. Without loss of generality, let $\mathfrak{H} = \{Q(x) : x \in \bar{Q}_0\}$, where $Q(x)$ is an open cube centred at x; write $a_0 = \sup\{\delta(Q(x)) : x \in \bar{Q}_0\}$. If $a_0 = \infty$, there exists $x \in \bar{Q}_0$ such that $Q(x) \supset \bar{Q}_0$. Suppose that $a_0 < \infty$. Choose $Q(x^{(1)}) \in \mathfrak{H}$ such that $\delta(Q(x^{(1)})) > \frac{1}{2}a_0$ and put $a_1 = \sup\{\delta(Q(x)) : x \in \bar{Q}_0 \backslash Q(x^{(1)})\}$; then choose $Q(x^{(2)})$, with $x^{(2)} \in \bar{Q}_0 \backslash Q(x^{(1)})$, such that $\delta(Q(x^{(2)})) > \frac{1}{2}a_1$; and so on, choosing $Q(x^{(m+1)})$ with $x^{(m+1)} \in \bar{Q}_0 \backslash \cup_{k=1}^{m} Q(x^{(k)})$ at the general stage. If $\bar{Q}_0 \backslash \cup_{k=1}^{m} Q(x^{(k)}) = \varnothing$ for some $m \in \mathbb{N}$, then the selection process stops and we have a finite subcover of $\bar{Q}_0$. Otherwise we obtain an infinite sequence $(Q(x^{(k)}) : k \in \mathbb{N})$ of cubes from the family $\mathfrak{H}$. Note that if $i, j \in \mathbb{N}$ and $i > j$, then $x^{(i)} \notin Q(x^{(j)})$ and $\delta(Q(x^{(j)})) > \frac{1}{2}\delta(Q(x^{(i)}))$; and so $\frac{1}{3}Q(x^{(i)}) \cap \frac{1}{3}Q(x^{(j)}) = \varnothing$, where $\frac{1}{3}Q$ is the cube concentric with Q and with sides one third the length of those of Q. To see that this is the case, suppose there is an x in $\frac{1}{3}Q(x^{(i)}) \cap \frac{1}{3}Q(x^{(j)})$; then for some $k \in \{1, 2, \ldots, n\}$, $|x_k^{(i)} - x_k^{(j)}| > \frac{1}{2}s_j$, where s_j is the side length of $Q(x^{(j)})$, while $|x_k - x_k^{(i)}| < \frac{1}{6}s_i < \frac{1}{3}s_j$ and $|x_k - x_k^{(j)}| < \frac{1}{6}s_j$, so that $|x_k^{(i)} - x_k^{(j)}| \leqslant |x_k^{(i)} - x_k| + |x_k - x_k^{(j)}| < \frac{1}{2}s_j$, which gives a contradiction.

For simplicity of notation, write $Q(x^{(j)}) = Q_j$ from now on. We claim that the Q_j form an open covering of $\bar{Q}_0$. To see this, suppose the contrary. Then there exists $x \in \bar{Q}_0 \setminus \cup_{k=1}^{\infty} Q_k$, and since the sets $\frac{1}{3}Q_k$ $(k \in \mathbb{N})$ are pairwise disjoint, there exists $k_0 \in \mathbb{N}$ such that $\delta(Q(x)) > 2\delta(Q_{k_0})$, which contradicts the process of selection of the $x^{(k)}$. Thus $\bar{Q}_0 \subset \cup_{k=1}^{\infty} Q_k$, and as $\bar{Q}_0$ is compact, there is a finite sub-cover $\mathfrak{H}_0$ of $\bar{Q}_0$, by cubes $Q_{k(1)}, Q_{k(2)}, \ldots, Q_{k(m)}$, say, with $k(i)$ monotonic increasing in i. We shall now prove that there exists $N \in \mathbb{N}$ depending only on n such that for any $i > N$, the cube $Q_{k(i)}$ is disjoint from one of its predecessors. First we show that each $z \in \mathbb{R}^n$ lies in at most c_n (a number depending only on n) of the cubes $Q_{k(i)}$. To do this we draw n hyperplanes through z, parallel to the coordinate hyperplanes, and consider the 2^n closed hyperquadrants through z determined by them. In each hyperquadrant the number of cubes $Q_{k(i)}$ with centres in the hyperquadrant is at most 2 since $x^{(i)} \in Q(x^{(j)})$ for $i > j$ and $\delta(Q^{(j)}) > \frac{1}{2}\delta(Q(x^{(i)}))$. Therefore z lies in at most $2^n \times 2 = 2^{n+1}$ of the cubes. We now claim that $N = 2^{n+1}(2^n + 2^n + 2n) + 1$ will do. Consider $Q_{k(i)}$ with $i > N$. Any cube $Q_{k(j)}$ with $j < i$ which meets $Q_{k(i)}$ must contain either a vertex, a centre point of an edge, or a centre of a face, since $\delta(Q_{k(j)}) > \frac{1}{2}\delta(Q_{k(i)})$. Hence from the argument above, $Q_{k(i)}$ can meet at most $2^{n+1}(2^n + 2^n + 2n)$ of its predecessors. It now follows that the covering has the claimed linkage. □

We can at last set about the proof of the existence of the covering $\mathfrak{H}$. We may suppose that $J(Q_0) = 1$, since otherwise we would simply have to change ε. Let $x \in \bar{Q}_0$ and let $Q(x)$ be a cube centred at x, with edges parallel to those of Q_0, and such that $J(Q(x) \cap \bar{Q}_0) = M^{-1}$; the existence of such a cube follows by continuity of J. Clearly $\{Q(x) : x \in \bar{Q}_0\}$ is a Besicovich covering of $\bar{Q}_0$. By Theorem 5.3, there is a sub-covering $\tilde{\mathfrak{H}}$ of this covering for which $K(\tilde{\mathfrak{H}}) \leqslant k_n$. Then

$$M^{-1} \;\#\{Q : Q \in \tilde{\mathfrak{H}}\} = \sum_{Q \in \tilde{\mathfrak{H}}} J(Q \cap \bar{Q}_0) \leqslant k_n J(\bar{Q}_0) = k_n,$$

and so

$$\#\{Q : Q \in \tilde{\mathfrak{H}}\} \leqslant M k_n.$$

The required covering $\mathfrak{H}$ is constructed from $\tilde{\mathfrak{H}}$ as follows. If $Q \in \tilde{\mathfrak{H}}$ and $Q \subset \bar{Q}_0$, let $Q \in \mathfrak{H}$. Now suppose that $Q \in \tilde{\mathfrak{H}}$ and that $Q \not\subset \bar{Q}_0$. Since the centre of Q lies in $\bar{Q}_0$, the ratio of the lengths of the longest and shortest edges of the rectangular box $Q \cap \bar{Q}_0$ is not greater than two, and so $Q \cap \bar{Q}_0$ can be covered by not more than 2^{n-1} cubes contained in it and open in $\bar{Q}_0$; all these cubes are also put in $\mathfrak{H}$, the description of which is now complete. It is now easy to see that properties (i), (ii), and (iii) hold for $\mathfrak{H}$, with $c_n = 2^{n-1}k_n$.

Theorem 5.4. Suppose that $n \geqslant 3$ and $q \in L^{n/2}(\mathbb{R}^n)$. Then

$$N(0, T, \mathbb{R}^n) \leqslant c_n \int_{\mathbb{R}^n} |q^-|^{n/2}. \qquad ■$$

Proof. Since $q \geqslant -q^-$, we have $N(0, -\Delta + q, \mathbb{R}^n) \leqslant N(0, -\Delta - q^-, \mathbb{R}^n)$ and hence we may suppose that $q \leqslant 0$. Let Q_0 be a cube so large that

$$\gamma_{n/2} \|q\|_{n/2, \mathbb{R}^n \setminus Q_0} < \tfrac{1}{8}, \tag{5.1}$$

and define

$$q_1(x) = \begin{cases} q(x) & \text{if } x \in Q_0, \\ 0 & \text{if } x \notin Q_0, \end{cases}$$

$q_2 = q - q_1$. By Proposition 5.2,

$$N(0, -\Delta + 2q_1, \mathbb{R}^n) \leqslant c_n \int_{\mathbb{R}^n} |q|^{n/2}. \tag{5.2}$$

From (5.1) and the fact that $q_2 = 0$ in Q_0, it follows that for any cube Q, with $g = q_2 - (q_2)_Q$,

$$\gamma_{n/2} \|g\|_{n/2, Q} \leqslant 2\gamma_{n/2} \|q_2\|_{n/2, Q} < \tfrac{1}{4},$$

and so

$$\gamma_{n/2} \|2g\|_{n/2, Q} < \tfrac{1}{2} < 1.$$

Hence, as in Proposition 5.1,

$$N(0, -\Delta + 2q_2, \mathbb{R}^n) = 0. \tag{5.3}$$

Now write $-\Delta + q = (-\tfrac{1}{2}\Delta + q_1) + (-\tfrac{1}{2}\Delta + q_2)$. The operators $-\tfrac{1}{2}\Delta + q_1$ and $-\tfrac{1}{2}\Delta + q_2$ are both self-adjoint and bounded below, and they have the same form domain $W^{1,2}(\mathbb{R}^n)$. We claim that

$$N(0, -\Delta + q, \mathbb{R}^n) \leqslant N(0, -\tfrac{1}{2}\Delta + q_1, \mathbb{R}^n) + N(0, -\tfrac{1}{2}\Delta + q_2, \mathbb{R}^n), \tag{5.4}$$

and granted this, the theorem follows immediately from (5.2) and (5.3).

All that remains is to prove (5.4). To do this, suppose that $-\tfrac{1}{2}\Delta + q_1$ has N negative eigenvalues, with eigenvectors $\psi_1, \psi_2, \ldots, \psi_N$. By (5.3), $-\tfrac{1}{2}\Delta + q_2$ has no negative eigenvalues, and so for all ϕ in $W^{1,2}(\mathbb{R}^n)$,

$$\tfrac{1}{2} \|\nabla\phi\|^2_{\mathbb{R}^n} + (q_2\phi, \phi)_{\mathbb{R}^n} \geqslant 0,$$

by the Max–Min Principle (Theorem 1.3). Moreover, if $\phi \in W^{1,2}(\mathbb{R}^n)$ is orthogonal to $\psi_1, \psi_2, \ldots, \psi_N$, then

$$\tfrac{1}{2} \|\nabla\phi\|^2_{\mathbb{R}^n} + (q_1\phi, \phi)_{\mathbb{R}^n} \geqslant 0,$$

again by the Max–Min Principle and since $-\tfrac{1}{2}\Delta + q_1$ has only N negative eigenvalues and 0 belongs to its essential spectrum (which coincides with $[0, \infty)$ since $q \in L^{n/2}(\mathbb{R}^n)$). Hence

$$\|\nabla\phi\|^2_{\mathbb{R}^n} + (q\phi, \phi)_{\mathbb{R}^n} = \tfrac{1}{2} \|\nabla\phi\|^2_{\mathbb{R}^n} + (q_1\phi, \phi)_{\mathbb{R}^n} + \tfrac{1}{2} \|\nabla\phi\|^2_{\mathbb{R}^n} + (q_2\phi, \phi)_{\mathbb{R}^n} \geqslant 0,$$

and so by the Max–Min Principle again the result follows. □

Remark 5.5. (i) Proposition 5.1 shows that there are no negative eigenvalues if $\|q^-\|_{n/2, \mathbb{R}^n} < 1/2\gamma_{n/2}$, where $\gamma_{n/2}$ is the norm of the embedding of $W^{1,2}((0,1)^n)$ in

$L^{2n/(n-2)}((0,1)^n)$ and $n \geqslant 3$. Sharper results have been obtained by Veling [2], using the optimal constant K for the Sobolev inequality

$$\|u\|_{2n/(n-2\theta), \mathbb{R}^n} \leqslant K \|\nabla v\|^{\theta}_{\mathbb{R}^n} \|v\|^{1-\theta}_{\mathbb{R}^n};$$

Weinstein [1] has also derived this optimal constant. Veling shows that there are no negative eigenvalues if

$$\|q^-\|_{n/2, \mathbb{R}^n} < \frac{\pi^{n/2} n (\tfrac{1}{2}n)^{n/2} (\tfrac{1}{2}n - 1)^{n/2} \Gamma(\tfrac{1}{2}) \Gamma(\tfrac{1}{2}n)}{\Gamma(1 + \tfrac{1}{2}n) \Gamma(\tfrac{1}{2}n + \tfrac{1}{2})} \qquad (n \geqslant 3).$$

(ii) Theorem 5.4 was first announced by Rosenbljum [1] and was independently proved by Cwikel [1] and Lieb [1, 2]. Lieb obtained the sharpest constant c_n and conjectured that the best possible value of c_n should be

$$C_n := \left[\left(\frac{n(n-2)}{4} \right)^{n/2} n\omega_n \frac{\Gamma(\tfrac{1}{2})\Gamma(\tfrac{1}{2}n)}{\Gamma[\tfrac{1}{2}(n+1)]} \right]^{-1}, \qquad \omega_n = \pi^{n/2}/\Gamma(1 + \tfrac{1}{2}n).$$

This conjecture remains open, but Li and Yau [1] showed that one could take c_n to be $e^{n/2} C_n$, and even though this is worse than Lieb's proven estimate we sketch their proof below because of its simplicity. Actually Li and Yau claim that their work improves that of Lieb but, as pointed out to us by Veling, there is a numerical error which invalidates their claim.

First let Ω be a bounded domain in $\mathbb{R}^n$ (with $n \geqslant 3$), let q be a positive function in $L^{n/2}(\Omega)$ and let μ_k be the kth eigenvalue (arranged in increasing order) of the problem

$$\Delta\psi = -\mu q \psi \text{ in } \Omega, \qquad \psi = 0 \text{ on } \partial\Omega, \tag{5.5}$$

with corresponding eigenfunction ψ_k ($k \in \mathbb{N}$) in $L^2(\Omega; q) := \{u : uq^{1/2} \in L^2(\Omega)\}$. We claim that

$$\mu_k^{n/2} \int_\Omega q^{n/2} \geqslant k \left(\frac{n(n-2)}{4e} \right)^{n/2} n\omega_n \, \Gamma(\tfrac{1}{2})\Gamma(\tfrac{1}{2}n)/\Gamma[\tfrac{1}{2}(n+1)]. \tag{5.6}$$

To prove this, suppose (without loss of generality) that the ψ_k form an orthonormal sequence in $L^2(\Omega; q)$, and put

$$H(x, y, t) = \sum_{k=1}^{\infty} e^{-\mu_k t} \psi_k(x) \psi_k(y)$$

for $x, y \in \bar{\Omega}$ and $t > 0$; H has the properties that in the weak sense,

$$\left(\frac{1}{q(y)} \Delta_y - \frac{\partial}{\partial t} \right) H(x,y,t) = 0, \qquad H(x,y,t) > 0 \quad \text{in } \Omega \times \Omega \times (0, \infty),$$

and

$$H(x, y, t) = 0 \quad \text{in } \partial\Omega \times \partial\Omega \times (0, \infty).$$

Put $h(t) = \sum_{k=1}^{\infty} e^{-2\mu_k t}$; clearly

$$h(t) = \int_\Omega \int_\Omega H^2(x, y, t)\, q(x) q(y)\, dx dy.$$

Moreover,

$$\begin{aligned}\frac{\partial h}{\partial t} &= 2\int_\Omega \int_\Omega H(x, y, t)\, q(x) q(y) \frac{\partial H}{\partial t}(x, y, t)\, dx dy \\ &= 2\int_\Omega \int_\Omega H(x, y, t) \Delta_y H(x, y, t) q(x)\, dy dx \\ &= -2\int_\Omega q(x) \int_\Omega |\nabla_y H(x, y, t)|^2\, dy dx. \end{aligned} \tag{5.7}$$

Also

$$\begin{aligned} h(t) &= \int_\Omega q(x) \int_\Omega H^2(x, y, t)\, q(y)\, dy dx \\ &\leqslant \int_\Omega q(x) \left(\int_\Omega H^{2n/(n-2)}(x, y, t)\, dy \right)^{(n-2)/(n+2)} \\ &\qquad \times \left(\int_\Omega H(x, y, t) q^{(n+2)/4}(y)\, dy \right)^{4/(n+2)} dx \\ &\leqslant \left[\int_\Omega q(x) \left(\int_\Omega H^{2n/(n-2)}(x, y, t)\, dy \right)^{(n-2)/n} dx \right]^{n/(n+2)} \\ &\qquad \times \left[\int_\Omega q(x) \left(\int_\Omega H(x, y, t) q^{(n+2)/4}(y)\, dy \right)^2 dx \right]^{2/(n+2)} \end{aligned} \tag{5.8}$$

Let

$$Q(x, t) = \int_\Omega H(x, y, t) q^{(n+2)/4}(y)\, dy.$$

Then in the weak sense,

$$\left(\frac{1}{q(x)} \Delta_x - \frac{\partial}{\partial t} \right) Q(x, t) = 0,$$

$Q(x,t) = 0$ on $\partial\Omega \times (0, \infty)$, and $Q(x, 0) = q^{(n-2)/4}(x)$, the last equality following since the ψ_k form an orthonormal basis of $L^2(\Omega; q)$. Also

$$\begin{aligned} \frac{\partial}{\partial t} \int_\Omega Q^2(x, t)\, q(x)\, dx &= 2\int_\Omega Q(x, t) \frac{\partial}{\partial t} Q(x, t)\, q(x)\, dx \\ &= 2\int_\Omega Q(x, t)\, \Delta_x Q(x, t)\, dx \\ &= -2\int_\Omega |\nabla_x Q(x, t)|^2\, dx \\ &\leqslant 0. \end{aligned}$$

Hence

$$\int_\Omega Q^2(x,t)\,q(x)\,dx \leqslant \int_\Omega Q^2(x,0)\,q(x)\,dx$$
$$= \int_\Omega q^{n/2}(x)\,dx,$$

which with (5.8) shows that

$$h^{(n+2)/n}(t)\left(\int_\Omega q^{n/2}(x)\,dx\right)^{-2/n} \leqslant \int_\Omega q(x)\left(\int_\Omega H^{2n/(n-2)}(x,y,t)\,dy\right)^{(n-2)/n} dx.$$

We now invoke Theorem V.3.6:

$$\int_{\mathbb{R}^n} |\nabla f|^2 \geqslant C\left(\int_{\mathbb{R}^n} |f|^{2n/(n-2)}\right)^{(n-2)/n}.$$

Application of this to $H(x, \bullet, t)$ and use of the best possible constant

$$C = \tfrac{1}{4}n(n-2)\,\{n\omega_n\,\Gamma(\tfrac{1}{2})\,\Gamma(\tfrac{1}{2}n)/\Gamma[\tfrac{1}{2}(n+1)]\}^{2/n}$$

(a result due to Aubin (cf. [1, p. 39]), Lieb and Thirring [1] and Talenti [1]) now gives, with (5.7) and with $\tilde{\omega}_{n-1} = n\omega_n\,\Gamma(\tfrac{1}{2})\,\Gamma(\tfrac{1}{2}n)/\Gamma[\tfrac{1}{2}(n+1)]$,

$$\frac{\partial h}{\partial t} \leqslant -\tfrac{1}{2}n(n-2)\,\tilde{\omega}_{n-1}^{2/n}\left(\int_\Omega q^{n/2}\right)^{-2/n} h^{(n+2)/n}(t).$$

Division of both sides of this by $h^{(n+2)/n}(t)$ and subsequent integration shows that

$$\sum_{i=1}^{\infty} e^{-2\mu_i t} = h(t) \leqslant (n-2)^{-n/2}\,\tilde{\omega}_{n-1}^{-1}\left(\int_\Omega q^{n/2}\right) t^{-n/2}.$$

With $t = n/(4\mu_k)$ this gives

$$(n-2)^{-n/2}\,\tilde{\omega}_{n-1}^{-1}\left(\int_\Omega q^{n/2}\right)\left(\frac{n}{4}\right)^{-n/2} \mu_k^{n/2} \geqslant \sum_{i=1}^{\infty} \exp\left(\frac{-n\mu_i}{2\mu_k}\right)$$
$$\geqslant k\,e^{-n/2},$$

which is (5.6).

We can now prove Theorem 5.4 in the form

$$N(0, -\Delta + q, \mathbb{R}^n) \leqslant \left(\frac{4e}{n(n-2)}\right)^{n/2} \tilde{\omega}_{n-1}^{-1} \int_{\mathbb{R}^n} |q^-|^{n/2}, \tag{5.9}$$

where $q \in L^{n/2}(\mathbb{R}^n)$ and $n \geqslant 3$. To do this, first observe that, as in the proof of

Theorem 5.4, it may be supposed that $q \leqslant 0$; and since we may approximate q by a sequence of negative functions, in the norm of $L^{n/2}(\mathbb{R}^n)$, we can assume that $q(x) < 0$ for all $x \in \mathbb{R}^n$. Moreover, as $\mathbb{R}^n$ may be expressed as the union of bounded domains, it is enough to prove (cf. Lemma 2.3 and Proposition 2.4) that

$$N(0, T_D, \Omega) \leqslant \left(\frac{4e}{n(n-2)}\right)^{n/2} \tilde{\omega}_{n-1}^{-1} \int_\Omega |q^-|^{n/2} \tag{5.10}$$

for every domain Ω in $\mathbb{R}^n$. We claim that for any such Ω, $N(0, T_D, \Omega)$ is equal to the number of eigenvalues less than 1 for the problem described by (5.5) and with q replaced by $-q$. To see this, note that the quadratic form associated with $T_{D,\Omega}$ is

$$\frac{\int_\Omega (|\nabla\phi|^2 + q|\phi|^2)}{\int_\Omega |\phi|^2} = \frac{\int_\Omega |q|\,|\phi|^2}{\int_\Omega |\phi|^2}\left(\frac{\int_\Omega |\nabla\phi|^2}{\int_\Omega |q|\,|\phi|^2} - 1\right) \tag{5.11}$$

Thus the subspace on which the left-hand side of (5.11) is non-positive has dimension equal to that of the subspace on which the quadratic form $(\int_\Omega |\nabla\phi|^2)/\int_\Omega |q|\phi^2$ is less than or equal to 1. As this latter quadratic form is that associated with the operator related to (5.5), the claim follows. The proof of (5.10) now follows on applying (5.6) with μ_k equal to the greatest eigenvalue less than or equal to 1. This gives

$$\int_\Omega |q|^{n/2} \geqslant \mu_k^{n/2} \int_\Omega |q|^{n/2} \geqslant k\left(\frac{n(n-2)}{4e}\right)^{n/2} \tilde{\omega}_{n-1}$$

$$\geqslant N(0, T_D, \Omega)\left(\frac{n(n-2)}{4e}\right)^{n/2} \tilde{\omega}_{n-1},$$

which is (5.10). This completes the proof of (5.9).

XII

Estimates for the singular values of $-\Delta + q$ when q is complex

1. Introduction

Unlike Chapter XI, in which the Schrödinger operator $-\Delta + q$ was considered with a real potential q and a self-adjoint map was induced, the present chapter deals with a complex potential, the corresponding operator being non-self-adjoint. Up to very recent times little detailed work appears to have been done on the spectral properties of the Schrödinger operator in this case and when the underlying open subset Ω of $\mathbb{R}^n$ is unbounded. Some information on the distribution of the singular values of this problem is given by Fleckinger [2], the central idea being the comparison of these singular values with the eigenvalues of a nearby self-adjoint problem (see also Ramm [1]). Fleckinger's work is typically concerned with the operator

$$-\Delta + (1 + |x|^2)^k + \mathrm{i}(1 + |x|^2)^{k/p},$$

with $p > 2$ and with zero Dirichlet conditions on the boundary of an unbounded open set in $\mathbb{R}^n$; it turns out that the spectrum of the corresponding operator is discrete and that as $\lambda \to \infty$,

$$\mathrm{M}(\lambda) := \sum_{s_j < \lambda} 1$$

(the s_j being the singular values, arranged in increasing order and repeated according to multiplicities) behaves like $\Sigma_{\mu_j < \lambda} 1$, where the μ_j are the eigenvalues of $-\Delta + (1 + |x|^2)^k$. Such a problem is included in the work of Robert [1] who is also able to determine the behaviour of $\Sigma_{\mathrm{re}\,\lambda_j < \lambda} 1$ or $\Sigma_{|\lambda_j| < \lambda} 1$ as $\lambda \to \infty$, the λ_j being the eigenvalues of the original problem.

In the present chapter we have objectives which are broader than those of Fleckinger and Robert in the papers referred to above, since we deal with general complex potentials q having non-negative real part and satisfying weak integrability conditions, and seek global estimates as well as asymptotic results. The price paid for this programme is that it is necessary to work with $\mathrm{M}(\lambda)$, the parameter λ being any real number lying below the essential spectrum of $|T|$, where T is the Dirichlet or Neumann operator associated with

$-\Delta+q$ on an open set Ω. In this we are in good company, for M is precisely the function used by Birman and Solomjak [1] in their study of non-self-adjoint problems. Ideally, of course, one would like to obtain estimates for $\Sigma_{\mathrm{re}\,\lambda_j<\lambda} 1$ or $\Sigma_{|\lambda_j|<\lambda} 1$, but such results are best obtained by the Carleman method, which involves a detailed analysis of the spectral kernel and requires more smoothness of q than we wish to allow. Moreover the spectral kernel technique gives rise to asymptotic results (and not the global estimates we seek) in which the contribution of $\mathrm{im}\, q$ features as a perturbation of the result for $-\Delta+\mathrm{re}\, q$; the significance of $\mathrm{im}\, q$ when it is not subordinate to $\mathrm{re}\, q$ is therefore undetected. We strive to deal with the problem as a genuine non-self-adjoint one, not as a perturbation of a self-adjoint one. We apply the standard technique of localization to cubes forming a covering of Ω, and with the use of the Max–Min Principle go some way towards obtaining an analogue of the general estimates in Chapter XI for the self-adjoint problem in which q is real. In the general circumstances under consideration it seems difficult to obtain results on eigenvalues directly in view of the lack of useful relations between individual eigenvalues and singular values of non-self-adjoint (and non-normal) maps. However we are able to say something (in §4) about the l^p classes to which the sequence of eigenvalues belong. Our treatment is based upon Edmunds, Evans and Fleckinger [1].

2. Bounds for $\mathrm{M}(\lambda, T_{\mathrm{D}}, \Omega)$ and $\mathrm{M}(\lambda, T_N, \Omega)$

2.1. *The Dirichlet and Neumann operators*

As always, Ω will stand for an open subset of $\mathbb{R}^n$. We shall be concerned with the Schrödinger operator

$$\tau := -\Delta+q, \tag{2.1}$$

where q is a complex-valued function on Ω which belongs to $\mathrm{L}^1_{\mathrm{loc}}(\Omega)$, and we shall extend q to the whole of $\mathbb{R}^n$ by setting it equal to zero outside Ω, this extension being still denoted by q. We decompose the extended q into its real and imaginary parts f and g, and assume that the real part is non-negative:

$$q = f+\mathrm{i}g, \qquad f \geqslant 0. \tag{2.2}$$

By $\mathfrak{F}$ we shall always mean a covering of Ω by a countable family of closed cubes Q with disjoint interiors and faces parallel to the coordinate axes. Under the assumption (2.2) and the supposition that the hypotheses of Theorem VII.1.15 hold, it has been shown in Theorem X.2.6 that the associated Dirichlet and Neumann operators $T_{\mathrm{D},\Omega}$ and $T_{\mathrm{N},\Omega}$ have essential spectra constrained by (X.2.11) and (X.2.12), while information about the location of their spectra is given by Theorem VII.1.15. We now claim that if, in addition,

$$\theta_0 := \inf_{Q\in\mathfrak{F}} [f_Q - \gamma_r \rho_r(f, Q)] > 0 \tag{2.3}$$

then

$$\sigma_e(T^*_{D,\Omega}T_{D,\Omega}) \subset [\phi_0^2, \infty), \tag{2.4}$$

where ϕ_0 is defined in Theorem X.2.6, that is,

$$\phi_0 = \lim_{N\to\infty} \inf_{Q\subset\Omega_N} [f_Q - \gamma_r \rho_r(f, Q)].$$

To establish this let $q_0 \in L^r(\Omega_N)$ and define

$$q_N(x) = \begin{cases} q(x) & \text{if } x\in\Omega_N, \\ q_0(x) & \text{if } x\in\Omega\backslash\Omega_N, \end{cases}$$

where the sets Ω_N $(N\in\mathbb{N})$ are defined just as in §2 of Chapter X. Choose q_0 so that q_N satisfies the hypothesis of Theorem VII.1.15 and with the infimum and suprema in III(i)–(iv) of that theorem attained for $Q\subset\Omega_N$; for instance choose $f_0 := \operatorname{re} q_0$ to be a sufficiently large and $g_0 := \operatorname{im} q_0$ a sufficiently small constant in each Q in Ω_N^c. By Theorem VII.1.15, there is an m-sectorial operator $S_{D,\Omega}$ associated with the sesquilinear form

$$\int_\Omega (\nabla u\cdot\nabla\bar{v} + q_N u\bar{v}).$$

Moreover, $T^{-1}_{D,\Omega}$ and $S^{-1}_{D,\Omega}$ both exist and are bounded on $L^2(\Omega)$, since the assumption (2.3) ensures that the numerical ranges of $T_{D,\Omega}$ and $S_{D,\Omega}$ exclude zero. By Theorem IV.5.1 (cf. §X.4), $T^{-1}_{D,\Omega} - S^{-1}_{D,\Omega}$ is compact, and so $\beta(T^{-1}_{D,\Omega}) = \beta(S^{-1}_{D,\Omega})$, where β is the measure of non-compactness (ball or set) of a map. Application of Theorem VII.1.15 shows that the numerical range of $S_{D,\Omega}$ lies in a set of complex numbers with real parts bounded below by $\theta_0^{(N)}$, where

$$\theta_0^{(N)} = \inf_{Q\subset\Omega_N} [f_Q - \gamma_r\rho_r(f, Q)].$$

Note that $\theta_0^{(N)} \geqslant \theta_0 > 0$. Thus, for any $u\in\mathscr{D}(S_{D,\Omega})$,

$$\begin{aligned}\theta_0^{(N)}\|u\|_\Omega^2 &\leqslant \operatorname{re}(S_{D,\Omega}u, u)_\Omega \\ &\leqslant \|S_{D,\Omega}u\|_\Omega\|u\|_\Omega,\end{aligned}$$

and so

$$\beta(S^{-1}_{D,\Omega}) \leqslant \|S^{-1}_{D,\Omega}\| \leqslant 1/\theta_0^{(N)}.$$

By Corollary I.2.15, Theorem I.2.19, Theorem I.4.10 and Corollary I.4.11,

$$\operatorname{re}[T^{-1}_{D,\Omega}(T^{-1}_{D,\Omega})^*] = \beta^2(T^{-1}_{D,\Omega}) = \beta^2(S^{-1}_{D,\Omega}) \leqslant (1/\theta_0^{(N)})^2.$$

Since this holds for arbitrary N it follows that

$$\operatorname{re}[T^{-1}_{D,\Omega}(T^{-1}_{D,\Omega})^*] \leqslant 1/\phi_0^2.$$

That (2.4) holds is now an immediate consequence of the spectral mapping result given by Theorem IX.2.3.

A similar proof may be employed to show that the same holds for $T_{N,\Omega}$ if $\mathfrak{F}$ is taken to be a Whitney covering of Ω.

We now put

$$M(\lambda, T_D, \Omega) = \sum_{s_m(T_{D,\Omega})<\lambda} 1,$$

where the $s_m(T_{D,\Omega})$ ($m\in\mathbb{N}$) are the singular values of $T_{D,\Omega}$, that is the eigenvalues of $|T_{D,\Omega}|$, the positive square root of $T^*_{D,\Omega}T_{D,\Omega}$, each repeated according to its multiplicity. To obtain a lower bound for M we first establish a lemma.

Lemma 2.1. Let Q be a cube in $\mathbb{R}^n$ and let $h\in L^r(Q)$ for some r such that

$$r\in[\tfrac{1}{2}n, \infty] \text{ if } n>4, \quad r\in(2,\infty] \text{ if } n=4, \; r\in[2,\infty] \text{ if } n\leqslant 3. \tag{2.5}$$

Put $I=(0,1)^n$ and

$$\alpha_r = \sup\{\|\phi\|_{2r/(r-2),I}/(\|\Delta\phi\|_I^2+\|\phi\|_I^2)^{\frac{1}{2}} : \phi\in W^{2,2}(I)\backslash\{0\}\}. \tag{2.6}$$

Then for all ϕ in $W^{2,2}(Q^0)$,

$$\|h\phi\|_Q^2 \leqslant \alpha_r^2|Q|^{-2/r+4/n}\|h\|_{r,Q}^2[\|\Delta\phi\|_Q^2+|Q|^{-4/n}\|\phi\|_Q^2]. \tag{2.7} ■$$

Proof. Without loss of generality we may assume that Q has faces parallel to the coordinate axes. By the Sobolev Embedding Theorem (Theorem V.4.13), $W^{2,2}(Q^\circ)$ is continuously embedded in $L^p(Q)$, where $\frac{1}{2}-2/n\leqslant 1/p\leqslant\frac{1}{2}$ if $n>4$, while $2\leqslant p<\infty$ if $n=4$, and where $W^{2,2}(Q^\circ)$ is continously embedded in $C(Q)$ if $n\leqslant 3$. Suppose that $n>4$. Then $2\leqslant p\leqslant 2n/(n-4)$ and, with $|Q| = t^n$ and $\tilde{\phi}(x)=\phi(a_Q+tx)$, where $y\in Q$ if and only if $y=a_Q+tx$ for some $x\in I$,

$$\|\phi\|_{p,Q}/(|Q|^{4/n}\|\Delta\phi\|_Q^2+\|\phi\|_Q^2)^{\frac{1}{2}} = t^{n/p-n/2}\|\tilde{\phi}\|_{p,I}/(\|\Delta\tilde{\phi}\|_I^2+\|\tilde{\phi}\|_I^2)^{\frac{1}{2}}.$$

Thus by Hölder's inequality, with p defined by $2/r+2/p=1$,

$$\begin{aligned}\int_Q |h|^2|\phi|^2 &\leqslant \left(\int_Q |h|^r\right)^{2/r}\left(\int_Q|\phi|^p\right)^{2/p}\\ &\leqslant \alpha_r^2\|h\|_{r,Q}^2|Q|^{-2/r}\{|Q|^{4/n}\|\Delta\phi\|_Q^2+\|\phi\|_Q^2\},\end{aligned}$$

and the result follows. The cases $n\leqslant 4$ are handled in a similar manner. □

2.2. *A lower bound for* $M(\lambda, T_D, \Omega)$

Theorem 2.2. Suppose that (2.2), (2.3) and the hypothesis of Theorem VII.1.15 hold with r satisfying (2.5) and assume that

$$\alpha := \alpha_r \sup_{Q\in\mathfrak{F}}(|Q|^{2/n-1/r}\|q-q_Q\|_{r,Q}) = \alpha_r \sup_{Q\in\mathfrak{F}}[|Q|^{2/n}\rho_r(q,Q)]<\infty. \tag{2.8}$$

Then if $\lambda < \phi_0$, where ϕ_0 is defined in Theorem X.2.6, we have

$$\mathrm{M}(\lambda, T_{\mathrm{D}}, \Omega) \geqslant (1+\alpha)^{-n/2}\omega_n(2\pi)^{-n}\Lambda_1 + O\big(\Lambda_1^{(n-1)/n}(N_1 + V_1\alpha^{n/4}\lambda^{n/2})^{1/n} + N_1 + V_1\alpha^{n/4}\lambda^{n/2}\big), \tag{2.9}$$

where

$$\Lambda_1 = \sum_{|q_Q|<\lambda} |Q|[(\lambda^2 - g_Q^2)^{\frac{1}{2}} - f_Q]^{n/2}, \tag{2.10}$$

$$V_1 = \sum_{|q_Q|<\lambda} |Q|, \tag{2.11}$$

$$N_1 = \#\{Q \in \mathfrak{F} : |q_Q| < \lambda\} \tag{2.12}$$

and ω_n is the volume of the unit ball in $\mathbb{R}^n$. ■

Proof. Since $q \in \mathrm{L}^r_{\mathrm{loc}}(\Omega)$ with r satisfying (2.5), it follows from Theorem VII.1.15 and Lemma VII.1.14 that the form domain $\mathfrak{Q}_{\mathrm{D},Q}$ of $T_{\mathrm{D},Q}$ is $\mathrm{W}_0^{1,2}(Q^\circ)$, for all Q in $\mathfrak{F}$; if $Q \not\subset \Omega$ we write, for simplicity, $T_{\mathrm{D},Q}$ for $T_{\mathrm{D},Q\cap\Omega}$ and $\mathfrak{Q}_{\mathrm{D},Q}$ for $\mathfrak{Q}_{\mathrm{D},Q\cap\Omega}$. Moreover, from Lemma 2.1, the density of $\mathrm{W}_0^{1,2}(Q^\circ) \cap \mathrm{W}^{2,2}(Q^\circ)$ in $\mathrm{W}_0^{1,2}(Q^\circ)$ and Theorem IV.2.8 we have

$$\mathscr{D}(T_{\mathrm{D},Q}) = \mathrm{W}_0^{1,2}(Q^\circ) \cap \mathrm{W}^{2,2}(Q^\circ). \tag{2.13}$$

Now suppose that $\lambda < \phi_0$. In view of (2.4) it follows that those points in $\sigma(T^*_{\mathrm{D},\Omega}T_{\mathrm{D},\Omega})$ which are to the left of λ^2 are eigenvalues of finite multiplicity; $\mathrm{M}(\lambda, T_{\mathrm{D}}, \Omega)$ is, of course the number of these eigenvalues.

From the Max–Min Principle (Theorem XI.1.2) and noting that $(T^*_{\mathrm{D},\Omega}T_{\mathrm{D},\Omega}f, f)_\Omega = \|T_{\mathrm{D},\Omega}f\|^2_\Omega$, we have, for all $m \in \mathbb{N}$,

$$\mathrm{s}_m^2(T_{\mathrm{D},\Omega}) = \max_{G \in G_{m-1}} \min_{f \perp G, f \in \mathscr{D}(T_{\mathrm{D},\Omega})} \|T_{\mathrm{D},\Omega}f\|^2_\Omega / \|f\|^2_\Omega \tag{2.14}$$

where G_{m-1} denotes the family of all linear subspaces of $\mathrm{L}^2(\Omega)$ of dimension $m-1$. The maximum is attained when G_{m-1} is the linear span of the eigenvectors of $T^*_{\mathrm{D},\Omega}T_{\mathrm{D},\Omega}$ corresponding to the first $m-1$ eigenvalues $\mathrm{s}_j^2(T_{\mathrm{D},\Omega})$, each being repeated according to its multiplicity. Given any open subset Ω' of Ω it follows that $\mathrm{s}_m^2(T_{\mathrm{D},\Omega}) \leqslant \mathrm{s}_m^2(T_{\mathrm{D},\Omega'})$ and hence

$$\mathrm{M}(\lambda, T_{\mathrm{D}}, \Omega') \leqslant \mathrm{M}(\lambda, T_{\mathrm{D}}, \Omega). \tag{2.15}$$

Let ($\mathfrak{F}_N$ be any finite subset of $\mathfrak{F}$ and put $\Omega_N^{\mathrm{c}} = (\cup_{Q\in\mathfrak{F}_N} Q \cap \Omega)^\circ$. Then (cf. (XI.2.20))

$$T_{\mathrm{D},\Omega_N^{\mathrm{c}}} = \bigoplus_{Q\in\mathfrak{F}_N} T_{\mathrm{D},Q}, \qquad T^*_{\mathrm{D},\Omega_N^{\mathrm{c}}}T_{\mathrm{D},\Omega_N^{\mathrm{c}}} = \bigoplus_{Q\in\mathfrak{F}_N} T^*_{\mathrm{D},Q}T_{\mathrm{D},Q}, \tag{2.16}$$

and

$$\mathrm{M}(\lambda, T_{\mathrm{D}}, \Omega_N^{\mathrm{c}}) = \sum_{Q\in\mathfrak{F}_N} \mathrm{M}(\lambda, T_{\mathrm{D}}, Q). \tag{2.17}$$

In fact the singular values of $T_{\mathrm{D},\Omega_N^c}$ are the singular values of the $T_{\mathrm{D},Q}$ and the eigenvectors of $T^*_{\mathrm{D},\Omega_N^c}T_{\mathrm{D},\Omega_N^c}$ are those of the $T^*_{\mathrm{D},Q}T_{\mathrm{D},Q}$. In view of (2.15) we therefore have

$$\mathrm{M}(\lambda, T_{\mathrm{D}}, \Omega) \geqslant \sum_{Q\in\mathfrak{F}_N} \mathrm{M}(\lambda, T_{\mathrm{D}}, Q). \tag{2.18}$$

This gives rise to a lower bound for $\mathrm{M}(\lambda, T_{\mathrm{D}}, \Omega)$ as follows. Define a function $\tilde{q}$ on $\bigcup_{Q\in\mathfrak{F}} Q$ by

$$\tilde{q}(x) = q_Q = f_Q + ig_Q \quad \text{for all } x\in Q^\circ \text{ and all } Q\in\mathfrak{F}.$$

By Lemma 2.1, with $h = q-\tilde{q}$ and $\phi\in \mathrm{W}^{2,2}(Q^\circ)$, we have

$$\|(q-\tilde{q})\phi\|_Q^2 \leqslant \alpha^2(\|\Delta\phi\|_Q^2 + |Q|^{-4/n}\|\phi\|_Q^2). \tag{2.19}$$

It follows that given any $\phi\in\mathscr{D}(T_{\mathrm{D},Q}) = \mathrm{W}_0^{1,2}(Q^\circ)\cap\mathrm{W}^{2,2}(Q^\circ)$ we have with $\tau_0 = -\Delta_{\mathrm{D},Q}+f_Q$ and $\tau_1 = -\Delta_{\mathrm{D},Q}+q_Q$,

$$\|\tau\phi\|_Q^2 = \|\tau_1\phi+(q-\tilde{q})\phi\|_Q^2 = \|\tau_1\phi\|_Q^2 + 2\,\mathrm{re}\,(\tau_1\phi, (q-\tilde{q})\phi)_Q + \|(q-\tilde{q})\phi\|_Q^2.$$

Hence, given any $\varepsilon>0$,

$$\begin{aligned} \big|\,\|\tau\phi\|_Q^2 - \|\tau_1\phi\|_Q^2\big| &\leqslant \varepsilon\|\tau_1\phi\|_Q^2 + (1+\varepsilon^{-1})\|(q-\tilde{q})\phi\|_Q^2 \\ &\leqslant \varepsilon\|\tau_1\phi\|_Q^2 + \alpha^2(1+\varepsilon^{-1})(\|\Delta\phi\|_Q^2+|Q|^{-4/n}\|\phi\|_Q^2), \end{aligned} \tag{2.20}$$

the second inequality following from (2.19). Also

$$\begin{aligned} \|\tau_1\phi\|_Q^2 &= \|\tau_0\phi\|_Q^2 + 2\,\mathrm{re}\,(\tau_0\phi, ig_Q\phi)_Q + \|ig_Q\phi\|_Q^2 \\ &= \|\tau_0\phi\|_Q^2 + g_Q^2\|\phi\|_Q^2, \end{aligned} \tag{2.21}$$

and

$$\begin{aligned} \|\tau_0\phi\|_Q^2 &= \|-\Delta\phi\|_Q^2 + 2\,\mathrm{re}\,(-\Delta\phi, f_Q\phi)_Q + f_Q^2\|\phi\|_Q^2 \\ &\geqslant \|\Delta\phi\|_Q^2 + f_Q^2\|\phi\|_Q^2, \end{aligned} \tag{2.22}$$

since $f_Q>0$, in view of our assumption that $\phi_0\geqslant\theta_0>0$, and $(-\Delta\phi,\phi)_Q\geqslant 0$ for all $\phi\in\mathrm{W}_0^{1,2}(Q^\circ)\cap\mathrm{W}^{2,2}(Q^\circ)$. Thus we have from (2.20), (2.21) and (2.22),

$$\begin{aligned} \big|\,\|\tau\phi\|_Q^2 - \|\tau_1\phi\|_Q^2\big| &\leqslant \varepsilon\|\tau_1\phi\|_Q^2 + \alpha^2(1+\varepsilon^{-1})(\|\tau_1\phi\|_Q^2 - |q_Q|^2\|\phi\|_Q^2 \\ &\qquad + |Q|^{-4/n}\|\phi\|_Q^2) \\ &= [\varepsilon+\alpha^2(1+\varepsilon^{-1})]\,\|\tau_1\phi\|_Q^2 - \alpha^2(1+\varepsilon^{-1})[|q_Q|^2 - |Q|^{-4/n}]\,\|\phi\|_Q^2. \end{aligned}$$

We now set $\varepsilon=\alpha$ and use (2.21) to obtain

$$\|\tau\phi\|_Q^2 \leqslant (1+\alpha)^2\,\|\tau_0\phi\|_Q^2 + (\alpha+1)[g_Q^2 - \alpha f_Q^2 + \alpha|Q|^{-4/n}]\,\|\phi\|_Q^2. \tag{2.23}$$

From (2.23) and (2.14),

$$\mathrm{s}_m^2(T_{\mathrm{D},Q}) \leqslant (1+\alpha)^2\mathrm{s}_m^2(-\Delta_{\mathrm{D},Q}+f_Q) + (\alpha+1)(g_Q^2 - \alpha f_Q^2 + \alpha|Q|^{-4/n})$$

and so

$$\sum_{s_m^2(T_{\mathrm{D},Q})<\lambda} 1 \geqslant \sum_{s_m^2(-\Delta_{\mathrm{D},Q}+f_Q)<\mu_Q^2} 1,$$

where

$$(1+\alpha)^2\mu_Q^2 = \lambda^2-(\alpha+1)(g_Q^2-\alpha f_Q^2+\alpha|Q|^{-4/n}). \tag{2.24}$$

Then

$$\begin{aligned} \mathrm{M}(\lambda, T_{\mathrm{D}}, Q) &= \sum_{s_m^2(T_{\mathrm{D},Q})<\lambda^2} 1 \geqslant \sum_{s_m^2(-\Delta_{\mathrm{D},Q}+f_Q)<\mu_Q^2} 1 \\ &= \mathrm{M}(\mu_Q, -\Delta_{\mathrm{D}}+f_Q, Q) = \mathrm{N}(\mu_Q, -\Delta_{\mathrm{D}}+f_Q, Q) \\ &= \mathrm{N}(\mu_Q-f_Q, -\Delta_{\mathrm{D}}, Q) := \sum_{\lambda_m(-\Delta_{\mathrm{D},Q})<\mu_Q-f_Q} 1. \end{aligned} \tag{2.25}$$

Then from (2.18) we conclude that for any finite subset $\mathfrak{F}_N$ of $\mathfrak{F}$,

$$\mathrm{M}(\lambda, T_{\mathrm{D}}, \Omega) \geqslant \sum_{Q\in\mathfrak{F}_N} \mathrm{N}(\mu_Q-f_Q, -\Delta_{\mathrm{D}}, Q). \tag{2.26}$$

Since $-\Delta_{\mathrm{D},Q}>0$, we need only consider those cubes $Q\in\mathfrak{F}$ for which

$$\mu_Q^2-f_Q^2 = (\alpha+1)^{-2}[\lambda^2-(\alpha+1)|q_Q|^2-\alpha(\alpha+1)|Q|^{-4/n}]>0, \tag{2.27}$$

and in view of the inequality $\lambda<\phi_0$ there are only finitely many such cubes, since every $Q\in\mathfrak{F}$ for which (2.27) holds must also be such that

$$\lambda > f_Q > f_Q-\gamma_r\rho_r(f, Q).$$

These remarks lead us to define

$$\mathfrak{F}_N = I_1 := \{Q\in\mathfrak{F}: \lambda^2-(\alpha+1)(|q_Q|^2+\alpha|Q|^{-4/n})>0\}, \tag{2.28}$$

and we now set about the estimation of the sum appearing in (2.26), with this particular family $\mathfrak{F}_N$.

First notice that from (2.24),

$$(\alpha+1)^2\mu_Q^2 = \lambda^2-g_Q^2+O(\alpha|g_Q|^2)+O(\alpha|Q|^{-4/n}), \tag{2.29}$$

where the O terms are bounded by a constant multiple of the quantities shown, the constant being independent of Q and of λ; notice that $\alpha=O(1)$ and that in our estimations we are guided by the ultimate need to make α small for the asymptotic results when $\lambda\to\infty$. From (2.29) we see that if $Q\in I_1$,

$$\mu_Q \geqslant (\alpha+1)^{-1}(\lambda^2-g_Q^2)^{\frac{1}{2}}-O(\alpha^{\frac{1}{2}}|q_Q|)-O(\alpha^{\frac{1}{2}}|Q|^{-2/n}), \tag{2.30}$$

and

$$\mu_Q-f_Q \geqslant (\alpha+1)^{-1}[(\lambda^2-g_Q^2)^{\frac{1}{2}}-f_Q]-O(\alpha^{\frac{1}{2}}|q_Q|)-O(\alpha^{\frac{1}{2}}|Q|^{-2/n}). \tag{2.31}$$

It also follows from (2.29) that

$$\mu_Q \leqslant (1+\alpha)^{-1}(\lambda^2 - g_Q^2)^{\frac{1}{2}} + O(\alpha^{\frac{1}{2}}|q_Q|) + O(\alpha^{\frac{1}{2}}|Q|^{-2/n}),$$

and

$$\mu_Q - f_Q \leqslant (1+\alpha)^{-1}[(\lambda^2 - g_Q^2)^{\frac{1}{2}} - f_Q] + O(\alpha^{\frac{1}{2}}|q_Q|) + O(\alpha^{\frac{1}{2}}|Q|^{-2/n}). \tag{2.32}$$

Note also that if $Q \in I_1$ then

$$\lambda^2 - g_Q^2 > (\alpha+1)f_Q^2 + \alpha g_Q^2 + \alpha(\alpha+1)|Q|^{-4/n} > f_Q^2,$$

so that

$$(\lambda^2 - g_Q^2)^{\frac{1}{2}} - f_Q > 0 \quad \text{for all } Q \in I_1. \tag{2.33}$$

The terms in the sum appearing in (2.26) are now estimated as follows (cf. Theorem XI.2.6):

$$\mathrm{N}(\mu_Q - f_Q, -\Delta_D, Q) \geqslant \omega_n(2\pi)^{-n}|Q|(\mu_Q - f_Q)^{n/2} - O(1 + [|Q|(\mu_Q - f_Q)^{n/2}]^{(n-1)/n}) \tag{2.34}$$

where the constants implicit in the O terms are independent of Q, μ_Q and f_Q. To proceed further we use the inequality (XI.2.37):

$$(A-B)^{n/2} \geqslant A^{n/2} - KB^{\frac{1}{2}}(A^{(n-1)/2} + B^{(n-1)/2})$$

valid for all A and B with $A \geqslant B \geqslant 0$, in which K is a constant depending only on n. Put

$$\Lambda_Q = |Q|[(\lambda^2 - g_Q^2)_+^{\frac{1}{2}} - f_Q]_+^{n/2}, \tag{2.35}$$

and note that if $Q \in I_1$,

$$\Lambda_Q = |Q|[(\lambda^2 - g_Q^2)^{\frac{1}{2}} - f_Q]^{n/2}.$$

Then from (2.31) and (XI.2.37),

$$|Q|(\mu_Q - f_Q)^{n/2} \geqslant (\alpha+1)^{-n/2}\Lambda_Q - O(\alpha^{\frac{1}{4}}\Lambda_Q^{(n-1)/n}[(|Q||q_Q|^{n/2})^{1/n} + 1] + \alpha^{n/4}(|Q||q_Q|^{n/2} + 1)). \tag{2.36}$$

Also from (2.32) we have

$$|Q|(\mu_Q - f_Q)^{n/2} = O(\Lambda_Q + \alpha^{n/4}[|Q||q_Q|^{n/2} + 1]). \tag{2.37}$$

Substitution of (2.35) and (2.36) in (2.34) yields

$$\begin{aligned}\mathrm{N}(\mu_Q - f_Q, -\Delta_D, Q) &\geqslant \omega_n(2\pi)^{-n}(1+\alpha)^{-n/2}\Lambda_Q - O(\alpha^{\frac{1}{4}}\Lambda_Q^{(n-1)/n}[(|Q||q_Q|^{n/2})^{1/n} \\ &\quad + 1] + \alpha^{n/4}(|Q||q_Q|^{n/2} + 1) + 1 + \Lambda_Q^{(n-1)/n} + \alpha^{(n-1)/4}\{(|Q||q_Q|^{n/2})^{(n-1)/n} + 1\}) \\ &\geqslant \omega_n(2\pi)^{-n}(1+\alpha)^{-n/2}\Lambda_Q - O(\alpha^{\frac{1}{4}}\Lambda_Q^{(n-1)/n}(|Q||q_Q|^{n/2})^{1/n} \\ &\quad + \alpha^{n/4}(|Q||q_Q|^{n/2}) + \Lambda_Q^{(n-1)/n} + 1),\end{aligned} \tag{2.38}$$

the final step following because $\alpha = O(1)$ and, by Young's inequality,

$$A^{(n-1)/n} \leqslant (n-1)A/n + 1/n \qquad (A \geqslant 0).$$

It is clear from the definition of I_1 in (2.28) that $|q_Q| < \lambda$ if $Q \in I_1$. Thus

$$\sum_{Q \in I_1} \Lambda_Q = \sum_{|q_Q| < \lambda} \Lambda_Q - \sum_{Q \in \mathfrak{F}'} |Q|[(\lambda^2 - g_Q^2)^{\frac{1}{2}} - f_Q]^{n/2},$$

where $\mathfrak{F}'$ is the set of all cubes $Q \in \mathfrak{F}$ such that $\lambda^2 > f_Q^2 + g_Q^2 \geqslant (1+\alpha)^{-1}[\lambda^2 - \alpha(\alpha+1)|Q|^{-4/n}]$. Hence in $\mathfrak{F}'$,

$$\lambda^2 - g_Q^2 \leqslant (1+\alpha) f_Q^2 + \alpha g_Q^2 + \alpha(1+\alpha)|Q|^{-4/n},$$

and so

$$(\lambda^2 - g_Q^2)^{\frac{1}{2}} - f_Q = O(\alpha^{\frac{1}{2}} f_Q + \alpha^{\frac{1}{2}} g_Q + \alpha^{\frac{1}{2}}|Q|^{-2/n}) = O(\alpha^{\frac{1}{2}}\lambda + \alpha^{\frac{1}{2}}|Q|^{-2/n}).$$

We therefore see that

$$\begin{aligned}
\sum_{Q \in I_1} \Lambda_Q &\geqslant \sum_{|q_Q| < \lambda} \Lambda_Q - O\left(\sum_{Q \in \mathfrak{F}'} |Q|(\alpha^{\frac{1}{2}}\lambda + \alpha^{\frac{1}{2}}|Q|^{-2/n})^{n/2}\right) \\
&= \sum_{|q_Q| < \lambda} \Lambda_Q - O\left(\alpha^{n/4} \sum_{|q_Q| < \lambda} |Q|(\lambda^{n/2} + |Q|^{-1})\right) \\
&= \sum_{|q_Q| < \lambda} \Lambda_Q - O(\alpha^{n/4}(V_1 \lambda^{n/2} + N_1)),
\end{aligned}$$

V_1 and N_1 being as defined in (2.11) and (2.12). We therefore obtain

$$\begin{aligned}
\sum_{Q \in I_1} \mathrm{N}(\mu_Q - f_Q, -\Delta_{\mathrm{D}}, Q) &\geqslant \omega_n (2\pi)^{-n}(1+\alpha)^{-n/2} \sum_{|q_Q| < \lambda} \Lambda_Q - O\Bigg(\alpha^{n/4}(V_1 \lambda^{n/2} + N_1) \\
&\quad + \alpha^{\frac{1}{4}} \sum_{Q \in I_1} (|Q||q_Q|^{n/2})^{1/n} \Lambda_Q^{(n-1)/n} + \alpha^{n/4} \sum_{Q \in I_1} |Q||q_Q|^{n/2} + \sum_{Q \in I_1} (\Lambda_Q^{(n-1)/n} + 1)\Bigg) \\
&\geqslant \omega_n (2\pi)^{-n}(1+\alpha)^{-n/2} \sum_{|q_Q| < \lambda} \Lambda_Q - O\Bigg(\alpha^{n/4}(V_1 \lambda^{n/2} + N_1) + \\
&\quad \alpha^{\frac{1}{4}}\lambda^{\frac{1}{2}} \sum_{|q_Q| < \lambda} |Q|^{1/n} \Lambda_Q^{(n-1)/n} + \alpha^{n/4}\lambda^{n/2} \sum_{|q_Q| < \lambda} |Q| + \sum_{|q_Q| < \lambda} \Lambda_Q^{(n-1)/n} + N_1\Bigg).
\end{aligned}$$

The estimate (2.9) now follows on using Hölder's inequality and the proof is complete. □

2.3. *An upper bound for* $\mathrm{M}(\lambda, T_{\mathrm{D}}, \Omega)$

To obtain an upper bound for

$$\mathrm{N}(\lambda, T_{\mathrm{D}}, \Omega) := \sum_{\lambda_m(T_{\mathrm{D},\Omega}) < \lambda} 1$$

in the self-adjoint problem when q is real, one can use the result

$$\mathrm{N}(\lambda, T_{\mathrm{D}}, \Omega) \leqslant \sum \mathrm{N}(\lambda, T_{\mathrm{N}}, Q),$$

the sum being over an appropriate selection of cubes Q from the covering $\mathfrak{F}$. This inequality is a consequence of the Max–Min Principle (Theorem XI.1.3) and the fact that $\mathcal{Q}(T_{\mathrm{D},\Omega}) \subset \bigoplus \mathcal{Q}(T_{\mathrm{N},Q})$; one argues with the form of $T_{\mathrm{D},\Omega}$ and not with $T_{\mathrm{D},\Omega}$ itself since it is not clear how $\mathcal{D}(T_{\mathrm{D},\Omega})$ is related to $\bigoplus \mathcal{D}(T_{\mathrm{N},Q})$. The same problem presents itself when q is complex, but is not so readily overcome. Ideally one should work with the positive square root $|T_{\mathrm{D},\Omega}|$ of $T^*_{\mathrm{D},\Omega}T_{\mathrm{D},\Omega}$ if results completely analogous to those obtained for the self-adjoint problem are to be derived. However $|T_{\mathrm{D},\Omega}|$ is not a differential operator and an explicit representation of it does not seem to be available. We have therefore been forced to steer clear of $|T_{\mathrm{D},\Omega}|$, but in doing this we obtain a result which is undoubtedly weaker than the best possible.

Theorem 2.3. Suppose that (2.2), (2.3) and the hypothesis of Theorem VII.1.15 hold and that

$$f_Q - |g_Q| - \sqrt{2\gamma_r}\rho_r(q, Q) \geqslant 0 \quad \text{for all } Q \text{ in } \mathfrak{F}. \tag{2.39}$$

Suppose also that

$$\lambda < \phi_2 := \lim_{N\to\infty} \inf_{Q\in\mathfrak{F},\, Q\subset\Omega_N} [f_Q - |g_Q| - \sqrt{2\gamma_r}\rho_r(q, Q)]. \tag{2.40}$$

Then

$$\mathrm{M}(\lambda, T_{\mathrm{D}}, \Omega) \leqslant (1-\gamma)^{-n/2}\omega_n(2\pi)^{-n}\Lambda_2 + O(N_2 + N_2^{1/n}\Lambda_2^{(n-1)/n}), \tag{2.41}$$

where

$$I_2 = \{Q\in\mathfrak{F}: \lambda - f_Q + |g_Q| + \sqrt{2\gamma_r}\rho_r(q, Q) \geqslant 0\}, \tag{2.42}$$

$$\Lambda_2 = \sum_{Q\in I_2} |Q|(\lambda - f_Q + |g_Q|)_+^{n/2}, \tag{2.43}$$

$$N_2 = \# I_2, \tag{2.44}$$

and

$$\gamma = \sqrt{2\gamma_r} \sup_{Q\in\mathfrak{F}} |Q|^{2/n}\rho_r(q, Q) < 1. \tag{2.45}$$ ∎

Proof. Given any $u \in \mathcal{D}(T_{\mathrm{D},\Omega})$,

$$\begin{aligned}
|(T_{\mathrm{D},\Omega}u, u)_\Omega|^2 &= |((-\Delta_{\mathrm{D},\Omega} + f + \mathrm{i}g)u, u)_\Omega|^2 \\
&= |((-\Delta_{\mathrm{D},\Omega} + f)u, u)_\Omega|^2 + |(gu, u)_\Omega|^2 \\
&= \tfrac{1}{2}[((-\Delta_{\mathrm{D},\Omega} + f + g)u, u)_\Omega^2 + ((-\Delta_{\mathrm{D},\Omega} + f - g)u, u)_\Omega^2] \\
&\geqslant |((-\Delta_{\mathrm{D},\Omega} + f + g)u, u)_\Omega|\,|((-\Delta_{\mathrm{D},\Omega} + f - g)u, u)_\Omega|.
\end{aligned} \tag{2.46}$$

From (2.14) and (2.46) we see that for any $m \in \mathbb{N}$ and any $G \in G_{m-1}$,

$$\mathrm{s}_m^2(T_{D,\Omega}) \geqslant \min_{u \perp G,\, u \in \mathscr{D}(T_{D,\Omega}),\, u \neq 0} |(T_{D,\Omega}u, u)_\Omega|^2/\|u\|_\Omega^4$$

$$\geqslant \left(\min_{u \perp G,\, u \in \mathscr{D}(T_{D,\Omega}),\, u \neq 0} |\|\nabla u\|_\Omega^2 + ((f+g)u, u)_\Omega|/\|u\|_\Omega^2\right) \times$$

$$\left(\min_{u \perp G,\, u \in \mathscr{D}(T_{D,\Omega}),\, u \neq 0} |\|\nabla u\|_\Omega^2 + ((f-g)u, u)_\Omega|/\|u\|_\Omega^2\right). \tag{2.47}$$

By Lemma VII.1.14,

$$\|\nabla u\|_Q^2 + ([f \pm g]u, u)_Q = \|\nabla u\|_Q^2 + (f_Q \pm g_Q)\|u\|_Q^2 + ([f - f_Q \pm (g - g_Q)]u, u)_Q$$

$$\geqslant \|\nabla u\|_Q^2 + (f_Q \pm g_Q)\|u\|_Q^2 - \sqrt{2}\,(|g - g_Q|u, u)_Q$$

$$\geqslant (1-\gamma)\|\nabla u\|_Q^2 + [f_Q \pm g_Q - \sqrt{2}\gamma_r \rho_r(q, Q)]\,\|u\|_Q^2. \tag{2.48}$$

Now define functions $\tilde{q}_\pm$ on $\mathbb{R}^n$ by setting

$$\tilde{q}_\pm(x) = (1-\gamma)^{-1}[f_Q \pm g_Q - \sqrt{2}\gamma_r \rho_r(q, Q)]$$

for all x in Q and for all Q in $\mathfrak{F}$, and setting $\tilde{q}_\pm(x) = 0$ otherwise. Enumerate the cubes in $\mathfrak{F}$ as Q_l ($l \in \mathbb{N}$) so that $\Omega \subset \cup_{l=1}^\infty Q_l$, and introduce approximating forms $t_\pm$ defined by

$$t_\pm[\![u, u]\!] = \sum_{l=1}^\infty (\|\nabla u_l\|_{Q_l}^2 + (\tilde{q}_\pm u_l, u_l)_{Q_l})$$

with domains

$$\mathfrak{Q}_\pm = \{u : u_l = u\restriction_{Q_l} \in \mathrm{W}^{1,2}(Q_l^\circ) \text{ for all } l \in \mathbb{N} \text{ and } t_\pm[\![u, u]\!] < \infty\}.$$

If $u \in \mathscr{D}(T_{D,\Omega})$ then $u\restriction_{Q_l} \in \mathrm{W}^{1,2}(Q_l^\circ)$ and so

$$\mathscr{D}(T_{D,\Omega}) \subset \mathfrak{Q}_+ = \mathfrak{Q}_-. \tag{2.49}$$

The self-adjoint operators associated with the forms $t_\pm$ in $\mathrm{L}^2(\Omega)$ are $T_\pm = \bigoplus_{l=1}^\infty T_\pm^{(l)}$, where the $T_\pm^{(l)}$ are the positive self-adjoint operators associated with the forms

$$t_\pm^{(l)}[\![u_l, u_l]\!] = \|\nabla u_l\|_{Q_l}^2 + (\tilde{q}_\pm u_l, u_l)_{Q_l} \qquad (u_l \in \mathrm{W}^{1,2}(Q_l^\circ)).$$

The eigenvalues of $T_\pm$ are $\lambda_m^\pm(Q)$ ($Q \in \mathfrak{F}$, $m \in \mathbb{N}$), where

$$\lambda_m^\pm(Q) = \lambda_m(Q) + (1-\gamma)^{-1}[f_Q \pm g_Q - \sqrt{2}\gamma_r \rho_r(q, Q)],$$

and the $\lambda_m(Q)$ are the eigenvalues of $-\Delta_{N,Q}$, repeated according to multiplicity. Also the eigenvector of $T_\pm$ corresponding to the eigenvalue $\lambda_m^\pm(Q)$ is

$\psi_m(Q)$, where $-\Delta_{N,Q}\psi_m(Q)=\lambda_m(Q)\psi_m(Q)$. Put

$$\Phi_\lambda := \{\lambda_m(Q): Q\in\mathfrak{F},\ m\in\mathbb{N},\ \lambda_m(Q)+(1-\gamma)^{-1}[f_Q-|g_Q|-\sqrt{2\gamma_r\rho_r(q,Q)}] < \lambda(1-\gamma)^{-1}\},$$

$$N=\#\Phi_\lambda,$$

and let G_N be the linear span of the set of eigenvectors $\{\psi_m(Q):\lambda_m(Q)\in\Phi_\lambda\}$. From (2.47), (2.48) and (2.49), with $m=N+1$ and $G=G_N$, we therefore obtain

$$s_{N+1}^2(T_{D,\Omega})\geqslant(1-\gamma)^2\left(\min_{\substack{u\in\mathfrak{Q}_+\cap G_N^\perp\\ u\neq 0}} t_+[\![u,u]\!]/\|u\|_\Omega^2\right)\left(\min_{\substack{u\in\mathfrak{Q}_-\cap G_N^\perp\\ u\neq 0}} t_-[\![u,u]\!]/\|u\|_\Omega^2\right)$$
$$\geqslant\lambda^2.$$

Hence

$$\sum_{s_m^2(T_{D,\Omega})<\lambda^2}1\leqslant N,$$

so that

$$\begin{aligned}M(\lambda,T_D,\Omega)&\leqslant\sum_{\lambda_m(Q)<(1-\gamma)^{-1}[\lambda-f_Q+|g_Q|+\sqrt{2\gamma_r\rho_r(q,Q)}]}1\\&=\sum_{Q\in I_2}N((1-\gamma)^{-1}[\lambda-f_Q+|g_Q|+\sqrt{2\gamma_r\rho_r(q,Q)}],\ -\Delta_N,Q),\end{aligned}\tag{2.50}$$

where I_2 is defined in (2.42). Note that, in view of (2.40), I_2 is a finite set. Notice also that if λ satisfies (2.40) then $\lambda<\phi_0$, where ϕ_0 is defined in Theorem X.2.6; ideally (2.40) should be a consequence of this latter condition which expresses only the need to remain below the essential spectrum.

To carry out estimates of (2.50) we use Theorem XI.2.6: thus

$$N(\mu,-\Delta_N,Q)=\omega_n(2\pi)^{-n}|Q|\mu^{n/2}+O(1+[|Q|\mu^{n/2}]^{(n-1)/n}),\tag{2.51}$$

with $\mu=(1-\gamma)^{-1}[\lambda-f_Q+|g_Q|+\sqrt{2\gamma_r\rho_r(q,Q)}]$. As in (XI.2.37) we see that if $A, B\geqslant 0$, there is a positive constant K, depending only upon n, such that

$$(A+B)^{n/2}\leqslant A^{n/2}+KB^{\frac{1}{2}}(A^{(n-1)/2}+B^{(n-1)/2}).\tag{2.52}$$

Thus, since $\rho_r(q,Q)=O(\gamma|Q|^{-2/n})$,

$$\begin{aligned}|Q|(1-\gamma)^{-n/2}[\lambda-f_Q+|g_Q|+\sqrt{2\gamma_r\rho_r(q,Q)}]^{n/2}&\leqslant(1-\gamma)^{-n/2}\Gamma_Q\\&+K\gamma_r^{\frac{1}{4}}\rho_r(q,Q)^{\frac{1}{4}}\{|Q|^{1/n}\Gamma_Q^{(n-1)/n}+|Q|[\gamma_r\rho_r(q,Q)]^{(n-1)/2}\}\\&=(1-\gamma)^{-n/2}\Gamma_Q+O(\gamma^{\frac{1}{2}}\Gamma_Q^{(n-1)/n}+\gamma^{n/2}),\end{aligned}\tag{2.53}$$

where

$$\Gamma_Q=|Q|(\lambda-f_Q+|g_Q|)_+^{n/2}.\tag{2.54}$$

Use of this in (2.51) shows that

$$N((1-\gamma)^{-1}[\lambda - f_Q + |g_Q| + \sqrt{2\gamma_r \rho_r(q, Q)}], -\Delta_N, Q),$$
$$\leqslant \omega_n(2\pi)^{-n}(1-\gamma)^{-n/2}\Gamma_Q + O(\gamma^{\frac{1}{2}}\Gamma_Q^{(n-1)/n} + \gamma^{n/2} + 1 + \Gamma_Q^{(n-1)/n} + \gamma^{(n-1)/2})$$
$$= (1-\gamma)^{-n/2}\omega_n(2\pi)^{-n}\Gamma_Q + O(\Gamma_Q^{(n-1)/n} + 1).$$

Summation of this over all Q in I_2 gives the desired result (2.41), with the aid of (2.50). The proof is complete. □

Remark 2.4. Techniques similar to those used in §XI.2.4 may be used to replace the sum

$$\Lambda_1 = \sum_{q_Q < \lambda} |Q|[(\lambda^2 - g_Q^2)^{\frac{1}{2}} - f_Q]^{n/2},$$

which appears in Theorem 2.2 by an integral

$$\int [(\lambda^2 - g^2)^{\frac{1}{2}} - f]_+^{n/2}.$$

The same applies to Theorem 2.3.

Remark 2.5. With the systematic use of Whitney coverings of Ω and natural changes in the argument already given for the Dirichlet case, it can be seen that Theorems 2.2 and 2.3 hold for $T_{N,\Omega}$ as well as for $T_{D,\Omega}$.

3. Asymptotic results for $\lambda \to \infty$

Throughout this section we suppose that $\Omega = \mathbb{R}^n$ and write T for $T_{D,\mathbb{R}^n}$. Under suitable hypotheses on the growth and regularity of the potential q we are able to determine the asymptotic behaviour of $M(\lambda, T, \mathbb{R}^n)$ as $\lambda \to \infty$, the upper and lower bounds in §2 giving the same leading term.

To do this, let $\lambda_0 > 1$ and $a \geqslant 1$, let β be a positive number such that $1 - a^{-1} \leqslant 2\beta < 1$, and for any $\lambda > \lambda_0$ let $\mathfrak{F}_\beta = \mathfrak{F}_\beta(\lambda)$ be a tesselation of $\mathbb{R}^n$ by congruent cubes of side $\lambda^{-\beta}$. Exactly as in §3 of Chapter XI let $\mathfrak{F} = \mathfrak{F}(\lambda)$ be a family of cubes covering $\mathbb{R}^n$ and obtained from $\mathfrak{F}_\beta$ by subdivision of the cubes in $\mathfrak{F}_\beta$ when necessary so that for all $\lambda > \lambda_0$ and all Q in $\mathfrak{F}$, the centre of Q being denoted by x_Q,

$$|Q|^{2/n}|x_Q|^{a-1} = O(1). \tag{3.1}$$

Note that

$$|Q|^{1/n} = \lambda^{-\beta} \text{ for all } Q \text{ in } \mathfrak{F} \text{ with } |x_Q| \leqslant K\lambda^{1/a}. \tag{3.2}$$

We now assume that there are positive constants c_1, c_2, c_3, c_4, ε and a constant $\delta \in [0, 1)$ such that for any $\lambda > \lambda_0$ and any $Q \in \mathfrak{F}$,

$$c_1(|x_Q|^a+1) \leqslant f_Q \leqslant c_2(|x_Q|^a+1), \tag{3.3}$$

$$|g_Q| \leqslant c_3(|x_Q|^a+1)^\delta, \qquad c_3 < c_1, \tag{3.4}$$

$$\gamma_r\rho_r(q, Q) \leqslant c_4|Q|^{\varepsilon/n}(|x_Q|^{a-1}+1). \tag{3.5}$$

Here the number r is supposed to satisfy (2.5).

Our object is to use the estimates of §2, which naturally become sharper in this more precise setting. First notice that for λ large enough we have

$$\begin{aligned} f_Q-\gamma_r\rho_r(f, Q) &\geqslant c_1(|x_Q|^a+1)-c_4|Q|^{\varepsilon/n}(|x_Q|^{a-1}+1) \\ &\geqslant (c_1-K\lambda^{-\varepsilon\beta})(|x_Q|^a+1) \\ &\geqslant \tfrac{1}{2}c_1(|x_Q|^a+1) \end{aligned}$$

and

$$\begin{aligned} |g_Q|+\gamma_r\rho_r(g, Q) &\leqslant c_3(|x_Q|^a+1)+c_4\lambda^{-\varepsilon\beta}(|x_Q|^{a-1}+1) \\ &\leqslant 2c_3(|x_Q|^a+1). \end{aligned}$$

Hence III(ii) of Theorem VII.1.15 is satisfied. Also, by (3.1) and (3.5),

$$\begin{aligned} |Q|^{2/n}\gamma_r\rho_r(f, Q) &\leqslant c_4|Q|^{(2+\varepsilon)/n}(|x_Q|^{a-1}+1) \\ &= O(|Q|^{\varepsilon/n}) \\ &= O(\lambda^{-\varepsilon\beta}). \end{aligned}$$

Thus $\beta_0 = O(\lambda^{-\varepsilon\beta})$, so that III(iii) of Theorem VII.1.15 is satisfied. Similarly $\beta_1 = O(\lambda^{-\varepsilon\beta})$, and it is now clear that q satisfies the hypothesis of Theorem VII.1.15.

Next we estimate the quantities in Theorem 2.2. Observe that whatever sequence (Ω_N) is chosen, $\phi_0 = \infty$: thus $\sigma_e(T) = \varnothing$ and the spectrum of $T * T$ is wholly discrete. From (3.5) and (3.1)

$$\begin{aligned} |Q|^{2/n}\gamma_r\rho_r(q, Q) &= O\big(|Q|^{(2+\varepsilon)/n}(|x_Q|^{a-1}+1)\big) \\ &= O(|Q|^{\varepsilon/n}) = O(\lambda^{-\varepsilon\beta}), \end{aligned}$$

whence

$$\alpha = O(\lambda^{-\varepsilon\beta}) \tag{3.6}$$

To estimate N_1 and V_1 we begin with the observation that if $|q_Q| < \lambda$, then $|f_Q| < \lambda$ and thus $c_1(|x_Q|^a+1) < \lambda$, which in turn gives $|x_Q| < K\lambda^{1/a}$. The set $\{Q : |q_Q| < \lambda\}$ therefore lies in a ball with centre the origin and radius $K\lambda^{1/a}$, and in view of (3.2) each Q in this set satisfies $|Q|^{1/n} = \lambda^{-\beta}$. Thus $N_1\lambda^{-\beta n} = O(\lambda^{n/a})$, and hence

$$N_1 = O(\lambda^{n\beta+n/a}) \tag{3.7}$$

and also

$$V_1 = O(\lambda^{n/a}). \tag{3.8}$$

From (3.4), for $|q_Q| < \lambda$ and hence for $|x_Q| < K\lambda^{1/a}$, there follows

$$(\lambda^2 - g_Q^2)^{\frac{1}{2}} - f_Q = \lambda - f_Q + O(\lambda^{-1} g_Q^2)$$
$$= \lambda - f_Q + O(\lambda^{2\delta - 1}). \tag{3.9}$$

Thus for $|q_Q| < \lambda$ we have, using (XI.2.37),

$$\Lambda_Q = |Q|[(\lambda^2 - g_Q^2)^{\frac{1}{2}} - f_Q]^{n/2} \geqslant |Q|(\lambda - f_Q)^{n/2} - O(\lambda^{-n\beta + n/2}(\lambda^{\delta-1} + \lambda^{-n(1-\delta)})) \tag{3.10}$$

and so, since $\delta < 1$, this and (3.7) give

$$\Lambda_1 \geqslant \sum_{|q_Q| < \lambda} |Q|(\lambda - f_Q)^{n/2} - O(\lambda^{n/2 + n/a - 1 + \delta}). \tag{3.11}$$

Each Q in the set $\{Q: |q_Q| < \lambda\}$ is such that $|x_Q| < K\lambda^{1/a}$ and $|Q|^{1/n} = \lambda^{-\beta}$. Hence, from (3.3) and (3.4),

$$\{Q: |q_Q| < \lambda\} \supset \{Q: \lambda^2 > c_2^2(|x_Q|^a + 1)^2 + c_3^2(|x_Q|^a + 1)^{2\delta}\}$$
$$\supset \{Q: \lambda^2 > c_2^2(|x_Q|^a + 1)^2 + K\lambda^{2\delta}\} = \{Q: |x_Q| < v\}, \tag{3.12}$$

where

$$v^a = (\lambda^2 - K\lambda^{2\delta})^{\frac{1}{2}} c_2^{-1} - 1 = \lambda c_2^{-1} - 1 + O(\lambda^{2\delta - 1}). \tag{3.13}$$

Also from (3.3),

$$\lambda - f_Q \geqslant c_2(\lambda c_2^{-1} - 1 - |x_Q|^a) = c_2(v^a - |x_Q|^a) - O(\lambda^{2\delta - 1}),$$

and hence

$$|Q|(\lambda - f_Q)^{n/2} \geqslant c_2^{n/2}|Q|(v^a - |x_Q|^a)^{n/2} - O(\lambda^{n/2 - n\beta + \delta - 1}).$$

By (XI.2.37) we obtain

$$\Lambda_1 \geqslant \Lambda_3 - O(\lambda^{n/2 + n/a + \delta - 1}), \tag{3.14}$$

where

$$\Lambda_3 = c_2^{n/2} \sum_{|x_Q| < v} |Q|(v^a - |x_Q|^a)^{n/2}.$$

The term Λ_3 is bounded below by

$$c_2^{-n/a} \lambda^{n/2 + n/a} \omega_n [1 - O(\lambda^{-a/2 - \beta/2} + \lambda^{-2 + 2\delta})] \int_0^1 (1 - r^a)^{n/2} r^{n-1}\, dr, \tag{3.15}$$

the argument to establish this being identical to that used to derive (XI.3.7). Collecting together all the estimates obtained above, namely (3.6), (3.7), (3.8) and (3.14), we find on substitution in (2.9),

$$M(\lambda, T, \mathbb{R}^n) \geqslant \omega_n (2\pi)^{-n} \Lambda_3 [1 - O(\lambda^{\delta - 1} + \lambda^{\beta - \frac{1}{2}} + \lambda^{-\beta\varepsilon/4})], \tag{3.16}$$

where Λ_3 is bounded below by the expression given in (3.15).

We now apply the same argument to Theorem 2.3. In view of (3.3), (3.4) and (3.5) we have as before

$$\gamma = O(\lambda^{-\beta\varepsilon}), \tag{3.17}$$

$$N_2 = O(\lambda^{n\beta+n/a}). \tag{3.18}$$

We also have for each $Q \in \mathfrak{F}$,

$$\begin{aligned} f_Q - |g_Q| - \sqrt{2\gamma_r \rho_r(q, Q)} &\geqslant (c_1 - c_3)(|x_Q|^a + 1) - K|Q|^{\varepsilon/n}(|x_Q|^{a-1} + 1) \\ &\geqslant (c_1 - c_3 - K\lambda^{-\beta\varepsilon})(|x_Q|^a + 1). \end{aligned}$$

Since $c_3 < c_1$, (2.39) is satisfied for large enough λ. For $Q \in I_2$,

$$\begin{aligned} \lambda &\geqslant f_Q - |g_Q| - \sqrt{2\gamma_r \rho_r(q, Q)} \\ &\geqslant (c_1 - c_3 - K\lambda^{-\beta\varepsilon})(|x_Q|^a + 1), \end{aligned} \tag{3.19}$$

and hence $|x_Q| = O(\lambda^{1/a})$. Consequently, for $Q \in I_2$,

$$\lambda \geqslant c_1(|x_Q|^a + 1) - K\lambda^{\delta} - K\lambda^{-\beta\varepsilon+1}$$

and

$$|x_Q|^a \leqslant \mu^a = \lambda c_1^{-1}[1 + O(\lambda^{\delta-1} + \lambda^{-\beta\varepsilon})]. \tag{3.20}$$

Also, if Q is such that (3.20) holds,

$$\begin{aligned} \Gamma_Q &= |Q|(\lambda - f_Q + |g_Q|)_+^{n/2} \\ &\leqslant |Q|[\lambda - c_1(1 + |x_Q|^a) + c_3(1 + |x_Q|^a)^{\delta}]_+^{n/2} \\ &\leqslant c_1^{n/2}|Q|[\mu^a - |x_Q|^a + O(\lambda^{\delta} + \lambda^{1-\beta\varepsilon})]_+^{n/2} \\ &\leqslant c_1^{n/2}|Q|(\mu^a - |x_Q|^a)^{n/2} + O(\lambda^{n/2 - n\beta + \eta/2 - \frac{1}{2}}), \end{aligned} \tag{3.21}$$

where $\eta = \max\{\delta, 1 - \beta\varepsilon\}$. Since, as for (3.17), we may show that

$$\#\{Q\colon |x_Q| \leqslant \mu\} = O(\lambda^{n/a + n\beta}),$$

we have

$$\Lambda_2 \leqslant \Lambda_4 + O(\lambda^{n/2 + n/a + (\eta-1)/2}), \tag{3.22}$$

where

$$\Lambda_4 = c_1^{n/2} \sum_{|x_Q| \leqslant \mu} |Q|\,(\mu^a - |x_Q|^a)^{n/2}.$$

The term Λ_4 can be estimated as in

$$\Lambda_4 \leqslant c_1^{-n/a}\lambda^{n/2+n/a}\omega_n[1 + O(\lambda^{-a/2-\beta/2} + \lambda^{\delta-1} + \lambda^{-\beta\varepsilon})]\int_0^1 (1 - r^a)^{n/2} r^{n-1}\,dr. \tag{3.23}$$

Hence (2.41) yields

$$M(\lambda, T, \mathbb{R}^n) \leqslant \omega_n(2\pi)^{-n}\Lambda_4[1 + O(\lambda^{-\beta\varepsilon/2} + \lambda^{-(1-\delta)/2} + \lambda^{(2\beta-1)/2})], \tag{3.24}$$

where Λ_4 satisfies (3.23).

We summarize our conclusions as follows:

Theorem 3.1. Let $a \geqslant 1$ and let β be a positive number such that $1 - a^{-1} \leqslant 2\beta < 1$, let r satisfy (2.5) and suppose that (3.3), (3.4) and (3.5) hold for all cubes Q in a tesselation of $\mathbb{R}^n$ satisfying (3.1) and (3.2) and arising out of a tesselation $\mathfrak{F}_\beta$ of $\mathbb{R}^n$ by cubes of side $\lambda^{-\beta}$, for all large enough λ. Then $\sigma_e(T) = \varnothing$ and $\sigma(T^*T)$ consists of a sequence of non-negative eigenvalues; moreover, as $\lambda \to \infty$,

$$M(\lambda, T, \mathbb{R}^n) = \omega_n^2(2\pi)^{-n}\Lambda[1 + O(\lambda^{-\beta\varepsilon/4} + \lambda^{-(1-\delta)/2} + \lambda^{-(\frac{1}{2}-\beta)})], \tag{3.25}$$

where

$$c_2^{-n/a} J[1 + O(\lambda^{-a/2-\beta/2} + \lambda^{-2(1-\delta)})] \leqslant \Lambda/\lambda^{n/a+n/2}$$
$$\leqslant c_1^{-n/a} J[1 + O(\lambda^{-a/2-\beta/2} + \lambda^{-1+\delta} + \lambda^{-\beta\varepsilon})], \tag{3.26}$$

with

$$J = \int_0^1 (1 - r^a)^{n/2} r^{n-1}\,\mathrm{d}r = a^{-1}\,\mathrm{B}(n/a, 1 + \tfrac{1}{2}n). \tag{3.27}$$ ■

To illustrate this result we consider the following example. Let $n \geqslant 2$ and $a \geqslant 1$, and suppose that the potential q is such that there are positive constants a_1, a_2, a_3, a_4, with $a_3 < a_1$, and a constant $\delta \in [0, 1)$ such that, for all $x \in \mathbb{R}^n$,

$$a_1(1 + |x|^a) \leqslant f(x) \leqslant a_2(1 + |x|^a), \tag{3.28}$$
$$|g(x)| \leqslant a_3(1 + |x|^a)^\delta, \tag{3.29}$$
$$|q(x) - q(y)| \leqslant a_4|x - y|(\max\{|x|, |y|\})^{a-1}. \tag{3.30}$$

We proceed to verify that the hypothesis in Theorem 3.1 is satisfied. First note that if $\mathfrak{F}$ is a tesselation as in that theorem then

$$|q_Q - q(x_Q)| = |Q|^{-1}\left|\int_Q [q(y) - q(x_Q)]\,\mathrm{d}y\right|$$
$$= O(|Q|^{1/n}(1 + |x_Q|)^{a-1}). \tag{3.31}$$

Hence

$$f_Q \geqslant f(x_Q) - |f(x_Q) - f_Q|$$
$$\geqslant [a_1 - O(\lambda^{-\beta})](1 + |x_Q|^a),$$

and similarly

$$f_Q \leqslant [a_2 + O(\lambda^{-\beta})](1 + |x_Q|^a).$$

Thus (3.3) is satisfied with $c_1 \sim a_1$ and $c_2 \sim a_2$ as $\lambda \to \infty$. Also, from (3.31) and

(3.29),

$$\begin{aligned}|g_Q| &\leqslant |g(x_Q)|+|g_Q-g(x_Q)| \\ &\leqslant a_3(1+|x_Q|^a)^\delta+O(\lambda^{-\beta}(1+|x_Q|^a)^{1-1/a}) \\ &\leqslant [a_3+O(\lambda^{-\beta})](1+|x_Q|^a)^{\delta_1},\end{aligned}$$

where $\delta_1=\max\{\delta, 1-a^{-1}\}$. This means that (3.4) holds for large enough λ, with δ replaced by δ_1. Finally (3.31) shows that

$$\gamma_r\rho_r(q, Q)=O(|Q|^{1/n}(1+|x_Q|^{a-1})),$$

and so (3.5) is satisfied with $\varepsilon=1$.

It now follows exactly as in §XI.3 that as $\lambda\to\infty$,

$$\mathrm{M}(\lambda, T, \mathbb{R}^n)=\omega_n(2\pi)^{-n}\int_{\mathbb{R}^n}(\lambda-f)_+^{n/2}[1+o(1)], \tag{3.32}$$

and that the integral in (3.32) satisfies the inequality (3.26) for $\omega_n\Lambda$. In other words, $\mathrm{M}(\lambda, T, \mathbb{R}^n)$ is asymptotically equal to $\mathrm{N}(\lambda, -\Delta+f, \mathbb{R}^n)$.

4. The l^p classes of the singular values and eigenvalues

The results of this chapter have so far been presented in terms of the singular values of the operator but it is possible to say something about eigenvalues, despite their lack of connection in any direct way with the singular values. Under the assumptions of Theorem 3.1 we know that for all large enough λ we have $\mathrm{M}(\lambda, T, \mathbb{R}^n)\asymp\lambda^x$, with $x=n/a+\frac{1}{2}n$, and we now claim that this is true if, and only if, $s_m\asymp m^{1/x}$, where s_m is the mth singular value of T. Of course, this is clear if the multiplicity of each s_m is one, but in general more work is needed. Accepting our claim for the moment, it follows that the sequence $(s_m^{-1})\in l^p$ if, and only if, $p>x$ and that by Weyl's inequality (Theorem II.5.13) $(\lambda_m^{-1})\in l^p$ if $p>x$, the λ_m being the eigenvalues of T in increasing order of magnitude and repeated according to their algebraic multiplicities. It remains to justify our claim which we put in the form of a proposition.

Proposition 4.1. With the above notation, $\mathrm{M}(\lambda, T, \mathbb{R}^n)\asymp\lambda^x$ if, and only if, $s_m\asymp m^{1/x}$. ■

Proof. Suppose that $M(\lambda):=\mathrm{M}(\lambda, T, \mathbb{R}^n)\asymp\lambda^x$. Then there are positive constants c_1 and c_2 such that for large enough λ,

$$c_1\leqslant M(\lambda)\lambda^{-x}\leqslant c_2.$$

It will in fact be enough to assume that there are positive functions C_1 and C_2 such that for all $\lambda>0$,

$$C_1(\lambda)\leqslant M(\lambda)\lambda^{-x}\leqslant C_2(\lambda),$$

where $C_1(\lambda) \to c_1$ and $C_2(\lambda) \to c_2$ as $\lambda \to \infty$, and we shall do so throughout. Since the singular values s_m have finite multiplicity, there is a sequence $(s_j)_{j\in J}$ of singular values, J being an infinite subset of $\mathbb{N}$, such that, given any $j \in J$, there is a non-negative integer k (depending on j) such that $s_{j-1} < s_j = s_{j+k} < s_{j+k+1}$. Given $j \in J$, choose λ such that $s_j > \lambda > \max\{s_{j-1}, s_j - 1\}$: thus $M(\lambda) = j-1$ and $j-1 \asymp \lambda^x$. In fact,

$$[(j-1)/C_2(\lambda)]^{1/x} \leqslant \lambda \leqslant [(j-1)/C_1(\lambda)]^{1/x},$$

and so

$$[(j-1)/C_2(\lambda)]^{1/x} < s_j < [(j-1)/C_1(\lambda)]^{1/x} + 1,$$

which implies the existence of positive functions k_1 and k_2 such that $k_1(j) \to c_1$ and $k_2(j) \to c_2$ as $j \to \infty$ through J, and

$$[j/k_2(j)]^{1/x} < s_j < [j/k_1(j)]^{1/x} \qquad (j \in J). \tag{4.1}$$

Similarly, choice of λ in the interval $(s_{j+k}, \min\{s_{j+k+1}, s_{j+k}+1\})$ leads to the inequalities

$$[(j+k)/d_2(j)]^{1/x} < s_{j+k} < [(j+k)/d_1(j)]^{1/x}, \tag{4.2}$$

where $d_i(j) \to c_i$ as $j \to \infty$ $(i = 1,2)$. From (4.1) and (4.2) we see that for some positive functions K_1 and K_2 such that $K_i(j) \to c_i$ as $j \to \infty$,

$$[j/K_2(j)]^{1/x} < s_j < [j/K_1(j)]^{1/x}$$

and

$$[(j+k)/K_2(j)]^{1/x} < s_{j+k} < [(j+k)/K_1(j)]^{1/x}.$$

Thus since $s_j = s_{j+k}$,

$$(j+k)/K_2(j) < j/K_1(j), \tag{4.3}$$

which shows that

$$\limsup_{\substack{j\to\infty \\ j\in J}} k(j)/j \leqslant c_2/c_1 - 1. \tag{4.4}$$

For $l = 1,2,\cdots,k(j)$, since $s_{j+l} = s_j$, we have

$$[(j+l)/K_2(j)]^{1/x} \leqslant [(j+k)/K_2(j)]^{1/x} < s_{j+l}$$
$$< [j/K_1(j)]^{1/x} \leqslant [(j+l)/K_1(j)]^{1/x}.$$

It follows that $s_m \asymp m^{1/x}$ as required.

For the converse, suppose that for all $m \geqslant m_0$, say, s_m satisfies an inequality of the form

$$p_1(m) \leqslant s_m m^{-1/x} \leqslant p_2(m),$$

where $p_i(m) \to p_i > 0$ as $m \to \infty$ $(i = 1,2)$. Then

$$M(\lambda) = \sum_{s_m < \lambda} 1 \leqslant \sum_{p_1(m)m^{1/\varkappa} < \lambda} 1 = \sum_{m < [\lambda/p_1(m)]^\varkappa} 1 \sim (\lambda/p_1)^\varkappa$$

as $\lambda \to \infty$. Similarly,

$$M(\lambda) \geqslant \sum_{p_2(m)m^{1/\varkappa} < \lambda} 1 \sim (\lambda/p_2)^\varkappa.$$

Thus $M(\lambda) \asymp \lambda^\varkappa$ and our proof is complete. □

An examination of the proof of Proposition 4.1 shows, by consideration of the cases $c_1 = c_2$ and $p_1 = p_2$, that the following holds.

Corollary 4.2. With $M(\lambda) = \mathrm{M}(\lambda, T, \mathbb{R}^n)$ we have that $\lim_{\lambda \to \infty} M(\lambda)/\lambda^\varkappa = c > 0$ if, and only if,

$$\lim_{m \to \infty} s_m/m^{1/\varkappa} = c^{-1/\varkappa}.$$ ■

A final point is that the proof of Proposition 4.1 also shows that the multiplicity $k(j)$ of s_j is $O(j)$ in the general case with $c_1 \neq c_2$ (cf. (4.4)), while if $c_1 = c_2$, (4.3) shows that $k(j) = o(j)$.

5. Perturbation results

The annoying condition $c_3 < c_1$ in (3.4), and the corresponding condition $a_3 < a_1$ in §4, can be removed by using procedures similar to those used by Ramm [1], our basic conclusion $M(\lambda) \asymp \lambda^\varkappa$ remaining intact. To do this it is enough to show that the perturbation of the potential by an arbitrary imaginary constant leaves that conclusion unaltered. Thus let k be a real constant, let $P: \phi \mapsto ik\phi$, and let $H = T + P$. Note that H is a sectorial operator with numerical range in $\{z \in \mathbb{C}\colon \operatorname{re} z \geqslant \theta_0 > 0\}$, and so H^{-1} and T^{-1} are bounded linear maps of $\mathrm{L}^2(\mathbb{R}^n)$ into itself. Since $H = (I + PT^{-1})T$, we have $I + PT^{-1} = HT^{-1}$, and so $(I + PT^{-1})^{-1} = TH^{-1}$. The boundedness of P on $\mathrm{L}^2(\mathbb{R}^n)$ shows that $\|H\phi\| \leqslant \text{const.}(\|T\phi\| + \|\phi\|)$ for all $\phi \in \mathscr{D}(T)$, and hence that

$$\|HT^{-1}u\| \leqslant \text{const.}(\|u\| + \|T^{-1}u\|) \leqslant \text{const.}\|u\|;$$

that is, $HT^{-1} \in \mathscr{B}(\mathrm{L}^2(\mathbb{R}^n))$. Similarly $TH^{-1} \in \mathscr{B}(\mathrm{L}^2(\mathbb{R}^n))$. We now write

$$H^{-1} = T^{-1}(I + PT^{-1})^{-1} = T^{-1}(I + Q),$$

where

$$Q = (I + PT^{-1})^{-1} - I = -(I + PT^{-1})^{-1}PT^{-1}.$$

Put $A = T^{-1}$ and $B = H^{-1}$, so that $B = A(I + Q)$; the operators B, A and Q are all compact since T^{-1} and H^{-1} are compact, the compactness of H^{-1}

following from that of T^{-1} and the relation $H^{-1} = T^{-1}(I+Q)$. Note that for all $m \in \mathbb{N}$,

$$s_m^2(A) = \lambda_m(A^*A) = \lambda_m\left((TT^*)^{-1}\right) = \lambda_m^{-1}(TT^*) = \lambda_m^{-1}(T^*T) = s_m^{-2}(T),$$

and similarly $s_m(B) = s_m^{-1}(H)$. Since under the assumptions of §4, $M(\lambda, T, \mathbb{R}^n) \asymp \lambda^x$, with $x = n(1/a + \frac{1}{2})$, it follows from Proposition 4.1 that $s_m^{-1}(A) = s_m(T) \asymp m^{1/x}$. Moreover,

$$\|Qu\| = \|(I+PT^{-1})^{-1}PT^{-1}u\| \leqslant \text{const.}\|T^{-1}u\| \leqslant \text{const.}\|Au\|;$$

and $I+Q = (I+PT^{-1})^{-1}$ is evidently invertible.

Let $U = A^*A$ and $V = B^*B = (I+Q^*)A^*A(I+Q)$, let L_m be the linear span of the eigenfunctions associated with the first m eigenvalues of U, put $v = (I+Q)u$, and set $M_m = (I+Q^*)L_m$. Note that $v \perp L_m$ (in $L^2(\mathbb{R}^n)$) if, and only if, $u \perp M_m$. By the Max–Min Principle, and since $u \perp L_m$ and $u \perp M_k$ implies that $u \perp (L_m + M_k)$ and $\dim(L_m + M_k) \leqslant m+k$, we have

$$\begin{aligned} s_{k+1+m}^2(B) = \lambda_{k+1+m}(V) &\leqslant \sup_{u \perp L_m, u \perp M_k, u \neq 0} (Vu,u)/\|u\|^2 \\ &\leqslant \left(\sup_{u \perp M_k, u \neq 0} (Uv,v)/\|v\|^2\right)\left(\sup_{u \perp L_m, u \neq 0} \|v\|^2/\|u\|^2\right) \\ &\leqslant s_{k+1}^2(A)\left(1 + \sup_{u \perp L_m, u \neq 0} (2\|Qu\|/\|u\| + \|Qu\|^2/\|u\|^2)\right) \\ &\leqslant s_{k+1}^2(A)\,[1 + O(s_m(A))], \end{aligned} \tag{5.1}$$

the final step following because $\|Qu\| \leqslant \text{const.}\|Au\|$ and $s_m(A) \to 0$ as $m \to \infty$. Moreover, $U = (I+S^*)V(I+S)$, where $I+S = (I+Q)^{-1}$, $S = -(I+Q)^{-1}Q$, and thus as above,

$$s_{k+1+2m}^2(A) \leqslant s_{k+1+m}^2(B)\,[1 + O(s_m(B))]. \tag{5.2}$$

From (5.1) and (5.2) it is clear that $\lim_{k \to \infty} s_k(B)/s_k(A) = 1$, and we conclude that

$$s_m^{-1}(B) \asymp m^{1/x}.$$

By Proposition 4.1 again, $M(\lambda, H, \mathbb{R}^n) \asymp \lambda^x$. The constants implicit in the symbol $\asymp$ are, in general, different from those connected with the statement $M(\lambda, T, \mathbb{R}^n) \asymp \lambda^x$; and of course we lose the integral representation formula. Nevertheless, if $M(\lambda, T, \mathbb{R}^n) \sim c\lambda^x$, then we can conclude that $M(\lambda, H, \mathbb{R}^n) \sim c\lambda^x$.

Bibliography

ADAMS, R. A.

1 *Sobolev spaces.* Academic Press, New York, 1975.

AGMON, S.

1 *Lectures on elliptic boundary-value problems.* Van Nostrand, Princeton, 1965.

AKHIEZER, N. I., and I. M. GLAZMAN.

1 *Theory of linear operators in Hilbert space* (English translation), volumes I and II. Pitman, London, 1981.

ALLEGRETTO, W.

1 On the equivalence of two types of oscillation for elliptic operators. *Pacific J. Math.* 55 (1974), 319–328.

AMICK, C. J.

1 Some remarks on Rellich's theorem and the Poincaré inequality. *J. Lond. Math. Soc.* (2) 18 (1978), 81–93.

AUBIN, T.

1 *Non-linear analysis on manifolds. Monge–Ampère equations.* Springer-Verlag, Berlin, Heidelberg, New York, 1982.

BENCI, V., and D. FORTUNATO.

1 Discreteness conditions of the spectrum of Schrödinger operators. *J. Math. Anal. Appl.* 64 (1978), 695–700.

BERGER, M. S., and M. SCHECHTER.

1 Embedding theorems and quasi-linear elliptic boundary-value problems for unbounded domains. *Trans. Amer. Math. Soc.* 172 (1972), 261–278.

BIRMAN, M. S., and M. Z. SOLOMJAK.

1 Estimates of singular numbers of integral operators. *Uspekhi Mat. Nauk.* 32 (1977), 17–84. Transl.: *Russian Math. Surveys* 32 (1977), 15–89.

2 Quantitative analysis in Sobolev imbedding theorems and applications to spectral theory. *Amer. Math. Soc. Transl.* (2) 114 (1980), 1–132.

BROWDER, F. E.

1 On the spectral theory of elliptic differential operators. *Math. Ann.* 142 (1961), 22–130.

BROWNELL, F. E., and C. W. CLARK.

1 Asymptotic distribution of the eigenvalues of the lower part of Schrödinger operator spectrum. *J. Math. Mech.* 10 (1961), 31–73; Addendum, ibid. 10 (1961), 525–528.

CALKIN, J. W.

1 Symmetric transformations in Hilbert space. *Duke Math. J.* 7 (1940), 504–508.

CARADUS, S. R., PFAFFENBERGER, W. E., and B. YOOD.

1 *Calkin algebras of operators on Banach spaces.* Marcel Dekker, New York, 1974.

CARL, B.

1 Entropy numbers, s-numbers and eigenvalue problems. *J. Funct. Anal.* 41 (1981), 290–306.

2 A remark on a paper by Edmunds and Triebel. *Math. Nachr.* 103 (1981), 39–43.

3 Inequalities of Bernstein-Jackson type and the degree of compactness of operators in Banach spaces. *Ann. Inst. Fourier* 35 (1985), 79–118.

CARL, B., and H. TRIEBEL.

1 Inequalities between eigenvalues, entropy numbers and related quantities in Banach spaces. *Math. Ann.* 251 (1980), 129–133.

CARLEMAN, T.

1 Über die asymptotische Verteilung der Eigenverte partieller Differentialgleichungen. *Ber. Sachs. Akad. Wiss. Math-Phys. Klasse* 88 (1936), 483–496.

CODDINGTON, E. A., and N. LEVINSON.

1 *Theory of ordinary differential equations.* McGraw-Hill, New York, 1955.

COURANT, R., and D. HILBERT.

1 *Methoden der mathematischen Physik,* vol. II. Springer-Verlag, Berlin, 1937.

CWIKEL, M.

1 Weak type estimates for singular values and the number of bound states of Schrödinger operators. *Annals of Math.* 106 (1977), 93–100.

DARBO, G.

1 Punti uniti in transformazioni a codominio non compatto. *Rend. Sem. Mat. Univ. Padova* 24 (1955), 84–92.

DAVIES, E. B.

1 Some norm bounds and quadratic form inequalities for Schrödinger operators (I). *J. Operator Theory* 9 (1983), 147–162.

2 Ditto (part II). *J. Operator Theory* 12 (1984), 177–196.

3 Trace properties of the Dirichlet Laplacian. *Math. Zeit.* 188 (1985), 245–251.

DIEUDONNE, J.

1 *Foundations of modern analysis.* Academic Press, New York, 1969.

EASTHAM, M. S. P.

1 Semi-bounded second-order differential operators. *Proc. Roy. Soc. Edin.* (A) 72 (1973), 9–16.

EASTHAM, M. S. P., EVANS, W. D., and J. B. McLEOD.

1 The essential self-adjointness of Schrödinger type operators. *Arch, Rat. Mech. Anal.* 60 (1976), 185–204.

EDMUNDS, D. E.

1 Embeddings of Sobolev spaces. Proc. Spring School 'Nonlinear analysis, function spaces and applications' *Teubner-Texte Math.* 19, 38–58, Teubner, Leipzig, 1979.

EDMUNDS, D. E., and R. M. EDMUNDS.

1 Entropy and approximation numbers of embeddings in Orlicz spaces. *J. Lond. Math. Soc.* 32 (1985), 528–538.

2 Entropy numbers of compact operators, *Bull. Lond. Math. Soc.* 18 (1986), 392–394.

EDMUNDS, D. E., and W. D. EVANS.

1 Elliptic and degenerate elliptic operators in unbounded domains. *Ann. Scuola Norm. Sup. Pisa* 27 (1973), 591–640.

2 Orlicz and Sobolev spaces on unbounded domains. *Proc. Roy. Soc. Lond.* A 342 (1975), 373–400.

3 Spectral theory and embeddings of Sobolev spaces. *Quart. J. Math. Oxford* (2) 30 (1979), 431–453.

4 On the distribution of eigenvalues of Schrödinger operators. *Arch. Rat. Mech. Anal.* 89 (1985), 135–167.

EDMUNDS, D. E., EVANS, W. D., and J. FLECKINGER.

1 On the spectrum and the distribution of singular values of Schrödinger operators with a complex potential. *Proc. Roy. Soc. Lond.* A 388 (1983), 195–218.

EDMUNDS, D. E., and V. B. MOSCATELLI.

1 Fourier approximations and embeddings of Sobolev spaces. *Dissertationes Mathematicae* 145 (1977), 1–50.

EDMUNDS, D. E., and H. TRIEBEL.

1 Entropy numbers for non-compact self-adjoint operators in Hilbert spaces. *Math Nachr.* 100 (1981), 213–219.

EDMUNDS, D. E., and H-O TYLLI.

1 On the entropy numbers of an operator and its adjoint. *Math. Nachr.* 126 (1986), 231–239.

ENFLO, P.

1 A counterexample to the approximation problem in Banach spaces. *Acta Math.* 130 (1973), 309–317.

EVANS, W. D.

1 On the essential self-adjointness of powers of Schrödinger-type operators. *Proc. Roy. Soc. Edin.* 79 A (1977), 61–77.

2 On the spectra of Schrödinger operators with a complex potential. *Math. Ann.* 255 (1981), 57–76.

3 On the spectra of non-self-adjoint realisations of second-order elliptic operators. *Proc. Roy. Soc. Edin.* 90 A (1981), 71–105.

4 Regularly solvable extensions of non-self-adjoint ordinary differential operators. *Proc. Roy. Soc. Edin.* 97 A (1984), 79–95.

EVANS, W. D., and D. J. HARRIS.

1 Sobolev embeddings for ridged domains. *Proc. Lond. Math. Soc.*, (3) 54 (1987), 141–176.

EVANS, W. D., KWONG, M., and A. ZETTL.

1 Lower bounds for the spectrum of ordinary differential operators. *J. Diff. Equs.* 48 (1983), 123–155.

EVANS, W. D., and A. ZETTL.

1 Dirichlet and separation results for Schrödinger-type operators. *Proc. Roy. Soc. Edin.* 80 A (1978), 151–162.

EVERITT, W. N.

1 On the spectrum of a second-order linear differential equation with a *p*-integrable coefficient. *Appl. Anal.* 2 (1972), 143–160.

FARIS, W. G.

1 *Self-adjoint operators.* Lecture Notes in Mathematics 433. Springer-Verlag, Berlin, Heidelberg, New York, 1975.

FEIGIN, V. I.

1 On spectral asymptotics for boundary-value problems and the asymptotics of the negative spectrum. *Soviet Math. Dokl.* 18 (1977), 255–259.

FLECKINGER, J.

1 Théorie spectrale des opérateurs uniformément elliptiques sur quelques ouverts irréguliers. *Sem. Anal. Num., Toulouse,* 1973.

2 Estimate of the number of eigenvalues for an operator of Schrödinger type. *Proc. Roy. Soc. Edin.* 89 A (1981), 355–361.
3 On the singular values of non-self-adjoint operators of Schrödinger type. Proc. Dundee Conf. Ordinary and Partial Differential Equations. *Lecture Notes in Mathematics* 964, Springer-Verlag, Berlin, Heidelberg, New York, 1982.

FORTUNATO, D.
1 Remarks on the spectrum of the non-self-adjoint Schrödinger operator. *Comment. Math. Univ. Carolin.* 20 (1979), 79–93.

FRAENKEL, L. E.
1 Formulae for high derivatives of composite functions. *Math. Proc. Camb. Phil. Soc.* 83 (1978), 159–165.
2 On regularised distance and related functions. *Proc. Roy. Soc. Edin.* 83 A (1979), 115–122.
3 On regularity of the boundary in the theory of Sobolev spaces. *Proc. Lond. Math. Soc.* (3) 39 (1979), 385–427.
4 On the embedding of $C^1(\bar{\Omega})$ in $C^{0,\alpha}(\bar{\Omega})$. *J. Lond. Math. Soc.* (2) 26 (1982), 290–298.

FRIEDMAN, A.
1 *Partial differential equations.* Holt, Rinehart and Winston, New York, 1969.

FUČIK, S., JOHN, O., and A. KUFNER.
1 *Function spaces.* Academia, Prague, 1977.

GALINDO, A.
1 On the existence of *J*-self-adjoint extensions of *J*-symmetric operators with adjoint. *Comm. Pure and Appl. Math.* 15 (1962), 423–425.

GILBARG, D., and N. S. TRUDINGER.
1 *Elliptic partial differential equations of second-order.* Springer-Verlag, Berlin, Heidelberg, New York, 1977.

GLAZMAN, I. M.
1 *Direct methods of qualitative spectral analysis of singular differential operators.* Israel Program for Scientific Translation, Jerusalem, 1965.

GOHBERG, I. C., and M. G. KREIN.
1 *Introduction to the theory of linear non-self-adjoint operators.* Translation of Mathematical Monographs volume 18. Amer. Math. Soc., Providence, R. I. 1969.

GOLDBERG, S.
1 *Unbounded linear operators.* McGraw-Hill, New York, 1966.

GOLDSTEIN, L. S., and A. S. MARKUS.
1 On a measure of non-compactness of bounded sets and linear operators. *Studies in Algebra and Mathematical Analysis.* Kishinev, 1965, 45–54.

GORDON, Y. KÖNIG, H., and C. SCHÜTT.
1 Geometric and probabilistic estimates for entropy and approximation numbers of operators. *J. Approx Theory* (1986).

GUSTAFSON, K., and J. WEIDMANN.
1 On the essential spectrum. *J. Math. Anal. Appl.* 25 (1969), 121–127.

GUZMAN, M. de.
1 *Differentiation of integrals in* $\mathbb{R}^n$. Lecture Notes in Mathematics 481. Springer-Verlag, Berlin, Heidelberg, New York, 1975.

HARDY, G. H., LITTLEWOOD, J. E., and G. POLYA.
1 *Inequalities.* Cambridge University Press, Cambridge, 1959.

HAUSDORFF, F.

1 *Mengenlehre*, 3 Aufl. W. de Gruyter, Berlin and Leipzig, 1935.

HERSTEIN, I. N.

1 *Topics in algebra*. Blaisdell, New York, Toronto, London, 1964.

HINTON, D. B., and R. T. LEWIS.

1 Discrete spectra criteria for singular differential operators with middle terms. *Math. Proc. Camb. Phil. Soc.* 77 (1975), 337–347.

2. Singular differential operators with spectra discrete and bounded below. *Proc. Roy. Soc. Edin.* 84A (1979), 117–134.

HÖRMANDER, L.

1 *The analysis of linear partial differential operators*, vol. I. Springer-Verlag, Berlin, Heidelberg, New York, 1983.

HÖRMANDER, L., and J. L. LIONS.

1 Sur la completion par rapport à une integral de Dirichlet. *Math. Scand.* 4 (3) (1956), 259–270.

IKEBE, T., and T. KATO.

1 Uniqueness of the self-adjoint extension of singular elliptic operators. *Arch. Rat. Mech. Anal.* 9 (1962), 77–92.

JÖRGENS, K.

1 Wesentliche Selbstadjungiertheit singulärer elliptischer Differentialoperatoren zweiter Ordnung in C_0^∞ (G). *Math. Scand.* 15 (1964), 5–17.

JÖRGENS, K., and J. WEIDMANN.

1 *Spectral properties of Hamiltonian operators*. Lecture Notes in Mathematics 313. Springer-Verlag, Berlin, Heidelberg, New York, 1973.

KADLEC, J., and A. KUFNER.

1 Characterisation of functions with zero traces by integrals with weight functions. *Časopis Pěst. Mat.* 91 (1966), 463–471.

KALF, H.

1 On the characterisation of the Friedrichs extension of ordinary or elliptic differential operators with a strongly singular potential. *J. Funct. Anal.* 10 (1972), 230–250.

2 Self-adjointness for strongly singular potentials with a $-|x|^2$ fall-off at infinity. *Math. Z.* 133 (1973), 249–255.

3 Gauss's theorem and the self-adjointness of Schrödinger operators. *Arkiv för Mat.* (1) 18 (1980), 19–47.

KALF, H and J. WALTER.

1 Strongly singular potentials and essential self-adjointness of singular elliptic operators in $C_0^\infty(\mathbb{R}^n \setminus \{0\})$. *J. Funct. Anal.* 10 (1972), 114–130.

2 Note on a paper of Simon on essentially self-adjoint Schrödinger operators with singular potentials. *Arch. Rat. Mech. Anal.* 52 (1973), 258–260.

KALF, H., SCHMINCKE, U-W., WALTER, J., AND R. WÜST.

1 On the spectral theory of Schrödinger and Dirac operators with strongly singular potentials. Proc. Dundee Symp. 1974. *Lecture Notes in Mathematics* 448, 182–226. Springer-Verlag, Berlin, Heidelberg, New York, 1975.

KATO, T.

1 *Perturbation theory for linear operators*, 2nd edition. Springer-Verlag, Berlin, Heidelberg, New York, 1976.

2 Schrödinger operators with singular potentials. *Israel J. Math.* 13 (1972), 135–148.

3 A second look at the essential self-adjointness of the Schrödinger operators. *Physical Reality and Mathematical Description*. pp. 193–201. D. Reidel, Dordecht, 1974.
4 On some Schrödinger operators with a singular complex potential. *Ann. Scuola Norm. Pisa* IV 5 (1978), 105–114.
5 Remarks on Schrödinger operators with vector potential. *Integral Equs. and Operator Theory* 1 (1978), 103–113.
6 Remarks on the self-adjointness and related problems for differential operators. *Spectral theory of differential operators* (Birmingham, Ala., 1981), pp. 253–266. North-Holland Math. Studies 55. North-Holland, Amsterdam, New York, 1981.

KELLOGG, O. D.
1 *Foundations of potential theory*. Springer-Verlag, Berlin, 1929.

KNOWLES, I.
1 On *J*-self-adjoint extensions of *J*-symmetric operators. *Proc. Amer. Math. Soc.* 79 (1980), 42–44.
2 On the boundary conditions characterising *J*-self-adjoint extensions of *J*-symmetric operators. *J. Diff. Equs.* 40 (1981), 193–216.

KÖNIG, H.
1 Operator properties of Sobolev imbeddings over unbounded domains. *J. Funct. Anal.* 24 (1977), 32–51.
2 Approximation numbers of Sobolev imbeddings over unbounded domains. *J. Funct. Anal.* 29 (1978), 74–87.
3 A formula for the eigenvalues of a compact operator. *Studia Math.* 65 (1979), 141–146.
4 *S*-numbers, eigenvalues and the trace theorem in Banach spaces. *Studia Math.* 67 (1980), 157–172.
5 Some inequalities for the eigenvalues of a compact operator. Proc. Oberwolfach Conf. Inequalities, 1983. Birkhäuser (1984), 213–219.

KURATOWSKI, K.
1 Sur les espaces complets. *Fund. Math.* 15 (1930), 301–309.

LANG, S.
1 *Analysis* II. Addison-Wesley, Reading, Massachusetts, 1969.

LEBOW, A. and M. SCHECHTER.
1 Semigroups of operators and measures of non-compactness. *J. Funct. Anal.*, 7 (1971), 1–26.

LEINFELDER, H., and C. G. SIMADER.
1 Schrödinger operators with singular magnetic vector potentials. *Math. Z.* 176 (1981), 1–19.

LEVINSON, N.
1 Criteria for the limit point case for second-order linear differential operators. *Časopis Pěst. Mat. Fys.* 74 (1949), 17–20.

LEWIS, R. T.
1 Singular elliptic operators of second order with purely discrete spectrum. *Trans. Amer. Math. Soc.* 271,2 (1982), 653–666.

LEWY, H., and G. STAMPACCHIA.
1 On the regularity of the solution of a variational inequality. *Comm. Pure and Appl. Math.* 22 (1969), 153–188.

LI. P., and S-T.YAU.

1 On the Schrödinger equation and the eigenvalue problem. *Comm. Math. Phys.* 88 (1983), 309–318.

LIEB, E.

1 Bounds on the eigenvalues of the Laplace and Schrödinger operators. *Bull. Amer. Math. Soc.* 82 (1976), 751–753.

2 The number of bound states of one-body Schrödinger operators and the Weyl problem. *Proc. Symp. Pure Math.* 36 (1980), 241–252.

LIEB, E., and W. THIRRING.

1 Inequalities for the moments of the eigenvalues of the Schrödinger equation and their relations to Sobolev inequalities. *Studies in Math. Phys.: Essays in honor of V. Bargmann.* Princeton Univ. Press, Princeton, N. J., 1976.

LIONS, J-L.

1 *Problèmes aux limites dans les équations aux derivées partielles.* University of Montreal Press, 1965.

LLOYD, N. G.

1 *Degree theory.* Cambridge University Press, Cambridge, 1978.

MARTINS, J. A. S.

1 Embeddings of Sobolev spaces in unbounded domains. *Ann. Mat. Pura Appl.* 95 (1977), 271–294.

MAZ'JA, V. G.

1 On (p, l)-capacity, embedding theorems, and the spectrum of a selfadjoint elliptic operator. *Isv. Akad. Nauk SSSR Ser. Mat.* 37 (1973), 356–385. *Transl. Math USSR IZV.* 7 (1973) 357–387.

2 *Einbettungssätze für Sobolewsche Räume*, Teil I und II. Teubner, Leipzig, 1979/80.

3 *Zur Theorie Sobolewscher Räume.* Teubner, Leipzig, 1981.

McLEOD, J. B.

1 The distribution of the eigenvalues for the hydrogen atom and similar cases. *Proc. Lond. Math. Soc.* (3) 11 (1961), 139–158.

MEYERS, N. G., and J. SERRIN.

1 $H = W$. *Proc. Nat. Acad. Sci. USA* 51 (1964), 1055–1056.

MICHLIN, S. G., and PRÖSSDORF, S.

1 *Singulare Integraloperatoren.* Akademie-Verlag, Berlin, 1980.

MITJAGIN, B. S., and A. PELCZYNSKI.

1 Nuclear operators and approximation dimension. Proc. I. C. M., Moscow, 1966, 366–372.

MOLCANOV, A. M.

1 Conditions for the discreteness of the spectrum of self-adjoint second-order differential equations. *Trudy Moskov Mat. Obšč.* 2 (1953), 169–200.

NAIMARK, M. A.

1 *Linear differential operators*, part II. Harrap, London, 1968.

NAMASIVAYAM, M.

1 Estimates of entropy numbers of embeddings of Sobolev spaces on unbounded domains. *Quart. J. Math. Oxford* (2) 36 (1985), 231–242.

NÉCAS, J.

1 *Les methodes directes en theorie des équations elliptiques.* Masson, Paris, 1967.

NUSSBAUM, R. D.

1 The radius of the essential spectrum. *Duke Math. J.* 37 (1970), 473–479.

PERSSON, A.
1 Bounds for the discrete part of the spectrum of a semi-bounded Schrödinger operator. *Math. Scand.* 8 (1960), 143–153.
2 Compact linear mappings between interpolation spaces. *Arkiv Mat.* 5 (1964), 215–219.
PHAM THÉ LAI.
1 Comportement asymptotique du noyau de la résolvante et des valeurs propres d'une classe d'opérateurs elliptiques dégénérés non-nécessairement auto-adjoint. *J. Math. Pures et Appl.* 55 (1976), 379–420.
PIETSCH, A.
1 *Operator ideals.* VEB Deutscher Verlag der Wissenschaften, Berlin, 1978.
2 Weyl numbers and eigenvalues of operators in Banach spaces. *Math. Ann.* 247 (1980), 149–168.
3 Eigenvalues of integral operators I. *Math. Ann.* 247 (1980), 169–178.
RACE, D.
1 The theory of *J*-self-adjoint extensions of *J*-symmetric operators. *J. Diff. Equs.* 57 (1985), 258–274.
RAMM, A.
1 Spectral properties of some non-self-adjoint operators and some applications. *Mathematical Studies* vol. 55. pp. 349–354, North Holland, Amsterdam, 1981.
RAUCH, J., and M. REED.
1 Two examples illustrating the difference between classical and quantum mechanics. *Comm. Math. Phys.* 29 (1973), 105–111.
REED, M., and B. SIMON.
1 *Methods of modern mathematical physics*, vol. II: *Fourier Analysis, self-adjointness.* Academic Press, New York, 1975.
2 Ditto vol. IV: *Analysis of operators.* Academic Press, New York, 1978.
ROBERT, D.
1 Propriétés spectrales d'opérateurs pseudodifférentielles. *Comm. Partial Diff. Equs.* 3 (1978), 755–826.
ROBINSON, D.
1 Scattering theory with singular potentials I: the two-body problem. *Ann. Inst. H. Poincaré* 21 (1974), 185–215.
ROSENBLJUM, G. V.
1 The distribution of the discrete spectrum for singular differential operators. *Soviet Math. Dokl.* 13 (1972), 245–249.
2 Estimates of the spectrum of the Schrödinger operator. *Problemy Mat. Anal.* 5 (1975), 152–166.
3 Asymptotics of the eigenvalues of the Schrödinger operator. *Mat. Sb.* 93 (1974), 347–367 (Russian). English translation: *Math. USSR Sb.* 22 (1975), 349–371.
4 An asymptotic of the negative discrete spectrum of the Schrödinger operator. *Math. Notes* 21 (1977), 222–227.
SCHECHTER, M.
1 On the essential spectrum of an arbitrary operator I. *J. Math. Anal. Appl.* 13 (1966), 205–215.
2 *Spectra of partial differential operators.* North-Holland, Amsterdam, 1971.
3 Essential self-adjointness of the Schrödinger operator with magnetic vector potentials. *J. Funct. Anal.* 20 (1975), 93–104.

SCHMINCKE, U-W.

1 Essential self-adjointness of a Schrödinger operator with a strongly singular potential. *Math. Z.* 124 (1972), 47–50.

2 The lower spectrum of Schrödinger operators. *Arch. Rat. Mech. Anal.* 75 (1981), 147–155.

SHORTLEY, G. H.

1 The inverse-cube central force field in quantum mechanics. *Phys. Rev.* 38 (1931), 120–127.

SIMADER, C. G.

1 Bemerkungen uber Schrödinger-operatoren mit stark singularen Potentialen. *Math. Z.* 138 (1974), 53–70.

SIMON, B.

1 Essential self-adjointness of Schrödinger operators with singular potentials. *Arch. Rat. Mech. Anal.* 52 (1973), 44–48.

2 Schrödinger operators with singular magnetic vector potentials. *Math. Z.* 131 (1973), 361–370.

3 *Trace ideals and their applications.* Lond. Math. Soc. Lecture Note series 35. Cambridge University Press, Cambridge, 1979.

4 Maximal and minimal Schrödinger forms. *J. Operator Theory* 1 (1979), 37–47.

5 Hardy and Rellich inequalities in non-integral dimensions. *J. Operator Theory* 9 (1983), 143–146.

SIMS, A. R.

1 Secondary conditions for linear differential operators of the second order. *J. Math. Mech.* 6 (1957), 247–285.

STAMPACCHIA, G.

1 *Equations elliptiques du second ordre à coefficients discontinus.* University of Montreal Press, 1966.

STEIN, E. M.

1 *Singular integrals and differentiability properties of functions.* Princeton University Press, 1970.

STUART, C.

1 Self-adjoint square roots of positive self-adjoint bounded linear operators. *Proc. Edin. Math. Soc.* 18 (1972), 77–99.

STUMMEL, F.

1 Singuläre elliptische Differentialoperatoren in Hilbertschen Raümen. *Math. Ann.* 132 (1956), 150–176.

TALENTI, G.

1 Best constant in Sobolev inequality. *Ann. Mat. Pura Appl.* 110 (1976), 353–372.

TAMURA, H.

1 The asymptotic distribution of discrete eigenvalues for Schrödinger operators. *J. Math. Soc. Japan* 29 (1977), 189–218.

2 Asymptotic formulas with sharp remainder estimates for bound states of Shrödinger operators, *I. J. d'Anal. Math.* 40 (1981), 166–182.

3 Asymptotic formulae with sharp remainder estimates for eigenvalues of elliptic operators of second-order. *Duke Math. J.*, 49 (1982), 87–119.

TAYLOR, A.

1 *Introduction to functional analysis.* Wiley, New York, 1958.

TEIXEIRA, M. F.

1 Entropy numbers and interpolation. *Math. Nachr.* 107 (1982), 131–137.

2 Weyl's inequality in Banach spaces. *Bull. Lond. Math. Soc.* 16 (1984), 1–7.

TITCHMARSH, E. C.
1 *Eigenfunction expansions associated with second-order differential equations.* Oxford University Press, Oxford. Part I (2nd edition) 1962, Part II 1958.
TRÈVES, F.
1 *Basic linear partial differential equations.* Academic Press, New York, 1975.
TRIEBEL, H.
1 Interpolationseigenschaften von Entropie und Durchmesseridealen kompacter Operatoren. *Studia Math.* 34 (1970), 89–107.
2 *Interpolation theory, function spaces, differential operators.* North-Holland, Amsterdam, 1978.
TRUDINGER, N. S.
1 On imbeddings into Orlicz spaces and some applications. *J. Math. Mech.* 17 (1967), 473–483.
TYLLI, H-O.
1 Nonkompacthelmsmått, semiFredholm operatorer och någa Riesz ideal. Dissertation, Helsinki, 1982.
2 On the asymptotic behaviour of some quantities related to semi-Fredholm operators. *J. Lond. Math. Soc.* (2) 31 (1985), 340–348.
VALENTINE, F. A.
1 A Lipschitz condition preserving extension for a vector function. *Amer. J. Math.* 67 (1945), 83–93.
VELING, E. J. M.
1 Optimal lower bounds for the spectrum of a second-order linear differential equation with a *p*-integrable coefficient. *Proc. Roy. Soc. Edin.* 92A (1982), 95–101.
2 A relation between the infimum of the spectrum of the Schrödinger equation in $\mathbb{R}^n$ and the Sobolev inequalities. To appear, preprint.
VISIK, M. I.
1 On general boundary problems for elliptic differential equations. *Amer. Math. Soc. Transl.* (2) 24 (1963), 107–172.
WEBB, J. R. L.
1 Remarks on *k*-set contractions. *Boll. Un. Mat. Ital.* (4) 4 (1971), 614–629.
2 On seminorms of operators. *J. Lond. Math. Soc.* 7 (1973), 339–342.
WEIDMANN, J.
1 *Linear operators in Hilbert spaces.* Springer-Verlag, Berlin, Heidelberg, New York, 1980.
WEINSTEIN, M. I.
1 Non-linear Schrödinger equations and sharp interpolation estimates. *Comm. Math. Phys.* 87 (1983), 567–576.
WET, J. S. de, and F. MANDL.
1 On the asymptotic distribution of eigenvalues. *Proc. Roy. Soc. Lond.* A 200 (1950), 572–580.
WEYL, H.
1 Über beschränkte quadratische Formen, deren Differenz vollstetig ist. *Rend. Circ. Mat. Palermo* 27 (1909), 373–392.
2 Über gewöhnliche Differentialgleichungen mit Singularitäten und die zugehörigen Entwicklungen wilkkürlicher Funktionen. *Math. Ann.* 68 (1910), 220–269.
3 Das asymptotische Verteilungsgesetz der Eigenwerte linearer partieller Differentialgleichungen. *Math. Ann.* 71 (1912), 441–479.
4 Inequalities between the two kinds of eigenvalues of a linear transformation. *Proc. Nat. Acad. Sci. USA* 35 (1949), 408–411.

WOLF, F.

1. On the essential spectrum of partial differential boundary problems. *Comm. Pure Appl. Math.* 12 (1959), 211–228.

WÜST, R.

1 Generalisations of Rellich's theorem on perturbations of (essentially) self-adjoint operators. *Math. Z.* 119 (1971), 276–280.

2 Holomorphic operator families and stability of self-adjointness. *Math. Z.* 125 (1972), 349–358.

YOSIDA, K.

1 *Functional analysis*. Springer-Verlag, Berlin, Heidelberg, New York, 1965.

ZEMÂNEK, J.

1 The essential spectral radius and the Riesz part of the spectrum. Functions, series, operators. Proc. Int. Conf. Budapest 1980. *Colloq. Math. Soc. Janos Bolyai* 35, 1275–1289.

2 Generalisations of the spectral radius formula. *Proc. Roy. Irish Acad.* 81 A (1) (1981), 29–35.

3 Geometric interpretations of the essential minimum modulus. Invariant subspaces and other topics (G. R. Arsene, Ed.) (Timisoara/Herculane 1981), pp. 225–227. *Operator theory*: Adv. Appl. 6, Birkhäuser, Basel, 1982.

4 The semi-Fredholm radius of a linear operator. *Bull. Polish Acad. Math.* To appear, preprint.

5 Geometric characterisation of semi-Fredholm operators and their asymptotic behaviour. To appear, preprint.

ZHIKHAR, N. A.

1 The theory of extensions of *J*-symmetric operators. *Ukrain. Mat. Z. XI* (4), (1959), 352–364.

Notation index

Author index

Subject index